Handbook of Experimental Pharmacology

Continuation of Handbuch der experimentellen Pharmakologie

Vol. 58/I

Cyclic Nucleotides

Part I: Biochemistry

Contributors

M.M. Appleman · M.A. Ariano · J.A. Beavo · L. Birnbaumer
G. Brooker · W.Y. Cheung · P. Greengard · T.K. Harden
J.F. Harper · R. Iyengar · E.M. Johnson · C.L. Kapoor
J.F. Kuo · C.K. Mittal · M.C. Mumby · F. Murad · J.P. Perkins
T.W. Rall · G.R. Revankar · R.K. Robins · H. Schulman
M. Shoji · A. Steiner · D.R. Storm · D.J. Takemoto · U. Walter
R.H. Whitson

Editors

J.A. Nathanson and J.W. Kebabian

Springer-Verlag Berlin Heidelberg New York 1982

Professor Dr. J. A. NATHANSON
Neuropharmacology Research Laboratory, Departments of Neurology and Pharmacology, Harvard Medical School, Massachusetts General Hospital, Boston, MA 02114/USA

Dr. J. W. KEBABIAN
Biochemical Neuropharmacology Unit, NINCDS National Institutes of Health, Bethesda, MD 20205/USA

With 69 Figures

ISBN 3-540-10786-X Springer-Verlag Berlin Heidelberg New York
ISBN 0-387-10786-X Springer-Verlag New York Heidelberg Berlin

Library of Congress Cataloging in Publication Data. Main entry under title: Biochemistry of cyclic nucleotides. (Handbook of experimental pharmacology; v. 58/I). Includes bibliographical references and indexes. 1. Cyclic nucleotides. I. Appleman, M. M. II. Nathanson, J. A. (James A.), 1947–. III. Kebabian, J. W. (John W.) IV. Series. [DNLM: 1. Nucleotides, Cyclic. W1 HA51L v. 58 pt. 1 etc./QU 58 B615] QP905.H3 vol. 58/I [QP625.N89] 615′.ls.81-9060. ISBN 0-387-10786-X (U.S.) [612′.01579] AACR2.

Printed in Germany

This book was edited by JOHN W. KEBABIAN, Ph.D., in his private capacity. No official endorsement or support by the U.S. Government is intended or should be inferred.

Typesetting, printing, and bookbinding: Brühlsche Universitätsdruckerei Giessen. 2122/3130-543210

Preface

The purpose of the present volume, the first of two on the pharmacology, biochemistry, and physiology of cyclic nucleotides, is to provide a comprehensive and up-to-date anthology on the nature and role of these important chemical regulators. Each of the chapters is the work of internationally known researchers who present a lucid and detailed review of their subject and not merely a single laboratory's viewpoint. The chapters emphasize critical assessments of the field rather than mere listings of experimental findings. By so doing, the contributors present the role of cyclic nucleotides in relationship to other intracellular regulators. Each chapter begins with a detailed summary to allow the reader to obtain a rapid overview of subsequent material. In addition, there are extensive bibliographies and a detailed subject index. Wherever pertinent, the chapters contain sections on drug mechanisms, physiological relevance, and disease processes.

The Volume is divided into two sections, each beginning with an overview written by Professors T. W. RALL and P. GREENGARD, respectively. The first section focuses on the detailed pharmacology and chemistry of cyclic nucleotides, including their formation, degradation, measurement, and interaction with various modulatory agents, such as receptors and calcium. The second section is concerned with the biochemistry of protein phosphorylation, a process which appears to be one of the most important mechanisms for the intracellular expression of cyclic nucleotide action in eukaryotic cells. The chapters cover such topics as cyclic AMP-dependent, cyclic GMP-dependent, and calcium-dependent phosphorylation, photoaffinity labeling, and the role of phosphorylation in gene expression.

We wish to express our gratitude to the contributors who so generously gave of their time. We wish also to extend very special thanks to Dr. PAUL GREENGARD – teacher, colleague, and friend.

Boston and Bethesda J. A. NATHANSON and J. W. KEBABIAN

List of Contributors

M. M. Appleman, Department of Biological Sciences, University of Southern California, Los Angeles, CA 90007/USA

M. A. Ariano, Department of Biological Sciences, University of Southern California, Los Angeles, CA 90007/USA

J. A. Beavo, Department of Pharmacology, University of Washington, Seattle, WA 98195/USA

L. Birnbaumer, Department of Cell Biology, Baylor College of Medicine, Houston, TX 77030/USA

G. Brooker, Department of Pharmacology, University of Virginia, Charlottesville, VA 22908/USA

W. Y. Cheung, Department of Biochemistry, St. Jude's Children's Research Hospital, Memphis, TN 38101/USA

P. Greengard, Department of Pharmacology, Yale Medical School, New Haven, CT 06510/USA

T. K. Harden, Department of Pharmacology, University of North Carolina, Chapel Hill, NC 27514/USA

J. F. Harper, Department of Medicine, University of Texas Medical Center, Houston, TX 77030/USA

R. Iyengar, Department of Cell Biology, Baylor College of Medicine, Houston, TX 77030/USA

E. M. Johnson, Rockefeller University, New York, NY 10021/USA

C. L. Kapoor, Laboratory of Pathophysiology, National Cancer Institute, National Institutes of Health, Bethesda, MD 20205/USA

J. F. Kuo, Departments of Pharmacology and Medicine (Hematology and Oncology), Emory University School of Medicine, Atlanta, GA, 30322/USA

C. K. Mittal, Division Pharmacology, Peoria School of Medicine, University of Illinois College of Medicine, Peoria, IL 61656/USA

M. C. Mumby, Department of Pharmacology, University of Washington, Seattle, WA 98193/USA

F. Murad, Department of Medicine, University of Virginia, Charlottesville, VA 22901/USA

J. P. PERKINS, Department of Pharmacology, North Carolina School of Medicine, Chapel Hill, NC 27514/USA

T. W. RALL, Department of Pharmacology, University of Virginia School of Medicine, Charlottesville, VA 22908/USA

G. R. REVANKAR, Cancer Research Center, Department of Chemistry, Brigham Young University, Provo, UT 84602/USA

R. K. ROBINS, Cancer Research Center, Department of Chemistry, Brigham Young University, Provo, UT 84602/USA

H. SCHULMAN, Department of Pharmacology, Stanford University School of Medicine, Stanford, CA 94305/USA

M. SHOJI, Departments of Pharmacology and Medicine (Hematology and Oncology), Emory University School of Medicine, Atlanta, GA 30322/USA

A. STEINER, Departments of Medicine and Pharmacology, North Carolina School of Medicine, Chapel Hill, NC 27514/USA

D. R. STORM, Department of Pharmacology, University of Washington School of Medicine, Seattle, WA 98195/USA

D. J. TAKEMOTO, Department of Biological Sciences, University of Southern California, Los Angeles, CA 90007/USA

U. WALTER, Department of Pharmacology, Yale Medical School, New Haven, CT 06510/USA

R. H. WHITSON, Department of Biological Sciences, University of Southern California, Los Angeles, CA 90007/USA

Contents

Section I: Biochemistry of Cyclic Nucleotides

CHAPTER 5

Guanylate Cyclase: Regulation of Cyclic GMP Metabolism.

C. K. MITTAL and F. MURAD. With 1 Figure

Section II: Biochemistry of Protein Phosphorylation

CHAPTER 13

Calcium-Dependent Protein Phosphorylation. H. SCHULMAN. With 10 Figures

CHAPTER 14

Photoaffinity Labeling of Cyclic AMP-Dependent and Cyclic GMP-Dependent Protein Kinases. U. WALTER and P. GREENGARD. With 11 Figures

CHAPTER 15

Nuclear Protein Phosphorylation and the Regulation of Gene Expression.
E. M. JOHNSON. With 2 Figures

Contents of Companion Volume 58, Part II
Cyclic Nucleotides: Physiology and Pharmacology

Section I
Biochemistry of Cyclic Nucleotides

CHAPTER 1

Formation and Degradation of Cyclic Nucleotides: An Overview

T. W. RALL

A. Introduction

This chapter provides some background and historical information about cyclic nucleotides as well as a brief overview of the section on their formation and degradation. Detailed and comprehensive treatment of the issues raised in this chapter will occur in the later chapters. This chapter will focus primarily on cyclic AMP; also, it will reflect the personal interest and biases of the author more so than will be the case in other chapters. However, it is hoped that this approach will be both informative and interesting for the reader.

B. Historical Review

I. Discovery of Cyclic AMP

Cyclic AMP was discovered as a chemical entity and as a biological entity virtually simultaneously as a result of two independent lines of investigation. In St. Louis, a group of chemists examining the alkaline hydrolysis of ATP in the presence of barium ions found that between 5% and 10% of the precursor was converted to a nucleotide which was readily distinguishable from the major product, 5′-AMP, most notably by its relative resistance to attack by snake venom. Meanwhile, in Cleveland, a group of biochemists were studying the mechanisms by which glucagon and epinephrine promoted glycogenolysis in liver. Their analysis had progressed from the correlation of hormone-induced glycogenolysis with increased recovery of glycogen phosphorylase from treated tissue (SUTHERLAND and CORI 1951) to the important discovery that a non-proteolytic enzyme released a characteristic amount of inorganic phosphate as it brought about decreased enzymic activity of phosphorylase (SUTHERLAND and WOSILAIT 1955). This was soon followed by the detection of the phosphorylase kinase reaction, whereby the dephosphorylated, relatively inactive product could be reactivated by the transfer of phosphate from ATP (FISCHER and KREBS 1955; RALL et al. 1956). It was hypothesized that the hormones acted primarily by accelerating the phosphorylase kinase reaction, rather than by inhibiting the phosphorylase phosphatase reaction. This was because the addition of radioactive phosphate to slices of dog liver, after hormone-induced accumulation of phosphorylase had reached a maximum, resulted in a high rate of incorporation of label into phosphorylase relative to that observed when NaF, a known inhibitor of phosphorylase phosphatase, was used to produce similar levels of phosphorylase. However, the phosphorylase kinase activity of even the crudest preparations was not affected by the addition of hormones.

Eventually, experiments were performed employing homogenates of slices of dog liver that had been incubated to reduce the level of active phosphorylase, and either omitting or reducing the concentration of NaF customarily used to inhibit phosphorylase phosphatase in the assay of phosphorylase kinase activity. Under these conditions, epinephrine and glucagon induced a marked increase in the rate of active phosphorylase accumulation that was dependent upon the presence of some portion of the particulate fraction sedimentable by low-speed contrifugation (RALL et al. 1957). This led to two-stage experiments, in which particulate material was first incubated with the hormones, then boiled, and finally incubated with supernatant fractions that were not responsive to the hormones. Fortunately, for reasons of simplicity more than anything else, ATP and magnesium ions were included in both stages of the experiment. Thus, evidence was produced that the hormones generated a "heat-stable factor" that could accelerate the phosphorylase kinase reaction catalyzed by supernatant fractions of liver homogenates (RALL et al. 1957). Curiously, even the crudest preparations of phosphorylase kinase, obtained in the customary fashion as side-fractions during the purification of phosphorylase (RALL et al. 1956) failed to respond to the "heat-stable factor," and its effect upon supernatant fractions of liver homogenates was used as the basis of assay procedures during isolation of the factor as well as for more than a decade thereafter (RALL and SUTHERLAND 1962a). On the other hand, it proved to be quite difficult to prepare phosphorylase *b* kinase from skeletal muscle that was *not* sensitive to cyclic AMP, and it required more than a decade to define the component that interacted with the cyclic nucleotide (see LANGAN 1973).

It was soon appreciated that ATP was essential for the production of the factor, suggesting the possible presence of a charged group; this in turn led to the use of ion-exchange chromatography in the initial attempts in its purification. This fortunate guess, combined with two serendipitous findings (described below) that permitted greatly increased accumulations of the factor, led to the isolation and crystallization of the active material less than a year after the first successful experiment showing effects of hormones in a broken-cell preparation. Chemical analysis and the ultraviolet spectrum suggested the presence of adenine, ribose, and phosphate in equimolar amounts (RALL et al. 1957). However, the factor was not destroyed by treatment with various commercially available phosphatases or nucleotidases and could be readily distinguished from the adenosine monophosphates by relative activity in the assay system and by separation during ion-exchange chromatography.

It was at this point that, within a span of a few weeks, both the Cleveland and St. Louis groups wrote to Dr. LEON HEPPEL, describing their respective substances and requesting some spleen phosphodiesterase in the hope that it might provide some further structural clues. Dr. HEPPEL sent the enzyme, but more importantly, suggested to both groups that the two substances might be identical. Following exchange of samples, this possibility was rapidly confirmed and the findings of the two groups were published simultaneously (COOK et al. 1957; SUTHERLAND and RALL 1957). The conclusion that the substance contained phosphate linked to both the 3′- and 5′-hydroxyl functions of ribose was inferred from the ability of snake venom to degrade the compound slowly to a mixture of adenosine-3′-phosphate and adenosine, the latter presumably arising from the action of 5′-nucleotidase upon adenosone-5′-phosphate. This was supported by the ability of an enzyme purified from cardiac tissue to convert the substance quantitatively to adenosine-5′-phosphate (SUTHERLAND and RALL 1958). Initially, it was proposed that the compound was a dinucleotide, consisting of two molecules of adenosine-5′-phosphate linked by phosphodiester bonds at the 3′-positions (COOK et al. 1957), leading some irreverent individuals to dub it "bicyclic AMP." The determination of its molecular weight and other evidence soon forced the assignment of a mononucleotide structure with an internal 3′–5′ phosphodiester ring (LIPKIN et al. 1959). The degree of puckering of the ribofuranose ring necessary to permit closure of the phosphodiester ring was later confirmed by proton magnetic resonance (JARDETZKY 1962).

II. Early Characterization of Adenylate Cyclase and Discovery of Cyclic Nucleotide Phosphodiesterase

The enzyme system responsible for the formation of cyclic AMP when cell particulate material was incubated with ATP and Mg^{++} ions was originally termed "adenyl cyclase" (SUTHERLAND et al. 1962); subsequently, the trivial designations, "adenylyl cyclase" and "adenylate cyclase" were promulgated, the latter probably being the most frequently used name. In any event, the initial two-stage experiments described above suggested that 1) while ATP was required for detectable accumulation of cyclic AMP, the hormones were not required; and 2) while cyclic AMP was chemically quite stable, it was biochemically very labile in that prolonging the first stage from the arbitrarily chosen 5 min period to 10 min resulted in nearly complete loss of assayable activity. Perhaps the most important consequence of the latter observation was that it later prompted a search for degradative enzymes that might provide insight into the structure of the active material. However, initially, the evanescent accumulation of product suggested the possibility that degradation of ATP might limit the duration and thus extent of the reaction. Accordingly, various experiments were performed including varying ATP concentrations, addition of metabolic intermediates, and inclusion of NaF (RALL and SUTHERLAND 1958). It was found that NaF markedly increased the accumulation of product, which originally was thought to result from reduced hydrolysis of ATP. However, when the amounts of particulate material were systematically varied in the attempt to establish conditions for assay of adenylate cyclase activity, it became clear that the effects of NaF could be explained only to a very limited extent by the preservation of ATP (SUTHERLAND et al. 1962). Stimulation to varying degrees by F^- ions has since been found to be a characteristic of the adenylate cyclase systems of virtually all animal cells (see PERKINS 1973). Despite intense investigation, the underlying mechanisms are relatively poorly understood. A detailed discussion of recent advances in the dissection of the adenylate cyclase system which bear upon this issue are beyond the scope of this chapter. For the moment, it is only important to point out that this action of NaF greatly aided investigation of the structure of cyclic AMP (by increasing amounts of product available for purification) and of the distribution and purification of adenylate cyclase (by facilitating assay of the enzyme system even when hormone-responsiveness was absent).

It is necessary to digress from this narrative for a moment in order to point out another important serendipitous observation. Soon after the initial demonstration of hormone effects in liver homogenates, an attempt was made to devise a method for the bioassay of glucagon based on these observations (BERTHET et al. 1957). It was noted that the magnitude of the hormone response decreased after freezing and storage of homogenates. This decrease was associated with an increase in the rate of the phosphorylase kinase reaction observed in the absence of glucagon. One possibility was that phosphorylase phosphatase activity had decayed, providing less opposition to the kinase reaction. Since it was known that caffeine accelerated the phosphorylase phosphatase reaction (SUTHERLAND and WOSILAIT 1955), probably by an action upon the substrate protein, this methylxanthine was added in the hope that it would reduce the "blank" rate of phosphorylase accumulation. This hope was not realized; but caffeine substantially increased the effect of low concentrations of glucagon, shifting the dose-response relationship to the left by a factor of about three. It was soon found that caffeine increased the accumulation of active material in hormone or NaF-treated particulate preparations, as well as the sensitivity of liver supernatant fractions to the factor. It was therefore hypothesized that both soluble and insoluble cell fractions contained degradative enzymes sensitive to caffeine. As preparations of purified "factor" became available, extracts of various tissues were explored for their ability to inactivate it, and those derived from brain and heart seemed especially powerful. Even without establishing precise assay conditions, it was possible to work out a preliminary fractionation procedure that yielded preparations free of nucleotidase activity if cardiac tissue was used as a source (SUTHERLAND and RALL 1958). These preparations were used to demonstrate the quantitative conversion of cyclic AMP to 5′-AMP, which was important in defining both the structure of the nucleotide and the 3′,5′ cyclic ribonucleotide phosphodiesterase reaction. Eventually, the cardiac enzyme was examined in more detail (BUTCHER and SUTHERLAND 1962), and the widespread occurence of cyclic nucleotide phosphodiesterase activity as well as the inhibitory effects of methylxanthines upon both soluble and insoluble forms of the enzyme were confirmed and documented. Later studies of the brain enzyme provided the first indications of multiple entities catalyzing this reaction and of potential regulation of some of these entities by Ca^{++} ions and a protein factor that ultimately proved to be calmodulin (see APPLEMAN et al. 1973; also, Chaps. 6 and 7 of this volume).

With the aid of the methylxanthines, an assay procedure for adenylate cyclase activity was established based upon the accumulation of product as judged by its ability to accelerate the phosphorylase kinase reaction (SUTHERLAND et al. 1962). Adenylate cyclase activity was detected in virtually every mammalian tissue tested and in tissues from a variety of other vertebrate species. However, these early experiments failed to detect adenylate cyclase activity in higher plants and in bacteria (E. coli). Subsequent studies have failed to provide clear evidence for its presence in higher plants (see LIN 1974) but have demonstrated both adenylate cyclase and important functions of cyclic AMP in bacteria (see PETERKOFSKY 1976). The insolubility of adenylate cyclase was of course obvious from the beginning, but establishment of its principal location in the plasma membrane was more difficult and involved both demonstrating its relative absence from major organelles (e.g., mitochondria, nuclei, lysosomes, "microsomes") and its relative enrichment in fractions containing "markers" for plasma membranes (e.g., 5′-nucleotidase, Na^+-K^+-ATPase).

III. Discovery of Cyclic GMP and Guanylate Cyclase

Soon after the structure of cyclic AMP had been established, a number of purine and pyrimidine analogues of this compound were chemically synthesized by several different approaches. Included in this group was cyclic GMP. This compound, along with cyclic IMP and cyclic UMP, was found to be about two orders of magnitude less potent than cyclic AMP in stimulating the phosphorylase kinase reaction in crude liver extracts (RALL and SUTHERLAND 1962b). A few years later, an investigation of the urinary excretion products following administration of radioactive inorganic phosphate to rats led to the identification of both cyclic AMP and cyclic GMP as the principle phosphate-containing substances in the urine (ASHMAN et al. 1963). Despite the obvious possibility that cyclic GMP was enzymatically derived from GTP, about six years elapsed before the first reports on guanylate cyclase appeared (HARDMAN and SUTHERLAND 1969; SCHULTZ et al. 1969; WHITE and AURBACH 1969). This interval primarily reflects the difficulty in establishing sensitive and reliable assay procedures for cyclic GMP. While this delay may seem remarkable to the younger investigators, it should be pointed out that it was only during the same period of time (i.e., 1968–1970) that procedures providing usable alternatives to the phosphorylase activation assay for cyclic AMP and adenylate cyclase were first reported.

Cyclic GMP and guanylate cyclase are distributed in nature in a fashion similar to cyclic AMP and adenylate cyclase, including their apparent absence from higher plants. In most, but certainly not all systems, the levels of cyclic GMP are about an order of magnitude lower than those of cyclic AMP. In animal tissues, a substantial fraction of the guanylate cyclase activity is either soluble or readily dissociable from membranes and other insoluble fractions of homogenates (see MURAD et al. 1979 and Chap. 5 of this volume). While a number of hormones can increase cyclic GMP levels in a variety of tissues, there is no uncontested instance wherein these or other hormones have been shown to stimulate guanylate cyclase activity in a cell-free system in a "specific," receptor-mediated fashion. These observations, coupled with nearly universal finding that extracellular Ca^{++} ions are required for hormonal effects on cyclic GMP accumulation, have led to the belief that hormone-receptor interactions are events relatively far removed from those regulating guanylate cyclase. Certain oxidizing agents are capable of stimulating cyclic GMP accumulation both in cell-free and intact cell preparation, and it is possible that hormonal influence on guanylate cyclase activity is a consequence of increased phospholipid metabolism (MURAD et al. 1979). In any event, the apparent complexity of the regulation of guanylate cyclase activity and of cyclic GMP accumulation is one of several factors that have made it difficult to establish that cyclic GMP plays an important role in a given hormone or autacoid action, and hopefully the realization of this complexity will help remedy this situation.

C. Regulation of Cyclic AMP Accumulation in Tissues

I. Introduction

The correlation of cyclic AMP accumulation with other consequences of hormone application has been one important line of evidence in attempts to define a "second messenger" role for the nucleotide in a particular hormone-tissue interaction. This line has also included characterization of the participating receptors as pharmacologically identical, potentiation of hormonal responses by inhibitors of phosphodiesterase, and mimicing the hormonal responses by producing cyclic AMP accumulation through means other than those involving the hormone receptors in question. The usual working hypothesis generated by "positive" results from one or more aspects of this line of investigation has been that the hormone-receptor interaction leads to an increased rate of cyclic AMP synthesis sufficient to overcome the opposition of phosphodiesterase, which in turn leads to the accumulation of cyclic AMP sufficient to activate cyclic AMP-dependent protein kinase and to bring about phosphorylation (and altered function) of key regulatory proteins. The strategy for investigating the mechanisms responsible for a particular hormone-tissue interaction then becomes "simplified" into searching for the relevant phosphoproteins and defining their functional roles. Some of these phosphoproteins are considered to be relatively tissue-specific, reflecting some of the cellular components that characterize a given differentiated cell. On the other hand, the hormone-receptor mechanisms leading to cyclic AMP accumulation are thought to be less variable, a given differentiated cell being characterized principally by the expression of an array of hormone recognition sites. These views have been useful in structuring investigations in the past, and probably will continue to be useful in the future. However, these and other generalizations will undoubtedly require modification as various systems are analyzed in detail. In what follows, a number of issues will be raised that may indicate a need for such modifications. These discussions will be neither detailed or extensively documented, and the reader is refered to subsequent chapters for most of the pertinent literature citations.

II. Regulation of Cyclic AMP Formation

1. Basal and Hormone-Induced Accumulation of Cyclic AMP in Intact Cells

The ability of a hormone to increase the accumulation of cyclic AMP in an intact-cell preparation has often, but not always, paralleled detection of hormonal stimulation of adenylate cyclase activity in cell-free preparations. Adenylate cyclase preparations from neural tissue are notable for frequent failure to respond to relevant agents. These and similar observations in certain other tissues are perhaps explicable on the basis of the following: (1) instability of the enzyme; (2) subtraction of essential components during preparation and assay; and (3) comingling of stimulating substances originally segregated from adenylate cyclase which might eliminate any hormone effects. In a few instances, the reverse has been true, in that hormone-sensitive adenylate cyclase can readily be prepared from tissues in which even maximal concentrations of hormone fail to produce detectable increases in cyclic AMP levels without the assistance of phosphodiesterase inhibitors.

It has been observed with far greater frequency that concentrations of hormone producing substantial or even nearly maximal physiological effects fail to increase cyclic AMP accumulation significantly in intact cell preparations. In part, this may be due to the relatively high levels of cyclic AMP in tissues in the absence of hormones, usually between 0.2 and 1.0 μmole/kg wet weight. While these concentrations probably are compatible with the unstimulated state of cyclic AMP-dependent protein kinase in tissues under control conditions, it is not at all clear that they are generated by the balance between the "basal" rate of the adenylate cyclase reaction and phosphodiesterase activity. This problem is particularly dramatic in the case of brain tissue, in which the phosphodiesterase activity of homogenates (1 μM cyclic AMP, no calcium, no cyclic GMP) is of the order of 100 pmoles cyclic AMP hydrolyzed min/mg protein (SINGER and RALL, unpublished observations). A comparably great basal rate of adenylate cyclase activity is difficult to reconcile with the relatively modest effect of most inhibitors of phosphodiesterase upon the accumulation of cyclic AMP in brain slices (up to a 3- to 4-fold increase). Thus, in addition to being an analytical nuisance, the basal levels of cyclic AMP generate questions as to their origin, locale, and potential functional significance, particularly in view of the numerous examples of receptor-mediated reduction of adenylate cyclase activity (e.g., acetylcholine, prostaglandins, adenosine, α-adrenergic agonists). However, despite earlier speculations to the contrary, there is no convincing evidence that hormonal inhibition of adenylate cyclase leads to reduction of cyclic AMP accumulation or an appreciable physiological response in the absence of agents that would otherwise stimulate adenylate cyclase activity and produce other characteristic effects.

Other speculations have viewed basal cyclic AMP as a potential reservoir that could be mobilized as one consequence of hormone action; this action, combined with stimulation of adenylate cyclase, could lead to an increase in the amount of cyclic AMP available for activation of protein kinase and for degradation without necessarily resulting in net accumulation of the nucleotide. While this view is attractive, it is also very difficult to substantiate. One group of observations that might be taken as evidence for hormone-induced "mobilization" of cyclic AMP has to do with the ability of at least some animal cells or tissues to export the nucleotide when exposed to stimuli that lead to increased cyclic AMP formation. This phenomenon was first described in 1963 (DAVOREN and SUTHERLAND 1963), in pigeon erythrocytes treated with β-adrenergic agonists. Subsequently, it has been observed that prominent cyclic AMP export occurred in a wide variety of cells incubated as dispersed suspensions or cultured in monolayers (see CHLAPOWSKI et al. 1975). In addition, certain tissues, notably liver and kidney tubule epithelium, have been found to release cyclic AMP into the blood and/or urine in situ and in vivo when presented with appropriate hormonal stimuli. Such export may perform some kind of excretory function, serving to dispose of "excess" cyclic AMP and thereby participating in the regulation of cyclic AMP accumulation. However, there are a number of instances where increased export of cyclic AMP can be detected at concentrations of hormone which fail to produce significant increases in the intracellular accumulation of cyclic AMP. One instructive example of the latter is provided by cultured adrenocortical tumor cells that respond to very low concentrations of ACTH by an increased elaboration of certain steroids and by increased accumulation of cyclic AMP in the medium without a detectable rise in cellular levels of the nucleotide (SCHIMMER and ZIMMERMANN 1976). At least in this case, the export of cyclic AMP is the most sensitive index of altered cyclic AMP metabolism that can be correlated with other indices of hormone action. The functional significance as well as the underlying mechanisms of cyclic AMP export are only poorly understood; however, this phenomenon deserves greater consideration in investigations aimed at understanding the regulation of cyclic AMP accumulation and/or the mode of cyclic AMP action.

2. Effects of Guanyl Nucleotides and Cholera Toxin

During the past decade, several important clues have been unearthed which have led to marked advances in our understanding of the mechanisms involved in the regulation of adenylate cyclase activity by hormones. Two of the most significant findings, the stimulatory effect of guanyl nucleotides (RODBELL et al. 1971 a) or cholera toxin (SHARP and HYNIE 1971) on adenylate cyclase activity, were reported within a few months of each other. These observations developed into more or less independent lines of investigation which recently have converged dramatically (see GILL 1977; MAGUIRE et al. 1977). The guanyl nucleotide story emanated from attempts to reconcile the kinetic parameters of glucagon binding with the kinetics of the stimulation by glucagon of adenylate cyclase in membrane preparations from rat liver (RODBELL et al. 1971 b). The ATP contained in the adenylate cyclase reaction mixtures markedly accelerated the dissociation of glucagon, there by leading to a reduction in the apparent affinity constant of the hormone. Other nucleotides were also effective, but GTP was more than 1,000 times as potent than other nucleoside triphosphates. This effect has proven to have general validity; guanyl nucleotides reduce the apparent affinity of agonist ligands (but not of antagonist ligands) in a wide variety of systems (see MAGUIRE et al. 1977 and Chap. 3 of this volume). The effects of GTP on hormone binding led to the establishment of adenylate cyclase assay conditions with limited amounts of nucleotides and the elimination of contaminating guanyl nucleotides; under such assay conditions, hormonal stimulation of adenylate cyclase activity depended upon the presence of very low amounts of GTP (RODBELL et al. 1971 a). This observation has proven to have general validity; GTP is required for both the stimulation and inhibition of adenylate cyclase activity induced by hormone-receptor activation in every membrane system that has been carefully examined (see ROSS and GILMAN 1980).

Further understanding of the mechanisms involved in hormonal regulation of adenylate cyclase activity and the relationship of guanyl nucleotides thereto has been aided enormously by the generation of variant cell lines that are virtually incapable of synthesizing cyclic AMP (see JOHNSON et al. 1980). Cholera toxin was an important tool in these efforts. Examination of these variants has provided important evidence for the existance of at least three membrane proteins that comprise the adenylate cyclase system: (1) the hormone-binding component ("receptor"); (2) the catalytic component (capable of forming cyclic AMP from ATP in the presence of manganous ions); and (3) a coupling component ("G/F factor") (see ROSS and GILMAN 1980). The latter component: (1) permits the utilization of Mg-ATP as substrate for the catalytic component; (2) is required for the GTP-dependent effects of liganded receptor upon the catalytic component and for the effects of guanyl nucleotides upon the binding of agonists to the receptor; and (3) is required for the stimulatory effects of NaF upon cyclic AMP formation. The G/F factor is susceptible to apparently irreversible modification by nonhydrolyzable analogues of GTP leading to a state of persistant activation of the adenylate cyclase system in which activity is no longer subject to alteration by hormones, NaF, or treatment with cholera toxin. This component also appears to be an important target for cholera toxin, in that one or more of its subunits can be ADP-ribosylated by the active subunit of the toxin in the presence of NAD^+ (see ROSS and GILMAN 1980); one consequence of this action is the generation of a persistant state of activation of adenylate cyclase in the presence of GTP. In addition to providing new insights, the results of these and related investigations afford important opportunities for the examination of the functioning of adenylate cyclase and perhaps other membrane systems. Among the numerous questions to be answered in the future are: (1) whether the G/F factor has GTP-ase activity that can be abolished by ADP-ribosylation; (2) the mechanism of action of NaF and its relation to the actions of the guanyl nucleotides; (3) the relationship of the reversible hormone-induced activation of the adenylate cyclase system to the irreversible (or slowly reversible) states of activation induced by NaF, cholera toxin; or non-hydrolyzable analogues of GTP; (4) whether the G/F factor functions in the regulation of receptor-effector systems other than adenylate cyclase; (5) whether modification of G/F factor could be an important facet of the phenomenon of refractoriness in those adenylate cyclase systems where major alterations in receptor number or affinity do not occur; and (6) whether there are other membrane or cytosolic components that have essential or modulatory functions in the regulation of cyclic AMP formation and/or accumulation. There is one known example relating to the last question. In adenylate cyclase systems derived from neural tissue, calmodulin (calcium dependent regulatory protein) has been found to mediate prominent stimulatory and inhibitory effects of calcium ions at concentrations below 1 μM (see WOLFF and BROSTROM 1979; PIASCIK et al. 1980). The potential role of calcium ions in regulating the accumulation of cyclic AMP in brain tissue will be discussed in a later section.

III. Regulation of Cyclic AMP Degradation

Compared to adenylate cyclase, much less attention has been paid to the potential role of the cyclic nucleotide phosphodiesterase in the regulation of cyclic AMP accumulation in tissues. This probably results both from the many documented examples of hormonal regulation of adenylate cyclase that survive cell disruption and from the relatively limited avenues for regulation of phosphodiesterase activity that have been uncovered so far (see WELLS and HARDMAN 1977 and Chap. 7 of this volume). In a few cases, hormone treatment of cells has produced small increases in phosphodiesterase activity upon subsequent assay of homogenates or fractions therefrom, both by mechanisms which are cyclic AMP-dependent (e.g., S-49 lymphoma cells) and cyclic AMP-independent (e.g., insulin in adipocytes). However, in a wide variety of tissues, far more dramatic stimulation of phosphodiesterase activity can be achieved by calcium ions (plus calmodulin) or by cyclic GMP. In the former instance, a species of enzyme is involved that is characterized by a high capacity for hydrolysis of cyclic AMP, by a greater affinity for cyclic GMP than for cyclic AMP, and by a predominantly cytosolic localization. In the latter case, a species of enzyme is involved that is characterized by a marked preference for cyclic AMP and by complex kinetic properties that include allosteric activation by high concentrations of cyclic AMP or by low concentrations of cyclic GMP. In view of the low concentrations of cyclic GMP and of cytosolic calcium ions prevailing under basal conditions in most tissues, all of the examples outlined above would appear to provide potential mechanisms only for the attenuation of cyclic AMP accumulation under certain circumstances.

This state of affairs is another reason why attention usually is riveted on the adenylate cyclase system in attempts to understand how increases in cyclic AMP accumulation occur. However, this view does not provide adequate quantitative explanations for the amount of cyclic AMP accumulation occuring under many circumstances. The difficulty in accounting for basal levels of cyclic AMP was discussed above. It is also difficult to account for the very large accumulation of cyclic AMP that can be observed in some instances without the aid of inhibitors of phosphodiesterase. One instructive example is provided by guinea pig cerebral cortical tissue. Exposure of slices to a combination of histamine and adenosine leads to the accumulation of around 800 pmoles of cyclic AMP per mg protein within 5 min (SATTIN and RALL 1970); this would correspond to an average cytosolic concentration of between 100 and 200 μM. Assays of homogenates at 30° in the absence of calcium and cyclic GMP revealed the rate of cyclic AMP hydrolysis to be about 1 nmole/min/mg protein at 5 μM substrate, and nearly 20 nmoles/min/mg protein at 100 μM cyclic AMP (SINGER and RALL, unpublished observations). Assays of adenylate cyclase activity in homogenates of the same tissue yielded maximum values of about 0.6 nmole/min/mg protein at 30°; these values resulted from incubation in the presence of histamine and GMPP(NH)P. These data would require either that less than 5% of the adenylate cyclase activity was recovered in homogenates or that phosphodiesterase activity was largely occult in the intact cell system. The former possibility would also require a substantial turnover of ATP and the consumption of an unreasonable amount of metabolic energy. Thus, the latter alternative would appear to be the more likely. Only three categories of potential explanations for these data come to mind: (1) the various phosphodiesterase enzymes exist largely as inactive molecular species in the cell and are accidently activated during homogenization and assay; (2) there exist powerful unidentified negative effectors that are diluted and/or destroyed during preparation and assay; and (3) the product of the adenylate cyclase reaction is largely segregated from phosphodiesterases in some kind of compartment. Hopefully, clues relating to one or more of these possibilities will be forthcoming from future investigations.

IV. Potentiative Interactions and the Potential Role of Calcium

Another puzzling phenomenon related to the regulation of cyclic AMP accumulation involves the observation that under some circumstances the stimulatory effects produced by the combination of two dissimilar hormone agonists are two- to threefold greater than the sum of the effects produced by the agents acting individually. So far, this phenomenon has been found only using slices of brain tissue. Guinea pig cerebral cortical tissue displays an unusually large number of such interactions; these include interactions between adenosine and either norepinephrine or histamine (SATTIN and RALL 1970) and between the latter two biogenic amines (HUANG et al. 1971). Quantitatively, the interactions are actually more dramatic than stated above in view of the fact that the α_1-adrenergic and H_1-histaminergic receptors involved do not have any appreciable impact on cyclic AMP accumulation when adenosine on H_2 receptors are blocked (SATTIN et al. 1975; DALY et al. 1979; DALY et al. 1980). Interestingly, the interaction of norepinephrine and adenosine in slices of rat cerebral cortex can be abolished by the removal of free calcium ions from incubation media (SCHWABE and DALY 1977). This observation, combined with the known involvement of α_1 and H_1 receptors in the mobilization of calcium ions in other tissues and the calmodulin-mediated stimulatory effects of calcium ions on neural adenylate cyclase (see WOLFE and BROSTROM 1979), has led to the hypothesis that these receptors can indirectly regulate adenylate cyclase activity by the alteration of cytosolic calcium. Presumably, the activated state of the adenylate cyclase system would be required in order for calcium plus calmodulin to have any effect. While this is a very attractive hypothesis, there are a number of considerations which may limit its general applicability: (1) calcium deprivation has less drastic effects on potentiative interactions when adult guinea pig cerebral cortical tissue is used and has no effect in fetal tissue (RALL 1979); (2) increases in cytosolic calcium would also be expected to stimulate cyclic AMP degradation, especially at the levels of cyclic AMP generated; and (3) at least in guinea-pig cerebral cortical preparations, calmodulin appears to mediate effects of calcium on adenylate cyclase activity that are both stimulatory (at 0.1 μM) and inhibitory (at 0.3 μM) (PIASCIK et al. 1980). The extreme cellular heterogeneity of brain tissue will make it very difficult to gather definitive evidence either supporting or refuting this hypothesis, especially if the putative alterations in calcium concentration must be circumscribed in amount and/or location. Although many other alternate hypotheses can be constructed to explain the observations (e.g., calcium-dependent release of unknown mediators), no substantial evidence in their favor has been found. Thus, for the time being, potentiative interactions in brain tissue may be considered as an example of indirect regulation of cyclic AMP formation in a fashion not too dissimilar from the apparent regulation of cyclic GMP formation.

D. Concluding Remarks

Twenty years ago, it was the opinion of this writer (and of many others) that our understanding of the mechanism of action of cyclic AMP would progress far more rapidly than that of the mechanisms underlying the hormonal regulation of adenylate cyclase. This attitude was forged during frustrating months spent in a cold room watching adenylate cyclase activity perish and was strengthened by observations that straight-forward protein fractionation procedures could lead to preparations enriched in phosphorylase kinase activity which were still responsive to cyclic AMP. Initially, this prediction was quite true, and the decade of the 1960's saw the characterization of the cyclic AMP-dependent protein kinases and the definition of their role in the regulation of glycogen metabolism and lipolysis. In the past decade, the analysis of adenylate cyclase systems had progressed quite rapidly, especially during the last five years. Meanwhile, our understanding of how cyclic AMP functions in the regulation of numerous cell processes (e.g., steroidogenesis, gluconeogenesis, thyroglobulin secretion, salivary amylase secretion, etc.) has improved only marginally. To be sure, these processes are complex; but the earlier expectation that cyclic AMP would be a powerful tool in dissecting these processes has to a large extent not been fulfilled.

The example provided by the regulation of adrenal steroidogenesis by ACTH is especially puzzling and instructive. The accumulation of evidence for the involvement of cyclic AMP began even before the nucleotide was chemically identified (HAYNES 1958). At present, the case for the *necessary* participation of cyclic AMP and cyclic AMP-dependent protein kinase in this action of ACTH is probably stronger than for any other hormone-tissue interaction in that it includes genetic evidence provided by variant cell lines (see SCHIMMER 1980). Furthermore, a great deal of information on the enzymology of the conversion of cholesterol to corticosteroids is now available, and the rate-determining reactions have been identified. Nevertheless, it has not been possible to construct a cell-free system demonstrating an obviously relevant response to cyclic AMP, much less to delineate the sequence of events involved. While there could be a host of specific explanations for this state of affairs in this system or in other systems, there is one general possibility worth considering, especially as it might relate to some of the puzzling features of the regulation of cyclic AMP accumulation and export discussed above. The hypothesis would have to do with a possible requirement for the interaction of cyclic AMP with a second cellular component in addition to the cyclic AMP-dependent protein kinase. This putative component might be contained within the same compartment postulated to be involved in the segregation of cyclic AMP from phosphodiesterase and permitting accumulation of cyclic AMP under apparently unfavorable conditions. The efficacy of exogenous cyclic AMP would require that both the product of the adenylate cyclase reaction and exogenous cyclic AMP would have access to this compartment in the intact cell. The lack of efficacy of cyclic AMP in cell-free systems might reflect the inaccessibility and/or the instability of the putative component under the conditions used. As stated, this hypothesis may seem outrageous and no more plausible than countless others that could be formulated, but it does have the virtue of attempting to link together several mysterious phenomena. In any event, it is necessary to face up to the fact that even after more than twenty years of work by numerous investigators our understanding of how cyclic AMP accumulates and exerts its actions in most systems is very limited.

Acknowledgement. The author wishes to acknowledge the support by a grant from the USPHS (NS 12764).

References

Appleman MM, Thompson WJ, Russell TR (1973) Cyclic nucleotide phosphodiesterases. Adv Cyclic Nucleotide Res 3:65–98

Ashman DF, Lipton R, Melicow MM, Price TD (1963) Isolation of adenosine 3′,5′-monophosphate and guanosine 3′,5′-monophosphate from rat urine. Biochem biophys Res Commun 11:330–334

Berthet J, Sutherland EW, Rall TW (1957) The assay of glucagon and epinephrine with use of liver homogenates. J Biol Chem 229:351–361

Butcher RW, Sutherland EW (1962) Adenosine 3′,5′-phosphate in biological materials. I. Purification and properties of cyclic 3′,5′-nucleotide phosphodiesterase and the use of this enzyme to characterize adenosine 3′,5′-phosphate in human urine. J Biol Chem 237:1244–1250

Chlapowski FJ, Kelly LA, Butcher RW (1975) Cyclic nucleotides in cultured cells. Adv Cyclic Nucleotide Res 6:245–338

Cook WH, Lipkin D, Markham R (1957) The formation of cyclic dianhydrodiadenylic acid by the alkaline degradation of adenosine-5′-triphosphoric acid. J Am Chem Soc 79:3607–3608

Daly JW, McNeal ET, Creveling CR (1979) Accumulation of cyclic AMP in brain tissue: Role of H_1- and H_2-histamine receptors. In: Histamine receptors. Spectrum Publications, Jamaica, New York pp 229–323

Daly JW, Padgett W, Nimitkitpaisan Y, Creveling CR, Cantacuzene D, Kirk KL (1980) Fluoronorepinephrines: Specific agonists for the activation of *alpha* and *beta* adrenergic-sensitive cyclic AMP-generating systems in brain slices. J Pharmacol Exp Ther 212:382–389

Davoren PR, Sutherland EW (1963) The cellular location of adenyl cyclase in the pigeon erythrocyte. J Biol Chem 238:3016–3023

Fischer EH, Krebs EG (1955) Conversion of phosphorylase *b* to phosphorylase *a* in muscle extracts. J Biol Chem 216:121–132

Gill DM (1977) Mechanism of action of cholera toxin. Adv Cyclic Nucleotide Res 8:85–118

Hardman JG, Sutherland EW (1969) Guanyl cyclase, an enzyme catalyzing the formation of guanosine 3′,5′-monophosphate from guanosine triphosphate. J Biol Chem 244:6363–6370

Haynes RC (1958) The activation of adrenal phosphorylase by the adrenocorticotropic hormone. J Biol Chem 233:1220–1222

Huang M, Shimizu H, Daly J (1971) Regulation of adenosine cyclic 3′,5′-phosphate formation in cerebral cortical slices: Interaction among norepinephrine, histamine, and serotonin. Mol Pharmacol 7:155–162

Jardetzky CD (1962) Proton magnetic resonance of nucleotides IV. Ribose conformation. J Am Chem Soc 84:62–66

Johnson GL, Kaslow HR, Farfel Z, Bourne HR (1980) Genetic analysis of hormone-sensitive adenylate cyclase. Adv Cyclic Nucleotide Res 13:1–37

Langan TA (1973) Protein kinases and protein kinase substrates. Adv Cyclic Nucleotide Res 3:99–153

Lin PP-C (1974) Cyclic nucleotides in higher plants? Adv Cyclic Nucleotide Res 4:439–461

Lipkin D, Cook WH, Markham R (1959) Adenosine-3′:5′-phosphoric acid: A proof of structure. J Am Chem Soc 81:6198–6203

Maguire ME, Ross EM, Gilman AG (1977) β-Adrenergic receptor: Ligand binding properties and the interaction with adenylyl cyclase. Adv Cyclic Nucleotide Res 8:1–83

Murad F, Arnold WP, Mittal CK, Braughler JM (1979) Properties and regulation of guanylate cyclase and some proposed functions for cyclic GMP. Adv Cyclic Nucleotide Res 11:175–204

Perkins JP (1973) Adenyl cyclase. Adv Cyclic Nucleotide Res 3:1–64

Peterkofsky A (1976) Cyclic nucleotides in bacteria. Adv Cyclic Nucleotide Res 7:1–48

Piascik MT, Wisler PL, Johnson CL, Potter JD (1980) Ca^{2+}-dependent regulation of guinea pig brain adenylate cyclase. J Biol Chem 255:4176–4181

Rall TW, Sutherland EW, Wosilait WD (1956) The relationship of epinephrine and glucagon to liver phosphorylase. III. Reactivation of liver phosphorylase in slices and in extracts. J Biol Chem 218:483–495

Rall TW, Sutherland EW, Berthet J (1957) The relationship of epinephrine and glucagon to liver phosphorylase. IV. The effect of epinephrine and glucagon on the reactivation of phosphorylase in liver homogenates. J Biol Chem 224:463–475

Rall TW, Sutherland EW (1958) Formation of a cyclic adenine ribonucleotide by tissue particles. J Biol Chem 232:1065–1076

Rall TW, Sutherland EW (1962a) Enzymes concerned with interconversion of liver phosphorylases. In: Methods in enzymology, vol. V. Academic Press, New York pp 217–226

Rall TW, Sutherland EW (1962b) Adenyl cyclase. II. The enzymatically catalyzed formation of adenosine 3′,5′-phosphate and inorganic pyrophosphate from adenosine triphosphate. J Biol Chem 237:1228–1232

Rall TW (1979) Regulation of cyclic adenosine monophosphate accumulation in brain tissue: Interactions of adenosine with other agonists. In: Physiological and regulatory functions of adenosine and adenine nucleotides. Raven Press, New York pp 217–226

Rodbell M, Birnbaumer L, Pohl SL, Krans HMJ (1971a) The glucagon-sensitive adenyl cyclase system in plasma membrane of rat liver. V. An obligatory role of guanyl nucleotides in glucagon action. J Biol Chem 246:1877–1882
Rodbell M, Krans HMJ, Pohl SL, Birnbaumer L (1971b) The glucagon-sensitive adenyl cyclase system in plasma membranes of rat liver. IV. Effects of guanyl nucleotides on binding of ^{125}I-glucagon. J Biol Chem 246:1872–1876
Ross EM, Gilman AG (1980) Biochemical properties of hormone-sensitive adenylate cyclase. Ann Rev Biochem 49:533–564
Sattin A, Rall TW (1970) The effect of adenosine and adenine nucleotide on the cyclic adenosine 3′,5′-phosphate content of guinea pig cerebral cortex slices. Mol Pharmacol 6:13–23
Sattin A, Rall TW, Zanella J (1975) Regulation of cyclic adenosine 3′,5′-monophosphate levels in guinea pig cerebral cortex by interaction of *alpha* adrenergic and adenosine receptor activity. J Pharmacol Exp Ther 192:22–32
Schimmer BP, Zimmerman AE (1976) Steroidogenesis and extracellular cAMP accumulation in adrenal tumor cell cultures. Mol Cell Endocrinol 4:263–270
Schimmer BP (1980) Cyclic nucleotides in hormonal regulation of adrenocortical function. Adv Cyclic Nucleotide Res 13:181–214
Schultz G, Böhme E, Munske K (1969) Guanyl cyclase. Determination of enzyme activity. Life Sci 8:1323–1332
Schwabe V, Daly JW (1977) The role of calcium ions in accumulations of cyclic AMP elicited by α- and β-adrenergic agonists in rat brain slices. J Pharmacol Exp Ther 202:134–143
Sharp GWG, Hynie S (1971) Stimulation of intestinal adenyl cyclase by cholera toxin. Nature 229:266–269
Sutherland EW, Cori CF (1951) Effect of hyperglycemic-glycogenolytic factor and epinephrine on liver phosphorylase. J Biol Chem 188:531–543
Sutherland EW, Wosilait WD (1955) Inactivation and activation of liver phosphorylase. Nature 175:169–170
Sutherland EW, Rall TW (1957) The properties of an adenine ribonucleotide produced with cellular particles, ATP, Mg^{++} and epinephrine or glucagon. J Am Chem Soc 79:3608
Sutherland EW, Rall TW (1958) Fractionation and characterization of a cyclic adenine ribonucleotide formed by tissue particles. J Biol Chem 232:1077–1091
Sutherland EW, Rall TW, Menon T (1962) Adenyl cyclase. I. Distribution, preparation, and properties. J Biol Chem 237:1220–1227
Wells JN, Hardman JG (1977) Cyclic nucleotide phosphodiesterases. Adv Cyclic Nucleotide Res 8:119–143
White AA, Aurbach GD (1969) Detection of guanyl cyclase in mammalian tissues. Biochim Biophys Acta 191:686–697
Wolff DJ, Brostrom CO (1979) Properties and functions of the calcium-dependent regulator protein. Adv Cyclic Nucleotide Res 11:27–88

CHAPTER 2

Chemistry of Cyclic Nucleotides and Cyclic Nucleotide Analogs

G. R. REVANKAR and R. K. ROBINS

Overview

Since the isolation and characterization of *c*AMP in early 1957, several enzymatic and chemical synthetic procedures have been developed to provide *c*AMP and various cyclic nucleotide analogs. The synthesis of *c*AMP and 2′-deoxy-*c*AMP by bacterial fermentation is of current commercial interest as a low cost source of nucleoside 3′,5′-cyclic phosphates. In this chapter, the improvements made over the years on the original synthesis of *c*AMP (*via* DCC mediated cyclization of AMP) by employing different coupling reagents, transesterification of "active" phosphates of AMP and isolation techniques are reviewed. By far the largest number of cyclic nuceotide derivatives have been prepared by the chemical transformation of parent *c*AMP and *c*GMP itself. A substantial number of *c*AMP derivates have been synthesized in which there are substituent modifications of the purine base, and structural modifications of the carbohydrate moiety and the cyclic phosphate moiety. Representative examples of cyclic nucleotide analogs related to *c*AMP, such as 1-deazapurine, 3-deazapurine, 7-deazapurine, 2-azapurine, 8-azapurine, formycin, and 1,N^6-ethenoadenosine cyclic phosphates, are treated in some detail. Description of some of the more important methods currently being used for the preparation of cyclic nucleotides related to *c*GMP and pyrimidine cyclic nucleotides are presented. Detailed information about the hydrolytic and spectral properties of cyclic nucleotides are also included. Due to the recent availability of an enormous number of synthetic cyclic nucleotide analogs, considerable success has been achieved in obtaining potent phosphodiesterase (PDE) inhibitors with greater tissue specificity. We now have cyclic nucleotides which are more resistant toward cellular PDE than the parent *c*AMP and, consequently, survive intracellularly long enough to provide an increased physiological response. An effort has been made, in the present chapter, to show that future cyclic nucleotide chemistry has the potential to provide highly potent cyclic nucleotide derivatives, which may have a significant effect on uncontrolled cellular proliferation, the immune response, asthma, the central nervous system, gastrointestinal function and other physiological responses characteristic of the natural cyclic nucleotides.

A. Introduction to the Chemistry of Cyclic Nucleotides

The present chapter is essentially an update of our earlier effort to organize the published reports on the chemistry of nucleoside cyclic phosphates (SIMON et al.

1973). Considerable material has appeared since 1973 and an attempt has been made to cover the literature until April 1980. An effort has been made to emphasize the chemistry and biochemistry which should be of interest to those working in the field of cyclic nucleotides in general.

The potential direct utilization of certain cyclic nucleotide derivatives as medicinal agents is especially emphasized. Although great hopes were expressed by many in the early days of cyclic nucleotide research for applying this new knowledge to major medical problems, progress in this direction has been rather slow. The major approach by medicinal chemists has been to search for new and more potent phosphodiesterase (PDE) inhibitors with greater tissue specificity. Our Laboratory and others have concentrated on the chemical syntheses of derivatives and analogs of adenosine 3′,5′-cyclic phosphate (*c*AMP) and guanosine 3′,5′-cyclic phosphate (*c*GMP) [for a review, see MILLER and ROBINS 1976], with a view of gaining greater tissue specificity and longer lasting potency over either *c*AMP or *c*GMP. Considerable success has been achieved in this effort and it is now clear that a cyclic nucleotide may be designed to be resistant toward cellular phosphodiesterases which will allow the cyclic nucleotide analog to survive intracellularly long enough to provide the desired physiological response. One of the more interesting derivatives of *c*AMP prepared in our Laboratory (MILLER et al. 1973a; MILLER et al. 1978) is 8-p-chlorophenylthio-*c*AMP (1) which showed a stimulation of bovine brain *c*AMP

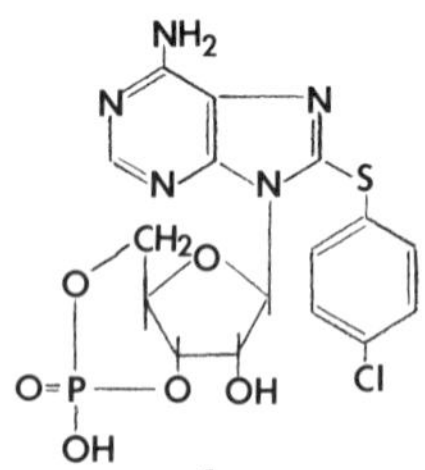

1

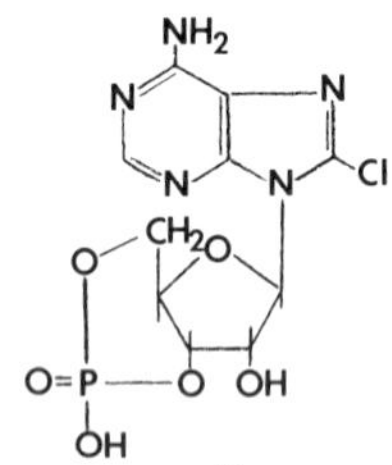

2

dependent protein kinase with an apparent $K_{a'}$ of 18 μM as compared to 1 μM for *c*AMP (MILLER et al. 1973a). 8-p-Chlorophenylthio-*c*AMP (1) is 50 times more potent than dibutyryl-*c*AMP as an inducer of tyrosine aminotransferase in rat liver (MILLER et al. 1975a). Compound (1) inhibited the spontaneous firing of 92% of the rat cerebellar Purkinje cells (SIGGINS and HENRIKSEN 1975). HALL et al. (1977) have recently reported that (1) mimics the effect of vasopressin in mammalian kidney tissue and is 100 times more potent in this system than *c*AMP, probably due to its resistance toward phosphodiesterase and its greater permeability across cellular membranes. More recently, STADEL and GOODMAN (1978) have shown that (1) mimics ADH (antidiuretic hormone) in the toad urinary bladder. The hydroosmotic response is equal in magnitude to oxytocin. Thus, the in vivo potency of (1) correlates well with the high degree of activation of *c*AMP dependent protein kinase (MILLER et al. 1973a).

8-Chloro-*c*AMP (2) (MUNEYAMA et al. 1974) has recently been shown to mimic the adrenal steroidogenic action of ACTH in vivo (FREE and PAIK 1977). Compound (2) elicited the secretion of corticosterone at normal plasma levels in both betamethasone-treated and hypophysectomized rats.

Early studies from our Laboratory (MUNEYAMA et al. 1971) indicated that substitution at the 8-position in *c*AMP usually increased activation with protein kinase and prevented degradation with PDE. Also, studies involving alkyl substitution at the N^6-amino group of *c*AMP gave products which possessed greater potency than *c*AMP (MEYER Jr et al. 1972; MILLER et al. 1978) but were readily cleaved by PDE. In an effort to prepare *c*AMP derivatives with superior potency and stability, a series of N^6-substituted 8-alkylthio-*c*AMP analogs were prepared in several steps from *c*AMP (MILLER et al. 1978; BOSWELL et al. 1975; ROBINS et al. 1977b). One of the most interesting compounds which emerged from this study is 6-*n*-butylamino-8-benzylthio-9-(*β*-D-ribofuranosyl)purine 3′,5′-cyclic phosphate (3). Com-

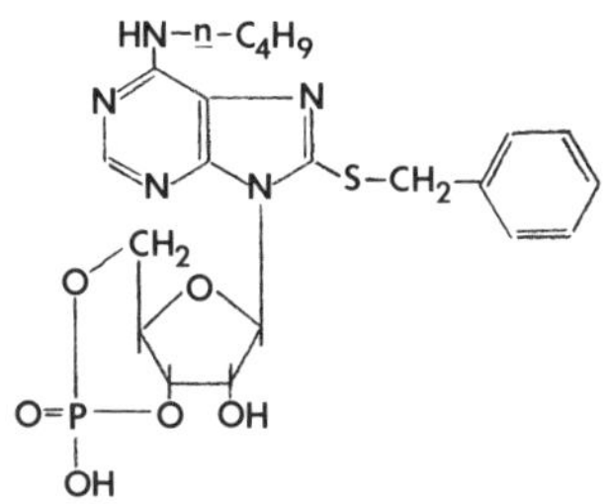

3

pound (3) has an apparent $K_{a'}$ of 1.8 μ*M* (bovine brain protein kinase) and is not degraded to any appreciable extent by PDE (MILLER et al. 1978). Compound (3) increases contractile force of excised cat ventricular papillary muscle and shows considerable stimulation of myocardial contractility in the anesthetized dog (EVANS et al. 1976). There is an increase in coronary blood flow and very little increase in heart rate. These effects were accompanied by some increase in blood glucose in vivo. Compound (3) showed an increase in mesenteric, renal and carotid arterial blood flow in dogs at doses as low as 1.0 mg/kg i.v. The positive inotropic effect was maintained in dogs *after* beta-receptor blockage (EVANS et al. 1976). IMAI and co-workers (1974) have shown that in vitro dibutyryl *c*AMP produced positive inotropic effects associated with an increase in the coronary flow. It was concluded by these workers (IMAI et al. 1974) that dibutyryl-*c*AMP passed the surface membrane and exerted its effect after being converted to *c*AMP within the cell. Thus, like *c*AMP, (3) acts within the cell unlike the beta-receptor agonists, which act on the cell membrane, and therefore (3) is effective even in the presence of beta-receptor blockade. Indeed, in a recent review of cyclic nucleotides and heart function, NAYLER (1977) points out that the beta-receptor on the cell surface is activated by catecholamines which results in the activation of adenyl cyclase and the formation of *c*AMP which, in turn, activates a protein kinase thereby resulting in the inotropic response within the cardiac cell. Compound (3) acts *directly* to stimulate protein kinase and give the desired inotropic response. Intravenous doses of 1 mg/kg of (3) produced a positive inotropic effect in the heart of dogs compromised by the intracoronary injection of microspheres. The contractility of these hearts was increased to levels that existed prior to insult of the myocardium. All these features make (3) an attractive candidate for the management of acute clinical myocardial failure (MILLER et al. 1980a).

The synthesis of 8-bromo-*c*GMP (4) was reported from our Laboratory (MILLER et al. 1973a) by direct bromination of *c*GMP. Compound (4) has been shown by Van Winkle of our Laboratory (MILLER 1977) to reduce the cutaneous anaphylactic response by 85% in rats at 200 mg/kg. Similarly, (4) protected guinea pigs against collapse from histamine-induced bronchoconstriction (MILLER 1977). SZADUYKIS-SZADURSKI and BERTI (1972) have reported that 8-bromo-*c*GMP (4) shows a greater relaxing activity on guinea pig trachea than theophylline, in vitro and in vivo. NAWRATH (1977, 1978) has shown that (4) reduced the force of contraction

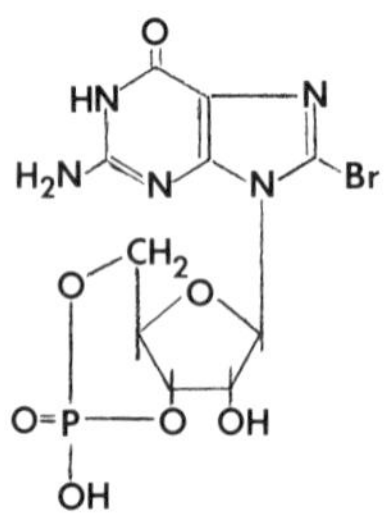

4

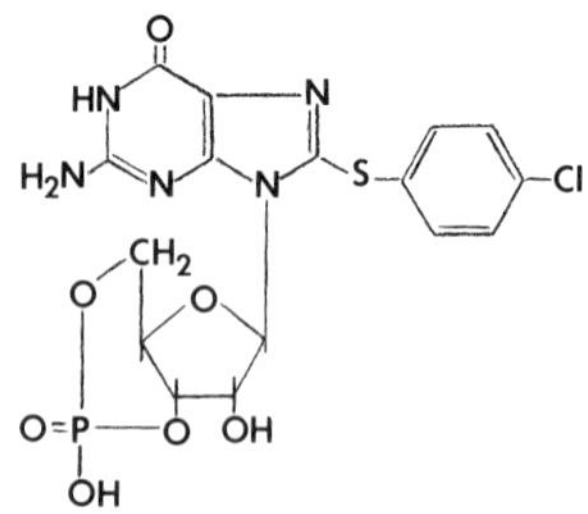

5

in the rat auricle to 40% of the control. 8-Bromo-*c*GMP also exerted a dose-dependent *negative* inotropic effect on isolated guinea pig auricles. NAWRATH (1977) postulates that *c*AMP and *c*GMP exert opposite effects on myocardial contraction. The apparent Ka′ for (4) activation of *c*GMP-dependent protein kinase from lobster muscle is 4.0 μ*M* as compared to 1.0 μ*M* for *c*GMP (MILLER et al. 1973a). Similarly 8-p-chlorophenylthio-*c*GMP (5) gives a value of 5.2 μ*M*, the most potent of the *c*GMP analogs tested so far (MILLER et al. 1973a). KNEER and co-workers (1974) have shown that (5) has the greatest activity among several *c*GMP derivatives in mimicking the action of epinephrine in stimulating gluconeogenesis in rat hepatocytes.

The isolation of cytidine 3′,5′-cyclic phosphate (*c*CMP, 6) by BLOCH (1974) from extracts of L-1210 cells and the subsequent growth initiation shown by (6) in vitro (BLOCH et al. 1974) has stimulated considerable interest in 3′,5′-cyclic nucleotides related to (6). It has recently been demonstrated that a cytidylate cyclase in mammalian cells converts CTP to *c*CMP. A phosphodiesterase has recently been

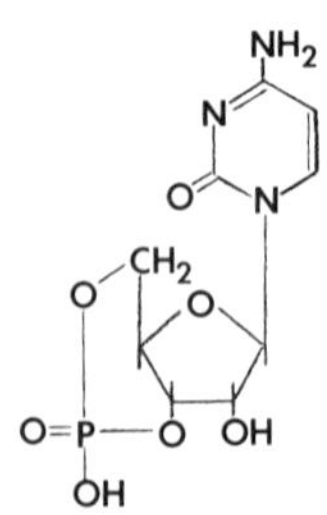

6

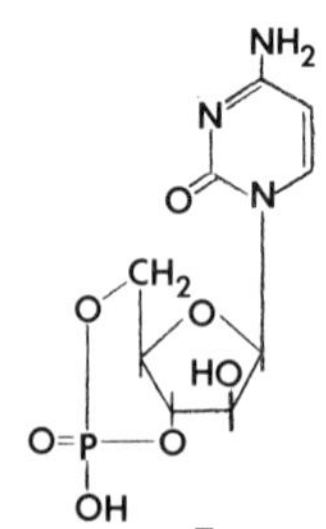

7

isolated from rat liver (KUO et al. 1978) and L-1210 cells (CHENG and BLOCH 1978) which preferentially hydrolyzes (6) to CMP. We reported on the synthesis of *ara*-*c*CMP (7) in 1972 from 1-(β-D-arabinofuranosyl)cytosine 3′-phosphate (LONG et al. 1972). In vitro studies showed (LONG et al. 1972) that (7) was hydrolyzed slowly to *ara*-CMP. The antileukemic activity of (7) suggests its use as a mechanism to get *ara*-CMP intracellularly as a precursor to *ara*-CTP, the active form of the clinically useful antileukemic agent *ara*-C. Indeed (7) should be active against *ara*-C-resistant leukemia strains, since (7) is not deaminated (SIDWELL et al. 1973) as is *ara*-C and could prove to be a much more useful form of the drug. Compound (7) has been shown (SIDWELL et al. 1973) to be considerably more effective than *ara*-C as an antiviral agent against herpes in the rabbit eye and vaccinia encephalitis infections in mice. It is now abundantly clear that future cyclic nucleotide chemistry will provide novel and potent cyclic nucleotide derivatives which will have a significant effect on uncontrolled cellular proliferation, the immune response, asthma, the central nervous system, gastrointestinal function and many other significant modern medical problems.

B. Synthesis of Cyclic Nucleotides and Cyclic Nucleotide Analogs

I. Cyclic Nucleotides Related to Adenosine 3′,5′-Cyclic Phosphate

In 1957, RALL, SUTHERLAND, and BERTHET published the first scientific paper on the discovery of the formation of a heat-stable nucleotide by tissue particles which mediates the glycolytic action of the peptide hormone glucagon. This heat-stable nucleotide (SUTHERLAND and RALL 1957; RALL and SUTHERLAND 1958), which is distributed very widely in nature, was subsequently shown to be identical to an adenine ribonucleotide isolated in low yield (ca. 5%–10%) from the barium hydroxide hydrolysis of adenosine 5′-triphosphate (8) (SUTHERLAND and RALL 1957; COOK et al. 1957). The adenine ribonucleotide was first assigned a cyclic dinucleotide structure (COOK et al. 1957; RALL and SUTHERLAND 1958) which was subsequently corrected to the monomeric cyclic phosphate, adenosine 3′,5′-cyclic phosphate (9, *c*AMP) (SUTHERLAND and RALL 1958; LIPKIN et al. 1959a, b).

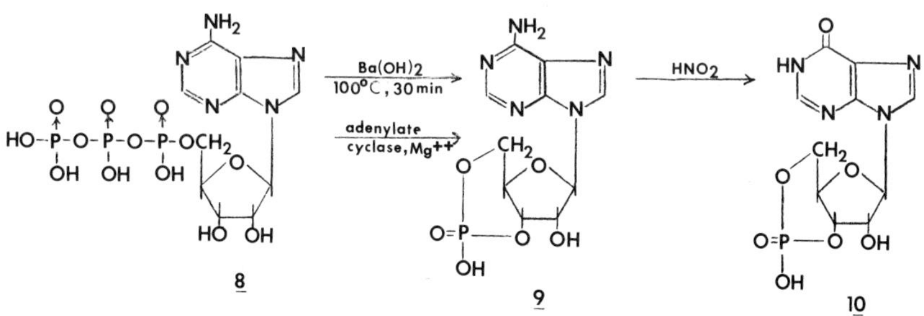

*c*AMP, which is the second messenger of hormone action (ROBISON and SUTHERLAND 1967; SUTHERLAND et al. 1968; ROBISON 1970), occurs in milk (KOBATA et al. 1961), urine (BUTCHER and SUTHERLAND 1962; HARDMAN et al. 1966), and most mammalian tissues. However, it does not accur in mammalian red cells, but does occur in avian erythrocytes (DAVOREN et al. 1963) and human platelets (ROBISON et al. 1969).

Deamination of *c*AMP with aqueous nitrous acid or barium hydroxide hydrolysis of inosine 5′-triphosphate readily gave inosine 3′,5′-cyclic phosphate (10, *c*IMP) (Lipkin et al. 1959a; Meyer Jr et al. 1972). The enzymatic preparation of adenosine-, inosine-, guanosine-, 2′-deoxyadenosine- and 2′-deoxyguanosine 3′,5′-cyclic phosphates from the corresponding nucleoside 5′-triphosphates is well recognized (Hardman et al. 1971; Ishiyama et al. 1974; Garbers et al. 1975; Sasaki et al. 1976). The synthesis of *c*AMP and 2′-deoxyadenosine 3′,5′-cyclic phosphate by bacterial fermentation is of current commercial interest as a low cost source of nucleoside 3′,5′-cyclic phosphate (Kikkoman Shoyu Co. 1971a, b; Ishiyama 1976).

An elegant improvement in the chemical synthesis of *c*AMP was achieved (Smith et al. 1961; Drummond et al. 1964) by the cyclization of adenosine 5′-monophosphate (11) in refluxing pyridine solution in the presence of N,N′-dicyclohexylcarbodiimide (DCC) as a dehydrating agent to form the new 3′,5′-phosphodiester linkage. This technique has been very successfully used by Khorana's group in the synthesis of phosphodiesters. However, for the intranucleotide linkage of the cyclic phosphates it is necessary to perform the reaction under high dilution conditions. Thus, only small-scale preparation of the desired 3′,5′-cyclic nucleotide derivatives is feasible. Subsequently, several alternative procedures (Borden and Smith 1966b; Marumoto et al. 1975), coupling reagents (e.g. triphenylphosphine and 2,2′-dipyridyl disulfide, TPP-PDS) (Mukaiyama and Hashimoto 1972), and isolation techniques (Rosett et al. 1970; Woods and Waitzman 1970; Kura and Ohashi 1971) have been reported for the preparation of 3′,5′-cyclic nucleotides from the corresponding 5′-phosphates. Cyclohexyl isocyanate (CIC) has also been used as a dehydrating agent for the preparation of *c*AMP; however, prior protection of the exocyclic amino function was required to prevent the formation of an N^6-cyclohexylcarbamoyl derivative (Naito and Sano 1965). Ynamines of type (12)

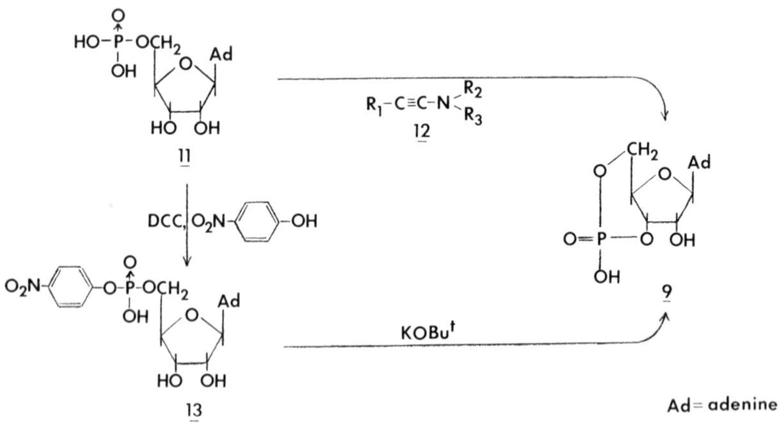

have also been used as dehydrating agents at a relatively lower solvent dilution than that used in the DCC or CIC procedures for the synthesis of *c*AMP (Fujimoto and Naruse 1968). Base-catalyzed transesterification of "active" phosphates of adenosine 5′-phosphate, e.g. adenosine 5′-(p-nitrophenyl)-phosphate (13) or a phosphotriester (van Boom et al. 1973, 1974, 1975, 1976), readily occurred in anhydrous dimethylsulfoxide at room temperature in high yields to give *c*AMP (Borden and Smith 1966 b). A similar active phosphate procedure for the synthesis of 3′,5′-cyclic phosphates (16) has been described (Eto et al. 1974) in which ribonucleoside 5′-methylthiophosphorothiolates (14) were selectively synthesized directly from borate complexes of ribonucleosides and 2-methylthio-4H-1,3,2-benzodioxaphosphorin-2-oxide in the presence of cyclohexylamine (Eto et al. 1978).

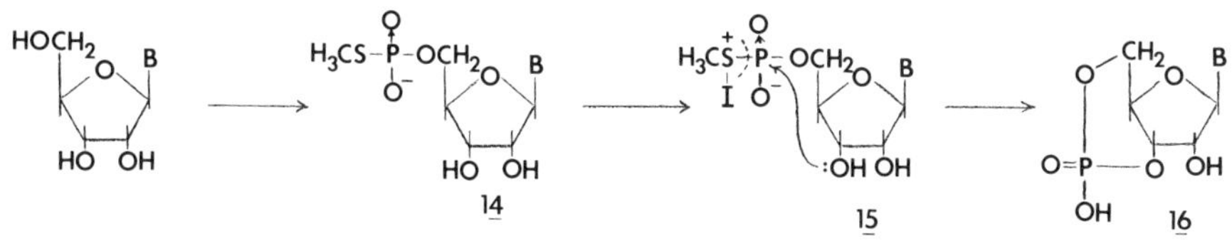

B = adenine, guanine, hypoxanthine, uracil, cytosine

Under anhydrous conditions, the phosphorothiolates (14) were cyclized to give ribonucleoside 3′,5′-cyclic phosphates (16) by iodine oxidation. Several other active phosphates for the synthesis of nucleoside 3′,5′-cyclic phosphates have been reported recently (Taguchi and Mushika 1975; Mushika and Taguchi 1976). The condensation of unprotected nucleosides with 2-(N,N-dimethylamino)-4-nitrophenyl phosphate in the presence of DCC in dimethylformamide gave the intermediate nucleoside-5′-2-(N,N-dimethylamino)-4-nitrophenyl phosphate (17) in good yields

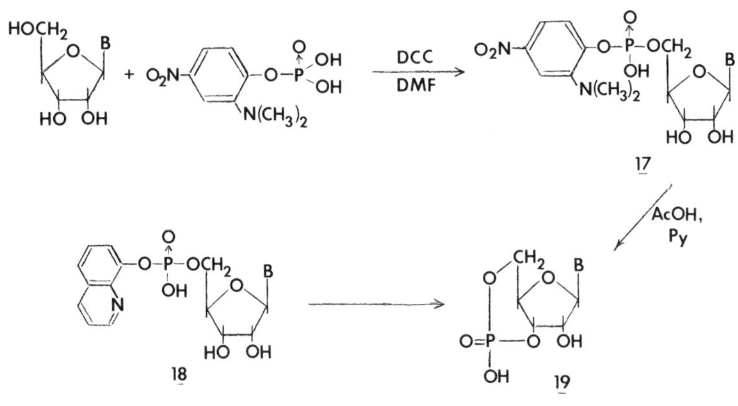

(Taguchi et al. 1976), which on treatment with acetic acid in boiling pyridine under high dilution conditions gave nucleoside 3′,5′-cyclic phosphates (19) in fairly good yields. A small amount of nucleoside 2′,3′-cyclic phosphate is also obtained by the reaction in dimethylformamide at 100 °C (Taguchi and Mushika 1975). Intramolecular cyclization of 8-quinolyl nucleoside 5′-phosphate (18) with cupric chloride in pyridine has recently been reported (Takaku et al. 1978) to give nucleoside 3′,5′-cyclic phosphate (19). Although the DCC and the "activated" ester methods are used extensively in the synthesis of nucleoside 3′,5′-cyclic phosphates, by far the largest number of analogs have been prepared by the chemical transformation of commercially available parent *c*AMP or *c*GMP. The DCC/pyridine ring closure procedure has been used for the preparation of ^{32}P-*c*AMP from ^{32}P-AMP (Symons 1970, 1973, 1974).

Adenosine 3′,5′-cyclic phosphate (cAMP) acts not only as a second messenger for the effects of almost all hormones (SUTHERLAND and ROBISON 1966), but also, together with the other natural cyclic nucleotide, guanosine 3′,5′-cyclic phosphate (cGMP), appears to be intimately involved in the regulation of nearly all facets of cellular activity (ROBISON et al. 1968, 1971 a, b; ROBISON and SUTHERLAND 1971; HARDMAN et al. 1971; ROBISON 1972; GREENGARD et al. 1972). Abnormal levels of cAMP and/or cGMP are implicated (ROBISON 1975) with a number of disease states, notably cancer (uncontrolled cellular proliferation), hypertension, asthma, cholera, diabetes (ROBISON and PARK 1970), a large number of endocrinopathies, diuresis, gastrointestinal function, and possibly aging (AMER and MCKINNEY 1973). There is a unique opportunity, therefore, for the medicinal and biochemist to design and study the interaction of synthetic cyclic nucleotide analogs as potential medicinal agents. These synthetic analogs provide excellent molecular probes to study in-depth the allosteric requirements for activation of a protein kinase of a given cell or tissue which in turn phosphorylates certain cellular proteins which give rise to a specific physiological response. Today we have a much clearer picture as to how such cyclic nucleotide analogs can be designed to transport across cellular membranes, how they may be designed to resist enzymatic degradation by phosphodiesterase and how they may be made to bind more tightly to the allosteric site of a given protein kinase.

The use of cyclic nucleotide analogs and other agents that act by way of systems that utilize cyclic nucleotides for the above mentioned disease states have recently been reviewed (AMER and MCKINNEY 1973, 1974, 1975). Further guides for the study of possible therapeutic uses of such cyclic nucleotide analogs may be found in a series of reviews on cyclic nucleotides in disease states (WEISS 1975) and on clinical cyclic nucleotide research (MURAD 1973; BROADUS 1977). Some chemical aspects of cyclic nucleotides have also been recently reviewed (POSTERNAK 1971; SIMON et al. 1973; MEYER Jr and MILLER 1974; MILLER and ROBINS 1976; MILLER 1977; MEYER Jr 1979; ROBINS 1979).

1. Substituent Modification of the Purine Base

a) N^1-Substituted Adenosine 3′,5′-Cyclic Phosphate Derivatives

The synthesis of adenosine 3′,5′-cyclic phosphate substituted at N^1-position has been achieved starting from cAMP itself. Treatment of cAMP with an ethereal so-

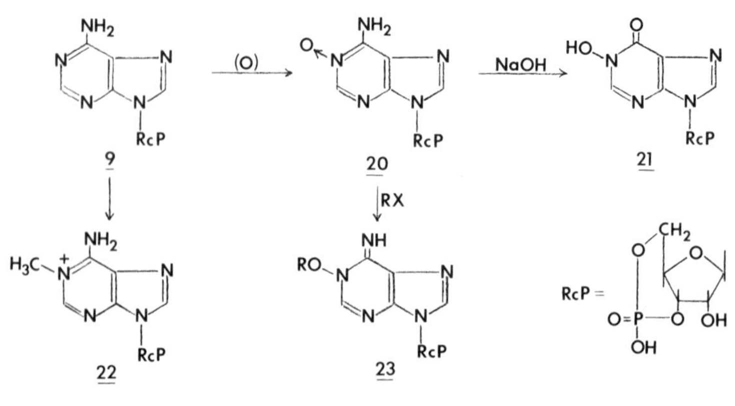

lution of perphthalic acid in phosphate buffer (POSTERNAK et al. 1962; FALBRIARD et al. 1967; JASTORFF and FREIST 1972) or m-chloroperbenzoic acid in a buffered two-phase system (MEYER Jr et al. 1973b; SHUMAN and MEYER Jr 1975) gave crystalline *c*AMP-N^1-oxide (20). Alkaline or nitrous acid deamination of (20) provided inosine 3′,5′-cyclic phosphate N^1-oxide (*c*IMP-N^1-oxide, 21) (MEYER Jr et al. 1973b). Alkylation of (20) with various alkylhalides gave the corresponding N^1-alkoxy-*c*AMP derivatives (23). The alkylation was conveniently performed in dimethyl sulfoxide solution using the soluble 1,5-diazabicyclo[5.4.0]-5-undecene (DBU) salt of (20). N^1-Methyladenosine 3′,5′-cyclic phosphate (22) was readily obtained from the DBU salt of *c*AMP and methyliodide in dimethyl sulfoxide (MEYER Jr et al. 1973b).

b) 2-Substituted Adenosine 3′,5′-Cyclic Phosphate Derivatives

Introduction of a substituent into the 2-position of a preformed purine cyclic nucleotide presented a synthetic challenge. Presently three methods are available. In the first method (MEYER Jr et al. 1974a), the key intermediate 5-amino-1-(β-D-ribofuranosyl)imidazole-4-carboxamidine 3′,5′-cyclic phosphate (25) (MEYER Jr et al. 1973b) was used as a precursor for the synthesis of 2-substituted *c*AMP derivatives. The reaction of (25) with aldehydes under oxidative conditions gave a multitude of 2-alkyl and 2-aryl substituted *c*AMP derivatives. This elegant method

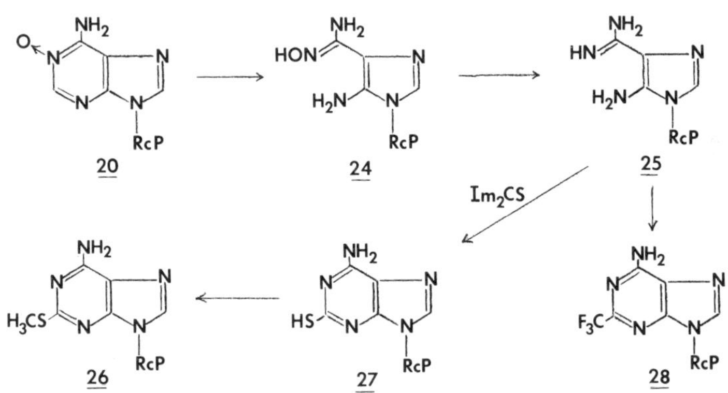

provides a direct route to the required *c*AMP derivatives and avoids the lengthy route of purine→nucleoside→5′-nucleotide→3′,5′cyclic nucleotide for each compound (SMITH et al. 1961; POSTERNAK et al. 1969; HONJO et al. 1973; MARUMOTO et al. 1975).

Treatment of (20) with refluxing 2N sodium hydroxide for ten minutes gave 5-amino-1-(β-D-ribofuranosyl)imidazole-4-carboxamidoxime 3′,5′-cyclic phosphate (24) in 43% yield and the Dimroth rearrangement product N-hydroxyadenosine 3′,5′-cyclic phosphate in minor amounts (MEYER Jr et al. 1973b). Reduction of (24) with hydrogen in the presence of sponge nickel catalyst gave the key intermediate (25). Compound (25) was also obtained by alkaline ring opening of 1-alkoxy-*c*AMP (23) to *O*-alkyl-5-amino-1-(β-D-ribofuranosyl)imidazole-4-carboxamidoxime 3′,5′-cyclic phosphate followed by Raney nickel reduction (MEYER Jr et al. 1973b). The intermediate (25) was quite insoluble in nonaqueous solvents and it was necessary to form its salt with DBU or 1,5-diazabicyclo[4.3.0]non-5-ene (DBN) to improve its solubility. Condensation of the DBU salt of (25) with trifluoroacetamide at 135 °C in N,N,N′,N′-tetramethylurea solution gave 2-trifluoromethyladenosine 3′,5′-cyclic phosphate (28) (MEYER Jr et al. 1974a). However, this procedure has its limitations since it cannot be used to obtain ring closure with strongly electron deficient carboxamides. The DBU or DBN salt of (25), upon treatment with 1,1′-thiocarbonyldiimidazole in dimethylsulfoxide solution, gave 2-thio-*c*AMP (27) (MEYER Jr et al. 1974a), alkylation of which, with alkyl halides in aqueous alkali furnished 2-alkylthio adenosine 3′,5′-cyclic phosphates (26) (MEYER Jr et al. 1975).

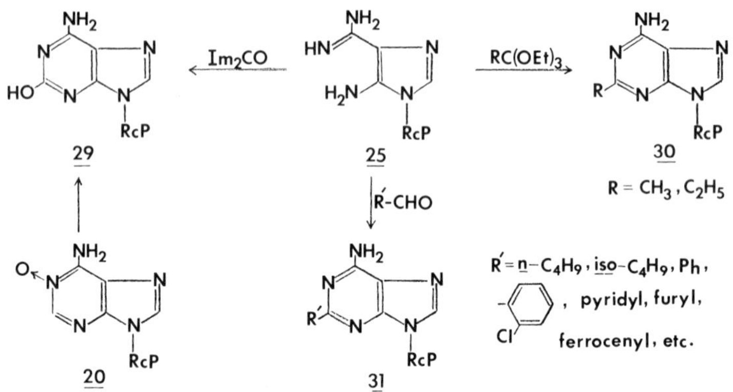

A similar reaction of (25) with 1,1′-carbonyldiimidazole gave 2-hydroxy-*c*AMP (isoguanosine 3′,5′-cyclic phosphate, 29) (MEYER Jr et al. 1974a), which has also been prepared by the photochemical transformation of (20) (KAZIMIERCZUK and SHUGAR 1973). Treatment of (25) with a wide variety of aldehydes under mild oxidative conditions has provided a series of unusual 2-alkyl- and 2-aryl-adenosine 3′,5′-cyclic phosphates (31), *via* the 2,3-dihydropurine intermediate (MEYER Jr et al. 1974a; MEYER Jr and SHUMAN 1975). Similarly, reaction of (25) with trialkyl orthoesters gave 2-alkyl-*c*AMP derivatives (30) (MEYER Jr et al. 1974a), which have also recently been prepared from corresponding 2-substituted adenosine *via* stepwise synthesis (MARUMOTO et al. 1979). Selective phosphorylation of unprotected nucleoside at the 5′-position with pyrophosphoryl chloride in acetonitrile (IMAI et al. 1969) or with phosphorus oxychloride in trialkyl phosphate (YOSHIKAWA et al. 1967) gave the 5′-nucleotide which then cyclized with DCC in boiling pyridine to the 3′,5′-cyclic phosphate (MARUMOTO et al. 1979).

2-Phenylinosine (32) (IMAI et al. 1971) was successfully converted to N^6-butyryl-2-phenyladenosine (37) *via* 6-chloro-2-phenyl-9-(β-D-ribofuranosyl)-purine (34), which was subsequently converted to the N^6-butyryl-2-phenyl-*c*AMP (35) (MARUMOTO et al. 1979). A few 2-alkyl and 2-alkylthio-*c*IMP derivatives (39) were also prepared by deamination of the corresponding 2-substituted-*c*AMP derivatives (38) (MEYER Jr et al. 1975a). These 2-substituted-*c*IMP derivatives could be prepared by the ring closure of 5-amino-1-(β-D-ribofuranosyl)imidazole-4-carboxamide 3′,5′-cyclic phosphate (40) produced by the ring opening of *c*IMP (MICHAL et al. 1974).

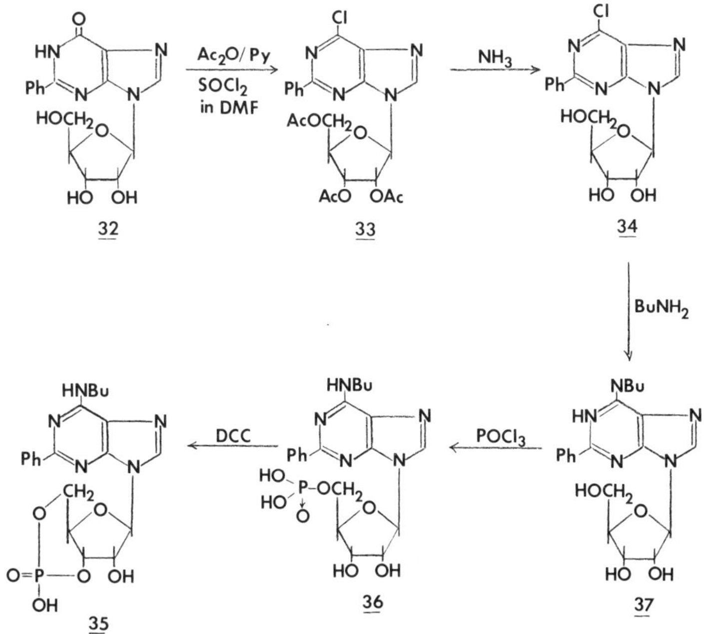

Several 2-substituted-*c*AMP derivatives were also prepared by an alternate method. Guanosine 3′,5′-cyclic phosphate (*c*GMP, 41) proved to be a very versatile starting material. *c*GMP occurs naturally and is formed from guanosine 5′-triphosphate by guanylate cyclase (HARDMAN et al. 1971). It has been chemically synthesized from guanosine 5′-p-nitrophenylphosphate (BORDEN and SMITH 1966b) and

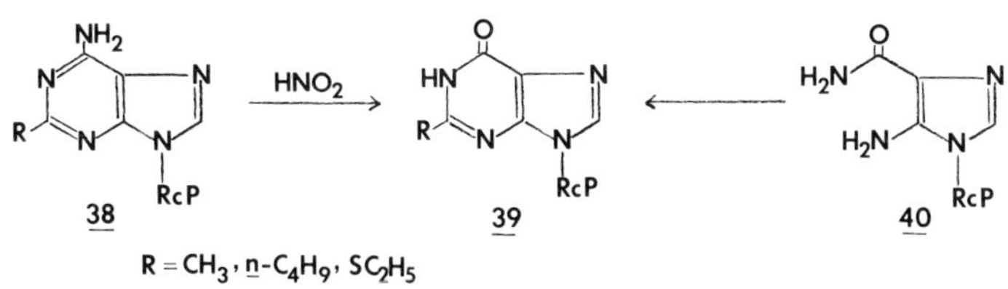

also by the cyclization of N^2-benzoylguanosine 5′-phosphate (SMITH et al. 1961; STRAUS and FRESCO 1965; FUJIMOTO and NARUSE 1968; POSTERNAK and FALBRIARD 1971) and guanosine 5′-phosphate (STRAUS and FRESCO 1965). Acetylation of *c*GMP to the 2′-*O*-acetyl derivative (41) followed by brief treatment with phosphoryl chloride in presence of N,N-diethylaniline provided 2-amino-6-chloro-9-(β-D-ribofuranosyl)purine 3′,5′-cyclic phosphate (42) (Boehringer Mannheim GmbH 1970a, 1971; MEYER Jr et al. 1975b; SHUMAN et al. 1975). Treatment of (42) with aqueous ammonia at 120 °C gave 2-aminoadenosine 3′,5′-cyclic phosphate (43) (MEYER Jr et al. 1975b). 2-Amino-6-chloro-9-(2,3,5-tri-*O*-acetyl-β-D-ribofuranosyl)purine (48) (GERSTER and ROBINS 1966) has als been used for the synthesis of (43) (POSTERNAK et al. 1971). Diazotization of (42) in cold concentrated hydrochloric acid, according to the method of GERSTER and ROBINS (1966) for the corresponding nucleoside, gave 2,6-dichloro-9-(β-D-ribofuranosyl)purine 3′,5′-

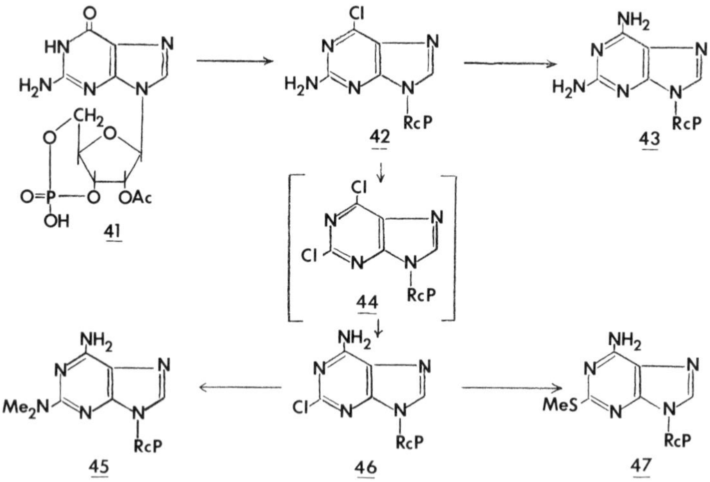

cyclic phosphate (44). Compound (44) was relatively unstable and often hydrolyzed to the parent hydroxypurine nucleotide. Treatment of (44) with concentrated ammonium hydroxide readily gave 2-chloroadenosine 3′,5′-cyclic phosphate (46) (Meyer Jr et al. 1975 b). Reaction of (46) with aqueous dimethylamine gave 2-(dimethylamino)adenosine 3′,5′-cyclic phosphate (45). 2-Chloro-*c*AMP (46) has also been prepared from 2-thioadenosine 3′,5′-cyclic phosphate (27) and chlorine gas in cold concentrated hydrochloric acid (Meyer Jr et al. 1975 b). The chlorine atom at C_2 in (46) is amenable to nucleophilic attack (Schaeffer and Thomas 1958) and was substituted by treatment with sodium methylmercaptide to give 2-methylthio derivative of *c*AMP (47) (Jastorff and Freist 1974). The synthesis of (47) has also been reported (Maguire et al. 1974) by the DCC in pyridine cycli-

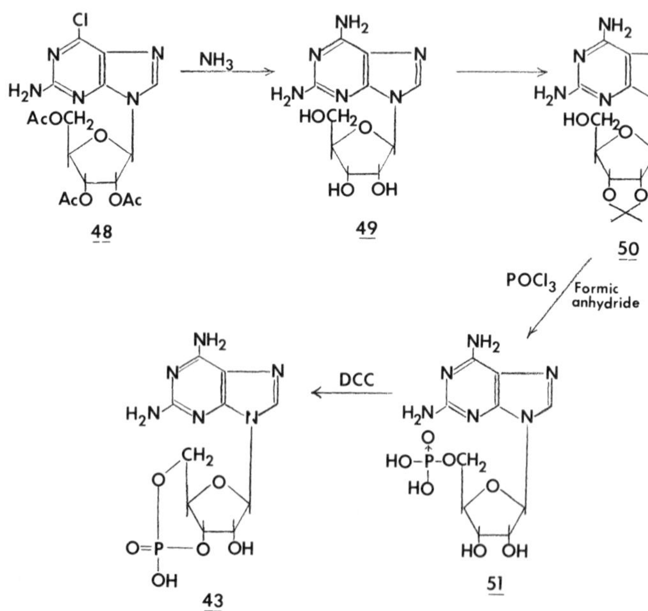

zation of 2-methylthioadenosine 5′-phosphate (MICHAL et al. 1969). Deacetylation of (48) with methanolic ammonia, followed by treatment with concentrated ammonium hydroxide, gave 2-aminoadenosine (49). Isopropylidenation of (42), followed by phosphorylation with phosphorus oxychloride in the presence of formic anhydride and pyridine, gave, after acid hydrolysis, 2-aminoadenosine 5′-phosphate (51). Cyclization of (51) by DCC in pyridine-dimethylformamide gave 2-aminoadenosine 3′,5′-cyclic phosphate (43) (POSTERNAK et al. 1971).

The synthesis of 2-chloro-, 2-bromo- (JASTORFF and FREIST 1972, 1974) and 2-fluoro- (ZIMMERMAN et al. 1976) adenosine 3′,5′-cyclic phosphates has been accomplished from the respective 2-haloadenosine by phosphorylation and subsequent cyclization with DCC. Oxidation of 2-chloroadenosine 3′,5′-cyclic phosphate (46) with m-chloroperbenzoic acid in a manner analogous to the preparation of (20)

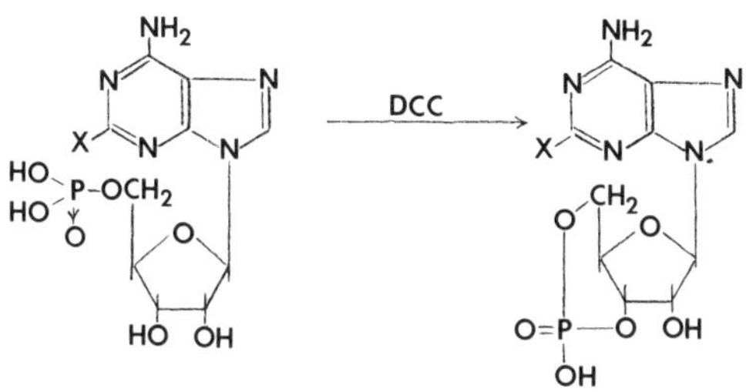

46

X = F, Cl or Br

gave 2-chloro-*c*AMP-N[1]-oxide (JASTORFF and FREIST 1972, 1974). Diazotization of *c*GMP Na-salt with nitrous acid and HBF_4 gave2-fluoro-*c*IMP, which when refluxed with benzylamine furnished 2-benzylamino-*c*IMP (BERGMEYER et al. 1972).

The third and the latest method utilizes easily accessible *c*AMP and provides a facile general route to 2-substituted-*c*AMP derivatives (YAMAJI et al. 1977e; YAMAJI and KATO 1978b). When 1,N[6]-ethenoadenosine 3′,5′-cyclic phosphate (52), obtained by reaction of chloroacetaldehyde with *c*AMP (SECRIST III et al. 1972a), was treated with sodium hydoxide (YIP and TSOU 1973), it gave 1-(β-D-ribo-

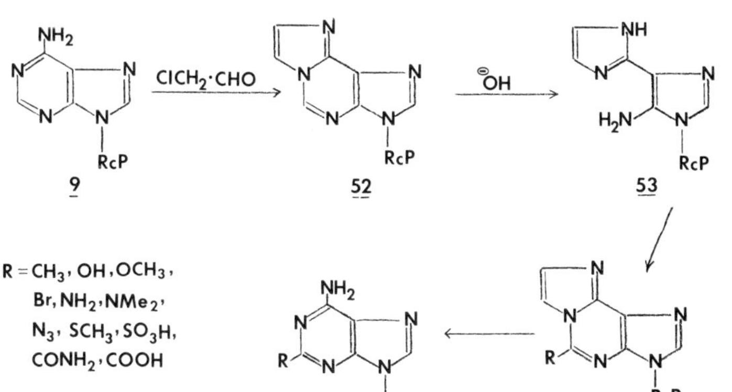

9 52 53 54 55

R = CH3, OH, OCH3, Br, NH2, NMe2, N3, SCH3, SO3H, CONH2, COOH furanosyl-5-amino-4-(imidazol-2-yl)imidazole 3′,5′-cyclic phosphate (53). Under relatively mild conditions (53) was ring closed to 2-substituted 1,N[6]-ethenoadenosine 3′,5′-cyclic phosphates (55) (YAMAJI et al. 1976b, c, 1980) with reagents like cyanogen bromide, triethyl orthoacetate, carbon disulfide, etc. Treatment of (55) with N-bromosuccinimide under mild conditions gave the corresponding 2-substituted-*c*AMP derivatives (54) in good yield (YAMAJI et al. 1977e).

c) N[6]- and 6-Substituted Adenosine 3′,5′-Cyclic Phosphate Derivatives

The first N[6]-substituted adenosine 3′,5′-cyclic phosphate derivatives were prepared (POSTERNAK et al. 1962) from *c*AMP and the appropriate acyl halide or a carboxylic acid anhydride in pyridine to yield (57) (FALBRIARD et al. 1967; STEINER et al. 1969; CAILLA and DELAAGE 1972; BROER et al. 1972; SEVERIN et al. 1975; MARUMOTO et al. 1979). These reagents also acylate the free 2′-sugar hydroxyl function.

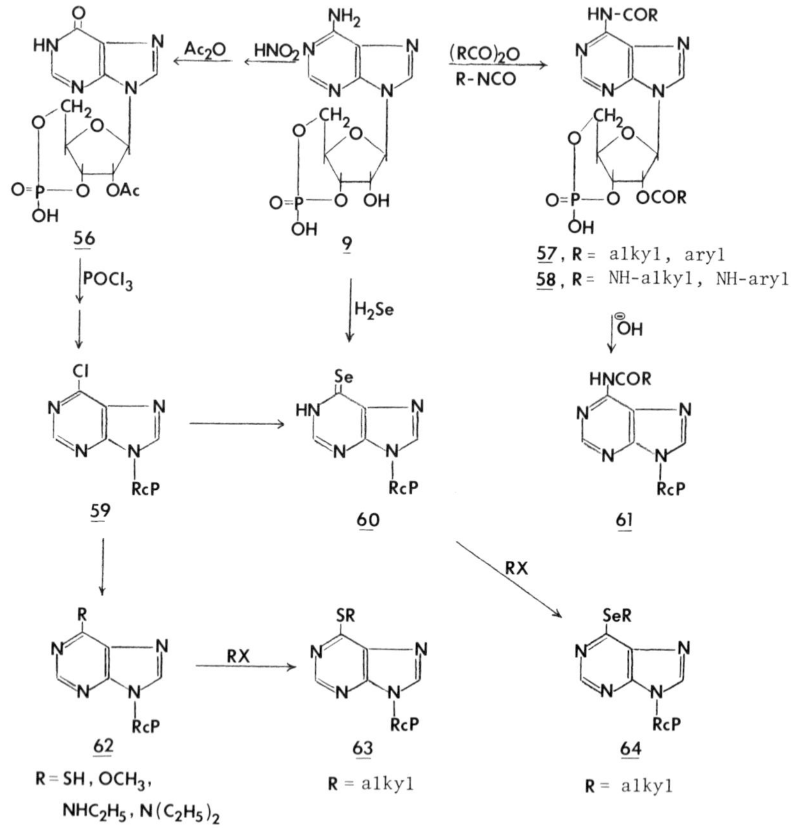

In this manner the N[6],2′-*O*-diacetyl, butyryl, hexanoyl, octanoyl, lauroyl, stearoyl, benzoyl, adamantoyl (FALBRIARD et al. 1967), palmitoyl, diazomalonyl (BRUNSWICK and COOPERMAN 1971) and aminocaproyl (WILCHEK et al. 1971) derivatives of *c*AMP have been synthesized from adenosine 3′,5′-cyclic phosphate (9). Since the 2′-*O*-acyl groups are very sensitive to hydrolysis by alkali or by *c*AMP phosphodiesterase (O'NEIL et al. 1975), they can be selectively removed quantitatively by brief treatment with dilute alkali at room temperature (POSTERNAK et al. 1962; MURAD et al. 1969). Similarly, the N^6-carbamoyl and alkoxycarbonyl derivatives (61) were prepared from *c*AMP and aryl or alkyl isocyanates or alkyl chloroformates (BOSWELL et al. 1973). The synthesis of several N^6-carbamoyl derivatives of *c*AMP has also been reported by DCC cyclization of corresponding 5′-phosphates (HONG et al. 1975).

The stability of N^6,2′-*O*-dibutyryladenosine 3′,5′-cyclic phosphate has been investigated (SWISLOCKI 1970; BLECHER et al. 1970). Dry N^6,2′-*O*-dibutyryladenosine 3′,5′-cyclic phosphate was stable for longer than four months at −15 °C, whereas aqueous solutions slowly decomposed to N^6-monobutyryladenosine 3′,5′-cyclic phosphate. Significant 2′-*O*-deacylation of N^6,2′-*O*-dibutyryladenosine 3′,5′-cyclic phosphate to predominantly N^6-monobutyryladenosine 3′,5′-cyclic phosphate occurred at 37 °C after one hour in basic media (0.15 *M* bicarbonate, pH 9.6) as well as in Krebs-Ringer bicarbonate (pH 7.4).

More stable 6-substituted derivatives of *c*AMP have been prepared (MEYER Jr et al. 1972; MICHAL et al. 1974). Adenosine 3′,5′-cyclic phosphate (9) was deaminated with aqueous nitrous acid and the resulting *c*IMP was acetylated with acetic anhydride to give 2′-*O*-acetylinosine 3′,5′-cyclic phosphate (56), which on treatment with refluxing phosphorus oxychloride in presence of N,N-dimethylaniline yielded the key intermediate 6-chloro-9-(*β*-D-ribofuranosyl)purine 3′,5′-cyclic phosphate (59) (MEYER Jr et al. 1972; WEIMANN et al. 1973). Nucleophilic displacement of the 6-chloro function of (59) occurred with various sulfur, oxygen and nitrogen nucleophiles to give the respective 6-substituted-9-(*β*-D-ribofuranosyl)purine 3′,5′-cyclic phosphates (62). 9-(*β*-D-Ribofuranosyl)purine-6-thione 3′,5′-cyclic phosphate (62, R=SH), prepared in good yield from (59) (MEYER Jr et al. 1972), has also been synthesized from the corresponding 5′-nucleotide by the DCC procedure in rather low yield (THOMAS and MONTGOMERY 1968), or by sulfhydrolysis of *c*AMP with liquid hydrogen sulfide in aqueous pyridine in good yield (MEYER Jr et al. 1978). Acylation of the 2′-hydroxy function of (62) (R=SH) with the carboxylic anhydrides in pyridine in the presence of the catalyst 4-dimethylaminopyridine (HÖFLE and STEGLICH 1972) gave 2′-*O*-acyl-6-thioinosinic 3′,5′cyclic phosphates (MEYER Jr et al. 1979). Alkylation of (62) (R=SH) with several alkylhalides gave 6-alkylthio-9-(*β*-D-ribofuranosyl)purine 3′,5′-cyclic phosphates (63). Adenosine 3′,5′-cyclic phosphate was also converted to 6-selenoxo-9-(*β*-D-ribofuranosyl)purine 3′,5′-cyclic phosphate (60) by treatment with hydrogen selenide in aqueous pyridine in 36% yield (SHIUE and CHU 1975). Alternately compound (60) was synthesized in rather low yield by allowing 6-chloro-9-(*β*-D-ribofuranosyl)purine 3′,5′-cyclic phosphate (59) to react with hydrogen selenide (SHIUE and CHU 1975). The seleno-cyclic phosphate (60) was transformed to 6-alkylseleno-9-(*β*-D-ribofuranosyl)purine 3′,5′-cyclic phosphate (64) with alkylhalides.

6-Alkylamino-9-(*β*-D-ribofuranosyl)purine 3′,5′-cyclic phosphates (68) have been synthesized by the classical procedure from 6-chloro-9-(2,3-*O*-isopropylidene-*β*-D-ribofuranosyl)purine (65) (POSTERNAK et al. 1969; CEHOVIC et al. 1971a; Agence Nationale de Valorisation de la Recherche 1972). Nucleophilic displacement of chlorine of (65) gave 6-alkylamino nucleoside (66), which on phosphorylation with phosphorus oxychloride and subsequent deisopropylidenation gave the

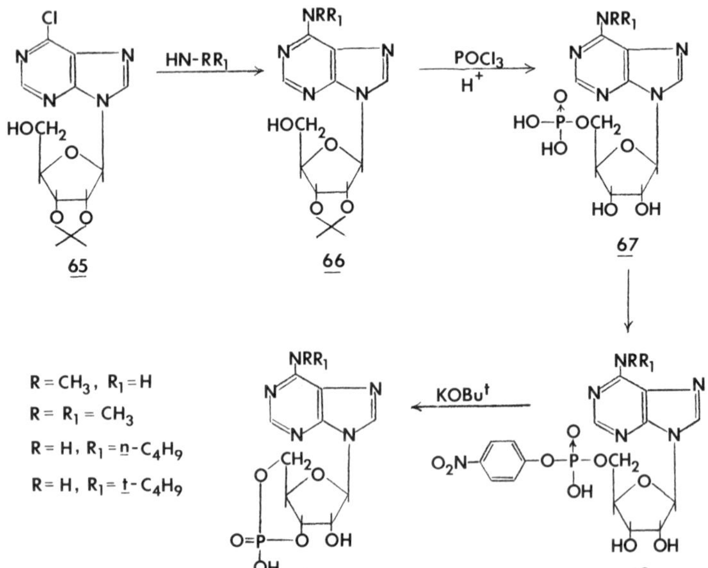

65 66 67 68 69

R = CH_3, R_1 = H
R = R_1 = CH_3
R = H, R_1 = n-C_4H_9
R = H, R_1 = t-C_4H_9

respective 6-alkylamino nucleotide (67). The 6-alkylamino 3′,5′-cyclic nucleotides (68) were then obtained by the base-catalyzed transesterification of their p-nitrophenyl esters (69) (Posternak et al. 1969; Cehovic et al. 1971a).

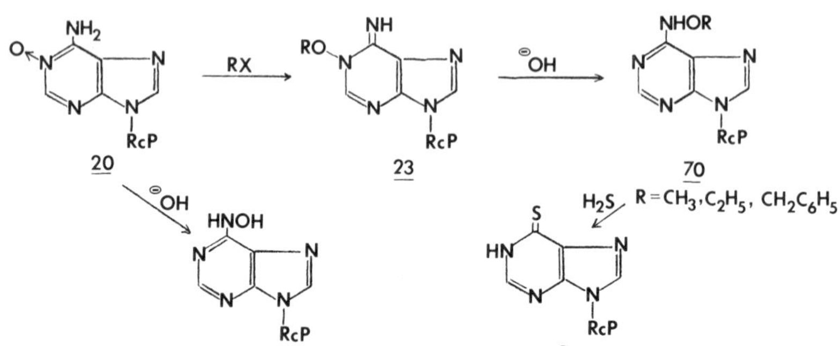

20 23 70 71 62

R = CH_3, C_2H_5, $CH_2C_6H_5$

The N[6]-alkoxyadenosine 3′,5′-cyclic phosphates (70) were prepared by *O*-alkylation of *c*AMP-N[1]-oxide (20) followed by Dimroth rearrangement in dilute sodium hydroxide solution at room temperature in low yield (~8%) (Meyer Jr et al. 1973b). However, the action of weaker alkaline media on N[1]-alkoxyadenosine 3′,5′-cyclic phosphates (23), like refluxing sodium bicarbonate solution, resulted in 53% yield of (70) (Meyer Jr et al. 1973b). Adenosine 3′,5′-cyclic phosphate N[1]-oxide itself, upon treatment with alkali, gave 12% yield of N[6]-hydroxy-*c*AMP (71). Sulfhydrolysis of (70) (R = CH_3) using liquid hydrogen sulfide under the conditions reported by Ueda and co-workers (1971), gave 9-(β-D-ribofuranosyl)purine-6-thione 3′,5′-cyclic phosphate (62) in good yield (Shuman et al. 1976).

d) 8-Substituted Adenosine 3′,5′-Cyclic Phosphate Derivatives

The syntheses of 8-substituted adenosine 3′,5′-cyclic phosphate derivatives has been extensively explored. 8-Bromoadenosine 3′,5′-cyclic phosphate (72) served as a key intermediate for the chemical synthesis of various 8-substituted derivatives of *c*AMP, and was readily prepared by direct bromination of *c*AMP in aqueous acetate buffer in over 80% yield (IKEHARA et al 1967; IKEHARA and UESUGI 1969; Boehringer Mannheim GmbH 1970b; MUNEYAMA et al. 1971; Boehringer Mannheim GmbH 1971; IKEHARA et al. 1978). 8-Bromo-*c*AMP (72) was also obtained by cyclization of 8-bromoadenosine 5′-phosphate (73) with DCC in pyridine (POSTERNAK et al. 1971). Facile nucleophilic displacement of the 8-bromo group of (72) occurred with thiourea to give 8-thioadenosine 3′,5′-cyclic phosphate (75) (Boehringer Mannheim GmbH 1970; MUNEYAMA et al 1971). Acylation of (75) with

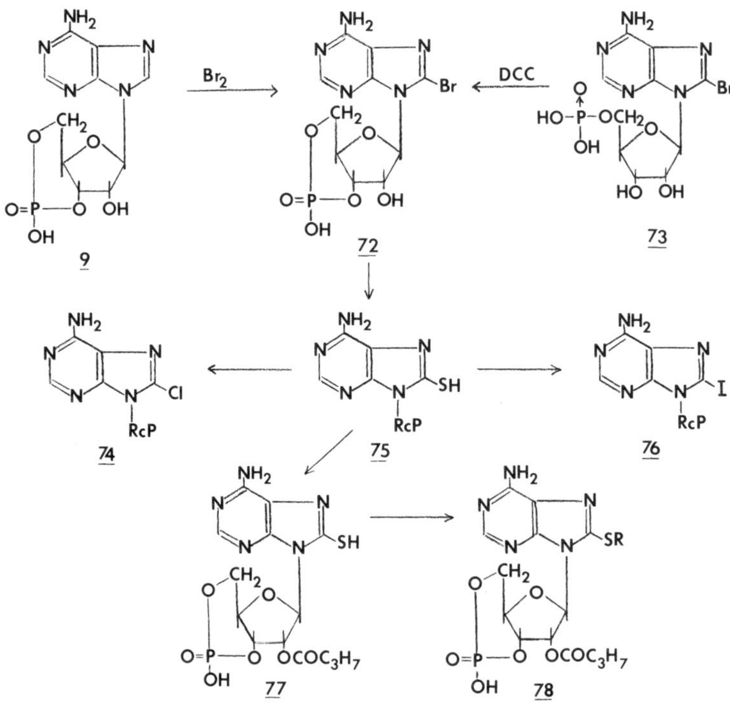

butyric anhydride in pyridine gave 2′-*O*-butyryl-8-thio-*c*AMP (77) which on alkylation provided 2′-*O*-butyryl-8-alkylthio-*c*AMP (78) (KHWAJA et al. 1975). 8-Chloroadenosine 3′,5′-cyclic phosphate (74) and 8-iodoadenosine 3′,5′-cyclic phosphate (76), which cannot be obtained readily by direct halogenation of *c*AMP, have been prepared from (75) (MUNEYAMA et al. 1974) by molecular halogen in the presence of excess halide ion (BEAMAN and ROBINS 1962; HOLMES and ROBINS 1964). When (75) was treated with chlorine gas in methanolic hydrochloric acid at −15 °C, (74) was formed in 75% yield (MUNEYAMA et al. 1974). Similarly, when (75) was treated with an aqueous iodine-potassium iodide solution, (76) and *c*AMP were formed and were separated by cellulose column chromatography to give 31% yield of 8-iodo-*c*AMP (76) (MUNEYAMA et al. 1974; SOWA et al. 1976).

Treatment of 8-bromo-*c*AMP (72) with sodium methoxide in presence of methanethiol in methanol gave 8-methylthioadenosine 3′,5′-cyclic phosphate (80, $R=CH_3$) in good yield (Muneyama et al. 1971). Similar nucleophilic displacements occurred with ethanetiol, p-chlorobenzenethiol, 2-mercaptoethanol, benzylmercaptan and other aryl or alkylthiol to give corresponding 8-substituted adenosine 3′,5′-cyclic phosphates (80) (Miller et al. 1973; Miller et al. 1980; Sasaki et al. 1976; Cehovic et al. 1976; Gulyaev et al. 1977). 8-Oxoadenosine 3′,5′-cyclic phosphate (79) was obtained by hydrogenation of 8-benzyloxyadenosine 3′,5′-cyclic phosphate (Boehringer Mannheim GmbH 1970, 1971) or

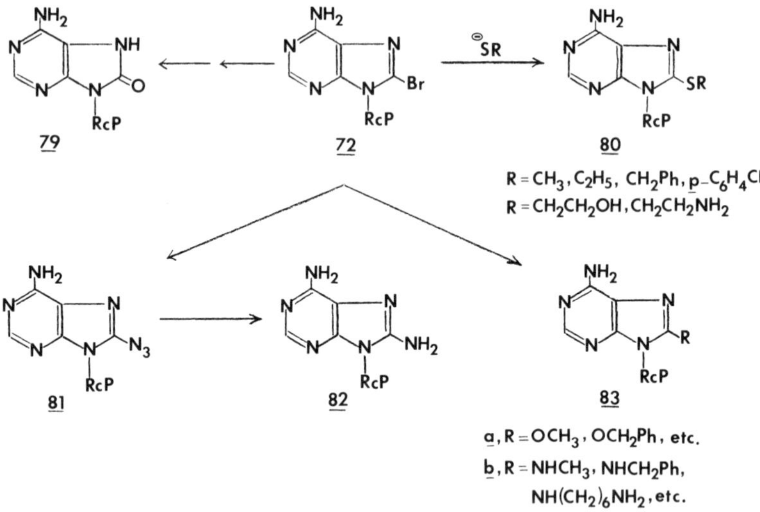

by treatment of (72) with a mixture of sodium acetate, acetic anhydride and glacial acetic acid at 100 °C to give N^6,2′-*O*-diacetyl-8-oxoadenosine 3′,5′-cyclic phosphate which was not isolated but was deacetylated with methanolic ammonia to the 8-oxo derivative (79) (Muneyama et al. 1971; Posternak et al. 1971). Azide displacement of the bromine of (72) gave 8-azidoadenosine 3′,5′-cyclic phosphate (81) which on hydrogenation gave 8-aminoadenosine 3′,5′-cyclic phosphate (82) (Muneyama et al. 1971). Compound (82) was also obtained by DCC cyclization of the corresponding 5′-phosphate (Posternak et al. 1971). Nucleophilic substitution of (72) with alkoxides gave several 8-alkoxy cyclic nucleotides (83a, $R=OCH_3$, $OCH_2C_6H_5$, etc.) (Muneyama et al. 1971; Boehringer Mannheim GmbH 1971). Treatment of (72) with refluxing aqueous methylamine, ethanolamine, furfurylamine and 1,6-diaminohexane, provided corresponding 8-substituted adenosine 3′,5′-cyclic phosphates (83b) (Muneyama et al. 1971; Jergil et al. 1974; Jergil and Mosbach 1974; Sasaki et al. 1976; Gulyaev et al. 1977). The 8-(*β*-aminoethylthio) derivative (80, $R=CH_2CH_2NH_2$) and 8-(6-aminohexyl)amino derivative [83, $R=NH(CH_2)_6NH_2$] have been used for the covalent binding of *c*AMP to Sepharose (Tesser et al. 1972; Jergil et al. 1974; Ramseyer et al. 1974; Guilford et al. 1972).

When 8-bromo-*c*AMP (72) was treated with sodium hydrogen selenide in refluxing methanol, 8-selenoadenosine 3′,5′-cyclic phosphate (84) was obtained,

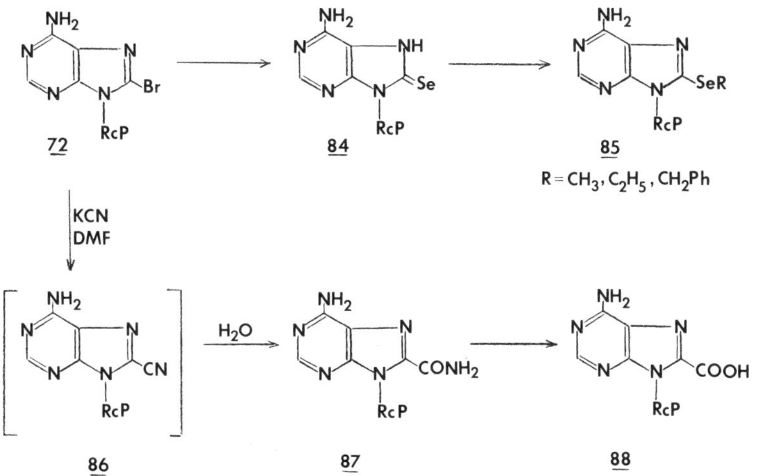

which with alkylhalides gave 8-alkylseleno-*c*AMP derivatives (85) (CHU et al. 1974). 8-Bromo-*c*AMP was also found to be a versatile intermediate for the synthesis of 8-carbamoyladenosine 3′,5′-cyclic phosphate (87). The reaction of (72) with potassium cyanide in hot dimethylformamide gave 8-carbamoyl-*c*AMP (87) *via* the intermediate (86) (NAKA and HONJO 1976a, b). This is probably the first example of successful introduction of a carbon nucleophile into the 8-position of adenine derivatives. Hydrolysis of (87) with aqueous sodium hydroxide gave 8-carboxyadenosine 3′,5′-cyclic phosphate (88) (NAKA and HONJO 1976a).

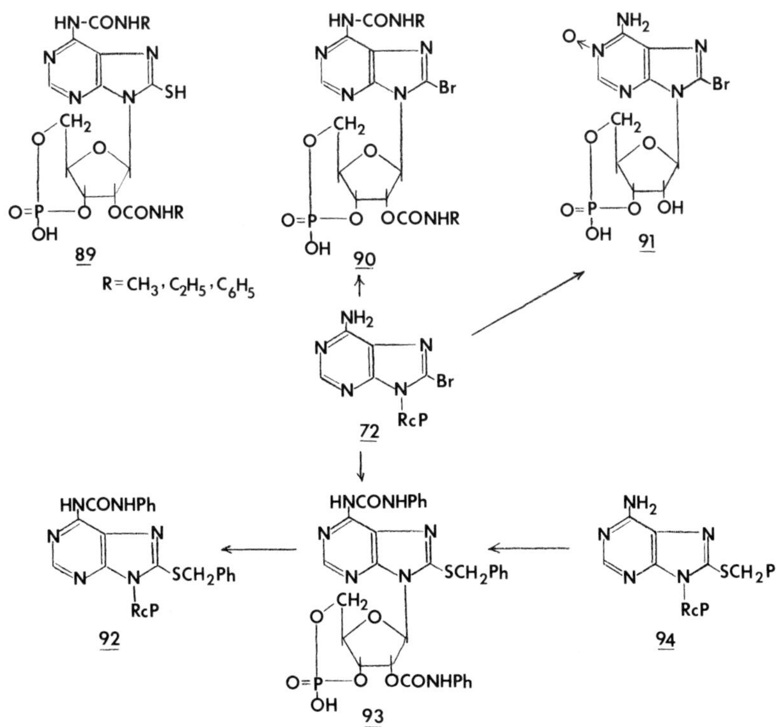

The N^6,2′-*O*-di(N-alkyl or arylcarbamoyl)-8-bromo-*c*AMP (90) and 8-benzylthio-N^6,2′-*O*-di(N-phenylcarbamoyl)-*c*AMP (93) were obtained by direct treatment of 8-bromo-*c*AMP (72) and 8-benzylthio-*c*AMP (94) with appropriate isocyanates (BOSWELL et al. 1973; ROBINS et al. 1977a). Alkaline hydrolysis of (93) gave 8-benzylthio-N^6-(N-phenylcarbamoyl)adenosine 3′,5′-cyclic phosphate (92). Oxidation of (72) with m-chloroperbenzoic acid in buffered two-phase system gave 8-bromoadenosine 3′,5′-cyclic phosphate N^1-oxide (91) (SHUMAN and MEYER Jr 1975; UNO et al. 1976).

e) 2,8-Disubstituted Adenosine 3′,5′-Cyclic Phosphate Derivatives

Relatively few 2,8-disubstituted adenosine 3′,5′-cyclic phosphate derivatives have been reported (UNO et al. 1976). Direct bromination of 2-methyl- (95, $R_2=CH_3$) and 2-butyl- (95, $R_2=C_4H_9$)adenosine 3′,5′-cyclic phosphate (MEYER Jr et al. 1974a) with bromine in aqueous sodium acetate buffer gave 8-bromo-2-methyl (or butyl)adenosine 3′,5′-cyclic phosphate (96). These 8-bromo nucleotides proved to

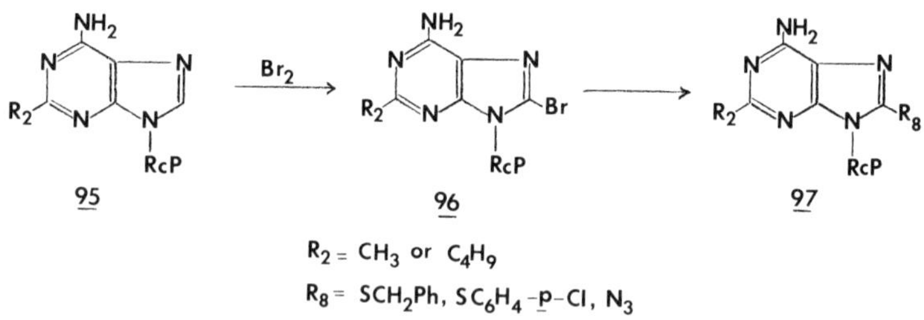

be convenient intermediates for the preparation of 2,8-disubstituted *c*AMP derivatives. Nucleophilic displacement of the 8-bromo function with sodium benzylmercaptide, sodium p-chlorothiophenolate and sodium azide gave 8-benzylthio, 8-p-chlorophenylthio, and 8-azido-2-alkyladenosine 3′,5′-cyclic phosphate derivatives (97), respectively (UNO et al. 1976).

The trichloromethylphosphorodichloridate procedure (MARUMOTO et al. 1975) has recently been used to prepare 2-chloro-8-bromoadenosine 3′,5′-cyclic phosphate (100). The intramolecular cyclization of the intermediate nucleoside 5′-trichloromethylphosphonate (99) with potassium *t*-butoxide in N,N-dimethylformamide led to the formation of (100) in 40% yield (MARUMOTO et al. 1975).

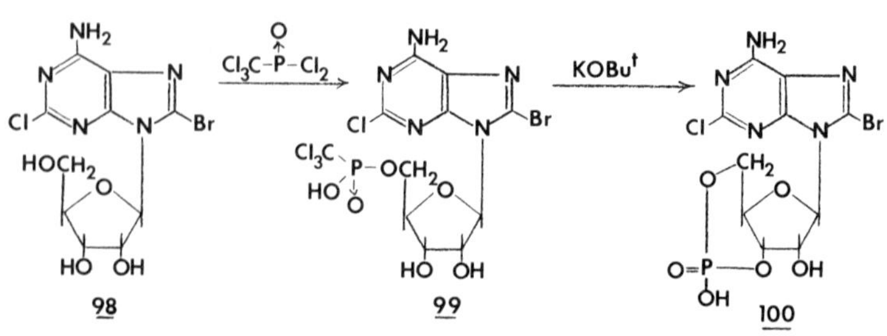

f) 6,8-Disubstituted and Other Structural Isomers of Adenosine 3′,5′-Cyclic Phosphate

A number of 6,8-disubstituted purine 3′,5′-cyclic phosphates have been synthesized using preformed 3′,5′-cyclic phosphate precursors. Three different approaches have been pursued in the preparation of these compounds: (i) the synthesis and subsequent reactions of 6,8-dichloro-9-(β-D-ribofuranosyl)purine 3′,5′-cyclic phosphate, (ii) the synthesis and reactions of 6-chloro-8-substituted-9-(β-D-ribofuranosyl)purine 3′,5′-cyclic phosphate, and (iii) N^1-alkylation followed by Dimroth rearrangement of 8-substituted adenosine 3′,5′-cyclic phosphates.

Deamination of 8-bromoadenosine 3′,5′-cyclic phosphate (72) gave 8-bromoinosine 3′,5′-cyclic phosphate (101) (MILLER et al. 1973a), which was also synthesized by bromination of *c*IMP (Boehringer Mannheim GmbH 1970b, 1971). Acetylation of (101) with acetic anhydride in pyridine led to 2′-*O*-acetyl-8-bromoinosine 3′,5′-cyclic phosphate (102) (MILLER et al. 1976). The use of 4-dimethylaminopyridine (HOFLE and STEGLICH 1972) as a catalyst in the acetylation of (101) with carboxylic anhydrides was found to be superior to the former procedures (MILLER et al. 1976). Treatment of (102) with boiling phosphorus oxychloride gave the key intermediate 6,8-dichloro-9-(2-*O*-acetyl-β-D-ribofuranosyl)purine 3′,5′-cyclic phosphate (105) (BOSWELL et al. 1975). Such a replacement of 8-bromo group by a chloro function has been observed with 2′,3′,5′-tri-*O*-acetyl-8-bromoguanosine (GERSTER et al. 1968). Treatment of (105) with

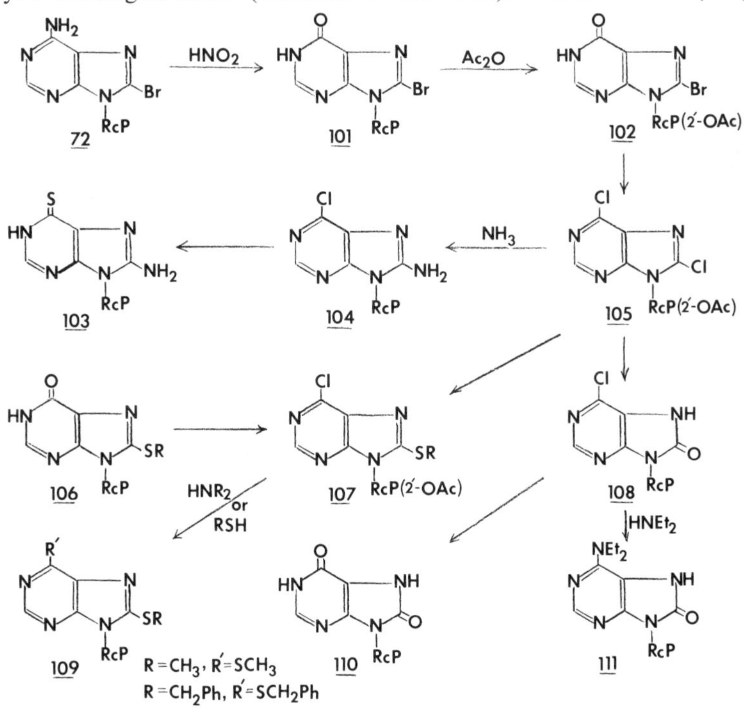

methanolic ammonia at 0 °C gave 8-amino-6-chloro-9-(β-D-ribofuranosyl)purine 3′,5′-cyclic phosphate (104) in 43% yield, which was converted to 8-amino-9-(β-D-ribofuranosyl)purine-6-thione 3′,5′-cyclic phosphate (103) with thiourea (BOS-

WELL et al. 1975; ROBINS et al. 1977). 6-Chloro-9-(β-D-ribofuranosyl)purin-8-one 3′,5′-cyclic phosphate (108) was isolated from the reaction of (105) with methanolic ammonia, or 1N sodium hydroxide at room temperature, and could be further converted to 6-diethylamino-9-(β-D-ribofuranosyl)purin-8-one 3′,5′-cyclic phosphate (111) with diethylamine. 8-Alkylthio-6-substituted -9-(β-D-ribofuranosyl)purine 3′,5′-cyclic phosphates (109) were formed by nucleophilic displacement of the 8-chloro function of (105) to give 8-alkylthio-6-chloro-9-(2-O-acetyl-β-D-ribofuranosyl)purine 3′,5′-cyclic phosphate (107), which was then subjected to attack by alkylthio and alkylamino nucleophiles to give (109). Compound (107) was also produced by treatment of 8-substituted inosine 3′,5′-cyclic phosphate (106) (MILLER et al. 1973a) with phosphorus oxychloride to yield (107). Vigorous hydrolysis of (108) with sodium acetate/acetic acid at 120 °C for three days gave 9-(β-D-ribofuranosyl)purin-6,8-dione 3′,5′-cyclic phosphate (110) (BOSWELL et al. 1975), identical with 8-hydroxyinosine 3′,5′-cyclic phosphate prepared by deamination of 8-hydroxy-cAMP (79) (MILLER et al. 1973a).

Application of Dimroth rearrangement upon 1-alkyl-8-substituted-cAMP (113) furnished an alternate method for the synthesis of 6,8-disubstituted-cAMP

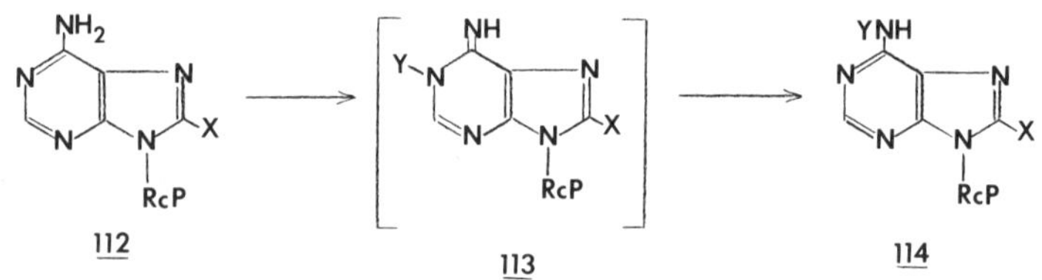

derivatives (114) (BOSWELL et al. 1975). Alkylation of 8-methylthio-(112, X = SCH_3) and 8-bromo-(112, X = Br)adenosine 3′,5′-cyclic phosphate with α-bromotoluene in dimethylsulfoxide gave the respective 1-alkyl-8-substituted nucleotides (113), which without isolation were rearranged by heating with aqueous bicarbonate to give (114) (BOSWELL et al. 1975). Recently this method has been used to prepare additional 6,8-disubstituted-cAMP derivatives (MILLER et al. 1980). Alkylation of cAMP with crotyl bromide at N^1-position (115) followed by

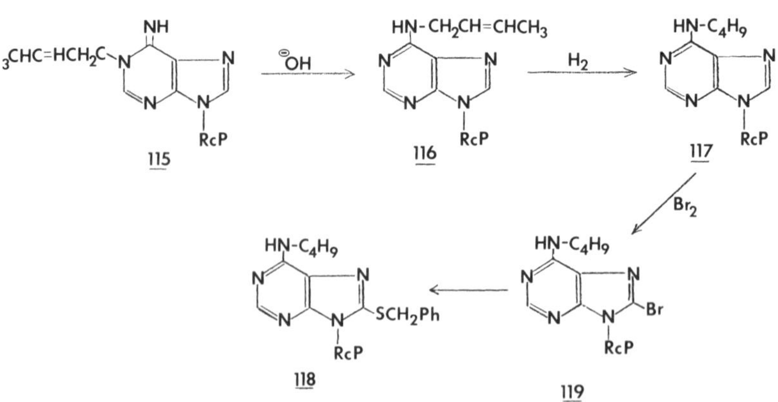

basic rearrangement gave N^6-crotyladenosine 3′,5′-cyclic phosphate (116). Hydrogenation of (116) afforded N^6-n-butyl-cAMP (117) which on bromination in a buffered media is reported to give N^6-n-butyl-8-bromoadenosine 3′,5′-cyclic phosphate (119). Treatment of (119) with benzylmercaptan at pH 12 gave N^6-*n*-butyl-8-benzylthioadenosine 3′,5′-cyclic phosphate (118) (MILLER et al. 1980).

Homolytic acylation of *c*IMP (10) with acetaldehyde or benzaldehyde in dilute mineral acid (0.3 *N* sulfuric acid) in presence of ammonium persulfate and ferrous sulfate gave 8-acetyl- or 8-benzoylinosine 3′,5′-cyclic phosphate (120), respectively,

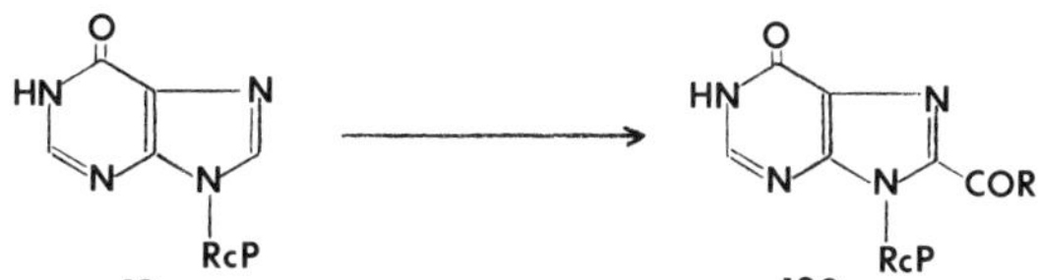

10 120

R = CH_3 or C_6H_5

in rather low yield (CHRISTENSEN et al. 1975; CHRISTENSEN and ROBINS 1976). When *c*AMP was irradiated in isopropanol with [^{60}Co] γ-rays (LEONOV and ELAD 1974), 8-(2-hydroxypropyl-2)adenosine 3′,5′-cyclic phosphate (121) was obtained (MARUMOTO et al. 1979).

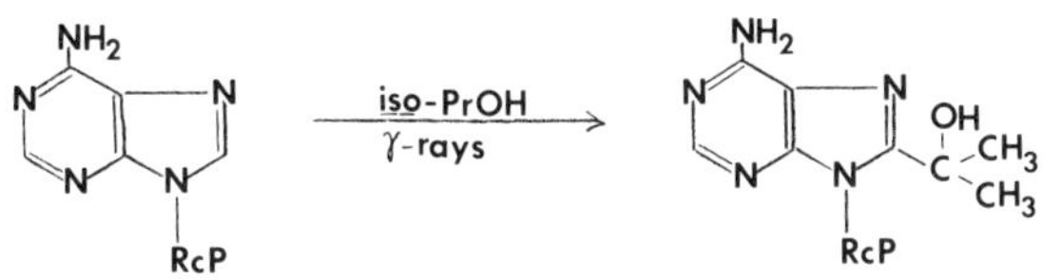

9 121

Some interesting structural isomers of adenosine 3′,5′-cyclic phosphate, like 3-(β-D-ribofuranosyl)adenine 3′,5′-cyclic phosphate(*iso*-*c*AMP, 122) have been synthesized (CEHOVIC et al. 1968; LAPINET et al. 1971; CEHOVIC et al. 1971b) from *iso*adenosine 5′-phosphate (LEONARD and LAURSEN 1965). 9-(β-L-Ribofuranosyl)-adenine 3′,5′-cyclic phosphate (123) has also been reported (JUROVCIK et al. 1971).

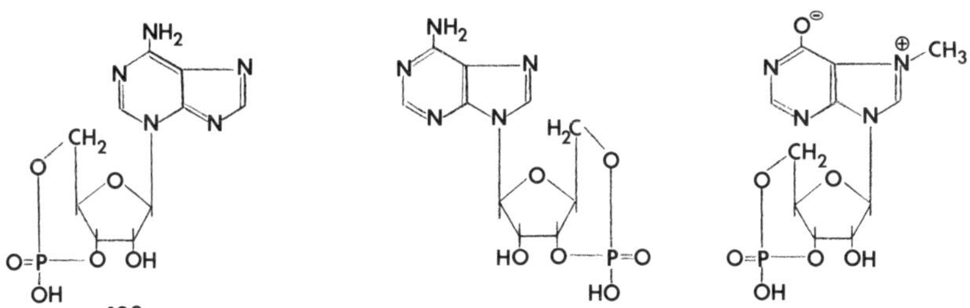

122 123 124

7-Methyl-*c*IMP (124) has been prepared by the methylation of *c*IMP (ANDERSON et al. 1972). 2-Amino-9-(β-D-ribofuranosyl)purine 3′,5′-cyclic phosphate (125) was prepared from 2-aminopurine nucleoside-5′-(4-nitrophenyl)phosphate (SCHEIT 1974). Compound (125) is readily hydrolyzed to 2-aminopurine nucleoside 5′-phosphate by PDE and this conversion leads to significant changes in both the absorption and fluorescence spectra (SCHEIT 1974). Another fluorescent "stretched-out" analog of *c*AMP, namely *linear* benzoadenosine 3′,5′-cyclic phosphate (126)

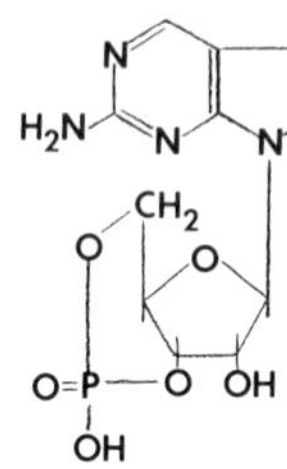

125

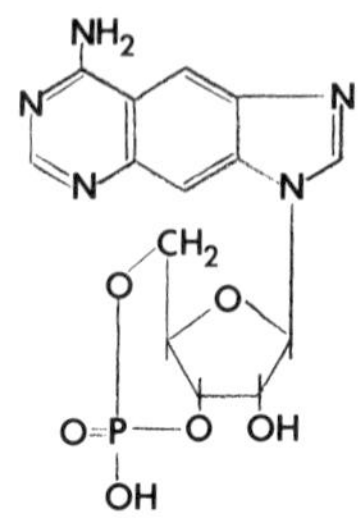

126

has been reported recently (SCHMIDT et al. 1978), and was synthesized by the base catalyzed cyclization (MARUMOTO et al. 1975) of *linear* benzoadenosine 5′-trichloromethyl phosphonate. It is of particular interest that compound (126) was only *slowly* hydrolyzed by the phosphodiesterases (SCOPES et al. 1977).

2. Structural Modification of the Carbohydrate Moiety

a) Arabinofuranosyl Cyclic Nucleotides

The first chemical synthesis of a nucleoside cyclic phosphate from an unnatural sugar (e.g. arabinofuranose) with the same phosphate ester stereochemistry as adenosine 3′,5′-cyclic phosphate was reported by GOODMAN and co-workers (HUBERT-HABART and GOODMAN 1969; LEE et al. 1971). The synthesis of such arabinofuranosyl 2′,5′- and 3′,5′-cyclic nucleotides has been pursued in several ways. The phosphorylation of 9-(β-D-arabinofuranosyl)adenine (*ara*-A, 127) (GLAUDEMANS and FLETCHER Jr 1963) with phosphorus oxychloride in trimethylphosphate gave

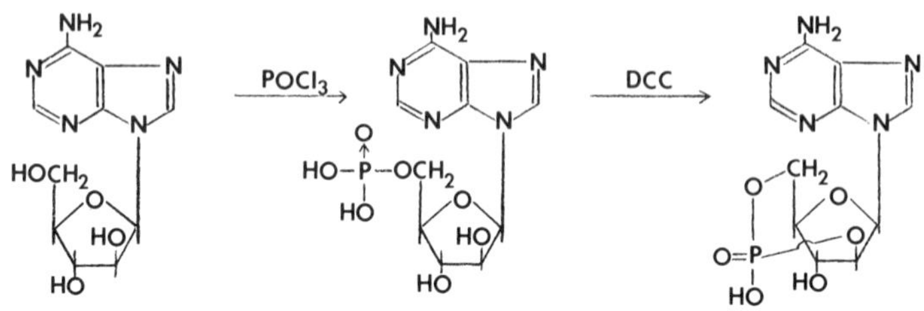

127 128 129

crystalline 9-(β-D-arabinofuranosyl)adenine 5′-phosphate (128) along with traces of 9-(β-D-arabinofuranosyl)adenine 2′,5′-cyclic phosphate (129) (HUBERT-HABART and GOODMAN 1969). However, direct cyclization of (128) with DCC in pyridine gave high yields of the 2′,5′-cyclic phosphate (129) (HUBERT-HABART and GOODMAN 1969), which was also obtained in low yield (11%) during the alkaline hydrolysis of the 5′-β-cyanoethylphosphate derivative of *ara*-A (FURTH and COHEN 1967).

Since the direct cyclization of *ara*-A 5′-phosphate (128) gave the 2′,5′-cyclic phosphate (129), other methods for the synthesis of the 3′,5′-cyclic phosphate (132) were explored. In addition to the possible enzymatic conversion of *ara*-ATP to the *trans* 3′,5′-cyclic *ara*-AMP (132) (ORTIZ 1972), two chemical methods for the synthesis of (132) have been developed. The first method (LEE et al. 1971) involves the selective blocking of the 2′-hydroxyl of *ara*-A. The synthesis of such a derivative,

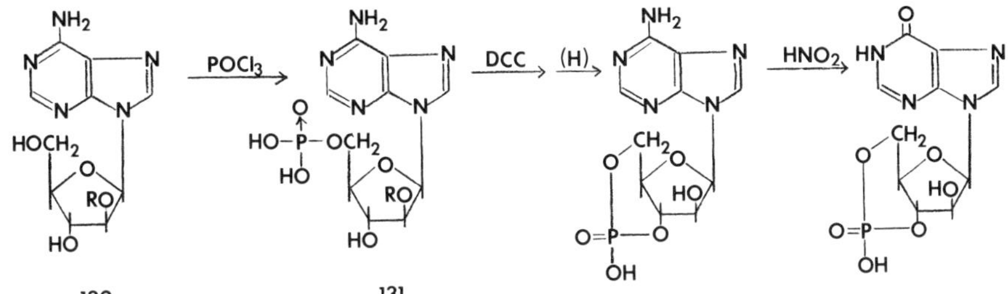

130 131 132 133

2′-*O*-benzylarabinofuranosyladenine (130) was achieved from 2-*O*-benzyl-3,5-di-*O*-p-nitrobenzoyl-α-D-arabinofuranosyl chloride and 6-benzamidopurine followed by subsequent deacylation with hot methanolic sodium methoxide. The 2′-*O*-benzylarabinofuranosyladenine (130) was then selectively phosphorylated to the 2′-*O*-benzylarabinofuranosyladenine 5′-phosphate (131) (LEE et al. 1971). Cyclization of (131) with DCC in hot pyridine followed by hydrogenolysis of the benzyl group gave 9-(β-D-arabinofuranosyl)adenine 3′,5′-cyclic phosphate (132). Deamination of (132) with aqueous nitrous acid gave 9-(β-D-arabinofuranosyl)hypoxanthine 3′,5′-cyclic phosphate (133) (REVANKAR et al. 1975). As with the ribonucleoside 3′,5′-cyclic phosphates, compound (132) was more stable to acidic hydrolysis than the corresponding 5′-nucleotide (128).

The second method involves the inversion of configuration of the 2′-hydroxyl of *c*AMP (KHWAJA et al. 1972; ROBINS and KHWAJA 1974). This procedure was also applied to *c*GMP to obtain 9-(β-D-arabinofuranosyl)guanine 3′,5′-cyclic phosphate (MIAN et al. 1974). The ability of nucleosides to undergo intramolecular cyclonucleoside formation was exploited as a general route to *arabino* 3′,5′-cyclic nucleotides from the more readily available *ribo* 3′,5′-cyclic nucleotides. 8-Bromo-2′-*O*-tosyladenosine 3′,5′-cyclic phosphate (135, $R=NH_2$, $Y=H$) (IKEHARA and UESUGI 1972), obtained by tosylation of 8-bromoadenosine 3′,5′-cyclic phosphate (134, $R=NH_2$, $Y=H$), was converted to 8-oxo-2′-*O*-tosyladenosine 3′,5′-cyclic phosphate (136, $R=NH_2$, $Y=H$) by treatment with refluxing acetic acid and

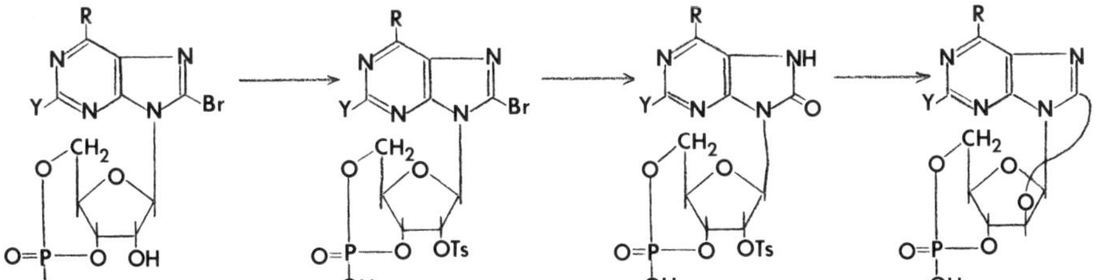

134 135 136 137

Ts = p-toluenesulfonyl

sodium acetate followed by methanolic ammonia at room temperature. The 8,2′-anhydro-9-(β-D-arabinofuranosyl)-8-hydroxyadenine 3′,5′-cyclic phosphate (137, $R=NH_2$, $Y=H$) was formed in 60% yield by heating (136) ($R=NH_2$, $Y=H$) in a solution of methanolic ammonia (KHWAJA et al. 1972).

The same procedure has also been applied to the synthesis of 8,2′-anhydro-9-(β-D-arabinofuranosyl)-8-hydroxyguanine 3′,5′-cyclic phosphate (137, R=OH, Y=NH_2). 8-Bromoguanosine 3′,5′-cyclic phosphate (134, R=OH, Y=NH_2) was tosylated with p-toluenesulfonylchloride to provide crystalline 8-bromo-2′-*O*-tosylguanosine 3′,5′-cyclic phosphate (135, R=OH, Y=NH_2) which when refluxed with sodium acetate in acetic acid gave 8-oxo-2′-*O*-tosylguanosine 3′,5′-cyclic phosphate (136, R=OH, Y=NH_2). Treatment of (136) (R=OH, Y=NH_2) with anhydrous sodium acetate in dimethylformamide resulted in intramolecular displacement of the 2′-*O*-tosyl group (IKEHARA and MUNEYAMA 1967) to give 8,2′-anhydro-8-hydroxy-9-(β-D-arabinofuranosyl)guanine 3′,5′-cyclic phosphate (137, R=OH, Y=NH_2) (MIAN et al. 1974).

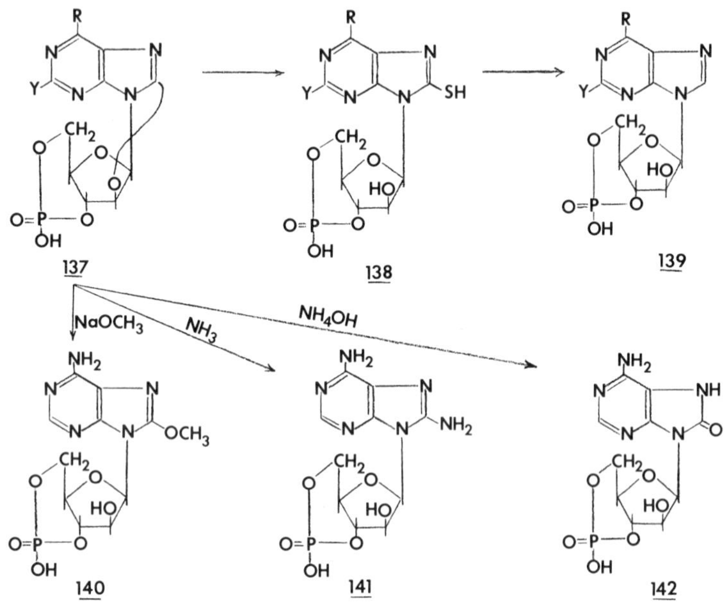

Reductive cleavage of the anhydro derivative (137) with hydrogen sulfide (MIAN et al. 1974; ROSOWSKY and GOHIL 1978), followed by Raney nickel desulfurization provided *ara*-cAMP (132) and 9-(β-D-arabinofuranosyl)guanine 3′,5′-cyclic phosphate (139, R=OH, Y=NH_2; *ara*-cGMP). Ring opening of the anhydronucleotide (137) (R=NH_2, Y=H) by nucleophilic attack at the 8-purinyl position with methoxide, ammonia or hydroxide gave 8-methoxy-(140), 8-amino-(141) and 8-oxo-(142)arabinofuranosyladenine 3′,5′-cyclic phosphate, respectively (MIAN et al. 1974).

b) Xylofuranosyl Cyclic Nucleotides

The importance of the stereochemical position of hydroxyl groups in the ease of cyclic phosphate formation is well illustrated by the direct formation of 9-(β-D-xylofuranosyl)adenine 3′,5′-cyclic phosphate (144) (HUBERT-HABART and GOODMAN 1969) and 9-(β-D-xylofuranosyl)guanine 3′,5′-cyclic phosphate (147) (REVANKAR et al. 1976) from 9-(β-D-xylofuranosyl)adenine (143) and 9-(β-D-xylofuranosyl)guanine (146), respectively, with phosphorus oxychloride in trimethylphosphate. Similar phosphorylation conditions gave only a trace amount of a 2′,5′-cyclic phosphate from 9-(β-D-arabinofuranosyl)adenine (HUBERT-HABART and GOODMAN 1969) and no identifiable formation of a 3′,5′-cyclic phosphate from either adenosine or guanosine (YOSHIKAWA et al. 1969). Similar results were also obtained in the phosphorylation of adenosine with pyrophosphoryl chloride in

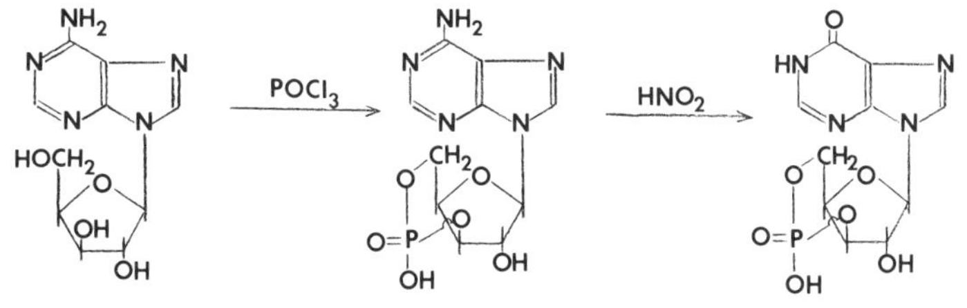

143 144 145

acetonitrile, nitrobenzene or ethyl acetate (IMAI et al. 1969). However, direct phosphorylation of (143) with pyrophosphoryl chloride in acetonitrile gave a mixture of 5′-phosphate and 3′,5′-cyclic phosphate (144), which was difficult to separate (IMAI et al. 1969). After deamination of the mixture with aqueous nitrous acid, the products were separated by ion-exchange chromatography and identified as 9-(β-D-xylofuranosyl)hypoxanthine 5′-phosphate and 9-(β-D-xylofuranosyl)hypoxanthine 3′,5′-cyclic phosphate (145) in a 1:2 ratio. Alkaline hydrolysis of 2′,3′-di-*O*-acetyl-9-(β-D-xylofuranosyl)adenine 5′-phosphoromorpholidate has also been reported to give, exclusively, the 3′,5′-cyclic phosphate (144) (IKEHARA and OHTSUKA 1963c).

Although the enzymatic phosphorylation of 9-(β-D-xylofuranosyl)guanine (146) using nucleoside phosphotransferase (MITSUGI et al. 1964) and p-nitrophenyl phosphate as the phosphate donor has been reported to yield 9-(β-D-xylofuranosyl)guanine 5′-phosphate (148) (SUZAKI et al. 1970), the direct chemical

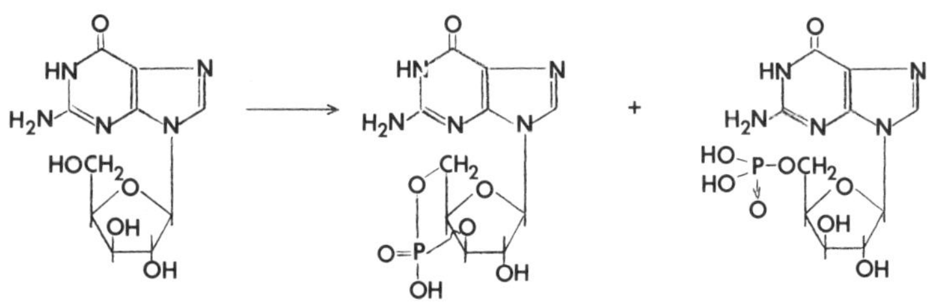

146 147 148

phosphorylation of (146) using phosphorus oxychloride in trimethylphosphate gave a mixture of (147) and (148), which were separated by ion-exchange chromatography (REVANKAR et al. 1976).

c) 2′-Modified Ribofuranosyl Cyclic Nucleotides

The structural modification at the 2′-position of the carbohydrate moiety of 3′,5′-cyclic nucleotides has been extensively studied by SUTHERLAND and co-workers

(FALBRIARD et al. 1967). A myriad of 2′-*O*-monoacyl derivatives of adenosine 3′,5′-cyclic phosphate (149) have been prepared by acylation of *c*AMP with an acyl halide or an acyl anhydride (FALBRIARD et al. 1967; BROER et al. 1972; MARUMOTO et al. 1979). These 2′-*O*-acyl derivatives can be further acylated to N^6,2′-*O*-diacyl-adenosine 3′,5′-cyclic phosphate (FALBRIARD et al. 1967). The acylation of tritiated

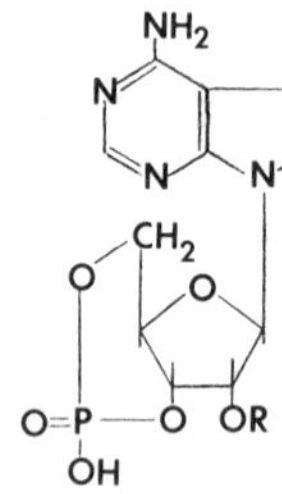

149

R = acetyl, butyryl, hexanoyl, octanoyl, lauroyl, stearoyl, benzoyl, adamantoyl, diazomalonyl, amino caproyl, etc.

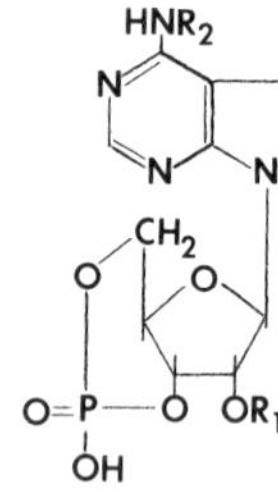

150

$R_1 = COCN_2CO_2Et$, $R_2 = H$

$R_1 = R_2 = COCN_2CO_2Et$

$R_1 = H$, $R_2 = COCN_2CO_2Et$

*c*AMP with diazomalonyl chloride has provided photo-affinity labeled *c*AMP (150) (BRUNSWICK and COOPERMAN 1971). These derivatives (150) have been used to study enzyme binding sites by generation of a reactive carbene upon photolysis leading to a covalent bond between the enzyme and *c*AMP.

A similar acylation of 3′,5′-cyclic nucleotides with succinic anhydride provided 2′-*O*-succinyl derivatives (151) (CAILLA and DELAAGE 1972; STEINER et al. 1972), which were then coupled with tyrosine methyl ester *via* mixed carboxylic-carbonic acid reaction to give the tyrosine methyl ester derivative (152). Iodination of (152) with $^{125}I_2$ gave ^{125}I-succinyl-cyclic nucleotide tyrosine methyl esters (153). These

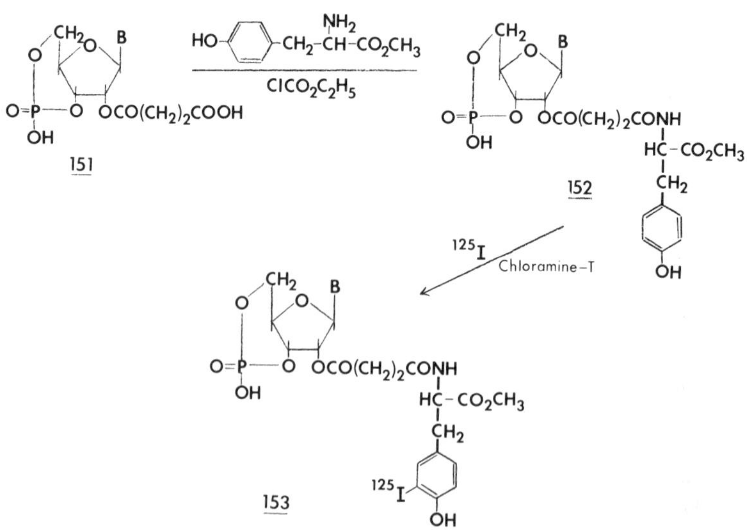

151 152 153

B = adenine, guanine, hypoxanthine or uracil

radioactive tyrosine cyclic nucleotides are extensively used in radioimmuno assay for the measurement of adenosine, guanosine, inosine and uridine 3′,5′-cyclic phosphates (STEINER et al. 1972a, b).

Reaction of guanosine 3′,5′-cyclic phosphate with a large excess of dihydropyran in dimethyl sulfoxide gave 2′-*O*-tetrahydropyranylguanosine 3′,5′-cyclic

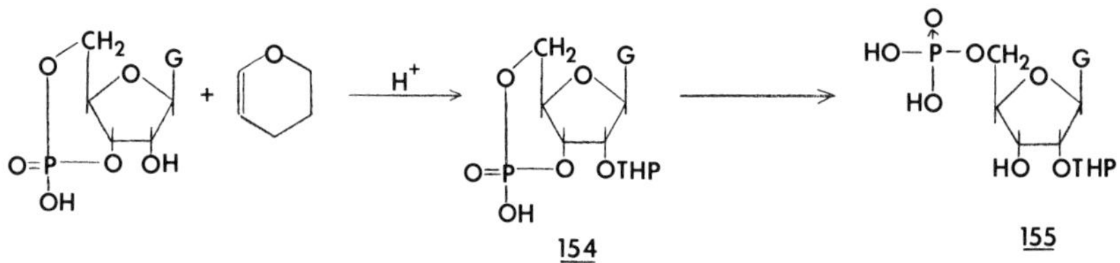

phosphate (154) (STRAUS and FRESCO 1965). Enzymatic hydrolysis of (154) gave exclusively the 5′-phosphate (155) which could be polymerized by the DCC procedure to obtain $C_{3'} \rightarrow C_{5'}$ phosphodiester linkages. A similar treatment of dihydropyran in anhydrous dioxane with uridine 3′,5′-cyclic phosphate (156) gave 2′-*O*-tetrahydropyranyl derivative (157), alkaline hydrolysis of which gave a mixture

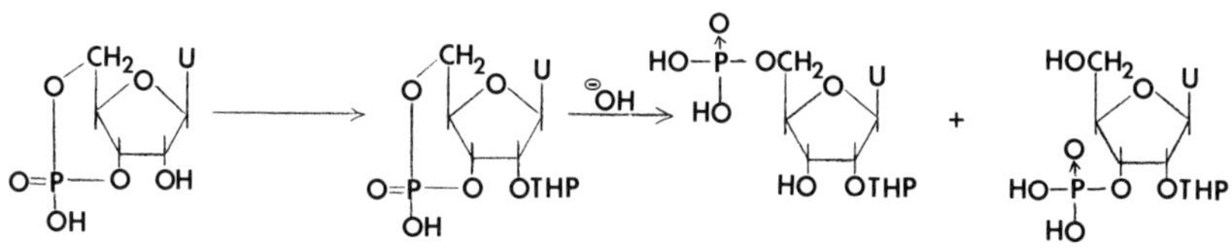

of 2′-*O*-tetrahydropyranyluridine 5′- and 3′-phosphates, (158) and (159) (SMITH and KHORANA 1959).

2′-*O*-Alkylation of nucleoside 3′,5′-cyclic phosphates with alkyl halides has provided a general route to 2′-*O*-alkylated adenine, hypoxanthine, cytosine and uracil 5′-nucleotides. When adenosine 3′,5′-cyclic phosphate was alkylated with methyl (or ethyl) iodide under alkaline conditions in dimethylformamide, 2′-*O*-methyl (or ethyl)adenosine 3′,5′-cyclic phosphate (160) was obtained (TAZAWA et al. 1972). Compound (160) was also prepared from 2′-*O*-methyladenosine (MILLER et al. 1973c). Phosphorylation of 2′-*O*-methyladenosine (BROOM and ROBINS 1965) gave 2′-*O*-methyladenosine 5′-phosphate (161) which on intramolecular cyclization with DCC in pyridine gave (160) (MILLER et al. 1973c; SHUMAN and ROBINS 1974a). Deamination of (160) with aqueous nitrous acid gave 2′-*O*-methylinosine 3′,5′-cyclic phosphate (163) (TAZAWA et al. 1972). The 2′-*O*-methyl-5′-nucleotides (161) and (164) were also prepared from (160) and (163), respectively, by phosphodiesterase (PDE) hydrolysis (TAZAWA et al. 1972). When *c*AMP was treated with 2,4-dinitrofluorobenzene in N,N-dimethylformamide in the presence of a base (like triethylamine or di-isopropylethylamine), the 2′-*O*-modified analog, 2′-*O*-(2,4-dinitrophenyl)adenosine 3′,5′-cyclic phosphate (162, R = H) was obtained, which was very stable under physiological conditions (JASTORFF and FREIST 1972, 1974; MILLER et al. 1973c). A similar treatment of 2-chloroadenosine 3′,5′-cyclic phosphate with 2,4-dinitrofluorobenzene in the presence of Hünig base (di-isopropyl-

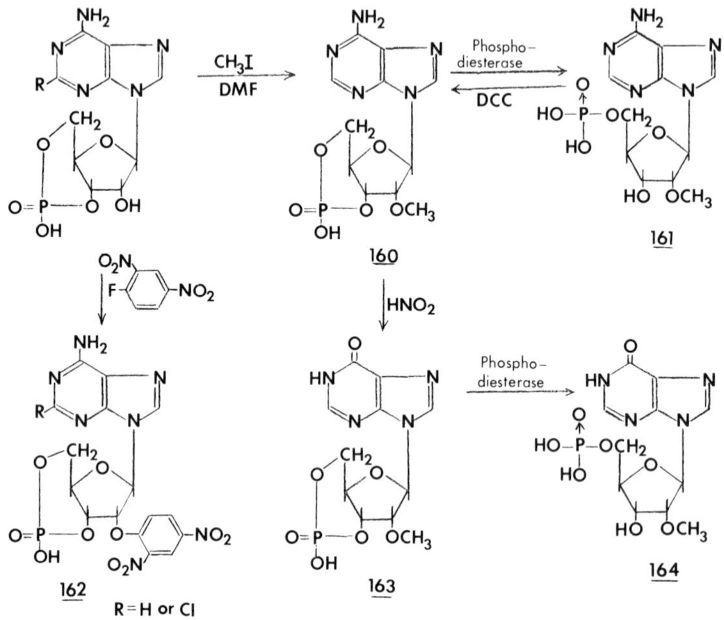

ethylamine) gave 2-chloro-2′-*O*-(2,4-dinitrophenyl)adenosine 3′,5′-cyclic phosphate (162, R=Cl) (JASTORFF and FREIST 1974). Although adenosine 5′-phosphorofluoridate has been prepared (BORDEN and SMITH 1966a) by the reaction of adenosine 5′-phosphate with 2,4-dinitrofluorobenzene, a similar attempt to prepare adenosine 3′,5′-cyclic phosphorofluoridate, however, gave only 2′-*O*-(2,4-dinitrophenyl)adenosine 3′,5′-cyclic phosphate (162, R=H) (KHWAJA and ROBINS 1979).

Other modifications of the glycosyl moiety of 3′,5′-cyclic nucleotides constitute the synthesis of 2′-deoxyribofuranosyl 3′,5′-cyclic nucleotides. The 2′-deoxy 3′,5′-cyclic phosphates of adenosine, inosine, guanosine, cytidine, uridine, and thymidine have been conveniently prepared by the cyclization of corresponding 5′-phosphates by the DCC (DRUMMOND et al. 1964) and ynamine (FUJIMOTO and NARUSE 1968) procedures or by the base-catalyzed "active phosphate" procedure (SMITH 1964; BORDEN and SMITH 1966b). Raney nickel desulfurization of 8,2′-anhydro-8-thio-9-(β-D-arabinofuranosyl)adenine 3′,5′-cyclic phosphate (165) (IKEHARA and UESUGI 1972; MIAN et al. 1974), obtained by treatment of 8,2′-anhydro-9-(β-D-arabinofuranosyl)-8-hydroxyadenine 3′,5′-cyclic phosphate (137) with

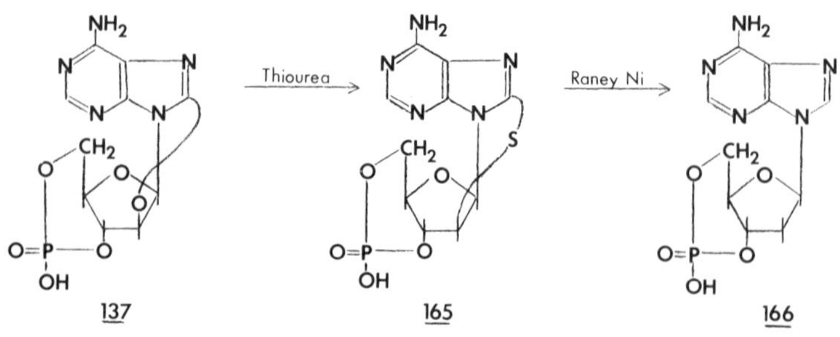

thiourea, readily gave 2′-deoxyadenosine 3′,5′-cyclic phosphate (166) (MILLER et al. 1976). Compound (166) has also been prepared enzymatically (HIRATA and HAYAISHI 1966; Kikkoman Shoyu Co. 1971 b; ISHIYAMA 1976).

Bromination of 2′-deoxyadenosine 3′,5′-cyclic phosphate gave 8-bromo-2′-deoxyadenosine 3′,5′-cyclic phosphate (Boehringer Mannheim GmbH 1970 b, 1971) in which the 8-bromo function is available for various nucleophilic displacement reactions. Reaction of 2′-deoxyinosine 3′,5′-cyclic phosphate (167) with

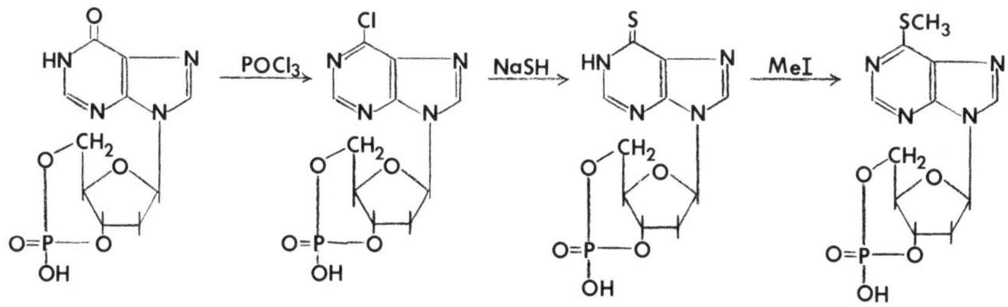

167 168 169 170

phosphorus oxychloride has been reported to give 6-chloro-9-(2-deoxy-β-D-ribofuranosyl)purine 3′,5′-cyclic phosphate (168) (HONJO et al. 1968). Nucleophilic displacement of the 6-chloro function of (168) with sodium hydrosulfide gave 9-(2-deoxy-β-D-ribofuranosyl)purine-6-thione 3′,5′-cyclic phosphate (169) which was subsequently methylated with methyliodide to 6-methylthio-9-(2-deoxy-β-D-ribofuranosyl)purine 3′,5′-cyclic phosphate (170) (HONJO et al. 1968).

d) 4′-Thioadenosine 3′,5′-Cyclic Phosphate and Aristeromycin 3′,6′-Cyclic Phosphate

Substitution of the oxygen atom in the D-ribofuranosyl ring in nucleoside 3′,5′-cyclic phosphate by sulfur has been reported by WHISTLER and co-workers (ANISUZZAMAN et al. 1973). Certain nucleoside and nucleotide analogs having the ring oxygen replaced by sulfur have shown interesting biological properties (BOBEK et al. 1970; HOFFMAN and WHISTLER 1970) among which is relatively low toxicity (HOFFMAN and WHISTLER 1968). Phosphorylation of 9-(2,3-*O*-isopropylidene-4-thio-β-D-ribofuranosyl)adenine (HOFFMAN and WHISTLER 1970) and subsequent removal of the blocking isopropylidene group gave 9-(4-thio-β-D-ribofuranosyl)-adenine 5′-phosphate (171) which on intramolecular cyclization with DCC in pyridine afforded 9-(4-thio-β-D-ribofuranosyl)adenine 3′,5′-cyclic phosphate (172) (ANISUZZAMAN et al. 1973). Treatment of (172) with butyric anhydride gave N^6,2′-*O*-dibutyryl-9-(4-thio-β-D-ribofuranosyl)adenine 3′,5′-cyclic phosphate (173) in 50% yield. However, when 6-chloro-9-(2,3,5-tri-*O*-acetyl-4-thio-β-D-ribofuranosyl)purine (BOBEK et al. 1970) was allowed to react with *n*-butylamine, N^6-(*n*-butyryl)-9-(4-thio-β-D-ribofuranosyl)adenine (174) was obtained. Phosphorylation of (174) *via* the 2′,3′-*O*-isopropylidene intermediate and subsequent hydrolysis gave N^6-(*n*-butyryl)-9-(4-thio-β-D-ribofuranosyl)adenine 5′-phosphate (175) which on cyclization with DCC in hot pyridine provided N^6-(*n*-butyryl)-9-(4-thio-β-D-ribofuranosyl)adenine 3′,5′-cyclic phosphate (176) (ANISUZZAMAN et al. 1973).

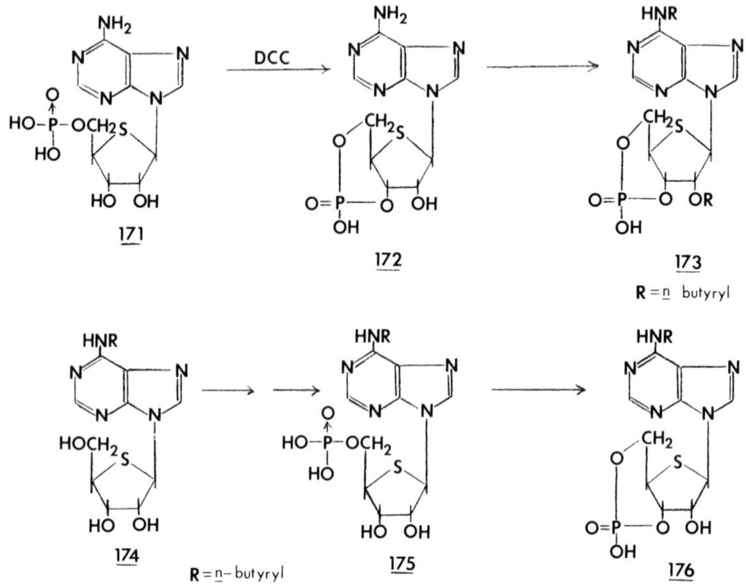

The carbocyclic analog of adenosine 3′,5′-cyclic phosphate, aristeromycin 3′,5′-cyclic phosphate (179), has been prepared from aristeromycin (177) by the base catalyzed intramolecular cyclization of nucleoside 5′-trichloromethylphosphonate (Marumoto et al. 1975). Treatment of aristeromycin (Kishi et al. 1967) with

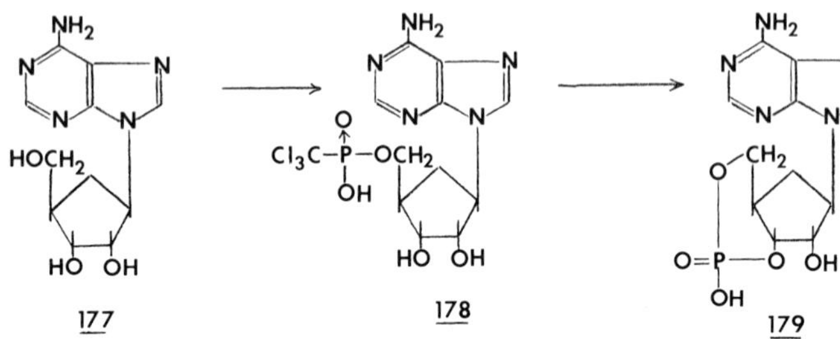

trichloromethylphosphorodichloridate in triethyl phosphate gave aristeromycin 6′-trichloromethyl phosphonate (178) in 70% yield. The reaction of (178) with KO-Bu^t in anhydrous dimethylsulfoxide afforded aristeromycin 3′,5′-cyclic phosphate (179) (Marumoto et al. 1975; Marumoto and Honjo 1975).

e) Hexopyranosyl Cyclic Nucleotides

Historically the first formation of a 4′,6′-cyclic phosphate of 7-(β-D-glucopyranosyl)theophylline has been suggested by Baddiley and co-workers (1954) in the reaction of 7-(β-D-glucopyranosyl)theophylline with phosphoryl chloride in pyridine (Fischer 1914). Stereochemically, the formation of 4′,6′-cyclic phosphates of hexopyranosyl derivatives is favorable as demonstrated by cyclization of 9-(β-D-glucopyranosyl)adenine 6′-phosphate (185) to 9-(β-D-glucopyranosyl)-adenine 4′,6′-cyclic phosphate (182) with trifluoroacetic anhydride at room tem-

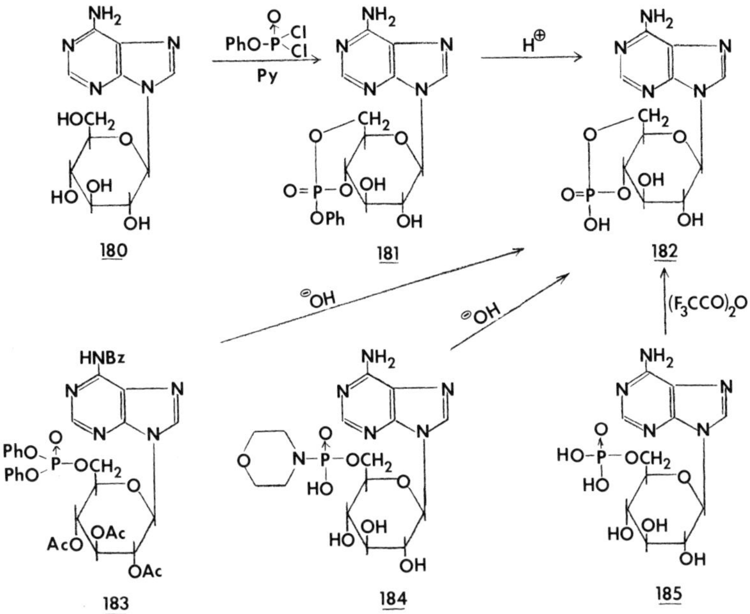

perature (BARKER and FOLL 1957). The 4′,6′-cyclic phosphate (182) was also obtained *via* the direct cyclic phosphate formation of 9-(β-D-glucopyranosyl)adenine (180) with phenyl phosphorodichloridate (BARKER and FOLL 1957) or by alkaline hydrolysis of 6-benzamido-9-(2,3,4-tri-*O*-acetyl-6-diphenylphosphoryl-β-D-glucopyranosyl)purine (183) (NOHARA et al. 1966), and from 9-(β-D-glucopyranosyl)-adenine 6′-phosphoromorpholidate (184) (IKEHARA and TAZAWA 1966).

f) Alditol Cyclic Nucleotides

9-D-Erythrityladenine 3′,4′-cyclic phosphate (186) has been reported to be formed by alkaline hydrolysis of 9-D-erythrityladenine 5′-triphosphate (IKEHARA and OHTSUKA 1963a) and by alkaline hydrolysis of erythrityladenine phosphoromorpholidate (IKEHARA and OHTSUKA 1963b).

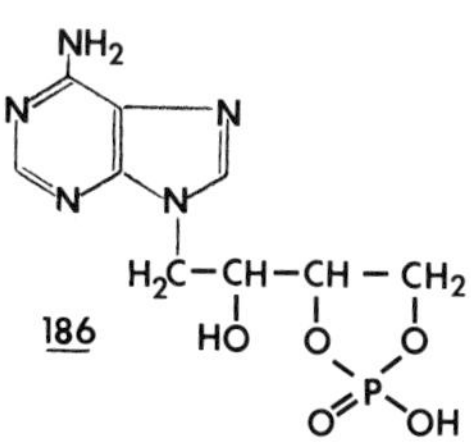

3. Structural Modification of the Cyclic Phosphate Moiety

a) Nucleoside 3′,5′-Cyclic Phosphorothioates

Perhaps one of the most critical structural features within the purine 3′,5′-cyclic nucleotide is the 3′,5′-cyclic phosphate ring. As a means of probing the importance of the cyclic phosphate ring, substitution of the phosphate oxygens by sulfur has been accomplished by several methods. The greater nucleophilic character toward carbon by sulfur as compared to oxygen with thiophosphates (AKERFELDT 1962) has been utilized for the synthesis of nucleoside 3′,5′-cyclic phosphorothioates,

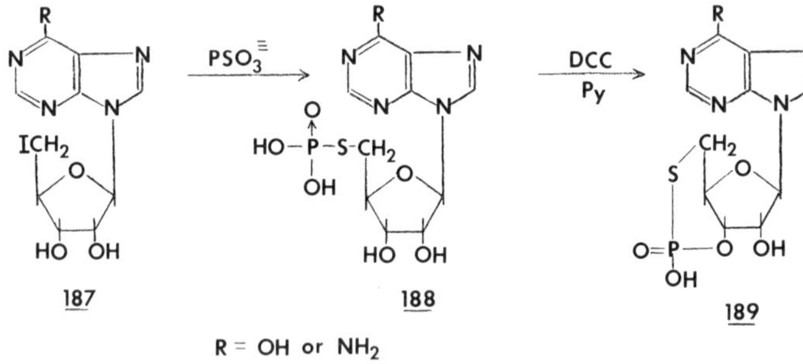

where sulfur has been incorporated within the cyclic phosphate ring. The acid-labile 5′-deoxy-5′-thioinosine 5′-phosphorothioate (188, R=OH) (HAMPTON et al. 1969; HAGA et al. 1971) and 5′-deoxy-5′-thioadenosine 5′-phosphorothioate (188, $R=NH_2$) (SHUMAN et al. 1973) were conveniently prepared by nucleophilic displacement of the 5′-iodo function of (187) by thiophosphate anion. When the 5′-phosphorothioates (188) were subjected to intramolecular cyclization with DCC in refluxing pyridine, 5′-deoxy-5′-thioinosine 3′,5′-cyclic phosphorothioate (189, R=OH) and 5′-deoxy-5′-thioadenosine 3′,5′-cyclic phosphorothioate (189, $R=NH_2$) were obtained in rather low yield (SHUMAN et al. 1973; SHUMAN and ROBINS 1974b).

Dreiding molecular models show that the cyclic phosphorothioate ring of (189) is slightly larger and less strained than the corresponding 3′,5′-cyclic phosphate ring. This observation is reflected in the ease of formation of the 2′,5′-dideoxy-5′-thionucleoside 3′,5′-cyclic phosphorothioates (192) by aqueous base-catalyzed intramolecular displacement of the 5′-*O*-tosyl nucleoside 3′-phosphorothioates (190) (CHLADEK and NAGYVARY 1972). The 2′,5′-dideoxy-5′-thionucleoside 3′,5′-cyclic phosphorothioates (192) were also readily formed at room temperature from the corresponding cyanoethyl esters (191). Although the pyrimidine 3′,5′-cyclic phos-

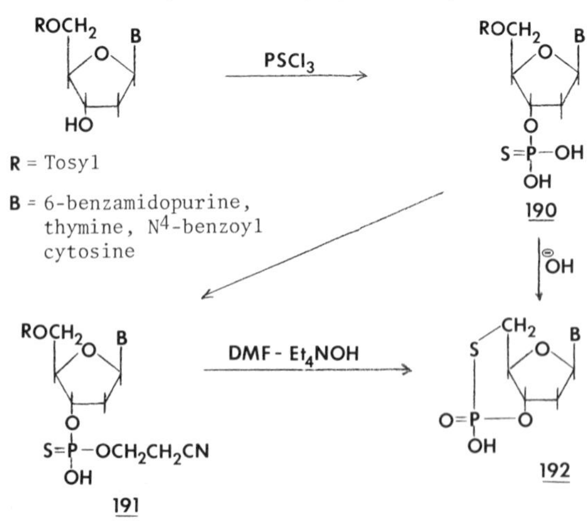

phorothioates (192, B = thymine or cytosine) were resistant to snake venom phosphodiesterase, 2′,5′-dideoxy-5′-thioadenosine 3′,5′-cyclic phosphorothioate (192, B = adenine) is readily hydrolyzed. This behavior parallels that of the natural nucleoside 3′,5′-cyclic phosphate (Drummond et al. 1964). The nucleoside 3′,5′-cyclic phosphorothioates (192), however, were stable to 0.1 *N* KOH at 20 °C for 24 h in the absence of oxygen (Chladek and Nagyvary 1972). The 5′-deoxy-5′-thionucleoside 3′,5′-cyclic phosphorothioates (189) were much more stable to acidic hydrolysis than the corresponding 5′-deoxy-5′-thionucleoside 5′-phosphorothioates (188). The cyclic phosphorothioates (189) were stable for 6 h at 50 °C to pH 2, 3, 7 or 9 buffers (Shuman et al. 1973).

Since ring closure of nucleoside 5′-phosphorothioate with DCC in pyridine usually leads to nucleoside 5′-phosphate through desulfurization, an alternate cyclizing agent was employed for the synthesis of adenosine 3′,5′-cyclic phosphorothioate (194) (Eckstein 1970). Activation of the phosphorothioate moiety of N^6-dimethylaminomethyleneadenosine 5′-phosphorothioate (193) with triisopropylbenzenesulfonyl chloride, followed by deprotection of the N^6-amino function gave 50% yield of a nucleotide identified as adenosine 5′-S-methylphosphorothioate (Eckstein et al. 1974). No formation of adenosine 3′,5′-cyclic phos-

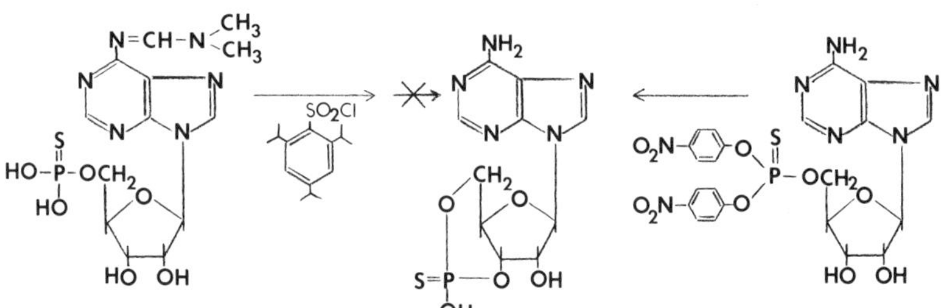

193 194 195

phate was detected in this reaction. However, the cyclization of adenosine 5′-*O*,*O*-bis(p-nitrophenyl)phosphorothioate (195) with potassium *tert*-butoxide in dimethylformamide gave the required adenosine 3′,5′-cyclic phosphorothioate (194) (Eckstein et al. 1974). As the P atom in (194) is asymmetric, adenosine 3′,5′-cyclic phosphorothioate exist as two diastereomers, the separation of which has not as yet been achieved (Eckstein et al. 1974).

The synthesis of adenosine 3′,5′-cyclic sulfate (197) has recently been reported (Gulyaev et al. 1976). Treatment of N^6-benzoyl-2′,3′-*O*-isopropylideneadenosine with sulfuryl chloride gave the 5′-*O*-sulfochloride derivative (196), which on hydrolysis and ring closure in the presence of DCC gave (197) (Gulyaev et al. 1976).

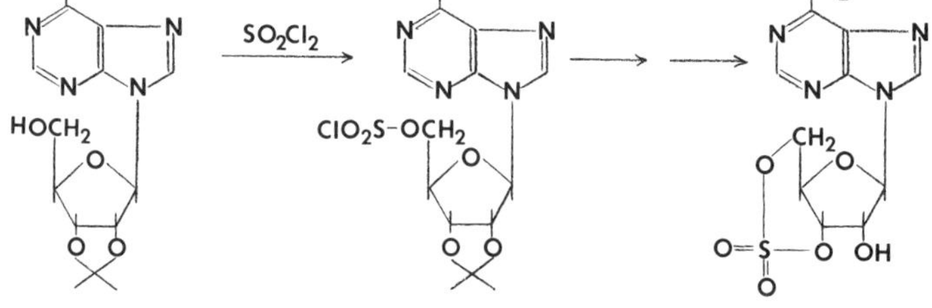

196 197

b) Nucleoside 3′,5′-Cyclic Phosphoramidates

Substitution of phosphoryl oxygen and phosphate oxygen of nucleoside 3′,5′-cyclic phosphates by NH, NHR or NR_2 has been described. The replacement of an exocyclic oxygen by NR_2 to obtain nucleoside 3′,5′-cyclic phosphoramidates has been accomplished from nucleoside 3′,5′-cyclic phosphate (MEYER Jr et al. 1973a) and also by cyclization of a nucleoside 5′-phosphoramide by nucleophilic attack on carbon (RUSSELL and MOFFATT 1969). In the former procedure (MEYER Jr et al. 1973a, b), the synthesis of nucleoside 3′,5′-cyclic phosphoramidates was accomplished by activation of nucleoside 3′,5′-cyclic phosphates to nucleophilic attack

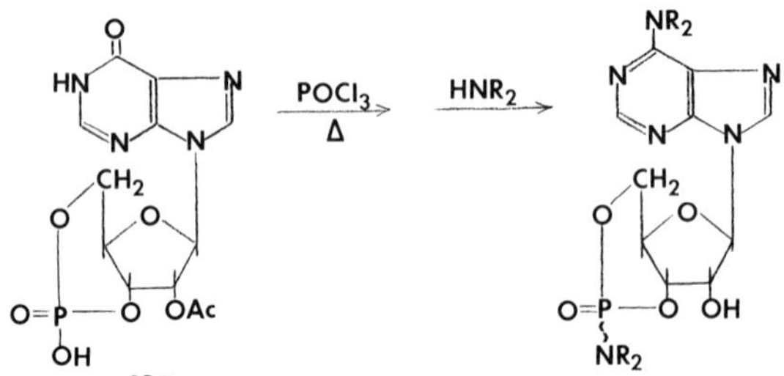

198 199

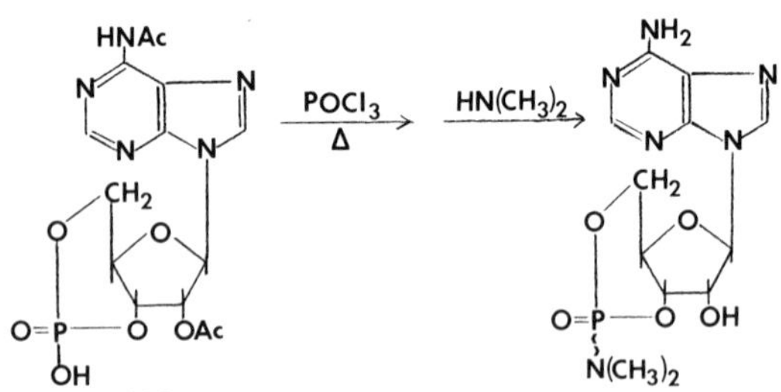

200 201

by reaction with phosphorus oxychloride (presumably by formation of a 3′,5′-cyclic phosphorochloridate or a dichlorophosphoric anhydride). Heating 2′-*O*-acetylinosine 3′,5′-cyclic phosphate (198) (MEYER Jr et al. 1972) and N^6,2′-*O*-diacetyladenosine 3′,5′-cyclic phosphate (200) (FALBRIARD et al. 1967) with phosphorus oxychloride followed by treatment with an alkylamine gave the nucleoside 3′,5′-cyclic phosphoramidates (199) and (201), respectively (MEYER Jr et al. 1973a, b). However, the question of the asymmetric configuration about phosphorus was not resolved (MEYER Jr et al. 1973a). Acidic hydrolysis of (201) in pH 1 buffer at 37 °C for 3 h gave approximately 50% hydrolysis to mainly adenosine 3′,5′-cyclic phosphate.

Because of the instability of 5′-amino-5′-deoxyadenosine 5′-phosphoramidate (202), the direct synthesis of 5′-amino-5′-deoxyadenosine 3′,5′-cyclic phosphoramidate (204, R = H) from (202) was unsuccessful (SHEN 1970). However, the neutral 5′-amino-5′-deoxyadenosine 3′,5′-cyclic phosphoramidates (204) were synthesized by transesterification of the "activated" bis-(p-nitrophenyl)phosphoramidate esters (203) (MURAYAMA et al. 1970, 1971; JASTORFF and KREBS 1972). The isolation

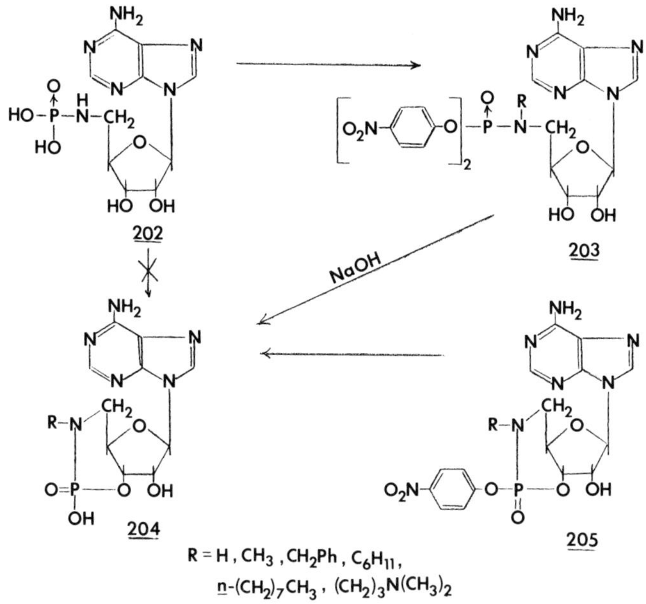

of the unusual cyclic p-nitrophenyl phosphoramidate diester (205) from the reaction suggests that the direct synthesis of the cyclic phosphoramidates (204) occurred *via* (205). The isolation of the diastereomers of (205) represents one of the few reported nucleoside 3′,5′-cyclic phosphate diastereomers to be characterized (MURAYAMA et al. 1973). The cyclic phosphoramidates (204) are much less stable to acidic hydrolysis than *c*AMP with significant P–N hydrolysis occurring after 5 h at 37 °C in buffer solutions with pH <7.

Replacement of 3′-oxygen by an amino group in *c*AMP gave 3′-amido-3′-deoxyadenosine 3′,5′-cyclic phosphate (208), the synthesis of which has recently been reported (MORR et al. 1974, 1975; MORR and ERNST 1978). The chemical or enzymatic phosphorylation of 3′-amino-3′-deoxyadenosine gave 3′-amino-3′-deoxyadenosine 5′-phosphate (206, R=H) which on intramolecular cyclization with DCC in dimethylformamide containing *tert*-butanol and water gave 3′-amido-3′-deoxyadenosine 3′,5′-cyclic phosphate (208) (MORR et al. 1974). The use

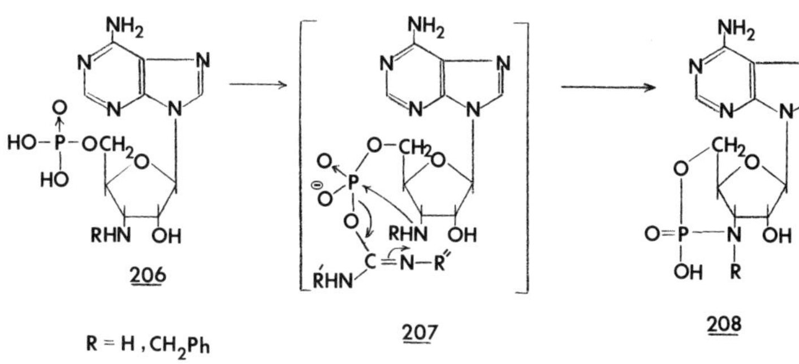

of water-soluble carbodiimides such as 1-ethyl-3-(3-dimethylaminopropyl)carbodiimide hydrochloride or bicarbonate (EDC · HCl/HCO_3) was found to be much superior to DCC and react under very mild conditions (in aqueous solution at pH 7.2–7.5 and 37 °C for 2–5 h) giving yields of (208) as high as 95% with no detectable side products (MORR et al. 1975; MORR and ERNST 1978). Nucleophilic attack of the 3′-amino group on the intermediate *O*-acylurea may proceed *via* a cyclic structure as formulated in (207), and this probably accounts for the rather high reaction rates and for the fact that intramolecular cyclization occurs almost exclusively. When 3′-amino-3′-deoxyadenosine 2′,5′-diphosphate (209) was treated with EDC · HCO_3 (see below) followed by alkaline phosphatase to remove the residual free phosphate group of (210), the 3′,5′-cyclic derivative (208) was formed rather than the 2′,3′-cyclic phosphoramidate derivative (MORR et al. 1975).

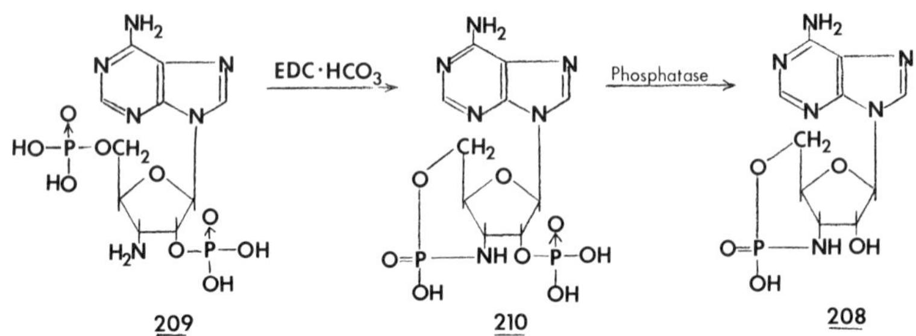

209 210 208

2′,3′-Bis(2-chloroethyl)aminophosphoryl-3′-amino-3′-deoxyadenosine (212) has recently been synthesized by treatment of 3′-amino-3′-deoxyadenosine with bis(2-chloroethyl)aminophosphorodichloridate in triethyl phosphate, which gave the isomeric (212) presumably through the intermediate (211) (OKRUSZEK and VERKADE 1979). The isomers were subsequently separated by column chromatography.

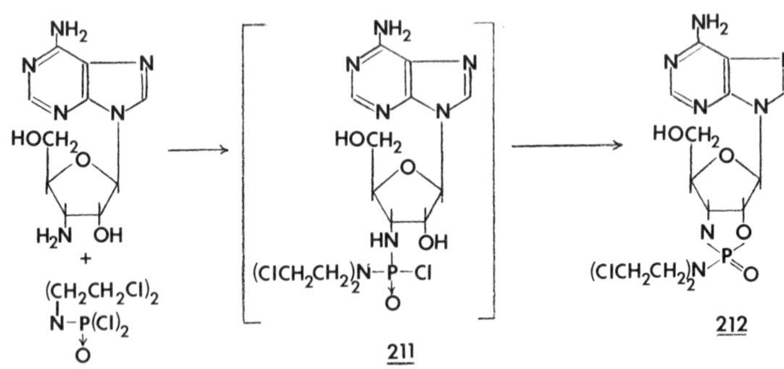

211 212

The 3′-N–P bond in (208) is considerably more stable (no hydrolysis after 5 h at pH 7 and 37 °C) than the 5′-N–P bond in (204) (R = H) (MORR et al. 1974). However, substitution of hydrogen by the benzyl group in the 3′-amido position (208, $R = CH_2C_6H_5$) makes it more susceptible to acidic hydrolysis of the P–N bond than the unsubstituted (208) (R = H) (MORR et al. 1975).

When *xylo*-3′-amino-3′-deoxyadenosine (213) is allowed to react with phosphorus oxychloride or with thiophosphoryl chloride, *xylo*-adenosine 3′,5′-cyclic phosphates (215) (X = O and X = S) are formed in high yields (MORR and MENGEL 1977).

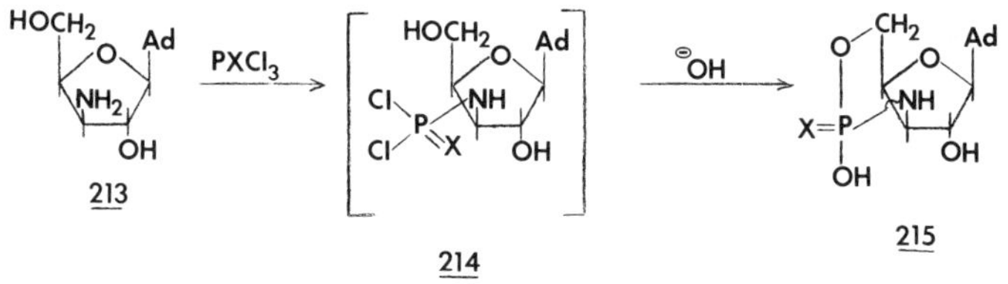

X = O or S

c) Nucleoside 3′,5′-Cyclic Thiophosphoramidates

The synthesis of 5′-deoxy-5′-aminoadenosine 3′,5′-cyclic thiophosphoramidates (217) requires much stronger alkaline conditions and higher temperatures than for analogous 5′-deoxy-5′-aminonucleoside 3′,5′-cyclic phosphoramidates (204) (JASTORFF and KREBS 1972, 1974), and the P–N bond of (217) is much more resistant to acidic hydrolysis. The relative stabilities found for the N-octyl derivatives (204) and (217) [$R = n\text{-}(CH_2)_7CH_3$] to acidic hydrolysis were attributed to steric effect.

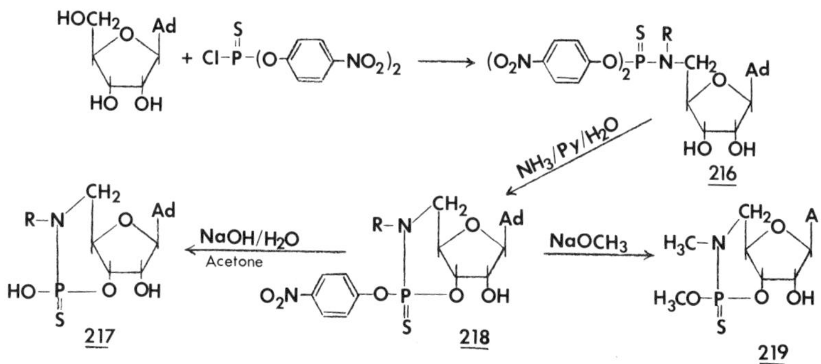

$R = CH_3$, $n\text{-}(CH_2)_7\text{-}CH_3$

In contrast to the P=O phosphoramidates (204), diastereomers of the P=S diesters (217) could not be detected by chromatography or ^{31}P spectral determinations. Treatment of the p-nitrophenyl diester (218) with sodium methoxide gave quantitative yield of methyl thiophosphoramidate (219) (JASTORFF and KREBS 1972).

The water soluble 1-ethyl-3-(3-dimethylaminopropyl)carbodiimide bicarbonate ($EDC \cdot HCO_3$) has also been used as the cyclizing agent to obtain 3′-amino-5′-thio-3′,5′-dideoxyadenosine 3′,5′-cyclic phosphate (221). Ring closure of 3′-amino-5′-thio-3′,5′-dideoxyadenosine phosphate (220) in presence of $EDC \cdot HCO_3$ at 27 °C for 7 h gave 28% yield of (221) (MORR 1976).

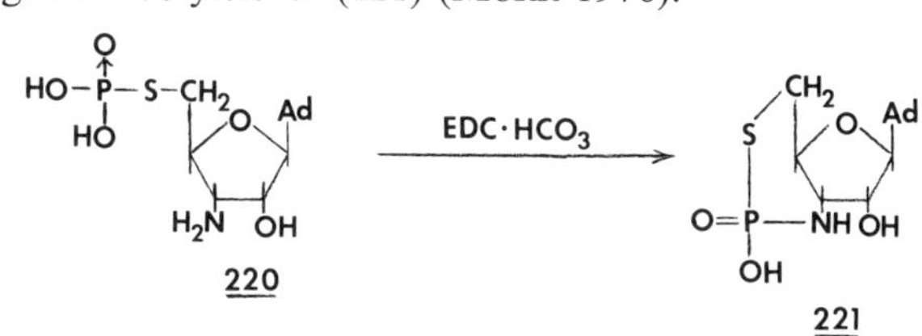

The reactions of aminonucleosides with thiophosphorylating reagents has recently been described (MORR and ERNST 1978). When 3′-alkylamino-3′-deoxyadenosine (222) was treated with thiophosphoryl chloride in alkaline media gave 3′-alkylamino-3′-deoxyadenosine 3′,5′-cyclic thiophosphoramidate (223). A similar reaction of 3′,5′-diamino-3′,5′-dideoxyadenosine with thiophosphoryl chloride gave 3′-amino-3′,5′-dideoxyadenosine 3′,5′-cyclic amidothiophosphor-

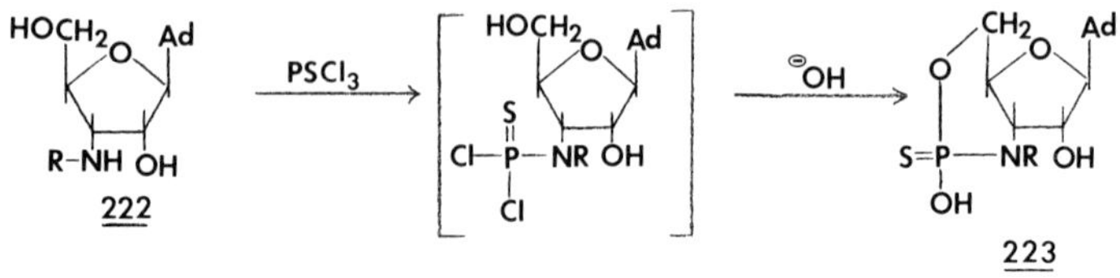

R = CH_2Ph, CH_2CH_2Ph, $CH(CH_3)_2$, $(CH_2)_7CH_3$

amidate (225). Diazomethane methylation of (225) in aqueous methanol yielded the corresponding methylthio derivative (226). However, when 3′-amino-3′-deoxyadenosine (227) was treated with thiophosphoryl chloride in alkaline media

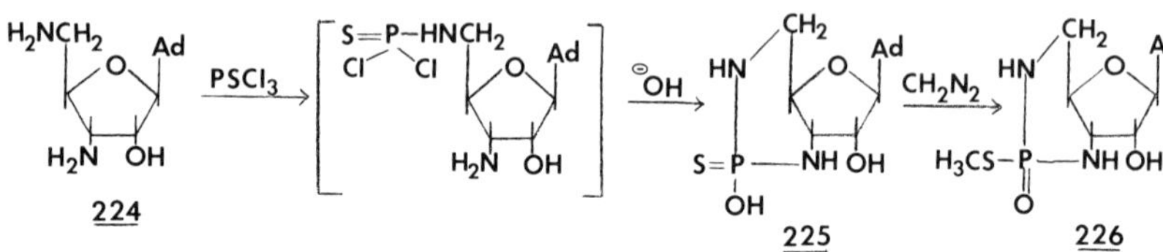

3′-amino-3′-deoxyadenosine 2′,3′-cyclic thiophosphoramidate (228) was formed (MORR and ERNST 1978). These cyclic thiophosphoramidates were formed in good yields, and as mixtures of diastereomers which were separated by DEAE-sephadex ion exchange chromotography into pure isomers (MORR and ERNST 1978). The P–N bonds of these cyclic phosphoramidates are less susceptible to acidolysis than the corresponding compounds having P=O groups.

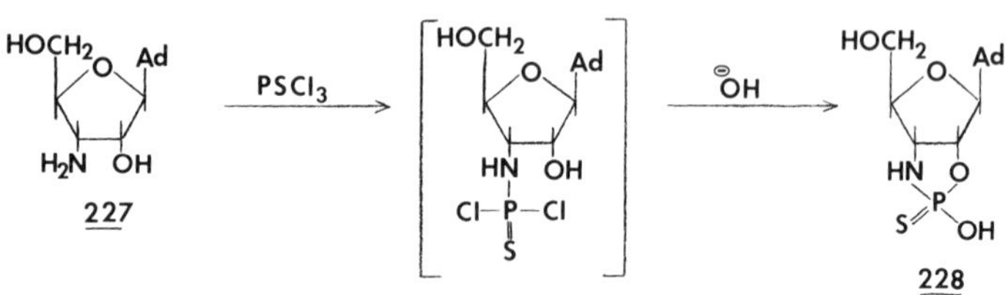

d) Nucleoside 3′,5′-Cyclic Alkylphosphates

In order to overcome the low cell membrane penetrating ability of exogenously added adenosine 3′,5′-cyclic phosphate (POSTERNAK 1974), the negative charge has been removed from *c*AMP by its conversion to esters. The formation of the methyl and ethyl esters of *c*AMP (230, R = CH_3 or C_2H_5) from the silver salt of adenosine 3′,5′-cyclic phosphate (229) and methyl or ethyl iodide was first reported by FALBRIARD and co-workers (1967). More recently, a series of neutral alkyl phosphotriesters of *c*AMP (230) have been described (GOHIL et al. 1974). Treatment of *c*AMP with triisopropylbenzenesulfonyl chloride in cold gave the reactive intermediate (231) without much side reaction, which when subjected to alcoholysis furnished the diastereomeric mixture of *c*AMP phosphotriesters in over 50% yield

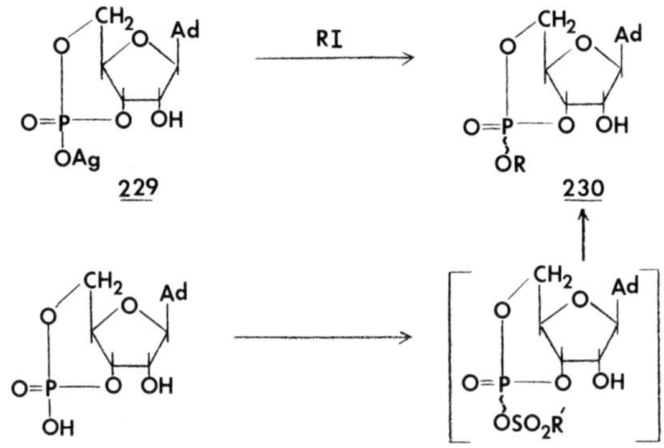

229 230 231

R = CH_3, C_2H_5, C_3H_7, C_4H_9, $(CH_2)_{15}CH_3$

(GOHIL et al. 1974). The ratio of two isomers, axial (endo) and equatorial (exo), may well depend on the size of the alkyl groups. Ethyl-*c*AMP (230, $R = C_2H_5$) in crystals exists in the axial form as it was determined by x-ray crystallography (COTTON et al. 1975).

Several triesters of *c*AMP have been prepared (ENGELS and SCHLAEGER 1977) according to the simple and straightforward method of TODD and co-workers (BROWN et al. 1955), through direct treatment of *c*AMP with diazaalkanes. This single-step reaction gave diastereomeric mixtures which were chromatographically

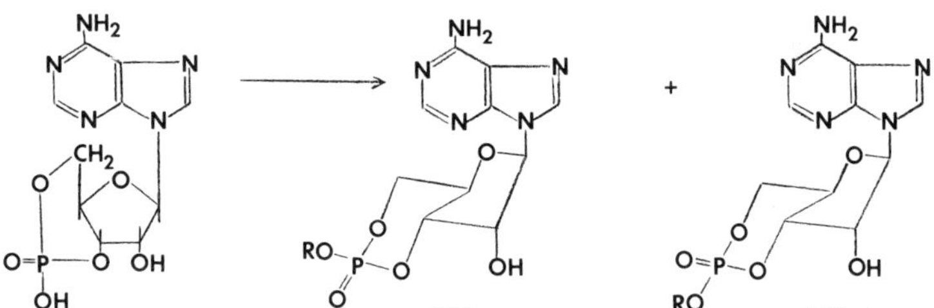

232 axial

233 equatorial

separated into the axial (232) and equatorial (233) isomers. The thermodynamically more stable axial isomer usually predominates (ENGELS and SCHLAEGER 1977). Treatment of ethylene oxide with triethylammonium salt of *c*AMP in aqueous solution gave only the axial isomer of adenosine 3′,5′-cyclic phosphate P-(2-hydroxyethyl) ester (234) in 32% yield (ZIELINKSI et al. 1978). The stabilization

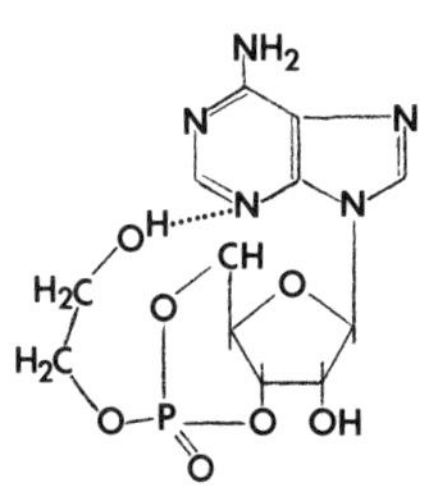

234

of such a structure by intramolecular hydrogen bonding between hydroxyl function of *β*-hydroxyethyl residue and N-3 nitrogen of the aglycon which adopts the preferred *syn* conformation (LEE and SARMA 1976) is quite conceivable.

The synthesis of N^6,2′-*O*-diacetyladenosine 3′,5′-cyclic phosphate methyl and ethyl esters (235) has been reported (NAGYVARY et al. 1973). Triesterification of the tri-*n*-butylammonium salt of N^6,2′-*O*-diacetyl-*c*AMP with anhydrous methanol or

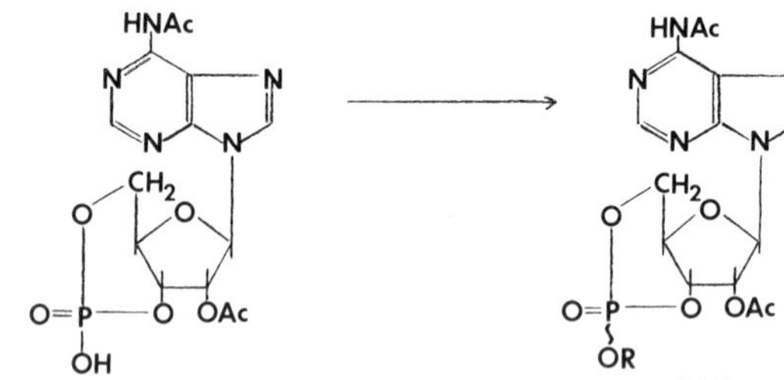

235

R = CH_3, C_2H_5

ethanol in presence of methanesulfonyl chloride gave the corresponding diastereomeric mixture of methyl or ethyl ester of N^6,2′-*O*-diacetyladenosine 3′,5′-cyclic phosphate (235) (NAGYVARY et al. 1973).

5′-Mono- and dialkyl analogs of *c*AMP have also been prepared (RANGANATHAN et al. 1974; JONES et al. 1975). Phosphorylation of the pure isomers of protected 6-deoxyhexofuranosyl nucleosides with the *β*-D-allo (236) and *α*-L-talo (237) configurations or a mixture of both, with bis(2,2,2-trichloroethyl)phosphorochloridate, followed by removal of the protecting groups, gave the corresponding epimeric 6-deoxyhexofuranosyladenine 5′-phosphates, (238) and (239) (RANGANATHAN et al. 1974; JONES et al. 1975). Intramolecular cyclization of both

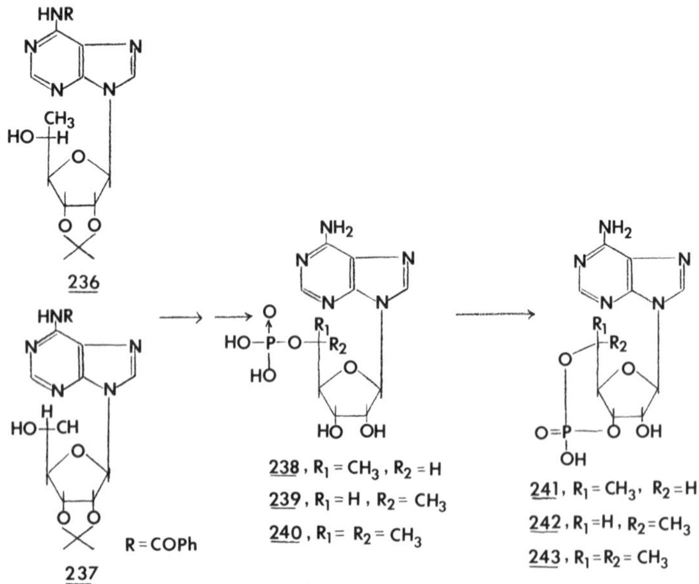

236

237

R = COPh

238, $R_1 = CH_3$, $R_2 = H$
239, $R_1 = H$, $R_2 = CH_3$
240, $R_1 = R_2 = CH_3$

241, $R_1 = CH_3$, $R_2 = H$
242, $R_1 = H$, $R_2 = CH_3$
243, $R_1 = R_2 = CH_3$

(238) and (239) using DCC in pyridine gave essentially quantitative yield of 3′,5′-cyclic phosphates (241) and (242), respectively, as crystalline free acids. Phosphorylation of 5′,5′-dimethyladenosine (NUTT and WALTON 1968), followed by cyclization gave 5′,5′-dimethyladenosine 3′,5′-cyclic phosphate (243) (RANGANATHAN et al. 1974).

Treatment of 9-(6-deoxy-β-D-allofuranosyl)adenine 3′,5′-cyclic phosphate (241) with bromine in a buffered aqueous solution gave the crystalline 8-bromo derivative (244a) which on subsequent treatment with benzyl mercaptan in the presence of sodium methoxide led to the formation of 8-benzylthio-9-(6-deoxy-β-D-allofuranosyl)adenine 3′,5′-cyclic phosphate (244b) (RANGANATHAN et al. 1974).

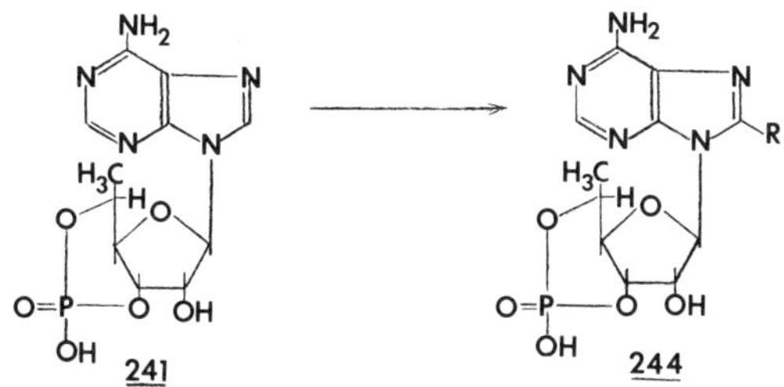

241

244

a, R = Br

b, R = SCH_2Ph

In view of the reported instability of several cyclic triesters of phosphoric acid (WESTHEIMER 1968), the cyclic phosphate triester moiety of nucleoside 3′,5′-cyclic alkylphosphates would be expected to be much more labile to hydrolysis than the cyclic phosphate ring in *c*AMP. In contrast to nucleoside 2′,3′-cyclic phosphate triesters (BROWN et al. 1955), which are too labile to be isolated, the adenosine 3′,5′-cyclic phosphate triesters (230) are relatively stable. Although the enthalpy of hydrolysis of *c*AMP was greater than that of adenosine 2′,3′-cyclic phosphate (GREENGARD 1971), the relative stability of *c*AMP triesters was somewhat unexpected. However, alkaline hydrolysis of these *c*AMP triesters (230) resulted predominantly in ring opening (GOHIL et al. 1974). On the other hand, nucleophilic attack by thiourea led to the formation of *c*AMP as the main product (GOHIL et al. 1974). Both alkaline and acidic hydrolysis of the benzyl ester gave rise mainly to *c*AMP (ENGELS and SCHLAEGER 1977).

e) Nucleoside 3′,5′-Cyclic Phosphonates

The synthesis of isosteric phosphonate analogs of both adenosine 5′-phosphate (245) (JONES and MOFFATT 1968) and adenosine 3′-phosphate (247) (ALBRECHT et al. 1970) has provided a route to the synthesis of adenosine 3′-5′-cyclic phosphonates in which the phosphate oxygen has been replaced by a methylene group (JONES and MOFFATT 1969; JONES et al. 1970). The facile intramolecular, high-dilution cyclization of the nucleoside 5′-phosphonate (245a) to adenosine 3′,6′-cyclic phosphonate (246a) with DCC occurred in hot pyridine. The cyclization of (245a) to (246a) was much more facile than that of adenosine 5′-phosphate and could be carried out even in aqueous pyridine in high yields (JONES et al. 1970). Adenosine 3′,5′-cyclic phosphonates (246a) were also obtained by hydrolysis of the intermediate aryl adenosine 3′,6′-cyclic phosphonates (246b) which were formed by base-

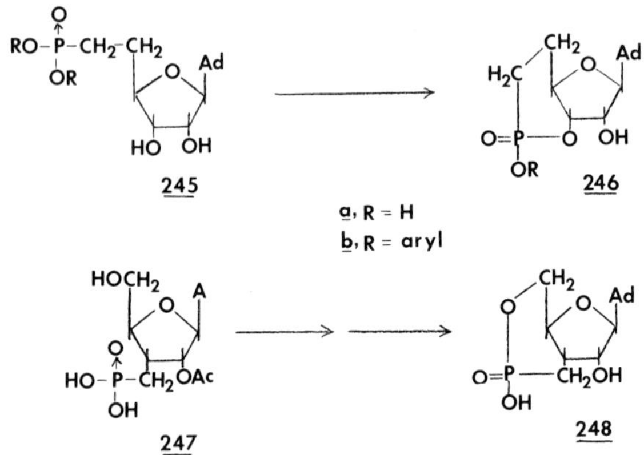

245 246

a, R = H
b, R = aryl

247 248

catalyzed transesterification of the aryl adenosine 5′-phosphonate (245b). However, prior protection of the 2′-hydroxy group of adenosine 3′-phosphonate (247) was necessary for cyclization of (247) to adenosine 3′,5′-cyclic phosphonate (248) (JONES et al. 1970).

The reaction of triethylammonium salt of adenosine 3′-phosphonate (249) with DCC in *tert*-butanol-dimethylformamide mixture at 80 °C gave quantitative yield of adenosine 2′,3′-cyclic phosphonate (250) (JONES et al. 1970). Adenosine 2′,3′-cyclic phosphonate (250) had a stability comparable to adenosine 2′,3′-cyclic phosphate, being 50% hydrolyzed by 0.1 *M* hydrochloric acid in 35 min at 23 °C (JONES et al. 1970).

249 250

f) Nucleoside 3′,5′-Cyclic Pyrophosphates

A further modification of the cyclic phosphate moiety is illustrated by the 2′,5′- and 3′,5′-cyclic pyrophosphates of adenosine (251 and 252) formed during barium hydroxide hydrolysis of adenosine 5′-triphosphate (LIPKIN et al. 1959a, b). Adenosine 3′,5′-cyclic pyrophosphate (252) has also been isolated from red algae *Porphyra perforata* (SU and HASSID 1962).

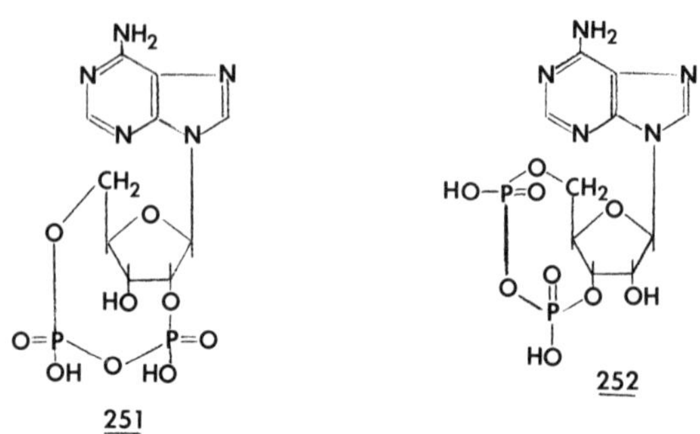

251 252

II. Cyclic Nucleotide Analogs Related to *c*AMP

1. 1-Deazapurine Cyclic Phosphates

Cyclic phosphates of imidazo[4,5-b]pyridine (1-deazapurine), an analog of purine with a ring nitrogen substituted by less electronegative carbon, has recently been described (MIZUNO et al. 1975). The first chemical synthesis of 3-(β-D-ribofuranosyl)imidazo[4,5-b]pyridine 5′-phosphate (1-deazanebularin 5′-phosphate, 253) by the conventional 2-cyanoethyl phosphate procedure is reported by WOENCKHAUS and PFLEIDERER (1965). However, the synthesis of 1-deazaadenosine 5′-phosphate

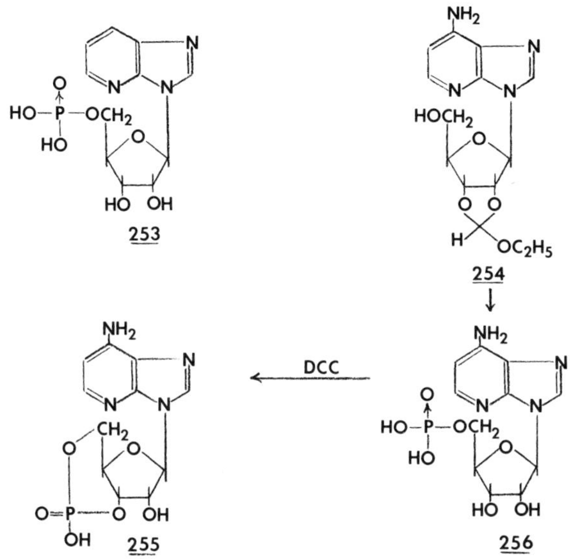

(256), which is a prerequisite for the preparation of the corresponding 3′,5′-cyclic phosphate (255), was prepared by the treatment of 2′,3′-*O*-ethoxymethylene-1-deazaadenosine (254) with pyrophosphoryl tetrachloride. Since 1-deazaadenosine is more acid-labile than 3-deazaadenosine with respect to N-glycosyl bond cleavage, substitution of 2′,3′-*O*-isopropylidene-1-deazaadenosine for (254), removal of the protecting groups under acidic conditions always accompanied by considerable cleavage of N-glycosyl bond. Cyclization of (256) with DCC in pyridine gave 1-deazaadenosine 3′,5′-cyclic phosphate (255). Hydrolysis of (255) with barium hydroxide gave (256) and 1-deazaadenosine 3′-phosphate in a ratio of 27:73 (MIZUNO et al. 1975).

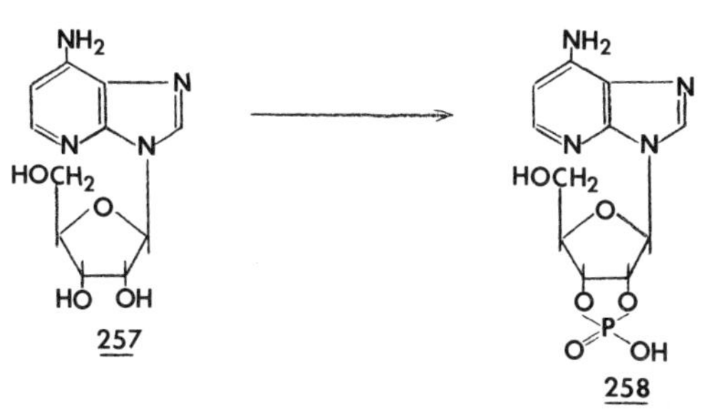

1-Deazaadenosine 2′,3′-cyclic phosphate (258) was also prepared (MIZUNO et al. 1975) from 1-deazaadenosine (257) employing pyrophosphate-dimethylformamide procedure (UEDA and KAWAI 1970). The 2′,3′-cyclic phosphate (258) was hydrolyzed to the corresponding nucleoside 3′-phosphate by RNase M.

2. 3-Deazapurine Cyclic Phosphates

A further modification of the aglycon moiety of adenosine 3′,5′-cyclic phosphate is illustrated in the synthesis of 3-deaza-*c*AMP. 4-Amino-1-(β-D-ribofuranosyl)-imidazo[4,5-*c*]pyridine 5′-phosphate (3-deaza-AMP, 260a) was prepared from 3-deazaadenosine (259) by phosphorylation with phosphorus oxychloride (MIZUNO et al. 1968) or by treatment of 2′,3′-*O*-isopropylidene-3-deazaadenosine with pyrophosphoryl tetrachloride in acetonitrile, followed by hydrolysis (MIZUNO et al. 1975). The general procedure of intramolecular cyclization of (260a) using DCC in pyridine, however, failed to give satisfactory results. By this approach the yield

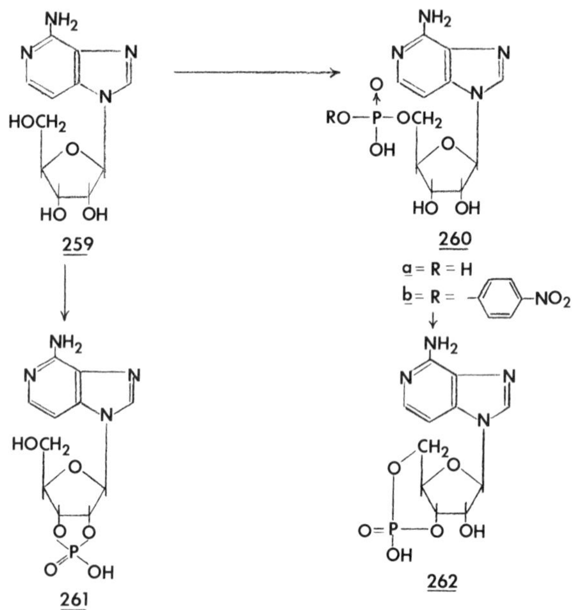

of 4-amino-1-(β-D-ribofuranosyl)imidazo[4,5-*c*]pyridine 3′,5′-cyclic phosphate (3-deaza-*c*AMP, 262) was very low and the product was always contaminated with 3-deazaadenosine 2′,3′-cyclic phosphate (261) (MIZUNO et al. 1975). However, the cyclization of the corresponding 5′-p-nitrophenyl ester (260b) in presence of potassium *tert* butoxide in dimethyl sulfoxide gave good yield of (262). By using tri-*n*-butylammonium pyrophosphate in dimethylformamide, 3-deazaadenosine (259) was converted to 3-deazaadenosine 2′,3′-cyclic phosphate (261), but only in 3% yield (MIZUNO et al. 1975). The pyrophosphoric acid in boiling dimethylformamide method was found to be convenient (over 50% yield) for the preparation of 4-chloro-1-(β-D-ribofuranosyl)imidazo[4,5-*c*]pyridine 2′,3′-cyclic phosphate (264) (UEDA and KAWAI 1970).

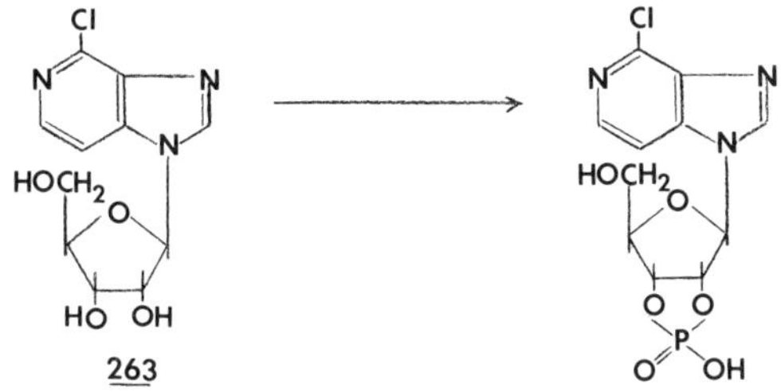

263 264

3. 7-Deazapurine Cyclic Phosphates

A few nucleoside cyclic phosphates with heterocyclic ring system related to purine, which had been modified by replacement of a ring nitrogen with carbon at position 7, have been synthesized. For the synthesis of the analogous tubercidin 3′,5′-cyclic phosphate (4-amino-7-β-D-ribofuranosylpyrrolo[2,3-*d*]pyrimidine or 7-deazaadenosine 3′,5′-cyclic phosphate, 266), tubercidin 5′-phosphate (265) (PIKE et al. 1964) served as a useful starting material. Reaction of (265) with DCC in

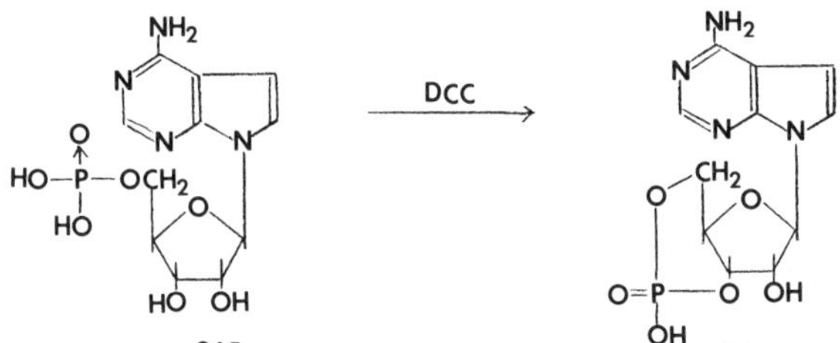

265 266

pyridine in presence of 4-morpholine-N,N′-dicyclohexylcarboxamidine according to the procedure of SMITH and co-workers (1961) gave crystalline tubercidin 3′,5′-cyclic phosphate (266) (HANZE 1967, 1968). A similar intramolecular cyclization of toyocamycin 5′-phosphate (267) with DCC in pyridine/N,N-dimethylformamide (1:1) gave toyocamycin 3′,5′-cyclic phosphate (268) (MILLER et al. 1973b). Alkaline hydrolysis of (268) with ammonical hydrogen peroxide gave sangivamycin 3′,5′-cyclic phosphate (269) (MILLER et al. 1973b).

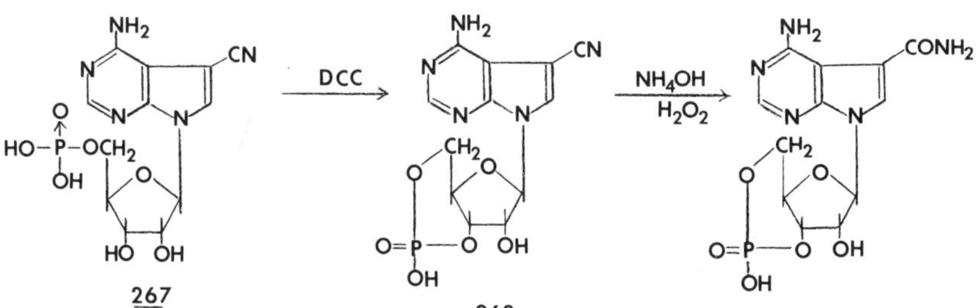

267 268 269

Cyclization of 4-substituted-7-(β-D-ribofuranosyl)pyrrolo[2,3-*d*]pyrimidine 5′-phosphate (270) with DCC and N,N′-dicyclohexylcarboxamidine in pyridine gave the corresponding 3′,5′-cyclic phosphates (271) (HANZE 1967). Tubercidin 2′,3′-

cyclic phosphate (272) was prepared directly from tubercidin in 50% yield, employing pyrophosphoric acid and tri-*n*-butylamine in N,N-dimethylformamide (HASHIMOTO et al. 1970). Treatment of 5′-monomethoxytrityltubercidin with P[1]-diphenyl-P[2]-morpholinopyrophosphorochloridate also gave (272) in rather low yield (1.8%) (IKEHARA et al. 1966).

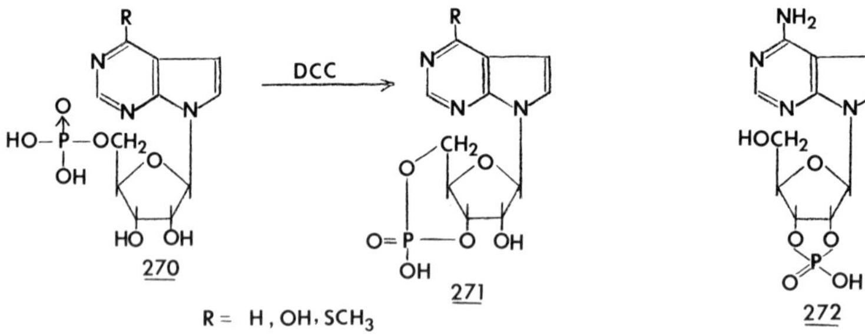

270 271 272

R = H, OH, SCH_3

4. 2-Azapurine Cyclic Phosphates

The purine ring carbon atom at position 2 of adenosine 3′,5′-cyclic phosphate has been replaced with nitrogen to obtain 2-azaadenosine 3′,5′-cyclic phosphate (273). 5-Amino-1-(β-D-ribofuranosyl)imidazole-4-carboxamidine 3′,5′-cyclic phosphate (25, X=NH) (MEYER Jr et al. 1973b, 1974a) was found to be a very useful intermediate for the synthesis of (273). Thus, treatment of (25) (X=NH) with sodium

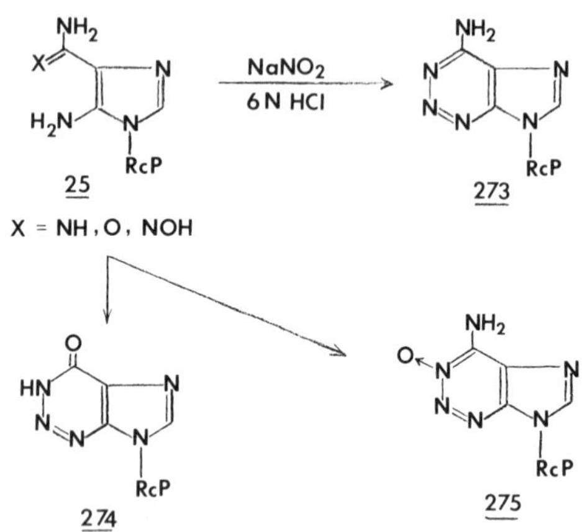

25 273

X = NH, O, NOH

274 275

nitrite under strongly acidic conditions (cold 6 N hydrochloric acid) (KAWANA et al. 1972) gave 4-amino-7-(β-D-ribofuranosyl)imidazo[4,5-*d*]-*v*-triazine 3′,5′-cyclic phosphate (2-aza-*c*AMP, 273) in 76% yield (MEYER Jr et al. 1974a; MEYER Jr and SHUMAN 1975). Similarly, treatment of 5-amino-1-(β-D-ribofuranosyl)imidazole-4-carboxamide/carboxamidoxime 3′,5′-cyclic phosphate (25, X=O or NOH) under the same conditions provided 2-azainosine 3′,5′-cyclic phosphate (274) and 2-azaadenosine 3′,5′-cyclic phosphate N[1]-oxide (275), respectively (MEYER Jr and SHUMAN 1975).

2-Azaadenosine 3′,5′-cyclic phosphate (273) was also conveniently prepared from 1,N^6-ethenoadenosine 3′,5′-cyclic phosphate. When 1-(β-D-ribofuranosyl)-5-amino-4-(imidazol-2-yl)imidazole 3′,5′-cyclic phosphate (53), obtained by alkaline

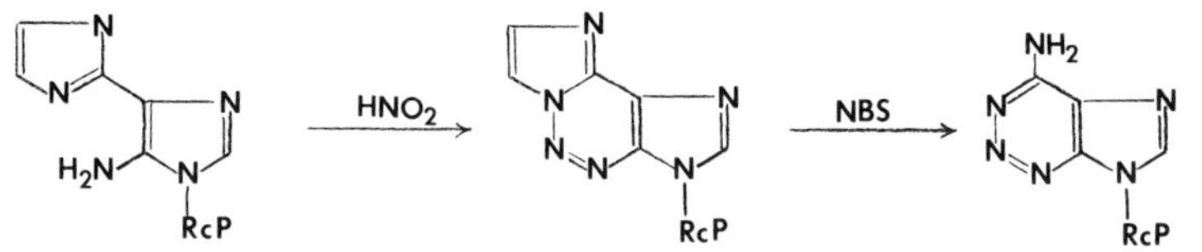

53 276 273

hydrolysis of 1,N^6-ethenoadenosine 3′,5′-cyclic phosphate (YAMAJI et al. 1977e; YAMAJI and KATO 1978a), was treated with aqueous nitrous acid at room temperature, 94% yield of crystalline 2-aza-1,N^6-ethenoadenosine 3′,5′-cyclic phosphate (276) was obtained. Treatment of (276) with N-bromosuccinimide or with bromine in aqueous dimethylformamide under mild conditions gave 2-azaadenosine 3′,5′-cyclic phosphate (273) in good yield (YAMAJI and KATO 1975; YAMAJI et al. 1976a).

5. 8-Azapurine Cyclic Phosphates

8-Azaadenosine 3′,5′-cyclic phosphate (279) was synthesized from 8-azaadenosine (7-amino-3-β-D-ribofuranosyl-*v*-triazolo[4,5-*d*]pyrimidine, 277) by phosphorylation and intramolecular cyclization (MILLER et al. 1978). Phosphorylation of 8-azaadenosine with phosphorus oxychloride in triethylphosphate gave 8-azaadenosine 5′-phosphate (278) which on subsequent cyclization with DCC and 4-morpholino-N,N′-dicyclohexylcarboxamidine in boiling pyridine afforded 8-azaadenosine 3′,5′-cyclic phosphate (279) (MILLER et al. 1978).

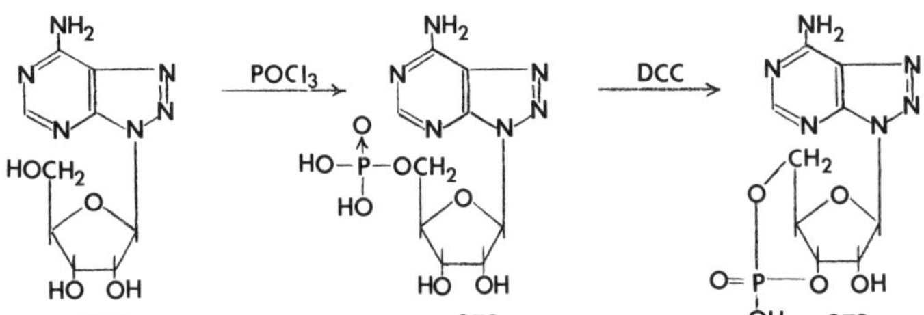

277 278 279

Although the synthesis of 8-azaguanosine 5′-phosphate (5-amino-3-β-D-ribofuranosyl-*v*-triazolo[4,5-*d*]pyrimidin-7-one 5′-phosphate) has been reported as early as 1961 (MONTGOMERY and THOMAS 1961), the preparation of 8-azaguanosine 3′,5′-cyclic phosphate has to date not been realized. However, the synthesis of 8-azaguanosine 2′,3′-cyclic phosphate (281) has been accomplished by the cyclization of 2′(3′)-phosphate of 8-azaguanosine (280) with DCC in *tert*-butanol under alkaline conditions (THOMAS et al. 1962). Hydrolysis of (281) with 0.1 *N* hydrochloric acid converted it to the 2′(3′)-phosphates (280). Alternatively (281) was also prepared from 8-azaguanosine 2′(3′)-phosphite (282) by the action of hexachloroacetone and DCC (HOLY and SMRT 1966).

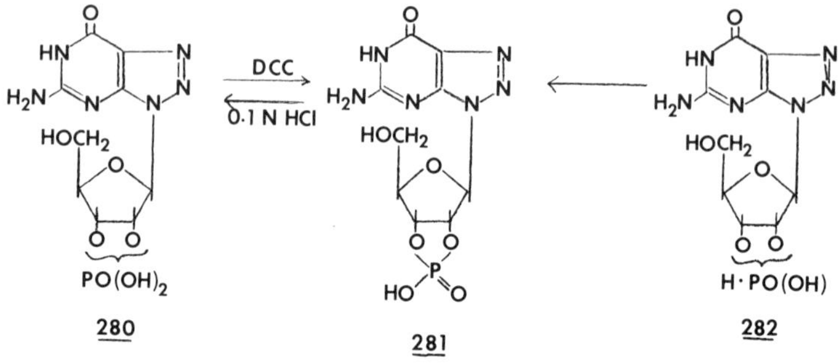

6. Formycin Cyclic Phosphates

Application of the general methods of synthesis of adenosine 3′,5′-cyclic phosphates to the preparation of formycin cyclic phosphates, proved to be unsatisfactory (MAKABE et al. 1978). However, the trichloromethyl phosphorodichloridate procedure (MARUMOTO et al. 1975) was found to be fruitful. Reaction of formycin (7-amino-3-*β*-D-ribofuranosylpyrazolo[4,3-*d*]pyrimidine, 283) with trichloromethyl phosphorodichloridate (TPD) in triethyl phosphate gave formycin 5′-trichloromethyl phosphonate (284) in 72% yield. Hydrolysis of (284) by treatment

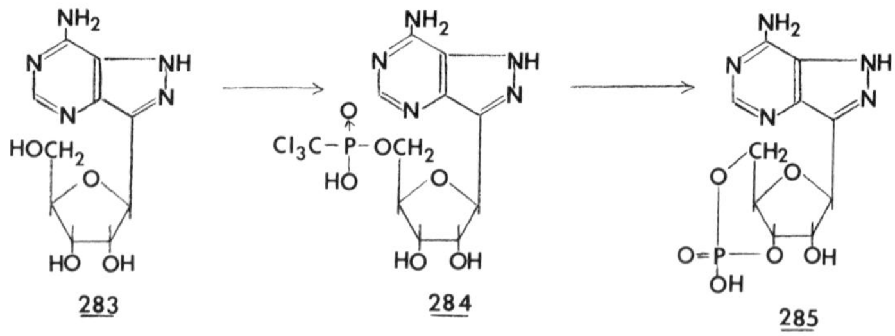

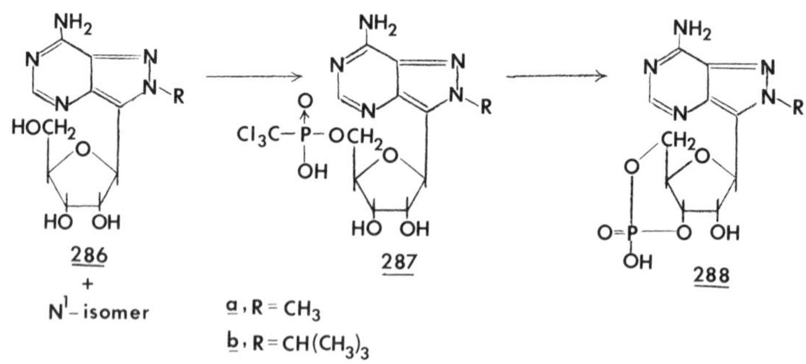

with potassium *tert*-butoxide gave formycin 3′,5′-cyclic phosphate (285) in good yield (MAKABE et al. 1978). Alkylation of formycin with methyl iodide and isopropyl iodide gave N-2 alkyl formycin (286) as the major product along with minor amounts of N-1 alkyl formycin. Phosphorylation of (286) with TPD, followed by alkaline cyclization gave 2-methyl- and 2-isopropylformycin 3′,5′-cyclic phosphates (288a, b) (MAKABE et al. 1978).

Formycin 2′,3′-cyclic phosphates were also synthesized. Treatment of 2′,3′-*O*-isopropylideneformycin with benzoyl chloride in pyridine gave the tetrabenzoyl derivative, which on deisopropylidenation furnished (289). Phosphorylation of (289) with phosphorus oxychloride in pyridine followed by hydrolysis with methanolic ammonia gave formycin 2′,3′-cyclic phosphate (290) (MAKABE et al. 1978). 2′,3′-Cyclic phosphates of 2-methyl- and 2-isopropylformycin were also similarly prepared. Isopropylidenation, benzoylation and deisopropylidenation of (286) gave tribenzoyl derivative (291) which on subsequent phosphorylation and hydrolysis afforded 2-alkylformycin 2′,3′-cyclic phosphates (292) (MAKABE et al. 1978).

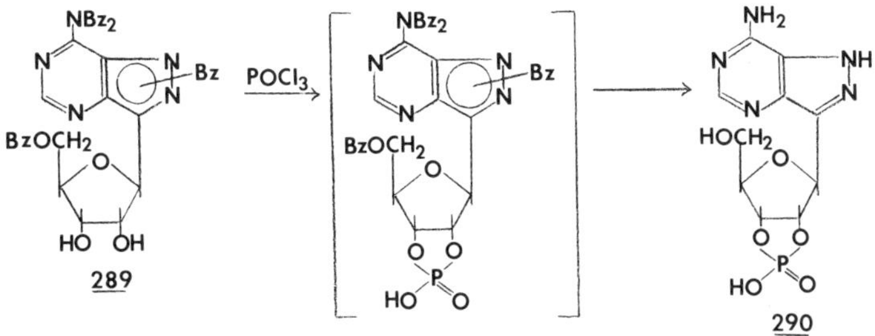

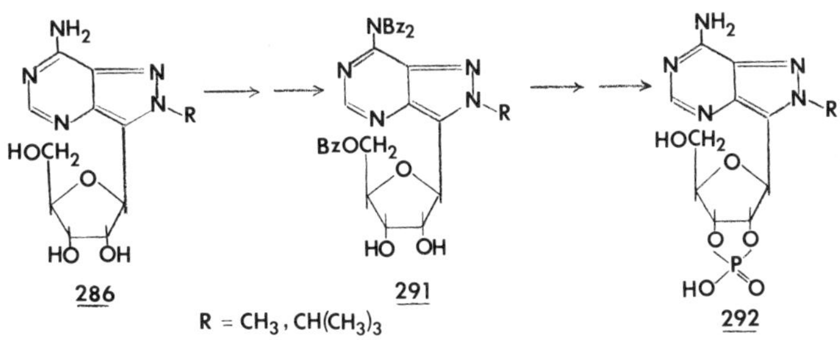

Hydrolysis of formycin 3′,5′-cyclic phosphate with 1 *N* barium hydroxide at 100 °C for 15 min gave formycin 3′- and 5′-monophosphates, whereas hydrolysis of formycin 2′,3′-cyclic phosphate with 1 *N* hydrochloric acid at 100 °C for 15 min or 0.5 *N* sodium hydroxide at 35 °C for 12 h gave formycin 2′- and 3′-monophosphates (MAKABE et al. 1978).

7. 1,N⁶-Ethenoadenosine Cyclic Phosphates

In view of the interesting enzymatic and fluorescent properties of 1,N⁶-ethenoadenosine triphosphate (3-β-D-ribofuranosylimidazo[2,1-*i*]purine 5′-triphosphate, εATP) (SECRIST III et al. 1972a; STEINER 1972; SHAHAK et al. 1973), several 1,N⁶-ethenoadenosine 3′,5′-cyclic phosphates (ε-*c*AMP) have been synthesized. The reaction of chloroacetaldehyde with adenosine 3′,5′-cyclic phosphate gave the highly fluorescent compound, 3-β-D-ribofuranosylimidazo[2,1-*i*]purine 3′,5′-cyclic phosphate (ε-*c*AMP, 294a) (SECRIST III et al. 1972b; SECRIST III 1974). Similar reactions starting with 8-bromo-, 8-methylthio- and 8-benzylthio

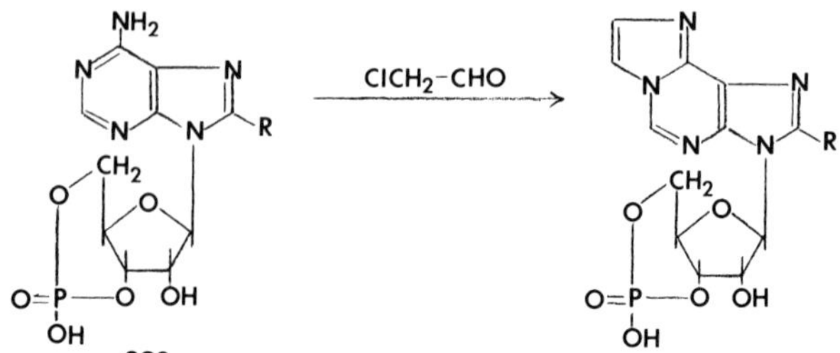

293 294

a, R = H
b, R = Br
c, R = SCH_3
d, R = SCH_2Ph

adenosine 3′,5′-cyclic phosphates led to the corresponding crystalline 1,N[6]-ethenoadenosine 3′,5′-cyclic phosphates (294) (Jones et al. 1973). Treatment of

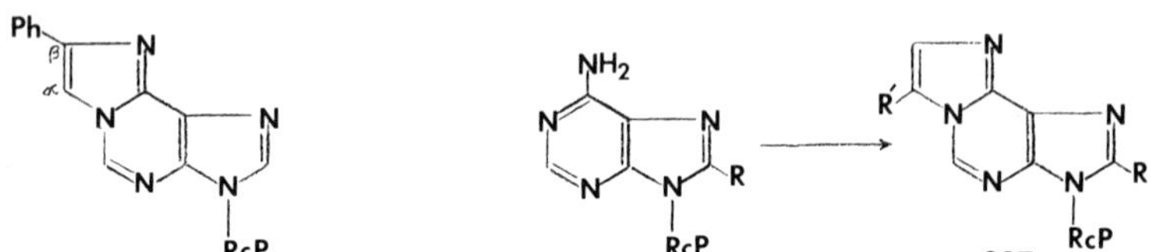

295 296 297

R = H, SCH_3
R′ = Ph, n-C_3H_7

*c*AMP with α-bromoacetophenone gave the β-phenyl derivative of 1,N[6]-ethenoadenosine 3′,5′-cyclic phosphate (295) (Jones et al. 1973; Meyer Jr et al. 1973b). On the other hand, treatment of *c*AMP or 8-methylthio-*c*AMP with 2-bromovaleraldehyde or reaction of *c*AMP with 2- bromo-2-phenylacetaldehyde gave the corresponding α-substituted derivatives (297) (Jones et al. 1973, 1975).

The synthesis of 2-substituted 1,N[6]-ethenoadenosine 3′,5′-cyclic phosphate derivatives was approached by the reaction of 2-substituted *c*AMP derivatives with excess chloroacetaldehyde (Miller et al. 1980b) or by ring opening of 1,N[6]-ethenoadenosine 3′,5′-cyclic phosphate in alkali to give 1-(β-D-ribofuranosyl)-5-amino-4-(imidazol-2-yl)imidazole 3′,5′-cyclic phosphate (53) (Yip and Tsou 1973;

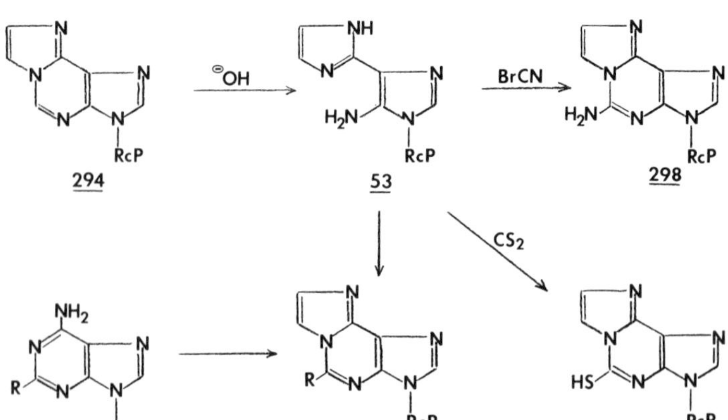

294 53 298 299 300 301

R = OH, CH_3

Yamaji and Kato 1975) and subsequent ring closure (Yamaji et al. 1976c, Miller et al. 1980b). Treatment of (53) with cyanogen bromide in the presence of triethylamine gave 2-amino-1,N^6-ethenoadenosine 3′,5′-cyclic phosphate (298) (Yamaji et al. 1976c, 1977b). Ring closure of (53) with 1,1′-carbonyldiimidazole gave 2-hydroxy-1,N^6-etheno-*c*AMP (300, R=OH) (Yamaji et al. 1977d) and with triethyl orthoacetate gave 2-methyl-1,N^6-etheno-*c*AMP (300, $R=CH_3$) (Yamaji et al. 1976a, b). Carbon disulfide and (53) in presence of anhydrous potassium carbonate in dimethylformamide afforded 2-mercapto-1,N^6-etheno-*c*AMP (301) (Yamaji et al. 1977a, e).

Various 2-alkylthio- and 2-benzylthio-1,N^6-ethenoadenosine 3′,5′-cyclic phosphates (302) were prepared by alkylation of (301) with alkyl halides in aqueous alkaline solution (Yamaji et al. 1976c, 1977c, 1978). The reaction of (301) with

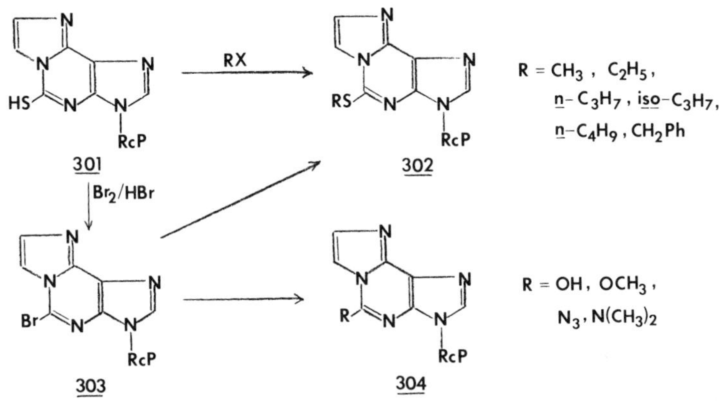

bromine in concentrated hydrobromic acid gave the versatile intermediate 2-bromo-1,N^6-ethenoadenosine 3′,5′-cyclic phosphate (303), the bromo group of which was available for various nucleophilic substitution reactions. 2-Azido-, 2-dimethylamino- and 2-methoxy-1,N^6-etheno-*c*AMP derivatives (304) were synthesized from (303) with sodium azide in dimethylformamide or methanolic dimethylamine or sodium methoxide in methanol, respectively (Yamaji et al. 1976a, b; Yamaji and Kato 1978a).

1-β-D-Ribofuranosyl-5-amino-4-(imidazol-2-yl)imidazole 3′,5′-cyclic phosphate (53) was also found to be a key intermediate for the synthesis of 1,N^6-etheno-2-azaadenosine 3′,5′-cyclic phosphate (305). As in the case of 2-aza-1,N^6-ethenoadenosine (Tsou et al. 1974), when (53) was allowed to react with sodium

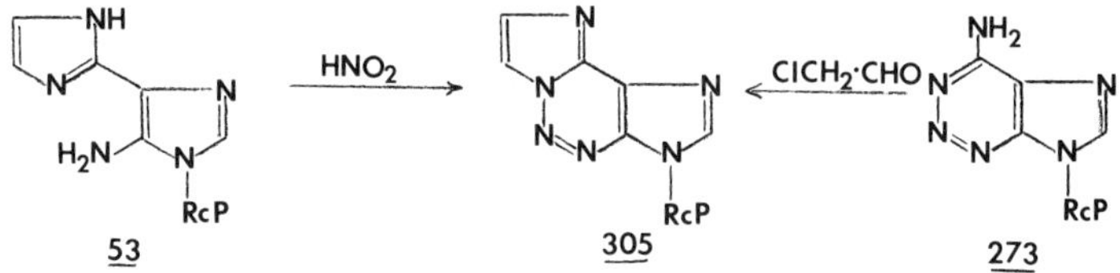

nitrite in acetic acid, 3-β-D-ribofuranosyl-3*H*-diimidazo[1,2-*c*:4′,5′-*e*]triazin3′,5′-cyclic phosphate (1,N^6-etheno-2-aza-*c*AMP, 305) was obtained (Yip and Tsou 1973; Yamaji and Kato 1975; Miller et al. 1980b). Treatment of 2-aza-*c*AMP

(273) with chloroacetaldehyde also gave 1,N^6-etheno-2-aza-*c*AMP (305) (YAMAJI and KATO 1975).

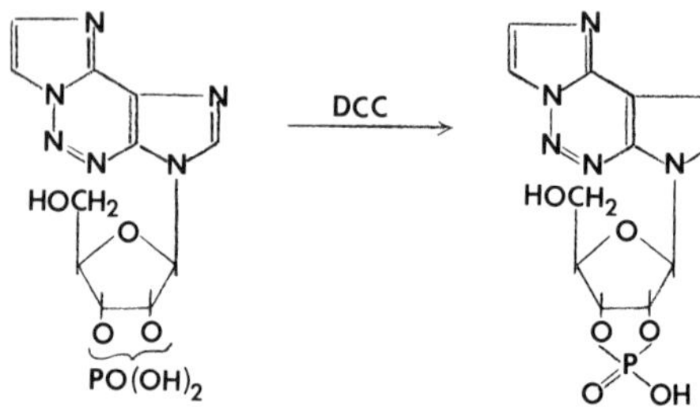

1,N^6-Etheno-2-azaadenosine 2′,3′-cyclic phosphate (307) was prepared by the intramolecular cyclization of 1,N^6-etheno-2-azaadenosine 2′(3′)-monophosphate (306) with DCC in pyridine (LO et al. 1975). When ten equivalents of chloroacetaldehyde in 0.1 *M* sodium acetate buffer (pH 5) at 50 °C was allowed to react with 3′-cyclic ester of 9-(5,6-dideoxy-6-dihydroxyphosphinyl-*β*-D-*ribo*-hexopyranosyl)-adenine and 9-(6-deoxy-*β*-D-*allo*furanosyl)adenine 3′,5′-cyclic phosphate, the 1,N^6-etheno-*c*AMP derivatives (308) and (310) were obtained in good yields (JONES et al. 1973; JONES and MOFFATT 1975; KAWANABE and MACHIDA 1976). The synthesis of (309) has also been reported from the corresponding adenosine cyclic phosphate derivative (JONES and MOFFATT 1974, 1975).

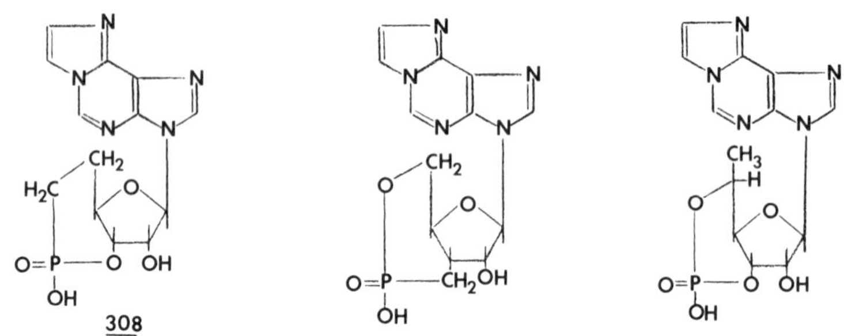

III. Cyclic Nucleotides Related to Guanosine Cyclic Phosphate

In 1963, the isolation of a second cyclic nucleotide, guanosine 3′,5′-cyclic phosphate (*c*GMP, 312) from rat urine was reported (ASHMAN et al. 1963). Subsequently *c*GMP has been detected in all mammalian tissues studied (GOLDBERG et al. 1969, 1973; ISHIKAWA et al. 1969; STEINER et al. 1972b). Like *c*AMP, guanosine 3′,5′-cyclic phosphate levels in tissues are extremely low and only recently have sensitive methods been available for detection of *c*GMP (HARDMAN et al. 1966, 1969; PRICE et al. 1967; ISHIKAWA et al. 1969; SHIBUYA et al. 1975; BROOKER et al. 1979). *c*GMP levels found in various tissues of rat have generally been between 10^{-8} and 10^{-7} moles/kg, and at least an order of magnitude lower than *c*AMP levels (GOLDBERG et al. 1969; ISHIKAWA et al. 1969). Relatively less is known about the biological role of *c*GMP than about that of *c*AMP, but increasing evidence is accumulating that in many cases *c*GMP acts antagonistically to *c*AMP (POGSON 1974; GOLDBERG et al. 1975b; GOLDBERG and HADDOX 1977). *c*GMP (312) is formed from guanosine 5′-triphosphate (311) by guanylate cyclase (HARDMAN and SUTHERLAND 1969; WHITE et al. 1969; KIMURA and MURAD 1974), and hydrolyzed by *c*GMP phosphodiesterase (ROSEN 1970; CLARK and BERNLOHR 1972; HWANG et al. 1974) to guanosine 5′-phosphate (313).

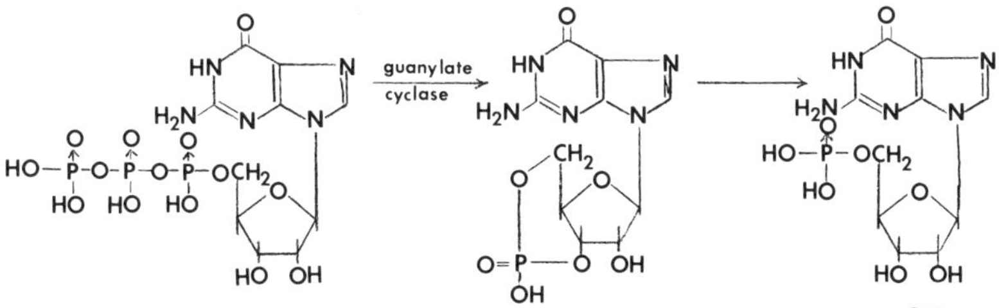

311 312 313

The chemical synthesis of guanosine 3′,5′-cyclic phosphate was accomplished (SMITH et al. 1961) before the natural occurrence of *c*GMP was reported. However, the synthesis encountered certain difficulties. Unlike *c*AMP, base-catalyzed transesterification of "active" guanosine 5′-p-nitrophenyl phosphate (313) in an aprotic solvent gave poor yield of *c*GMP (312) (BORDEN and SMITH 1966b). Because of the poor solubility of guanosine 5′-phosphate (313a), the direct intramolecular cyclization in pyridine in the presence of DCC was accomplished with difficulty (STRAUS and FRESCO 1965). The use of N^2-benzoylguanosine 5′-phosphate (313b)

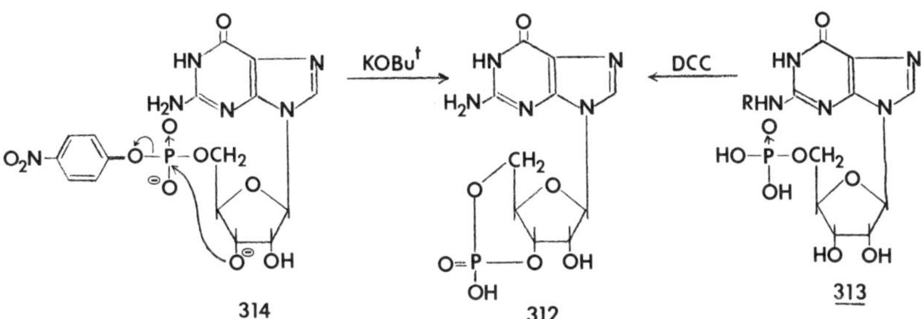

314 312 313

a, R = H

b, R = Bz

in the cyclization reaction was found to be beneficial, but the yield of (312) obtained was rather low (SMITH et al. 1961; FUJIMOTO and NARUSE 1968; POSTERNAK and FALBRIARD 1971). The low cost enzymatic formation of *c*GMP from guanosine 5′-triphosphate is now well known (HARDMAN et al. 1971; GARBERS et al. 1975; SASAKI et al. 1976). Although the DCC method with (313b) can be used for the synthesis of *c*GMP derivatives, by far they have been prepared by the chemical transformation of commercial *c*GMP.

Several 8-substituted guanosine 3′,5′-cyclic phosphate derivatives have been synthesized from *c*GMP (MILLER et al. 1973a; PAOLETTI et al. 1973). Bromination of *c*GMP with bromine in aqueous solution gave 8-bromo-*c*GMP (316) (Boehringer Mannheim GmbH 1970b, 1971; MILLER et al. 1973a). Metathesis of the bromo group of (316) occurred in good yield and with minimum side reactions when (316) was treated with nitrogen, sulfur, and oxygen containing nucleophiles at elevated temperatures. In this manner, treatment of a methanolic solution of (316) with dimethylamine, methylamine, benzylamine, p-chlorobenzenethiol and aminoethanethiol hydrochloride gave the respective 8-substituted *c*GMP derivatives (315) (Miller et al. 1973a; Dills et al. 1976, 1979). 8-Hydroxyguanosine 3′,5′-cyclic phosphate (315, R = OH) was prepared by the reaction of (316) in refluxing sodium acetate in glacial acetic acid (Miller et al. 1973a). Thiourea and

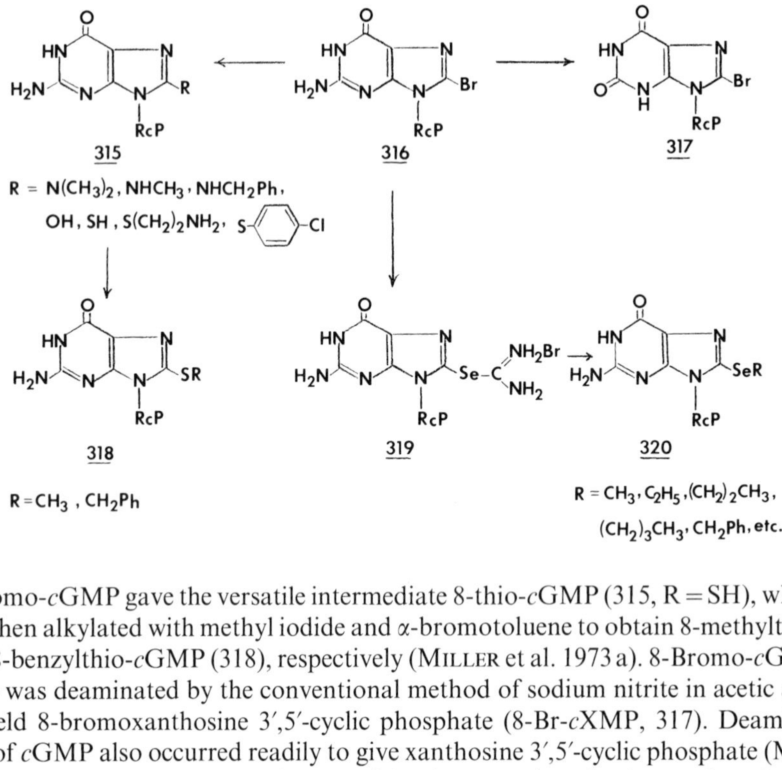

8-bromo-*c*GMP gave the versatile intermediate 8-thio-*c*GMP (315, R = SH), which was then alkylated with methyl iodide and α-bromotoluene to obtain 8-methylthio- and 8-benzylthio-*c*GMP (318), respectively (Miller et al. 1973a). 8-Bromo-*c*GMP (316) was deaminated by the conventional method of sodium nitrite in acetic acid to yield 8-bromoxanthosine 3′,5′-cyclic phosphate (8-Br-*c*XMP, 317). Deamination of *c*GMP also occurred readily to give xanthosine 3′,5′-cyclic phosphate (Miller et al. 1973a). Treatment of (316) with selenourea in refluxing methanol gave guanosine 3′,5′-cyclic phosphate 8-isoselenouronium hydrobromide (319) (Chu et al. 1975). Alyklation of (319) with various alkylhalides gave the corresponding 8-alkylseleno-*c*GMP derivatives (320) (Chu et al. 1975). It is of particular interest that 8-seleno-*c*GMP (320, R = H) could not be synthesized from 8-bromo-*c*GMP and sodium hydrogen selenide (Chu et al. 1975).

8-Bromo-*c*GMP (316) proved to be a versatile starting material for the introduction of an arylthio function in position 8. Reaction of (316) and p-chlorophenylthiol gave 8-p-chlorophenylthio-*c*GMP (312). Acetylation followed by chlorination with phosphoryl chloride afforded 2-amino-6-chloro-8-p-chlorophenylthio-9-(2-*O*-acetyl-*β*-D-ribofuranosyl)purine 3′,5′-cyclic phosphate (322). Replacement of the 6-chloro group of (322) with piperidine gave 2-amino-6-piperidino-8-p-chlorophenylthio-9-(*β*-D-ribofuranosyl)purine 3′,5′-cyclic phosphate (323) (Boswell and Robins 1979).

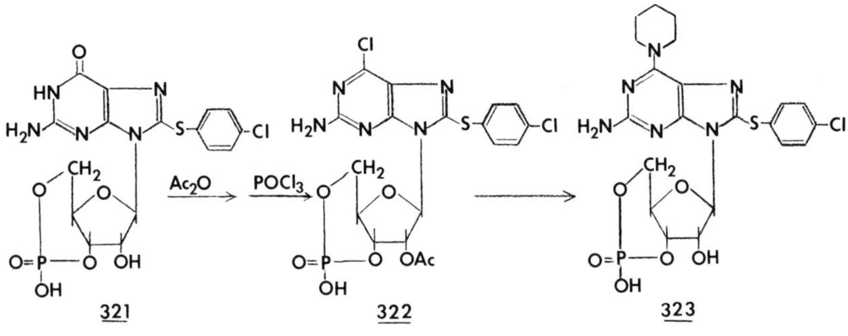

The enzymatic preparation of [^{32}P]-*c*GMP has recently been reported (JOHNSON and WALSETH 1979), which on bromination gave 8-bromo-[^{32}P]-*c*GMP. Nucleophilic displacement of the 8-bromo group from 8-bromo-[^{32}P]-*c*GMP with large excess of triethylammonium azide provided a route to the synthesis of 8-azido-[^{32}P]-*c*GMP, which should prove to be a useful probe for the localization of *c*GMP receptors (GEAHLEN et al. 1979).

Several 2-substituted adenosine 3′,5′-cyclic phosphates, like 2-amino-*c*AMP and 2-chloro-*c*AMP have been prepared by the chemical transformation of *c*GMP (MEYER Jr et al. 1975) (see Sect. B.I.1.b). A number of 8-alkyl and 8-acyl derivatives of *c*GMP have also been synthesized by homolytic alkylation or acylation of parent *c*GMP *via* free radicals generated in situ (CHRISTENSEN et al. 1975). When *c*GMP was treated with ferrous sulfate and *tert*-butyl hydroperoxide in 0.3 *N* sulfuric acid, 8-methylguanosine 3′,5′-cyclic phosphate (325, R′=CH_3) was obtained in 75% yield. Treatment of *c*GMP with trimethylhydroperoxypentane under similar conditions gave 8-neopentylguanosine 3′,5′-cyclic phosphate (325, R′=CH_2CMe_3). 8-Benzyl-*c*GMP (325, R′=$CH_2C_6H_5$) was prepared from *c*GMP and toluene in aqueous acetic acid containing ferrous sulfate and ammonium persulfate (CHRISTENSEN et al. 1975).

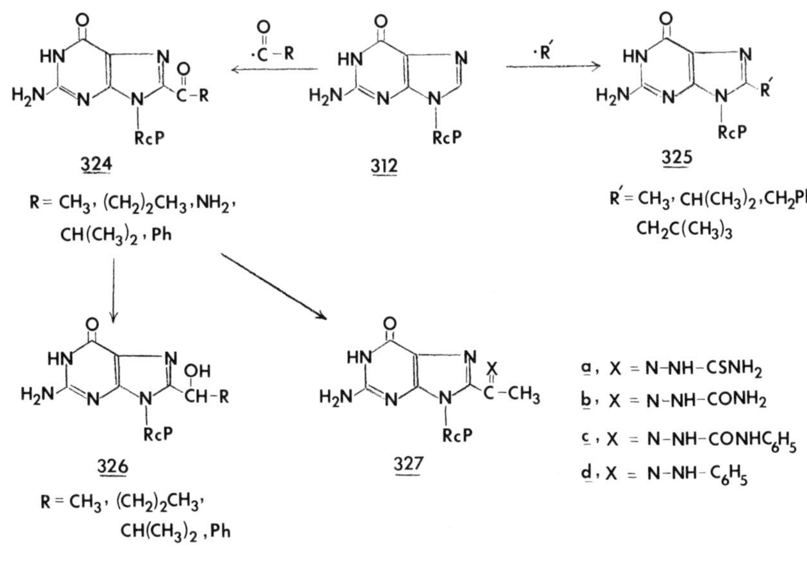

Acetaldehyde and *c*GMP in presence of ammonium persulfate and ferrous sulfate in 0.3 *N* sulfuric acid gave 8-acetylguanosine 3′,5′-cyclic phosphate (324, R = CH_3) in 43% yield (CHRISTENSEN et al. 1975). Using appropriate aldehydes under similar conditions, a series of 8-acyl-*c*GMP derivatives were prepared (324). Formamide and *c*GMP under similar conditions gave 8-carbamoylguanosine 3′,5′-cyclic phosphate (324, R = NH_2). Although the side reactions were minimal in the homolytic acylation reaction, isobutyraldehyde and *c*GMP gave, in addition to 47% yield of (324) (R = $CHMe_2$), 8-isopropylguanosine 3′,5′-cyclic phosphate (325, R′ = $CHMe_2$) in 22% yield (CHRISTENSEN et al. 1975). Reduction of these novel 8-acyl-*c*GMP derivatives (324) with sodium borohydride in water readily gave 8-(1-hydroxyalkyl)guanosine 3′,5′-cyclic phosphate derivatives (326). 8-Acetylguanosine 3′,5′-cyclic phosphate (324, R = CH_3) also yielded thiosemicarbazone (327a), semicarbazone (327b), 4-phenylsemicarbazone (327c) and phenylhydrazone (327d) upon treatment with the appropriate reagent in water (CHRISTENSEN et al. 1975).

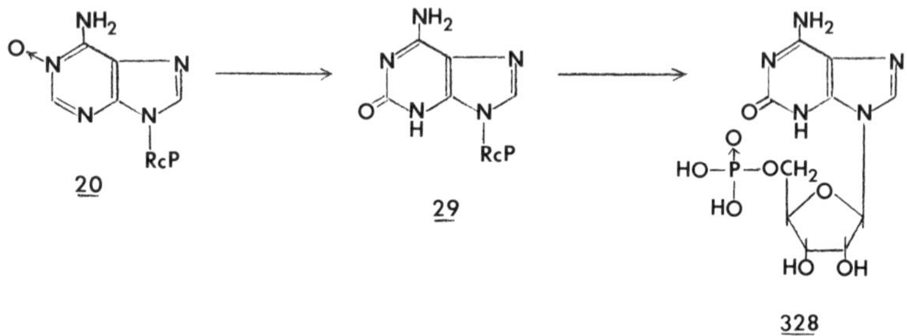

Adenosine 3′,5′-cyclic phosphate N^1-oxide (20) was found to undergo photochemical transformation to isoguanosine 3′,5′-cyclic phosphate (2-oxo-*c*AMP, 29)

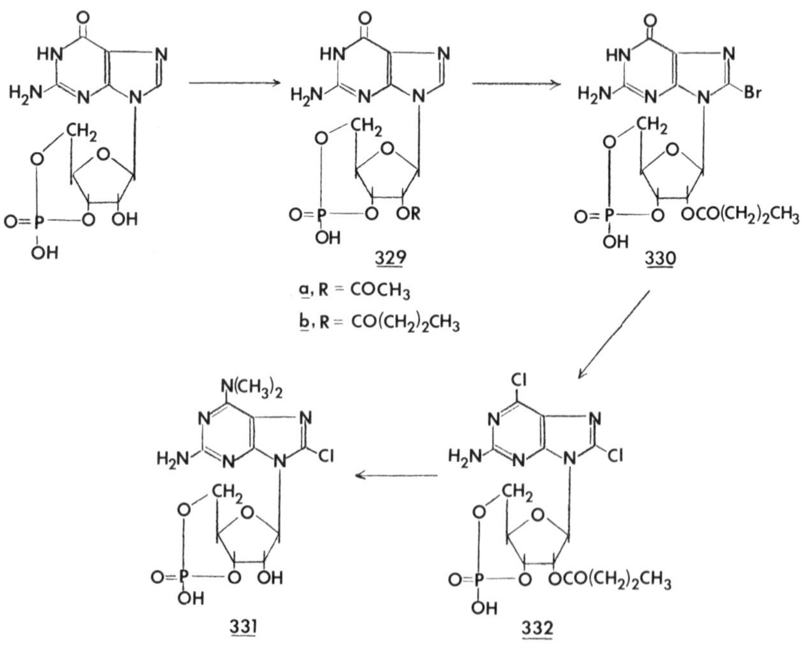

(Kazimierczuk and Shugar 1973; Fikus et al. 1974). 2-Oxo-*c*AMP is susceptible to hydrolysis by phosphodiesterase to isoguanosine 5′-phosphate (328), the rate of enzymatic hydrolysis being 50% that for *c*AMP (Fikus et al. 1974).

The customary procedure of acylation using acetic or butyric anhydride with sodium salt of *c*GMP in conjunction with 4-dimethylaminopyridine as a catalyst readily gave 2′-*O*-acetyl-*c*GMP (329a) and 2′-*O*-butyryl-*c*GMP (329b), respectively (Miller et al. 1976; Boswell and Robins 1979). Bromination of (329b) with bromine in water gave 8-bromo-2′-*O*-butyrylguanosine 3′,5′-cyclic phosphate (330) (Miller et al. 1976), which on subsequent chlorination with phosphorus oxychloride in presence of 2,6-lutidine yielded 2-amino-6,8-dichloro-9-(2-*O*-butyryl-*β*-D-ribofuranosyl)purine 3′,5′-cyclic phosphate (332). It is of interest to note the replacement of 8-bromo group of (330) by chloro group in this reaction (Boswell and Robins 1979). Treatment of (332) directly with N,N-dimethylamine in boiling ethanol gave 2-amino-8-chloro-6-dimethylamino-9-(*β*-D-ribofuranosyl)-purine 3′,5′-cyclic phosphate (331) (Boswell and Robins 1979).

The synthesis of 2′-deoxyguanosine 3′,5′-cyclic phosphate (d*c*GMP, 334) by DCC mediated cyclization of the corresponding 5′-phosphate (333) has been reported (Drummond et al. 1964). Raney nickel desulfurization of 8,2′-anhydro-8-thio-9-(*β*-D-arabinofuranosyl)guanine 3′,5′-cyclic phosphate (335) also furnished

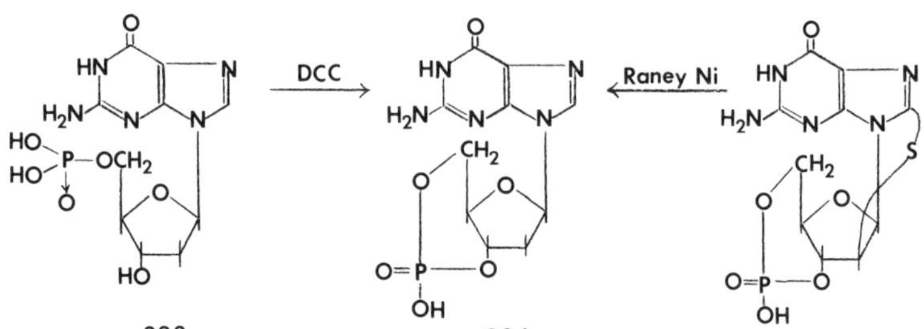

333 334 335

d*c*GMP (Miller et al. 1976). Treatment of 8-bromo-2′-*O*-tosyl-*c*GMP with thiourea in butanol gave 8-mercapto-8,2′-anhydro-*ara*-*c*GMP (335), which was then converted to (334) with Raney nickel. Similarly, reductive cleavage of the anhydro derivative (336) with hydrogen sulfide, followed by Raney nickel desulfurization provided 9-(*β*-D-arabinofuranosyl)guanine 3′,5′-cyclic phosphate (338) (Mian et al. 1974).

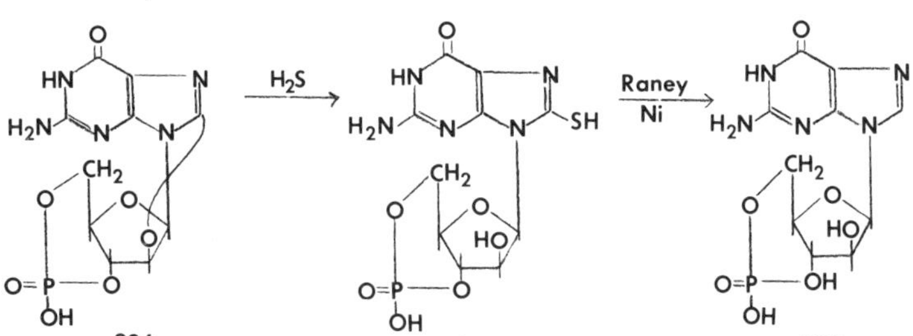

336 337 338

The importance of the stereochemical position of hydroxyl groups in the ease of cyclic phosphate formation is demonstrated (Revankar et al. 1976) by the direct formation of 9-(β-D-xylofuranosyl)guanine 3′,5′-cyclic phosphate (*xylo-c*GMP, 147) from 9-(β-D-xylofuranosyl)guanine (*xylo*-G, 146) upon treatment of phosphorus oxychloride (see Sect. B.I.2.b):

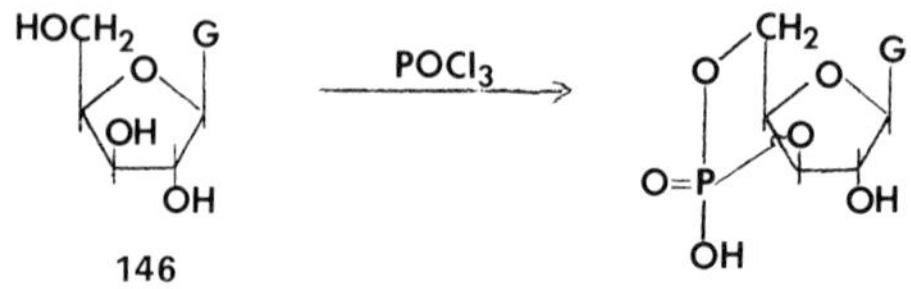

146 147

Guanosine 3′,5′-cyclic phosphoramidate (341) with an exocyclic amino group has been synthesized from *c*GMP (MEYER Jr et al. 1973a). The synthesis of (341) was achieved by activation of (339) to nucleophilic attack by reaction with phosphorus oxychloride. Heating N^2,2′-*O*-diacetylguanosine 3′,5′-cyclic phosphate (339) with phosphorus oxychloride followed by treatment with an alkylamine gave the intermediate acyl guanosine 3′,5′-cyclic phosphoramidate (340). Further alkaline hydrolysis gave (341) (MEYER Jr et al. 1973a, 1974b). Acidic hydrolysis of (341) in pH 1 buffer at 37 °C for 3 h gave approximately 50% hydrolysis to mainly *c*GMP (MEYER Jr et al. 1973a).

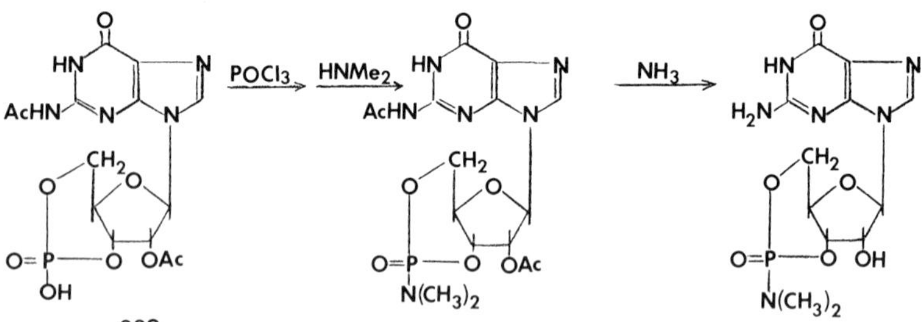

339 340 341

Since esterification of nucleoside 3′,5′-cyclic phosphates facilitates membrane-transportation and use of the benzyl group allows a subsequent selective hydrolytic cleavage (ENGELS and SCHLAEGER 1977), several acyl/alkyl esters of *c*GMP (342) have been prepared (ENGELS 1979), by reaction of *c*GMP with appropriate

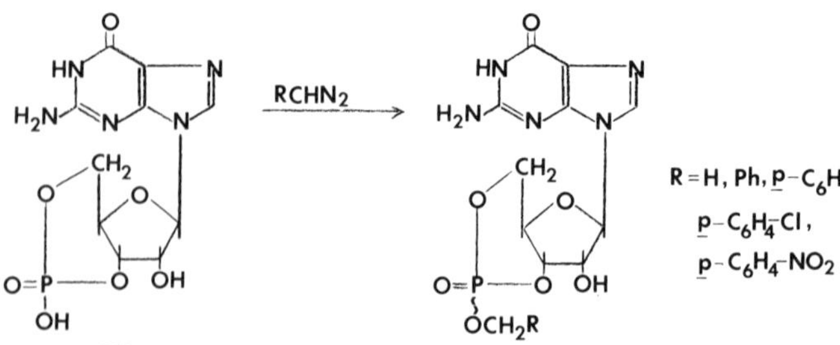

312 342

diazoalkanes in hexamethylphosphoric triamide. The alkylation of the aglycon also occurs as a side reaction, decreasing in extent, however, in the series diazomethane, phenyldiazomethane and substituted phenyldiazomethane. The side products obtained with diazomethane are identified as O^6-methyl- (343) and N^1-methyl- (344) derivatives (ENGELS 1979). The triesters (342), (343), and (344), all occur as diastereomeric mixtures which were chromatographically separated into pure axial and equatorial isomers.

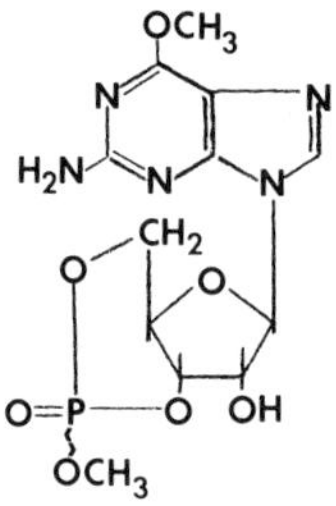

343

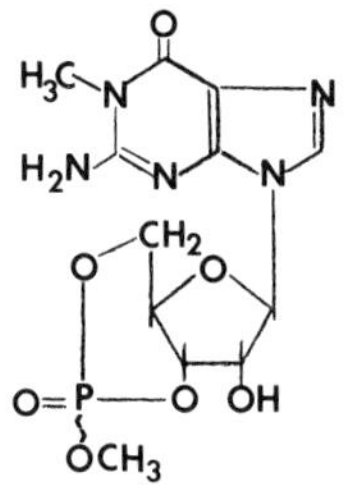

344

Facile 2′,3′-cyclic phosphate formation by the DCC procedure occurred with guanosine 2′(3′)-phosphate (345) to give guanosine 2′,3′-cyclic phosphate (346) in high yield (SMITH et al. 1958; SMRT 1967; HOLY' 1970a). The synthesis of various 5′-*O*-substituted derivatives of guanosine 2′,3′-cyclic phosphate (347), which are prerequisite for the preparation of guanylyl-(3′→5′)-uridine, has been reported (HOLY' 1969a). The acetylation of (346) with acetic anhydride in the presence of triethylamine (or pyridine) gave 5′-*O*-acetylguanosine 2′,3′-cyclic phosphate (347a) (MICHELSON 1959; HOLY' 1969a; SMRT 1974) along with N^2,5′-*O*-diacetylguanosine 2′,3′-cyclic phosphate (SMRT 1974). Treatment of triethylammonium salt of (346) with triphenylphosphite in presence of triethylamine afforded guanosine 2′,3′-cyclic phosphate 5′-phosphite (347b).

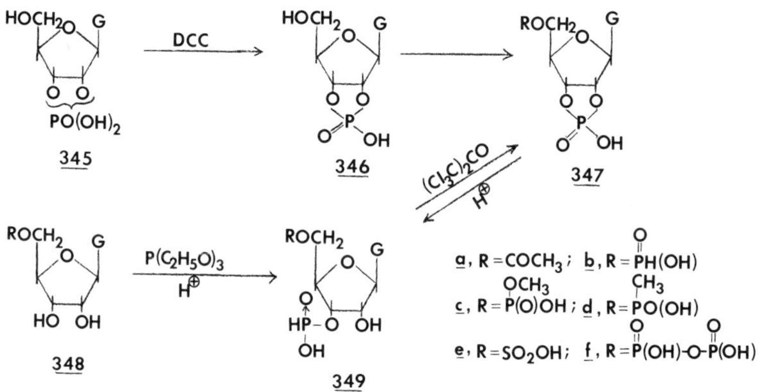

The 5′-*O*-substituted guanosine 2′,3′-cyclic phosphates (347c–f) were prepared by an alternate route. The reaction between guanosine 5′-phosphate methyl ester [348, R, $\overset{\mathrm{O}}{\overset{\|}{\mathrm{P}}}$—(OCH$_3$)OH] and triethylphosphite in dimethylformamide containing 6 *N* hydrochloride acid occurred slowly to give nucleotide phosphite [349, R, $\overset{\mathrm{O}}{\overset{\|}{\mathrm{P}}}$—(OCH$_3$)OH] which, upon treatment with hexachloroacetone, gave guanosine 5′-phosphate 2′,3′-cyclic phosphate P^5′-methylester (347c) (Holý 1969a).

Similarly, 5′-methanephosphonate-(347d), 5′-sulfate-(347e) and 5′-pyrophosphate-(347f) derivatives of guanosine 2′,3′-cyclic phosphate have been prepared (Holy' 1969a). The structure of these cyclic nucleotides (347b–e) was confirmed on the bases of their ease of hydrolysis in 50% aqueous acetic acid at 50 °C or ribonuclease T1 to the corresponding 2′(3′)-phosphates (349) (Holy' 1969a).

The synthesis of the mixture of two diastereomers of guanosine 2′,3′-cyclic phosphorothioate (351) is reported by the reaction of 5′-acetylguanosine (350) with triimidazolyl-1-phosphine sulfide (Eckstein et al. 1972). Alkaline hydrolysis of (351) gave guanosine 2′(3′)-phosphorothioate, and only the axial isomer is hydrolyzed by ribonuclease T1 to guanosine 3′-phosphorothioate, without loss of sulfur (Eckstein et al. 1972).

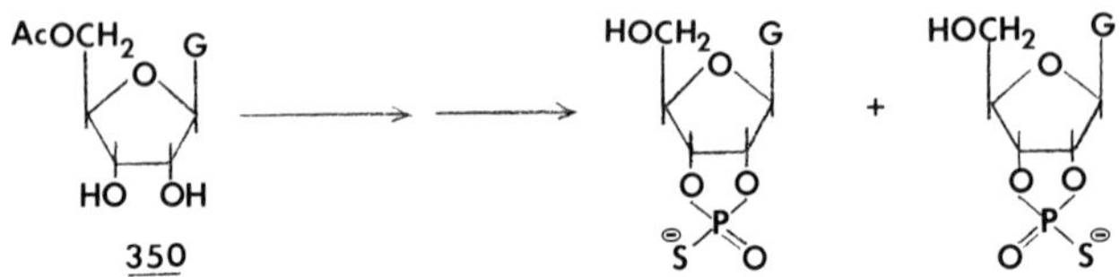

350 351

IV. Pyrimidine Cyclic Nucleotides

Whereas the presence of *c*AMP and *c*GMP in most mammalian tissues has long been established, the natural occurrence of pyrimidine cyclic nucleotides was not demonstrated until 1974, although such nucleotides had been prepared chemically as early as 1961 (Smith et al. 1961). It was Alexander Bloch who first isolated cytidine 3′,5′-cyclic phosphate (*c*CMP) from the extracts of leukemia L-1210 cells (Bloch 1974; Bloch et al. 1974) and demonstrated that addition of *c*CMP causes the initiation of growth of the leukemic cells in culture (Bloch et al. 1974; Bloch 1975a). The concentration of *c*CMP in regenerating liver is higher than in normal cells, and its concentration in leukemic cells is much higher than in regenerating liver (Bloch 1975a). Although *c*CMP has been isolated from the urine of human leukemia patients, it has not been detected in normal subjects (Bloch et al. 1975). Subsequently, uridine 3′,5′-cyclic phosphate (*c*UMP) was also isolated from rat liver extracts (Bloch 1975b). The isolation of both *c*CMP and *c*UMP from bacterial culture fluids has recently been reported (Ishiyama 1975). Cytidine 3′,5′-cyclic phosphate (*c*CMP, 353) is formed from cytidine 5′-triphosphate (352) by

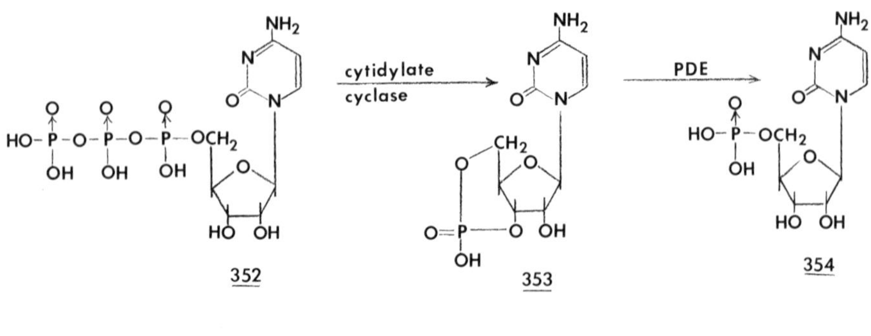

352 353 354

cytidylate cyclase in the presence of manganese or iron at neutral pH (CECH et al. 1976; CECH and IGNARRO 1977, 1978), and hydrolyzed by *c*CMP phosphodiesterase to cytidine 5′-phosphate (354) (KUO et al. 1978; CHENG and BLOCH 1978). The hydrolyzing activity of *c*CMP phosphodiesterase appears to be quite different from that reported for *c*UMP phosphodiesterase (HARDMAN and SUTHERLAND 1965). The growth lag (BLOCH et al. 1974) observed with *c*CMP and particularly with *c*UMP may lead to the elucidation of yet another "yin-yang" mechanism by which cell processes are controlled.

The chemical synthesis of cytidine 3′,5′-cyclic phosphate (*c*CMP, 353) and uridine 3′,5′-cyclic phosphate (*c*UMP, 355) has been accomplished by procedures similar to those used for synthesis of purine 3′,5′-cyclic nucleotides. Intramolecular cyclization of the requisite 5′-nucleotide by DCC under high dilution conditions (SMITH and KHORANA 1959; SMITH et al. 1961) or by the ynamine procedure (FUJIMOTO and NARUSE 1968), or preferably, base-catalyzed cyclization of "active" phosphate derivative (BORDEN and SMITH 1966b) gave good yields of the pyrimidine 3′,5′-cyclic phosphates (353) and (355). In the case of (353) the N^4-benzoyl derivative of cytidine 5′-phosphate was required for solubilization in pyridine (SMITH et al. 1961). DCC mediated cyclization of 4-morpholino-N,N′-dicyclohexylcarboxamidine salt of cytidine 5′-phosphate with an unprotected amino group in pyridine-dimethylformamide, however, gave (353) in rather low yield (ZIELINSKI et al. 1974; WIERENGA and WOLTERSOM 1977). Treatment of (353) with butyric anhy-

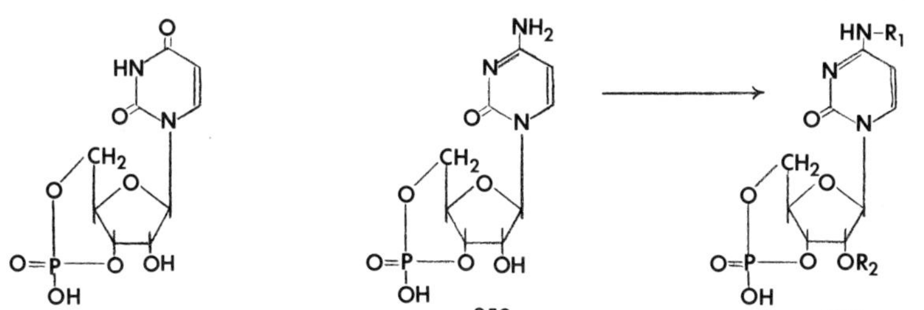

355 353 356

a, $R_1 = R_2 = CO(CH_2)_2CH_3$
b, $R_1 = H$; $R_2 = CO(CH_2)_2CO$

dride in pyridine gave N^4,2′-*O*-dibutyryl cytidine 3′,5′-cyclic phosphate (356a) (WIERENGA and WOLTERSOM 1977). Selective 2′-*O*-succinylation of (353) with succinic anhydride gave 2′-*O*-succinyl-*c*CMP (356b), which hydrolyzed readily to *c*CMP with dilute alkali (WIERENGA and WOLTERSOM 1977). Reaction of *c*UMP with N-bromosuccinimide in glacial acetic acid produced 5-bromouridine 3′,5′-

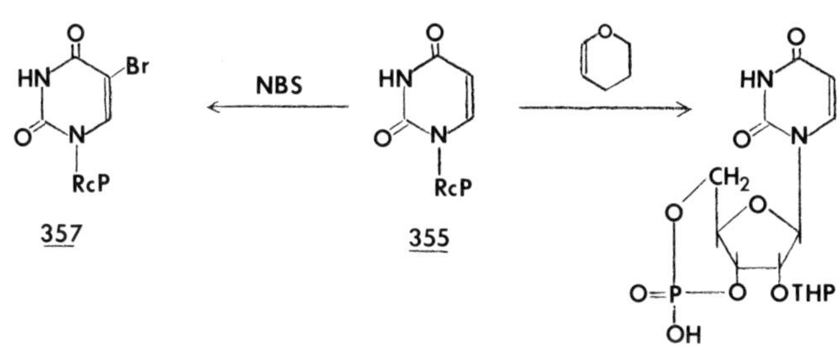

357 355 358

cyclic phosphate (357) in good yield (LONG and ROBINS 1978). The selective introduction of a protecting group at 2′-position of (355) was achieved by the reaction of dihydropyran with *c*UMP to obtain 2′-*O*-tetrahydropyranyluridine 3′,5′-cyclic phosphate (358) (SMITH and KHORANA 1959). Alkaline hydrolysis of (358) gave a mixture of 2′-*O*-tetrahydropyranyluridine 3′- and 5′-phosphates (SMITH and KHORANA 1959).

Several triesters of *c*UMP (359) were also prepared by direct treatment of (355) with diazoalkanes (ENGELS and PFLEIDERER 1975). This single step reaction gave diastereomeric mixtures, which were chromatographically separated into the axial and equatorial isomers.

359

R = H, Ac, Piv.

R′ = CH_3, C_2H_5, CH_2Ph

Activation of the phosphate of *c*UMP with diphenylphosphorochloridate gave the mixed anhydride (360), which was used for the synthesis of uridine 3′,5′-cyclic (methyl phenylalanyl)-phosphoramidate (361) (PREOBRAZHENSKAYA et al. 1967). Acid and alkaline hydrolysis, with cleavage of the phosphate ester bonds, of the phosphoramidate (361) was faster than for uridine 3′,5′-cyclic phosphate.

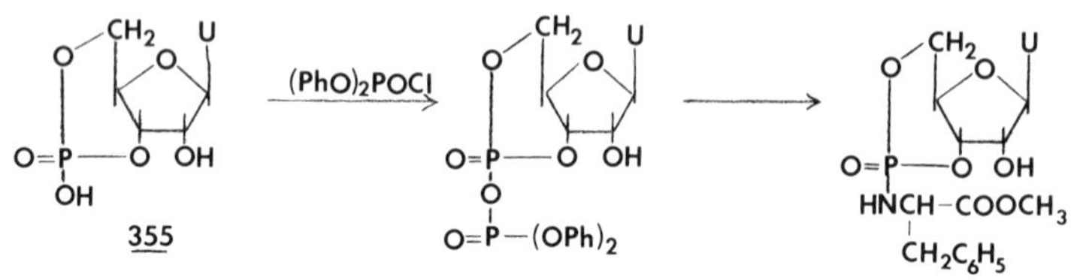

355 360 361

The 2′-deoxy-3′,5′-cyclic phosphates of uridine and cytidine have been conveniently synthesized as in the case of the *ribo* derivatives by the DCC (DRUMMOND et al. 1964) or by the base-catalyzed "active phosphate" methods (BORDEN and SMITH 1966b). Other "active phosphate" nucleotides, such as nucleoside 5′-phosphorofluoridate, 5′-(2,4-dinitrophenyl)phosphate and 5′-(P^1,P^1-diphenyl)pyrophosphates, were also cyclized to nucleoside 3′,5′-cyclic phosphates in good yield when treated with potassium *t*-butoxide in dimethylsulfoxide (BORDEN and SMITH 1966b).

The formation of thymidine 3′,5′-cyclic phosphate (363) as a minor product was observed during the polymerization of thymidine 5′-phosphate to oligonucleotides (TENER et al. 1958). Subsequently a convenient synthesis of (363) was accomplished by reaction of thymidine 5′-phosphate (362) with DCC in dilute pyridine solution (TENER et al. 1958; DRUMMOND et al. 1964). Thymidine 3′,5′-cyclic phosphate was also formed when the p-nitrophenyl esters of thymidine 3′-

phosphate (364) or thymidine 5′-phosphates were treated with potassium *t*-butoxide in dimethylsulfoxide (TURNER and KHORANA 1959; BORDEN and SMITH 1966 b). The 3′-p-nitrophenyl ester (364) reacts most readily (SMITH 1964; BORDEN and SMITH 1966 b). Alkaline or enzymatic hydrolysis of (363) afforded a mixture of thymidine 3′- and 5′-phosphates (TENER et al. 1958). The glycosyl linkage of (363) is much more labile to acid hydrolysis than is the corresponding linkage in thymidine or in thymidine 5′-phosphate (362).

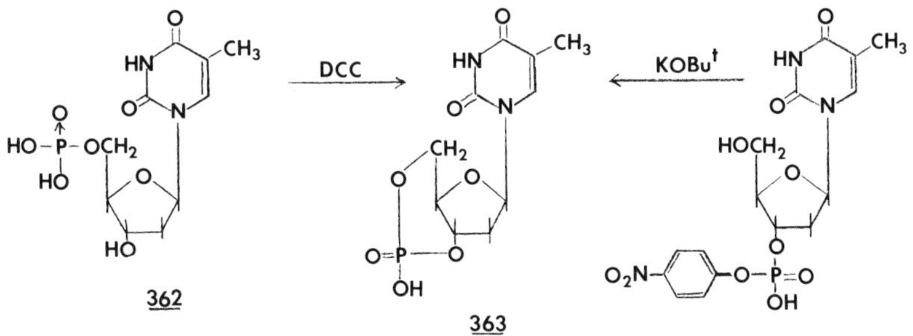

362 363 364

During the selective iodination of thymidine with methyltriphenoxyphosphonium iodide, a low yield of thymidine 3′,5′-cyclic methylphosphonate (367) was isolated (VERHEYDEN and MOFFATT 1970). The mechanism of this reaction was suggested to occur by intramolecular attack of the 3′-hydroxyl upon the phosphorus atom of the 5′-diphenoxymethylphosphonium intermediate (365) giving a cyclic phosphonium salt (366) or related phosphorane which decomposed to thymidine 3′,5′-cyclic methylphosphonate (367).

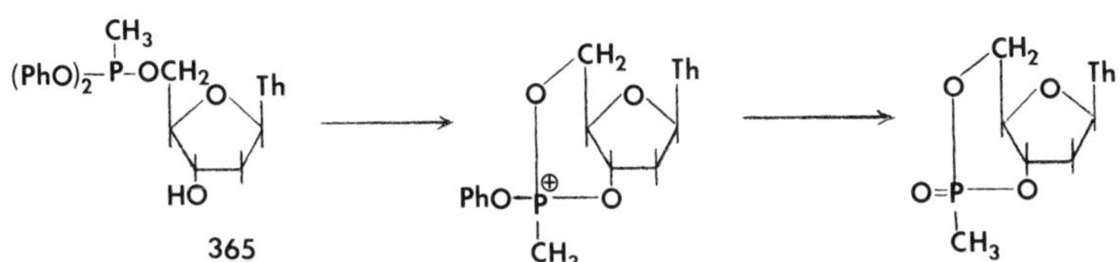

365 366 367

1-(2-Deoxy-β-D-*threo*-pentofuranosyl)thymine 3′,5′-cyclic phosphate (369) has been synthesized by intramolecular nucleophilic displacement of iodide from 3′-deoxy-3′-iodothymidine 5′-phosphate (368) by the phosphate anion (RUSSELL and MOFFATT 1969). Treatment of 1-(2-deoxy-β-D-*threo*-pentofuranosyl)thymine 5′-diphenylphosphate (370) with 1,5-diazabicyclo[4.3.0]nonene-5 in dioxane gave the separable diastereomers (371) and (372). Alkaline hydrolysis of the diastereomers (371) and (372) or the diphenylphosphate (370) also gave 1-(2-deoxy-β-D-*threo*-pentofuranosyl)thymine 3′,5′-cyclic phosphate (369). Compound (369) was considerably more stable to alkaline hydrolysis than thymidine 3′,5′-cyclic phosphate and was not hydrolyzed by barium hydroxide under conditions known to hydrolyze thymidine 3′,5′-cyclic phosphate (RUSSELL and MOFFATT 1969).

Intramolecular nucleophilic displacement of the 3′-iodo function by the phosphate anion of 3′-deoxy-3′-iodothymidine 5′-phosphoromorpholidate (373) gave 1-(2-deoxy-β-D-*threo*-pentofuranosyl)thymine 3′,5′-cyclic phosphoromorpholidate (374) (RUSSELL and MOFFATT 1969). In contrast to the nucleoside 3′,5′-cyclic

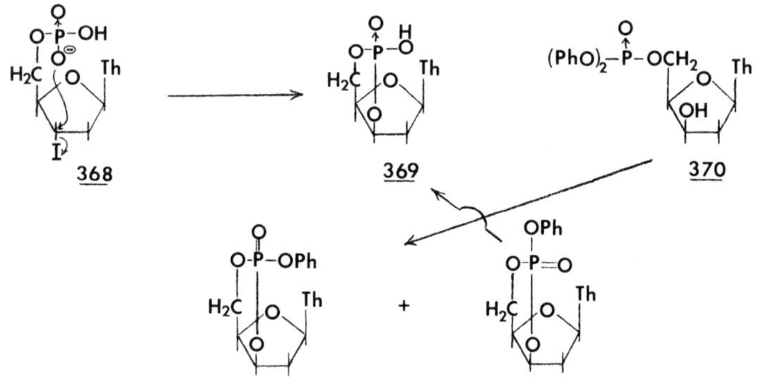

368 369 370 371 372

phosphoramidates (199) and (201), the P–N bond of (374) could not be selectively hydrolyzed under the conditions tried.

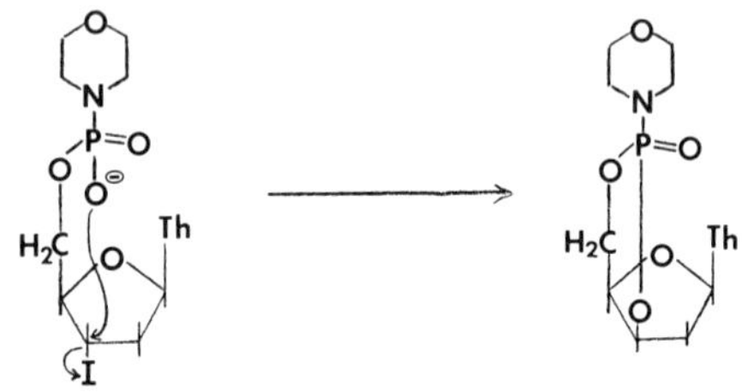

373 374

Several novel thymidine 3′,5′-cyclic phosphate derivatives have been prepared from thymidine by way of its 3′,5′-cyclic phosphite (375). When thymidine was treated with tris(dimethylamino)phosphane in dioxane at 80 °C or in presence of carbon dioxide at 20°–25 °C, thymidine 3′,5′-cyclic dimethylaminophosphite (375) was obtained (BASCHANG and KVITA 1973). The cyclic phosphite (375) was found to be very sensitive to hydrolysis. On oxidation with potassium permanganate, hydroperoxides or peroxy acids, (375) gave thymidine 3′,5′-cyclic dimethylaminophosphate (377). Replacement of dimethylamino group of (375) by an alkoxy function on reaction with p-acetoxyphenol afforded (376) which, on oxidation with *tert*-butyl hydroperoxide gave thymidine 3′,5′-cyclic p-acetoxyphenylphosphate (378) (BASCHANG and KVITA 1973). Treatment of thymidine 3′,5′-cyclic phosphate with ethereal diazomethane in anhydrous methanol gave diastereomeric mixture of thymidine 3′,5′-cyclic methylphosphate (379) (BLACKBURN et al. 1964).

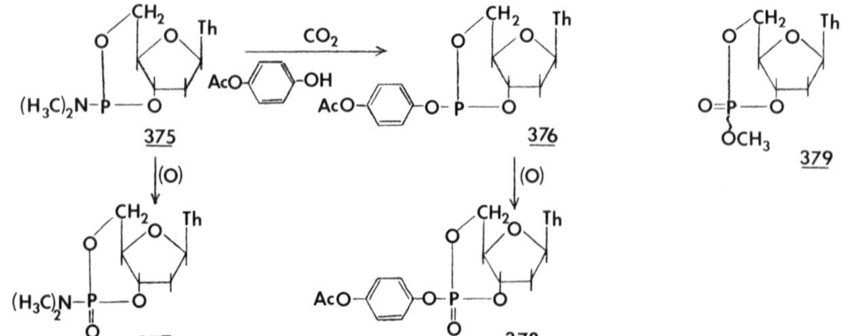

375 376 379 377 378

Facile 2′,5′-cyclic phosphate formation by the DCC procedure occurred with either N[4]-benzoyl-1-(β-D-arabinofuranosyl)cytosine 5′-phosphate (380) or N[4]-benzoyl-1-(β-D-arabinofuranosyl)cytosine 2′-phosphate (382) to give crystalline 1-(β-D-arabinofuranosyl)cytosine 2′,5′-cyclic phosphate (381) in high yield (WECHTER 1969). Typical of seven-membered cyclic phosphates the 2′,5′-cyclic phosphate ring of (381) was stable to acid and alkali and to conditions which normally hydrolyzed the D-*ribo*-3′,5′-cyclic nucleotides.

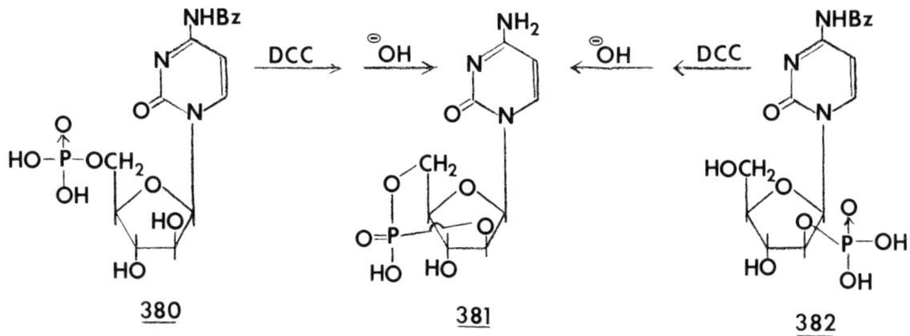

In order to obtain 1-(β-D-arabinofuranosyl)cytosine 3′,5′-cyclic phosphate (*ara-c*CMP, 387) the 1-(β-D-arabinofuranosyl)cytosine 3′-phosphate (385) was synthesized (WECHTER 1967) from cytosine 2′,3′-cyclic phosphate (383) (NAGYVARY 1969). Ring closure of (385) with DCC in pyridine and dimethylformamide gave *ara-c*CMP (387) in 35% yield (LONG et al. 1972; KREIS and WECHTER 1972). Deamination of (387) with aqueous nitrous acid gave quantitative yield of 1-(β-D-arabinofuranosyl)uracil 3′,5′-cyclic phosphate (*ara-c*UMP, 388) (LONG et al. 1972). Sulfhydrolysis of *ara-c*CMP with hydrogen sulfide in aqueous pyridine gave 1-(β-D-arabinofuranosyl)-4-thiouracil 3′,5′-cyclic phosphate (386, R=H), which was subsequently methylated with methyl iodide in dilute alkali to 1-(β-D-arabinofuranosyl)-4-methylthio-2-pyrimidone 3′,5′-cyclic phosphate (386, $R=CH_3$) (LONG et al. 1972).

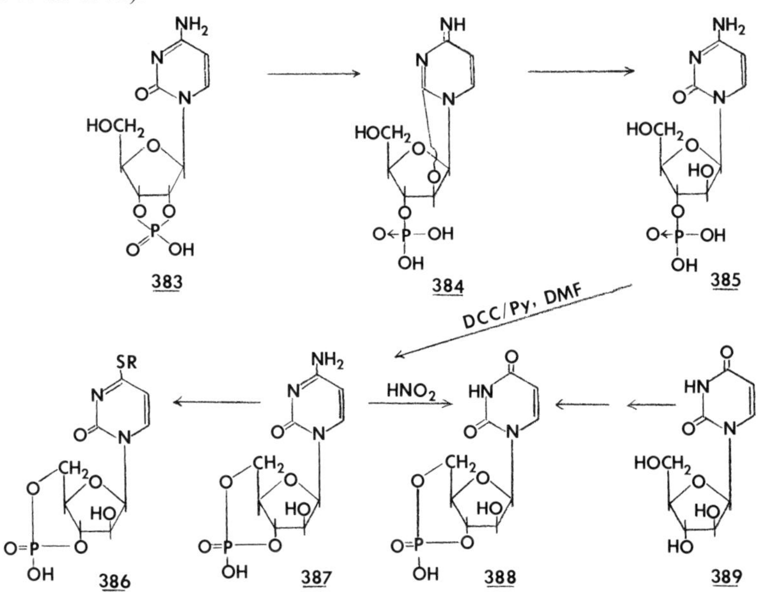

The reaction between 1-(β-D-arabinofuranosyl)uracil (389) and triethyl phosphite occurred slowly to give a low yield of nucleoside phosphites which, upon treatment with hexachloroacetone, gave a single cyclic nucleotide. The structure of this nucleotide was suggested to be *ara-c*UMP (388) based on its ease of hydrolysis in 0.1 *N* hydrochloric acid and 0.2 *M* sodium hydroxide at 50 °C (HOLY' and ŠORM 1969). The formation of (388) from an unprotected arabinonucleoside is in contrast to the formation of the 2′,5′-cyclic phosphate (381).

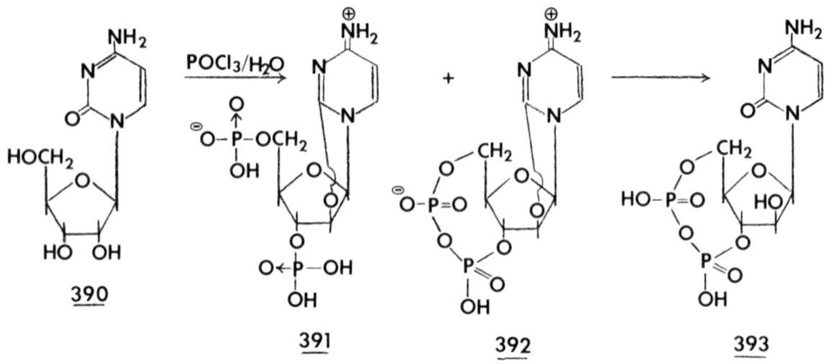

When cytidine (390) was heated at 60 °C with partially hydrolyzed phosphorus oxychloride (4.5 ml $POCl_3$ with 0.8 ml H_2O), two phosphorylated O^2,2′-cyclocytidines, (391) and (392) were obtained (KANAI et al. 1971). Further hydrolysis of (392) with 1 *M* triethylammonium bicarbonate at 50 °C gave the interesting *ara*-cytidine 3′,5′-cyclic pyrophosphate (393) (KANAI et al. 1971).

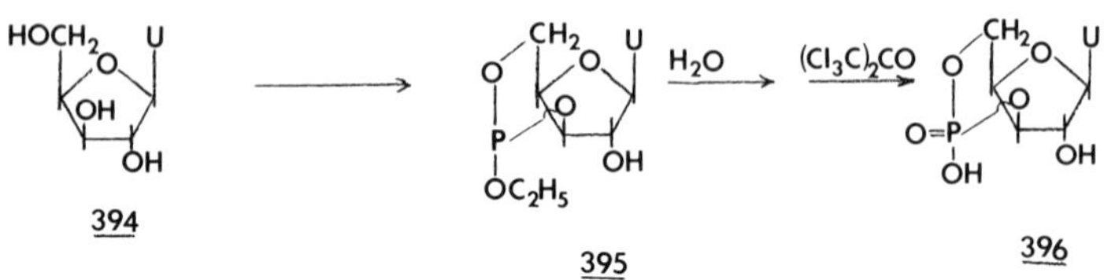

Triethyl phosphite undergoes transesterification with 1-(β-D-xylofuranosyl)-uracil (394) to give initially the cyclic triester (395). Hydrolysis of the triester (395) and oxidative cyclization of the resulting nucleoside phosphite with hexachloroacetone gave 1-(β-D-xylofuranosyl)uracil 3′,5′-cyclic phosphate (396) (HOLY' and ŠORM 1969).

The formation of three lyxofuranosyl cyclic nucleotides is stereochemically feasible from lyxofuranosyl nucleosides. The DCC cyclization of the 2′(3′)phosphate (397a) or the oxidative cyclization of the 2′(3′)phosphite (397b) of 1-(β-D-lyxofuranosyl)uracil gave only 1-(β-D-lyxofuranosyl)uracil 2′,3′-cyclic phosphate (398) (UKITA and HAYATSU 1961; HOLÝ and ŠORM 1969). The cyclization of the 5′-phosphate (399a) or the 5′-phosphite (399b), however, gave a mixture of 1-(β-D-lyxofuranosyl)uracil 3′,5′-cyclic phosphate (400) and 1-(β-D-lyxofuranosyl)uracil 2′,5′-cyclic phosphate (401). The acidic and alkaline stability of the 3′,5′-cyclic phosphate (400) and the 2′,5′-cyclic phosphate (401) are in accordance with other six- and seven-membered cyclic phosphates, in that both (400) and (401) were more stable to hydrolysis than the five-membered cyclic phosphate (398) (UKITA and

(+2′-isomer)
397
a, R = PO_3H_2
b, R = PO_2H_2

398

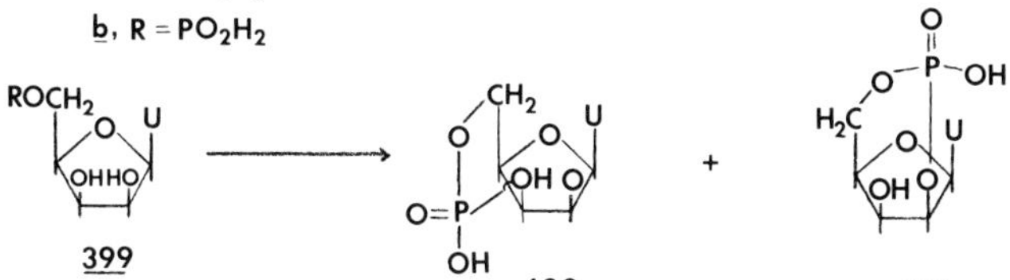

399
a, R = PO_3H_2
b, R = PO_2H_2

400

401

HAYATSU 1961). The 2′,5′-, 3′,5′- and 2′,3′-cyclic phosphates of 1-(α-L- and α-D-lyxofuranosyl)thymine were similarly prepared from the respective 2′, 3′- and 5′-phosphates (HOLÝ and ŠORM 1969).

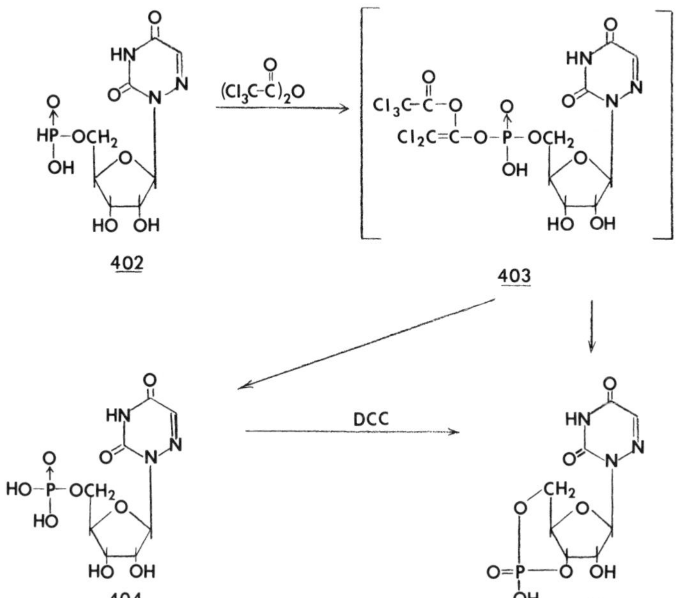

402 403 404 405

The oxidation of 6-azauridine 5′-phosphite (402) with trichloroacetic anhydride gave 6-azauridine 5′-phosphate (404) as well as 6-azauridine 3′,5′-cyclic phosphate (405), and P^1,P^2-di(6-azauridine-5′-yl)pyrophosphate presumably *via* the intermediate (403). The cyclic phosphate (405) was also prepared in high yield from the 5′-phosphate (404) by the DCC procedure (HOLÝ et al. 1965).

During hydrolysis of ribonucleic acid with pancreatic ribonuclease or with barium carbonate or with liquid ammonia, the 2′,3′-cyclic phosphates of pyrimidine nucleosides were formed (TANAKA 1960) which then undergo fission in two directions to give both 2′- and 3′-phosphates of pyrimidine nucleosides (MARKHAM and SMITH 1951, 1952; BROWN and TODD 1952; LIPKIN and TALBERT 1955). The isolation of these 2′- and 3′-phosphates from hydrolyzates of ribonucleic acid led to a chemical understanding of along-known phenomenon that ribonucleic acid is broken down to its mononucleotides with mild alkali, whereas 2′-deoxyribonucleic

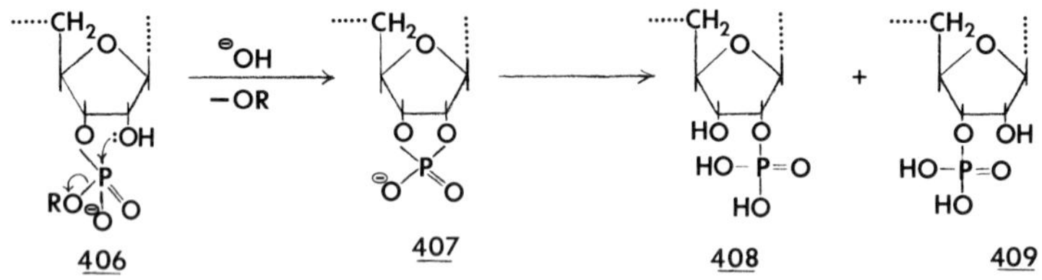

acid is not hydrolyzed under these conditions. The phospho-diester bonds of ribonucleic acid (406) undergo an S_{N^2} attack on phosphorus by the *cis*-vicinal hydroxyl group, with elimination of −OR (transesterification) to form 2′,3′-cyclic phosphates (407), which are then hydrolyzed to nucleoside 2′- and 3′-phosphates (408) and (409). 2′-Deoxyribonucleic acid, which lacks the vicinal hydroxyl groups cannot undergo this type of fission (UEDA and FOX 1967; HUTCHINSON 1971).

The chemical cyclization of a nucleoside phosphate monoester leading to *cis*-fused cyclic phosphate is relatively easy. Quantitative yields of pyrimidine nucleoside 2′,3′-cyclic phosphates can be obtained under mild conditions with acid chlorides (MICHELSON 1958, 1959; GASSEN and WITZEL 1967), acid anhydrides (BROWN et al. 1952), dimethylformamide acetals (HOLÝ et al. 1969) or carbodiimides (DEKKER and KHORANA 1954; SMITH et al. 1958; LOHRMAN and KHO-

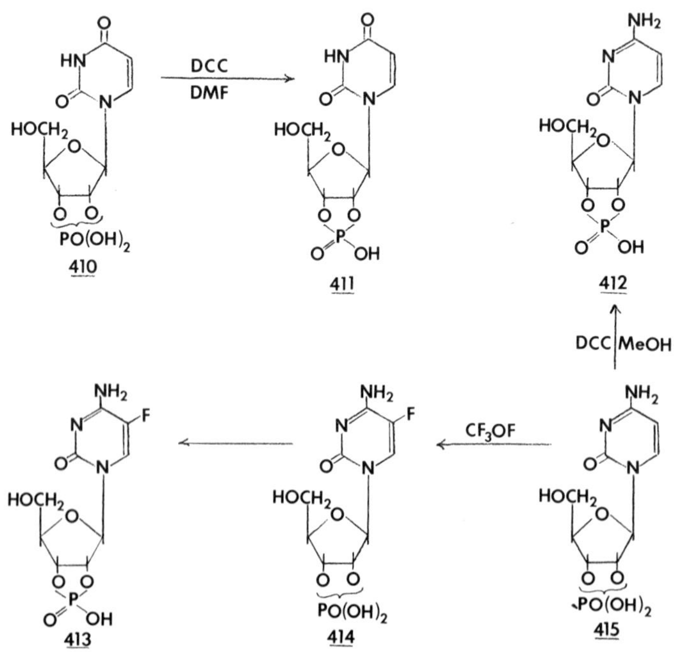

RANA 1964; NAYLOR and GILHAM 1966). The dimethylformamide acetal procedure was preferred to the acid chloride or carbodiimide methods for the preparation of pyrimidine 2′,3′-cyclic phosphates. The presence of a 5′-phosphate group [e.g. cytidine 2′(3′),5′-diphosphate] was found not to interfere with the dimethylformamide acetal cyclization to the 2′,3′-cyclic nucleotide (HOLÝ et al. 1969).

A simple preparative procedure for uridine 2′,3′-cyclic phosphate (411) or cytidine 2′,3′-cyclic phosphate (412) has been described (SHUGAR and WIERZCHOWSKI 1958; SZER and SHUGAR 1963) which involves DCC mediated intramolecular cyclization of corresponding 2′(3′)-phosphates (410) and (415) in dimethylformamide or methanol. The yields by this method are essentially quantitative. Uridine 2′,3′-cyclic phosphate has also been prepared by phosphorylation of 5′-*O*-benzoyluridine with P[1]-diphenyl P[2]-morpholinopyrophosphorochloridate, followed by treatment with dilute ammonium hydroxide (IKEHARA and TAZAWA 1966). Electrophilic fluorination of cytidine 2′(3′)-phosphate (415) with trifluoromethyl hypofluorite produced 5′-fluorocytidine 2′(3′)-phosphate (414) which on treatment with ethyl chloroformate at ambient temperature gave 5-fluorocytidine 2′,3′-cyclic phosphate (413) (MIAN et al. 1979). Compound (413) was found to be a key intermediate for the synthesis of antiviral agent 1-(β-D-arabinofuranosyl)-2-amino-1,4(2H)-4-imino-5-fluoropyrimidine 3′-phosphate (MIAN et al. 1979).

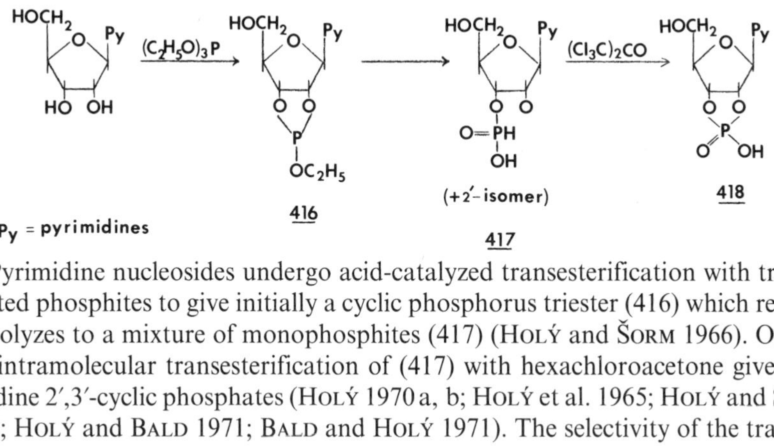

Py = pyrimidines

416 417 418

Pyrimidine nucleosides undergo acid-catalyzed transesterification with trisubstituted phosphites to give initially a cyclic phosphorus triester (416) which readily hydrolyzes to a mixture of monophosphites (417) (HOLÝ and ŠORM 1966). Oxidative intramolecular transesterification of (417) with hexachloroacetone gives pyrimidine 2′,3′-cyclic phosphates (HOLÝ 1970a, b; HOLÝ et al. 1965; HOLÝ and SMRT 1966; HOLÝ and BALD 1971; BALD and HOLÝ 1971). The selectivity of the transes-

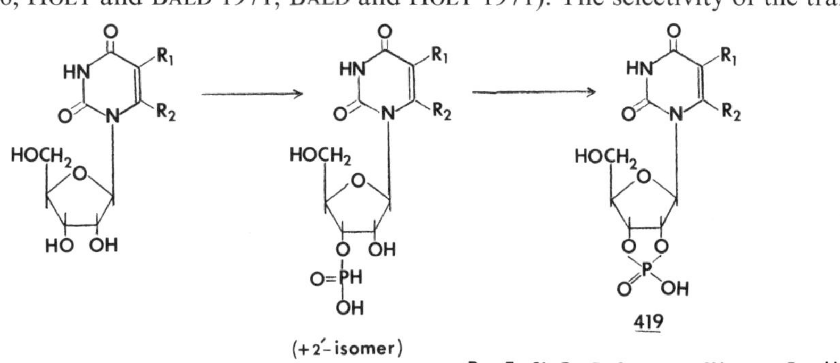

419

R_1 = F, Cl, Br, I, OH, NH_2, CH_3, etc. R_2 = H

R_1 = H ; R_2 = CH_3, COOH

R_1 = R_2 = CH_3, etc.

terification is a function of the stereochemistry of the diol and the phosphite. Unprotected ribonucleosides react selectively and quantitatively with triethylphosphite at the *cis*-2′,3′-position, resulting ultimately in high yields of the 2′,3′-cyclic nucleotides. Using this elegant phosphite-phosphate procedure, several 2′,3′-cyclic phosphates of base-substituted pyrimidine *ribo*-nucleosides (419) and other related compounds have been synthesized (Holý and Bald 1971; Bald and Holý 1971). The synthesis of other nucleoside 2′,3′-cyclic phosphates related to (419) has also been performed analogously. Isocytidine-(420), 5-methyl-6-azauridine-(421), 6-azauridine-(422), N^3-(2-hydroxyethyl)uridine-(423) and C-1′,5-dimethyluridine (424) 2′,3′-cyclic phosphates were obtained by this method in good yields (Holý and Bald 1971; Holý et al. 1971). The 2′,3′-cyclic phosphates of 2-thiouridine

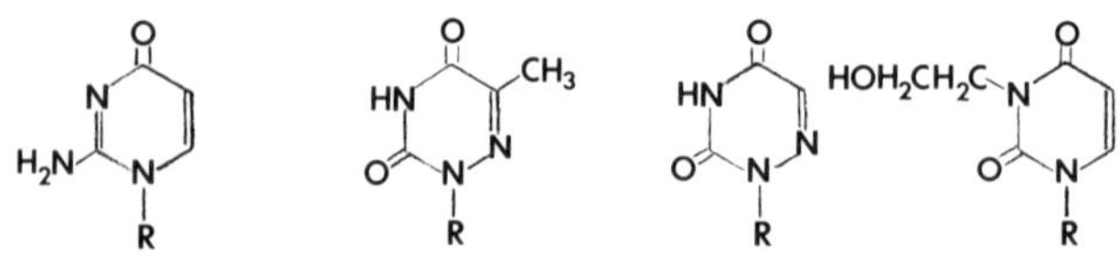

420 421 422 423

R = β-D-ribofuranosyl 2',3'-cyclic phosphate

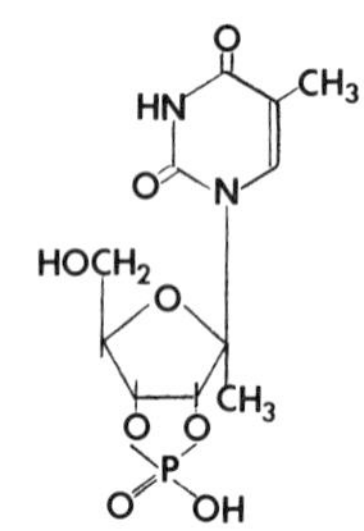

424

(425), 2-thio-6-azauridine (426) and 4-thio-6-azauridine (427) have been synthesized from unprotected nucleosides with triethylammonium phosphate in the presence of trichloroacetonitrile (Pischel and Holý 1970).

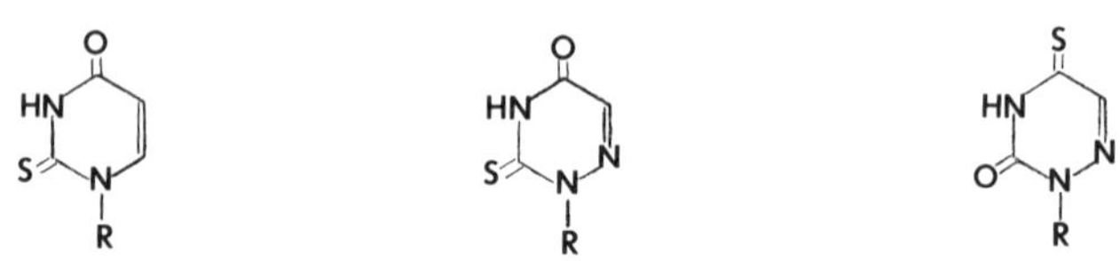

425 426 427

R = β-D-ribofuranosyl 2',3'-cyclic phosphate

Since the phosphite-phosphate procedure failed to give 5-nitrouridine 2′,3′-cyclic phosphate, an alternate procedure was used. 5-Nitrouridine was transformed by reaction with substituted trityl chloride to the 5′-tritylated derivative (428), the treatment of which with 2-cyanoethyl phosphate and DCC followed by removal of protecting groups first in alkaline and then in acidic media, gave 5-ni-

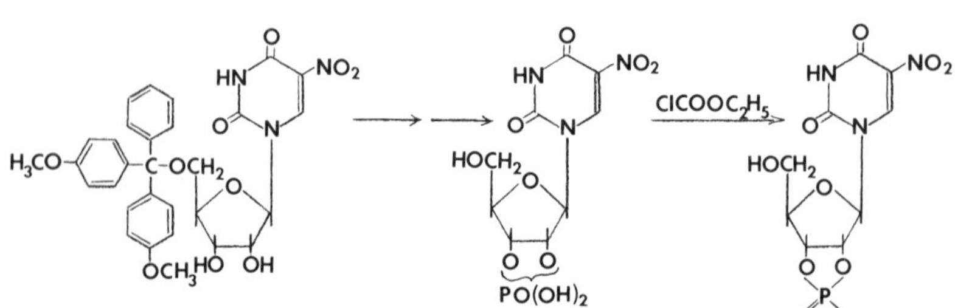

428 429 430

trouridine 2′(3′)-phosphate (429). Treatment of (429) with ethyl chloroformate gave 5-nitrouridine 2′,3′-cyclic phosphate (430) (HOLÝ and BALD 1971).

1-(β-D-Ribofuranosyl)-4-methyl-2-pyrimidone (431) (KLEIN et al. 1970), when phosphorylated with triethylammonium phosphate in presence of trichloroacetonitrile, 1-(β-D-ribofuranosyl)-4-methyl-2-pyrimidone 2′,3′-cyclic phosphate (432) was obtained (BALD and HOLÝ 1971). Hydrolysis of (432) with 50% aqueous acetic acid gave the 2′(3′)-phosphate (433) and the bacterial phosphatase degradation of (432) afforded quantitatively the starting nucleoside (431) (BALD and HOLÝ 1971).

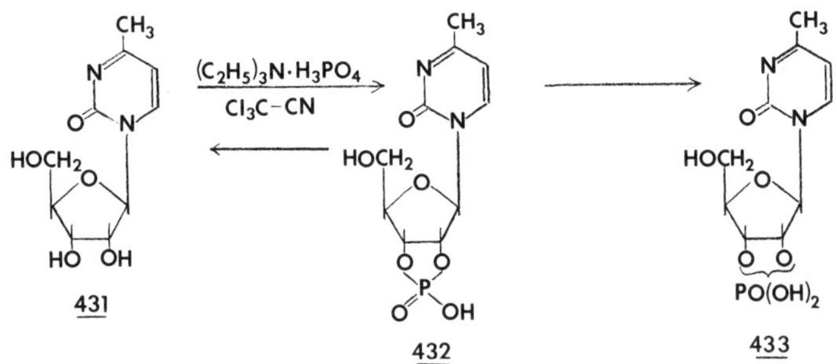

Several 3-(β-D-ribofuranosyl)uracil 2′,3′-cyclic phosphate derivatives (436) have been prepared from the corresponding nucleosides (434) (WINKLEY and ROBINS 1968; PRYSTÁS and ŠORM 1969), by the phosphite-phosphate procedure (BALD and HOLÝ 1971). Compounds (436) were cleaved with 50% aqueous acetic acid to the corresponding 2′(3′)-phosphates and were dephosphorylated to the starting nucleosides (434) by the action of alkaline bacterial phosphates (BALD and HOLÝ 1971).

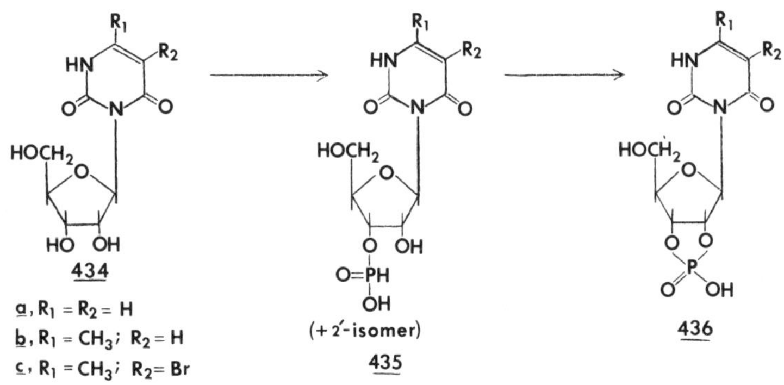

Some interesting structural isomers of uridine 2′,3′-cyclic phosphates, like pseudouridine 2′,3′-cyclic phosphate (437), 1-(β-D-ribofuranosyl)-3,6-dioxopyridazine 2′,3′-cyclic phosphate (438) and 1-(β-D-ribofuranosyl)-2-pyridon 2′,3′-cyclic phosphate (439) have been prepared (GASSEN and WITZEL 1967). D-1-(2,3-Dihydroxypropyl)uracil 2′,3′-cyclic phosphate (440 a) and D-1-(2,3-dihydroxypropyl)thymine 2′,3′-cyclic phosphate (440 b) have been reported to be formed by neutral buffered hydrolysis or by DCC mediated cyclization of the corresponding 3′-phosphates (HOLÝ and IVANOVA 1974).

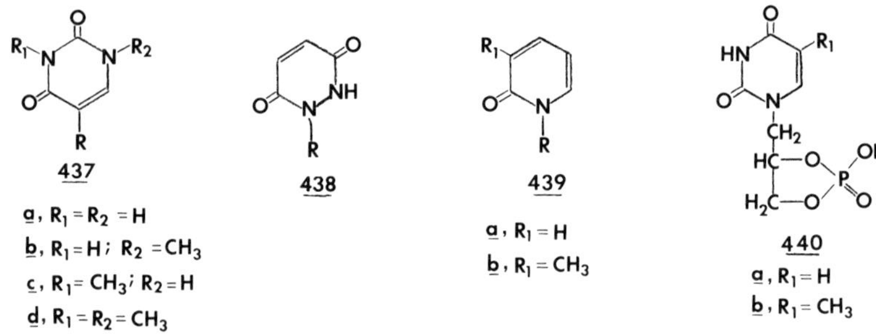

R = β-D-ribofuranosyl 2',3'-cyclic phosphate

When 5′-*O*-acetyluridine (441) was treated with triimidazolyl-1-phosphinsulfide (442) in anhydrous pyridine, (441) was smoothly cyclized into uridine 2′,3′-cyclic phosphorothioate (443), and the possible two diastereomers were separated by crystallization (ECKSTEIN and GINDL 1968). The absolute configuration of (443) was established by x-ray analysis (SAENGER and ECKSTEIN 1969). Whereas the alkaline hydrolysis of (443) is lower than that of uridine 2′,3′-cyclic phosphate by a factor of 6, the acid hydrolysis is slower by a factor of 200 (ECKSTEIN 1968). Treatment

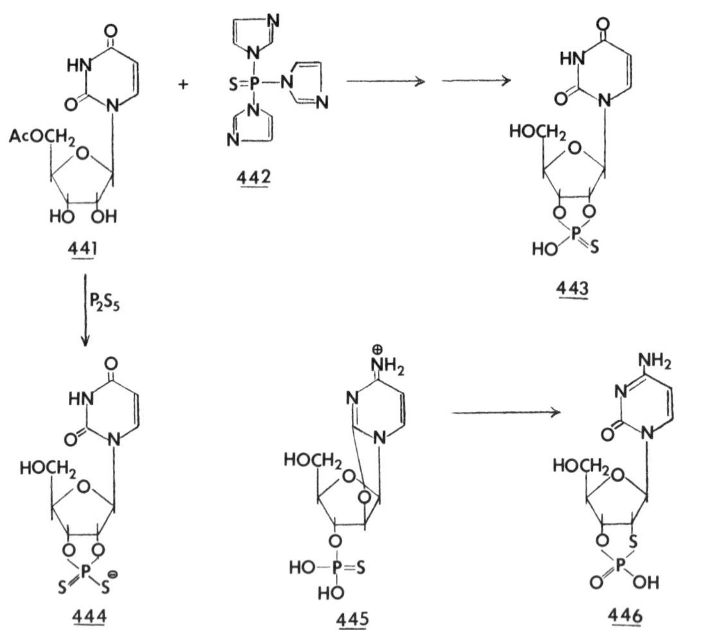

of 5′-*O*-acetyluridine with P_2S_5 at ambient temperature in pyridine, followed by brief reaction with ammonia, gave uridine 2′,3′-cyclic phosphorodithioate (444) (ECKSTEIN 1970). The phosphorodithioate (444) was a substrate for pancreatic ribonuclease and was hydrolyzed to presumably uridine 3′-phosphorodithioate without loss of sulfur (ECKSTEIN 1970). The preparation of 2′-thio-2′-deoxycytidine 2′,3′-cyclic phosphorothioate (446) has recently been accomplished by the intramolecular displacement reaction of anhydro-*ara*-cytidine 3′-*O*-phosphorothioate (445) in 90% aqueous pyridine (BRADBURY and NAGYVARY 1976).

Like pyrimidine nucleoside 2′,3′-cyclic phosphates, the synthesis of cyclic nucleotides with pyranose sugars has been extensively studied (HOLÝ 1969b). The ease of cyclic phosphate formation from glucosyl phosphates was exemplified by the ready formation of cyclic phosphates from 1-(β-D-glucopyranosyl)uracil 2′,3′,4′ or 6′-monophosphate under the mild conditions of DCC and aqueous pyridine at room temperature (ZMUDZKA et al. 1962). Transesterification of

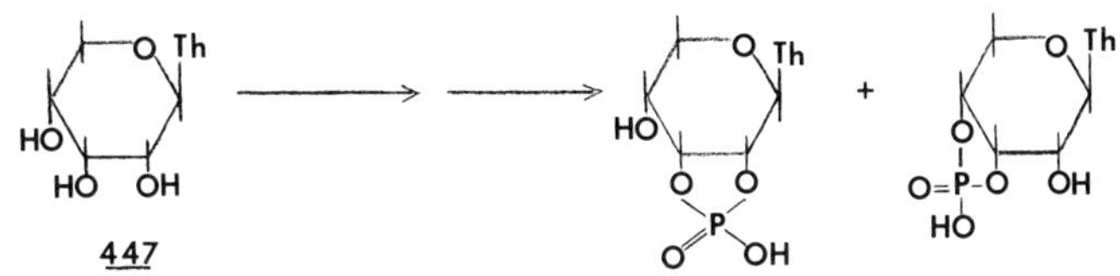

447 448 449

pyranosyl derivatives possessing *cis*-hydroxyls like 1-(β-D-ribopyranosyl)thymine (447) with triethylphosphite gave high yields of nucleoside phosphites which readily formed cyclic phosphates (448) and (449) by oxidative cyclization. The transesterification of 1-(α-D-mannopyranosyl)thymine (450) was, however, complicated and gave mono- and diphosphite derivatives (HOLÝ 1969b). Oxidative cyclization of these phosphites gave the 2′,3′-cyclic phosphate (451) and the *bis*(cyclic phosphate) (452). Acetic acid hydrolysis of the *bis*(cyclic phosphate) (452) gave the 2′(3′)-phosphate 4′,6′-cyclic phosphate which was further hydrolyzed to the 4′,6′-cyclic phosphate (453) with *E. coli* alkaline phosphatase.

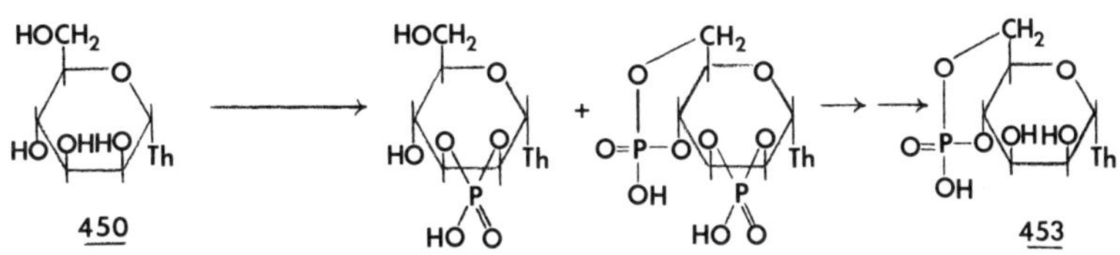

450 451 452 453

V. Miscellaneous Cyclic Nucleotides

5-Amino-1-(β-D-ribofuranosyl)imidazole-4-carboxamide 5′-phosphate (AICA-Ribotide, 454) plays an important role as central intermediate in the biosynthesis of purine nucleotides (LITCHFIELD and SHAW 1971; LEHNINGER 1975). Its unique structure shows possibilities of intra- and intermolecular interactions which are important from chemical and biochemical points of view (YAMAZAKI and OKUTSU 1978). The chemical synthesis of 5-amino-1-(β-D-ribofuranosyl)imidazole-4-carboxamide 3′,5′-cyclic phosphate (40) has been pursued by two ways (BERGMEYER et al. 1971). In the first procedure, the DCC mediated intramolecular cyclization of (454) gave a low yield (18%) of (40). Ring opening of *c*IMP or N[1]-methoxymethyl-2′-*O*-acetylinosine 3′,5′-cyclic phosphate (455) with hot 2 *N* potassium hydroxide, however, gave good yield of (40) (BERGMEYER et al. 1971; MICHAL et al.

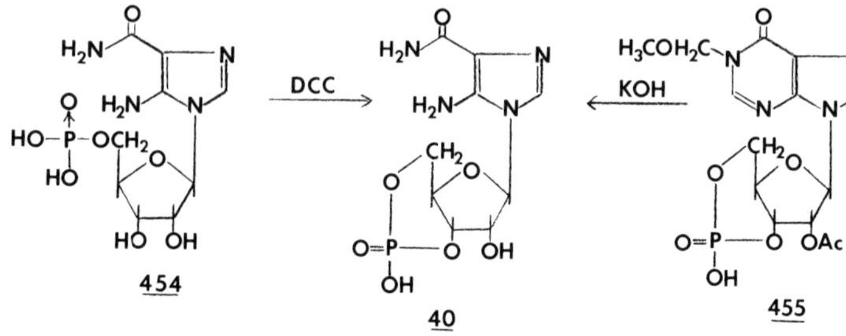

1974). A similar alkaline hydrolysis of *c*AMP-N[1]-oxide (20) gave 5-amino-1-(β-D-ribofuranosyl)imidazole-4-carboxamidoxime 3′,5′-cyclic phosphate (24) along with other minor products (see Sect. B.I.1.b). Raney nickel reduction of (24) gave 5-amino-1-(β-D-ribofuranosyl)imidazole-4-carboxamidine 3′,5′-cyclic phosphate (25) (MEYER Jr et al. 1973b). Compound (25) was also obtained by alkaline ring opening of 1-alkoxy-*c*AMP (23) to *O*-alkyl-5-amino-1-(β-D-ribofuranosyl)imid-

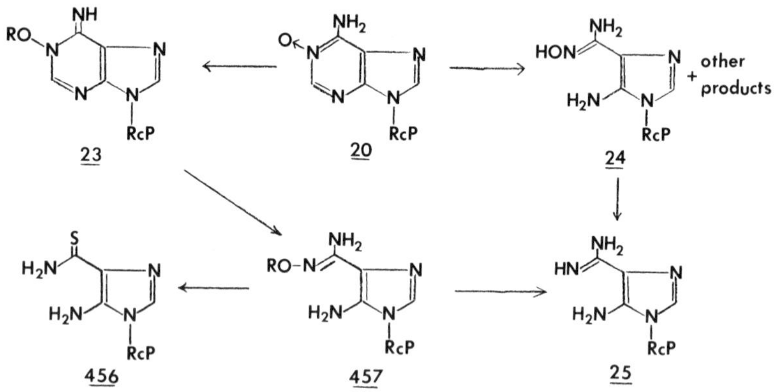

azole-4-carboxamidoxime 3′,5′-cyclic phosphate (457) (MEYER Jr et al. 1973b; SUEHIRO et al. 1976), followed by Raney nickel reduction. Treatment of (457) with liquid hydrogen sulfide in aqueous pyridine at elevated temperature gave 5-amino-1-(β-D-ribofuranosyl)imidazole-4-thiocarboxamide 3′,5′-cyclic phosphate (456). The 5-amino-1-(β-D-ribofuranosyl)imidazole 3′,5′-cyclic phosphate derivatives are poor activators of *c*AMP-dependent protein kinase (MEYER Jr et al. 1973b).

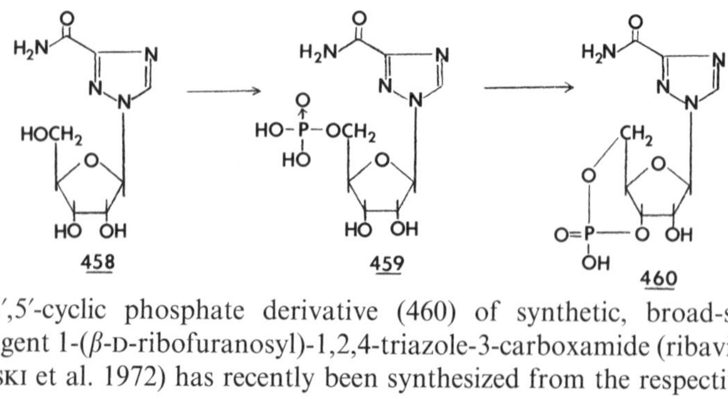

The 3′,5′-cyclic phosphate derivative (460) of synthetic, broad-spectrum antiviral agent 1-(β-D-ribofuranosyl)-1,2,4-triazole-3-carboxamide (ribavirin, 458) (WITKOWSKI et al. 1972) has recently been synthesized from the respective 5′-nucleotide (459) (ALLEN et al. 1978). 5,6-Dimethyl-1-(α-D-ribofuranosyl)benzimidazole 3′,5′-cyclic phosphate (461) was also obtained by intramolecular cyclization of the corresponding 5′-phosphate (FRIEDRICH 1963). Riboflavin 4′,5′-cyclic phosphate (462) has been prepared directly from riboflavin and phosphorus oxychloride (FORREST and TODD 1950). 2′,3′-Di-*O*-butyrylriboflavin 4′,5′-cyclic phosphate was obtained, interestingly, by treatment of riboflavin 5′-phosphate with butyric anhydride or a butyryl halide (SUZUKI et al. 1972).

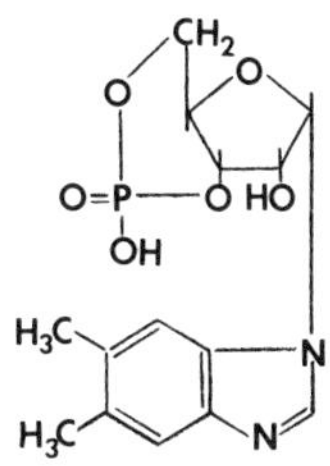

461

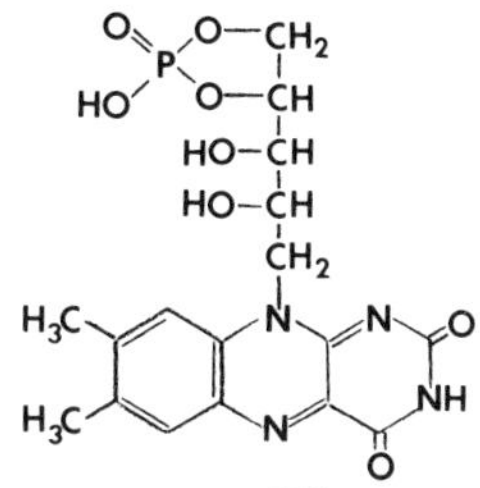

462

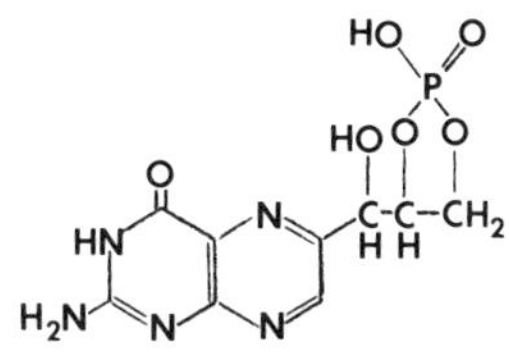

463

VISCONTINI and FURUTA (1973) have reported the synthesis of D-neopterin 2′,3′-cyclic phosphate (463) by DCC cyclization of D-neopterin 3′-phosphate. The preparation of 2′,3′-*O*-adamantylphosphonates of purine nucleosides has recently been reported (NEDOREZOVA et al. 1978). By the reaction of adenosine (464, $R=NH_2$) and 6-methylthio-9-(β-D-ribofuranosyl)purine (464, $R=SCH_3$) or their 5′-*O*-acyl derivatives with 1-adamantylphosphonedichloride, the appropriate 2′,3′-*O*-adamantylphosphonates were obtained. The diastereomers (465) and (466) were separated.

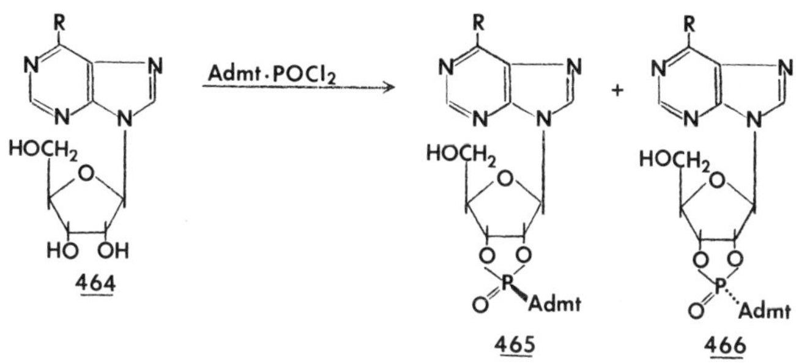

464 465 466

C. Chemical and Physical Properties of Cyclic Nucleotides

I. Hydrolytic Studies

The cyclic phosphates of ethylene glycol (467), propane-1,3-diol (468), and butane-1,4-diol (469) are the simple examples of 5-, 6-, and 7-membered cyclic phosphates. In alkali, (467) hydrolyzes some 10^6–10^7 times faster than the cyclic diester, dimethyl phosphate (KUMAMOTO et al. 1956; HAAKE and WESTHEIMER 1961) or methyl ethylene cyclic phosphate (470) (WESTHEIMER 1968). The 6-membered analog (468) is much more stable than (467) and hydrolyzes only approximately 10 times faster than dimethyl phosphate (KHORANA et al. 1957; CHERBULIEZ et al. 1959; KHORANA 1961a). The 7-membered cyclic phosphate (469) is even more stable than (468) and hydrolyzes approximately as fast as dimethyl phosphate

(KHORANA et al. 1957; CHERBULIEZ et al. 1959). The same order of stability is also found for these compounds in acidic conditions. Thus, the stability of these cyclic phosphates is a function of ring size and also phosphate ring configuration (PIZER and BALLOU 1959).

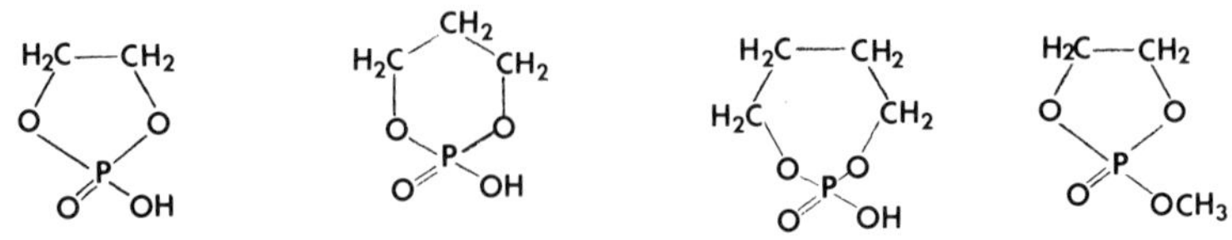

467 468 469 470

The lability of 5-membered cyclic phosphate is attributed to considerable ring strain (COX et al. 1959; WESTHEIMER 1957), and x-ray studies of methyl ethylene cyclic phosphate (470) have shown the existence of some deformations, particularly in the angles between the phosphorus and the two ring oxygen atoms (USHER et al. 1965; STEITZ and LIPSCOMB 1965). The instability of these simple cyclic phosphates has been correlated not only with the amount of ring strain but also with the degree of $d_\pi p_\pi$ bonding in the phosphate bonds (COLLIN 1966; BOYD 1969). Different hydrolysis rates do exist within each group where additional ring strain is imposed by *trans*- and *cis*-fused rings. In such cases, *cis*-fused rings were generally found to be more stable to hydrolysis than *trans*-fused ring systems (PIZER and BALLOU 1959; KHORANA 1961 a). Differences in fused ring size (e.g. in the 6-membered cyclic phosphates pyranose 4,6-cyclic phosphate and furanose 3,5-cyclic phosphate), ring conformation, or the presence of other substituents, would be expected to influence the stability of cyclic phosphates, thus drawing general conclusions rather difficult.

Ribonucleoside 2′,3′-cyclic phosphates are more labile toward acid and alkali than are noncyclic phosphodiesters. The hydrolysis of nucleoside 2′,3′-cyclic phosphate with acid or alkali is a simple phenomenon; facile ring opening of the cyclic phosphate ring occurs before glycosyl bond cleavage at approximately the same rate of hydrolysis of other 5-membered cyclic phosphates (KOCHETKOV et al. 1972). In 0.1 *N* hydrochloric acid at 25 °C, the hydrolysis of a nucleoside 2′,3′-cyclic phosphate is complete in 4 h producing a mixture of approximately equal amounts of 2′- and 3′-phosphates (MARKHAM 1957). Acid and alkali catalyzed alcoholysis of nucleoside 2′,3′-cyclic phosphates also produces a mixture of 2′- and 3′-alkyl phosphates (CHERBULIEZ et al. 1959; TENER and KHORANA 1955; BARKER et al. 1957). This reaction is apparently reversible (KHORANA et al. 1957) and this behavior is a rather general characteristic of 5-membered cyclic phosphates (KHORANA 1961 a). Phosphodiesterases have also been employed for the preparation of specific nucleoside 2′- or 3′-phosphates by selective cleavage of a particular O–P linkage in nucleoside 2′,3′-cyclic phosphates (KHORANA 1961 b; EGAMI et al. 1964). The rates of hydrolysis of uridine and cytidine 2′,3′-cyclic phosphates, however, differ. The cytidine phosphodiester is hydrolyzed to cytidine 3′-phosphate at twice the rate of uridine phosphodiester in the presence of ribonuclease (WIEKER and WITZEL 1967). This rate difference has been attributed to the conformational feature and the chemical nature of the bases in the context of the mechanism of ribonuclease catalysis (COULTER 1973; DEL ROSARIO and HAMMES 1970).

Table 1. Alkaline hydrolysis of ribonucleoside 3′,5′-cyclic phosphates and ribonucleoside 5′-phosphates in 1 *N* NaOH at 100 °C

Compound	Half-life (min)	Products
Adenosine 3′,5′-cyclic phosphate	36[a]	Adenine, transient formation of adenosine
Adenosine 5′-phosphate	10% after 2 h[b]	Adenine, two minor compounds
Uridine 3′,5′-cyclic phosphate	90[b]	Uridine 5′-phosphate (trace), Uridine 3′-phosphate (major), Uracil
Uridine 5′-phosphate	Unaffected after 2 h[a]	
Thymidine 3′,5′-cyclic phosphate	120[c]	Thymidine 3′-phosphate (80%), Thymidine 5′-phosphate (20%)

[a] LIPKIN et al. 1959a, b [b] SMITH et al. 1961 [c] TENER et al. 1958

The nucleoside 3′,5′-cyclic phosphate (with a 6-membered cyclic phosphate ring) are generally more stable to hydrolysis than the corresponding 2′,3′-cyclic phosphates. Although adenosine 3′,5′-cyclic phosphate is more stable than the corresponding 2′,3′-cyclic phosphate, its hydrolysis in alkaline medium is much faster than that of trimethylene phosphate. In 1 *N* sodium hydroxide at 100 °C, 50% of *c*AMP was hydrolyzed in 36 min, whereas trimethylene phosphate requires about 20 h (Table 1). The greater lability of *c*AMP is presumably due to the extra strain resulting from the *trans*-fusion of the phosphate ring with the furanose moiety.

The nucleoside 3′,5′-cyclic phosphates are less labile to hydrolysis than the corresponding 5′-nucleotides with glycosyl cleavage observed with *c*AMP and *c*UMP. The catalytic effect of polyvalent cations on the rate of hydrolysis of diesters of phosphoric acid (DIMROTH et al. 1959; BADDILEY et al. 1954; SZABO and SZABO 1958), nucleotides (KOCHETKOV et al. 1972), and 5- and 6-membered cyclic phosphates (LIPKIN et al. 1959; WESTHEIMER 1957; BADDILEY et al. 1954; SZABO and SZABO 1958) has been well documented. The catalytic effect of barium ions on hydrolysis of the phosphate ring of nucleoside 3′,5′-cyclic phosphates is shown by their complete hydrolysis in 30 min with 0.2 *M* barium hydroxide at 100 °C to the respective nucleoside 5′- and 3′-phosphates in the ratio of 1:5 (LIPKIN et al. 1959a; SMITH et al. 1961). The barium hydroxide hydrolysis of cytidine 3′,5′-cyclic phosphate resulted in concomitant deamination to uracil nucleotides, and the 3′- and 5′-phosphates of both uridine and cytidine were produced. In addition, both the uridine and cytidine 3′,5′-cyclic phosphates yielded on hydrolysis small amounts (<5%) of new nucleotides. Although these new nucleotides have not been fully characterized, it has been suggested (SMITH et al. 1961; KHORANA et al. 1961) that these nucleosides and nucleotides may have the D-*xylo* configuration (475, 476), and probably arising through anhydronucleotide intermediate (471) (YUNG and FOX 1961). The minor nucleotide also obtained by the alkaline hydrolysis of 2′-deoxypyrimidine 3′,5′-cyclic phosphate, for similar reasoning, has been postulated to be 1-(2-deoxy-β-D-*threo*-pentofuranosyl)pyrimidine 5′-phosphate (KHORANA et al. 1961; KHORANA and VIZSOLYI 1961).

MECHANISM A

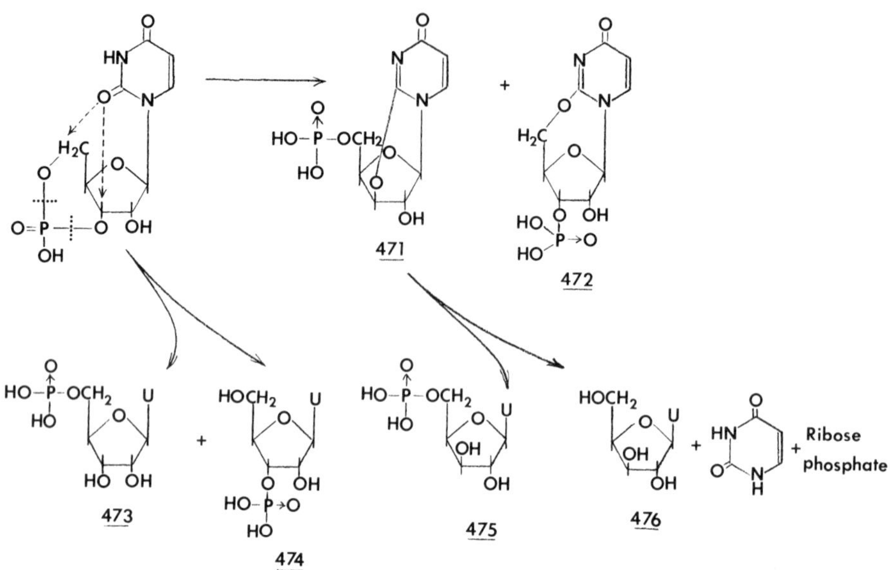

Far more interesting are the results obtained on acid hydrolysis of nucleoside 3′,5′-cyclic phosphates. The ribonucleoside 3′,5′-cyclic phosphates are unusually stable to acid hydrolysis (SUTHERLAND and RALL 1957, 1958; SMITH et al. 1961), whereas nucleosides and their 5′-phosphates usually hydrolyzed in acid in the order purine > pyrimidine. The presence of the respective 3′,5′-cyclic phosphates reversed this order. The D-ribo- and 2′-deoxy-D-ribo 3′,5′-cyclic phosphates of adenosine and guanosine are much more stable to acidic hydrolysis than their respective nucleotides. However, the reverse was found to be true for uridine and thymidine 3′,5′-cyclic phosphates (Table 2).

The kinetic studies on acid-catalyzed hydrolysis of purine and pyrimidine nucleosides suggest that hydrolysis proceeds *via* the *Mechanism B* (ZOLTEWICZ et al. 1970; ZOLTEWICZ and CLARK 1972; PANZICA et al. 1972; HEVESI et al. 1972). The

MECHANISM B

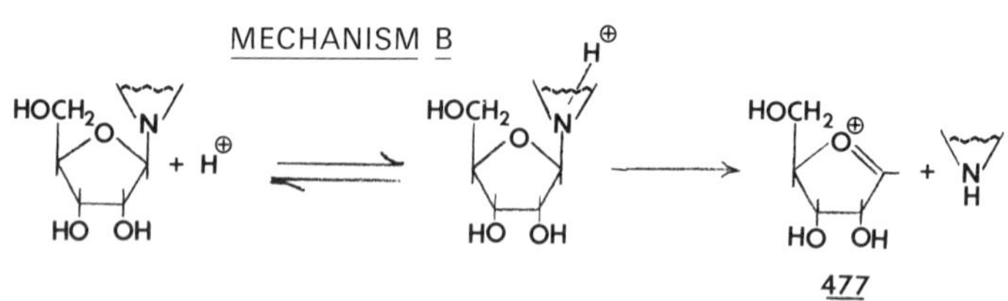

first event will be fixation of a proton and the subsequent rate determining step will be the cleavage of the nitrogen-glycosyl bond to give the aglycon and a carboxonium ion (477). If hydrolysis of a purine 3′,5′-cyclic phosphate occurs *via* a carboxonium ion, then the formation of the carboxonium ion could be hindered in the transition state by the rigid nonplanar *trans*-fused ring system, resulting in an increased stability to hydrolysis. Hydrolysis of the phosphate ring may thus be prerequisite for glycosidic cleavage. However, this does not explain the unexpected

Table 2. Acidic hydrolysis of ribo(deoxyribo)nucleoside 3′,5′-cyclic phosphates and ribo(deoxyribo)nucleoside (2′)(3′)(5′) phosphates in 1 *N* HCl at elevated temperatures

Compound	Half-life (min)	Temperature (°C)	Products
Adenosine 3′,5′-cyclic phosphate	30[a]	100	Adenine, transient adenosine, orthophosphoric acid, ribose, trace of ribose phosphates
Adenosine 5′ or 2′(3′) phosphate	2–4[a]	100	Adenine
2′-Deoxyadenosine 3′,5′-cyclic phosphate	Unaffected after 2 h[b]	50	
2′-Deoxyadenosine 3′,5′-cyclic phosphate	3[b]	100	Adenine
2′-Deoxyadenosine 5′-phosphate	5[b]	50	Adenine
Guanosine 3′,5′-cyclic phosphate	28[a]	100	Guanine, transient guanosine phosphates
Guanosine 2′(3′)-phosphate	1.5[a]	100	Guanine
2′-Deoxyguanoside 3′,5′-cyclic phosphate	Unaffected after 2 h[b]	50	
2′-Deoxyguanosine 5′-phosphate	5[b]	50	Guanine
Cytidine 3′,5′-cyclic phosphate	26[a]	100	Cytosine 9%, cytidine 5′-phosphate 11%, cytidine 2′(3′) phosphate 70%
Cytidine 2′(3′)-phosphate	[a]	100	9% Cytosine after 2 h
2′-Deoxycytidine 3′,5′-cyclic phosphate	Unaffected after 2 h[b]	50	
2′-Deoxycytidine 3′,5′-cyclic phosphate	22[b]	100	Cytosine
2′-Deoxycytidine 5′-phosphate	Unaffected after 2 h[b]	50	
2′-Deoxycytidine 5′-phosphate	110[b]	100	Cytosine
Thymidine 3′,5′-cyclic phosphate	13[b]	50	Thymine
Thymidine 5′-phosphate	Unaffected after 2 h[b]	50	
Uridine 3′,5′-cyclic phosphate	8[a]	100	Uracil (67%), uridine 5′-phosphate (6%), uridine 2′(3′)-phosphate (27%), ribose, orthophosphate
Uridine 5′-phosphate	Unaffected after 2 h[b]	100	

[a] Smith et al. 1961 [b] Drummond et al. 1964

acid-lability of the glycosidic bond of the respective 3′,5′-cyclic phosphate of uridine and thymidine, as compared with cytidine or their parent 5′-phosphates. The alternate *Mechanism A*, as described above *via* the anhydro intermediates (471) and/or (472) (Fox and Miller 1963), is consistent with these results (Ueda and Fox 1967). Thymidine 2,3′- or 2,5′-anhydronucleosides are also known to be rapidly hydrolyzed under similar acidic conditions (Ueda and Fox 1967).

Examination of the pKa's of adenosine, guanosine, uridine, thymidine, and cytidine lend further support to these mechanisms for hydrolysis. Protonation of adenosine, guanosine, and cytidine occurs in 1 *N* hydrochloric acid (pH 0.1) (DUNN and HALL 1970). The protonation of the aglycon would reduce the electron density at N-3 (purine) and O-2 (cytidine), thereby decreasing the potential of anhydro nucleotide formation. Uridine (pKa = −3.38) (SHAPIRO and DANZIG 1972) is not completely protonated under these conditions, and anhydro formation is still possible. The different rates of hydrolysis of uridine 3′,5′-cyclic phosphate and cytidine 3′,5′-cyclic phosphate can then be explained by the protonation of the cytidine base where *Mechanism B* is favored over *Mechanism A*. The fact that cytidine 3′,5′-cyclic phosphate hydrolyzed without cleavage of the glycosyl bond and at a rate similar to *c*AMP and *c*GMP suggests that all three cyclic nucleotides hydrolyze by a similar mechanism, with the rate determining step most probably the hydrolysis of the phosphate ring. However, more proof is needed for the validity of these mechanisms.

The stability of several nucleoside 3′,5′-cyclic phosphates to enzymatic hydrolysis has also been reported. 2-Deoxyadenosine 3′,5′-cyclic phosphate was hydrolyzed about 30% faster than adenosine 3′,5′-cyclic phosphate, while inosine- and guanosine-3′,5′-cyclic phosphates are cleaved at 60% and 33% the rate of *c*AMP, respectively (DRUMMOND et al. 1964, 1966; NAIR 1966). Tubercidin 3′,5′-cyclic phosphate hydrolyzed almost three times more rapidly than *c*AMP (DRUMMOND and POWELL 1970). The pyrimidine cyclic nucleotides, *c*UMP and *c*CMP are relatively stable to hydrolysis, using a dog heart phosphodiesterase preparation (DRUMMOND et al. 1964), whereas the hydrolysis of pyrimidine nucleoside 2′,5′-cyclic phosphates was catalyzed by pancreatic ribonuclease (TIKHOMIROVA-SIDOROVA et al. 1971, 1973; KAVUNENKO and HOLÝ 1978). Adenosine 3′,5′-phosphorothioate was resistant to diesterase action (ECKSTEIN and BÄR 1969) and 2′-thio-2′-deoxycytidine 2′,3′-cyclic phosphorothioate was not hydrolyzed by ribonuclease A (LEE et al. 1979).

Probably with the exception of hydrolytic studies of nucleoside cyclic phosphates, the studies of the chemical properties of this important class of phosphate diesters have been limited to the thermodynamic parameters associated with the enzymatic hydrolysis of the phosphate ester bonds (BAHR et al. 1965; GREENGARD et al. 1969; RUDOLPH et al. 1971; HAYAISHI et al. 1971). The free energy of hydrolysis of adenosine 3′,5′-cyclic phosphate was found to be −11.9 kcal/mole (pH 7.3, pMg 3, 25 °C) (HAYAISHI et al. 1971). Under the same conditions the free energy of hydrolysis of adenosine 5′-triphosphate to yield adenosine 5′-diphosphate and inorganic phosphate has been determined to be −8.9 kcal/mole (ALBERTY 1969; GREENGARD 1971). The high free energy of 3 kcal/mole associated with the hydrolysis of *c*AMP to adenosine 5′-monophosphate indicates the thermodynamic possibility that *c*AMP might interact with its receptor by formation of a covalent bond (GREENGARD 1971). The possibility of reversible phosphorylation of enzymes by nucleoside 3′,5′-cyclic phosphates is supported by the reversible phosphorylation of α-chymotrypsin with catechol cyclic phosphate (KAISER et al. 1971). The enthalpy of hydrolysis of *c*AMP has been determined to be -11.1 ± 0.2 kcal/mole (GERLT et al. 1975) and the entropy change for the hydrolysis is calculated to be −7.4 cal/mole deg. (GREENGARD et al. 1969). Thus, the free energy of hydrolysis

Table 3. Enthalpy of hydrolysis of cyclic nucleotides[a]

3′ Bond of 3′,5′-cyclic nucleotide	ΔH, kcal mole^{-1}
*c*AMP	−14.1
*c*dAMP	−13.0
*c*IMP	−13.4
*c*GMP	−10.5
*c*UMP	−12.0
2′ Bond of 2′,3′-cyclic nucleotide	
*c*AMP	− 9.4
*c*GMP	− 9.5
*c*CMP	− 8.1
*c*UMP	− 7.8

[a] RUDOLPH et al. 1971

is dominated by a large exothermic enthalpy. Similar enthalpies (as determined by calorimetric methods) were found for the hydrolysis of other nucleoside 3′,5′- and 2′,3′-cyclic phosphates (Table 3) (RUDOLPH et al. 1971). The data in Table 3 reveals several interesting chemical features of these cyclic nucleotides: a) the bond hydrolyzed was of a high-energy nature; b) the enthalpy of hydrolysis of 3′,5′-cyclic nucleotides was greater than that of the corresponding 2′,3′-cyclic nucleotides; and c) changing the nature of the aglycon of the 2′,3′-cyclic nucleotides had only a slight effect on the measured enthalpy of hydrolysis; in contrast, the enthalpy of hydrolysis of the 3′,5′-cyclic nucleotides was significantly affected by the nature of the purine or pyrimidine base (GREENGARD 1971).

The enthalpies of hydrolysis of several cyclic phosphate diesters, which can be considered as structural analogs of the *trans*-fused trimethylene phosphate-ribofuranoside ring system of *c*AMP, have been determined recently by microcalorimetric techniques using the metal-dependent phosphohydrolase from *Enterobacter aerogenes* as catalyst (GERLT et al. 1980a). The results indicate that the more exothermic enthalpies of hydrolysis of 3′,5′-cyclic nucleotides relative to trimethylene phosphate could be attributed to intramolecular geometric distortion, and to an unexpected and unexplained effect which appears to be caused by the presence of the endocyclic oxygen atom of the ribofuranoside ring (GERLT et al. 1980a). The results of ab initio and molecular mechanics calculations indicate that the effect of the oxygen atom is not due to introduction of additional geometric distortion (MARSH et al. 1980). The results of ^{1}H NMR studies on the solution conformations of trimethylene phosphate substituted at the 5-position with alkyl and alkoxy groups implicate differential solvation of the reactants and products of the hydrolysis reaction as the source of the unaccounted for enthalpy of hydrolysis in *c*AMP (GERLT et al. 1980b).

In addition to the high free energy of hydrolysis, *c*AMP has the highest enthalpy of hydrolysis of any of the cyclic nucleotides investigated to date. It is thus possible that the high energetic properties of the cyclic phosphate moiety of *c*AMP were an important aspect of the central role that this compound came to assume in the course of the evolution of mechanisms for the regulation of metabolism (GREENGARD 1971).

II. Spectral Measurements

The conformational properties of a number of nucleoside and nucleotide derivatives have been extensively studied by x-ray crystallographic (SUNDARALINGAM 1975) and ^{1}H NMR methods (TS'O 1974). Determination of precise conformation of nucleoside cyclic phosphates is of great importance for an understanding of the factors responsible for their chemical properties and unusual biological activity. The nucleoside 3′,5′-cyclic phosphates contain a 6-membered phosphate ring fused *trans* (1,2) to a 5-membered furanose ring, producing a rigid system that also has been extensively studied by x-ray and ^{1}H NMR methods.

Single-crystal x-ray analysis (Fig. 1) of adenosine 3′,5′-cyclic phosphate shows two molecules of *c*AMP per unit cell, one in the *anti* conformation, the other *syn*, about the glycosidic bond (WATENPAUGH et al. 1968; BRYAN and OLAFSSON 1973). The 3′,5′-*trans*-fused phosphate of ribonucleoside 3′,5′-cyclic phosphates restricts the conformation of the furanose ring to twist C(3′)-*endo*-C(4′)-*exo* ($^3T_4$) with normal bond lengths and angles. The phosphate rings in both molecules are locked in the chair conformation. Unlike 6-membered cyclic organic phosphates (HAQUE et al. 1970), the conformation of a 3′,5′-cyclic phosphodiester leads to appreciable strain, which is why their synthesis requires fairly drastic conditions like heating in boiling pyridine in the presence of DCC for several hours.

Adenosine 3′,6′-cyclic phosphonate (246a), a structural analog of *c*AMP where the 5′-*O*- has been substituted by a methylene, also has a *syn*, $^3T_4$ conformational structure in the solid state (SUNDARALINGAM and ABOLA 1972a, b), which is similar to that for one of the molecules of *c*AMP itself. The 6-membered 9-(β-D-xylofuranosyl)adenine 3′,5′-cyclic phosphate (144) having a *cis*(1,2) ring-fusion is relatively strain-free and the sugar conformation of which is ^3E (MACCOSS et al. 1978). The 7-membered 9-(β-D-arabinofuranosyl)adenine 2′,5′-cyclic phosphate (129) with a *cis*(1,3) ring-fusion has the sugar conformation ^2E and large distortions from normal bond-angles have been observed in the phosphate ring of (129) (KUNG et al. 1977; MACCOSS et al. 1978). The phosphate ring conformation in both (129) and (144) are in chair form.

Single-crystal x-ray diffraction analysis of *c*GMP shows *syn* conformation (Fig. 2). The cyclization of the phosphate group forces the furanose ring into the $^3T_4$ conformation, characteristic of the nucleoside 3′,5′-cyclic phosphates (CHWANG and SUNDARALINGAM 1974). The phosphate ring is locked into a chair conformation, being puckered mostly about the C(3′)–C(4′) bond and flattened at the phosphate end.

The rigidity of cyclic phosphate and furanose rings in nucleoside 3′,5′-cyclic phosphates and their α-anomers (MACCOSS et al. 1977) has also been confirmed in solution by ^{1}H NMR and ^{13}C NMR studies. The ^{13}C, ^{31}P, and ^{1}H chemical shifts and coupling constants have been reported on some *ribo* (JARDETZKY 1962; SMITH et al. 1973; LAPPER et al. 1972, 1973; SMITH and JARDETZKY 1968; BLACKBURN et al. 1973; SCHWEIZER and ROBINS 1973; LEE and SARMA 1976; ROBINS and MACCOSS 1977), and *arabino* 3′,5′-cyclic nucleotides (ROBINS and MACCOSS 1977), and various 8-substituted adenosine 3′,5′-cyclic phosphates (SCHWEIZER and ROBINS 1973; BOBRUSKIN et al. 1979; UESUGI et al. 1979). The ^{1}H and ^{13}C NMR spectra are consistent with $^3T_4$ conformations. This conformation of the ribose ring causes the

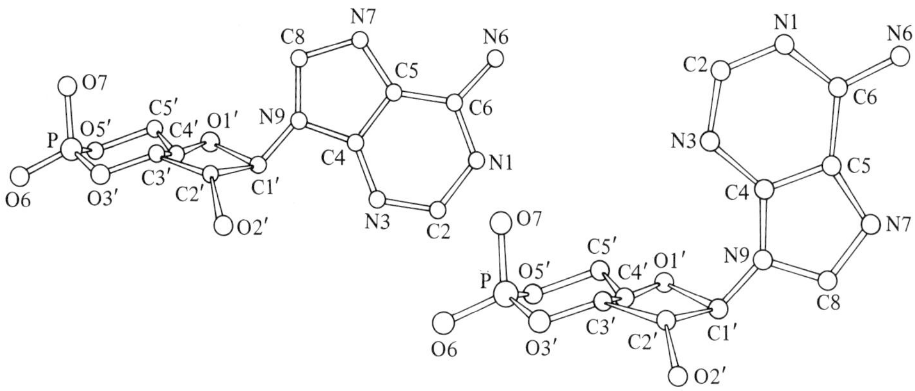

Fig. 1. A schematic view of *c*AMP showing *syn-anti* conformations about the glycosidic linkage (WATENPAUGH et al. 1968)

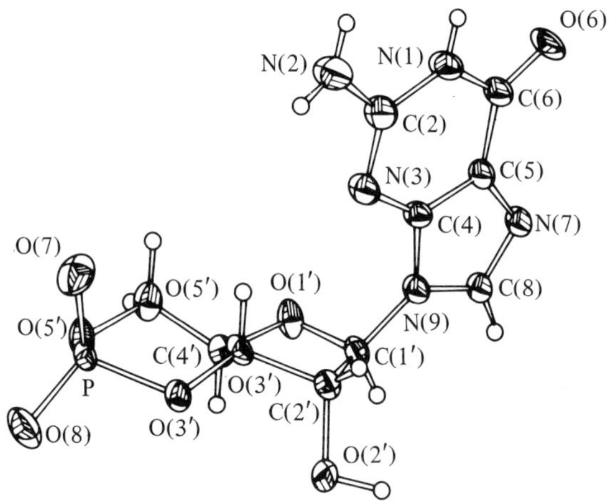

Fig. 2. Molecular conformation of *c*GMP (SUNDARALINGAM 1975)

phosphorus atom to be too far removed in ribonucleoside 3′,5′-cyclic phosphates for deshielding of base protons as observed in the ^{1}H NMR of 5′-nucleotides (SCHWEIZER et al. 1968; TS'O et al. 1969) and *arabino*-nucleoside 2′,5′-cyclic phosphates (LEE et al. 1971). Additionally, the coupling $J_{1'-2'}$ is observed to be less than 1 Hz (sharp singlet) for β-anomers of ribonucleoside 3′,5′-cyclic phosphates, whereas the corresponding ^{1}H NMR signal for α-anomers appears as a doublet ($J_{1'-2'} \gtrsim 3.5$ Hz) (ISHIYAMA 1976; ROBINS and MACCOSS 1977). Similarly, small $J_{1'-2'}$ *(trans)* and large $J_{1'-2''}$ *(cis)* vacinal proton coupling constants were observed for the corresponding *ribo* and *arabino* 2′-deuterio-2′-deoxynucleoside 3′,5′-cyclic phosphates, respectively (MACCOSS et al. 1977). These observations have a diagnostic importance in structural determinations of cyclic nucleotides.

High resolution ^{1}H NMR and ^{31}P chemical shifts of *c*AMP and its analogs in solution at room temperature may be interpreted in terms of either the chair or twist conformation for the phosphate ring (SCHWEIZER and ROBINS 1973). A comparison of the chemical shifts of H-2′ and H-3′ of various *ribo* and *arabino* nucleoside 3′,5′-cyclic phosphates indicates an *anti* conformation for *c*AMP in solution, whereas *c*GMP and various 8-substituted *c*AMP derivatives are probably *syn*. Ara-*c*AMP probably exists preferentially as the *anti* conformer (SCHWEIZR and ROBINS 1973). Ultrasonic relaxation evidence indicates that in aqueous solution (pH 8.0) *c*AMP coexists in two glycosyl conformations which rapidly equilibrate (HEMMES et al. 1976). Unusually large differences in optical rotatory dispersion (ORD) amplitudes between adenosine and *c*AMP in dilute aqueous solution, however, have been interpreted as the latter having predominantly *syn* conformation (KLEE and MUDD 1967). Examination of the aqueous solution conformation of gadolinium complexes of purine 3′,5′-cyclic phosphates indicates a rapid *syn-anti* equilibrium with a slight preference for the *anti* forms. The presence of a bulky substituent at position 8, such as 8-methylthio-*c*AMP, however, drives this equilibrium towards the *syn* configuration (FAZAKERLEY et al. 1975, 1977). Thus, in spite of the variety of experimental techniques employed, no consistent conclusion has yet been arrived at the conformation of these molecules in solution. However, the ^{1}H NMR analysis of the conformation of nucleoside 3′,5′-cyclic phosphates employing lanthanide ions as shift reagents was found to be consistent with the structure in the crystalline state (KAINOSHO and AJISAKA 1975; LAVALLEE and ZELTMANN 1974; BARRY et al. 1974; INAGAKI et al. 1976).

A number of theoretical conformational calculation studies have also been utilized for the determination of the conformation of cyclic nucleotides. Conformational energy calculations predict that *c*GMP and *c*IMP favor *syn* conformation to the *anti* by 95:5 and 70:30, respectively, while *c*AMP shows a preference for *anti* conformation to *syn* by 70:30, indicating purines show a greater probability for the *syn* conformation than the pyrimidines in 3′,5′-cyclic nucleotides (YATHINDRA and SUNDARALINGAM 1974) (Table 4). The 8-substituted *c*AMP derivatives, however, exist primarily in *syn* conformation (SHELDRICK and RIEKE 1978; BOBRUSKIN et al. 1979).

Table 4. Population of the *Anti* and *Syn* conformers in purine and pyrimidine 3′,5′-cyclic nucleotides[a]

	Anti	Syn
3′,5′-*c*UMP (I)	96 (69)	4 (31)
3′,5′-*c*UMP (II)	99 (89)	1 (11)
3′,5′-*c*CMP	99 (95)	1 (5)
3′,5′-*c*TMP	99 (87)	1 (13)
3′,5′-*c*AMP	69 (28)	31 (72)
3′,5′-*c*IMP	36 (31)	64 (69)
3′,5′-*c*GMP	8 (27)	92 (73)

[a] Values in parentheses correspond to those obtained by considering van der Waals interaction energy alone (YATHINDRA and SUNDARALINGAM 1974)

The quantum mechanical perturbative configuration interaction using localized orbitals (PCILO) method (MASSON et al. 1970; PULLMAN and SARAN 1976) has been utilized for the determination of the conformational properties of nucleoside 2′,3′- and 3′,5′-cyclic phosphates (SARAN et al. 1973). The pyrimidine 2′,3′-cyclic

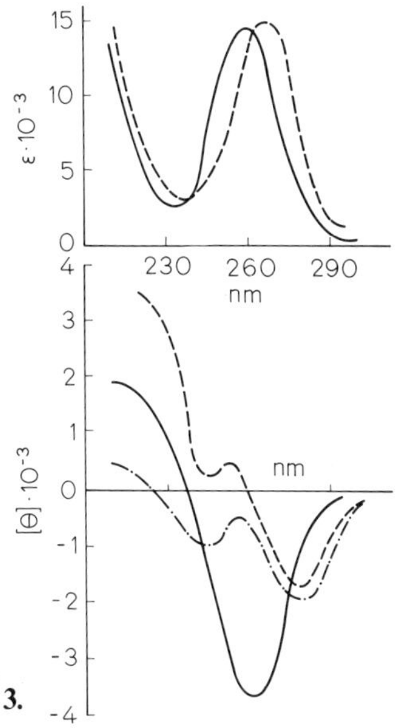

Fig. 3.

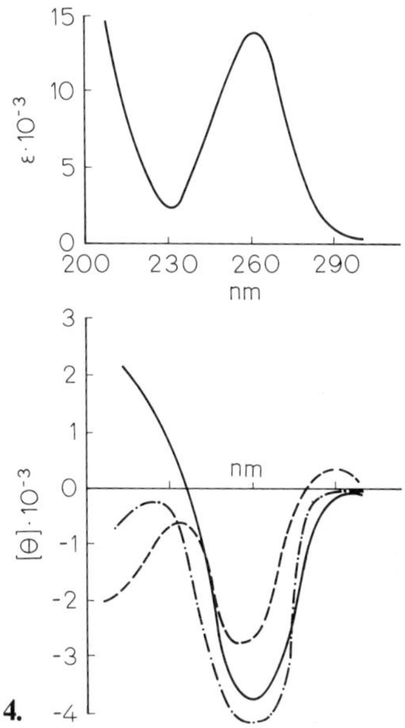

Fig. 4.

Fig. 3. Absorption and CD spectra of adenosine (1), 5′-AMP (2), and *c*AMP (3) at pH 7

Fig. 4. Absorption and CD spectra of adenosine (1), 8-bromoadenosine (2), and 8-bromo-*c*AMP (3) at pH 7 (TUNITSKAYA et al. 1977)

phosphates show a preference for a *syn* configuration about the glycosidic linkage, while the purine 2′,3′-cyclic phosphates are at the borderline between *syn* and *anti* conformers. On the other hand, the pyrimidine 3′,5′-cyclic phosphates show a strong preference for the *anti* conformation, whereas the purine 3′,5′-cyclic phosphates are again at the borderline between *syn* and *anti* conformers about the glycosyl bond (SARAN et al. 1973). The PCILO method also indicates a C(2′)-*exo*-C(3′)-*endo* ($_2T^3$) pucker for α-nucleoside 3′,5′-cyclic phosphates in contrast with $^3T_4$ form found in the β-anomers (TEWARI and DANYLUK 1978). The extended Hückel theory (EHT) has also been used as a quantum mechanical approach for the determination of the conformation of *c*AMP (JORDAN 1973; LESPINASSE and VASILESCU 1974), *c*GMP (CHWANG and SUNDARALINGAM 1973), and adenosine 3′,6′-cyclic phosphonate (JORDAN 1973). The EHT calculations confirmed the $^3T_4$ conformation for *c*AMP (LESPINASSE and VASILESCU 1974). Circular dichroic (CD) spectra indicate that substitution at N^6 of *c*AMP does not markedly influence the conformational equilibrium in *c*AMP about the glycosidic bond, whereas 8-substituted *c*AMP derivatives assume a *syn* conformation (HOPPE and WAGNER 1974; TUNITSKAYA et al. 1977) (Figs. 3 and 4).

Water soluble magnesium, calcium, and manganous salts of gold (III)-adenosine 3′,5′-cyclic phosphate have been prepared. The solubility of these heavy metal complexes has been suggested as potentially useful for cytochemical localizations of enzymes by electron microscopy (GIBSON et al. 1971).

^1H NMR of copper (II) complexes of adenosine (2′) (3′) (5′)-monophosphates have been compared with the spectra of adenosine 2′,3′- and 3′,5′-cyclic phospha-

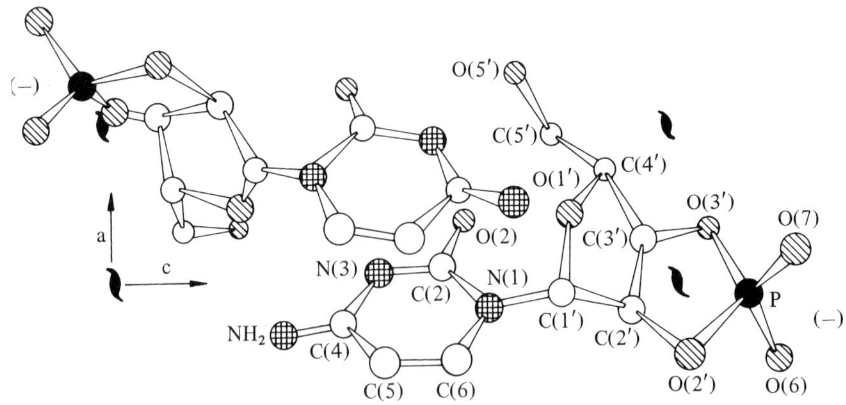

Fig. 5. A schematic view of the 2′,5′-*c*CMP anions down the *b* axis (COULTER and GREAVES 1970)

tes. A comparison of the chemical shifts and broadening effects with varying copper (II) concentrations suggested that greater base stacking and no phosphate-base deshielding occurred in the cyclic nucleotides relative to the nucleoside monophosphates (BERGER and EICHHORN 1971).

The mass spectra of *c*AMP has been determined. In addition to the important peaks observed with 5′-nucleotides, the *c*AMP spectrum exhibits two peaks apparently unique to the 3′,5′-cyclic structure. An increased stability of the cyclic phosphate was shown in the greater abundance of *M* ion (molecular ion) and *M*-15 ion (LAWSON et al. 1971).

The molecular structure of monoclinic crystals of sodium/ammonium salt of cytidine 2′,3′-cyclic phosphate (2′,3′-*c*CMP) has been studied by x-ray crystallography (COULTER and GREAVES 1970; REDDY and VISWAMITRA 1971; COULTER 1973). Like *c*AMP crystals, 2′,3′-*c*CMP crystals were also found to contain two nucleotides per asymmetric unit, but both in the *syn* conformation about the *β*-linkage (Fig. 5). The conformation of the ribose ring of one nucleotide is planar, whereas the other has oxygen-1′ puckered toward carbon-5′. The phosphorus atoms in the 5-membered phosphodiester rings are puckered toward the sugars (LAPPER and SMITH 1973). The strain in both these conformations was shown by the O(2′)-P–O(3′) angles of 96.2° and 96.1°, which are smaller than the usual 103° found in less strained esters (COULTER 1968, 1969). Further, the C(2′)–C(3′)–C(4′) angles are both 107°, which is larger than normally observed (100°), and the C(3′)–O(3′) distances are larger than expected (SASISEKHARAN et al. 1967). The dihedral angles with O(1′) are 43.5° and 61° and with C(3′) are 161° and 54°; the projected O(5′) position falls inside the sugar ring between O(1′) and C(3′) in the second case, and outside the ring in the first anion. Both 2′,3′-*c*CMP and 2′,3′-*c*UMP are in the *syn* conformation in solution (LAVALLEE and COULTER 1973; KARPEISKII and YAKOVLEV 1975; LAVELLEE and MYERS 1978).

The molecular conformation of uridine 3′,5′-cyclic phosphate (*c*UMP) in crystalline state is quite different from 2′,3′-*c*UMP (COULTER 1968), and the conformations of the uridine and cytidine 2′,3′-cyclic phosphates in solution are appreciably different from those found in the solid state (LAPPER and SMITH 1973; LAVAL-

LEE and COULTER 1973; LAVALLEE and MYERS 1978). *c*UMP also crystallizes with two molecules per unit cell and in contrast to the conformational mobility of ribonucleoside 2′,3′-cyclic phosphates (LAPPER and SMITH 1973), the 3′,5′-*trans* fused phosphate of ribonucleoside 3′,5′-cyclic phosphates restricts the conformation of the furanose ring to twist C(3′)-*endo*-C(4′)-*exo* with normal bond lengths and angles. The two independent uracil rings are planar and in the keto form. The base-sugar torsion angles are in the *anti* range. The ribose conformation is C(4′)-*endo* for both anions in the asymmetric unit, and the conformation about the C(4′)–C(5′) bond is trans-gauche (COULTER 1968, 1970).

The conformation of *ara*-cytidine 2′,5′-cyclic phosphate in crystal (KUNG et al. 1977) and in solution (WECHTER 1969; KAINOSHO and AJISAKA 1975) has been studied in some detail. An extremely large vicinal ^{3}JP,H coupling constant of 31.0 Hz has been noted (KAINOSHO and AJISAKA 1975), which is probably due to bond-angle distortions in the 7-membered phosphate ring (KUNG et al. 1977).

The lanthanide-ion probe procedure has been used to determine the molecular conformation of thymidine 3′,5′-cyclic phosphate (3′,5′-*c*TMP) in solution (INAGAKI et al. 1976). The deoxyribose and cyclic phosphate rings 3′,5′-*c*TMP were shown to be locked in a rigid conformation.

The ^{13}C, ^{31}P, and ^{1}H NMR chemical shifts and coupling constants have been measured for several pyrimidine nucleoside 2′,3′-cyclic phosphates (JARDETZKY 1962; SMITH and JARDETZKY 1968; LAPPER et al. 1972; SMITH et al. 1973; GORENSTEIN 1975). As expected from molecular models and x-ray analysis, conformational mobility of the ribose ring is possible and is observed by comparison of the ^{1}H–^{31}P couplings in the furanose rings. A rapid interconversion between puckered forms occurs with little preference for any other form (LAPPER et al. 1972; SMITH et al. 1973). Conformational energy calculations indicate that the pyrimidine bases show a strong energetic preference for the *anti* conformation in 3′,5′-cyclic nucleotides (YATHINDRA and SUNDARALINGAM 1974), whereas in 2′,3′-cyclic nucleotides *syn* conformation (LIPKIND 1976).

X-ray structural analysis of *O*-ethyladenosine 3′,5′-cyclic phosphate (230) has recently been studied (COTTON et al. 1975). The base ring has the *anti* conformation. The ethyl group is *endo* to the base ring and is axial in the flattened chair-conformer 6-membered ring formed by the 3′,5′ O–P–O cyclization (COTTON et al. 1975). The absolute configuration at phosphorus in diastereoisomeric adenosine 3′,5′-cyclic phosphorothioates has recently been assigned (BARANIAK et al. 1979).

X-ray structural analysis, ^{13}C and ^{1}H NMR studies of uridine 2′,3′-cyclic phosphorothioate have been reported (SAENGER and ECKSTEIN 1970; LAPPER et al. 1972). In solid state the base is in *anti* position with respect to the furanose. The conformation about the C(4′)–C(5′) bond is trans-gauche (SAENGER and ECKSTEIN 1970). ^{1}H NMR spectral data for 5′-deoxy-5′-thiothymidine 3′,5′-cyclic phosphorothioate, 5′-thio-2′,5′-dideoxyadenosine 3′,5′-cyclic phosphorothioate, and 5′-thio-2′,5′-dideoxycytidine 3′,5′-cyclic phosphorothioate have also been reported (CHLADEK and NAGYVARY 1972). As in all 5′-deoxy-5′-thio compounds, the 5′-protons exhibited a characteristic upfield shift. The anomeric protons in these cyclic phosphorothioates appear as the usual triplet with large coupling constant (J = 8 Hz). There is no particular strain in these cyclic phosphorothioates (CHLADEK and NAGYVARY 1972).

III. Ultraviolet Spectrophotometric Measurements

Ultraviolet absorption spectra of nucleoside cyclic phosphates have been determined in increasing numbers in recent years. The parent compound, adenosine 3′,5′-cyclic phosphate, has essentially the same ultraviolet absorption spectra as those of adenosine and other adenosine nucleotides at various pH's. Introduction of different functionalities at C_2, C_6, and C_8 position of the aglycon moiety results in considerable change of ultraviolet absorption peaks. The following Tables 5–7, in which the ultraviolet absorption spectra of numerous representative nucleoside cyclic phosphates in both acidic and alkaline pH's have been described, do not provide any broad correlation between the structure and absorption characteristics.

D. Structure and Enzymic Activity Relationships of Certain Nucleoside 3′,5′-Cyclic Phosphates

I. Structural Requirements for the Stimulation of Various Protein Kinases

The cellular function of *c*AMP and *c*GMP is dependent on the type of cell such as adipose, hepatic, cardiac, mast cell, etc. and also the specific nature and structure of the *c*AMP or *c*GMP-dependent protein kinase present (Type I or Type II). It is generally accepted that a majority (if not all) of the known effects of *c*AMP in eukaryotes are mediated by *c*AMP-dependent protein kinases (KUO and GREENGARD 1969; KUO et al. 1970). The mechanism by which *c*AMP activates protein kinases has been extensively studied (RUBIN and ROSEN 1975). Cyclic-AMP activated protein kinases are composed of two dissimilar subunits; the regulatory subunit (R) and the catalytic subunit (C). The R subunit binds *c*AMP whereas the C subunit catalyzes the transfer of phosphate from ATP to the protein substrate. In the presence of *c*AMP, the R unit binds the cyclic nucleotide and dissociates from the C subunit; the latter is then enzymatically active. Removal of *c*AMP permits the reassociation of the two subunits and the protein kinase returns to an inactive state (SANDS and RICKENBERG 1978).

Since the discovery of *c*AMP-dependent protein kinases in 1968 (WALSH et al. 1968), it has been established that there are at least two native forms of the enzyme in tissues. These forms are usually referred to as Type I and Type II *c*AMP-dependent protein kinases depending on their order of elution during ion-exchange chromatography using DEAE-cellulose. The C subunits show considerable homology while the R subunits are quite different (ZOLLER et al. 1979). OWENS and HALEY (1978), using radiolabelled 8-azido-*c*AMP (MUNEYAMA et al. 1971) and γ-[^{32}P]-ATP in a photoaffinity study, obtained data supporting a tightly bound regulatory subunit R and a catalytic subunit C that is released from the membrane in the presence of *c*AMP. These results were obtained with the Type I protein kinase (OWENS and HALEY 1978). The Type I *c*AMP protein seems to be membrane bound or compartmentalized as compared to the more soluble Type II enzyme.

It would appear that two identical *c*GMP kinase subunits containing separate regulatory and catalytic domains are arranged in an antiparallel manner to each other (LINCOLN et al. 1978). Similarities in the mode of action of *c*GMP kinase to

Table 5. Ultraviolet absorption spectra of substituted purine 3′,5′-cyclic nucleotides

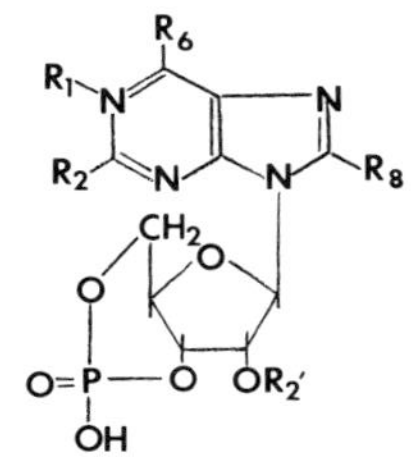

R_1	R_2	R_6	R_8	λ max (mμ) (ε × 10^{-3})		Reference
				pH 1	pH 11	
OCH_3	H	NH	H	257 (12.4)	257 (17.5) 285 sh (3.5)	MEYER Jr et al. (1973b)
H	NH_2	O	H	258 (11.3)	262 (12.4)	ASHMAN et al. (1963)
	CF_3	NH_2	H	258 (12.9)	258 (13.3)	MEYER Jr et al. (1974a)
	CH_3	NH_2	H	256 (13.5)	261 (14.4)	MEYER Jr et al. (1974a)
	Ph	NH_2	H	270 (16.5), 287 sh (13.7)	238 (24.3), 268 sh (14.4)	MEYER Jr et al. (1974a)
	C_6H_4–2–Cl	NH_2	H	263 (17.6) 283 sh (10.3)	262 (15.9)	MEYER Jr et al. (1974a)
	NH_2	NH_2	H	252 (12.1), 292 (10.2)	255 (9.8), 280 (9.8)	MEYER Jr et al. (1975b)
	Cl	NH_2	H	262 (13.7)	262 (14.3)	MEYER Jr et al. (1975b)
	OH	NH_2	H	234 (6.7), 280 (12.7)	252 (7.3), 283 (10.9)	MEYER Jr et al. (1974a)
	SH	NH_2	H	231 (15.1), 288 (21.6)	241 (21.4), 283 (16.7)	MEYER Jr et al. (1974a)
	SCH_3	NH_2	H	268 (16.7)	234 (21.8), 272 (14.7)	MEYER Jr et al. (1974a)
H	NH_2	S	H	262 (7.75), 342 (19.6)	250 (10.9), 321 (14.0)	MEYER Jr et al. (1975b)
H	CH_3	O	H	249 (12.7)	253 (13.0)	MEYER Jr et al. (1975a)

Table 5 (continued)

R_1	R_2	R_6	R_8	λ max (mμ) (ε × 10⁻³) pH 1	λ max (mμ) (ε × 10⁻³) pH 11	Reference
H	SC_2H_5	O	H	264 (14.0)	262 (14.0)	MEYER Jr et al. (1975a)
	H	NH_2	H	257 (14.5)	261	ASHMAN et al. (1963)
	H	NHOH	H	263 (17.1)	270 (9.6), 291 sh (8.1)	MEYER Jr et al. (1973b)
	H	NHC_2H_5	H	262 (18.7)	266 (17.5)	MEYER Jr et al. (1972)
	H	$N(C_2H_5)_2$	H	268 (17.7)	277 (17.2)	MEYER Jr et al. (1972)
	H	Cl	H	265		MEYER Jr et al. (1972)
H	H	O	H	247.5 (11.8)	252 (12.8)	MEYER Jr et al. (1972)
	H	OCH_3	H	248 (11.3)	248 (10.8)	MEYER Jr et al. (1972)
H	H	S	H	321 (23.7)	310 (22.2)	MEYER Jr et al. (1972)
	H	SCH_3	H	292 (16.4)	288 (18.0)	MEYER Jr et al. (1972)
	H	SCH_2Ph	H	292 (17.0)	291 (18.3)	MEYER Jr et al. (1972)
	H	$NHCONHCH_3$	H	276 (22.9), 285 sh (17.2)	267 (21.8), 275 sh (18.5)	BOSWELL et al. (1973)
	H	NHCONHPh	H	285 (28.3)	277 (28.6)	BOSWELL et al. (1973)
	H	NHCONHPh	SCH_2Ph	308 (27.9)	296 (30.3)	BOSWELL et al. (1973)
	H	NH_2	Cl	259.5 (15.6)	261 (14.2)	MUNEYAMA et al. (1974)
	H	NH_2	Br	263	265	IKEHARA and UESUGI (1969)
	H	NH_2	I	268 (18.5), 281 (19.7)	266 (16.3), 278 (18.5)	MUNEYAMA et al. (1974)
	H	NH_2	NH_2	266 (14.1)	268 (16.7)	MUNEYAMA et al. (1974)
	H	NH_2	$NHCH_3$	273 (14.6)	274 (16.0)	MUNEYAMA et al. (1971)
	H	NH_2	$N(CH_3)_2$	278 (17.8)	273 (18.4)	MUNEYAMA et al. (1971)
	H	NH_2	N_3	280 (17.7)	281 (13.3)	MUNEYAMA et al. (1971)
	H	NH_2	OH	262 (11.2) 284 (10.2)	278 (15.3)	MUNEYAMA et al. (1971)
	H	NH_2	OCH_3	259 (12.9)	256 (13.6)	MUNEYAMA et al. (1971)
	H	NH_2	SH	220 (13.4), 244 (9.9), 308 (25.2)	224 (17.4), 291 (22.4)	MUNEYAMA et al. (1971)

	H	NH_2	SCH_3	281 (19.7)	278 (18.5)	MUNEYAMA et al. (1971)
	H	NH_2	SCH_2Ph	283 (19.2)	283 (16.2)	MUNEYAMA et al. (1971)
H	NH_2	O	Br	259 (21.2), 256 sh (19.4)	260 (19.2)	MILLER et al. (1973b)
H	OH	O	H	234 (8.5), 260 (9.9)	247 (10.9), 276 (9.7)	MILLER et al. (1973b)
H	OH	O	Br	250 (12.3), 263 (12.5)	262 (13.5), 280 sh (12.5)	MILLER et al. (1973b)
H	NH_2	O	$NHCH_3$	253 (16.1), 290 (9.0)	260 (14.7)	MILLER et al. (1973b)
H	NH_2	O	$N(CH_3)_2$	262 (18,2), 290 sh (11.9)	265 (16.2)	MILLER et al. (1973b)
H	NH_2	O	OH	246 (11.4), 294 (9.5)	248 (10.0), 278 (8.7)	MILLER et al. (1973b)
H	NH_2	O	SCH_3	273 (17.7)	283 (14.9)	MILLER et al. (1973b)
H	NH_2	O	SCH_2Ph	273 (13.8), 290 sh (11.4)	293 (13.6)	MILLER et al. (1973b)
H	NH_2	O	$S-C_6H_4-Cl$	275 (21.5)	296 (21.5)	MILLER et al. (1973b)
	H	NH_2	$S-C_6H_4-Cl$	281 (15.9)	283 (13.7)	MILLER et al. (1973b)
H	H	O	Br	253 (13.9)	258 (12.6)	MILLER et al. (1973b)
H	H	O	OH	253 (11.0), 283 sh (5.4)	263 (11.7)	MILLER et al. (1973b)
H	H	O	N_3	273 (15.7)	280 (13.0)	MILLER et al. (1973b)
H	H	O	NH_2	253 (13.4), 283 sh (4.2)	261 (14.3)	MILLER et al. (1973b)
H	H	O	SH	291 (13.4)	288 (17.5)	MILLER et al. (1973b)
H	H	O	SCH_3	268 (12.2)	275 (12.0)	MILLER et al. (1973b)
H	H	O	SCH_2Ph	276 (13.1)	283 (13.5)	MILLER et al. (1973b)
H	H	O	$S-C_6H_4-Cl$	256 (15.6), 275 sh (15.0)	244 (12.9), 282 (15.8)	MILLER et al. (1973b)
H	H	O	Cl	251	255	BOSWELL et al. (1975)
	H	Cl	NH_2	238 (4.9), 277 (13.4)	257 (8.0), 286 (12.2)	BOSWELL et al. (1975)

Table 5 (continued)

R_1	R_2	R_6	R_8	λ max (mμ) (ε × 10^{-3}) pH 1	pH 11	Reference
	H	Cl	OH	239 (3.8), 277 (11.1)	261 (6.5) 292 (10.2)	Boswell et al. (1975)
	H	$N(C_2H_5)_2$	OH	274, 300 sh	290	Boswell et al. (1975)
	H	$N(C_2H_5)_2$	Cl	272 (21.1)	276 (19.4), 286 sh (17.9)	Boswell et al. (1975)
	H	Cl	$N(C_2H_5)_2$	290 (13.8)	296 (13.1), 265 sh (6.7)	Boswell et al. (1975)
	H	$N(C_2H_5)_2$	$N(C_2H_5)_2$	289 (18.9)	286 (20.0)	Boswell et al. (1975)
	H	SCH_3	SCH_3	246 (13.7), 300 sh (14.5), 309 (16.5), 330 sh (11.6)	245 (15.6), 301 sh (21.6), 307 (22.2)	Boswell et al. (1975)
	H	SCH_2Ph	SCH_2Ph	313 (13.4), 340 sh (7.6)	309 (20.9)	Boswell et al. (1975)
	H	$NHCH_2Ph$	Br	269 (24.3)	271 (21.7)	Boswell et al. (1975)
	H	$NHCH_2Ph$	SCH_3	289 (24.6)	286 (23.4)	Boswell et al. (1975)
H	H	S	NH_2	234 (13.7), 297 sh (10.0), 331 (17.5)	239 (17.0), 312 (22.9)	Boswell et al. (1975)
H	NH_2	O	CH_3	260 (13.3)	257 (13.2)	Christensen et al. (1975)
H	NH_2	O	CH_2Ph	262 (16.2)	262 (16.2)	Christensen et al. (1975)
H	NH_2	O	$COCH_3$	276 (8.1), 330 (16.2)	349 (17.6)	Christensen et al. (1975)
H	NH_2	O	COPh	271 (13.4), 351 (14.9)	253 (16.1), 375 (16.1)	Christensen et al. (1975)
H	NH_2	O	$CONH_2$	275 (14.8), 300 (14.2)	312 (15.1)	Christensen et al. (1975)
H	NH_2	O	$CHOHCH_3$	261 (14.8)	262 (14.3)	Christensen et al. (1975)
H	NH_2	O	$N.NHCSNH_2$	282 (12.5), 335 (32.2)	275 (9.5), 347 (29.2)	Christensen et al. (1975)

H	NH_2	O	$N.NHCONH_2$	317 (23.0)	324 (22.7)	CHRISTENSEN et al. (1975)
H	H	O	$COCH_3$	297 (7.7)	321 (7.0)	CHRISTENSEN et al. (1975)
O	H	NH_2	Br	236 (12.5), 262 (13.2)	236 (35.4), 263 (9,7), 293 sh (2.5)	UNO et al. (1976)
	CH_3	NH_2	Br	262 (18.0)	267 (16.8)	UNO et al. (1976)
	CH_3	NH_2	N_3	276 (14.9)	275 (11.5)	UNO et al. (1976)
	CH_3	NH_2	SCH_2Ph	283 (18.7)	282 (16.3)	UNO et al. (1976)
CH_3	H	NH	SCH_2Ph	285 (18.5)	285 (12.5)	UNO et al. (1976)

R_6	R_8	$R_{2'}$	λ max (mμ) (ε × 10^{-3})		Reference
			pH 1	pH 11	
NH_2	H	CH_3	256 (14.7)	258 (15.1)	MILLER et al. (1973a)
$NHCONHCH_3$	H	$CONHCH_3$	276 (23.1), 285 sh (17.6)	267 (21.9), 275 sh (18.6)	BOSWELL et al. (1973)
$NHCONHCH_3$	SH	$CONHCH_3$	244 (19.8), 313 (33,8)	233 (17.5), 311 (29.6)	BOSWELL et al. (1973)
NHCONHPh	SCH_2Ph	CONHPh	309 (17.5)	298 (24.8)	BOSWELL et al. (1973)
OH	Br	$COCH_3$	252 (14.7)	257 (13.4)	BOSWELL et al. (1975)
Cl	Cl	$COCH_3$	248 (6.6), 266 (10.5), 274 sh (8.8)	248 (6.6), 66 (10.2), 274 sh (8.6)	BOSWELL et al. (1975)
Cl	SCH_2Ph	$COCH_3$	294 (15.4)	294 (15.4)	BOSWELL et al. (1975)
Cl	$S-C_6H_4-Cl$	$COCH_3$	247 (11.0), 293 (13.8)	247 (11.0), 293 (14.0)	BOSWELL et al. (1975)

Table 6. Ultraviolet absorption spectra of aza- and deazaadenosine 3′,5′-cyclic phosphates

3′,5′-Cyclic Phosphate	λ max (mμ) ($\varepsilon \times 10^{-3}$)		Reference
	pH 1	pH 11	
2-Azaadenosine	253 (7.1), 281 (3.4)	255 (7.1), 296 (5.2)	MEYER Jr et al. (1974)
8-Azaadenosine	262 (12.2)	278 (11.4)	MILLER et al. (1978)
1-Deazaadenosine	262 (13.9)		MIZUNO et al. (1975)
3-Deazaadenosine	263 (10.2)		MIZUNO et al. (1975)
7-Deazaadenosine	226 (20.8), 270 (9.4)		HANZE (1968)
Formycin	233 (10.0), 295 (10.5)	234 (16.3), 303 (8.1)	MAKABE et al. (1978)
N_2-Methylformycin	230 (11.9), 260 (4.9), 271 (4.9), 303 (10.2)	238 (5.36), 295 (11.2), 305 (12.8), 317 (8.4)	MAKABE et al. (1978)

Table 7. Ultraviolet absorption spectra of pyrimidine 3′,5′-cyclic phosphates

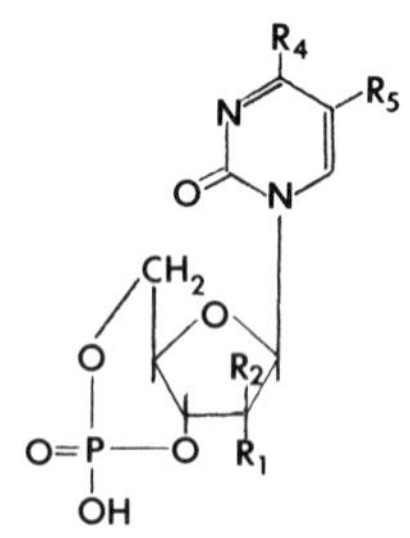

R_1	R_2	R_4	R_5	λ max (mμ) ($\varepsilon \times 10^{-3}$)			Reference
				pH 1	pH 7	pH 11	
OH	H	OH	H		261 (9.9)	260 (7.7)	SMITH et al. (1961)
OH	H	NH_2	H	279 (12.4)	272 (9.3)		SMITH et al. (1961)
H	H	OH	CH_3	264.5		264.5	TENER et al. (1958)
OH	H	OH	Br	277 (9.5)		232 (7.4) 275 (7.3)	LONG and ROBINS (1978)
H	OH	OH	H	262 (7.2)		260 (6.0)	LONG et al. (1972)
H	OH	NH_2	H	277 (14.2)		236 268 (10.4)	LONG et al. (1972)
H	OH	SH	H	330 (21.8)		317 (21.6)	LONG et al. (1972)
H	OH	SCH_3	H			303 (14.9)	Long et al. (1972)

that of *c*AMP-dependent protein kinase are suggested by self-phosphorylation and inhibition of catalytic activity by the regulatory subunit of both enzymes, and by the fact that the two enzymes catalyze phosphorylation of each other (LINCOLN et al. 1978). The structural requirements for stimulation of *c*AMP-dependent protein kinases have largely been summarized as a result of studies from our Laboratories.

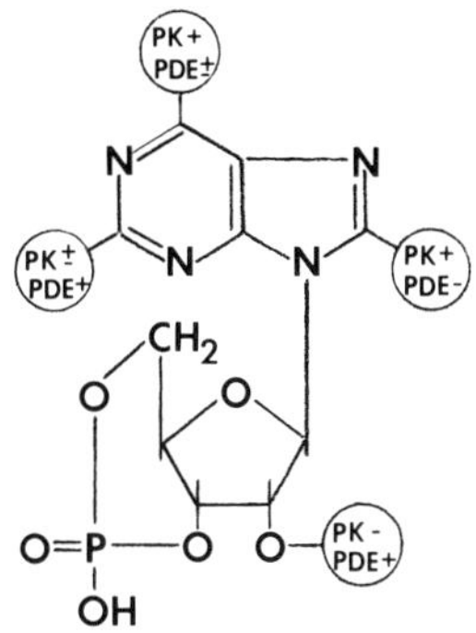

Fig. 6. Summary of structure-activity relationships. PK = Protein Kinase Activation, + = Good, PDE = Hydrolysis by Phosphodiesterase, − = Resists

These studies have been reviewed elsewhere (MEYER Jr and MILLER 1974; MILLER 1977; MILLER and ROBINS 1976; MEYER Jr 1979).

The substitutions at positions 2, 6, and 8 also effect the enzymatic degradation by phosphodiesterase. These studies have also been similarly reviewed (MEYER Jr and MILLER 1974; MILLER 1977; MILLER and ROBINS 1976; MEYER Jr 1979).

A summary of these results can be presented diagramatically (Fig. 6), where various substituents at the 2′-*O*-position and/or at positions 2, 6, and 8 in general affect protein kinase and inhibit phosphodiesterase as indicated.

Studies have indicated that 7-deaza-*c*AMP (266) and 1-deaza-*c*AMP (255) are more potent activators of bovine brain, bovine heart and rat liver protein kinases than *c*AMP itself (MILLER et al. 1978). The presence of N^3 in *c*AMP has been postulated as a necessary binding site for Type II protein kinases (MILLER et al. 1978).

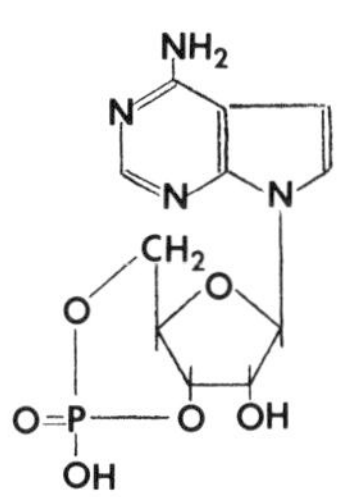

266

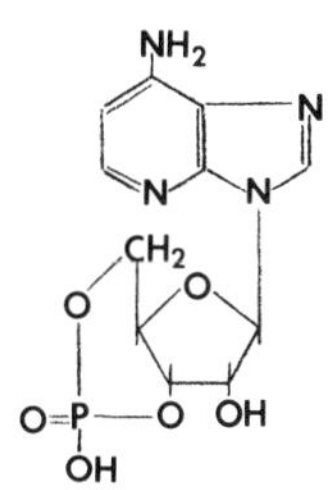

255

In contrast, JASTORFF and co-workers (1979) have recently shown that Type I *c*AMP-dependent protein kinases bind the adenine moiety of *c*AMP in a hydrophobic cleft without any hydrogen bond interactions. These workers (JASTORFF et al. 1979) have proposed that Type I cAMP-dependent protein kinases bind *c*AMP *via* its 3′- and 5′-oxygens and the 2′-hydroxyl. The negative charge on the phosphate oxygen is also believed to be bound to the protein kinase in a fixed position (JASTORFF et al. 1979). Substitution of nitrogen or sulfur for the 3′- and 5′-oxygens in *c*AMP gives analogs which reduce the binding affinity to *c*AMP-dependent protein kinases (SHUMAN et al. 1973; PANITZ et al. 1975; JASTORFF et al. 1979). Further evidence for the necessity of the presence of an ionic oxygen is found in that adenosine 3′,5′-cyclic sulfate (197) has a 25,000 times lowered affinity for porcine brain protein kinase as compared to *c*AMP (SEVERIN et al. 1976). The esterification of the phosphate moiety in *c*AMP provided compounds which were poor competitors for the *c*AMP binding site with beef heart protein kinase (GILLEN and NAGYVARY 1976). The benzyl ester of *c*AMP did not activate a *c*AMP-dependent protein kinase from bovine brain (ENGELS and SCHLAEGER 1977).

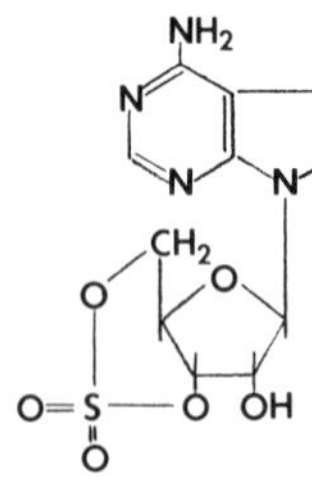

197

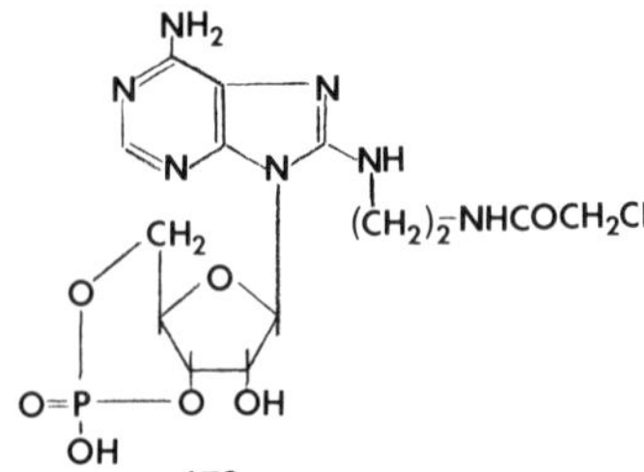

478

SEVERIN and co-workers (GULYAEV et al. 1977) have employed 8-(N-chloroacetylaminoethylamino)-*c*AMP (478) in the study of protein kinase from pig brain. This analog, (478), interacts irreversibly with the enzyme via a covalent bond through the chloroacetyl group, and it is suggested that (478) is bound to the regulatory site of the protein kinase in the *syn*-conformation. SEVERIN and co-workers (1979a, b) have recently summarized their results using pig brain *c*AMP-dependent protein kinase and it is concluded that the presence of the negative charge on the phosphate anion is an indispensable requirement of activation and the 2′-hydroxyl must be unsubstituted. These workers postulate that a proton *donor locus* must be present on the enzyme near the N^6-exocyclic amino group which is also important in binding *c*AMP (SEVERIN et al. 1979a, b). Studies on Type I and Type II *c*AMP-dependent protein kinases indicate that these enzymes may be activated by different mechanisms (HOFMANN et al. 1979). Indeed, MILLER and co-workers (YAGURA et al. 1980) have recently demonstrated that protein kinase Type I from rabbit skeletal muscle and protein kinase Type II from bovine heart have different structural requirements for binding *c*AMP analogs. It would appear there is significantly more steric interference at position 2 for binding to protein kinase Type I than there is for protein kinase Type II (YAGURA et al. 1980). 2-Chloro-*c*AMP and 2-trifluoromethyl-*c*AMP, both of which exert a relatively strong electron withdrawing effect on the purine ring, were more potent activators of protein kinase I than of protein kinase II (YAGURA et al. 1980). This data is consistent with lowering the electron density at N^3, a proposed binding site for Type II protein kinase (MILLER et al. 1978). It is conceivable that *c*AMP-dependent protein kinases from different organs and organisms bind *c*AMP in different ways. For example, the protein kinase of *Dictyostelium* has a binding site involving two specific hydrogen bonds with N^6 as a donor and N^7 as an acceptor which are essential for chemotactic activity (MATO et al. 1978). It has been postulated that in this instance the *anti* conformation of *c*AMP is involved in binding, whereas the *syn* conformation is preferred with Type II protein kinase (MATO et al. 1978). Thus the 8-substituents would favor the *syn* conformation and activation of Type II protein kinase.

II. Structural Requirements for Stability Against Cyclic Nucleotide Phosphodiesterases

Resistance of a *c*-AMP analog toward *c*AMP phosphodiesterase (PDE) may be induced by an appropriate substitution at C_6 or C_8 of the purine ring. Most interesting is the fact that N^6-(N,N)diethyl-*c*AMP is resistant to PDE from pig brain or rabbit kidney (MEYER Jr et al. 1972). The attachment of a carbamoyl substituent at N^6 of *c*AMP almost completely inhibits PDE activity while retaining significant activation of protein kinase (BOSWELL et al. 1973). Introduction of a substituent at 8-position of the purine ring in *c*AMP larger than amino group, resulted in considerable resistance to *c*AMP-PDE (MUNEYAMA et al. 1971). Similarly, a bulky 8-substituent in *c*AMP confers the same resistance to *c*GMP-PDE (MILLER et al. 1973a). In this instance, the 8-substituent should be larger than methyl or amino (MILLER et al. 1973a; CHRISTENSEN et al. 1975). The structure-activity relationships of various cyclic nucleotides with regard to cleavage by PDE has been reviewed (MEYER Jr and MILLER 1974; MILLER and ROBINS 1976; MILLER 1977) and are summarized in Fig. 6. Aza and deazapurine analogs of *c*AMP without additional ring substituents are rather readily hydrolyzed by *c*AMP-phosphodiesterase (MILLER et al. 1978). Recently, a number of N^6-alkyl-8-alkylthio derivatives of *c*AMP have been prepared and studied in detail with regard to degradation with PDE (MILLER et al. 1980). In most instances, the 8-alkylthio substituent prevents PDE degradation. 8-Alkylamino and 8-alkylthio *c*AMP derivatives were found to be competitive inhibitors of the *c*AMP high Km PDE from hog brain (SASAKI et al. 1976). Thus, it is reasonable to assume that these 8-substituted derivatives bind to the same site as *c*AMP (SASAKI et al. 1976). It has been suggested that the enzyme from hog brain cortex requires the *syn* conformation for hydrolysis of *c*AMP (SASAKI et al. 1978). Alkylphosphotriesters of *c*AMP are not substrates but are inhibitors of *c*AMP-phosphodiesterase isolated from beef heart (GILLEN and NAGYVARY 1976). SEVERIN and co-workers (1979) studied the interaction of *c*AMP analogs in a partially purified rat liver phosphodiesterase preparation. The high affinity enzyme with 8-substituted derivatives of *c*AMP showed strong inhibition. Unlike the case with *c*AMP-dependent protein kinase, the presence of a free hydroxyl at the 2′-position is not essential for binding to the PDE enzyme (SEVERIN et al. 1979). 8-(β-Aminoethylthio)-*c*AMP (80) was the most potent inhibitor. Although the presence of a negative oxygen anion was found crucial to stimulate protein kinase, the absence of this charge did not affect the binding affinity to PDE (SEVERIN et al. 1979). It is postulated that *anti* conformation is necessary for PDE hydrolysis of *c*AMP analogs (SEVERIN et al. 1979). It is also postulated that bonding of the

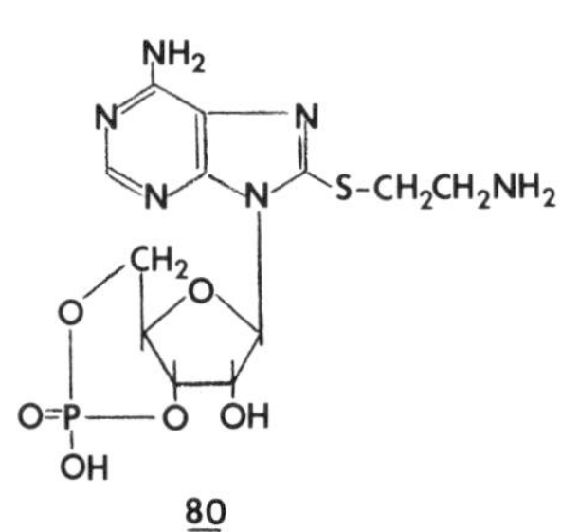

80

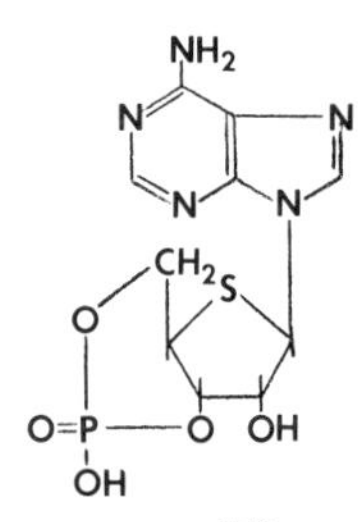

172

PDE enzyme to N^6 and the purine moiety is involved in hydrophobic interaction with an aromatic residue of an aromatic amino acid of the PDE enzyme. An electron pair of the N^6 amino group is bound with a proton donor group of the enzyme near the hydrophobic site. Another proton donor group interacts with the 5′-phosphodiester oxygen since 5′-deoxy-5′-methylene-*c*AMP is not bound by this PDE (SEVERIN et al. 1979).

Of interest is the fact that adenosine 3′,5′-cyclic sulfate (197) bound strongly to rabbit skeletal muscle PDE and was hydrolyzed at a rate of 23% that of *c*AMP (SEVERIN et al. 1976). This is further substantiation of the fact that an ionic oxygen is unnecessary for the binding and action of PDE.

Phosphodiesterase from bovine heart and rabbit brain hydrolyzed 5′-amido-*c*AMP and PDE from bovine heart also hydrolyzed 5′-thio-*c*AMP (JASTORFF and BAR 1973; PANITZ et al. 1975). Rabbit kidney *c*AMP phosphodiesterase hydrolyzed 5′-thio-*c*AMP at 29% the rate of *c*AMP (SHUMAN et al. 1973). 4′-Thio-*c*AMP (172) is hydrolyzed by PDE from bovine heart muscle at 95% of the rate of *c*AMP (ANISUZZAMAN et al. 1973).

III. Cyclic Nucleotides and Inhibition of Cellular Proliferation

In 1971, IRA PASTAN and co-workers (JOHNSON et al. 1971) reported the restoration of several morphological characteristics of normal fibroblasts in sarcoma cells by the treatment of these cells with adenosine 3′,5′-cyclic phosphate administered as the N^6,2′-*O*-dibutyryl derivative. Almost simultaneously, HSIE and PUCK (1971 a, b) observed the morphological conversion in vitro of transformed Chinese hamster ovary cells to a fibroblastic form by the treatment of a combination of dibutyryl-*c*AMP and testosterone. The changes in the following characteristics were consistent with the conversion from a malignant to a normal fibroblastic state (HSIE and PUCK 1971 b).

(1) Acquisition of contact inhibition;
(2) Change from a random growth pattern to a parallel pattern;
(3) Decrease in the ability to be agglutinated;
(4) Disappearance of knob-like structures around the cell periphery;
(5) Return of anchorage dependence.

These properties of the *c*AMP treated malignant cell have been confirmed and extended by use of the electron microscope (PORTER et al. 1974). PASTAN and co-workers (JOHNSON et al. 1971) suggested that *c*AMP might be an important factor in the morphology of normal fibroblasts and this function may be altered or lost during the transformation process. As illustrated by these experiments, *c*AMP has been shown to play an important role in cell growth and differentiation (PASTAN et al. 1975), which represents one of the most exciting areas of cancer research today. It has been postulated (PUCK 1977) that the microtubular-microfilamentous structure of the cell conveys growth regulatory information from the cell membrane via *c*AMP to the nucleus and its disorganization leads to malignant growth. Recently, it has been found that 8-bromo-*c*AMP is highly active in producing reverse transformation of Chinese hamster ovary cells (PUCK 1979).

The malignant cell is characterized by a cell cycle which is apparently identical to that of normal cells; the major difference between the two cell types is the loss of growth regulatory mechanisms which allow transformed cells to enter or remain in the G_0 phase or quiescent state. Considerable evidence is accumulating which indicates that *c*GMP levels fall markedly in cultured fibroblasts as they enter quiescence. Furthermore, a dramatic transient rise in *c*GMP is observed after release from quiescence by serum re-addition (RUDLAND et al. 1974; MOENS et al. 1975). If *c*GMP is the mediator of the G+ state, then addition of *c*GMP or its analogs to quiescent cultures should cause cells to enter DNA synthesis and cell division. SEIFERT and RUDLAND (1974) have reported such experiments and note definite success using *c*GMP with a 15% stimulation of DNA synthesis. Thus, it now appears that cell proliferation may be stimulated by either or both of the following (HALPRIN 1976; MILLER et al. 1975b):

a) A fall in *c*AMP levels.

b) A rise in *c*GMP levels.

A reverse of (a) and (b) inhibits cell proliferation (MILLER et al. 1975; HALPRIN 1976). The simple concept that *c*AMP acts as a negative and *c*GMP acts as a positive signal for cellular growth is receiving continual support from various laboratories (GOLDBERG et al. 1975a; DE RUBERTIS et al. 1976; FRIEDMAN 1976; DE RUBERTIS and CRAVEN 1977).

The *c*AMP content of fibroblasts transformed by viruses is decreased to about 50% of normal. The lower level of *c*AMP is thought to be responsible for most of the properties of transformed cells (FRIEDMAN 1976), including their altered morphology. WILLINGHAM and PASTAN (1975) have presented electron microscopic evidence that dibutyryl-*c*AMP promotes formation and altered distribution of cytoplasmic microtubules and microfilaments which are responsible for changes in morphology, mobility and adhesiveness (FRIEDMAN 1976).

One of the most striking studies of the effect of *c*AMP on transformed cells is that of POLLACK and co-workers (OEY et al. 1974). The dependence on anchorage for growth is lost in transformed cells. (Tumor cells can be grown in semi-solid media but normal cells cannot.) It was shown that there was a correlation between revertant lines that had regained their normal high *c*AMP levels and those that regained anchorage dependence. Thus, loss in anchorage dependence correlated well with in vivo tumorigenicity (OEY et al. 1974; SHIN et al. 1975).

Recently, ZEILIG and GOLDBERG (1977) reported that in a fast growing rat hepatoma cell line, *c*GMP levels vary independently and reciprocally with *c*AMP levels during the cell cycle and *c*GMP levels increase up to ten-fold at the onset of mitosis. These results are consistent with possible regulatory roles for both *c*AMP and *c*GMP in cell proliferation (for a detailed treatment of this subject, see ABOU-SABE 1976). JOHNSON and HADDEN (1977) have shown that *c*GMP increased the activity of RNA polymerase I and III by apparently specifically binding to these enzymes. It has recently been shown that levels of *c*GMP in plasma and urine were considerably increased in patients with disseminated cancer over those of control patients with other diseases (CHAWLA et al. 1979).

As early as 1969, GERICKE and CHANDRA (1969) reported that *c*AMP definitely inhibited the growth of transplanted NKL-lymphosarcoma in mice. Dibutyryl-*c*AMP, 8-methylthio-*c*AMP and 8-bromo-*c*AMP significantly inhibited MTW9 and Walker 256 mammary carcinomas and 5123 hepatoma in experimental animals (CHO-CHUNG 1974). Growth inhibition was dose-dependent. 8-Methylthio-

*c*AMP inhibited human prostatic ephthelial cell replication (NILES et al. 1976). More recently, 8-chloro-*c*AMP (74) has been shown to inhibit significantly human carcinoma in cell culture (NILES et al. 1979). 9-(β-D-Ribofuranosyl)-6-thiopurine 3′,5′-cyclic phosphate and 9-(β-D-ribofuranosyl)-6-methylthiopurine 3′,5′-cyclic phosphate (63) penetrated cells intact, and they may be viewed as an internal ester which is less polar than the 5′-nucleotide. *Ara-c*AMP not only penetrated the tumor cells, but was cleaved by a phosphodiesterase to the desired 5′-nucleotide intracellularly (LE PAGE and HERSH 1972). Thus the synthesis of the 3′,5′-cyclic nucleotide may be an excellent way to transport the active form of the nucleoside (5′-phosphate) inside the tumor cell. This could well circumvent the resistance due to the lack of nucleoside kinase or bypass the requirement of high kinase levels in order for the cell to be susceptible to inhibition. *Ara-c*CMP (LONG et al. 1972) was shown to exhibit activity in L-1210 leukemia in mice approximately equivalent to *ara*-C and to be definitely active against an *ara*-C resistant strain (KREIS and WECHTER 1972). Tests by KREIS and WECHTER (1972) indicate that *ara-c*CMP penetrated the cell membrane and was most likely hydrolyzed in vivo to the 5′-nucleotide in the neoplastic cell. In vitro studies with partially purified *c*AMP phosphodiesterase showed (LONG et al. 1972) that *ara-c*CMP was slowly hydrolyzed to the 5′-phosphate. More recent studies with 6-thio and 6-methylthiopurine ribonucleoside 3′,5′-cyclic phosphates (KOONTZ and WICKS 1977) have shown that the cytotoxicity of both of these derivatives results principally from intracellular conversion to the corresponding 5′-nucleotide. Evindence has recently been presented (KOONTZ and WICKS 1977) which would indicate that 6-methylthiopurineribonucleoside 3′,5′-cyclic phosphate (63) may indeed exert antitumor effect as the unmetabolized cyclic nucleotide analog. The selective action of these cyclic nucleotides on transformed *vs* normal cells is noteworthy (KOONTZ and WICKS 1977). We have studied 9-(2′-*O*-butyryl-β-D-ribofuranosyl)-6-mercaptopurine 3′,5′-cyclic phosphate (KHWAJA et al. 1974) in a comparison with the nonbutyrylated 6-mercaptopurine cyclic ribonucleotide against L-1210 leukemia in mice. The butyrylated analog appears to be substantially superior (KHWAJA et al. 1974) (Table 8), which may be due to better cellular penetration. This may be a good example of the prodrug concept which is supported by recent cell culture data (MEYER Jr et al. 1979).

X-ray induced adenocarcinomas in the bowel of rats contained a significantly lower concentration of *c*AMP than did normal rat intestinal tissue (STEVENS et al. 1976). The concentration of *c*AMP was found to be low in normal rat kidney cells transformed by Kirsten sarcoma virus (CARCHMAN et al. 1974). Using the technique of somatic cell hybridization, TISDALE and PHILLIPS (1974) were able to correlate intracellular *c*AMP levels to the malignancy of cell lines. Cells of high malignancy had low levels of *c*AMP, while cells of low malignancy had a high level of *c*AMP.

Cyclic AMP has been shown to suppress the tumorigenicity of CELO virus-transformed hamster skin cells in vivo. This was brought about by pretreating the transformed cells with dibutyryl-*c*AMP before inoculation into the animal (REDDI and CONSTANTINIDES 1972). Treatment of the human lymphocytic cell line RPMI 1788 with dibutyryl-*c*AMP caused attachment of the cells to the surface of the culture vessels and their conversion to the normal fibroblastlike forms (SMITH et al. 1974). SHIELDS (1974) has shown that dibutyryl-*c*AMP greatly increased the adhe-

Table 8. Activity of 2′-*O*-Butyryl-6-mercaptopurine cyclic nucleotide against leukemia L-1210 (I.P. inoculation – i.p. treatment) in Bf_1 female mice (KHWAJA et al. 1974)

	(i.p.) Treatment mg/kg × days	Δ Wt. (gm.)	T/C
9-(β-D-Ribofuranosyl)-6-mercapto-purine 3′,5′-cyclic phosphate	150 × 7	+1.80	1.24
	200 × 7	+2.38	1.25
	250 × 7	+1.66	1.42
	300 × 7	+0.75	1.30
9-(2′-*O*-Butyryl-β-D-ribofuranosyl)-6-mercaptopurine 3′,5′-cyclic phosphate	50 × 7	+3.30	1.23
	100 × 7	+1.23	1.35
	150 × 7	+0.18	1.56
	200 × 6	+0.00	1.75
	250 × 7	−0.96	1.67
	250 × 7	+0.98	1.75
6-Mercaptopurine	20 × 7	+1.71	1.18
	35+7	+1.26	1.58

sion of cells to a culture dish. This suggests that *c*AMP might be useful to inhibit proliferation of metastasizing tumor cells (TISDALE 1979). It is of interest that dibutyryl-*c*AMP (NSC 143108) is currently being developed by the United States National Cancer Institute for Phase I clinical trials (see LOMAX and NARAYANAN 1979). TISDALE (1979) has pointed out that the great potential of *c*AMP in cancer chemotherapy may lie in its ability to cause redifferentiation of certain types of tumors. Dibutyryl-*c*AMP induces irreversible morphological differentiation of mouse neuroblastoma cells in culture to a form characteristic of mature neurons (PRASAD and SHEPPARD 1972).

PARDEE et al. (1974) have postulated that uncontrolled growth leading to malignancy is brought about by permanent transformation of the cell surface (caused by carcinogenic chemicals or tumor viruses). The changes in the membrane are such as to inhibit the membrane bound adenylate cyclase which in turn lowers the *c*AMP concentration inside the cell. There is an increase in *c*GMP and a resultant rapid cellular proliferation (PARDEE et al. 1974).

Chick embryo fibroblasts transformed by Rous sarcoma virus have decreased adenylate cyclase activity (ANDERSON et al. 1973 a, b). This decreased enzymatic activity which leads to lowering of *c*AMP would appear to be responsible for most of the abnormal properties of transformed cells, including an increased rate of cellular proliferation (ANDERSON et al. 1973 b). In lymphocytes of patients with chronic lymphatic leukemia, adenylate cyclase activity is decreased as compared to normal human lymphocytes (POLGAR et al. 1973). The altered adenylate cyclase activity has more recently been correlated with in vitro viral transformation of various cell lines (PASTAN et al. 1975 b).

The methyl and ethyl phosphate esters of *c*AMP exhibited a significant inhibition of Ehrlich ascites carcinoma in vivo while *c*AMP was not active under these conditions (NAGYVARY et al. 1973). Presumably this is another example of a form of *c*AMP which penetrates cells more readily and could be considered a prodrug form of *c*AMP.

ZIMMERMAN (1979) has recently pointed out that a number of adenine and adenosine analogs and nucleoside antibiotics may owe their biological activity to the fact that these compounds are converted enzymatically to the corresponding 5′-triphosphate which then may be converted by adenylcyclase to the 3′,5′-*c*AMP analog which could be the actual form of the drug contributing to the antitumor or chemotherapeutic effect.

IV. Cyclic Nucleotides and Antiviral Effects

Within the last two years considerable progress has been made in the elucidation of the mechanism of action of interferon (for a review, see MARX 1979; BAGLIONI 1979). Interferon is a high molecular weight glycoprotein which binds to the surface of the cellular plasma membrane (FRIEDMAN 1978b; BESANCON and ANKEL 1974; KOBLET et al. 1978) and effectively reduces cellular proliferation. Just how this is accomplished is a subject under active investigation in many laboratories (BAGLIONI 1979). The involvement of *c*AMP as the mediator of the action of interferon was first suggested by FRIEDMAN (1977). However, as early as 1971 our Laboratory (SIDWELL et al. 1972) had shown that *c*AMP and certain 8-substituted derivatives of *c*AMP exhibit significant antiviral activity in vitro against both DNA and RNA viruses. As an extension of this study, certain 6-substituted-9-(β-D-ribofuranosyl)purine 3′,5′-cyclic phosphates related to *c*AMP were studied as antiviral agents (SIDWELL et al. 1974) against Type 1 and Type 2 herpes, cytomegalo, vaccinia and certain rhinoviruses in vitro. 6-Methylthio-9-(β-D-ribofuranosyl)purine 3′,5′-cyclic phosphate exhibited the best antiviral spectrum and was active against all of these viruses (SIDWELL et al. 1974). FRIEDMAN and PASTAN (1969) showed that *c*AMP potentiated the antiviral activity of interferon. This observation was confirmed and extended by WEBER and STEWART (1975) who demonstrated that interferon elevated *c*AMP levels in L cells. Dibutyryl-*c*AMP inhibited adenovirus Type 2 in three cell lines (BIRON and RASKA 1973). Dibutyryl-*c*AMP similarly reduced the yields of herpes virus in cultures (STANWICK and NAHMIAS 1979). Recently (ROBBINS and RAPP 1979) *c*AMP has been shown to substantially (99.9%) reduce the production of measles virus when added to cell cultures. The rise in *c*AMP levels in interferon treated cells is suggested (KOBLET et al. 1979) to be due to stimulation of adenylate cyclase (membrane bound) by interferon binding. Human KB-3 cells, although they bind mouse interferon, do not exhibit *c*AMP elevation (MELDOLESI et al. 1977) and show no antiviral effect. This can be interpreted in that the interferon from another species does *not* activate human adenylate cyclase. Studies by FUSE and KUWATA (1978) showed that human leucocyte interferon inhibits the proliferation of virus-transformed human embryonic cells. These workers also noted that extracellularly added dibutyryl-*c*AMP showed an inhibition of proliferation to almost the same degree as interferon in the infected cells. A decrease in *c*AMP and G_1-phase arrested cells by serum addition was prevented by pretreatment with interferon. It was concluded that intracellular *c*AMP *may* mediate the inhibitory effect of interferon on cell growth (FUSE and KUWATA 1978). FRIEDMAN and co-workers (GROLLMAN et al 1978) have summarized the similarities between the structure and function of the interferon receptor and other glycoprotein hormones. Data is presented to show that the interferon receptor, like the thyrotropin receptor, is a two-component system (glycoprotein and ganglioside) receptor, that both components are necessary for message transmission and that membrane changes as a consequence of interferon binding involve *c*AMP and activation of protein kinase as mediators of the response to interferon. Cholera toxin, a glycoprotein enterotoxin, exerts its effects on cells by binding to the cellular plasma membrane and activating adenylate cyclase (for a review, see Moss and Vaughan 1979) in a similar manner. These similarities have been noted by several investigators (Friedman 1978b; Coffino and Gray 1978). Coffino and Gray (1978) report that cholera toxin treated S49 lymphoma cells arrest in the G_1 phase. Mutant cells deficient in adenylate cyclase are not inhibited by cholera toxin (Coffino and Gray 1978). The potential involvement of *c*AMP as a mediator of the action of interferon (Baglioni et al. 1978) has been reviewed by Friedman (1978a) and summarized by Robins (1979).

The broad spectrum antiviral activity of 6-methylthio-9-(β-D-ribofuranosyl)-purine 3′,5′-cyclic phosphate (63) was considered unique and was studied further. It was found that (63) in mouse L cells enhanced interferon activity as much as 32-fold over that of *c*AMP where certain 8-substituted *c*AMP derivatives exhibited up to a 4-fold enhancement (Allen et al. 1974). This amazing enhancement of the action of interferon took place despite the known fact that (63) is readily cleaved by PDE (Meyer Jr et al. 1972; Koontz and Wicks 1977) to the corresponding 5′-phosphate. It is interesting, however, that (63) is also cytotoxic to rat hepatoma cells (Koontz and Wicks 1977) and appears to exert an effect in part, independent of the corresponding nucleoside. Thus, the interferon enhancement by (63) may be a general effect. Indeed, (63) has been shown to be active in tumor cells resistant to the corresponding 6-methylthiopurineribonucleoside (Epps et al. 1975).

Weber and Stewart (1975) have studied in detail the interferon potentiation of the antiviral activity of *c*AMP in L cells and found that the antiviral activity of interferon alone (24%) was enhanced 98% with *c*AMP. This potentiation was achieved due to an increased intracellular level of *c*AMP and that interferon acts to increase levels of *c*AMP during early stages of viral infection (Weber and Stewart 1975). Endogenous *c*AMP does indeed cross cellular membranes. Tritium labelled *c*AMP administered intravenously to rats was recovered in 20% yield in the urine (Brenner and Gorin 1978).

Tovey and co-workers (Tovey et al. 1979) have pointed out that interferon exerts a number of effects on cells. These may be summarized as follows:

(1) Inhibition of cell multiplication
(2) Modification of the cellular surface
(3) Antiviral effect
(4) Marked antitumor effect
(5) Enhancement of specialized cellular function (differentiation).

These effects have been reviewed by Gresser (1977). Although there has been some effort to explain these varied biological effects as probably due to "*impure*" interferon, Gresser and co-workers (Gresser et al. 1979) have recently produced electrophoretically pure mouse interferon and have concluded that these and related effects attributed to interferon are due to interferon itself and *not* to protein contamination. These authors state: "We are obliged therefore to pose the question: how can one substance (interferon) exert such a variety of effects on cells? It seems likely that interferon triggers a cascade of events." Indeed, *c*AMP is best known for its cascade mechanism of action and for its multiple effects in the cell.

It is quite probable that the multiple biological effects of interferon are related to the various side effects of the clinical use of human leukocyte interferon (MARX 1979; INGEMARSSON et al. 1979). Interferon has recently been shown to enhance the excitability of cultured neurons, which increased the frequency of spontaneous firing (CALVET and GRESSER 1979). These changes began approximately within one-half hour and lasted several hours (CALVET and GRESSER 1979). This type of response is known to be mediated by *c*AMP (NATHANSON and GREENGARD 1977). More recently, BLALOCK and STANTON (1980) have pointed out the similarity of the action of interferon with other types of hormone action. The exposure of myocardial cells to interferon caused an increased beat frequency in myocardial cells, a response also known to be caused by *c*AMP (BLALOCK and STANTON 1980). It would be most useful and gratifying if one could separate the various biological effects of interferon as desired. Can the antiviral effects be separated from the antitumor effects? If the effects of interferon *are* indeed modulated by *c*AMP, it is readily seen why these effects are so diverse. A possible approach to this problem is the direct employment of a *c*AMP analog which would only activate certain cellular protein kinases to achieve a much more specific antiviral or antitumor effect.

V. Cyclic Nucleotide Analogs and Cardiovascular Effects

In 1970 KUKOVETZ and POECH studied the effects of *c*AMP, N^6,O^2-dibutyryl-*c*AMP and N^6,O^2-dihexanoyl-*c*AMP in isolated perfused hearts of guinea pigs, rats, and rabbits. In all cases, strong and long lasting effects in the rate, amplitude and coronary flow were noted. Theophylline augmented the inotropic effects. Several recent reviews have appeared dealing with the role of *c*AMP and its regulatory function in contractility (NAYLER 1977; TSIEN 1977; DRUMMOND and SEVERSON 1979), which has now become rather firmly established. A systematic search for more specific *c*AMP analogs exhibiting greater potency as inotropic agents has resulted in the synthesis and study of N^6-*n*-butyryl-8-benzylthio-*c*AMP (479), an agent which in dog increased cardiac output by 84% without an increase in heart rate. The action of (479) was quite specific in that the only other physiological effects were a release of insulin followed by an increase in plasma glucose levels (MILLER et al. 1980). A strong case has been made for (479) as an attractive candidate for the clinical management of acute myocardial failure (MILLER et al. 1980).

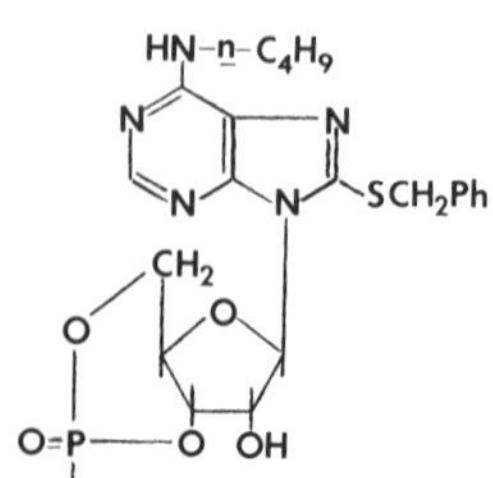

479

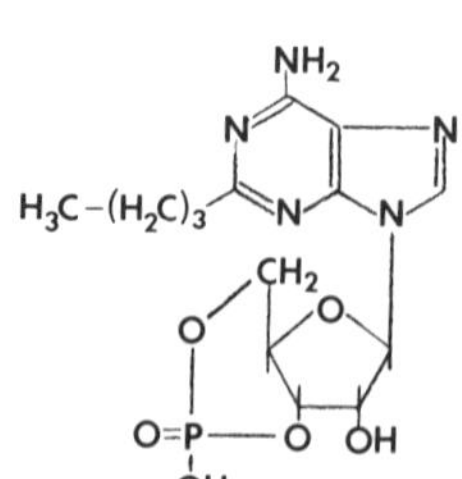

480

AMER (1975) has suggested that an increased $\frac{c\mathrm{GMP}}{c\mathrm{AMP}}$ ratio in the vasculature results in elevated vascular smooth muscle tone which causes an increased peripheral resistance and a resultant hypertensive state. An increase in *c*AMP which lowers the $\frac{c\mathrm{GMP}}{c\mathrm{AMP}}$ ratio should result in reducing hypertension (AMER, 1975). Indeed, it should be possible to prepare a *c*AMP analog which would be much more superior to *c*AMP itself in reducing hypertension. Such a derivative of *c*AMP has been prepared in our Laboratory, 2-*n*-butyryl-*c*AMP (480) (MEYER JR et al. 1974a). 2-*n*-Butyryl-*c*AMP (480) is orally effective and superior to aldomet in the hypertensive rat (Giarrusso and Robins, unpublished results). The cyclic nucleotide (480) exhibits a good dose response when given orally to dogs. A 10–30 mg/Kg dose results in a 50% drop in blood pressure with a 10-hour duration and no tachycardia or significant increase in heart rate noted. It is interesting to note that (480) was the most potent of a series of 2-substituted-*c*AMP derivatives in activating *c*AMP-dependent protein kinase (MEYER JR et al. 1974a). The action of (480) may be due in part to the fact that this cyclic nucleotide is a powerful inhibitor of *c*AMP-dependent phosphodiesterase isolated from beef heart (MEYER JR et al. 1975a). Certainly these studies suggest further work with additional cyclic nucleotides in the field of cardiovascular pharmacology.

The chemistry, biochemistry and molecular biology of nucleoside 3′,5′-cyclic phosphates has opened a new era of understanding of many fundamental biological processes and responses. It is now abundantly clear that new analogue derivatives can act as molecular probes to more clearly define and focus the activation of certain cyclic nucleotide protein kinases on a rather specific basis within the given cell type. It is possible to design cyclic nucleotides which are resistant to PDE degradation, and consequently would have a considerably prolonged cellular effect. Application of new cyclic nucleotide chemistry to major medical problems should create a fruitful area of medicinal chemistry of considerable importance over the next few decades.

Acknowledgements. We wish to thank Mr. HOWARD B. COTTAM for proofreading of the manuscript and Ms. EDITH JOHNSON for very competent typing of this manuscript

E. List of Abbreviations Used

Ac = acetyl
ACTH = adrenocorticotropic hormone
Ad = adenine
ADH = antidiuretic hormone
AICAR = 5-amino-1-(β-D-ribofuranosyl)-imidazole-4-carboxamide
5′-AMP = adenosine 5′-monophosphate
ara-A = 9-(β-D-arabinofuranosyl)adenine
ara-C = 1-(β-D-arabinofuranosyl)cytosine
ara-*c*AMP = 9-(β-D-arabinofuranosyl)-adenine 3′,5′-cyclic phosphate
ara-CMP = 1-(β-D-arabinofuranosyl)-cytosine 5′-phosphate
ara-*c*CMP = 1-(β-D-arabinofuranosyl)-cytosine 3′,5′-cyclic phosphate
ara-*c*GMP = 9-(β-D-arabinofuranosyl)-guanine 3′,5′-cyclic phosphate
ara-CTP = 1-(β-D-arabinofuranosyl)-cytosine 5′-triphosphate
ara-*c*UMP = 1-(β-D-arabinofuranosyl)-uracil 3′,5′-cyclic phosphate
ara-U = 1-(β-D-arabinofuranosyl)uracil
ATP = adenosine 5′-triphosphate
2-Aza-*c*AMP = 4-amino-7-(β-D-ribofuranosyl)imidazo[4,5-*d*]-*v*-triazine 3′,5′-cyclic phosphate

8-Aza-*c*AMP = 7-amino-3-(β-D-ribofuranosyl)-*v*-triazolo[4,5-*d*]pyrimidine 3′,5′-cyclic phosphate
Bz = benzoyl
*c*AMP = adenosine 3′,5′-cyclic phosphate
*c*CMP = cytidine 3′,5′-cyclic phosphate
*c*dAMP = 2′-deoxyadenosine 3′,5′-cyclic phosphate
*c*GMP = guanosine 3′,5′-cyclic phosphate
CIC = cyclohexyl isocyanate
*c*IMP = inosine 3′,5′-cyclic phosphate
*c*TMP = thymidine 3′,5′-cyclic phosphate
CTP = cytidine 5′-triphosphate
*c*UMP = uridine 3′,5′-cyclic phosphate
DBN = 1,5-diazabicyclo[4.3.0]non-5-ene
DBU = 1,5-diazabicyclo[5.4.0]-5-undecene
DCC = N,N′-dicyclohexylcarbodiimide
1-Deaza-*c*AMP = 7-amino-3-(β-D-ribofuranosyl)imidazo[4,5-*b*]pyridine 3′,5′-cyclic phosphate
3-Deaza-*c*AMP = 4-Amino-1-(β-D-ribofuranosyl)imidazo[4,5-*c*]pyridine 3′,5′-cyclic phosphate
DMF = N,N-dimethylformamide
EDC = 1-ethyl-3-(3-dimethylaminopropyl)carbodiimide
Formycin = 7-amino-3-(β-D-ribofuranosyl)pyrazolo[4,3-*d*]pyrimidine
G = guanine
iso-*c*AMP = 6-amino-3-(β-D-ribofuranosyl)purine 3′,5′-cyclic phosphate
$KOBu^t$ = potassium *tert*-butoxide
Neopterin = 2-amino-4-hydroxy-6-trihydroxypropylpteridine
PDE = phosphodiesterase
PDS = 2,2′-dipyridyl disulfide
Ph = phenyl
PK = protein kinase
Py = pyridine
ReP = β-D-ribofuranosyl 3′,5′-cyclic phosphate
Th = thymine
THP = tetrahydropyranyl
TPP = triphenylphosphine
Ts = *p*-toluenesulfonyl (tosyl)
Tu-*c*AMP = tubercidin 3′,5′-cyclic phosphate (4-amino-7-β-D-ribofuranosylpyrrolo[2,3-*d*]pyrimidine 3′,5′-cyclic phosphate)
U = uracil
xylo-*c*AMP = 9-(β-D-xylofuranosyl)-adenine 3′,5′-cyclic phosphate
xylo-*c*GMP = 9-(β-D-xylofuranosyl)-guanine 3′,5′-cyclic phosphate
ε-*c*AMP = 3-(β-D-ribofuranosyl)imidazo-[2,1-*i*]purine 3′,5′-cyclic phosphate

References

Abou-Sabe M (1976) Cyclic nucleotides and the regulation of cell growth. Dowden, Hutchinson and Ross, Inc. Stroudsburg PA

Agence Nationale de Valorisation de la Recherche (1972) Cyclic (iso) adenosine-3′,5′-monophosphoric acid. French Patent 2,112,084

Akerfeldt S (1962) Further studies on S-substituted phosphorothioic acids. II. Synthesis and certain properties of some potential antiradiation drugs. Acta Chim Scand 16:1897–1907

Alberty RA (1969) Standard Gibbs free energy, enthalpy, and entropy changes as a function of pH and pMg for several reactions involving adenosine phosphates. J Biol Chem 244:3290–3302

Albrecht HP, Jones GH, Moffat JG (1970) 3′-Deoxy-3′-(dihydroxyphosphinylmethyl)nucleosides. Isosteric phosphonate analogs of nucleoside 3′-phosphates. J Am Chem Soc 92:5511–5513

Allen LB, Boswell KH, Khwaja TA, Meyer RB Jr, Sidwell RW, Witkowski JT, Christensen LF, Robins RK (1978) Synthesis and antiviral activity of some phosphates of the broad-spectrum antiviral nucleoside, 1-β-D-ribofuranosyl-1,2,4-triazole-3-carboxamide (Ribavirin). J Med Chem 21:742–746

Allen LB, Eagle NC, Huffman JH, Shuman DA, Meyer RB Jr, Sidwell RW (1974) Enhancement of interferon antiviral action in L-cells by cyclic nucleotides. Proc Soc Exp Biol Med 146:580–584

Amer MS (1975) Cyclic nucleotides in disease; on the biochemical etiology of hypertension. Life Sci 17:1021–1038

Amer MS, McKinney GR (1973) Possibilities for drug development based on the cyclic AMP system. Life Sci 13:753–767

Amer MS, McKinney GR (1974) Cyclic nucleotides and drug discovery. Annu Rept Med Chem 9:203–212
Amer MS, McKinney GR (1975) Cyclic nucleotides as mediators of drug action. Annu Rept Med Chem 10:192–201
Anderson WB, Johnson GS, Pastan IH (1973a) Transformation of chick-embryo fibroblasts by wild-type and temperature-sensitive Rous Sarcoma virus alters adenylate cyclase activity. Proc Natl Acad Sci USA 70:1055–1059
Anderson WB, Lovelace E, Pastan I (1973b) Adenylate cyclase activity is decreased in chick embryo fibroblasts transformed by wild type and temperature sensitive Schmidt-ruppin rous sarcoma virus. Biochem Biophys Res Commun 52:1293–1299
Anderson WB, Perlman RL, Pastan I (1972) Effect of adenosine 3′,5′-monophosphate analogs on the activity of the cyclic adenosine 3′,5′-monophosphate receptor in *Escherichia coli*. J Biol Chem 247:2717–2722
Anisuzzaman AKM, Lake WC, Whistler RL (1973) 4′-Thioadenosine 3′,5′-cyclic phosphate and derivatives. Chemical synthesis and hydrolysis by phosphodiesterase. Biochemistry 12:2041–2045
Ashman DF, Lipton R, Melicow MM, Price TD (1963) Isolation of adenosine 3′,5′-monophosphate and guanosine 3′,5′-monophosphate from rat urine. Biochem Biophys Res Commun 11:330–334
Baddiley J, Buchanan JG, Szabo L (1954) Sugar phosphates. Part I. Derivatives of glucose 4:6-(hydrogen phosphate). J Chem Soc 3826–3839
Baglioni C (1979) Interferon-induced enzymatic activities and their role in the antiviral state. Cell 17:255–264
Baglioni C, Minks MA, Maroney PA (1978) Interferon action may be mediated by activation of a nuclease by pppA2′A2′A. Nature 273:684–687
Bahr JT, Cathou RE, Hammes GG (1965) A thermodynamic study of the hydrolysis of cytidine 2′,3′-cyclic phosphate. J Biol Chem 240:3372–3378
Bald RW, Holý A (1971) Nucleic acid components and their analogues. CXLII. Preparation of 3-(β-D-ribofuranosyl)uracil 2′,3′-cyclic phosphate and related compounds and their behaviour towards pancreatic ribonuclease and ribonuclease T2. Coll Czech Chem Commun 36:3657–3669
Baraniak J, Lesiak K, Stec WJ (1979) Assignment of absolute configuration at phosphorus in diastereoisomeric adenosine 3′,5′-cyclic phosphorothioates (cAMPS). Pol J Chem 53:1387–1391
Barker GR, Montague MD, Moss RJ, Parsons MA (1957) Biosynthesis of polynucleotides. Part I. The mode of action of ribonuclease. J Chem Soc 3786–3793
Barker GR, Foll GE (1957) Biosynthesis of polynucleotides. Part II. The synthesis and properties of phosphoryl derivatives of adenine glucoside. J Chem Soc 3794–3798
Barry CD, Martin DR, Williams RJP, Xavier AV (1974) Quantitative determination of the conformation of cyclic 3′,5′-adenosine monophosphate in solution using lanthanide ions as nuclear magnetic resonance probes. J Mol Biol 84:491–502
Baschang G, Kvita V (1973) New derivatives of thymidine 3′,5′-cyclophosphate. Angew Chem (Engl) 12:71–72
Beaman AG, Robins RK (1962) Potential purine antagonists. XXXIII. Synthesis of chloropurines. J Appl Chem 12:432–437
Berger NA, Eichhorn GL (1971) Interaction of metal ions with polynucleotides and related compounds. XIV. Nuclear magnetic resonance studies of the binding of copper (II) to adenine nucleotides. Biochemistry 10:1847–1857
Bergmeyer HU, Michal G, Nelboeck-Hochstetter M, Stork H, Weimann G, (Boehringer Mannheim GmbH) (1971) Antilipolytic cyclophosphate derivatives of 5-amino-1-β-D-ribosylimidazole-4-carboxamides. German Patent 2,026,040
Bergmeyer HU, Nelboeck-Hochstetter M, Michal G, Muehlegger K, Weimann G, Roesch E (1972) Substituted purine ribonucleoside 3′,5′-cyclic phosphates. Ger. Offen. 2,125,077
Besancon F, Ankel H (1974) Binding of interferon to gangliosides. Nature 252:478–480
Biron KK, Raska K Jr (1973) Effects of dibutyryl-cAMP on adenovirus replication in different cell lines. Virology 56:383–385

Blackburn BJ, Lapper RD, Smith ICP (1973) A proton magnetic resonance study of the conformations of 3′,5′-cyclic nucleotides. J Am Chem Soc 95:2873–2878
Blackburn GM, Cohen JS, Todd Lord (1964) Cyclic phosphate and phosphate triesters – A P^{31} N.M.R. study. Tetrahedron Lett 2873–2879
Blalock JE, Stanton JD (1980) Common pathways of interferon and hormonal action. Nature 283:406–408
Blecher M, Ro'Ane JT, Flynn PD (1970) Metabolism of dibutyryl cyclic adenosine 3′,5′-monophosphate during its regulation of lipolysis and glucose oxidation in isolated rat epididymal adipocytes. J Biol Chem 245:1867–1870
Bloch A, private communication
Bloch A (1974) Cytidine 3′,5′-monophosphate (cyclic CMP). I. Isolation from extracts of leukemia L-1210 cells. Biochem Biophys Res Commun 58:652–659
Bloch A (1975a) Isolation of cytidine 3′,5′-monophosphate from mammalian tissues and body fluids and its effects on leukemia L-1210 cell growth in culture. Adv Cyclic Nucleotide Res 5:331–338
Bloch A (1975b) Uridine 3′,5′-monophosphate (cyclic UMP). I. Isolation from rat liver extracts. Biochem Biophys Res Commun 64:210–218
Bloch A, Dutschman G, Maue R (1974) Cytidine 3′,5′-monophosphate (cyclic CMP). II. Initiation of leukemia L-1210 cell growth in vivo. Biochem Biophys Res Commun 59:955–959
Bloch A, Hromchak R, Henderson ES (1975) Isolation of cytidine 3′,5′-monophosphate (cyclic CMP) from the urine of leukemic patients. Proc Am Assoc Cancer Res 16:191
Bobek M, Whistler RL, Bloch A (1970) Preparation and activity of the 4′-thio derivatives of some 6-substituted purine nucleosides. J Med Chem 13:411–413
Bobruskin ID, Gulyaev NN, Kirpichnikov MP, Severin ES, Tunitskaya VA, Florent'ev VL (1979) Nuclear magnetic resonance study of conformation of nucleotides, oligonucleotides, and their analogs in solution. I. Conformation of adenosine-3′,5′-cyclophosphate and its analogs in aqueous solutions. Mol Biol (Mosk) 13:87–95
Boehringer Mannheim GmbH (1970a) Cyclofosfaten en werkwijze voor hun bereiding. Nederland patent 6,913,671
Boehringer Mannheim GmbH (1970b) Werkwijze voor het bereiden van de op 8-planta gemodificocorde purineribofuranoside 3′,5′-cyclofosfaten. Nederland Patent 7,003,222
Boehringer Mannheim GmbH (1971) Purineribofuranoside 3′,5′-cyclophosphates. British Patent 1,257,546
Borden RK, Smith M (1966a) Nucleotide synthesis. II. Nucleotide *p*-nitrophenyl and 2,4-dinitrophenyl esters. J Org Chem 31:3241–3246
Borden RK, Smith M (1966b) Nucleotide synthesis. III. Preparation of nucleoside-3′,5′-cyclic phosphates in strong base. J Org Chem 31:3247–3253
Boswell KH, Christensen LF, Shuman DA, Robins RK (1975) Synthesis of 6,8-disubstituted-9-*β*-D-ribofuranosylpurine 3′,5′-cyclic phosphates. J Heterocycl Chem 12:1–9
Boswell KH, Miller JP, Shuman DA, Sidwell RW, Simon LN, Robins RK (1973) Synthesis and biological activity of certain carbamoyl and alkoxycarbonyl derivatives of adenosine 3′,5′-cyclic phosphate. J Med Chem 16:1075–1079
Boswell KH, Robins RK (1979) The synthesis of 2-amino-6,8-disubstituted purine-9-*β*-D-ribofuranosyl-3′,5′-cyclic monophosphates from guanosine-3′,5′-cyclic monophosphate (c-GMP). J Carbohydr Nucleosides Nucleotides 6:359–370
Boyd DB (1969) Mechanism of hydrolysis of cyclic phosphate esters. J Am Chem Soc 91:1200–1205
Bradbury E, Nagyvary J (1976) Preparation of 2′-thio-2′-deoxycytidine 2′,3′-phosphorothioate. Nucleic Acids Res 3:2437–2442
Brenner T, Gorin E (1978) Metabolic fate of exogenous 5′-AMP, cyclic AMP and cyclic GMP in rats. Biochem Med 20:160–166
Broadus AE (1977) Clinical cyclic nucleotide research. Adv Cyclic Nucleotide Res 8:509–548
Broer Y, Fouchereau M, Rosselin G (1972) Dosages radio-immunologiques de l'AMPc et du GMPc. Préparation, purification et contrôle du 2′0 succinyl AMP-3′,5′-cyclique et du 2′0 succinyl-GMP-3′,5′-cyclique. C.R. Acad Sci Paris 275D:619–622

Brooker G, Harper JF, Teraski WL, Moylan RD (1979) Radioimmunoassay of cyclic AMP and cyclic GMP. Adv Cyclic Nucleotide Res 10:1–33
Broom AD, Robins RK (1965) The direct preparation of 2′-*O*-methyladenosine from adenosine. J Am Chem Soc 87:1145–1146
Brown DM, Magrath DI, Todd AR (1952) Nucleosides. Part XII. The preparation of cyclic 2′,3′-phosphates of adenosine, cytidine, and uridine. J Chem Soc 2708–2714
Brown DM, Magrath DI, Sir Todd AR (1955) Nucleotides. Part XXXIV. The dydrolysis of dialkyl esters of uridine-3′ phosphate and its relevance to the question of phosphotriester linkages in ribonucleic acids. J Chem Soc 4396–4401
Brown DM, Todd AR (1952) Nucleotides. Part X. Some observations on the structure and chemical behaviour of the nucleic acids. J Chem Soc 52–58
Brunswick DJ, Cooperman BS (1971) Photo-affinity labels for adenosine 3′,5′-cyclic monophosphate. Proc Natl Acad Sci USA 68:1801–1804
Bryan AM, Olafsson PG (1973) Calorimetric investigations of crystalline 3′,5′-cyclic nucleotides. Biopolymers 12:229–235
Butcher RW, Sutherland EW (1962) Adenosine 3′,5′-phosphate in biological materials. I. Purification and properties of cyclic 3′,5′-nucleotide phosphodiesterase and use of this enzyme to characterise adenosine 3′,5′-phosphate in human urine. J Biol Chem 237:1244–1250
Cailla H, Delaage M (1972) Succinyl derivatives of adenosine 3′,5′-cyclic monophosphate: Synthesis and purification. Anal Biochem 48:62–72
Calvet M-C, Gresser I (1979) Interferon enhances the excitability of cultured neurones. Nature 278:558–560
Carchman RA, Johnson GS, Pastan I, Scolnick EM (1974) Studies on the levels of cyclic AMP in cells transformed by wild-type and temperature-sensitive Kirsten sarcoma virus. Cell 1:59–64
Cech SY, Gross RA, Ignarro LJ (1976) Pharmacologist 18:223
Cech SY, Ignarro LJ (1977) Cytidine 3′,5′-monophosphate (cyclic CMP) formation in mammalian tissues. Science 198:1063–1065
Cech SY, Ignarro LJ (1978) Cytidine 3′,5′-monophosphate (cyclic CMP) formation by homogenates of mouse liver. Biochem Biophys Res Commun 80:119–125
Cehovic G, Bayer M, Giao N-B (1976) Synthesis of new cyclic nucleotides and their differential stimulatory effects on thyroid function in mice. J Med Chem 19:899–903
Cehovic G, Gabbal A, Marcus I, Posternak T (1971 a) Dérivés d'acide adénosine-3′,5′-phosphorique cyclique et leur préparation. French patent 6,934,747
Cehovic G, Marcus I, Posternak T (1971 b) Iso-adenosine-3′,5′-monophosphorsäure und ihre Salze. German Patent 1,944,000
Cehovic G, Marcus I, Vengadabady S, Posternak T (1968) Sur la préparation de l'acide iso-adénosine-3′,5′-phosphorique (iso-AMP cyclique) et sur certaincs de ses propriétés biologiques. CR Soc Phys Hist Nat 3:135–139
Chawla RK, Nixon DW, Shoji M, Rudman D (1979) Plasma and urine cyclic guanosine 3′,5′-monophosphate in disseminated cancer. Ann Int Med 91:862–864
Cheng YC, Bloch A (1978) Demonstration in leukemia L-1210 cells, of a phosphodiesterase acting on 3′,5′-cyclic CMP but not on 3′,5′-cyclic AMP or 3′,5′-cyclic GMP. J Biol Chem 253:2522–2524
Cherbuliez E, Probst H, Rabinowitz J (1959) Recherches sur la formation et la transformation des esters XVII. Sur la formation d'esters phosphoriques cycliques et sur la vitesse d'hydrolyse de ces esters et des esters dialcoylphosphoriques. Helv Chim Acta 42:1377–1384
Chládek S, Nagyvary J (1972) Nucleophilic reactions of some nucleoside phosphorothioates. J Am Chem Soc 94:2079–2085
Cho-Chung YS (1974) In vivo inhibition of tumor growth by cyclic adenosine 3′,5′-monophosphate derivatives. Cancer Res 34:3492–3496
Christensen LF, Meyer RB Jr, Miller JP, Simon LN, Robins RK (1975) Synthesis and enzymic activity of 8-acyl and 8-alkyl derivatives of guanosine 3′,5′-cyclic phosphate. Biochemistry 14:1490–1496

Christensen LF, Robins RK (1976) 8-Substituted cyclic nucleotides by free radical alkylation and acylation. US Patent 3,968,101
Chu SH, Shiue CY, Chu MY (1974) Synthesis and biological activity of some 8-substituted seleno cyclic nucleotides and related compounds. J Med Chem 17:406–409
Chu SH, Shiue CY, Chu MY (1975) Synthesis and biological activity of some 8-substituted selenoguanosine cyclic 3′,5′-phosphates and related compounds. J Med Chem 18:559–564
Chwang AK, Sundaralingam M (1973) Molecular conformation of guanosine 3′,5′-cyclic monophosphate. Nature New Biol 244:136–137
Chwang AK, Sundaralingam M (1974) The crystal and molecular structure of guanosine 3′,5′-cyclic monophosphate (cyclic GMP) sodium tetrahydrate. Acta Cryst 30B:1233–1240
Clark VL, Bernlohr RW (1972) Guanyl cyclase of *Bacillus licheniformis*. Biochem Biophys Res Commun 46:1570–1575
Coffino P, Gray JW (1978) Regulation of S49 lymphoma cell growth by cyclic adenosine 3′,5′-monophosphate. Cancer Res 38:4285–4288
Collin RL (1966) The electronic structure of phosphate esters. J Am Chem Soc 88:3281–3287
Cook WH, Lipkin D, Markham R (1957) The formation of a cyclic dianhydrodiadenylic acid (I) by the alkaline degradation of adenosine-5′-triphosphoric acid (II). J Am Chem Soc 79:3607–3608
Cotton FA, Gillen RG, Gohil RN, Hazen EE Jr, Kirchner CR, Nagyvary J, Rouse JP, Stanislowski AG, Stevens JD, Tucker PW (1975) Tumor-inhibiting properties of the neutral *P–O*-ethyl ester of adenosine 3′,5′-monophosphate in correlation with its crystal and molecular structure. Proc Natl Acad Sci USA 72:1335–1339
Coulter CL (1968) Cyclic uridine-3′,5′-phosphate: Molecular Structure. Science 159:888–889
Coulter CL (1969) The crystal and molecular structure of the triethylammonium salt of cyclic uridine-3′,5′-phosphate. Acta Cryst 25B:2055–2065
Coulter CL (1970) The crystal and molecular structure of the triethylammonium salt of the cyclic uridine-3′,5′-phosphate. Corrigendum. Acta Cryst 26B:441
Coulter CL (1973) Structural chemistry of cyclic nucleotides. II. Crystal and molecular structure of sodium *β*-cytidine 2′,3′-cyclic phosphate. J Am Chem Soc 95:570–575
Coulter CL, Greaves ML (1970) Cyclic cytidine 2′,3′-phosphate: Molecular structure. Science 169:1097–1098
Cox JR Jr, Wall RE, Westheimer FH (1959) Thermochemical demonstration of strain in a cyclic phosphate. Chem Ind (Lond) 929
Davoren PR, Sutherland EW, Maxwell AM (1963) The cellular location of adenyl cyclase in the pigeon erythrocyte. J Biol Chem 238:3016–3023
Dekker CA, Khorana HG (1954) Carbodiimides. VI. The reaction of dicyclohexylcarbodiimide with yeast adenylic acid. A new method for the preparation of monoesters of ribonucleoside 2′- and 3′-phosphates. J Am Chem Soc 76:3522–3527
DeRubertis FR, Craven P (1977) Increased guanylate cyclase activity and guanosine 3′,5′-monophosphate content of ethionone-induced hepatomas. Cancer Res 37:15–21
DeRubertis FR, Chayoth R, Field JB (1976) The content and metabolism of cyclic adenosine 3′,5′-monophosphate and cyclic guanosine 3′,5′-monophosphate in adenocarcinoma of the human colon. J Clin Invest 57:641–649
Dills WL Jr, Beavo JA, Bechtel PJ, Meyers KR, Sakai LJ, Krebs EG (1976) Binding of adenosine 3′,5′-monophosphate dependent protein kinase regulatory subunit to immobilized cyclic nucleotide derivatives. Biochemistry 15:3724–3731
Dills WL Jr, Goodwin CD, Lincoln TM, Beavo JA, Bechtel PJ, Corbin JD, Krebs EG (1979) Purification of cyclic nucleotide receptor proteins by cyclic nucleotide affinity chromatography. Adv Cyclic Nucleotide Res 10:199–217
Dimroth K, Witzel H, Hülsen W, Mirbach H (1959) Über die Hydrolyse von Ribonucleinsäuren in Gegenwart von Metallhydroxyden. Ann Chem 620:94–132
Drummond GI, Gilgan MW, Reiner EJ, Smith M (1964) Deoxyribonucleoside-3′,5′-cyclic phosphates. Synthesis and acid-catalyzed and enzymic hydrolysis. J Am Chem Soc 86:1626–1630

Drummond GI, Duncan L, Hertzman E (1966) Effect of epinephrine on phosphorylase b kinase in perfused rat hearts. J Biol Chem 241:5899–5903

Drummond GI, Powell CA (1970) Analogues of adenosine 3′,5′-cyclic phosphate as activators of phosphorylase b kinase and as substrates for cyclic 3′,5′-nucleotide phosphodiesterase. Mol Pharmacol 6:24–30

Drummond GI, Severson DL (1979) Cyclic nucleotides and cardiac function. Circ Res 44:145–153

Dunn DB, Hall RH (1970) Purines, pyrimidines, nucleosides and nucleotides: physical constants and spectral properties. In: Handbook of biochemistry, 2nd ed, Edit. Sober, Chemical Rubber Co., Cleveland, p G-1-238

Eckstein F (1968) Uridine 2′,3′-*O,O*-cyclophosphorothioate as substrate for pancreatic ribonuclease. FEBS Lett 2:85–86

Eckstein F (1970) Nucleoside phosphorothioates. J Am Chem Soc 92:4718–4723

Eckstein F, Gindl H (1968) Uridin-2′,3′-*O,O*-cyclothiophosphat. Chem Ber 101:1670–1673

Eckstein F, Bär H-P (1969) Enzymatic hydrolysis of adenosine 3′,5′-cyclic phosphorothioate. Biochim Biophys Acta 191:316–321

Eckstein F, Schulz HH, Rüterjans H, Haar W, Maurer W (1972) Stereochimistry of the transesterification step of ribonuclease T_1. Biochemistry 11:3507–3512

Eckstein F, Simonson LP, Bär H-P (1974) Adenosine 3′,5′-cyclic phosphorothioate: Synthesis and biological properties. Biochemistry 13:3806–3810

Egami F, Takahashi K, Uehida T (1964) Ribonuclease in taka-diastase: Properties, chemical nature, and applications. In: Davidson, Cohn (eds) Progress in nucleic acid research and molecular biology, Academic Press Inc, New York, vol 3, pp 59–101

Engels J (1979) Synthesis and reactivity of guanosine 3′,5′-phosphoric acid alkyl esters. Bioorg Chem 8:9–16

Engels J, Pfleiderer W (1975) Nucleotide, V: Synthese und Eigenschaften von Uridin-3′,5′-cyclophosphat-triestern. Tetrahedron Lett 1661–1664

Engels J, Schlaeger E-J (1977) Synthesis, structure, and reactivity of adenosine cyclic 3′,5′-phosphate benzyl triesters. J Med Chem 20:907–911

Epps D, Chang IM, Sherwood E, Kimball AP (1975) Feedback inhibition by 6-methylthioinosine 3′,5′-cyclic monophosphate in tumor cells resistant to the nucleoside. Proc Soc Exp Biol Med 150:578–580

Eto M, Iio M, Kobayashi Y, Omura H, Eto M (1974) Syntheses of ribonucleoside 5′-S-methyl phosphorothiolates and ribonucleoside 3′,5′-cyclic phosphates from nucleosides applying a new phosphorylating agent, MTBO. Agr Biol Chem 38:2081–2092

Eto M, Iio M, Omura H, Eto M (1978) Direct synthesis of ribonucleoside 2′,3′-cyclic phosphate from ribonucleoside. Agr Biol Chem 42:199–200

Evans DB, Parham CS, Schenck MT, Laffan RJ (1976) Stimulation of myocardial contractility by a new cyclic nucleotide analog, 8-(benzylthio)-N^6-*n*-butyladenosine cyclic 3′,5′-phosphate (SQ80122). J Cyclic Nucleotide Res 2:307–319

Falbriard J-G, Posternak T, Sutherland EW (1967) Preparation of derivatives of adenosine 3′,5′-phosphate. Biochim Biophys Acta 148:99–105

Fazakerley GV, Russell JC, Wolfe M (1975) Conformation of the gadolinium complexes of adenosine 3′,5′-cyclic monophosphoric acid and inosine 3′,5′-cyclic monophosphoric acid. J Chem Soc Chem Commun 527

Fazakerley GV, Russell JC, Wolfe MA (1977) Determination of the *syn-anti* equilibrium of some purine 3′:5′-nucleotides by nuclear-magnetic-relaxation perturbation in the presence of a lanthanide-ion probe. Eur J Biochem 76:601–605

Fikus M, Welfeld JK, Kazimierczuk Z, Shugar D (1974) Biochemical studies on some new analogues of adenosine-3′,5′-cyclic phosphate, including isoguanosine-3′,5′-cyclic phosphate. Acta Biochim Pol 21:465–474

Fischer E (1914) Über Phosphorsäureester des Methyl-glucosids und Theophyllin-glucosids. Chem Ber 47:3193–3205

Forrest HS, Todd AR (1950) Nucleotides. Part V. Riboflavin 5′-phosphate. J Chem Soc 3295–3299

Fox JJ, Miller NC (1963) Nucleosides. XVI. Further studies of anhydronucleosides. J Org Chem 28:936–941

Free CA, Paik VS (1977) Adrenal steroidogenic actions of cyclic nucleotide derivatives in the rat. Endocrinology 100:1287–1293
Friedman DL (1976) Role of cyclic nucleotides in cell growth and differentiation. Physiol Rev 56:652–708
Friedman RM (1977) Antiviral activity of interferons. Bacteriol Rev 41:543–567
Friedman RM (1978a) Interferons and cancer. J Natl Cancer Inst 60:1191–1194
Friedman RM (1978b) Interferon action and the cell surface. Pharmacol Ther pt A, 2:425–438
Friedman RM, Pastan I (1969) Interferon and cyclic-3′,5′-adenosine monophosphate: potentiation of antiviral activity. Biochem Biophys Res Commun 36:735–740
Friedrich VW (1963) Zur Kenntnis der Phosphorsäureester der Benzimidazolriboside. Z Naturforsch 186:455–462
Fujimoto Y, Naruse M (1968) Nucleoside 3′,5′-cyclic phosphates. Japanese Patent 68 16,988
Furth JJ, Cohen SS (1967) Inhibition of mammalian DNA polymerase by the 5′-triphosphate of 9-β-D-arabinofuranosyladenine. Cancer Res 27:1528–1533
Fuse A, Kuwata T (1978) Inhibition of DNA synthesis and alteration of cyclic adenosine 3′,5′-monophosphate levels in RSa cells by human leukocyte interferon. J Natl Cancer Inst 60:1227–1232
Garbers DL, Suddath JL, Hardman JG (1975) Enzymatic formation of inosine 3′,5′-monophosphate and of 2′-deoxyguanosine 3′,5′-monophosphate. Inosinate and deoxyguanylate cyclase activity. Biochim Biophys Acta 377:174–185
Gassen HG, Witzel H (1967) Zum Mechanismus der Ribonuclease-Reaktion. 1. Die Aufgabe der Pyrimidinbase bei der Reaktion. Eur J Biochem 1:36–45
Geahlen RL, Haley BE, Krebs EG (1979) Synthesis and use of 8-azidoguanosine 3′,5′-cyclic monophosphate as a photoaffinity label for cyclic GMP-dependent protein kinase. Proc Natl Acad Sci USA 76:2213–2217
Gericke D, Chandra P (1969) Inhibition of tumor growth by nucleoside cyclic 3′,5′-monophosphates. Hoppe-Seylers Z Physiol Chem 350:1469–1471
Gerlt JA, Gutterson NI, Datta P, Belleau B, Penney CL (1980a) Thermochemical identification of the structural factors responsible for the thermodynamic instability of 3′,5′-cyclic nucleotides. J Am Chem Soc 102:1655–1660
Gerlt JA, Gutterson NI, Drews RE, Sokolow JA (1980b) Conformational properties of 5-alkoxy and 5-alkyl substituted trimethylene phosphates in solution. J Am Chem Soc 102:1665–1670
Gerlt JA, Westheimer FH, Sturtevant JM (1975) The enthalpies of hydrolysis of acyclic, monocyclic, and glycoside cyclic phosphate diesters. J Biol Chem 250:5059–5067
Gerster JF, Hinshaw BC, Robins RK, Townsend LB (1968) Purine nucleosides. XIX. The synthesis of certain 8-chloropurine nucleosides and related derivatives. J Org Chem 33:1070–1073
Gerster JF, Robins RK (1966) Purine nucleosides XIII. The synthesis of 2-fluoro and 2-chloroinosine and certain derived purine nucleosides. J Org Chem 31:3258–3262
Giarrusso FF, Robins RK, unpublished results
Gillen RG, Nagyvary J (1976) Some biochemical properties of alkyl phosphotriesters of cyclic AMP. Biochem Biophys Res Commun 68:836–840
Gibson DW, Beer M, Barrnett RJ (1971) Gold (III) complexes of adenine nucleotides. Biochemistry 10:3669–3679
Glaudemans CPJ, Fletcher HG Jr (1963) Syntheses with partially benzylated sugars. III. A simple pathway to a "*cis*-nucleoside," 9-β-D-arabinofuranosyladenine (spongoadenosine). J Org Chem 28:3004–3006
Gohil RN, Gillen RG, Nagyvary J (1974) Synthesis and properties of some cyclic AMP alkyl phosphotriesters. Nucleic Acids Res 1:1691–1701
Goldberg ML, Burke GC, Morris HP (1975a) Cyclic AMP and cyclic GMP content and binding in malignancy. Biochem Biophys Res Commun 62:320–327
Goldberg ND, Dietz SB, O'Toole AG (1969) Cyclic guanosine 3′,5′-monophosphate in mammalian tissues and urine. J Biol Chem 244:4458–4466
Goldberg ND, Haddox MK (1977) Cyclic GMP metabolism and involvement in biological regulation. Annu Rev Biochem 46:823–896

Goldberg ND, Haddox MK, Nicol SE, Glass DB, Sanford CH, Kuehl FA Jr, Estensen R (1975) Biologic regulation through opposing influences of cyclic GMP and cyclic AMP: The yin yang hypothesis. Adv Cyclic Nucleotide Res 5:307–330

Goldberg ND, O'Dea RF, Haddox MK (1973) Cyclic GMP. Adv Cyclic Nucleotide Res 3:155–223

Gorenstein DG (1975) Dependence of ^{31}P chemical shifts on oxygen-phosphorus-oxygen bond angles in phosphate esters. J Am Chem Soc 97:898–900

Greengard P (1971) On the reactivity and mechanism of action of cyclic nucleotides. Ann NY Acad Sci 185:18–26

Greengard P, Paoletti R, Robison GA (1972) Adv Cyclic Nucleotide Res 1, 2

Greengard P, Rudolph SA, Sturtevant JM (1969) Enthalpy of hydrolysis of the 3′ bond of adenosine 3′,5′-monophosphate and guanosine 3′,5′-monophosphate. J Biol Chem 244:4798–4800

Gresser I (1977) On the varied biologic effect of interferon. Cell Immunol 34:406–415

Gresser I, DeMaeyer-Guignard J, Tovey MG, DeMaeyer E (1979) Electrophoretically pure mouse interferon exerts multiple biologic effects. Proc Natl Acad Sci USA 76:5308–5312

Grollman EF, Lee G, Ramos S, Lazo PS, Kaback HR, Friedman RM, Kohn LD (1978) Relationships of the structure and function of the interferon receptor to hormone receptors and establishment of the antiviral state. Cancer Res 38:4172–4185

Guilford H, Larsson P-O, Mosbach K (1972) On adenine nucleotides for affinity chromatography. Chimica Scripta 2:165–170

Gulyaev NN, Baranova LA, Severin ES (1976) Adenosine-3′,5′-cyclosulfate for selective regulation of the level of adenosine-3′,5′-cyclophosphate in cell. U.S.S.R. Patent 502,887. Chem Abstr 84:136012x

Gulyaev NN, Tunitskaya VL, Nesterova MV, Mazurova LA, Murtuzaev IM, Severin ES (1977) Interaction of 8-substituted derivatives and adenosine-3′,5′-cyclophosphate esters with protein kinase from pig brain. Biokhimiya (Moscow) 42:2071–2078

Haake PC, Westheimer FH (1961) Hydrolysis and exchange in esters of phosphoric acid. J Am Chem Soc 83:1102–1113

Haga K, Kainosho M, Yoshikawa M (1971) Studies of phosphorylation. V. The synthesis of inosine-5′-thiophosphates. Bull Chem Soc (Japan) 44:460–463

Hall DA, Barnes LD, Dousa TP (1977) Cyclic AMP in action of antidiuretic hormone: effects of exogenous cyclic AMP and its new analogue. Am J Physiol 232F:368–376

Halprin KM (1976) Cyclic nucleotides and epidermal cell proliferation. J Invest Dermatol 66:339–343

Hampton A, Brox LW, Bayer M (1969) Analogs of inosine 5′-phosphate with phosphorus-nitrogen and phosphorus-sulfur bonds. Binding and kinetic studies with inosine 5′-phosphate dehydrogenase. Biochemistry 8:2303–2311

Hanze AR (1967) Deazapurine riboside cyclic 3′,5′-phosphates and process therefor. US Patent 3,300,479

Hanze AR (1968) Nucleic Acids. V. Nucleotide derivatives of tubercidin (7-deazaadenosine). Biochemistry 7:932–939

Haque M, Caughlan CN, Moats WL (1970) The crystal and molecular structure of propane-1,3-diol cyclic phosphate ($C_3H_7PO_4$). J Org Chem 35:1446–1448

Hardman JG, Davis JW, Sutherland EW (1966) Measurement of guanosine 3′,5′-monophosphate and other cyclic nucleotides. J Biol Chem 241:4812–4815

Hardman JG, Davis JW, Sutherland EW (1969) Effects of some hormonal and other factors on the excretion of guanosine 3′,5′-monophosphate and adenosine 3′,5′-monophosphate in rat urine. J Biol Chem 244:6354–6362

Hardman JG, Robison GA, Sutherland EW (1971) Cyclic nucleotides. Annu Rev Physiol 33:311–336

Hardman JG, Sutherland EW (1965) A cyclic 3′,5′-nucleotide phosphodiesterase from heart with specificity for uridine 3′,5′-phosphate. J Biol Chem 240:3704–3705

Hardman JG, Sutherland EW (1969) Guanyl cyclase, an enzyme catalyzing the formation of guanosine 3′,5′-monophosphate from guanosine triphosphate. J Biol Chem 244:6363–6370

Hashimoto J, Uchida T, Egami F (1970) Action of ribonucleases T_1, T_2, and U_2 on dinucleoside monophosphates containing 7-deazapurine base. Biochim Biophys Acta 199:535–536

Hayaishi O, Greengard P, Colowick SP (1971) On the equilibrium of the adenylate cyclase reaction. J Biol Chem 246:5840–5843

Hemmes P, Oppenheimer L, Jordan F (1976) Ultrasonic relaxation evidence for a two-state glycosyl conformational equilibrium in aqueous solution of adenosine 3′,5′-cyclic monophosphate. J Chem Soc Chem Commun 929–930

Hevesi L, Davidson EW, Nagy JB, Nagy OB, Bruglants A (1972) Contribution to the mechanism of the acid-catalyzed hydrolysis of purine nucleosides. J Am Chem Soc 94:4715–4720

Hirata M, Hayaishi O (1966) Enzymatic formation of deoxyadenosine 3′,5′-phosphate. Biochem Biophys Res Commun 24:360–364

Hoffman DJ, Whistler RL (1968) Diabetogenic action of 5-thio-D-glucopyranose in rats. Biochemistry 7:4479–4482

Hoffman DJ, Whistler RL (1970) Synthesis and properties of nucleotides containing 4-thio-D-ribofuranose. Biochemistry 9:2367–2372

Hofmann F, Flockerzi V, Schwechheimer K, Speichermann N (1979) Krause EG, Pinna L, Wollenberger A (eds) In: Studies on the activation mechanism of cyclic nucleotide-dependent protein kinases. FEBS 12th Meeting, Dresden, 1978, vol 54, Symposium 7: Cyclic Nucleotides and Protein Phosphorylation in Cell Regulation. Pergamon Press, London, New York, p 115

Höfle G, Steglich W (1972) 4-Dialkylaminopyridines as acylation catalysts; III. Acylation of sterically hindered alcohols. Synthesis 619–621

Holmes RE, Robins RK (1964) Purine nucleosides. VII. Direct bromination of adenosine, deoxyadenosine, guanosine, and related purine nucleosides. J Am Chem Soc 86:1242–1245

Holý A (1969a) Oligonucleotidic compounds. XXX. Synthesis of some 5′-*O*-substituted derivatives of guanosine 2′,3′-cyclic phosphate and guanylyl-(3′→5′)-uridine. Coll Czech Chem Commun 34:1261–1277

Holý A (1969b) Nucleic acid components and their analogues. CXXVIII. Phosphorylation of some thymine 1-pentopyranosyl and 1-hexopyranosyl derivatives by the reaction with triethyl phosphite and hexachloroacetone. Coll Czech Chem Commun 34:3510–3522

Holý A (1970a) Oligonucleotidic compounds. XXXVIII. Synthesis of diribonucleoside phosphates and triribonucleoside diphosphates derived from adenosine 3′-phosphate and guanosine 3′-phosphate: A general approach to the synthesis of oligonucleotides. Coll Czech Chem Commun 35:3686–3711

Holý A (1970b) Nucleic acid components and their analogues. CXXX. Preparation of nucleotide derivatives of 1′-homouridine and their behaviour towards some nucleolytic enzymes. Coll Czech Chem Commun 35:81–88

Holý A, Bald RW (1971) Nucleic acid components and their analogues. CXXXVIII. Synthesis of 2′,3′-cyclic phosphates derived from some pyrimidine ribonucleosides and their behaviour towards pancreatic ribonuclease and ribonuclease T_2. Coll Czech Chem Commun 36:2809–2823

Holý A, Bald RW, Hong NgD (1971) Nucleic acid components and their analogues. CXXXVII. Preparation and properties of some N-(2-hydroxyethyl) derivatives of ribonucleosides and nucleotides. Coll Czech Chem Commun 36:2658–2676

Holý A, Chládek S, Žemlicka J (1969) Oligonucleotidic compounds. XXIX. Reactions of ribonucleoside 2′(3′)-phosphates with dimethylformamide acetals. Coll Czech Chem Commun 34:253–271

Holý A, Smrt J (1966) Oligonucleotide compounds. XI. Synthesis of ribonucleoside-2′,3′ cyclophosphates from nucleosides *via* nucleoside 2′,3′-phosphites. Coll Czech Chem Commun 31:1528–1534

Holý A, Smrt J, Šorm F (1965) Nucleic acids components and their analogues. LXXI. Oxidation of nucleoside 5′-phosphates on treatment with halo acid derivatives and hexachloroacetone. Coll Czech Chem Commun 30:3309–3319

Holý A, Šorm F (1966) Nucleic acids components and their analogues. LXXXI. A selective synthesis of ribonucleoside-2′(3′) phosphites. Coll Czech Chem Commun 31:1562–1568

Holý A, Šorm F (1969) Oligonucleotidic compounds. XXXII. Phosphorylation of 1-lyxofuranosyl, 1-xylofuranosyl and 1-arabinofuranosyl derivatives of uracil and thymine with triethyl phosphite and hexachloroacetone. Coll Czech Chem Commun 34:1929–1953

Holý H, Ivanova GS (1974) Aliphatic analogues of nucleotides: synthesis and affinity towards nucleases. Nucleic Acids Res 1:19–34

Hong CI, Tritsch GL, Mittelman A, Hebborn P, Chheda GP (1975) Synthesis and antitumor activity of 5′-phosphates and cyclic 3′,5′-phosphates derived from biologically active nucleosides. J Med Chem 18:465–473

Honjo M, Furukawa Y, Yoshioka Y, Imada A, Fujii S, Ootsu K, Kimura T, Komeda T, Matsumoto T (1968) Synthesis of 6-mercaptopurine 2′-deoxyribonucleoside and its related compounds and their biological activities. Ann Rep Takeda Res Lab 27:1–19

Honjo M, Marumoto R, Yoshioka Y, Takatsuki S (1973) 2-Substituted adenosine 3′,5′-cyclomonophosphates. Ger Offen 2,324,130

Hoppe J, Wagner KG (1974) Synthesis and properties of N^6,C-8 and C-2 spin-labelled derivatives of adenosine cyclic 3′,5′-monophosphate. Eur J Biochem 48:519–525

Hsie AW, Puck TT (1971a) Morphological transformation of Chinese hamster cells by dibutyryl adenosine cyclic 3′,5′-monophosphate and testosterone. Proc Natl Acad Sci USA 68:358–361

Hsie AW, Janes C, Puck TT (1971b) Further changes in differentiation state accompanying the conversion of Chinese hamster cells to fibroblastic form by dibutyryl adenosine cyclic 3′:5′-monophosphate and hormones. Proc Natl Acad Sci USA 68:1648–1652

Hubert-Habart M, Goodman L (1969) The direct formation of a 3′,5′-cyclic mononucleotide from an adenine nucleoside. J Chem Soc Chem Commun 740–741

Hutchinson DW (1971) Phosphates and phosphonates of biochemical interest. In: Trippett (ed) Organophosphorus chemistry, vol 2. John Wright and Sons Ltd, Bristol, pp 119–155

Hwang TC, Clark VL, Bernlohr RW (1974) Guanosine 3′,5′-cyclic monophosphate phosphodiesterase activity of *Bacillus Licheniformis*. Biochem Biophys Res Commun 58:707–713

Ikehara M, Harada F, Ohtsuka E (1966) Polynucleotides. III. Synthesis of four trinucleoside diphosphates containing tubercidin (7-deazaadenosine) and N^6-dimethyladenosine. Chem Pharm Bull (Tokyo) 14:1338–1346

Ikehara M, Muneyama K (1967) Studies of nucleosides and nucleotides. XXXIV. Purine cyclonucleosides. 4. Synthesis of a cyclonucleoside having an *O* cyclo linhage derived from guanosine. J Org Chem 32:3039–3042

Ikehara M, Ohtsuka E (1963a) Studies on coenzymc analogs. XVI. Synthesis of 9-D-erythritvladenine and its phosphates. Chem Pharm Bull (Tokyo) 11:1095–1101

Ikehara M, Ohtsuka E (1963b) Studies on coenzyme analogs. XIX. Further investigations of phosphorylation using morpholinophosphorodicholoridate and P^1-diphenyl P^2-morpholino pyrophosphorochloridate. Chem Pharm Bull (Tokyo) 11:1353–1358

Ikehara M, Ohtsuka E (1963c) Studies of nucleosides and nucleotides. XXI. A new synthesis of thymidine 5′-triphosphate and the use of P^1,P^2-di-(2-cyanoethyl)-pyrophosphate in the nucleoside triphosphate synthesis. Chem Pharm Bull (Tokyo) 11:1358–1363

Ikehara M, Tazawa I (1966) Studies of nucleosides and nucleotides. XXIX. Direct synthesis of nucleoside-2′,3′-cyclic phosphates. J Org Chem 31:819–821

Ikehara M, Uesugi S (1969) Studies of nucleosides and nucleotides. XXXVIII. Synthesis of 8-bromoadenosine nucleotides. Chem Pharm Bull (Tokyo) 17:348–354

Ikehara M, Uesugi S (1972) Studies of nucleosides and nucleotides. LIII. Purine cyclonucleosides -18. Selective tosylation of adenine nucleotides. Synthesis of 8,2′-anhydro-8-mercapto-9-β-D-arabinofuranosyladenine 5′- and 3′,5′-cyclic phosphate. Tetrahedron 28:3687–3694

Ikehara M, Uesugi S, Kaneko M (1967) Bromination of adenine nucleoside and nucleotide. J Chem Soc Chem Commun 17–18

Ikehara M, Uesugi S, Kaneko M (1978) 8-Bromoadenosine and its 5′-phosphate. Nucleic acid chem., improved and new synthetic procedures, methods, and techniques. Townsend, Tipson (eds) John Wiley and Sons, New York, pt 2 p 837–841

Imai K, Fujii S, Takanohashi K, Furukawa Y, Masuda T, Honjo M (1969) Studies on phosphorylation. IV. Selective phosphorylation of the primary hydroxyl group in nucleosides. J Org Chem 34:1547–1550

Imai K, Marumoto R, Kobayashi K, Yoshioka Y, Toda J, Honjo M (1971) Synthesis of compounds related to inosine 5′-phosphate and their flavor enhancing activity. IV. 2-Substituted inosine 5′-phosphates. Chem Pharm Bull (Tokyo) 19:576–586

Imai S, Otorii T, Takeda K, Katano Y, Horii D (1974) Effects of cyclic AMP and dibutyryl cyclic AMP on the heart and coronary circulation. Jpn J Pharmacol 24:499–510

Inagaki F, Takahashi S, Tasumi M, Miyazawa T (1976) Conformation of thymidine 3′,5′-cyclic phosphate in aqueous solution. Compatibility of the information from the lanthanide-ion probe and the spin-coupling constants. Bull Chem Soc (Jpn) 49:611–616

Ingemarsson S, Cantell K, Strander H (1979) J Infect Dis 140:560

Ishikawa E, Ishikawa S, Davis JW, Sutherland EW (1969) Determination of guanosine 3′,5′-monophosphate in tissues and of guanyl cyclase in rat intestine. J Biol Chem 244:6371–6376

Ishiyama J (1975) Isolation of cyclic 3′,5′-pyrimidine mononucleotides from bacterial culture fluids. Biochem Biophys Res Commun 65:286–292

Ishiyama J (1976) Isolation of cyclic deoxyadenosine 3′,5′-monophosphate from the culture fluid of *Corynebacterium murisepticum* No 7. J Biol Chem 251:438–440

Ishiyama J, Yokotsuka T, Saito N (1974) Cyclic-AMP production by *Corynebacterium murisepticum* No 7 (ATCC 21374) and *Microbacterium sp.* No 205 (ATCC 21376). Agr Biol Chem 38:507–514

Jardetzky CD (1962) Proton magnetic resonance of nucleotides. IV. Ribose conformation. J Am Chem Soc 84:62–66

Jastorff B, Bär H-P (1973) Effects of 5′-amido analogues of adenosine 3′,5′-monophosphate and adenosine 3′,5′-monophosphothioate on protein kinase, binding protein and phosphodiesterases. Eur J Biochem 37:497–504

Jastorff B, Freist W (1972) New analogs of adenosine-3′,5′-cyclophosphate. Angew Chem (Engl) 11:713

Jastorff B, Freist W (1974) Synthesis and biological activities of cyclic AMP analogs modified in the 1,2 and 2′-positions. Bioorg Chem 3:103–113

Jastorff B, Hoppe J, Morr M (1979) A model for the chemical interactions of adenosine 3′,5′-monophosphate with the R subunit of protein kinase type I. Refinement of the cyclic phosphate binding moiety of protein kinase type I. Eur J Biochem 101:555–561

Jastorff B, Krebs T (1972) Analoge des Adenosin-(3′.5′)-cyclophosphats mit Stickstoff und Schwefelatomen im Phosphatring. Chem Ber 105:3192–3202

Jastorff B, Krebs T (1974) Amidocyclothiophosphates. Ger Offen 2,247,782

Jergil B, Guilford H, Mosbach K (1974) Biospecific affinity chromatography of an adenosine 3′,5′-cyclic monophosphate-stimulated protein kinase (Protamine kinase from trout testis) by using immobilized adenine nucleotides. Biochem J 139:441–448

Jergil B, Mosbach K (1974) Cyclic AMP: Purification of protamine kinase. Methods Enzymol 34:261–264

Johnson GS, Friedman RM, Pastan I (1971) Restoration of several morphological characteristics of normal fibroblasts in sarcoma cells treated with adenosine-3′,5′-cyclic monophosphate and its derivatives. Proc Natl Acad Sci USA 68:425–429

Johnson LD, Hadden JW (1977) Modification of human DNA-dependent RNA polymerase activity by cyclic GMP. Nucleic Acids Res 4:4007–4014

Johnson RA, Walseth TF (1979) The enzymatic preparation of [α-^{32}P]ATP, [α-^{32}P]GTP, [^{32}P]cAMP, and [^{32}P]cGMP, and their use in the assay of adenylate and guanylate cyclases and cyclic nucleotide phosphodiesterases. Adv Cyclic Nucleotide Res 10:135–167

Jones GH, Albrecht HP, Damodaran NP, Moffatt JG (1970) Synthesis of isosteric phosphonate analogs of some biologically important phosphodiesterases. J Am Chem Soc 92:5510–5511

Jones GH, Moffatt JG (1968) The synthesis of 6′-deoxy-homonucleoside-6′-phosphonic acids. J Am Chem 90:5337–5338
Jones GH, Moffatt JG (1969) 3′-Cyclic esters of 5′-deoxy-5′-(dihydroxyphosphinylmethyl)-nucleosides. US Patent 3,446,793
Jones GH, Moffatt JG (1974) Cyclic phosphoric acid ester of 1,N^6-ethenoadenosines. Ger Offen 2,350,608
Jones GH, Moffatt JG (1975) 1,N^6-Ethenoadenosine cyclophosphate compounds. US Patent 3,872,098
Jones GH, Murthy DVK, Tegg D, Golling R, Moffatt JG (1973) Analogs of adenosine 3′,5′-cyclic phosphate. II. Synthesis and enzymatic activity of derivatives of 1,N^6-ethenoadenosine 3′,5′-cyclic phosphate. Biochem Biophys Res Commun 53:1338–1343
Jones GH, Ranganathan RS, Moffatt JG (1975) Purine nucleoside 3′,5′-cyclic phosphates. US Patent 3,872,084
Jordan F (1973) The electronic structure of and conformational energy barrier to rotation around the C–N glycosidic linkage in adenosine-3′,5′-cyclicmonophosphate (cyclic AMP) and its phosphonate analog. J Theor Biol 41:23–40
Jurovcik M, Holý A, Šorm F (1971) The utilization of L-adenosine by mammalian tissues. FEBS Lett 18:274–276
Kainosho M, Ajisaka K (1975) Conformational study of cyclic nucleotides. Lanthanide ion assisted analysis of the hydrogen-1 nuclear magnetic resonance spectra. J Am Chem Soc 97:6839–6843
Kaiser ET, Lee TWS, Boer FP (1971) Structure and enzymatic reactivity of an aromatic five-membered cyclic phosphate diester. Biological implications. J Am Chem Soc 93:2351–2353
Kanai T, Ichino M, Hoshi A, Kuretani K (1971) Some phosphate esters of cyclocytidine and aracytidine. Tetrahedron Lett 1965–1968
Karpeiskii MYa, Yakovleve GI (1975) Nuclear Overhauser effect study on base and sugar relative arrangement in aqueous solutions of cytidine nucleotides. Bioorg Khim 1:749–757
Kavunenko AP, Holý A (1978) Acceptor activity of uracil nucleosides in the dinucleoside monophosphate synthesis catalyzed by pancreatic ribonuclease. Nucleic Acids Res Spe Publ 4:s143–146
Kawana M, Ivanovics GA, Rousseau RJ, Robins RK (1972) Azapurine nucleosides. 3. Synthesis of 7-(β-D-ribofuranosyl)imidazo[4,5-*d*]-v-triazin-4-one (2-azainosine) and related derivatives. J Med Chem 15:841–843
Kawanabe T, Machida M (1976) Cyclic phosphoric acid esters of 1,N^6-ethenoadenosines. Ger Offen 2,365,720
Kazimierczuk Z, Shugar D (1973) Preparative photochemical synthesis of isoguanosine ribo- and deoxyribonucleosides and nucleotides, and isoguanosine-3′,5′-cyclic phosphate, a new cAMP analogue. Acta Biochim Pol 20:395–402
Khorana HG (1961 a) Cyclic phosphate formation and its role in the chemistry of phosphate esters of biological interest. In: Some recent developments in the chemistry of phosphate esters of biological interest. John Wiley and Sons, Inc. New York, pp 44–68
Khorana HG (1961 b) Phosphodiesterases. In: Boyer PD, Lardy H, Myrbäck K (eds) Enzymes. Academic Press, New York, pp 79–94
Khorana HG, Vizsolyi JP (1961) Studies on polynucleotides. VIII. Experiments on the polymerization of mononucleotides. Improved preparation and separation of linear thymidine polynucleotides. Synthesis of corresponding members terminated in deoxycytidine residues. J Am Chem Soc 83:675–685
Khorana HG, Turner AF, Vizsolyi JP (1961) Studies on polynucleotides. IX. Experiments on the polymerization of mononucleotides. Certain protected derivatives of deoxycytidine-5′ phosphates and the synthesis of deoxycytidine polynucleotides. J Am Chem Soc 83:686–698
Khorana HG, Tener GM, Wright RS, Moffatt JG (1957) Cyclic phosphates. III. Some general observations on the formation and properties of five-, six- and seven-membered cyclic phosphate esters. J Am Chem Soc 79:430–436

Khwaja TA, Boswell KH, Robins RK, Miller JP (1975) 8-Substituted derivatives of adenosine 3′,5′-cyclic phosphate require an unsubstituted 2′-hydroxyl group in the ribo configuration for biological activity. Biochemistry 14:4238–4244

Khwaja TA, Harris R, Robins RK (1972) Synthesis of 9-β-D-arabinofuranosyladenine-3′,5′-cyclic phosphate from adenosine-3′,5′-cyclic phosphate. Tetrahedron Lett 4681–4684

Khwaja TA, Kigwana L, Robins RK (1974) Nucleoside 3′,5′-cyclic phosphates as antitumor agents. Proc Am Assoc Cancer Res 15:94. Abstr 373

Khwaja TA, Robins RK (1979) Personal communication

Kikkoman Shoyu Company (1971a) 3′,5′-Cyclic adenylic acid production. US Patent 3,630,842

Kikkoman Shoyu Company (1971b) 3′,5′-Cyclic deoxyadenylic acid. Japanese Patent 71 42,958

Kimura H, Murad F (1974) Nonenzymatic formation of guanosine 3′,5′-monophosphate from guanosine triphosphate. J Biol Chem 249:329–331

Kishi T, Muroi M, Kusaka T, Nishikawa M, Kamiya K, Mizuno K (1967) The structure of aristeromycin. J Chem Soc Chem Commun 852–853

Klee WA, Mudd SH (1967) The conformation of ribonucleosides in solution. The effect of structure on the orientation of the base. Biochemistry 6:988–998

Klein RS, Wempen I, Watanabe KA, Fox JJ (1970) Nucleosides. LXVII. The chemistry of 4-methyl-2-pyrimidinone ribonucleosides. J Org Chem 35:2330–2334

Kneer NM, Bosch AL, Clark MG, Lardy HA (1974) Glucose inhibition of epinephrine stimulation of hepatic gluconeogenesis by blockade of the α-receptor function. Proc Natl Acad Sci USA 71:4523–4527

Kobata A, Kida M, Ziro S (1961) Occurrence of 3′,5′-cyclic AMP in milk. J Biochem 50:275–276

Koblet H, Wyler R, Kohler U (1978) Kinetics of interferon action. Experientia 34:1164–1165

Koblet H, Wyler R, Kohler U (1979) Increase of interferon antiviral activity by exogenous cyclic adenosine-3′,5′-monophosphate (cAMP). Experientia 35:575–576

Kochetkov NK, Budovskii EI, Sverdlov ED, Simukova NA, Turchinskii MF, Shibaev VN (1972) Cleavage of phosphoester bonds and some other reactions of phosphate groups of nucleic acids and their components. In: Kochetkov, Budovskii (eds) Organic chemistry of nucleic acids, Part B, Plenum Press, London

Koontz JW, Wicks WD (1977) Inhibition of human granulocyte function by methotrexate. Cancer Res 37:651–655

Kreis W, Wechter WJ (1972) Studies on drug resistance. IV. Synthesis and biological activities of 1-β-D-arabinofuranosylcytosine 3′,5′-cyclic monophosphate and a derivative. Res Commun Chem Pathol Pharmacol 4:631–640

Kreis W, Wechter WJ (1972) Biological activity of 3′,5′-cyclic phosphates of 1-β-D-arabinofuranosylcytosine (arc-C). Proc Am Assoc Cancer Res 13:62

Kukovetz WR, Poch G (1970) Cardiostimulatory effects of cyclic 3′,5′-adenosine monophosphate and its acylated derivatives. Naunyn-Schmiedebergs Arch Pharmakol 266:236–254

Kumamoto J, Cox JR Jr, Westheimer FH (1956) Barium ethylene phosphate. J Am Chem Soc 78:4858–4860

Kung W, Marsh RE, Kainosho M (1977) Crystal and solution structure of 2′,5′-arabinosylcytidine monophosphate. Influence of POC bond angles on the protonphosphorus vicinal coupling constants in the P–O–C–H fragments. J Am Chem Soc 99:5471–5477

Kuo JF, Brackett NL, Shoji M, Tse J (1978) Cytidine 3′:5′-monophosphate phosphodiesterase in mammalian tissues. J Biol Chem 253:2518–2521

Kuo JF, Greengard P (1969) Cyclic nucleotide-dependent protein kinases. IV. Widespread occurrence of adenosine 3′,5′-monophosphate-dependent protein kinase in various tissues and phyla of the animal kingdom. Proc Natl Acad Sci USA 64:1349–1355

Kuo JF, Krueger BK, Sanes JR, Greengard P (1970) Cyclic nucleotide-dependent protein kinases. V. Preparation and properties of adenosine 3′,5′-monophosphate-dependent protein kinase from various bovine tissues. Biochim Biophys Acta 212:79–91

Kura G, Ohashi S (1971) Chromatographic separation of cyclic phosphates by means of an anion-exchange dextran gel. J Chromatogr 56:111–120
Lapinet E, Cehovic G, Posternak T (1971) New cosmetic compositions intended for the physiological care of human skin, hair and nails. French Patent 7,004,557
Lapper RD, Mantsch HH, Smith ICP (1972) A carbon-13 and hydrogen-1 nuclear magnetic resonance study of the conformations of 3′,5′- and 2′,3′-cyclic nucleotides. A demonstration of the angular dependence of three-bond spin-spin couplings between carbon and phosphorus. J Am Chem Soc 94:6243–6244
Lapper RD, Mantsch HH, Smith ICP (1973) A carbon-13 nuclear magnetic resonance study of the conformations of 3′,5′-cyclic nucleotides. J Am Chem Soc 95:2878–2880
Lapper RD, Smith ICP (1973) A ^{13}C and ^{1}H nuclear magnetic resonance study of the conformations of 2′,3′-cyclic nucleotides. J Am Chem Soc 95:2880–2884
Lavallee DK, Coulter CL (1973) Structural chemistry of cyclic nucleotides. III. Proton magnetic resonance studies of β-pyrimidine nucleotides. J Am Chem Soc 95:576–581
Lavallee DK, Myers RB (1978) Role of the pyrimidine base in ribonuclease A hydrolysis of RNA. Determination of the conformation of cyclic β-cytidine 2′,3′-phosphate and cyclic β-uridine 2′,3′-phosphate in solution. J Am Chem Soc 100:3907–3912
Lavallee DK, Zeltmann AH (1974) Conformation of cyclic β-adenosine 3′,5′-phosphate in solution using the lanthanide shift technique. J Am Chem Soc 96:5552–5556
Lawson AM, Stillwell RN, Tacker MM, Tsuboyama K, McCloskey JA (1971) Mass spectrometry of nucleic acid components. Trimethylsilyl derivatives of nucleotides. J Am Chem Soc 93:1014–1023
Lee WW, Fisher LV, Goodman L (1971) 9-(β-D-Arabinofuranosyl)adenine 3′,5′-cyclic phosphate. J Heterocycl Chem 8:179–180
Lee CC, Schrier WH, Nagyvary J (1979) The enzymatic hydrolysis of the phosphate ester bond in some thionucleotides. Biochim Biophys Acta 561:223–231
Lee C-H, Sarma RH (1976) Aqueous solution conformation of rigid nucleosides and nucleotides. J Am Chem Soc 98:3541–3548
Lehninger AL (1975) Biochemistry. 2nd edn. Worth Publishers, New York, p 729
Leonard NJ, Laursen RA (1965) Synthesis of 3-β-D-ribofuranosyladenine and (3-β-D-ribofuranosyladenine)-5′-phosphate. Biochemistry 4:354–365
Leonov D, Elad D (1974) Ultraviolet- and γ-ray-induced reactions of nucleic acid constituents. Reactions of purines with ethers and dioxolane. J Org Chem 39:1470–1473
LePage GA, Hersh EM (1972) Cyclic nucleotide analogs as carcinostatic agents. Biochem Biophys Res Commun 46:1918–1922
Lespinasse JN, Vasilescu D (1974) Conformational analysis of a cyclic nucleotide: 3′,5′-Adenosine monophosphate. A study based on the extended Hückel theory. Biopolymers 13:63–75
Lincoln TM, Flockhart DA, Corbin JD (1978) Studies on the structure and mechanism of activation of the guanosine 3′,5′-monophosphate-dependent protein kinase. J Biol Chem 253:6002–6009
Lipkin D, Cook WH, Markham R (1959a) Adenosine-3′:5′-phosphoric acid: A proof of structure. J Am Chem Soc 81:6198–6203
Lipkin D, Markham R, Cook WH (1959b) The degradation of adenosine-5′-triphosphoric acid (ATP) by means of aqueous barium hydroxide. J Am Chem Soc 81:6075–6080
Lipkin D, Talbert PT (1955) The base-catalysed conversion of yeast ribonucleic acid to cyclic phosphate esters. Chem Industry (Lond) 143
Lipkind GM (1976) Theoretical conformational analysis of cytidine 2′,3′-cyclophosphate. Mol Biol (Mosk) 10:490–493
Litchfield GJ, Shaw G (1971) Purines, pyrimidines, and imidazoles. Part XXXV. Potentiometric and spectroscopic studies of some imidazoles related to intermediates in the biosynthesis *de novo* of purine nucleotides. J Chem Soc C:817–820
Lo KW, Yip KF, Tsou KC (1975) Fluorometric assay of 2′,3′-cyclic adenosine monophosphate 3′-phosphohydrolase with 1,N^6-etheno-2-aza-adenosine 2′,3′-monophosphate. J Neurochem 25:181–183
Lohrmann R, Khorana HG (1964) Studies on polynucleotides. XXXIV. The specific synthesis of $C^{3'}$–$C^{5'}$-linked ribooligonucleotides. New protected derivatives of ribonucleo-

sides and ribonucleoside 3′-phosphates. Further synthesis of diribonucleoside phosphates. J Am Chem Soc 86:4188–4194

Lomax NR, Narayanan VL (1979) Chemical structures of interest to the Division of Cancer Treatment. National Cancer Institute, Drug Synthesis and Chemistry Branch, January, p 4

Long RA, Robins RK (1978) 5-Bromouridine 3′,5′-cyclic monophosphate. Nucleic Acid Chemistry, Improved and New Synthetic Procedures, Methods and Techniques. Townsend, Tipson (eds) pt 2, John Wiley and Sons, New York, pp 817–819

Long RA, Szekeres GL, Khwaja TA, Sidwell RW, Simon LN, Robins RK (1972) Synthesis and antitumor and antiviral activities of 1-β-D-arabinofuranosylpyrimidine 3′,5′-cyclic phosphates. J Med Chem 15:1215–1218

MacCoss M, Ezra FS, Robins MJ, Danyluk SS (1977) Conformational characteristics of rigid cyclic nucleotides. 2. The solution conformation of α-nucleoside 3′,5′-cyclic monophosphates and the role of the 2′-hydroxyl group. J Am Chem Soc 99:7495–7502

MacCoss M, Ezra FS, Robins MJ, Danyluk SS (1978) Proton magnetic resonance studies of 9-(β-D-xylofuranosyl)adenine 3′,5′-cyclic monophosphate and 9-(β-D-arabinofuranosyl)adenine 2′,5′-cyclic monophosphate. Carbohydr Res 62:203–212

Maguire MH, Caredes FP, Gough GR (1974) Effects of the 3′,5′-cyclic phosphates of 2-methylthioadenosine, 2-chloroadenosine and adenosine on platelet aggregation. Experientia 30:922–924

Makabe O, Miyadera A, Kinoshita M, Umezawa S (1978) Cyclic phosphates of formycin. J Antibiot 31:456–467

Markham R (1957) The preparation and assay of cyclic nucleotides. Methods Enzymol 3:805–810

Markham R, Smith JD (1951) Structure of ribonucleic acid. Nature 168:406–408

Markham R, Smith JD (1952) The structure of ribonucleic acids. 1. Cyclic nucleotides produced by ribonuclease and by alkaline hydrolysis. Biochem J 52:552–557

Marsh FJ, Weiner P, Douglas JE, Kollman PA, Kenyon GL, Gerlt JA (1980) Theoretical calculations on the geometric destabilization of 3′,5′- and 2′,3′-cyclic nucleotides. J Am Chem Soc 102:1660–1665

Marumoto R, Honjo M (1975) Aristeromycin-3′,6′-cyclic phosphoric acid. Japan Kokai 75 40,590

Marumoto R, Nishimura T, Honjo M (1975) A new method for synthesis of nucleoside 3′,5′-cyclic phosphates. Cyclization of nucleoside 5′-trichloromethyl phosphonates. Chem Pharm Bull (Tokyo) 23:2295–2300

Marumoto R, Yoshioka Y, Naka T, Shima S, Miyashita O, Maki Y, Suzuki T, Honjo M (1979) Synthesis and enzymatic activity of adenosine 3′,5′-cyclic phosphate analogs. Chem Pharm Bull (Tokyo) 27:990–1003

Marx JL (1979) Interferon (II): Learning about how it works. Recent research is clarifying how interferon inhibits viral reproduction but its other actions are less well understood. Science 204:1293–1295

Masson A, Levy B, Malrieu JP (1970) Formaldéhyde: calcul de l'energie dans l'état fondamental par une méthode de perturbation. Theoret Chim Acta 18:193–207

Mato JM, Jastorff B, Morr M, Konijn TM (1978) A model for cyclic AMP-chemoreceptor interaction in *Dictyostelium discoideum*. Biochim Biophys Acta 544:309–314

Meldolesi MF, Friedman RM, Kohn LD (1977) An interferon-induced increase in cyclic AMP levels precedes the establishment of the antiviral state. Biochem Biophys Res Commun 79:239–246

Meyer RB Jr (1979) Analogs of cyclic nucleotides. In: Wolff (ed) *Burger's medicinal chemistry*, 4th edn, part II, John Wiley and Sons Inc, New York, pp 1201–1224

Meyer RB Jr, Miller JP (1974) Analogs of cyclic AMP and cyclic GMP: General methods of synthesis and the relationship of structure to enzymic activity. Life Sci 14:1019–1040

Meyer RB Jr, Shuman DA (1975) 2-Substituted cyclic AMP derivatives. US Patent 3,917,583

Meyer RB Jr, Shuman DA, Robins RK (1973a) Synthesis of purine nucleoside 3′,5′-cyclic phosphoramidates. Tetrahedron Lett 269–272

Meyer RB Jr, Shuman DA, Robins RK (1974a) A new purine ring closure and the synthesis of 2-substituted derivatives of adenosine cyclic 3′,5′-phosphate. J Am Chem Soc 96:4962–4966
Meyer RB Jr, Shuman DA, Robins RK (1974b) Phosphoramidates of 3′,5′-cyclic purine nucleotides. US Patent 3,852,267
Meyer RB Jr, Shuman DA, Robins RK, Bauer RJ, Dimmitt MK, Simon LN (1972) Synthesis and biological activity of several 6-substituted 9-β-D-ribofuranosylpurine 3′,5′-cyclic phosphates. Biochemistry 11:2704–2709
Meyer RB Jr, Shuman DA, Robins RK, Miller JP, Simon LN (1973b) Synthesis and enzymic studies of 5-aminoimidazole and N^1- and N^6-substituted adenine ribonucleoside cyclic 3′,5′-phosphates prepared from adenosine cyclic 3′,5′-phosphate. J Med Chem 16:1319
Meyer RB Jr, Stone TE, Heinzel FP (1978) Direct sulfhydrolysis of cyclic AMP: A one-step synthesis of the cyclic ribonucleotide of 6-mercaptopurine. J Heterocycl Chem 15:1511–1512
Meyer RB Jr, Stone TE, Ullman B (1979) 2′-*O*-Acyl-6-thioinosine cyclic 3′,5′-phosphates as prodrugs of thioinosinic acid. J Med Chem 22:811–815
Meyer RB Jr, Uno H, Robins RK, Simon LN, Miller JP (1975a) 2-Substituted derivatives of adenosine and inosine cyclic 3′,5′-phosphates. Synthesis, enzymic activity, and analysis of the structural requirements of the binding locale of the 2-substituent on bovine brain protein kinase. Biochemistry 14:3315–3321
Meyer RB Jr, Uno H, Shuman DA, Robins RK, Simon LN, Miller JP (1975b) The synthesis of 2,6-disubstituted-9-β-D-ribofuranosylpurine cyclic 3′,5′-phosphates and the selectivity of cAMP and cGMP-specific enzymes to substituents in these positions. J Cyclic Nucleotide Res 1:159–167
Mian AM, Harris R, Sidwell RW, Robins RK, Khwaja TA (1974) Synthesis and biological activity of 9-β-D-arabinofuranosyladenine cyclic 3′,5′-phosphate and 9-β-D-arabinofuranosylguanine cyclic 3′,5′-phosphate. J Med Chem 17:259–263
Mian AM, Long RA, Allen LB, Sidwell RW, Robins RK, Khwaja TA (1979) Synthesis and antitumor and antiviral activities of 1-β-D-arabinofuranosyl-2-amino-1,4(2H)iminopyrimidine and its derivatives. J Med Chem 22:514–517
Michal F, Maguire MH, Gough G (1969) 2-Methylthioadenosine-5′-phosphate: a specific inhibitor of platelet aggregation. Nature 222:1073–1074
Michal G, Muehlegger K, Nelboeck M, Thiessen C, Weimann G (1974) Cyclophosphates. VI. Cyclophosphates as substrates and effectors of phosphodiesterase. Pharm Res Commun 6:203–252
Michelson AM (1958) Synthesis of nucleoside cyclic phosphates. Chem Ind 70–71
Michelson AM (1959) Polynucleotides. Part II. Homopolymers of cytidylic and pseudouridylic acid, copolymers with repeating subunits, and the stepwise synthesis of polyribonucleotides. J Chem Soc 3655–3669
Miller JP (1977) Cyclic nucleotide analogues. In: Cramer, Schultz (eds) Cyclic 3′,5′-nucleotides; mechanisms of action. John Wiley and Sons, New York, pp 77–105
Miller JP, Beck AH, Simon LN, Meyer RB Jr (1975a) Induction of hepatic tyrosine aminotransferase *in vivo* by derivatives of cyclic adenosine 3′,5′-monophosphate. J Biol Chem 250:426–431
Miller JP, Boswell KH, Meyer RB Jr, Christensen LF, Robins RK (1980a) Synthesis and enzymatic and inotropic activity of some new 8-substituted and 6,8-disubstituted derivatives of adenosine cyclic 3′,5′-monophosphates. J Med Chem 23:242–251
Miller JP, Boswell KH, Mian AM, Meyer RB Jr, Robins RK, Khwaja TA (1976) 2′-Derivatives of guanosine and inosine cyclic 3′,5′-phosphates. Synthesis, enzymic activity, and the effect of 8-substituents. Biochemistry 15:217–223
Miller JP, Boswell KH, Muneyama K, Simon LN, Robins RK, Shuman DA (1973a) Synthesis and biochemical studies of various 8-substituted derivatives of guanosine 3′,5′-cyclic phosphate, inosine 3′,5′-cyclic phosphate, and xanthosine 3′,5′-cyclic phosphate. Biochemistry 12:5310–5319
Miller JP, Boswell KH, Muneyama K, Tolman RL, Scholten MB, Robins RK, Simon LN, Shuman DA (1973b) Activity of tubercidin-, toyocamycin-, and sangivamycin-3′,5′-

cyclic phosphates and related compounds with some enzymes of adenosine-3′,5′-cyclic phosphate metabolism. Biochem Biophys Res Commun 55:843–849
Miller JP, Christensen LF, Andrea TA, Meyer RB Jr, Kitano S, Mizuno Y (1978) Interaction of "aza" and "deaza" analogs of adenosine cyclic 3′,5′-phosphate with some enzymes of adenosine cyclic 3′,5′-phosphate metabolism: Evidence that the lone pair electrons of N-3 are involved in the binding of adenosine cyclic 3′,5′-phosphate to type II adenosine cyclic 3′,5′-phosphate-dependent protein kinase. J Cyclic Nucleotide Res 4:133–144
Miller JP, Robins RK (1976) The chemical modification of cyclic AMP and cyclic GMP. Ann Rep Med Chem 11:291–300
Miller JP, Robins RK, Christensen LF, Boswell KH, Meyer RB Jr, submitted to J Cyclic Nucleotide Res
Miller JP, Shuman DA, Scholten MB, Dimmitt MK, Stewart CM, Khwaja TA, Robins RK, Simon LN (1973c) Synthesis and biological activity of some 2′-derivatives of adenosine 3′,5′-cyclic phosphate. Biochemistry 12:1010–1016
Miller JP, Yagura TS, Meyer RB Jr, Robins RK, Uno H (1980b) Synthesis and enzymic studies of 2-substituted and 2-modified derivatives of 1,N^6-ethenoadenosine cyclic 3′,5′-phosphate. J Carbohydr Nucleosides Nucleotides 7:167–184
Miller Z, Lovelace E, Gallo M, Pastan I (1975b) Cyclic guanosine monophosphate and cellular growth. Science 190:1213–1215
Mitsugi K, Komagata K, Takahashi M, Iizuka H, Katagiri H (1964) Bacterial synthesis of nucleotides. Part II. Distribution of nucleoside phosphotransferase in bacteria. Agr Biol Chem 28:586–600
Mizuno Y, Kitano S, Nomura A (1975) Nucleotides. III. Synthesis of deazaadenosine 3′,5′-cyclic phosphates and related nucleotides of biological interest. Chem Pharm Bull (Tokyo) 23:1664–1670
Mizuno Y, Tazawa S, Kageura K (1968) Synthetic studies of potential antimetabolites. XII. Synthesis of 4-substituted 1-(β-D-ribofuranosyl)-1H-imidazo[4,5-*c*]pyridines. Chem Pharm Bull (Tokyo) 16:2011–2017
Moens W, Vokaer A, Kram R (1975) Cyclic AMP and cyclic GMP concentrations in serum- and density-restricted fibroblast cultures. Proc Natl Acad Sci USA 72:1063
Montgomery JA, Thomas HJ (1961) Synthesis of potential anticancer agents. XXVII. The ribonucleotides of 6-mercaptopurine and 8-azaguanine. J Org Chem 26:1926–1929
Morr M (1976) Synthese des 5′-thio-3′-amido-5′,3′-didesoxyadenosin-3′,5′-cyclophosphats, ein cAMP-derivat mit s und N im Cyclophosphatring. Tetrahedron Lett 2127–2128
Morr M, Ernst L (1978) Zur Reaktion von Aminonucleosiden mit Thiophosphorylierungsreagenzien. Chem Ber 111:2152–2172
Morr M, Kula M-R, Ernst L (1975) Synthesis and ^{13}C NMR spectra of 3′-amido analogs of adenosine 3′,5′-cyclic monophosphate. Tetrahedron 31:1619–1622
Morr M, Kula M-R, Roesler G, Jastorff B (1974) Synthesis of 3′-amido-3′-deoxyadenosine 3′,5′-cyclophosphate. Angew Chem (Engl) 13:280–281
Morr M, Mengel R (1977) Synthese der cycloamide der *xylo*-3′-amino-3′-desoxyadenosin-5′-phosphorsäure und -5′-thionophosphorsäure, zweier neuer cAMP-analoga. Chem Ber 110:3947–3949
Moss J, Vaughan M (1979) Activation of adenylate cyclase by choleragen. Ann Rev Biochem 48:581–600
Mukaiyama T, Hashimoto M (1972) Synthesis of oligothymidylates and nucleoside cyclic phosphates by oxidation-reduction condensation. J Am Chem Soc 94:8528–8532
Muneyama K, Bauer RJ, Shuman DA, Robins RK, Simon LN (1971) Chemical synthesis and biological activity of 8-substituted adenosine 3′,5′-cyclic monophosphate derivatives. Biochemistry 10:2390–2395
Muneyama K, Shuman DA, Boswell KH, Robins RK, Simon LN, Miller JP (1974) Synthesis and biological activity of 8-haloadenosine 3′,5′-cyclic phosphates. J Carbohydr Nucleosides Nucleotides 1:55–60
Murad F (1973) Clinical studies and applications of cyclic nucleotides. Adv Cyclic Nucleotide Res 3:355–383

Murad F, Rall TW, Vaughan M (1969) Conditions for the formation, partial purification and assay of an inhibor of adenosine 3′,5′-monophosphate. Biochim Biophys Acta 192:430–445
Murayama A, Jastorff B, Cramer F, Hettler H (1971) 5′-Amido analogs of adenosine 3′,5′-cyclic monophosphate. J Org Chem 36:2029–3033
Murayama A, Jastorff B, Hettler H (1970) Synthesis of 5′-amido analogs of adenosine (3′-5′)-cyclophosphates. Angew Chem (Engl) 9:640–641
Murayama A, Jastorff B, Hettler H, Cramer F (1973) Die Stereoselektivität bei der Synthese von 5′-N-substituierten 5′-amido-analoga des Adenosin-3′,5′-cyclophosphat-*p*-nitrophenylesters. Chem Ber 106:3127–3131
Mushika Y, Taguchi Y (1976) Cyclic nucleotides. Japan Kokai 76 65,773
Nagyvary J (1969) Arabinonucleotides II. The synthesis of O^2, 2′-anhydrocytidine 3′-phosphate, a precursor of 1-β-D-arabinosylcytosine. J Am Chem Soc 91:5409–5410
Nagyvary J, Gohil RN, Kirchner CR, Stevens JD (1973) Studies on neutral esters of cyclic AMP. Biochem Biophys Res Commun 55:1072–1077
Nair KG (1966) Purification and properties of 3′,5′-cyclic nucleotide phosphodiesterase from dog heart. Biochemistry 5:150–157
Naito T, Sano M (1965) Adenosine cyclic 3′,5′-phosphate. Japanese Patent 9,063
Naka T, Honjo M (1976a) Synthesis of 8-carbamoyl- and 8-carboxyadenosine 3′,5′-cyclic phosphates. Chem Pharm Bull (Tokyo) 24:2052–2056
Naka T, Honjo M (1976b) 8-Substituted adenosine-3′,5′-cyclic phosphates. Japan Kokai 76 04,195
Nathanson JA, Greengard P (1977) Second messengers in the brain. Sci Am 237:108–119
Nawrath H (1977) Does cyclic GMP mediate the negative inotropic effect of acetylcholine in the heart? Nature 267:72–74
Nawrath H (1978) Rec Adv Stud Card Struct Metab 11:419
Nayler WG (1977) Cyclic nucleotides and the heart. Adv. Drug Res 12:39–51
Naylor R, Gilham PT (1966) Studies on some interactions and reactions of oligonucleotides in aqueous solution. Biochemistry 5:2722–2728
Nedorezova TP, Melnik SYa, Yartseva IV, Preobrazhenskaya MN (1978) (R)- and (S)-2′,3′-*O*-adamantylphosphonates of purine nucleosides. Bioorg Khim (USSR) 4:1058–1064
Niles RM, Ludwig KW, Makarski JS (1979) Differential growth inhibition in two human carcinoma cell lines by cyclic adenosine 5′-monophosphate analogs. J Natl Cancer Inst 63:909–911
Niles RM, Makarski JS, Kurtz MJ, Rutenburg AM (1976) Inhibition of human prostatic epithelial cell replication by cAMP and selcted analogs. Exp Cell Res 102:95–103
Nohara A, Imai K, Honjo M (1966) Synthesis of the glucose analogs of inosine-5′ phosphate. Chem Pharm Bull (Tokyo) 14:491–495
Nutt RF, Walton E (1968) Branched-chain sugar nucleosides. II. 5′,5′-Di-*C*-methyladenosine. J Med Chem 11:151–153
Oey J, Vogel A, Pollack R (1974) Intracellular cyclic AMP concentration responds specifically to growth regulation by serum. Proc Natl Acad Sci USA 71:694–698
Okruszek A, Verkade JG (1979) 2′,3′-Bis(2-chloroethyl)aminophosphoryl-3′-amino-3′-deoxyadenosine: A cyclic nucleotide with antitumor activity. J Med Chem 22:882–885
O'Neill JP, Schröder CH, Hsie AW (1975) Hydrolysis of butyryl derivatives of adenosine cyclic 3′,5′-monophosphate by Chinese hamster ovary cell extracts and characterization of the products. J Biol Chem 250:990–995
Ortiz PJ (1972) The inhibition of *E. coli* adenyl cyclase by *ara* ATP. Biochem Biophys Res Commun 46:1728–1733
Owens JR, Haley BE (1978) Use of photoaffinity nucleotide analogs to determine the mechanism of ATP regulation of a membrane-bound, cAMP-activated protein kinase. J Supramol Str 9:57–68
Panitz N, Reike E, Morr M, Wagner KG, Roesler G, Jastorff B (1975) The 3′-amido and 5′-amido analogues of adenosine 3′,5′-monophosphate; Interaction with cAMP-specific proteins. Eur J Biochem 55:415–422

Panzica RP, Rousseau RJ, Robins RK, Townsend LB (1972) A study on the relative stability and a quantitative approach to the reaction mechanism of the acid-catalyzed hydrolysis of certain 7- and 9-β-D-ribofuranosylpurines. J Am Chem Soc 94:4708–4714

Paoletti R, Berti F, Spano PF (1973) Cyclophosphates. V. *In vivo* metabolic and cardiovascular effects of new cyclophosphates. Pharmacol Res Commun 5:87–100

Pardee AB, de Asua LJ, Rozengurt E (1974) In: Clarkson B, Baserga R (eds) Control of Proliferation of Animal Cells, vol 1. Cold Spring Harbor Conferences on Cell Proliferation, p 547

Pastan IH, Johnson GS, Anderson WB (1975) Role of cyclic nucleotides in growth control. Ann Rev Biochem 44:491–522

Pike JE, Slechta L, Wiley PF (1964) Tubercidin and related compounds. J Heterocycl Chem 1:159–161

Pischel H, Holý A (1970) Oligonucleotid-verbindungen XXXVII. Synthese einiger nucleotid-derivate von 1-(β-D-ribofuranosyl)pyrimidon-(2) und 3-(β-D-ribofuranosyl)pyrimidon-(4). Coll Czech Chem Commun 35:3584–3596

Pizer FL, Ballou CE (1959) Studies on *myo*-inositol phosphates of natural origin. J Am Chem Soc 81:915–921

Pogson CI (1974) Guanine nucleotides and their significance in biochemical processes. Am J Clin Nutr 27:380–402

Polgar P, Vera JC, Kelley PR, Rutenburg AM (1973) Adenylate cyclase activity in normal and leukemic human leukocytes as determined by a radioimmunoassay for cyclic AMP. Biochim Biophys Acta 297:378–383

Porter KR, Puck TT, Hsie AW, Kelley D (1974) An electron microscope study of the effects of dibutyryl cyclic AMP on Chinese hamster ovary cells. Cell 2:145–162

Posternak T (1971) Chemistry of cyclic nucleoside phosphates and synthesis of analogs. In: Robison, Butcher, Sutherland (eds) Cyclic AMP. Academic Press, New York, pp 48–71

Posternak T (1974) Cyclic AMP and Cyclic GMP. Annu Rev Pharmacol 14:23–33

Posternak T, Falbriard JG (1971) Preparation of cyclic GMP. In: Robison, Butcher, Sutherland (eds) Cyclic AMP. Academic Press, New York

Posternak T, Marcus I, Cehovic G (1971) Préparation de nouveaux dérivés de l'AMPs (substitués en position C-8 et C-2) et étude de leur action sur la libération de l'hormone de croissance. CR Acad Sci Paris 272D:622–625

Posternak T, Marcus I, Gabbai A, Cehovic G (1969) Préparation et étude de quelques properiétés biologiques d'analogues de l'acide adénosine-3′,5′-phosphorique. CR Acad Sci Paris 269D:2409–2412

Posternak T, Sutherland EW, Henion WF (1962) Derivatives of cyclic 3′,5′-adenosine monophosphate. Biochim Biophys Acta 65:558–560

Prasad KN, Sheppard JR (1972) Inhibitors of cyclic-nucleotide phosphodiesterase induce morphological differentiation of mouse neuroblastoma cell culture. Exp Cell Res 73:436–440

Preobrazhenskaya NN, Shabarova ZA, Prokofev MA (1967) The synthesis and hydrolytic stability of amino acid derivatives of uridine cyclophosphates. Dokl Akad Nauk SSSR 174:100–103

Price TD, Ashman DF, Melicow MM (1967) Organophosphate of urine, including adenosine 3′,5′-monophosphate and guanosine 3′,5′-monophosphate. Biochim Biophys Acta 138:452–465

Prystás M, Šorm F (1969) Nucleic acid components and their analogues. CXXVII. Synthesis of uridine and cytidine 6-methyl and 5,6-dimethyl derivatives. Coll Czech Chem Commun 34:2316–2347

Puck TT (1977) Cyclic AMP, the microtubule-microfilament system, and cancer. Proc Natl Acad Sci USA 74: 4491–4495

Puck TT (1979) Studies on cell transformation. Somatic Cell Genet 5:973–990

Pullman B, Saran A (1976) Quantum-mechanical studies on the conformation of nucleic acids and their constituents. Progr Nucleic Acid Res Mol Biol 18:215–325

Rall TW, Sutherland EW (1958) Formation of a cyclic adenine ribonucleotide by tissue particles. J Biol Chem 232:1065–1076

Rall TW, Sutherland EW, Berthet J (1957) The relationship of epinephrine and glucagon to liver phosphorylase. IV. Effect of epinephrine and glucagon on the reaction of phosphorylase in liver homogenates. J Biol Chem 224:463–475

Ramseyer J, Kaslow HR, Gill GN (1974) Purification of the cAMP receptor protein by affinity chromatography. Biochem Biophys Res Commun 59:813–821

Ranganathan RS, Jones GH, Moffatt JG (1974) Novel Analogs of nucleoside 3′,5′-cyclic phosphates. I. 5′-Mono- and dimethyl analogs of adenosine 3′,5′-cyclic phosphate. J Org Chem 39:290–298

Reddi PK, Constantinides SM (1972) Partial suppression of tumor production by dibutyryl cyclic AMP and theophylline. Nature 238:286–287

Reddy BS, Viswamitra MA (1971) Crystal data of an ammonium salt of cyclic cytidine-2′,3′-phosphate. Current Sci 40:625

Revankar GR, Huffman JH, Allen LB, Sidwell RW, Robins RK, Tolman RL (1975) Synthesis and antiviral activity of certain 5′-monophosphates of 9-D-arabinofuranosyladenine and 9-D-arabinofuranosylhypoxanthine. J Med Chem 18:721–726

Revankar GR, Huffman JH, Sidwell RW, Tolman RL, Robins RK, Allen LB (1976) Synthesis and anti-DNA virus activity of the 5′-monophosphate and the cyclic 3′,5′-monophosphate of 9-(β-D-xylofuranosyl)guanine. J Med Chem 19:1026–1028

Robbins SJ, Rapp F (1979) Abstracts Mtgs Chem Soc for Microbiology, p 261, Abstract S 130, Los Angeles

Robins MJ, MacCoss M (1977) Nucleic acid related compounds. 26. A "geometry-only" method for determining the anomeric configuration of nucleosides based on the H-1′ NMR signal of cyclic α and β 3′,5′-mononucleotides. J Am Chem Soc 99:4654–4660

Robins MJ, MacCoss M, Wilson JS (1977) nucleic acid related compounds. 27. "Virtual Coupling" of the anomeric proton of cyclic 2′-deoxynucleoside 3′,5′-monophosphates. Reassessment of conformation using praseodymium shifts and assignment of H-2′,2″ signals by biomimetic deuteration at C-2. J Am Chem Soc 99:4660–4666

Robins RK (1979) The synthesis and structure activity relationship of analogues and derivatives of 3′,5′-cyclic nucleotides. In: Simkin (ed) Proc 6th Internatl Sym Med Chem, Brighton, UK. Cotswold Press Ltd, Oxford, UK, pp 127–135

Robins RK, Khwaja TA (1974) 3′,5′-Cyclic monophosphate nucleosides. US Patent 3,849,397

Robins RK, Shuman DA, Boswell KH (1977a) 6-Aminocarbonyl purine 3′,5′-cyclic nucleotides. US Patent 4,038,480

Robins RK, Shuman DA, Boswell KH (1977b) 6,8-Disubstituted purine derivatives of 9-β-D-ribofuranosyl 3′,5′-cyclic phosphate. US Patent 4,058,659

Robison GA (1970) Cyclic AMP as a second messenger. J Reprod Fert Suppl 10:55–74

Robison GA (1972) Cyclic AMP and hormone action. Am J Pharm Educ 36:723–733

Robison GA (1975) Cyclic AMP and disease: An overview. In: Good, Day, Yums (eds) Molecular pathology. CC Thomas Publ, Springfield, Ill, pp 394–404

Robison GA, Arnold A, Hartmann RC (1969) Divergent effects of epinephrine and prostaglandin E_1 on the level of cyclic AMP in human blood platelets. Pharm Res Commun 1:325–332

Robison GA, Butcher RW, Sutherland EW (1968) Cyclic AMP. Annu Rev Biochem 37:149–174

Robison GA, Butcher RW, Sutherland EW (1971a) Cyclic AMP. Academic Press, New York

Robison GA, Nahas GG, Triner L (1971b) Cyclic AMP and cell function. Ann NY Acad Sci 185:5–556

Robison GA, Park CR (1970) Cyclic adenylate in mammalian tissues. In: Ellenberg, Rifkin (eds) Diabetes mellitus: Theory and practise. McGraw-Hill, New York, pp 132–149

Robison GA, Sutherland EW (1967) Adenylic acid (cyclic). McGraw Hill Yearbook of Science and Technology, New York, pp 91–92

Robison GA, Sutherland EW (1971) On the relation of cyclic AMP to adrenergic receptors and sympathin. Adv Cytopharmacol 1:263–272

Rosario EJ del, Hammes GG (1970) Relaxation spectra of ribonuclease. VII. The interaction of ribonuclease with uridine 2′,3′-cyclic phosphate. J Am Chem Soc 92:1750–1753

Rosen OM (1970) Interaction of cyclic GMP and cyclic AMP with a cyclic nucleotide phosphodiesterase of the frog erythrocyte. Arch Biochem Biophys 139:447–449
Rosett T, Smith JG Jr, Matsuo I, Bailey PA, Smith DB, Surakiat S (1970) The quantitative separation of 3′,5′-cyclic adenosine monophosphate, adenosine-5′-monophosphate, adenosine-5′-diphosphate, and adenosine-5′-triphosphate by ion-exchange chromatography on diethylaminoethyl sephadex. J Chromatogr 49:308–312
Rosowsky A, Gohil RN (1978) 6-Amino-9-β-D-arabinofuranosylpurine-8-thiol 3′,5′-cyclic phosphate. Nucleic Acid Chem, Improved and New Synthetic Procedures, Methods, and Techniques, Townsend and Tipson (eds) pt 2. John Wiley and Sons, New York, pp 857–860
Rubin CS, Rosen OM (1975) Protein phosphorylation. Ann Rev Biochem 44:831–887
Rudland PS, Seeley M, Seifert W (1974) Cyclic GMP and cyclic AMP levels in normal and transformed fibroblasts. Nature 251:417–419
Rudolph SA, Johnson EM, Greengard P (1971) The enthalpy of hydrolysis of various 3′,5′- and 2′,3′-cyclic nucleotides. J Biol Chem 246:1271–1273
Russell AF, Moffatt JG (1969) Synthesis of some nucleotides derived from 3′-deoxythymidine. Biochemistry 8:4889–4896
Saenger W, Eckstein F (1969) Crystal structure of uridine 2′,3′-*O*,*O*-cyclophosphorothioate. Angew Chem (Engl) 8:595–596
Saenger W, Eckstein F (1970) Stereochemistry of a substrate for pancreatic ribonuclease. Crystal and molecular structure of the triethylammonium salt of uridine 2′,3′-*O*,*O*-cyclophosphorothioate. J Am Chem Soc 92;4712–4718
Sands H, Rickenberg HV (1978) Assessment of the role of cyclic nucleotides as hormonal mediators. In: Rickenberg HV (ed) Int rev biochim: Biochemistry and mode of action of hormones II, vol 20, University Park Press, Baltimore, Maryland, pp 45–80
Saran A, Berthod H, Pullman B (1973) Molecular orbital calculations on the conformation of nucleic acids and their constituents. VIII. Conformations of 2′,3′- and 3′,5′-cyclic nucleotides. Biochim Biophys Acta 331:154–164
Sasaki Y, Kodaira R, Nozawa R, Yokota T (1978) Inhibitory effect of 8-substituted adenosine derivatives on Ca^{++} and modulator protein-dependent phosphodiesterase activity. Biochem Biophys Res Commun 84:277–284
Sasaki Y, Suzuki N, Sowa T, Nozawa R, Yokota T (1976) Effects of 8-substituted adenosine 3′,5′-monophosphate derivatives on high K_m phosphodiesterase activity. Biochemistry 15:1408–1413
Sasisekharan V, Lakshminarayanan AV, Ramachandran GN (1967) In: Ramachandran (ed) Conformation of biopolymers. Academic Press, New York, pp 641–654
Schaeffer HJ, Thomas HJ (1958) Synthesis of potential anticancer agents. XIV. Ribosides of 2,6-disubstituted purines. J Am Chem Soc 80:3738–3742
Scheit KH (1974) Biological and spectroscopic properties of a fluorescent cyclic AMP analogue, 2-aminopurine nucleoside-5′.3′-cyclic phosphate. J Carbohydr Nucleosides Nucleotides 1:385–399
Schmidt MJ, Truex LL, Leonard NJ, Scopes DI, Barrio JR (1978) Effect of *lin*-benzoadenosine and *lin*-benzoadenosine 3′,5′-monophosphate on cyclic AMP-dependent protein kinase activity *in vitro*. J Cyclic Nucleotide Res 4:201–207
Schweizer MP, Broom AD, Ts'o POP, Hollis DP (1968) Studies of inter- and intramolecular interaction in mononucleotides by proton magnetic resonance. J Am Chem Soc 90:1042–1055
Schweizer MP, Robins RK (1973) NMR studies on the conformation of nucleosides and 3′,5′-cyclic nucleotides. Conformation of Biological molecules and polymers. In: Pullman, Bergmann (eds) Proceedings of the Fifth Jerusalem Symposium on Quantum Chemistry and Biochemistry. Academic Press, New York, pp 329–343
Scopes DIC, Barrio JR, Leonard NJ (1977) Defined dimensional changes in enzyme cofactors: Fluorescent "stretched-out" analogs of adenine nucleotides. Science 195:296–298
Secrist III JA (1974) The synthesis of 1,N^6-ethenoadenosine 3′,5′-monophosphate; A fluorescent analog of cyclic AMP. Methods Enzymol 38c:428–430

Secrist III JA, Barrio JR, Leonard NJ, Palasi CV, Gilman AG (1972a) Fluorescent modification of adenosine 3′,5′-monophosphate: Spectroscopic properties and activity in enzyme systems. Science 177:279–280
Secrist III JA, Barrio JR, Leonard NJ (1972b) A fluorescent modification of adenosine triphosphate with activity in enzyme systems: 1,N^6-Ethenoadenosine triphosphate. Science 175:646–647
Seifert WE, Rudland PS (1974) Possible involvement of cyclic GMP in growth control of cultured mouse cells. Nature 248:138–140
Severin ES, Gulyaev NN, Bulargina TV, Kochetkova MN (1979a) Specific inhibition of cyclic AMP-dependent protein kinase, adenylate cyclase and phosphodiesterase by ATP and cyclic AMP analogs. In: Weber G (ed) Adv. Enzyme Regulation, vol 17. Pergamon Press, New York, p 251
Severin ES, Nesterova MV, Sashchenko LP, Rasumova VV, Tunitskaya VL, Kochetkov SN, Gulyaev NN (1975) Investigation of the adenosine 3′,5′-cyclic phosphate binding site of pig brain histone kinase with the aid of some analogues of adenosine 3′,5′-cyclic phosphate. Biochim Biophys Acta 384:413–422
Severin ES, Sashchenko LP, Kochetkov SN, Gulyaev NN, Kurochkin SN, Trakht IN (1979b) Structure and mechanism of action of cyclic AMP-dependent protein kinase. In: Krause EG, Pinna L, Wollenberger A (eds) FEBS 12th Meeting, Dresden 1978, vol 54. Symposium 7, Cyclic Nucleotides and Protein Phosphorylation in Cell Regulation. Pergamon Press, London, p 101
Severin ES, Tkachuk VA, Gulyaev NN (1976) Interaction of adenosine-3′,5′-cyclosulfate with adenosine-3′,5′-cyclophosphate-dependent protein kinase and phosphodiesterase. Biokhimiya 41:384–388
Shahak Y, Chipman DM, Shavit N (1973) Photophosphorylation studies with fluorescent adenine nucleotide analogs. FEBS Lett 33:293–296
Shapiro R, Danzig M (1972) Acidic hydrolysis of deoxycytidine and deoxyuridine derivatives. The general mechanism of deoxyribonucleoside hydrolysis. Biochemistry 11:23–29
Shields R (1974) In: Dumont JE, Brown BL, Marshall NJ (eds) Eukaryotic cell function and growth. Plenum Press, London, p 747
Sheldrick WS, Reike E (1978) 8-[(2-Aminoethyl)amino]adenosine cyclic 3′,5′-monophosphate tetrahydrate. Acta Cryst 34B:2324–2327
Shen TY (1970) Nucleosides and nucleotides as potential therapeutic agents. Angew Chem (Engl) 9:678–688
Shibuya M, Arai K, Kaziro Y (1975) A novel method for the determination of guanosine 3′,5′-cyclic monophosphate (cyclic GMP). Biochem Biophys Res Commun 62:129–135
Shin S, Freedman VH, Risser R, Pollack R (1975) Tumorigenicity of virus-transformed cells in *nude* mice is correlated specifically with anchorage independent growth *in vitro*. Proc Natl Acad Sci USA 72:4435–4439
Shiue C-Y, Chu S-H (1975) A convenient one-step synthesis of 6-selenoxo-9-(β-D-ribofuranosyl)purine 3′,5′-cyclic phosphate and related compounds. J Heterocycl Chem 12:493–500
Shugar D, Wierzchowski KL (1958) Preparation of nucleoside 3′,5′-cyclic phosphates and some ribonucleotide isomers. Bull Acad Polon Sci, Ser Sci Biol 6:283
Shuman DA, Meyer RB Jr (1975) 6-Substituted purine nucleotides. US Patent 3,915,958
Shuman DA, Meyer RB Jr, Robins RK (1975) 2,6-Disubstituted purine cyclic nucleotides. US Patent 3,897,413
Shuman DA, Meyer RB Jr, Robins RK (1976) 6-Substituted purine 3′,5′-cyclic nucleotides. US Patent 3,948,886
Shuman DA, Miller JP, Scholten MB, Simon LN, Robins RK (1973) Synthesis and biological activity of some purine 5′-thio-5′-deoxynucleoside 3′,5′-cyclic phosphorothioates. Biochemistry 12:2781–2786
Shuman DA, Robins RK (1974a) 2′-*O*-Methyladenosine 3′,5′-cyclic phosphate. US Patent 3,816,400
Shuman DA, Robins RK (1974b) Nucleoside 3′,5′-cyclic phosphorothioates. US Patent 3,853,844

Sidwell RW, Huffman JH, Allen LB, Meyer RB Jr, Shuman DA, Simon LN, Robins RK (1974) *In vitro* antiviral activity of 6-substituted 9-β-D-ribofuranosylpurine 3′,5′-cyclic phosphates. Antimicrob Agents Chemother 5:652–657

Sidwell R, Huffman J, Shuman D, Muneyama K, Robins R (1972) Cyclic AMP derivatives as antiviral agents. Proceedings of the VIIth Int Cong of Chemotherapy, Prague, 1971, pp 313–314

Sidwell RW, Simon LN, Huffman JH, Allen LB, Long RA, Robins RK (1973) DNA virus inhibitory activity of 1-β-D-arabinofuranosylcytosine-3′,5′-cyclic phosphate. Nature New Biol 242:204–206

Siggins GR, Henriksen SJ (1975) Analogs of cyclic adenosine monophosphate: Correlation of inhibition of purkinje neurons with protein kinase activation. Science 189:559–560

Simon LN, Shuman DA, Robins RK (1973) The chemistry and biological properties of nucleotides related to nucleoside 3′,5′-cyclic phosphates. Adv Cyclic Nucleotide Res 3:225–353

Smith ICP, Mantsch HH, Lapper RD, Deslauriers R, Schleich T (1973) A study of the conformations of nucleic acids by carbon-13 and hydrogen nuclear magnetic resonance spectroscopy. In: Pullman, Bergman (ed) 5th Jerusalem Symposium on Quantum Chem Biochem. Academic Press, New York, pp 381–401

Smith M (1964) Synthesis of deoxyribonucleoside-3′,5′-cyclic phosphates by base-catalysed transesterification. J Am Chem Soc 86:3586

Smith M, Drummond GI, Khorana HG (1961) Cyclic phosphates. IV. Ribonucleoside-3′,5′-cyclic phosphates. A general method of synthesis and some properties. J Am Chem Soc 83:698–706

Smith M, Jardetzky CD (1968) The NMR spectra and conformation of the nucleotide 3′,5′-cyclic phosphates. J Mol Spectroscopy 28:70–80

Smith M, Khorana HG (1959) Specific synthesis of the $C_{5'}$–$C_{3'}$ interribonucleotide linkage: The synthesis of uridylyl-(5′→3′)-uridine. J Am Chem Soc 81:2911–2912

Smith M, Moffatt JG, Khorana HG (1958) Carbodiimides. VIII. Observations on the reactions of carbodiimides with acids and some new applications in the synthesis of phosphoric acid esters. J Am Chem Soc 80:6204–6212

Smith SW, Werthamer S, Artman M (1974) Differential effects of dibutyryl cyclic AMP on the growth and morphology of an established human lymphocyte line. *In Vitro* 10:225–229

Smrt J (1967) Oligonucleotidic compounds. XVI. Synthesis of 5,6-dihydrouridine analogues of guanylyl-(3′→5′)-uridylyl-(3′→5′)-uridine. Coll Czech Chem Commun 32:198–205

Smrt J (1974) A remark on the preparation of protected guanosine 3′-phosphate by means of a mixture of ribonucleases T1 and T2. Coll Czech Chem Commun 39:969–971

Sowa T, Suzuki N, Sasaki Y, Yokota T (1976) Cyclic adenosine monophosphate derivative. Japan Kokai 76 41,388

Stadel JM, Goodman DBP (1978) 8-p-Chlorophenylthio-cyclic AMP: A potent partial simulator of antidiuretic hormone action. J Cyclic Nucleotide Res 4:35–43

Stanwick TL, Nahmias AJ (1979) Abstracts, Mtgs Am Soc for Microbiology, p 290, Abstract No. S306 Los Angeles

Steiner AL, Ferrendelli JA, Kipnis DM (1972c) Radio-immunoassay for cyclic nucleotides. III: Effects of ischemia, changes during development and regional distribution of adenosine 3′,5′-monophosphate and guanosine 3′,5′-monophosphate in mouse brain. J Biol Chem 247:1121–1124

Steiner AL, Pagliara AS, Chase LR, Kipnis DM (1972b) Radioimmunoassay of cyclic nucleotides. II. Adenosine 3′,5′-monophosphate and guanosine 3′,5′-monophosphate in mammalian tissues and body fluids. J Biol Chem 247:1114–1120

Steiner AL, Parker CW, Kipnis DM (1972a) Radioimmunoassay for cyclic nucleotides. I. Preparation of antibodies and iodinated cyclic nucleotides. J Biol Chem 247:1106–1113

Steiner AL, Kipnis DM, Utiger R, Parker C (1969) Radioimmunoassay for the measurement of adenosine 3′,5′-cyclic phosphate. Proc Natl Acad Sci USA 64:367–373

Steiner RF (1972) The activity of ε-adenosine derivatives as allosteric modifiers of phosphorylase b. FEBS Lett 23:139–141

Steitz TA, Lipscomp WN (1965) Molecular structure of methyl ethylene phosphate. J Am Chem Soc 87:2488–2489
Stevens RH, Smith DD, Osborne JW, Oberley LW (1976) Adenosine 3′,5′-cyclic monophosphate levels in x-ray-induced small-bowel adenocarcinoma in the rat. J Natl Cancer Inst 57:43–45
Straus DB, Fresco JR (1965) Synthesis of N-benzoyl-2′-*O*-tetrahydropyranylguanosinephosphate, an intermediate in the chemical synthesis of polyriboguanylic acid. J Am Chem Soc 87:1364–1374
Su J-C, Hassid WZ (1962) Carbohydrates and nucleotides in the red alga *Porphyra perforate*. II. Separation and identification of nucleotides. Biochemistry 1:474–480
Suehiro H, Kikugawa K, Ichino M, Nakamura T (1976) 2-Substituted thioadenosine-5′-monophosphates. Japan Kokai 76 76,291
Sundaralingam M (1975) Structure and conformation of nucleosides and nucleotides and their analogs as determined by x-ray diffraction. Ann NY Acad Sci 255:3–42
Sundaralingam M, Abola J (1972a) Molecular conformation of adenosine 3′,5′-monophosphate monohydrate. Nature New Biol 235:244–245
Sundaralingam M, Abola J (1972b) Stereochemistry of nucleic acids and their constituents. XXVII. The crystal structure of 5′-methyleneadenosine 3′,5′-cyclic monophosphonate monohydrate, a biologically active analog of the secondary hormonal messenger cyclic adenosine 3′,5′-monophosphate. Conformational "rigidity" of the furanose ring in cyclic nucleotides. J Am Chem Soc 94:5070–5076
Sutherland EW, Rall TW (1957) The properties of an adenine ribonucleotide produced with cellular particles, ATP, Mg^{++}, and epinephrine or glucagon. J Am Chem Soc 79:3608
Sutherland EW, Rall TW (1958) Fractionation and characterization of a cyclic adenine ribonucleotide formed by tissue particles. J Biol Chem 232:1077–1091
Sutherland EW, Robison GA (1966) Metabolic effects of catecholamines. A. The role of cyclic-3′,5′-AMP in responses to catecholamines and other hormones. Pharmacol Rev 18:145–161
Sutherland EW, Robison GA, Butcher RW (1968) Some aspects of the biological role of adenosine 3′,5′-monophosphate (cyclic AMP). Circulation 37:279–306
Suzaki S, Yamazaki A, Kamimura A, Mitsugi K, Kumashiro I (1970) Synthesis of 9-β-D-xylofuranosyl-6-mercaptopurine and 9-β-D-xylofuranosylguanine 5′-phosphate. Chem Pharm Bull (Tokyo) 18:172–176
Suzuki T, Kambayashi M, Matsuishi N (1972) 2′,3′-Dibutyrylriboflavin cyclic-4′,5′-phosphate. Japan Patent 72 08,316
Swislocki NI (1970) Decomposition of dibutyryl cyclic AMP in aqueous buffers. Anal Biochem 38:260–269
Symons RH (1970) ^{32}P-3′,5′-cyclic AMP: A simple preparative procedure. Biochem Biophys Res Commun 38:807–870
Symons RH (1973) Improved synthesis of ^{32}P-labelled 3′,5′-cyclic AMP, 3′,5′-cyclic GMP and other 3′,5′-cyclic ribo- and deoxyribonucleotides of high specific activity. Biochim Biophys Acta 320:535–539
Symons RH (1974) The synthesis of [^{32}P]adenosine 3′,5′-cyclic phosphate and other ribo- and deoxyribonucleoside 3′,5′-cyclic phosphates. Methods Enzymol 38:410–420
Szabó P, Szabó L (1958) Alkaline hydrolysis of methyl α-D-glucoside 4,6-phosphate. Compt Rend 247:1748–1750
Szaduykis-Szadurski Berti F (1972) Smooth muscle relaxing activity of 8-bromoguanosine-3′,5′-monophosphate. Pharmacol Res Commun 4:53–61
Szer W, Shugar D (1963) Ribonucleoside 2′,3′-cyclic phosphates. Biochem Prep 10:139–144
Taguchi Y, Mushika Y (1975) Synthetic studies on phosphorylating reagents. IV. A novel synthesis of nucleoside-3′,5′-cyclic phosphates by the use of 2-(N,N-dimethylamino)-4-nitrophenyl phosphate. Bull Chem Soc (Japan) 48:1528–1532
Taguchi Y, Sishimura N, Kakimoto T, Mushika Y (1976) Synthetic studies on phosphorylating reagents. V. A convenient synthesis of 2′,3′-cyclic coenzyme A. Bull Chem Soc (Japan) 49:1122–1125
Takaku H, Kato M, Hata T (1978) Preparation of nucleoside 3′,5′-cyclic phosphates *via* 8-quinolyl nucleoside 5′-phosphates as a useful intermediate. Chemistry Lett 181–184

Tanaka K (1960) A simple preparative method for ribonucleoside-2′,3′-cyclic phosphates. J Biochem (Tokyo) 47:398–399
Tazawa I, Tazawa S, Alderfer JL, Ts'o POP (1972) A novel procedure for the synthesis of 2′-*O*-alkyl nucleotides. Biochemistry 11:4931–4937
Tener GM, Khorana HG (1955) Cyclic Phosphates. II. Further studies of ribonucleoside 2′,3′-cyclic phosphates. J Am Chem Soc 77:5349–5351
Tener GM, Khorana HG, Markham R, Pol EH (1958) Studies on polynucleotides. II. The synthesis and characterization of linear and cyclic thymidine oligonucleotides. J Am Chem Soc 80:6223–6230
Tesser GI, Fisch H, Schwyzer R (1972) Limitations of affinity chromatography: Sepharose-bound cyclic 3′,5′-adenosine monophosphate. FEBS Lett 23:56–58
Tewari R, Danyluk SS (1978) Configurational effects on conformational properties of cyclic nucleotides. I. Theoretical calculations of conformer preferences in α-nucleoside 3′,5′-cyclic monophosphates. Biopolymers 17:1181–1196
Thomas HJ, Hewson K, Montgomery JA (1962) The 2′(3′)-phosphates of 6-mercaptopurine ribonucleoside and 8-azaguanosine. J Org Chem 27:192–194
Thomas HJ, Montgomery JA (1968) Derivatives and analogs of 6-mercaptopurine ribonucleotide. J Med Chem 11:44–48
Tikhomirova-Sidorova NS, Kogan ÉM, Sysoev VA, Ustyuzhanin GE (1971) Alcoholysis of pyrimidine nucleoside 2′,3′-cyclophosphates with purine nucleosides in frozen solutions in presence of pancreatic ribonuclease. Zh Obshch Khim 41:2570–2575
Tikhomirova-Sidorova NS, Kogan ÉM, Ustyuzhanin GE (1973) Synthesis of pyrimidine oligonucleotides in frozen solutions in presence of pancreatic ribonuclease. Zh Obshch Khim 43:393–399
Tisdale MJ (1979) The significance of cyclic AMP and cyclic GMP in cancer treatment. Cancer Tratment Rev 6:1–15
Tisdale MJ, Phillips BJ (1974) Apparent correlation between adenosine 3′,5′-cyclic monophosphate levels and malignancy in somatic cell hybrids. Exp Cell Res 88:111–120
Tovey MG, Rochette-Egly C, Castagna M (1979) Effect of interferon on concentration of cyclic nucleotides in cultured cells. Proc Natl Acad Sci USA 76:3890–3893
Tsien RW (1977) Cyclic AMP and contractile activity in heart. Adv Cyclic Nucleotide Res 8:363–420
Ts'o POP (1974) Bases, nucleosides, and nucleotides. In: Ts'o POP (ed) Basic principles in nucleic acid chemistry, vol 1. Academic Press, Inc, New York, pp 453–584
Ts'o POP, Kondo NS, Schweizer MP, Hollis DP (1969) Studies of the conformation and interaction in dinucleoside mono- and diphosphates by proton magnetic resonance. Biochemistry 8:997–1029
Tsou KC, Yip KF, Miller EE, Lo KW (1974) Synthesis of 1,N^6-etheno-2-aza-adenosine (2-aza-ε-adenosine): a new cytotoxic fluorescent nucleoside. Nucleic Acids Res 1:531–547
Tunitskaya VL, Gulyaev NN, Poletaev AI, Severin ES (1977) Study of conformation of cyclic adenosine-3′,5′-monophosphate and some of its derivatives by the method of circular dichroism. Biokhimiya 42:746–753
Turner AF, Khorana HG (1959) Studies on polynucleotides. VI. Experiments on the chemical polymerization of mononucleotides. Oligonucleotides derived from thymidine-3′ phosphate. J Am Chem Soc 81:4651–4656
Ueda T, Fox JJ (1967) The mononucleotides. In: Wolfrom, Tipson (eds) Advances in carbohydrate chemistry, vol 22. Academic Press, New York, pp 307–419
Ueda T, Imazawa M, Miura K, Iwata R, Odajima K (1971) A facile conversion of amino to thiono group in certain nucleobases. Tetrahedron Lett 2507–2510
Ueda T, Kawai I (1970) A convenient synthesis of ribonucleoside 2′,3′-cyclic phosphates from ribonucleosides and ribonucleotides. Chem Pharm Bull (Tokyo) 18:2303–2308
Uesugi S, Tanaka S, Ikehara M (1979) Carbon-13 NMR spectra of nucleoside 3′,5′-cyclic phosphates. Proposition for revised signal assignments. Org Magnetic Resonance 12:143–145
Ukita T, Hayatsu H (1961) Organic phosphates. XVIII. Synthesis of lyxouridine 2′,3′-cyclic phosphate and related compounds. Chem Pharm Bull (Tokyo) 9:1000–1005

Uno H, Meyer RB Jr, Shuman DA, Robins RK, Simon LN, Miller JP (1976) Synthesis of some 1,8- and 2,8-disubstituted derivatives of adenosine cyclic 3′,5′-phosphate and their interaction with some enzymes of cAMP metabolism. J Med Chem 19:419–422

Usher DA, Dennis EA, Westheimer FH (1965) Calculation of the bond angles and conformation of methyl ethylene phosphate and related compounds. J Am Chem Soc 87:2320–2321

van Boom JH, Burgers PM, Deursen P (1974) Preparation of nucleoside 3′,5′-cyclic phosphates *via* phosphotriester intermediates. J Chem Soc Chem Commun 618–619

van Boom JH, Burgers PMJ, Haasnoot CAG (1976) Synthesis of 3′-phosphates of diribonucleoside monophosphates *via* phosphotriester intermediates. Nucleic Acids Res 3:2731–2747

von Boom JH, Haasnoot CAG (1975) Preparation of respectively 5′-phosphoryladenylyl-(3′-5′)-uridine 2′,3′-cyclic phosphate, 3′-phosphate and 2′/3′-phosphates exclusively *via* phosphotriester intermediates. Nucleic Acids Res. Sp Publ s117–120

van Boom JH, de Rooy JFM, Reese CB (1973) The synthesis of oligoribonucleotides. Part X. Preparation of 2′,3′-cyclic phosphates of ribonucleosides and diribonucleoside phosphates *via* phosphotriester intermediates. J Chem Soc, Perkin Trans 1:2513–2517

Verheyden JPH, Moffatt JG (1970) Halo sugar nucleosides. II. Iodination of secondary hydroxyl groups of nucleosides with methyltriphenoxyphosphonium iodide. J Org Chem 35:2868–2877

Viscontini M, Furuta Y (1973) Über Pterinchemie 43. Synthese von D-Neopterin-3′- phosphat und D-Neopterin-2′,3′-cyclophosphat. Betrachtungen über die Biopterinbiogenese. Helv Chim Acta 56:1819–1825

Walsh DA, Perkins JP, Krebs EG (1968) An adenosine 3′,5′-monophosphate-dependent protein kinase from rabbit skeletal muscle. J Biol Chem 243:3763–3765

Watenpaugh K, Dow J, Jensen LH, Furberg S (1968) Crystal and molecular structure of adenosine 3′,5′-cyclic phosphate. Science 159:206–207

Weber JM, Stewart RB (1975) Cyclic AMP potentiation of interferon antiviral activity and effect of interferon on cellular cyclic AMP levels. J Gen Virology 28:363–372

Wechter WJ (1967) Nucleic acids. I. The synthesis of nucleotides and dinucleoside phosphates containing *ara*-cytidine. J Med Chem 10:762–773

Wechter WJ (1969) Nucleic acids. VIII. Synthesis and chemistry of *ara*-cytidine 2′,5′-cyclic phosphate. Phosphate anisotropy. J Org Chem 34:244–247

Weimann G, Haid E, Mühlegger K, Bergmeyer HU, Dietmann K, Michal G, Hochstetter MN (1973) 6-Substituted purineribonucleoside 3′,5′-cyclophosphates. US Patent No 3,712,885

Weiss B (1975) Cyclic nuceotides and disease. University Park Press, Baltimore

Westheimer FH (1957) Studies of the solvolysis of some phosphate esters. Chem Soc (London), Spe Publ 8:1–15

Westheimer FH (1968) Pseudo-rotation in the hydrolysis of phosphate esters. Acc Chem Res 1:70–78

White AA, Aurbach GE, Carlson SJ (1969) Identification of guanyl cyclase in mammalian tissues. Fed Proc 28:473

Wieker H-J, Witzel H (1967) Zum Mechanismus der Ribonuclease-Reaktion 3. Zuordnung der kinetischen Parameter k_{+1}, k_{-1}, k_{+2} und interpretation von Km. Eur J Biochem 1:251–258

Wierenga W, Woltersom JA (1977) An efficient preparation of cyclic CMP and conversion to its dibutyryl and succinyl derivatives. J Carbohydr Nucleosides Nucleotides 4:189–198

Wilchek M, Salomon Y, Lowe M, Selinger Z (1971) Conversion of protein kinase to a cyclic AMP independent form by affinity chromatography on N^6-caproyl 3′,5′-cyclic adenosine monophosphate-sepharose. Biochem Biophys Res Commun 45:1117–1183

Willingham MC, Pastan I (1975) Cyclic AMP modulates microvillus formation and agglutinability in transformed and normal mouse fibroblasts. Proc Natl Acad Sci USA 72:1263–1267

Winkley MW, Robins RK (1968) Pyrimidine nucleoside. I. The synthesis of 6-methylcytidine, 6-methyluridine, and related 6-methylpyrimidine nucleosides. J Org Chem 33:2822–2827

Witkowski JT, Robins RK, Sidwell RW, Simon LN (1972) Design, synthesis, and broad spectrum antiviral activity of 1-β-D-ribofuranosyl-1,2,4-triazole-3-carboxamide and related nucleosides. J Med Chem 15:1150–1154

Woenckhaus C, Pfleiderer G (1965) Biochemische Eigenschaften der coenzymanaloga nicotinamid-purin-dinucleotid, nicotinamid-1-desazapurine-dinucleotid und nicotinamid-benzimidazol-dinucleotid. Biochem Z 341:495–501

Woods WD, Waitzman MB (1970) Isolation of cyclic 3′,5′-adenosine monophosphate on a neutral silicic acid-glass microfiber matrix. J Chromatogr 47:536–542

Yagura TS, Sigman CC, Sturm PA, Reist EJ, Johnson HL, Miller JP (1980) Mapping cyclic AMP binding sites on type I and type II cyclic AMP-dependent protein kinases using 2-substituted derivatives of cyclic AMP. Biochem Biophys Res Commun 92:463–469

Yamaji N, Ishiyama J, Kato M, Yoshida F (1977a) 1,N^6-Etheno-2-thio-adenosine-3′,5′-cyclic phosphoric acids and their salts. Japan Kokai 77 00,296

Yamaji N, Ishiyama J, Kato M, Yoshida F (1977b) 1,N^6-Etheno-2-aminoadenosine 3′,5′-cyclic phosphate. Japan Kokai 77 25,798

Yamaji N, Ishiyama J, Kato M, Yoshida F (1977c) 1,N^6-Etheno-2-thioadenosine 3′,5′-cyclic phosphate and its S-alkyl derivatives. Japan Kokai 77 53,897

Yamaji N, Ishiyama J, Kato M, Yoshida F (1977d) 1,N^6-Etheno-2-hydroxyadenosine 3′,5′-cyclic phosphate. Japan Kokai 77 77,094

Yamaji N, Ishiyama J, Kato M, Yoshida F (1978) 1,N^6-Etheno-2-substituted-adenosine-3′,5′-cyclic phosphoric acids. Japan Kokai 78 12,891

Yamaji N, Kato M (1975) The synthesis of 2-aza-adenosine-3′,5′-cyclic phosphate *via* 1,N^6-etheno-adenosine-3′,5′-cyclic phosphate. Chem Lett 311–314

Yamaji N, Kato M (1978a) 1,N^6-Etheno-2-substituted-adenosine-3′,5′-cyclic phosphoric acids. Japan Kokai 78 12,892

Yamaji N, Kato M (1978b) 2-Substituted adenosine 3′,5′-cyclic phosphoric acids. Japan Kakai 78 46,998

Yamaji N., Katoh M, Ishiyama J, Yoshida F (1976a) 2-Azapurines. Japan Kokai 76 70,793

Yamaji N, Suda K, Kato M (1976b) The synthesis of 2-substituted adenosine-3′,5′-cyclic phosphate *via* 1,N^6-etheno-adenosine-3′,5′-cyclic phosphate. Nucleic Acids Res, Sp Publ No. 2, s 59–62

Yamaji N, Suda K, Oneue Y, Kato M (1977e) Studies on the synthesis of compounds related to adenosine 3′,5′-cyclic phosphate. II. Removal of etheno group of 2-substituted 1,N^6-etheno-adenosine 3′,5′-cyclic phosphates. Chem Pharm Bull (Tokyo) 25:3239–3246

Yamaji N, Tahara K, Kato M (1980) Studies on the synthesis of compounds related to adenosine 3′,5′-cyclic phosphate. IV. The synthesis of 2-sulfo- and 2-carboxy-adenosine 3′,5′-cyclic phosphate. Chem Pharm Bull (Tokyo) 28:115–119

Yamaji N, Yuasa Y, Kato M (1976c) The synthesis of 2-substituted 1,N^6-etheno-adenosine-3′,5′-cyclic phosphate by ring reclosure of alkali-hydrolizate of 1,N^6-etheno-adenosine-3′,5′-cyclic phosphate. Chem Pharm Bull (Tokyo) 24:1561–1567

Yamazaki A, Okutsu M (1978) Cyclization of 5-amino-1-β-D-ribofuranosylimidazole-4-carboxamide (AICA-riboside): A review. J Heterocycl Chem 15:353–358

Yathindra N, Sundaralingam M (1974) Conformations of cyclic 3′,5′-nucleotides. Effect of the base on the *syn-anti* conformer distribution. Biochem Biophys Res Commun 56:119–126

Yip KF, Tsou KC (1973) Synthesis of fluorescent adenosine derivatives. Tetrahedron Lett 3087–3090

Yoshikawa M, Kato T, Takenishi T (1967) A novel method for phosphorylation of nucleosides to 5′-nucleotides. Tetrahedron Lett 5065–5068

Yoshikawa M, Kato T, Takenishi T (1969) Studies of phosphorylation. III. Selective phosphorylation of unprotected nucleosides. Bull Chem Soc (Japan) 42:3505–3508

Yung NC, Fox JJ (1961) Nucleosides. X. Anhydronucleosides and related compounds derived from 2′,5′-di-*O*-trityluridine. J Am Chem Soc 83:3060–3066

Zeilig CE, Goldberg ND (1977) Cell-cycle-related changes of 3′,5′-cyclic GMP levels in novikoff hepatoma cells. Proc Natl Acad Sci USA 74:1052–1956
Zielinski WS, Hynie S, Smrt J (1978) Synthesis of adenosine 3′,5′-cyclic phosphate P-(2-hydroxyethyl)ester and its action as activator of protein kinase. Coll Czech Chem Commun 43:1655–1659
Zielinski WS, Smrt J, Beránek J (1974) Novel preparation of cytidine 5′-phosphate and cytidine 3′,5′-cyclic phosphate. Coll Czech Chem Commun 39:3560–3563
Zimmerman TP (1979) Nucleoside 3′,5′-cyclic monophosphate metabolites of purine analogs. Biochem Pharmacol 28:2533–2539
Zimmerman TP, Rideout JL, Wolberg G, Duncan GS, Elion GB (1976) 2-Fluoroadenosine 3′,5′-monophosphate. J Biol Chem 251:6757–6766
Zmudzka B, Szer W, Shugar D (1962) Preparation and chemical and enzymic properties of phosphate esters of 1-(β-D-glucopyranosyl)-uracil and thymine. Acta Biochim Polonica 9:321–341
Zoller MJ, Kerlavage AR, Taylor SS (1979) Structural comparisons of *c*AMP-dependent protein kinases I and II from porcine skeletal muscle. J Biol Chem 254:2408–2412
Zoltewicz JA, Clark DF (1972) Kinetics and mechanism of the hydrolysis of guanosine and 7-methylguanosine nucleosides in perchloric acid. J Org Chem 37:1193–1197
Zoltewicz JA, Clark DF, Sharpless TW, Grahe G (1970) Kinetics and mechanism of the acid-catalyzed hydrolysis of some purine nucleosides. J Am Chem Soc 92:1741–1750

CHAPTER 3

Coupling of Receptors to Adenylate Cyclases

L. BIRNBAUMER and R. IYENGAR

Overview

Hormone- and neurotransmitter-sensitive adenylate cyclases are information-transfer systems that are dependent for their activities on Mg and are stimulated synergistically by guanine nucleotides and agonists. Three separate membranebound entities are responsible for the transmembrane signalling: receptor (R), the catalytic moiety of adenylate cyclase bearing the site that converts ATP to cAMP (C), and a guanine nucleotide-binding regulatory component (N) that affects receptor and catalytic component behavior, is a GTPase that hydrolyzes GTP to GDP, and acts as the actual coupler or transducer of the information transfer.

Studies on the kinetics of cAMP formation by isolated membrane particles containing functionally active hormone- and nucleotide-responsive adenylate cyclase systems have revealed that the basic readout system, i.e. the nucleotide-affected adenylate cyclase or CN complex, constitutes a hysteritic enzyme where binding of guanine nucleotide proceeds rapidly but the ensuing activation is slow, and that occupied hormone receptors are anti-hysteretic conferring to the system the capacity to respond rapidly to stimulatory nucleotides.

Kinetic studies on hormone and neurotransmitter binding to receptors have revealed, on the other hand, that guanine nucleotides promote a decrease in the apparent K_D of the receptor-agonist interaction, indicating that receptors can exist in both a low affinity and a high afinity state. The latter is seen under conditions where its anti-hysteretic action is not observed and the former appears concomitantly with its anti-hysteretic effect.

N components can be extracted from membranes with detergents and shown to interact with C in solution or with C and R after reinsertion into N-deficient membranes such as provided by the so-called *cyc*$^-$ variant of S49 lymphoma cells. Reconstitution studies of R and C regulation in *cyc*$^-$ membranes showed that guanine nucleotides including GTP, GDP, and analogs such as GMP–P(NH)P and GTPγS, not only affect C through interaction with N, but also affect the high to low affinity state transition of receptors through interaction with N. However, while GTP and its analogs stimulate basal adenylate cyclase activity and GDP and its analogs either do not affect it or inhibit it, both GTP and GTP-like as well as GDP and GDP-like compounds promote equally the appearance of so-called "low affinity" states of receptors.

This chapter describes and contrasts various schools of thought on the role that guanine nucleotides and the regulatory component N may have in regulating R and C. We discuss whether it is the high affinity or the low affinity form of the receptor that is the active one responsible for N-mediated activation of C, whether R does or does not form part of the hormone-affected activated adenylate cyclase complex and if so whether one or two N components need to form part of it, and whether the role of hormone-occupied receptors may be merely one of "facilitating" the activation of C by N.

A. Introduction

Hormone and neurotransmitter sensitive adenylate cyclases are hysteretic enzyme systems. As such they respond slowly both to guanine nucleotides, their main intracellular regulators, and to receptor occupancy by hormones, their extracellular regulators whose changes in concentration need to be transduced into changes in cAMP formation (RODBELL et al. 1971a, b; LONDOS et al. 1974; SALOMON et al. 1975; SCHRAMM and RODBELL 1975). Much work has been done to elucidate the mechanism by which hormones and guanine nucleotides regulate adenylate cyclases. It has been found that these systems are constituted of at least three components: the receptor, responsible for hormone binding and initiation of the hormonal response (ORLY and SCHRAMM 1976; LIMBIRD and LEFKOWITZ 1977; SCHRAMM 1979); the catalytic component, responsible for converting ATP to cAMP; and a nucleotide regulatory component, playing a central role in mediating the guanine nucleotide regulation of the system (ROSS and GILMAN 1977; ROSS et al. 1978; HOWLETT et al. 1979; STERNWEIS and GILMAN 1979; PFEUFFER 1977, 1979). Kinetic and biochemical studies have revealed that adenylate cyclase has associated with it a GTPase activity and that this GTPase activity is inhibited by cholera toxin (BLUME and FOSTER 1975; LEVINSON and BLUME 1977; CASSEL and SELINGER 1976, 1977a). It is thought that the catalytic GTPase site is the allosteric guanine nucleotide binding site of the nucleotide regulatory component and that the action of this GTPase leads to formation of an enzyme-GDP complex that is inactive. This led to the proposal, principally by Cassel and Selinger (CASSEL and SELINGER 1977a, 1978; LAMBERT et al. 1979), that in the presence of GTP the activity of adenylate cyclase is controlled by a turnover cycle involving the hydrolysis of GTP with formation of P_i and a relatively inactive enzyme-GDP complex, and that the known stimulatory effect of cholera toxin is due to inhibition of GTP hydrolysis at the regulatory site and accumulation of active enzyme-GTP complex. Given this basic make up, the following questions need to be asked:

(1) What is the nature of the "coupling" process by which receptors affect cAMP formation?

(2) What precisely is the role (or roles) of guanine nucleotides in this coupling process?

The answers to these questions will have to take into consideration the nature of the hysteretic response of adenylate cyclases to nucleotides, the effects nucleotides have on hormone-receptor binding, and the architectural make up of adenylate cyclases in as much as deciding how many nucleotides are simultaneously involved with a system when hormone receptor couples to it.

Since hormonal stimulation of adenylate cyclases is accompanied by a disappearance of hysteresis in their response to nucleotides, causing these systems to bind and respond rapidly to guanine nucleotides (SALOMON et al. 1975; SCHRAMM and RODBELL 1975; CASSEL and SELINGER, 1978; CUATRECASAS et al. 1975; LONDOS et al. 1977; IYENGAR et al. 1980; LEFKOWITZ and CARON 1975; RENDELL et al. 1977; TOLKOVSKY and LEFKOWITZ 1978a), the nature of this hysteretic response has been a center of attention in many laboratories. The fact that adenylate cyclases exhibit hysteresis was discovered in kinetic experiments that examined the mode of action of GMP–P(NH)P. This analog of GTP activates adenylate cyclases not only more than the natural effector GTP, but does so with a characteristic lag leading to activated states that have rather long half-lives (LONDOS et al. 1974; SALOMON et al. 1975; SCHRAMM and RODBELL 1975). Two independent and mutually exclusive explanations have been set forth to account for the hysteresis observed with GTP analogs, and under special circumstances, also with GTP. One comes from analyzing the system as an allosterically controlled enzyme. It proposes that lags in nucleotide activation of adenylate cyclase are due to a slow conversion or transition of an inactive enzyme-nucleotide complex to its active form (BIRNBAUMER et al. 1980a, b) and that the primary mechanism by which receptors affect adenylate cyclase is by accelerating the forward conversion of the inactive to the active state of the enzyme. Since this results not only in faster activation by the nucleotide but also in an increase in the proportion of the enzyme system in the active form, this explanation accounts for both the increased rate of nucleotide activation and the enhanced catalytic activity at steady state observed in the presence of hormones (RENDELL et al. 1977; IYENGAR et al. 1980a, b; ABRAMOWITZ et al. 1980). The other explanation comes from analyzing the system in terms of its GTPase-dependent turnover cycle. It assumes the enzyme-GTP complex to be stable and not susceptible to stimulation by hormone and proposes that the lags in progress curves with GTP analogs are due to slow release of GDP from the inactive enzyme-GDP complex. In this model, the primary action of receptor is to decrease the affinity of the system for inhibitory GDP (CASSEL and SELINGER 1978; ECKSTEIN et al. 1979; CASSEL et al. 1979). The reduced affinity for GDP is then assumed to facilitate its exchange for a stimulating nucleotide, a process that results in activation of the system. Close examination of these two proposals indicates that while the former is not dependent on presence of a GTPase activity at the regulatory site for explaining nucleotide and hormone effects, the latter is forcibly linked to existence of the GTPase turn-off mechanism and precludes the possibility of hormonal stimulation upon constant occupancy of the system by GDP.

The effects of guanine nucleotides on the interaction of receptor with hormone have also received much attention. Two forms of receptor can be defined: that seen in the absence of guanine nucleotides, and that seen in their presence. Two schools of thought exist as to which is the active form. One believes that the hormone-receptor complex seen in the absence of nucleotide is the active one (ECKSTEIN et al. 1979; CASSEL et al. 1979), the other tends to think that the nucleotide sensitized form of the hormone-receptor complex is the one that "couples" to adenylate cyclase activation (IYENGAR et al. 1980b). There are considerations that would indicate that both forms of receptor may be required and hence that neither is "inactive."

Since at least two functional aspects, cyclizing activity and hormone binding, are affected by guanine nucleotides, the obvious question arises as to whether both functions are mediated simultaneously by a common nucleotide binding site. Kinetic and biochemical evidence points to the existence of two types of guanine-nucleotide sites. One is associated with the basic adenylate cyclase and appears to have GTPase activity. The other is associated with receptor regulation; it is not known whether it is also a GTPase. On the other hand, unsuccessful complementation attempts between two S 49 cell lines: a cyc^- variant, deficient in nucleotide regulation of cyclizing activity and hormonal stimulation, and an UNC variant, defective in hormonal stimulation but competent in nucleotide stimulation, would indicate that only one nucleotide regulatory component, and hence one guanine nucleotide binding site is involved in the regulation of both cyclizing activity and hormonal stimulation.

This article presents our view of how all of the above may integrate into a coherent picture of the mode by which hormones and nucleotides interact to regulate adenylate cyclases. We will present opposing views and shall include the analysis of adenylate cyclase systems in terms of a two-state enzyme model whose state transitions are regulated by both nucleotides and receptors. This model is found to account for many nucleotide-dependent amplification and dampening effects that are observed in the transduction of hormone-triggered receptor signals into cAMP production.

B. Guanine Nucleotide Regulation of Adenylate Cyclases

Although the effects of guanine nucleotides were first discovered in relation to hormone binding in an hormonally responsive adenylate cyclase system (RODBELL et al. 1971 b) it is now well established that these nucleotides can regulate adenylate cyclase activity in the absence of any receptor-mediated event. This was evidenced by the finding that nucleotide regulation of cyclizing activity is present in unaltered form in two hormone unresponsive variants of S 49 cells: the so-called UNC variant that has both adenylate cyclase and hormone receptors but lacks a coupling mechanism (HAGA et al. 1977 a) and the so-called β_d variant that also has adenylate cyclase but lacks up to 90% of its hormone receptors (JOHNSON et al. 1979). Further, extraction of adenylate cyclases from membranes with non-ionic detergents results in loss of hormone effects but not of guanine nucleotide stimulation (LEFKOWITZ and CARON 1975; WELTON et al. 1977, 1978; NORTHUP et al. 1979).

The principal characteristics derived from studies on the effects of GTP, GTP analogs and cholera toxin on the rat liver adenylate cyclase in the absence of hormone (BIRNBAUMER et al. 1980 a; IYENGAR et al. 1980 a) are summarized in part in Fig. 1. These include the following observations: (1) The enzyme system is hysteretic in its response to all guanine nucleotides, being slower in its responses at lower the Mg ion concentrations than at higher Mg ion concentrations. (2) Different GTP analogs stimulate this system to different degrees, with GMP–P(NH)P and GTPγS stimulating more than GMP–P(CH_2)P which in turn stimulates somewhat more than GTP. (3) GTP and its analogs activate this system at different rates, with GTP being fastest followed by the much slower GTPγS which in turn is somewhat faster than GMP–P(NH)P and GMP–P(CH_2)P. (4) The rate of activation by nucleotides is independent of the extent to which they activate at steady state. (5) Cholera toxin only affects the extent to which GTP stimulates the system.

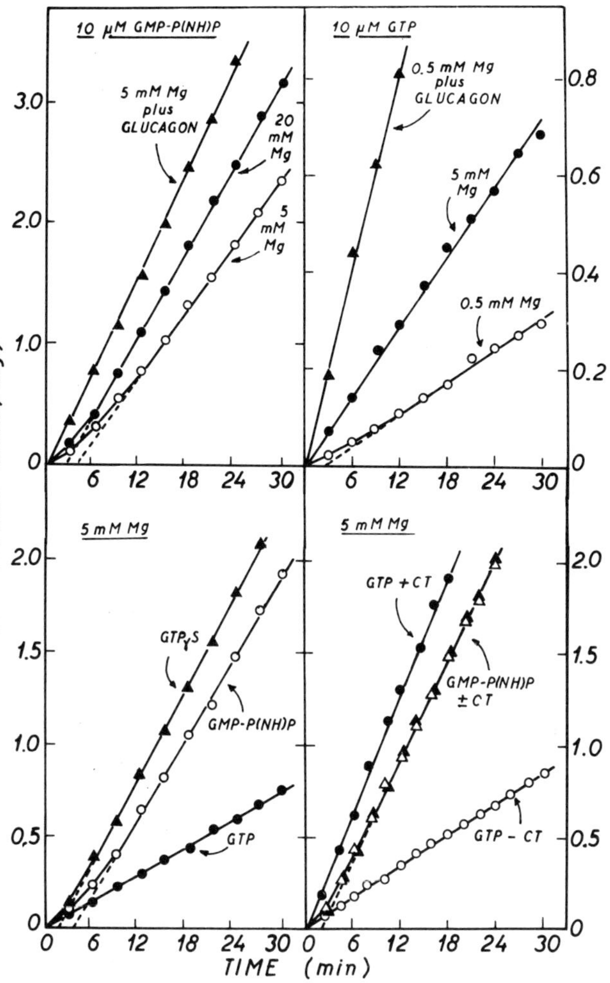

Fig. 1. Hysteretic nature of nucleotide activation of liver adenylate cyclase: Effects of Mg^{++} ion, hormonal stimulation and cholera toxin. *Upper Left.* Hysteresis in the response induced by GMP–P(NH)P: loss upon hormonal stimulation and decrease upon increase in Mg^{++} concentration in the assay. *Upper Right.* Hysteresis in the response induced by GTP: loss upon hormonal stimulation and decrease upon increasing the Mg^{++} concentration in the assay. *Lower Left.* Dependency of hysteresis on type of nucleotide. Assays carried out at 5 m*M* Mg^{++}. *Lower Right.* Selective enhancement of GTP response by cholera toxin and lack of effect on hysteretic nature of the response to the GTP analog GMP–P(NH)P. Mg^{++} in assays: 5 m*M*. Mg^{++} ion concentrations are expressed as m*M* in excess of added EDTA (1 m*M*) and ATP (0.1 m*M*). (For further details see BIRNBAUMER et al. 1980a)

(6) Cholera toxin does not alter the transients (lags) observed in the activation of this system by GTP analogs. (7) Other experiments show that the lags observed with GTP analogs can not be shortened by increasing their concentration (SALOMON et al. 1975; BIRNBAUMER et al. 1980a). These observations indicate clearly that in the rat liver adenylate cyclase system the rate limiting step in activation

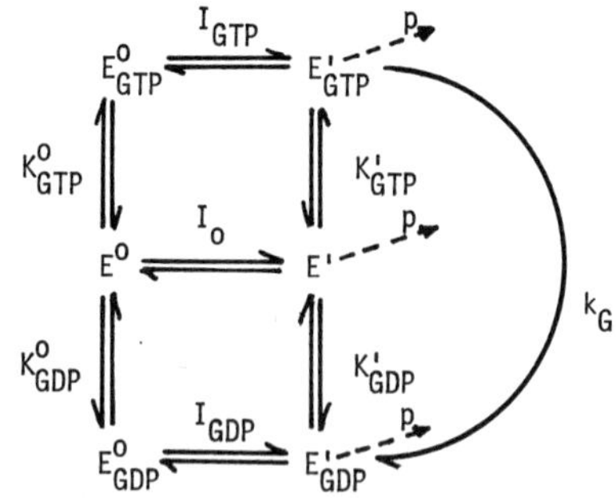

Fig. 2. Scheme of a two state enzyme model with the capacity to metabolize the regulatory activating ligand as applied to adenylate cyclase regulation by guanine nucleotides. The scheme is based on principles set forth by MONOD, CHANGEUX and collaborators (1963, 1965). The enzyme system is assumed to exist in either of two states: an inactive ^{0} (zero) and an active $'$ (prime) state and to be distributed into six compartments depending on its occupancy by GTP or GDP. K values are intrinsic equilibrium association constants for binding of ligands, GTP or GDP (subscripts) to the ^{0} or $'$ states (superscripts) of the enzyme. I values are equilibrium constants of the ^{0} to $'$ transitions in the absence (I_0) or presence of ligands (I_{GTP} and I_{GDP}), with $I_0 = E'/E^0$, $I_{GTP} = E'_{GTP}/E^0_{GTP}$ and $I_{GDP} = E'_{GDP}/E^0_{GDP}$. k_G is the first order rate constant describing E'_{GTP} to E'_{GDP} conversion. Since the $'$ state has $k_{cat} = p$ regardless of whether free (E') or occupied (E'_{GTP} or E'_{GDP}), fractional activation of this system v/V_{max} is given by the sum $E'/E_t + E'_{GTP}/E_t + E'_{GDP}/E_t$. Due to the circular arrangements arising from the two-state assumption, the following equalities need to hold: 1. $K^0_{GTP} \cdot I_{GTP} = I_0 \cdot K'_{GTP}$, and 2. $K^0_{GDP} \cdot I_{GDP} = I_0 \cdot K'_{GDP}$. It follows that I values in the absence of ligand or in the presence of an inactivating ligand such as GDP are lower than I values in the presence of activating ligands such as GTP or GTP analogs, i.e., that $I_0 < I_{GTP}$ and $I_{GDP} < I_{GTP}$. Thus, the $I^0/I_{Nucleotide}$ ratio will determine the $K^0_{Nuclotide}/K'_{Nucleotide}$ ratio. Since I_0/I_{GTP}, I_0/I_{GDP}, and $I_0/I_{GMP-P(NH)P}$ ratios are susceptible to experimental measurement by assaying the enzyme in the absence and presence of ligand, it is possible to assign relative values to the affinities of the inactive and active forms of the enzyme for each of the ligands. For example, measurement of activities of the basic adenylate cyclase system in liver membranes has shown that their basal activity ($E'/E^0 = I_0$ in the above model) is not altered by GDP addition. It follows that in the liver system $I_0 = I_{GDP}$ and hence that $K^0_{GDP} = K'_{GDP}$, i.e. that the microscopic equilibrium affinity constants for GDP binding are the same for the active and the inactive states of the liver enzyme system. On the other hand, since GTP and GTP analogs stimulate the system, it follows that I_{GTP} (or $I_{GTP\,analog}$) $> I_0$ and that $K^0_{GTP} < K'_{GTP}$, i.e. that activation of the system is associated with an increase in the intrinsic affinity of the system for GTP and its analogs. It should be realized, of course, that the actual overall affinity of the system for nucleotide is only one since it is given by $K^0_{GTP} \cdot I_{GTP} = I_0 \cdot K'_{GTP}$. It should be noted, however, that rates of exchance may differ upon enzyme activation. If an increase in K_{GTP} from K^0_{GTP} to K'_{GTP} is assumed to be due to a decrease in the first order rate constant of GTP dissociation from E' (l'_{-} on Figs. 3 and 4), this would result in a slowdown in the rate of relaxation of the active complex. Although this has not been observed for GTP, this is the main characteristic of action of GMP–P(NH)P on the liver adenylate cyclase system (LONDOS et al. 1977; BIRNBAUMER et al. 1980a)

by nucleotides is nucleotide specific and, that dissociation of enzyme-GDP complex to free enzyme plus GDP, a phenomenon that would be common to the effects of all nucleotides, cannot be the rate limiting step in the action of GTP and GMP–P(NH)P in this system. The effect of cholera toxin, increasing GTP-induced activity only, is consistent with the current view that the toxin acts by inhibition of GTP hydrolysis at the nucleotide regulatory site.

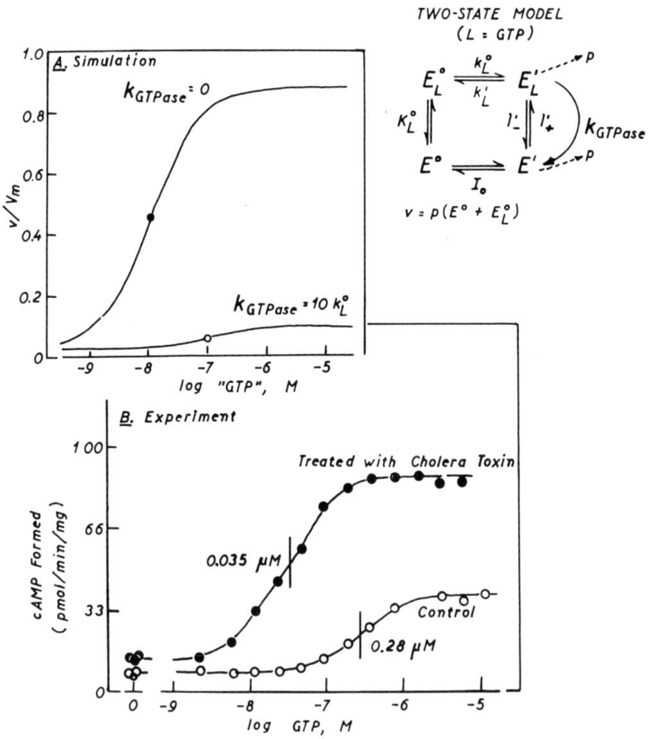

Fig. 3 A, B. Effect of cholera toxin treatment on GTP-stimulated adenylate cyclase in liver membranes. **A** Simulation of effect of presence of GTPase-like activity (K_{GTPase}) in the two-state enzyme model on height and position of concentration-effect curve for stimulating ligand L (GTP). Simulations were carried out according to analytical solutions to the two-state scheme shown on the upper right hand corner of the figure (BIRNBAUMER et al. 1980 b). **B** Effect of cholera toxin on the concentration-effect curve of the stimulatory action of GTP on liver membrane adenylate cyclase. (For details see BIRNBAUMER et al. 1980 a.) Note that the observed effect of cholera toxin is consistent with the assumption that it acts by inhibiting a GTPase intrinsic to the adenylate cyclase system

Based on the above findings, we analyzed the liver adenylate cyclase system in terms of a two-state enzyme model (MONOD et al. 1963, 1965), in which the binding of allosteric ligands is rapid and temporally isolated from the process or processes leading to conversion of an inactive enzyme-nucleotide complex to an active-enzyme nucleotide complex. Such a model, applied to adenylate cyclase is shown in Fig. 2. It depicts the system (E) to exist in either in inactive ^{0}(zero) state or in an active $'$(prime) state, the latter capable of forming cAMP from ATP with a catalytic rate $k_{cat} = p$. The model includes a GTPase feature which upon enzyme activation by GTP degrades enzyme-GTP to enzyme-GDP. Activation of a two-state enzyme by an allosteric ligand comes about by shifts in I values, i.e., in the active/inactive state ratios.

Figure 3 shows a simulation of the effect of cholera toxin on the GTPase-regulated basic two-state enzyme model and its positive correlation with the experimental finding in liver membranes. Since in the rat liver system GDP release is not rate

Table 1. Effects of guanine nucleotides and hormones on selected adenylate cyclase systems Turkey erythrocyte membranes, prepared by sucrose density gradient centrifugation of a Polytron homogenate, were subjected to a 30 min treatment at 32.5 °C with 10^{-3} *M* GMP and 10^{-4} *M* isoproterenol to clear the adenylate cyclase system of endogenous GDP and then were assayed in the presence of 0.75 m*M* AMP−P(NH)P, 10 µ*M* [α−^{32}P]ATP (15×10^6 cpm), 5 m*M* $MgCl_2$, 1.0 m*M* EGTA, 10 m*M* KCl, 1.0 m*M* [^{3}H]cAMP (ca. 10,000 cpm, 2 m*M* DTT and 10 m*M* Tris−HCl pH 7.5 for 25 min at 32.5 °C. When activities of membranes from rat liver, rabbit corpora lutea and NS−20 neuroblastoma cells were determined, adenylate cyclase assays were for 5 min at 32.5 °C in the presence of 0.5 m*M* AMP–P(NH)P and 10 µ*M* [α−^{32}P[ATP (50×10^5 cpm), 5 m*M* $MgCl_2$, 1.0 m*M* [^{3}H] cAMP (ca. 10,000 cpm), 1.0 m*M* EDTA, and 25 m*M* Tris−HCl pH 7.5. For determination of activities in membranes from S49 murine lymphoma cells $MgCl_2$ was 10 m*M* and the buffer was Na−HEPES pH 8.0. Final volume of assays was 50 µl. When present the nucleoside triphosphate regenerating system (RS) was composed of 20 m*M* creatine phosphate (charcoal treated to remove guanine nucleotide-like contaminants), 0.2 mg/ml creatine phosphokinase and 0.02 mg/ml myokinase. For further details see Iyengar et al. (1980b)

Nucleotide additions to assays	Adenylate cyclase activities[a] (cpm × 10^{-4}/5 min/mg)	
	No Stimulant	Stimulant Added
A. *Turkey erythrocyte membranes:*		
Pretrated with isoproterenol+GMP; 5.3 µg protein per assay; stimulant: 10^{-4} *M* isoproterenol		
1. *No RS*		
—	87.2±5.7	69.5± 1.6
100 µ*M* GDP	0.8±0.3	0.9± 0.2
100 µ*M* GMP−P (NH) P	141.2±6.4	152.7±25.0
2. *Plus RS*		
—	138.0±4.8	107.8± 4.7
100 µ*M* GTP	13.7±0.3	43.2± 0.1
100 µ*M* GMP−P (NH) P	213.8±2.1	239.0± 9.9
B. *Rat liver membranes:*		
0.5 µg protein per assay; stimulant: 10 µg/ml glucagon		
1. *No RS*		
—	22.6±3.2	43.9± 2.9
100 µ*M* GDP	18.6±1.4	123.7± 4.6
100 µ*M* GMP−P (NH) P	102.3±5.3	228.6± 3.6
2. *Plus RS*		
—	17.1±2.6	57.1± 3.6
100 µ*M* GTP	47.6±3.3	184.6±10.6
100 µ*M* GMP−P (NH) P	112.3±8.2	216.3± 1.4
C. *NS-neuroblastoma membranes:*[b]		
2.5 µg protein per assay; stimulant: 10 µg/ml PGE_1; no RS		
—	10.2±0.9	11.3± 1.2
100 µ*M* GDP	5.1±0.8	56.2± 4.5
100 µ*M* GTP	22.5±2.0	119.0± 8.7
100 µ*M* GMP−P (NH) P	86.5±0.5	126.9± 1.8

Table 1 (continued)

Nucleotide additions to assays	Adenylate cyclase activities[a] (cpm × 10^{-4}/5 min/mg)	
	No Stimulant	Stimulant Added
D. *Rabbit corpus luteum membranes:* 1.2 µg protein per assay; stimulant: 10 µg/ml PGE_1, no RS		
—	13.7 ± 1.2	10.1 ± 2.3
100 µ*M* GDP	14.6 ± 1.6	24.8 ± 1.0
100 µ*M* GTP	44.0 ± 4.3	168.3 ± 15.2
100 µ*M* GMP – P (NH) P	140.7 ± 2.6	205.3 ± 6.0
E. *S49 murine lymphoma cell membranes:*[c] 1.75 µg protein per assay; stimulant: 10^{-4} *M* isoproterenol; no RS		
—	7.8 ± 1.2	9.1 ± 0.8
100 µ*M* GDP	8.6 ± 0.9	61.8 ± 4.6
100 µ*M* GTP	19.5 ± 0.7	93.0 ± 1.6
100 µ*M* GMP – P (NH) P	60.6 ± 1.1	106.0 ± 2.8

[a] Activities are expressed in terms of cpm × 10^{-4} [α – ^{32}P] ATP converted in 5 min to [^{32}P] cAMP per mg membrane protein. If substrate efficacy of ATP and AMP – P (NH) P are assumed to be equal the values reported can also be taken as pmoles/min/mg. Values are mean ± standard deviation of triplicate determinations

[b] We thank Dr. Arthur Blume (Roche Institute of Molecular Biology) for providing us with the cells for preparation of these membranes

[c] We thank Drs. Henry Bourne (Department of Medicine, University of California at San Francisco) and Alfred G. Gilman (Department of Pharmacology, University of Virginia at Charlottesville for providing us with the starting stocks of S49 cells

limiting, the model used for the simulation was collapsed to the form shown on the figure. Both the model and the experimental system agree in that the presence of GTPase activity leads to a right shift in the concentration-effect curve for GTP to higher concentrations *and* in a decrease in the apparent V_{max} of the system. Other experiments showed that neither the apparent K_m for activation nor the V_{max} observed with GMP–P(NH)P, GTPγS or GMP–P(CH_2)P are affected by toxin treatment (*Birnbaumer* et al. 1980 a), thus providing support for the proposal of CASSEL and SELINGER that cholera toxin acts via inhibition of a GTPase associated with adenylate cyclase (CASSEL and SELINGER 1977 a).

By enzyme system E we refer to a complex of the two known components involved in nucleotide regulation of adenylate cyclases: the catalytic *C* component responsible for the cyclization reaction and the guanine nucleotide regulatory *N* component responsible for binding of nucleotide. The existence of these components was discovered independently by Gilman and collaborators (ROSS and GILMAN 1977; ROSS et al. 1978; HOWLETT et al. 1979; STERNWEIS and GILMAN 1979) exploring the biochemical lesion of the *cyc*$^-$ S 49 cell variant developed by BOURNE et al. (1975), and by PFEUFFER (1977, 1979) working on the separation by affinity chromatography of pigeon erythrocyte adenylate cyclase subunits. Their studies revealed that *C* in the absence of *N* only exhibits cyclizing activity when assayed with Mn^{++} ion and that *N* is responsible both for relaxing the divalent cation specificity so as to allow *C* to use more abundantly available Mg^{++} and for conferring nucleotide regulation to *C*. JOHNSON et al. (1978) and CASSEL and PFEUFFER (1978) showed that *N* is the site of cholera toxin action, the effect being due to NAD^+-dependent ADP-ribosylation of a 42,000 molecular weight subunit. PFEUFFER, using [^{32}P]ADP-ribosylated *N* and resolved *C* from pigeon erythrocytes showed that the GTP analog GTPγS promotes the association of *C* and *N* as seen in sucrose density gradient centrifugations by appearance of a Mg^{++}-dependent activity and co-migration of labeled *N* (PFEUFFER 1979). It is possible therefore, that after nucleotide binding, activation of adenylate cyclase involves first association of *C* and *N* followed by isomerization from the inactive to the active form. Recent experiments in our laboratory (BHAT et al. 1981a) using solubilized fractions from liver membranes, indicate that also in this tissue guanine nucleotides affect *C–N* interactions. This suggests strongly that slow transitions observed in time course studies need to be analyzed in terms of both changes in subunit interactions and subsequent isomerizations of state.

While it is clear that dissociation of GDP from the enzyme complex is not rate limiting in nucleotide activation of the liver adenylate cyclase, a study similar to that shown on the bottom left side of Fig. 1, using turkey erythrocyte membranes (ABRAMOWITZ et al. 1980), showed that activation by either GMP–P(NH)P or GTPγS proceedes very slowly in control membranes, and very rapidly in membranes pretreated with isoproterenol and GMP, a treatment known to cause dissociation of GDP from these membranes as seen in isotope labeling studies (CASSEL and SELINGER 1978). These results indicate that the rate limiting step in the activation of untreated turkey erythrocyte adenylate cyclase is the binding of nucleotide rather than nucleotide-induced activation. As illustrated in Table 1, these studies also showed that the basal activity of the treated membranes is rather high (50%–60% of maximum in this experiment and up to 80% of maximum in other experiments) and is strongly inhibited by GDP (ABRAMOWITZ et al. 1980). It follows that contrary to what happens in liver membranes, the microscopic equilibrium affinity constant for GDP in the turkey erythrocyte system (K_{GDP}) is lower in the active state than in the inactive state, i.e. $K'_{GDP} < K^{0}_{GDP}$ (see legend to Fig. 2). Table 1 shows other examples in which GDP either does not affect the basal state distribution (liver, S 49 cells, corpora lutea) or causes it to shift towards the inactive form by a factor of about 2 (NS-20 neuroblastoma cells).

The basic properties of guanine nucleotide regulation of adenylate cyclase activities, independent of hormonal control, can therefore be summarized as follows:

(1) Guanine nucleotide stimulation requires the interaction between *C* and *N* components followed by isomerization of the complex from an inactive form to one that has the capacity to cyclize ATP to cAMP.

(2) Full activation by guanine nucleotides can be, but need not be, obtained in the absence of hormone.

(3) Activation by guanine nucleotides is often slow, and this time-dependency is abolished by hormone.

(4) There is likely to exist an enzymatic activity associated with *N* which inactivates the system by converting N_{GTP} to N_{GDP}, and which maintains the system in a constant turnover in the presence of GTP.

(5) GDP has varying effects on adenylate cyclases that range from not affecting basal activity (liver, S 49 cells, corpus luteum), to inhibiting it partially (NS-20 neuroblastoma), to abolishing all detectable activity (turkey erythrocyte system). The more inhibitory GDP is in a given system the more tightly it is bound to *N*.

(6) The experimental data are well explained by a two-state enzyme model based on the principles set forth by Monod, Changuex, and collaborators set forth in 1963 and 1965.

C. Regulation of Coupling of Hormone Receptor to Adenylate Cyclase

Coupling comprises all of the events that intervene between hormone-receptor complex formation and modification of adenylate cyclase activity. At saturating hormone concentrations, stimulations develop rapidly (BIRNBAUMER et al. 1972; KAUMANN and BIRNBAUMER 1974; RODBELL et al. 1974; NAKAHARA and BIRNBAUMER 1974; MUKHERJEE et al. 1975; BIRNBAUMER and POHL 1973), and are reversible, as demonstrated by the fact that hormone effects can be washed off and blocked by post-addition of competitive blockers (BIRNBAUMER et al. 1972; KAUMANN and BIRNBAUMER 1974; BIRNBAUMER and POHL 1973).

The first evidence for GTP being involved in coupling of hormone receptors to adenylate cyclase was obtained by Rodbell and collaborators in 1971 using hepatic plasma membranes. Since then, in most systems examined with assay components free of GTP or "GTP-like" contaminants, there is an obligatory requirement for guanine nucleotides for coupling of the hormone receptor to adenylate cyclase. This has been observed, among others, for coupling of receptors for catecholamines, serotonin, parathyroid hormone, prostaglandins, and prostacyclin (ROSS et al. 1977; NORTHUP and MANSOUR 1978; SEVILLA et al. 1976; GOLTZMANN et al. 1978; KRISHNA et al. 1972; BIRNBAUMER and YANG 1974; ABRAMOWITZ and BIRNBAUMER 1979). In addition, evidence has been presented for a role of intracellular GTP in the regulation of isoproterenol and prostaglandin E_1-induced elevation of cAMP levels in intact cells (JOHNSON and MUKKU 1979; MUKKU et al. 1979), clearly indicating that these in vitro effects have physiological bearing.

Further evidence for a role of GTP in receptor-cyclase coupling comes from the studies of Gilman and co-workers (ROSS and GILMAN 1977; ROSS et al. 1978). Using membranes from the *cyc*$^-$ variant of S 49 cells, which contain β-adrenergic receptors and the *C* component but lack the regulatory *N* component, these authors demonstrated that functional reconstitution of hormone sensitivity requires *N* and a guanine nucleotide. Without added guanine nucleotide during reconstitution of *C* with *N* in *cyc*$^-$ membranes, catecholamine-sensitive adenylate cyclase activity was not restored even though nucleotide responsive adenylate cyclase activity could be obtained under these conditions. Interestingly, not only is *N* necessary for hormonal stimulation of adenylate cyclase, but its source can alter the kinetics of hormone activation of adenylate cyclase. *Kaslow* et al. (1979) reported that transfer of *N*, solubilized from turkey erythrocytes, into *cyc*$^-$ membranes results in recombinant membranes that respond to nucleotide and hormone as if the *cyc*$^-$ membranes were turkey erythrocyte membranes. Experiments by PIKE et al. (1979) showed that transfer of turkey erythrocyte β-receptors into frog erythrocytes in which their β-adrenergic receptors had been inactivated by treatment with dicyclohexyl-carbodiimide, resulted in coupling of β-receptors to the frog adenylate cyclase system with preservation of the patterns of specificity and order of potency of the donating β-turkey erythrocyte receptor. However, the pattern of partial agonism was that of the receptive frog effector system, presumably formed by frog *C* associated with frog *N*. These results clearly indicate that a given receptor signal is susceptible to "interpretation" by the effector system.

Cassel and Selinger, working with turkey erythrocytes (CASSEL and SELINGER 1978; ECKSTEIN et al. 1979; CASSEL et al. 1979), have proposed that catecholamines act by facilitating either the binding or the binding and the exchange of nucleotides at the level of the guanine nucleotide regulatory component according to one of the following reactions:

Enzyme–GDP + GTP-like Nucleotide $\rightleftharpoons$

Enzyme–GTP-like Nucleotide Complex + GDP

Enzyme + GTP-like Nucleotide $\rightleftharpoons$ Enzyme–GTP-like Nucleotide Complex

with the final activity of the enzyme system being that of the given enzyme-nucleotide complex whose formation is promoted by the active receptor. Although this model has been extrapolated to other adenylate cyclase systems (SVOBODA and CHRISTOPHE, 1979), it appears incomplete. It predicts that steady state activities obtained with non-hydrolyzable nucleotides should not be affected by hormone. There are, however, systems where activity stimulated fully by GMP–P(NH)P can be stimulated further by hormone (SALOMON et al. 1975; IYENGAR et al. 1979, 1980a; CLARK 1978). Further, in most if not all adenylate cyclase systems hormones cause at least a doubling in the activity given by GMP–P(CH_2)P (IYENGAR et al. 1980a). The nucleotide exchange model would also predict that the rate of nucleotide activation would be independent of nucleotide used, thereby reflecting release of GDP. This, while being true in the particular case of turkey erythrocytes, is not the case in liver as shown above. Finally, as proposed, the model of Cassel and Selinger does not make any provisions for the observations that (1) different nucleotides activate, at steady state, to different extents as reported for GMP–P(NH)P and GTPγS, which in the presence of isoproterenol activate the turkey enzyme to 200 and 300 pmol/min/mg, respectively (CASSEL and SELINGER 1977b), and (2) hormonal activation can be seen in several systems in the presence of saturating concentrations of GDP (Table 1) and its analog GDPβS (CASSEL et al. 1979; IYENGAR et al. 1980b).

However, by accepting that hormone receptors act on the rate limiting step of the two-state enzyme model used to account for nucleotide regulation of the liver and other mammalian adenylyl cyclases, all of the above features not accounted for by the nucleotide exchange model of Cassel and Selinger, can be accounted for. We have therefore proposed (IYENGAR et al. 1980a) that hormonal regulation of adenylate cyclase, i.e. coupling, results in a facilitation of the conversion of inactive state to the active state of the system shown in Fig. 2. Thermodynamic considerations requiring that $K^{\mathring{o}}_{GDP} I_{GDP} = K'_{GDP} I_O$, indicate that all isomerization reactions and hence all I values (I_O, I_{GTP}, and I_{GDP}) have to be affected. Thus, in the scheme shown in Fig. 2 describing the nucleotide regulation of the basic adenylate system, hormonal stimulation will cause the concerted changes of I_O, I_{GDP}, and I_{GDP} to ${}^{H}I_O$, ${}^{H}I_{GTP}$, and ${}^{H}I_{GDP}$, respectively, with ${}^{H}I$ values being related to I values by a given multiplication factor that describes quantitatively the hormone effect. We termed this factor *efficacy factor* or simply *efficacy* (IYENGAR et al. 1980b).

Simulations of hormonal effects as a function of hormone concentration (C_H), employing the simplified form of the model found adequate to describe the liver adenylate cyclase system, indicate that both the position and the height of the dose-response curves for hormone depend not only on the efficacy of the receptor, but also on the starting activity ratio of the system, i.e. on the quality of the nucleotide regulatory component *N* and the type and quantity of nucleotide present at the moment of hormone addition. This is illustrated in Fig. 4 in which the effect of "receptors" with efficacies of 10, 100, and 1,000 are shown to act on an adenylate cyclase system with very low basal activity ($I_0 = 0.001$) in the absence and the presence of nucleotide with varying degrees of effectiveness. Four main conclusions can be drawn from these simulations:

(1) A given, hormone-occupied, receptor will or will not stimulate the system maximally depending on the type and quantity of nucleotide used.

(2) The more active the nucleotide is in stimulating adenylate cyclase in the absence of hormone, the more potent a given hormone will appear in the presence of that nucleotide, i.e. the system generates a difference between K_{act} – the observed dose-response relationship – and K_D – the observed occupancy of receptor.

(3) Provided the efficacy of the receptor is high enough, the effector system (adenylate cyclase proper) need not be occupied by any nucleotide to show a response to hormone or may indeed be occupied by an inhibitory nucleotide such as GDP or its analog GDPβS.

(4) If a given effector system is affected by more than one hormone receptor, and these receptors have differing efficacies, nucleotides of high effectiveness will allow for expression of receptors with both high and low efficacy, while nucleotides of low effectiveness will allow for expression of only those receptors that have a high enough efficacy. This may result in a pronounced "damoning" or even a "silencing" of the expression of low efficacy receptors.

The data shown in Table 1 illustrate that indeed the change from an activating nucleotide (GTP) to a less active or inhibitory nucleotide (GDP) will result in varying degrees of dampening of the hormonal effect in accordance with the prediction of the two-state model. Figure 5 illustrates two points: (1) While in one system (rabbit corpus luteum) the dampening effect of GDP leads to almost total abolishment of the PGE_1 response, in NS-20 neuroblastoma cells the prostaglandin receptor has a high enough efficacy that it can overpower even an inhibitory effect of GDP; (2) In both instances the apparent K_a with which PGE_1 stimulates is higher with the less effective coupling nucleotide GDP than with the more effective GTP. Both types of results are accounted for by the above proposed model for hormonal stimulation of the nucleotide regulated adenylate cyclase. The fact that dose-response curves in the presence of more active nucleotides, especially GMP–P(NH)P, are to the left of dose-response curves to the same hormones obtained with less active nucleotides has been documented in many other systems (cf. IYENGAR et al. 1980a). Further, in accordance with predictions of the two-state model proposed here, it is known that in some instances hormonal responses are not seen with GTP

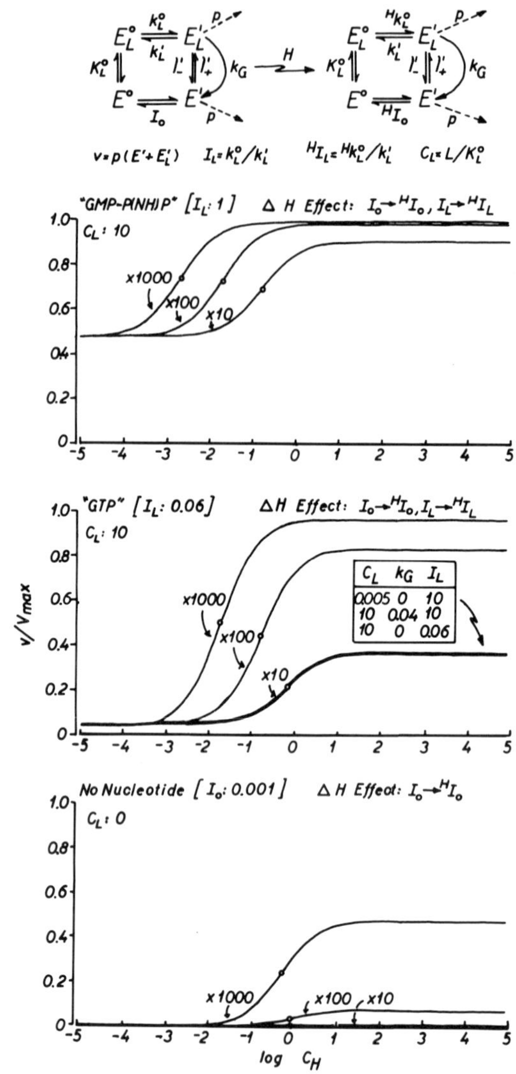

Fig. 4. Simulations of concentration-effect curves of hormone action on adenylate cyclase. It was assumed that: (1) the enzyme system behaves as a two-state enzyme according to the simplified scheme shown on the top of the figure; (2) the enzyme system is activated by nucleotides (L) that affect the transition equilibrium between the two states of the system according to $I_L = k_L^0/k'_L = E'_L/E_L^0$; (3) the enzyme system may be under control of a GTPase-type turnoff mechanism which leads from E'_L to E' with a rate constant k_G; and (4) hormones stimulate the system by altering specifically its transition equilibrium constants I_0 and I_L by a fixed *efficacy* factor in a manner that is linearly related to receptor occupancy converting them into ${}^{H}I_0$ and ${}^{H}I_L$. Three sets of simulations for hormones acting upon full occupancy of receptor with efficacies of 10, 100, and 1,000 are shown. *Top Panel:* Concentration-effect curves in the presence of an effective ligand with I_L of 1.0, i.e. capable of stimulating the system by itself to a fractional activity of 0.5 at saturation. GTPase value k_G is set at 0 to mimick the effect seen with "GMP–P(NH)P" or "GTP" after cholera toxin. A saturating concentration of L ($C_L = 10$) is assumed to be present throughout. C_L is an expression for concentration normalized with respect to its intrinsic binding affinity for the E^0 form of the enzyme according to $C_L = L \cdot K_L^0$ (see figure). *Middle Panel:* Concentration-effect curves for hormones in the presence of 10 C_L of ligand L with $I_L = 0.06$ acting on a system with $k_G = 0$,

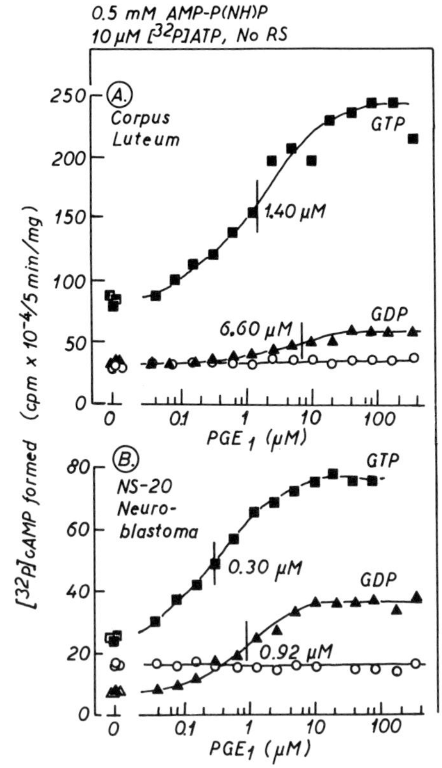

Fig. 5 A, B. Amplifying and dampening effects of GTP and GDP, respectively, on the responsiveness to prostaglandin E_1 (PGE_1) of corpus luteum and NS-20 neuroblastoma cell membrane adenylate cyclases. For details see (Iyengar et al. 1980b). Note that in both systems, the concentration effect curves for PGE_1 in the presence of stimulatory ATP is left-shifted with respect to that obtained with GDP which either does not stimulate (corpus luteum) or even inhibits (neuroplastoma) basal adenylate cyclase activity. These results are in accord with predictions from simulations such as shown in Fig. 4. It may be speculated that the corpus lutem PGE_1 receptor operates with a lower efficacy than that of neuroplastoma cells

---

i.e. in the presence of a system without GTPase affected by a guanine nucleotide with low effectiveness. Lower curve illustrates that the same results are obtained regardless of which combination of initial conditions is given, provided they lead to an identical initial fractional activation of the system onto which hormones exert their action. Three separate combinations of ligand concentration (C_L) ligand effectiveness (I_L) and GTPase activity (k_G) are shown. *Lower Panel:* Concentration-effect curve for hormones acting on the same enzyme system of $I_0 = 0.001$ in the absence of added ligand L, or in the presence of a hypothetical ligand that does not affect the transition equilibrium between states (e.g. as seen for GDP acting on liver, corpus luteum and S49 cell membrane adenylate cyclases). *Note:* (1) hormone can stimulate the system significantly provided their efficacy factor ($^{H}I/I$) is large enough; (2) the concentration-effect curves for hormones in the presence of various nucleotides are left-shifted in proportion to the effectiveness of the nucleotide used, in proportion to the efficacy of the hormone signal and, in the presence of a given nucleotide, in proportion to the concentration of this nucleotide; and (3) the GTPase-turnoff mechanism causes concentration-effect curves for hormones to become right-shifted. $C_H = H/K_H$, i.e. concentration of hormone normalized with respect to its affinity for the receptor. K_H: equilibrium dissociation constant of the receptor-hormone interaction. (For further details see Iyengar et al. 1980a) and that their detection requires the use of artificial but more effective GTP analogs such as GMP–P(NH)P. This is especially true if the hormone elicits only a weak signal from the receptor such as seen with partial agonists (KAUMANN et al. 1978). Since the degree of hormonal stimulation in response to a given receptor signal depends not only on the nucleotide, but also on the effector system to which the nucleotide binds, i.e. the type of nucleotide regulatory component *N*, the findings of PIKE et al. (1979) showing preservation of partial agonistic responses in the frog adenylate cyclase system in spite of having exchanged the receptor, and of KASLOW et al. (1979) showing that the S49 *cyc*$^{-}$ system acquires turkey erythrocyte characteristics upon complementation with turkey *N*, are also accounted for by the two-state model presented above.

The primary mode of action of hormones asumed in the two-state model provides an explanation for the finding by Cassel and Selinger that during hormonal stimulation the exchange of GDP for other nucleotides is facilitated. Thus, a receptor-mediated shift in the inactive to active transition probability is reasoned to lead concomitantly to a greater proportion of the enzyme in the active state with lowered affinity for GDP. Consequently, while there is no doubt that the dissociation of tightly bound GDP is a limiting factor in the activation of the turkey erythrocyte adenylate cyclase, this fact does not necessitate the hormone receptor to act via a primary alteration in the binding characteristics of nucleotide for the system.

TOLKOVSKY and LEVITZKI (1978a, b) studied the rate of activation of the turkey erythrocyte system by GMP–P(NH)P and found that a progressive decrease in membrane receptors results in a proportional slow-down of the rate of activation by GMP–P(NH)P without a decrease in the final activity achieved. They proposed receptors to act trough a "collision-coupling" process in which the receptor-activated state of the system is required to have a half-time that is longer than that of the receptor-enzyme complex, thus permitting one receptor to maintain concominantly active more than one adenylate cyclase complex. They did not, however, address the question as to what the actual kinetic structure of the basic nucleotide-regulated system might be and hence also did not address the question as to which kinetic parameter of the cyclizing system might be affected in a "lasting" manner by active, occupied receptor. By analyzing the kinetics of adenylate cyclase regulation in terms of a two-state enzyme model, it is possible to provide such a structure. Due to the nonlinearity in the coupling function that arises when occupied receptor affects I values of the two-state model, a "lasting" effect such as required in "collision-coupling" can be elicited. This requires, however, that the system be fully stimulated by the combination of receptor and nucleotide, as occurs in the presence of GMP–P(NH)P and catecholamines. For example, if a combination of nucleotide and active receptor leads to final activity ratio of ${}^{H}I_{L}=100$, it follows that after uncoupling, the system will show hysteresis. It can relax by as much as 90% to an activity ratio of 10 without detectable or significant decay in activity. Only after the system relaxes further will activity start falling detectably. It is at this point that the active receptor needs to couple again to "re-stimulate" the system. Since, in the meantime, receptors can interact with other systems, the two-state model asumed here allows "collision-coupling" to work. It then follows also that with active receptor combinations that do not stimulate the system supramaximally, no collision coupling effect will be seen even though they may still be acting through the same mechanism of facilitating the transition from the inactive to the active state of the system. The conclusions set forth by Levitzki and collaborators (HANSKI et al. 1979; BRAUN and LEVITZKI 1979) that adenosine affects adenylate cyclase through a basically different mechanism than the mechanism through which catecholamines exert their action may require re-investigation and redefinition.

The following proposals are therefore presented here to account for receptor-cyclase coupling and nucleotide regulation of adenylate cyclases:

(1) Hormonal activation, resulting from "coupling" of active, occupied receptors to the two-state adenylate cyclase system (Fig. 2) is due to an increase in the inactive to active state transition probability of all forms of the enzyme: the free form, the GTP-occupied form and the GDP-occupied form.

(2) Both these features account for changes in lag times in activation by GTP and its analogs, for the increased dissociation of inhibitory GDP as observed in the turkey erythrocyte system upon hormone addition, and for variations of position and height of dose-response curves to stimulatory hormones as observed in the presence of different guanine nucleotides.

D. Receptor Regulation and Receptor Forms

Throughout the above discussion on "coupling" we have often qualified receptors as active and/or occupied. Studies have revealed that these are both parameters that occur independently and that are under control of guanine nucleotides.

I. Effects of Guanine Nucleotides on Binding of Hormone to Receptor

The interaction of many hormones and neurotransmitters is affected by guanine nucleotides. The effect consists of an increase in both the association and the dissociation rate constants that characterize the interaction. It is observed only with so-called agonists, i.e. with receptor-activating hormones or neurotransmitters, be they natural or analogs, but not with competitive blockers which occupy receptors but do not elicit a response (MAGUIRE et al. 1976). In almost all instances, the *off* rate constant is affected more by guanine nucleotides than the *on* rate constant, leading to right-shifts in the concentration dependent occupancy curves of receptor by the stimulating hormone or neurotransmitter. Nucleotide effects are most readily detected in any one of three ways: (1) determination of occupancy by labeled effector as a function of concentration; (2) determination of concentration dependent displacement of a labeled receptor probe by unlabeled effector; and (3) determination of rate of dissociation of labeled effector. Of these, the first and last require an adequately labeled agonist, while the second can use any receptor probe, for it tests for the potency of the competing, unlabeled agonist. The effect of guanine nucleotides in hormone binding was first reported in 1971 by RODBELL et al. (1971 b), who studied the binding of ^{125}I-labeled glucagon to liver membranes. In these membranes, nucleotide addition resulted in an about 2-fold decrease in the affinity of this system for glucagon and an approximately 10-fold increase in the dissociation rate of prebound glucagon. More recently, using a ^{125}I-glucagon

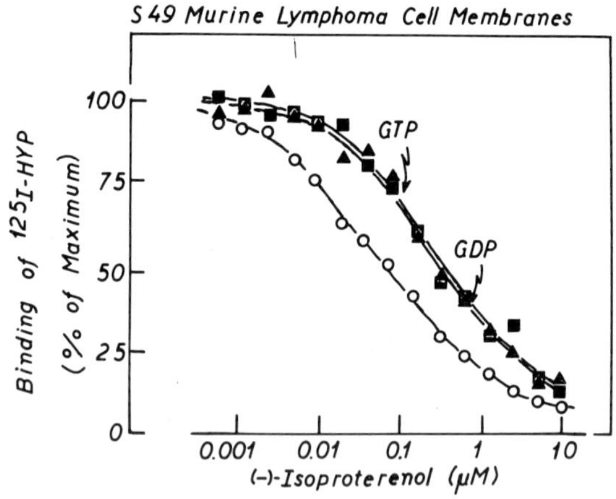

Fig. 6. Effects of GTP and GDP on agonist binding to β-receptor of S 49 lymphoma cells. (For details see IYENGAR et al. 1980b.) Note that both nucleotides are equally effective in promoting a right-shift in the concentration-dependent displacement of the β-receptor probe ^{125}I–HYP (an antagonist) by the agonist (–)-isoproterenol. This differs from their effect in these systems on basal adenylate cyclase activity which is stimulated 2- to 3-fold by GTP but is unaffected by GDP, and on isoproterenol-stimulated adenylate cyclase activity which in the presence of GDP is only about 70% of that seen in the presence of GTP (IYENGAR et al. 1980b). This and other considerations presented in the text led us to the notion that possibly more than one nucleotide binding site is involved in receptor-mediated activation of adenylate cylcases

probe purified by high pressure liquid chromatography and reagents especially freed of most if not all contaminating guanine nucleotides, we have observed nucleotide effects on glucagon binding that are much larger, yielding decreases in affinity due to GTP addition in the order of 60- to 100-fold (SWARTZ and BIRNBAUMER, unpublished). Increased dissociation rates and/or decreased affinity of hormone or neurotransmitter for its receptor under the influence of guanine nucleotides have now been observed, among others, for angiotensin binding to adrenal cortex (GLOSSMAN et al. 1974), catecholamine binding to S 49 cell membranes (MAGUIRE et al. 1976), binding of catecholamines (WILLIAMS and LEFKOWITZ 1977) and prostaglandins (LEFKOWITZ et al. 1977) to frog erythrocyte membranes, catecholamine binding to α-adrenergic receptors in calf brain membranes (U'PRICHARD and SNYDER 1978), vasoactive intestinal peptide binding to guinea pig brain membranes (ROBBERECHT et al. 1978), dopamine binding to rat corpus striatum membranes (CREESE et al. 1979), muscarinic ligand binding to rat myocardial membranes (BERRIE et al. 1979; ROSENBERGER et al. 1979) and opiate ligand binding to membranes from a hybrid neuroblastoma-glioma cell line (BLUME et al. 1979). Figure 6 illustrates a typical experiment showing the effect of guanine nucleotides on the displacement of the β-receptor probe ^{125}I-HYP by increasing concentrations of the β-agonist isoproterenol in "wild-type" S 49 cell membranes. Although most of the receptor systems listed above are known to be associated with either positive or negative modulation of adenylate cyclase, this may not be always the case. Thus, while at this time muscarinic receptors do interfere with cAMP production in the

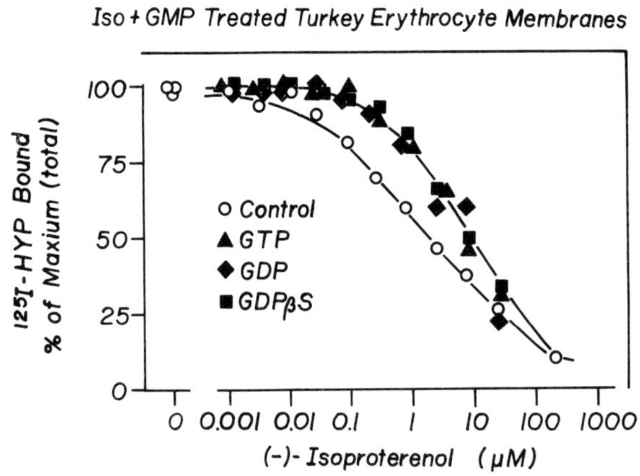

Fig. 7. Effect of guanine nucleotides on agonist binding to β-receptor in turkey erythrocyte membranes previously treated with isoproterenol and GMP. Note that GTP, GDP, and the GDP analog GDPβS affect agonist binding in an equivalent manner. (For details see IYENGAR et al. 1980b)

neuroblastoma-glioma cell line (LICHTSHTEIN et al. 1979) and in heart membranes (JAKOBS et al. 1980), it is not known, for example, whether α-adrenergic stimulation of heart muscle affects cAMP production. The possible role of nucleotide effects on binding of membrane receptors, seemingly not related to adenylate cyclase, is of course a center of current research.

Guanine nucleotide regulation of receptor binding requires its association with a nucleotide-binding regulatory component, much in the way that Mg^{++}-dependent cyclizing activity of the *C* component of adenylate cyclase requires association with *N*. Evidence for this comes from studies by Gilman's group (ROSS and GILMAN 1977; ROSS et al. 1978; STERNWEIS and GILMAN 1979), who showed that in both *cyc*⁻ and UNC S49 cell membranes, agonist binding is unaffected by guanine nucleotides, and that upon "reconstitution" with heated "wild-type" extracts the nucleotide effect on binding is restored. It should be noted that agonist binding in *cyc*⁻ and UNC membranes is of the low affinity type such as would be observed in the "wild-type" system upon nucleotide addition, and that reconstitution results in appearance of high affinity binding and capacity of nucleotides to induce the right shift in the agonist binding curve (HAGA et al. 1977a).

The nucleotide specificity for effects on hormone binding differs from that seen on basal adenylate cyclase activity. Thus, while GTP and GDP have widely differing effects on the cyclizing activity, with GDP either not affecting it or inhibiting it, both nucleotides affect receptor binding readily and to the same extent. This observation was made originally by RODBELL et al. (1971b) when the nucleotide effects on glucagon binding to liver membranes were studied, and is illustrated for the catecholamine interaction with S49 cell membranes in Fig. 6 and for turkey erythrocyte membranes in Fig. 7. The similarity in action of GTP, GDP, and the GDP analog GDPβS on binding appears to be of a relatively general nature, as evidenced by BLUME's work (1978) which described GDP effects on opiate binding to membranes of a neuroblastoma-glioma cell.

Further it has also been shown that cholera toxin treatment which has a marked effect on GTP stimulation of adenylate cyclase because it prevents GTP hydrolysis to the less active GDP and markedly reduces the K_{act} for hormone in the presence of GTP, has no effect on the K_D for hormone binding in the presence of GTP (ROSS et al. 1977).

II. Definition of an Active Form of the Receptor

Two alternative definitions of the active form of hormone-occupied receptor are possible: the nucleotide-free high affinity form, or (2) the nucleotide-sensitized low affinity form. Lefkowitz's laboratory used both an agonist and an antagonist to study the frog β_2-receptor system and correlated nucleotide effects with agonism (STADEL et al. 1980; KENT et al. 1980). Using the antagonist [^{3}H]dihydroalprenolol ([^{3}H]DHA) as a probe it was observed that while antagonists such as propranolol displace [^{3}H]DHA according to a simple binding isotherm, the agonist isoproterenol yields displacement curves that in the absence of nucleotide are best fitted by a two-site isotherm with a high (H) and a low (L) affinity site defined by the binding equilibrium constants K_H and K_L, respectively. The K_H/K_L ratio for isoproterenol was found to be in the order of 100. About 90% of the total sites appeared in the H form. Addition of GTP resulted in abolishment of high affinity binding due to a right shift of the isoproterenol displacement curve, which now showed 100% of the sites in the L form. Binding studies with the agonist [^{3}H]hydroxybenzylisoproterenol ([^{3}H]HBI) showed a binding isotherm corresponding to the H form of the receptor with the total number of sites measurable being about 90% of those detectable with [^{3}H]DHA. Addition to GTP abolished all measurable specific [^{3}H]HBI binding in agreement with the fact that an approximately 100-fold decrease in binding affinity for agonist would make its assay by direct binding experimentally impossible. Displacement of [^{3}H]DHA by unlabeled partial agonists, i.e. analogs that interact competitively with the frog receptor but stimulate adenylate cyclase to a lower extent than the full agonist isoproterenol, yielded curves that had a K_H/K_L ratio that was lower and that showed a lower proportion of the total sites in the high affinity state. In fact a positive correlation was established relating degree of agonism (stimulation of adenylate cyclase) either to the K_H/K_L ratio or to the proportion of total sites in the H form when measured in the absence of guanine nucleotide. Since in the absence of nucleotide agonists tend to interact (stabilize) a larger proportion of the receptor sites in the H state, Lefkowitz and collaborators have concluded that this state is the active state of the receptor, whose formation is required for the triggering of adenylate cyclase activation.

Although the conclusion that the high affinity form of receptor is the active one appears reasonable and correlates with the independent information that uncoupled and hence inactive receptors in *cyc*$^-$ and UNC membranes are in their low affinity form (L state), kinetic experiments on nucleotide dependency of receptor-mediated stimulation of adenylate cyclase, especially as seen in the presence of GDP, suggest otherwise. Of importance in this respect are the data shown in Table 1 and Fig. 5 on PGE_1 and GDP action on NS-20 neuroblastoma adenylate cyclase. As shown, this adenylate cyclase when assayed in the absence of hormone is inhibited by about 50% upon addition of GDP. In terms of the two-state model

(Fig. 2), this indicates that $I_0 > I_{GDP}$, a situation qualitatively similar to that of the turkey erythrocyte adenylate cyclase. Further, prostaglandin does not stimulate this system in the absence of nucleotide addition, but does so both in the presence of GTP as well as in the presence of GDP. If receptor were active in the absence of nucleotide and nucleotide were required at the adenylate cyclase level only, then prostaglandin should have stimulated either only in the presence of GTP, or in the presence of GTP and in the absence of nucleotide, or under all three conditions tested: presence of GTP, presence of GDP *and* absence of nucleotide. The fact that stimulation was observed only in the presence of guanine nucleotides and not in their absence indicates that nucleotide is required at the receptor level. Since this requirement leads to stimulation, it would follow that the role of nucleotide at the receptor level must be one of receptor activation. Since hormonal stimulation of adenylate cyclase does not occur in the absence of hormone, the final conclusion that can be drawn from these experiments is that receptors exist in at least two states: a resting or inactive state and an active state, the latter being stabilized (or induced) by the simultaneous interaction of hormone and nucleotide with receptor. Thus, currently there are two schools of thought, each based on kinetic data and each concluding that receptors exist in two states. But while one proposes the nucleotide and hormone-affected L state to be active, the other proposes this state to be inactive. One relies on a functional analysis of receptor effects (L state is active), the other relies on measurements of receptor interaction separately from receptor effects (L state is inactive). It is possible that depending on the mode of observation both interpretations would explain the observed data. For example, it is conceivable that "coupling" rather than being a static situation in which the receptor in a given state couples to adenylate cyclase and activates it, is a more dynamic affair in which a sequential set of reactions involves both states of the receptor in an obligatory manner. It is possible to envision the complex between receptor and nucleotide regulatory component as being in the absence of nucleotide in the H form readily picking up hormone molecules and hence becoming charged. Nucleotide binding can then stabilize the L form of the complex having two critical properties: (1) highly increased probability of "coupling" to the adenylate cyclase system causing its state transitions to change according to a two-state enzyme model such as discussed above; and (2) lowered affinity for both hormone and nucleotide, the latter due to symmetry considerations. It follows that the "coupled" system due to its low affinity for the ligands that induced it tends to fall apart into the uncoupled adenylate cyclase plus a complex between the nucleotide- and hormone-free receptor and the nucleotide regulatory component. This complex would then relax into its H state and be ready to re-initiate the cycle by binding hormone with high affinity. It should be easy to see that by assigning proper half-lives and rate constants to the reactions involved, *both* the H and the L forms of receptor would appear to be considered active since both are sequentially required in the overall binding and adenylate cyclase activation process.

III. Number of Sites Involved in Coupling

It remains to be determined if guanine nucleotide occupancy of a single site can affect regulation of hormone-receptor interactions, adenylate cyclase activity modulation and coupling. Current data indicate that more than one site needs to be occupied to achieve coupling of the hormone receptor to adenylate cyclase. The evidence on which this conclusion is based is as follows:

(1) Kinetic experiments on nucleotide and hormone stimulation of adenylate cyclase show that nucleotides cause left-shifts in the dose-response curves for hormones. Because of symmetry considerations, hormones should cause corresponding left-shifts in the dose-response curves for nucleotides. In fact, however, dose-response curves for nucleotides are not left-shifted by hormones (IYENGAR et al. 1979, 1980a, b). The data are consistent therefore with involvement of nucleotide not only at the adenylate cyclase level but additionally at the receptor level.

(2) Rodbell and co-workers prelabeled liver membranes with ^{125}I-glucagon and prepared a Lubrol PX extract which was subjected to gel filtration. Analysis of the column fractions for GMP–P(NH)P-stimulable adenylate cyclase and for guanine nucleotide stimulation of dissociation of ^{125}I-glucagon from macromolecular protein showed that the pattern of elution of nucleotide-sensitive receptor differed from that of nucleotide-sensitive enzyme (WELTON et al. 1977). This indicated that receptor and enzyme can interact independently with guanine nucleotides.

(3) ROSS et al. (1977) found that treatment of S 49 cell membranes with GMP–P(NH)P, followed by washing, resulted in complete reversal of the effects of the nucleotide on agonist binding but not of the effects on the enzyme which remained active. This strongly suggests that while GMP–P(NH)P dissociates very slowly from the site responsible for adenylate cyclase activation, it dissociates readily from the site responsible for receptor modulation.

(4) Preactivation of the liver membrane adenylate cyclase with GMP–P(NH)P followed by washing, leads to an active enzyme system that does not respond further to addition of GMP–P(NH)P alone or of glucagon alone, but which is stimulated by as much as 30% if glucagon is added in combination with GMP–P(NH)P, GTP or GDP (IYENGAR et al. 1979; IYENGAR and BIRNBAUMER 1979). These results suggest both that continued presence of free nucleotide is necessary for receptor activation and coupling, and that the characteristics of the nucleotide site associated with cyclase, causing wash-resistant "persistent" activation by GMP–P(NH)P, differ from those of the nucleotide site mediating the nucleotide effect on coupling of receptor to cyclase.

(5) CASSEL and SELINGER (1979) observed that the isoproterenol-induced release of [^{3}H]GDP from prelabeled turkey erythrocyte membranes could be enhanced by simultaneous addition of free guanine nucleotide. This "allosteric" effect of guanine nucleotide on guanine nucleotide dissociation from membranes is inconsistent with only one nucleotide site being involved in the receptor-mediated changes of nucleotide exchange rates and suggests the involvement of a second site, possibly in receptor activation.

(6) An inconsistent picture is obtained if symmetry and thermodynamic equilibrium laws are applied under the assumption that a single nucleotide binding site is mediating stimulating effects on both receptor and adenylate cyclase. Since GTP-like nucleotides stabilize an active adenylate cyclase complex, it follows that regardless of the structure of the complex, its active form must have a higher affinity for the activating nucleotide than the inactive form (MONOD et al. 1963, 1965).

Similarly, since nucleotides decrease the affinity of hormone for receptors, hormones are required to cause a decrease in the affinity for nucleotides. It is difficult, therefore, to reconcile how hormonal stimulation of adenylate cyclase in the presence of activating nucleotides can lead to *increased* nucleotide affinity at the adenylate cyclase level while at the same time leading to a *decrease* in nucleotide affinity at the receptor level, without invoking two sites.

Arguments have been set forth to oppose the above conclusion that more than one nucleotide site is involved in regulation of hormone action on adenylate cyclase. They are: a) Separation of guanine nucleotide-sensitive, bound form of ^{125}I-glucagon from nucleotide-sensitive adenylate cyclase is by no means total and excess nucleotide regulatory components over adenylate cyclase and receptors may exist; b) The reason for which GMP–P(NH)P-pretreated membranes "lose" their nucleotide effect on binding or coupling is that at the moment of addition of hormone the bound nucleotide dissociates giving rise to hormone-induced "clearing" of nucleotide from the enzyme. This would result in not enough free nucleotide to lead to significant occupancy of the nucleotide site and in the observed loss (or reversal) of the nucleotide effect on binding. However, this speculation does not withstand experimental testing. We have found that liver membranes pretreated with GMP–P(NH)P, subjected to incubation with glucagon and then washed, still have GMP–P(NH)P associated to their adenylate cyclase and still need presence of additional free nucleotide to show a hormonal response (IYENGAR, SWARTZ and BIRNBAUMER, unpublished). Furthermore, to overcome the thermodynamic considerations, it would be necessary to show that a true resting state of the liver membrane system, as well as of most of other mammalian systems, is one where enzyme is more active in the absence of GMP–P(NH)P than in its presence, i.e. where $I_0 > I_{GMP \hat{=} P(NH)P}$. Such a speculation does not seem warranted. On the other hand, STERNWEIS and GILMAN (1979) showed that *cyc*$^-$ extracts cannot promote coupling in UNC membranes under conditions that allowed extracts from "wild-type" S49 to reconsitute coupling. Although this suggests the existence of only one type of nucleotide regulatory component, it is necessary to remember that *cyc*$^-$ and UNC are variants and not deletion or amber-type point mutants. This unknown cause of their phenotypes raises the possibility that their defects overlap, with UNC missing a receptor-related regulatory coupling component and *cyc*$^-$ being deficient or defective in more than one component, specifically N and the putative coupling factor.

With the experimental evidence currently at hand, we feel inclined to analyze the mode of action of nucleotides by involving two sites, each associated with a separate nucleotide regulatory component: one binding to receptor and promoting coupling to the adenylate cyclase and the other binding to adenylate cyclase and promoting increased cyclization at ATP to cAMP. We hasten to add that this does not imply the obligatory existence of more than one kind of nucleotide regulatory component. The data are equally well interpreted if a single fully coupled system involves receptor R, catalytic component C and multiple identical N subunits bridging the "gap" between R and C, than if it involves more than one type of N component. However, a simple R–N–C complex with only one nucleotide binding site mediating nucleotide effects simultaneously on cyclizing activity and on hormone binding cannot account for the data.

E. Concluding Remarks and Future Outlooks

We hope to have presented a coherent view of how nucleotides and regulatory nucleotide-binding component(s) are involved kinetically in the transduction of the receptor signal into increased cAMP formation. The discussion was by no means exhaustive and was not meant to be complete. Rather, we wished to present a framework onto which the reader may add new information as it becomes available, hopefully without having to introduce major alterations.

I. Roles of Mg^{++}

This ion is known to affect adenylate cyclases allosterically and some of the nucleotide effects may indeed be mediated by regulating Mg^{++} effects on adenylate cyclases and receptors as evidence by the finding that hysteresis in the nucleotide effects (Fig. 1) can be decreased by increasing Mg^{++} ion. Of interest is the recent report by BIRD and MAGUIRE (1979) that not only the cyclizing reaction but also the binding of hormone to receptor is under Mg^{++} control. Studies with "wild-type" and *cyc*$^-$ S 49 cell membranes indicated that Mg^{++} augments the proportion of the S 49 cell β-receptor in the H state and that this Mg^{++} effect is absent in *cyc*$^-$ membranes. Thus, the regulatory *N* component is involved in regulation of divalent cation effects at both ends of the system. LIMBIRD et al. (1979) showed that Mn^{++}, the ion preferred by *C*, may disrupt the architectural arrangement of the system, for it leads to loss of hormonal response, i.e. uncoupling, without loss of the binding function of the receptor.

II. Inhibitory Effects of Hormones

Recent reports by several laboratories indicate that certain receptor-mediated effects are inhibitory (LICHTSHTEIN et al. 1979; SABLOL and NIRENBERG 1979; BLUME et al. 1979; COOPER and RODBELL 1979; JAKOBS and SCHULTZ 1980; JAKOBS et al. 1980; AKTORIES et al. 1980). Within the framework of the two state model developed above, two explanations come to mind. One is to assume that inhibitory receptors act by the same mechanism as stimulatory receptors, i.e. by affecting state transitions, but do so in the opposite sense. The other explanation is to assume that they act by stimulating the GTPase activity of the *N* component using a different coupling mechanism than stimulatory receptors. Since the latter explanation is dependent on GTP-mediated turnover of the system, it can be tested by assaying for the inhibitory effects both in the presence of GMP–P(NH)P and in the presence of GDP. Both nucleotides stop the turnover cycle while supporting stimulatory effects of hormones (IYENGAR and BIRNBAUMER 1979; IYENGAR et al. 1979, 1980b; KIMURA and NAGATA 1979; HANOUNE et al. 1975). They should abolish hormonal inhibition effects if these effects are due to stimulation of GTPase activity.

III. Is GTPase Indeed an Integral Part of all Adenylate Cyclases?

Although hormone-affected GTPase activities have been determined in turkey and frog erythrocytes (CASSEL and SELINGER 1976; PIKE and LEFKOWITZ 1980), as well as in a rat pancreatic membrane fraction (LAMBERT et al. 1979), such activities are difficult to measure and have not yet been reported in any other membrane, especially not in membranes whose adenylate cyclases have been extensively studied, e.g. liver, fat cell or S49 cell membranes. Research into detection of hormone responsive GTPase should prove fruitful. Although we feel it to be improbable, the possibility that effects of hormone and cholera toxin on GTPase are not causally related to changes of activity seen in the adenylate cyclase system cannot be ruled out on the basis of the kinetic data available (IYENGAR et al. 1980a). The final proof will have to come from studying GTPase activity in *N* purified to homogeneity.

IV. Are Levels of *N* Regulated?

The fact that basal Mg^{++}-dependent activity as well as hormonal responses are dependent on *N* component, clearly leads one to ask whether levels of the *N* component are constant or susceptible to regulation. An interesting observation was reported by LIN et al. (1979). They showed that addition of sodium butyrate to suspension cultures of the S3 HeLa cell strain, which under control conditions has rather high levels of basal and GTP-stimulated adenylate cyclase activities and no detectable β-receptors, results in appearance of β-receptors and, concommitant with appearance of coupling, in a lowering of both the basal Mg^{++}-dependent activity and its stimulability by GTP. It is possible that in the presence of butyrate, *N* component levels fell leading to an enzyme with lowered basal and nucleotide-stimulated activity which is now better suited to modulation by the newly induced receptor than if it would be in the presence of high levels of *N*.

V. Molecular Basis for Coupling: Size Studies and Subunit Interactions

A central question in future investigations on the molecular, rather than kinetic, basis for coupling of receptors to adenylate cyclase will be the elucidation of the architectural make up of the system under various conditions of regulation. This is already being studied in several laboratories from several angles: some laboratories are purifying the components involved, others are studying size patterns of [^{32}]ADP-ribosylated *N* component after crosslinking to neighboring proteins with bifunctional crosslinking reagents, still others are exploring size changes of adenylate cyclases in the presence of various regulators as seen in radiation inactivation studies. This last type of study has given new and rather unexpected results; SCHLEGEL et al. (1979) evaluated sizes of functional "units" that contribute to basal Mn^{++}-dependent activity, to basal Mg^{++}-dependent and hence *N*-modulated activity, to Mg^{++} activity assayed in the presence of GTP [or GMP–P(NH)P] and to the activity of the fully regulated glucagon plus GTP-stimulated enzyme system. Sizes were 150,000 for Mn^{++} activity, 230,000 for Mg^{++} activity stimulated by GMP–P(NH)P, and 350,000 for the fully stimulated GTP and glucagon affected enzyme. This indicated that the liver adenylate cyclase system increases in size with increasing complexity in regulation. Puzzling is the additional finding that basal Mg^{++} activity shows two size components: one of 240,000 molecular weight and another of 1,300,000 molecular weight. Thus one may speculate that liver *C* is of

150,000 daltons, liver *N* is of about 80,000–90,000 daltons and liver glucagon receptor is of 100,000–150,000 daltons. Not clear, and requiring further investigation, is the finding that maintenance of Mg^{++}-dependent activity is itself dependent on a 1,300,000 "pool" of presumably multimeric *N* or multimeric *CN*. In S49 cells, the protein complex responsible for Mg^{++}-dependent activity of the "wild-type" strain was reported to be of ca. 220,000–230,000 daltons (ROSS et al. 1978) and that responsible for Mn^{++}-dependent activity in *cyc*$^-$ of 190,000 daltons (HAGA et al. 1977b). The difference, presumably S49 cell *N* is in the order of 40,000, close to the size of one of the [^{32}P]ADP-ribosylated bands seen upon SDS gel electrophoretic analysis of toxin-treated S49 cell membranes (KASLOW et al. 1979). However, hydrodynamic analysis of the S49 cell *cyc*$^-$ reconstituting activity has a $S_{20,w}$ of 4.8 and a molecular weight of 126,000 (KASLOW et al. 1980). This indicates that the sizes that are being obtained cannot correspond to simple *C–N* segregations. Recently we discovered that 300,000 × *g* supernatants from various mammalian tissues (liver, heart, muscle), obtained from homogenates prepared without detergents, contain up to 30% of the total *N*-type *cyc*$^-$ reconstituting activity (BHAT et al. 1980b). This indicates that it is likely that *N* exists partitioned between membranes and cytosol. All these findings raise important questions as to what the stoichiometric *C* to *N* ratio might be and indeed whether *C* and *N* are associated in a constant ratio or whether this ratio may change with the degree of regulation. With the data at hand, the best approximation is that *C* interacts with *N* or a subunit of *N* with ca. 40,000 molecular weight, to give a *CN* complex regulated by guanine nucleotides in a manner well explained by a two-state enzyme model, and that the function of hormone receptors may be a dynamic one in which they "add" *N* subunits to the initial *CN* complex. These extra-numerary *N* subunits when associated with the basic *CN* complex would function to remove its intrinsically hysteretic behavior by increasing its state transition rates. They would tend to dissociate in the absence of receptor stimulation, thus accounting for some of the differential properties observed in the kinetics of nucleotide action on receptor and adenylate cyclase. It is clear that future investigations into the architectural make up of adenylate cyclase systems will give us more than one surprise. These surprises will be revealing and may offer new insight into the workings of these fascinating signal transducing enzyme systems.

Acknowledgements. We should like to thank Drs. Abramowitz and Bhat for carrying out some of the experiments commented on and for helpful reading and discussion of the manuscript. Thanks are due also to Mr. Mick Scheib for the secretarial assistance he provided. Parts of this work were supported by grants AM-19318 and AM-27542 from the National Institutes of Health. RI is recipient of USPHS Research Service Award AM-06066.

References

Abramowitz J, Birnbaumer L (1979) Prostacyclin activation of adenylyl cyclase in rabbit corpus luteum membranes: Comparison with 6-keto prostaglandin F_{1alpha} and prostaglandin E_1. Biol Reprod 21:609–616

Abramowitz J, Iyengar R, Birnbaumer L (1980) On the mode of action of catecholamines on the turkey erythrocyte adenylyl cyclase: Evaluation of basic activity states after removal of endogenous GDP and interpretation of nucleotide regulation and hormone activation in terms of a two-state model. J Biol Chem 255:8259–8265

Aktories K, Schultz G, Jakobs KH (1980) Inhibition of hamster fat cell adenylate cyclase by prostaglandin E_1 and epinephrine: Requirement for GTP and sodium ions. FEBS Letters (in press)

Berrie CP, Birdsall NJM, Burgen ASV, Hulme EC (1979) Guanine nucleotides modulate muscarin receptor binding in the heart. Biochem Biophys Res Commun 87:1000–1005

Bhat MK, Riser ME, Birnbaumer L (1981 a) Evidence for association-dissociation reactions between components of the rat liver adenylyl cyclase system: Effects of guanine nucleotides and glucagon. J Biol Chem (submitted)

Bhat MK, Iyengar R, Abramowitz J, Riser ME, Birnbaumer L (1980 b) Identification of naturally soluble component(s) which confer(s) guanine nucleotide and fluoride sensitivity to adenylyl cyclase. Proc Natl Acad Sci USA 77:3836–3840

Birnbaumer L, Pohl SL (1973) Relation of glucagon-specific binding sites to glucagon-dependent stimulation of adenylyl cyclase activity in plasma membranes of rat liver. J Biol Chem 248:2056–2061

Birnbaumer L, Yang PCh (1974) Studies on receptor-mediated activation of adenylyl cyclases. III. Regulation by purine nucleotides of the activation of adenylyl cyclases from target organs for prostaglandins, luteinizing hormone neurohypophyseal hormones and catecholamines. Tissue and hormone-dependent variations. J Biol Chem 249:7867–7873

Birnbaumer L, Pohl SL, Rodbell M, Sundby F (1972) The glucagon-sensitive adenylate cyclase system in plasma membranes of rat liver. VII. Hormonal stimulation: Reversibility and dependence on concentration of free hormone. J Biol Chem 247:2038–2043

Birnbaumer L, Swartz TL, Abramowitz J, Mintz PW, Iyengar R (1980 a) Transient and steady state kinetics of the interaction of nucleotides with the adenylyl cyclase system from rat liver plasma membranes: Interpretation in terms of a simple two-state model. J Biol Chem 255:3542–3551

Birnbaumer L, Bearer CF, Iyengar R (1980 b) A two-state model of an enzyme with an allosteric regulatory site capable of metabolizing the regulatory ligand: Simplified mathematical treatments of transient and steady state kinetics of an activator and its competitive inhibition as applied to adenylyl cyclases. J Biol Chem 255:3552–3557

Bird SJ, Maguire ME (1979) The agonist-specific effect of magnesium ion on binding by *beta*-adrenergic receptors in S 49 lymphoma cells. J Biol Chem 253:8826–8834

Blume AJ (1978) Interaction of ligands with the opiate receptors of brain membranes: Regulation by ions and nucleotides. Proc Natl Acad Sci USA 75:1713–1717

Blume AJ, Foster CJ (1975) Mouse neuroblastoma adenylate cyclase. J Biol Chem 250:5003–5008

Blume AJ, Lichtshtein D, Boone G (1979) Coupling of opiate receptors to adenylate cyclase: Requirement for Na^+ and GTP. Proc Natl Acad Sci USA 76:5626–5630

Bourne HR, Coffino P, Tomkins GM (1975) Selection of a variant lymphoma cell deficient in adenylate cyclase. Science 187:750

Braun S, Levitzki A (1979) Adenosine receptor permanently coupled to turkey erythrocyte adenylate cyclase. Biochemistry 18:2134–2138

Cassel D, Selinger Z (1976) Catecholamine-stimulated GTPase activity in turkey erythrocyte membranes. Biochim Biophys Acta 452:538–551

Cassel D, Selinger Z (1977 a) Mechanism of adenylate cyclase activation by cholera toxin: Inhibition of GTP hydrolysis at the regulatory site. Proc Natl Acad Sci USA 74:3307–3311

Cassel D, Selinger Z (1977 b) Activation of turkey erythrocyte adenylate cyclase and blocking of the catecholamine-stimulated GTPase by guanosine 5′-(γ-thio) triphosphate. Biochem Biophys Res Commun 77:868–873

Cassel D, Selinger Z (1978) Mechanism of adenylate cyclase activation through the *beta*-adrenergic receptor: Catecholamine-induced displacement of bound GDP by GTP. Proc Natl Acad Sci USA 75:4155–4159

Cassel D, Pfeuffer T (1978) Mechanism of cholera toxin action: Covalent modification of the guanyl nucleotide-binding protein of the adenylate cyclase system. Proc Natl Acad Sci USA 75:2669–2673

Cassel D, Eckstein F, Lowe M, Selinger Z (1979) Determination of the turn-off reaction for the hormone-activated adenylate cyclase. J Biol Chem 254:9835–9838
Clark RB (1978) Adenylylimidodiphosphate: Effect of contaminants on adenylate cyclase activity. J Cyclic Nucleotide Res 4:259–270
Creese I, Usdin TB, Snyder SH (1979) Dopamine receptor binding regulated by guanine nucleotides. Mol Pharmacol 16:69–76
Cooper DMF, Rodbell M (1979) ADP is a potent inhibitor of human platelet plasma membrane adenylate cyclase. Nature 282:517–518
Cuatrecasas P, Jacobs S, Bennett V (1975) Activation of adenylate cyclase by phosphoramidate and phosphonate analogs of GTP: Possible role of covalent enzyme-substrate intermediates in the mechanism of hormonal activation. Proc Natl Acad Sci USA 72:1739–1743
Eckstein F, Cassel D, Levkovitz H, Lowe M, Selinger Z (1979) Guanosine 5′-O-(2-thiodiphosphate). An inhibitor of adenylate cyclase stimulation by guanine nucleotides and fluoride ions. J Biol Chem 254:9829–9834
Glossmann H, Baukal AJ, Catt KJ (1974) Properties of angiotensin II receptors in the bovine and rat adrenal cortex. Modification of angiotensin II binding by guanyl nucleotides. J Biol Chem 249:825–834
Goltzman D, Callahan EN, Tregear GW, Potts JT Jr (1978) Influence of guanyl nucleotides on parathyroid hormone-stimulated adenylyl cyclase activity in renal cortical membranes. Endocrinology 103:1352–1360
Haga T, Ross EM, Anderson HJ, Gilman AG (1977a) Adenylate cyclase permanently uncoupled from hormone receptors in a novel variant of S 49 mouse lymphoma cells. Proc Natl Acad Sci USA 74:2016–2020
Haga T, Haga K, Gilman AG (1977b) Hydrodynamic properties of the *beta*-adrenergic receptor and adenylate cyclase from wild type and variant S 49 lymphoma cells. J Biol Chem 252:5776–5782
Hanoune J, Lacombe ML, Pecker F (1975) The epinephrine-sensitive adenylate cyclase of rat liver plasma membranes. Role of guanyl nucleotides. J Biol Chem 250:4569–4574
Hanski E, Rimon G, Levitzki A (1979) Adenylate cyclase activation by the *beta*-adrenergic receptors as a diffusion-controlled process. Biochemistry 18:846–853
Howlett AC, Sternweis PC, Macik BA, VanArsdale PM, Gilman AG (1979) Reconstitution of catecholamine-sensitive adenylate cyclase. Association of a regulatory component of the enzyme with membranes containing the catalytic protein and *beta*-adrenergic receptors. J Biol Chem 254:2287–2295
Iyengar R, Birnbaumer L (1979) GDP promotes coupling and activation of cyclizing activity in the glucagon-sensitive adenylyl cyclase system of rat liver plasma membranes. Evidence for two levels of regulation in adenylyl cyclase. Proc Natl Acad Sci USA 76:3189–3193
Iyengar R, Swartz TL, Birnbaumer L (1979) Coupling of glucagon receptor to adenylyl cyclase: Requirement of a receptor-related guanyl nucleotide binding site for coupling of receptor to enzyme. J Biol Chem 254:1119–1123
Iyengar R, Abramowitz J, Riser M, Birnbaumer L (1980a) Hormone receptor-mediated stimulation of the rat liver plasma membrane adenylyl cyclase system: Nucleotide effects and analysis in terms of a two-state model for the basic receptor-affected enzyme. J Biol Chem 255:3558–3564
Iyengar R, Abramowitz J, Bordelon-Riser ME, Blume AJ, Birnbaumer L (1980b) Regulation of hormone-receptor coupling to adenylyl cyclases. Role of GTP and GDP in amplification and dampening of receptor signal. J Biol Chem 255:10312–10321
Jakobs KH, Schultz G (1980) Different inhibitory effect of adrenaline on platelet adenylate cyclase in the presence of GTP plus cholera toxin and of stable GTP analogs. Naunyn-Schmiedebergs Arch Pharmacol (in press)
Jakobs KH, Aktories K, Schultz G (1980) Muscarimic receptor-mediated inhibition of myocardial adenylate cyclase. Naunyn-Schmiedebergs Arch Pharmacol 236:1–12
Johnson GL, Kaslow HR, Bourne HR (1978) Reconstitution of cholera toxin activated adenylate cyclase. Proc Natl Acad Sci USA 75:3113–3117

Johnson GL, Bourne HR, Gleason MK, Coffino P, Insel PA, Melmon KL (1979) Isolation and characterization of S 49 lymphoma cells deficient in *beta*-adrenergic receptors: Relation of receptor number to activation of adenylate cyclase. Mol Pharmacol 15:16–27
Johnson GS, Mukku VR (1979) Evidence in intact cells for an involvement of GTP in the activation of adenylate cyclase. J Biol Chem 254:95–100
Kaslow HR, Farfel Z, Johnson GL, Bourne HR (1979) Adenylate cyclase assembled *in vitro*: Cholera toxin substrates determines different patterns of regulation by isoproterenol and guanosine 5′-triphosphate. Mol Pharmacol 15:472–483
Kaslow HR, Johnson GL, Brothers VM, Bourne HR (1980) A regulatory component of adenylate cyclase from human erythrocyte membranes. J Biol Chem (in press)
Kaumann AJ, Birnbaumer L (1974) Studies on receptor-mediated activation of adenylyl cyclases. IV. Characteristics of the adrenergic receptor coupled to myocardial adenylyl cyclase: Stereospecificity for ligands and determination of apparent affinity constants for *beta*-blockers. J Biol Chem 249:7874–7885
Kaumann AJ, Birnbaumer L, Wittman R (1978) Heart *beta*-adrenoceptors. In: Receptors and hormone action, vol 3. O'Malley BW, Birnbaumer L (eds). Academic Press, New York, pp 134–177
Kent RS, DeLean A, Lefkowitz RJ (1980) A quantitative analysis of *beta*-adrenergic receptor interactions: Resolution of high and low affinity states of the receptor by computer modeling of ligand binding data. Mol Pharmacol 17 (in press)
Kimura N, Nagata N (1979) Mechanism of glucagon stimulation of adenylate cyclase in the presence of GDP in rat liver plasma membrane. J Biol Chem 254:3451–3457
Krishna G, Harwood JP, Barber AJ, Jamieson GA (1972) Requirement for guanosine triphosphate in the prostaglandin activation of adenylate cyclase of platelet membranes. J Biol Chem 247:2253–2254
Lambert M, Svoboda M, Chrystophe J (1979) Hormone-stimulated GTPase activity in rat pancreatic plasma membranes. FEBS Letters 99:303–307
Lefkowitz RJ, Caron MG (1975) Characteristics of 5′-guanylyl imidodiphosphate-activated adenylate cyclase. J Biol Chem 250:4418–4422
Lefkowitz RJ, Mullikin D, Wood CL, Gore TB, Mukherjee C (1977) Regulation of prostaglandin receptors by prostaglandins and guanine nucleotides in frog erythrocytes. J Biol Chem 252:5295–5303
Levinson SL, Blume AJ (1977) Altered guanine nucleotide hydrolysis as basis for increased adenylate cyclase activity after cholera toxin treatment. J Biol Chem 252:3766–3774
Lichtshtein D, Boone G, Blume A (1979) Muscarinic receptor regulation of Ng 108–15 adenylate cyclase requirement for Na^+ and GTP. J Cycl Nucl Res 5:367–375
Limbird LE, Lefkowitz RJ (1977) Resolution of *beta*-adrenergic receptor binding and adenylate cyclase activity by gel exclusion chromatography. J Biol Chem 252:799–802
Limbird LE, Hickey AR, Lefkowitz RJ (1979) Unique uncoupling of the frog erythrocyte adenylate cyclase system by manganese. Loss of hormone and guanine nucleotide-sensitive enzyme activities without loss of nucleotide-sensitive high affinity agonist binding. J Biol Chem 254:2677–2683
Lin MC, Lin CS, Whitlock JP Jr (1979) Reduction of GTP activation of adenylate cyclase system by its coupling to hormone receptor. J Biol Chem 254:4684–4688
Londos C, Salomon Y, Lin MC, Harwood JP, Schramm M, Wolff J, Rodbell M (1974) 5-Guanylylimidodiphosphate, a potent activator of adenylate cyclase systems in eukaryotic cells. Proc Natl Acad Sci USA 71:3087–3090
Londos C, Lin MC, Welton AF, Lad PM" Rodbell M (1977) Reversible activation of hepatic adenylate cyclase by guanyl-5′-yl-(α,β-methylane) diphosphonate and guanyl-5′-yl imidodiphosphate. J Biol Chem 252:5180–5182
Maguire ME, Van Arsdale PM, Gilman AG (1976) An agonist-specific effect of guanine nucleotides on binding to the *beta*-adrenergic receptor. Mol Pharmacol 12:335–339
Monod J, Changeux J-P, Jacob F (1963) Allosteric proteins and cellular control systems. J Mol Biol 6:306–329
Monod J, Wyman J, Changeux J-P (1965) On the nature of allosteric transitions: A plausible model. J Mol Biol 12:88–118

Mukherjee C, Caron MG, Coverstone M, Lefkowitz RJ (1975) Identification of adenylate cyclase-coupled *beta*-adrenergic receptors in frog erythrocytes with (–)-[^{3}H]alprenolol. J Biol Chem 250:4869–4876
Mukku VR, Anderson WB, Johnson GS (1979) Enhancement of hormonal stimulation in intact cells. Potentiation of GTP-dependent activation of adenylate cyclase. J Biol Chem 254:5588–5590
Nakahara T, Birnbaumer L (1974) Studies on receptor-mediated activation of adenylyl cyclases. V. Transient kinetics of the activation of beef renal medullary adenylyl cyclase by neurohypophyseal hormones. Estimation of apparent rate constants of the receptor hormone interaction. J Biol Chem 249:7886–7891
Northup JK, Mansour TE (1978) Adenylate cyclase from *fasciola hepatica*. 2. Role of guanine nucleotides in coupling adenylate cyclase and serotonin receptors. Mol Pharmacol 14:820–833
Northup JK, Renard MF, Grove JR, Mansour TE (1979) Serotonin-activated adenylate cyclase from *fasciola hepatica*. Properties of the solubilized enzyme. J Biol Chem 254:11861–11867
Orly J, Schramm M (1976) Coupling of catecholamine receptor from one cell with adenylate cyclase from another cell by cell fusion. Proc Natl Acad Sci USA 73:4410–4414
Pfeuffer T (1977) GTP-binding proteins in membranes and the control of adenylate cyclase activity. J Biol Chem 252:7224–7234
Pfeuffer T (1979) Guanine nucleotide-controlled interactions between components of adenylate cyclase. FEBS Letters 101:85–89
Pike LJ, Lefkowitz RJ (1980) Parallel activation and desensitization of *beta*-adrenergic receptor coupled GTPase and adenylate cyclase of frog and turkey erythrocyte membranes. J Biol Chem 255 (in press)
Pike LJ, Limbird LE, Lefkowitz RJ (1979) *Beta*-adrenoceptors determine affinity but not intrinsic activity of adenylate cyclase stimulants. Nature 280:502–504
Rendell MS, Rodbell M, Berman M (1977) Activation of hepatic adenylate cyclase by guanyl nucleotides. Modeling of the transient kinetics suggests an "excited" state of GTPase is a control component of the system. J Biol Chem 252:7909–7912
Rodbell M, Birnbaumer L, Pohl SL, Krans HMJ (1971a) The glucagon-sensitive adenylyl cyclase system in plasma membranes of rat liver. V. An obligatory role of guanylnucleotides in glucagon action. J Biol Chem 246:1877–1882
Rodbell M, Krans HMJ, Pohl SL, Birnbaumer L (1971b) The glucagon-sensitive adenylyl cyclase system in plasma membranes of rat liver. IV. Effects of guanylnucleotides on binding of ^{125}-glucagon. J Biol Chem 246:1872–1876
Rodbell M, Lin MC, Salomon Y (1974) Evidence for interdependent action of glucagon and nucleotides on the hepatic adenylate cyclase system. J Biol Chem 249:59–65
Robberecht P, DeNeef P, Lammens M, Deschodt-Lanckman M, Christophe JP (1978) Specific binding of vasoactive intestinal peptide to brain membranes from the guinea pig. Eur J Biochem 90:147–154
Rosenberger LB, Yamamura HI, Roeske WR (1980) Cardiac muscarinic cholinergic receptor binding is regulated by Na^+ and guanyl nucleotides. J Biol Chem 255:820–823
Ross EM, Gilman AG (1977) Resolution of some components of adenylate cyclase necessary for catalytic activity. J Biol Chem 252:6966–6969
Ross EM, Maguire ME, Sturgill TW, Biltonen RL, Gilman AG (1977) Relationship between the *beta*-adrenergic receptor and adenylate cyclase. Studies of ligand binding and enzyme activity in purified membranes of S49 lymphoma cells. J Biol Chem 252:5761–5775
Ross EM, Howlett AC, Ferguson KM, Gilman AG (1978) Reconstitution of hormone-sensitive adenylate cyclase activity with resolved components of the enzyme. J Biol Chem 253:6401–6412
Sabol SL, Nirenberg M (1979) Regulation of adenylate cyclase of neuroblastoma × glioma cells by *alpha*-adrenergic receptors. I. Inhibition of adenylate cyclase mediated by *alpha*-receptors. J Biol Chem 254:1913–1920
Salomon Y, Lin MC, Londos C, Rendell B, Rodbell M (1975) The hepatic adenylate cyclase system. I. Evidence for transition states and structural requirements for guanine nucleotide activation. J Biol Chem 250:4239–4245

Schlegel W, Kempner ES, Rodbell M (1979) Activation of adenylate cyclase in hepatic membranes involves interactions of the catalytic unit with multimeric complexes of regulatory proteins. J Biol Chem 254:5168–5176
Schramm M (1979) Transfer of glucagon receptor from liver membranes to a foreign adenylate cyclase by a membrane fusion procedure. Proc Natl Acad Sci USA 76:1174–1178
Schramm M, Rodbell M (1975) A persistent active state of the adenylate cyclase system produced by the combined actions of isoproterenol and guanylyl imidodiphosphate in frog erythrocyte membranes. J Biol Chem 250:2232–2237
Sevilla N, Steer ML, Levitzki A (1976) Synergistic activation of adenylate cyclase by guanylyl imidophosphate and epinephrine. Biochemistry 15:3493–3499
Stadel JM, DeLean A, Lefkowitz RJ (1980) A high affinity agonist – *beta*-adrenergic receptor complex is an intermediate for catecholamine stimulation of adenylate cyclase in frog and turkey erythrocyte membranes. J Biol Chem 255 (in press)
Sternweis PC, Gilman AG (1979) Reconstitution of catecholamine-sensitive adenylate cyclase. Reconstitution of the uncoupled variant of S49 lymphoma cell. J Biol Chem 254:3333–3340
Svoboda M, Christophe J (1979) Effect of hormone and guanyl nucleotide pretreatment on the activation energy of pancreatic adenylate cyclase. J Cyclic Nucleotide Res 5:377–384
Tolkovsky AM, Levitzki A (1978a) Mode of coupling between the *beta*-adrenergic receptor and adenylate cyclase in turkey erythrocytes. Biochemistry 17:3795–3810
Tolkovsky AM, Levitzki A (1978b) Coupling of a single adenylate cyclase to two receptors: adenosine and catecholamine. Biochemistry 17:3811–3817
U'Prichard DC, Snyder SH (1978) Guanyl nucleotide influences on ^{3}H-ligand binding to *alpha*-noradrenergic receptors in calf brain membranes. J Biol Chem 253:3444–3452
Welton AF, Lad PM, Newby AC, Yamamura H, Nicosia S, Rodbell M (1977) Solubilization and separation of glucagon receptor and adenylate cyclase in guanine nucleotide-sensitive states. J Biol Chem 252:5947–5950
Welton AF, Lad PM, Newby AC, Yamamura H, Nicosia S, Rodbell M (1978) The characteristics of lubrol-solubilized adenylate cyclase from rat liver plasma membranes. Biochim Biophys Acta 522:625–629
Williams LT, Lefkowitz RJ (1977) Slowly reversible binding of catecholamine to a nucleotide-sensitive state of the *beta*-adrenergic receptor. J Biol Chem 252:7202–7213

CHAPTER 4

Acute and Chronic Modulation of the Responsiveness of Receptor-Associated Adenylate Cyclases

J. P. PERKINS, T. K. HARDEN, and J. F. HARPER

Overview

Exposure of certain target cells to hormone agonists not only causes the activation of adenylate cyclase and the accumulation of cyclic AMP but sets in motion a complicated series of events designed to down-regulate the system if exposure to the agonist is extended in time. The development of refractoriness has been shown to result from a number of specifiable phenomena, each exhibiting unique kinetic properties. In a number of systems it has been demonstrated that upon exposure of cells to catecholamines there is a rapid uncoupling of β-adrenergic receptors from adenylate cyclase. This rapidly reversible event results in a selective loss of cellular responsiveness to catecholamines. With continued exposure of cells to catecholamine there occurs a slowly reversible loss of β-receptors from the cell. Evidence from several laboratories suggests that these phenomena are associated with changes in the membrane form of β-adrenergic receptors. In addition, it also has been shown that a nonspecific loss of responsiveness of adenylate cyclase can occur during incubation of cells with hormones that elevate cyclic AMP levels. At least in the C 6 rat glioma cell line, this phenomenon appears to occur as the result of a cyclic AMP-induced increase in the synthesis of a protein that confers on adenylate cyclase a reduced capacity to respond to all activators. The most detailed studies of agonist-induced refractoriness have been carried out with homogeneous cell populations maintained in vitro. However, there is widespread evidence for the occurrence of similar adaptive phenomena in vivo.

In addition to the adaptive phenomena that occur in hormone-stimulated adenylate cyclase systems, there is an increasing amount of evidence indicating that a number of neurohumoral agents can cause acute inhibition of adenylate cyclase. The fact that agonist-mediated inhibition is dependent on the presence of guanosine triphosphate suggests that these "inhibitory" receptors may directly interact with the same membrane components that are involved in activation of adenylate cyclase.

A. Introduction

I. Historical Perspective

The early formulations of the 2nd messenger concept of hormone action (SUTHERLAND et al. 1965) focused on the relation between changes in the concentration of hormone (the first messenger) and resultant alterations in the intracellular concentration of adenosine 3′,5′-monophosphate (cyclic AMP) (the 2nd messenger). Thus, a straightforward on-off mechanism was envisioned based on elevations or reductions in extracellular hormone concentrations. However, by the end of the 1960's it had become apparent that a more complex relationship exists between hormone concentration and cyclic AMP accumulation (ROBISON et al. 1971). The earliest studies in this regard reported that even in the continued presence of hormone many tissues or cell preparations exhibited a decline in cyclic AMP content (BUTCHER et al. 1965). KAKIUCHI and RALL (1968) were the first to publish experiments directly demonstrating a "tachyphylaxis" to the effect of freshly added 1st messenger on cyclic AMP formation. Subsequently, it has been reported that the effects of 1st messengers can be modulated by a variety of conditions. For example, cellular responsiveness to agonists has been shown to be altered by other hormones and neurohumors (BITENSKY et al. 1970; ROBISON et al. 1971; SHARMA et al. 1975; WOLFE et al. 1976; FOSTER and PERKINS 1978), during ontogenesis of hormone sensitive tissues (SCHMIDT et al. 1970; PERKINS and MOORE 1973; HARDEN et al. 1977a), during cell growth (HASLAM and GOLDSTEIN 1974; HARDEN et al. 1979a), etc. A variety of experimental approaches are being taken to elucidate what now appears to be an elaborate homeostatic regulatory system that functions to modulate the process through which hormones and neurotransmitters affect adenylate cyclase activity. Our present knowledge of a number of these regulatory mechanisms will be considered in this review.

II. Scope of the Review

The hormone-sensitive adenylate cyclase system is known to be composed of at least three separate moieties: a hormone receptor, a catalytic protein and a guanine nucleotide binding protein, the latter being involved in coupling the effects of hormone binding to enzyme activation. The response of adenylate cyclase to hormones could thus be regulated at any of the steps that are involved in the sequence of interaction of these proteins. For example, there is now good evidence that the *number* of hormone receptors can be regulated independently of the other components of the adenylate cyclase system. Such changes appear to involve chronic adaptive or developmental changes in cells and have been observed during ontogenesis and as a result of chronic alterations in the exposure of target cells or tissues to receptor agonists. In addition to changes in receptor number, changes in the functional properties of receptors theoretically could be expected to alter their capacity for interaction with the guanine nucleotide binding protein. Evidence that such modifications in receptor function occur will be presented.

The cellular response to hormones also could be regulated by alterations in the coupling moiety or the catalytic protein per se. There is evidence implicating both of these sites in the regulation of 1st messenger action. We will review examples of two types of such effects. First, a variety of agents, such as muscarinic, α-adrenergic, opiate, and adenosine receptor agonists all can acutely inhibit the stimulatory action of 1st messengers on adenylate cyclase activity. The observation that GTP is required for such inhibition suggests interaction with the GTP-binding component of adenylate cyclase. Second, certain cell types exhibit a tachyphylaxis to the effects of hormones that involves nonspecific changes in adenylate cyclase; under such a condition basal, NaF-stimulated, and guanine nucleotide-stimulated activities, as well as hormone-stimulated adenylate cyclase activity are all reduced. Losses in enzyme responsiveness that are nonspecific in nature suggest that changes occur in components of the adenylate cyclase system other than in hormone-specific receptor sites.

In addition to changes in the components of the adenylate cyclase system, changes in the activity state or in the amount of cyclic AMP-phosphodiesterase could alter the accumulation of cyclic AMP in response to a hormone challenge. Although some evidence exists in support of regulation at this level of the process of hormone action (MANGANIELLO and VAUGHAN 1972; SCHWARTZ and PASSONEAU 1974; NEWCOMBE et al 1975; BROWNING et al. 1976), these studies will not be considered in detail in this review.

Finally, if the measured response is not cyclic AMP accumulation but rather is some cellular function that is responsive to cyclic AMP, a number of additional potential regulatory sites must be considered. Unfortunately, little is known in most instances about the events leading to cellular responses, and regulation of hormonal effects at this level have yet to be described in sufficient detail to merit review.

In light of these considerations our discussion will focus primarily on two aspects of the regulation of hormone action: (1) the mechanisms of 1st messenger-induced decreases in the responsiveness of adenylate cyclase which include agonist-specific and non-specific processes; and (2) the GTP-dependent, acute inhibition of hormonal stimulation of adenylate cyclase by other hormones.

III. The Structure and Function of Adenylate Cyclase

Since the discussion will focus on the regulation of the response of adenylate cyclase to hormones, it will be useful to have in mind some idea of the structure and function of this complex enzyme system. Furthermore, the regulatory phenomena we will discuss involve relationships derived for the most part from studies of the catecholamine-sensitive adenylate cyclase; thus, regulation of the β-adrenergic receptor-linked adenylate cyclase system will be emphasized. Excellent and more exhaustive reviews of the molecular structure and function of adenylate cyclase have recently appeared (MAGUIRE et al. 1977; STADEL et al. 1980; ROSS and GILMAN 1980; also see Chap. 3 of this volume).

The identification and characterization of hormone receptors, a guanine nucleotide binding protein and a catalytic protein as specifiable moieties comprising the hormone-sensitive adenylate cyclase system has involved a variety of approaches. Using chromatographic and density gradient techniques it is possible to separate β-adrenergic receptors from adenylate cyclase and demonstrate that the separated proteins exhibit distinct hydrodynamic properties (HAGA et al. 1977a; LIMBIRD and LEFKOWITZ 1977; VAUQUELIN et al. 1977). Reconstitution studies have been utilized to further resolve the constituents of this enzyme system and to understand some of the specifics of the interaction of these components in a membrane environment (ROSS and GILMAN 1977; JOHNSON et al. 1978a; ROSS et al. 1978;

HOWLETT et al. 1979; STERNWEIS and GILMAN 1979). These studies have permitted the identification of a heat labile, N-ethylmaleimide-sensitive protein (the catalytic component, C) and a thermostable protein (the guanine nucleotide binding protein, N). Although evidence for more than one guanine nucleotide binding site exists in several systems (COOPER et al. 1979; IYENGAR and BIRNBAUMER 1979; IYENGAR et al. 1979), only a single species of N appears to be involved in the activation of the β-adrenergic receptor-linked adenylate cyclase (NORTHUP et al. 1980).

Several laboratories have demonstrated that agonists bind to β-adrenergic receptors in a high affinity, slowly dissociable complex (LEFKOWITZ et al. 1976; MAGUIRE et al. 1976; WILLIAMS and LEFKOWITZ 1977; KENT et al. 1980; STADEL et al. 1980; DE LEAN et al. 1980). This complex apparently involves an association of hormone (H), R and N, and serves as an obligatory intermediate through which hormones stimulate the rate of activation of adenylate cyclase by guanine nucleotides. The complex is responsible for the high affinity agonist binding observed in competition binding experiments with radiolabelled antagonists; in the presence of GTP there is a marked decrease in the apparent affinity of agonists (LEFKOWITZ et al. 1976; MAGUIRE et al. 1976). Similarly, the agonist [^{3}H]-hydroxybenzylisoproterenol ([^{3}H]-HBI) binds in a tight complex to the β-receptors of frog erythrocyte membranes; in the presence of guanine nucleotides this complex rapidly dissociates (WILLIAMS and LEFKOWITZ 1977; LEFKOWITZ and WILLIAMS 1978). The formation of a high affinity complex in washed membranes can be taken as a measure of the capacity for "coupling" between R and N. For example, the efficacies of a series of β-adrenergic receptor agonists for activation of adenylate cyclase are directly proportional to their capacity to form a high affinity binding complex (KENT et al. 1980). High affinity agonist binding is lost in β-receptor/adenylate cyclase systems that are genetically deficient in N or are chemically uncoupled (HAGA et al. 1977; ROSS et al. 1977; HOWLETT et al. 1979). The catalytic protein of the system appears to be unnecessary for the normal interaction of R and N since a cell line (HC-1 hepatoma cells) deficient in C still exhibits high affinity, GTP-sensitive agonist binding (ROSS et al. 1978).

LIMBIRD, LEFKOWITZ and coworkers have shown that incubation of erythrocyte membranes with [^{3}H]-HBI (but not antagonists) changes the apparent Stokes radius of the solubilized β-receptor as determined by gel exclusion chromatography (LIMBIRD and LEFKOWITZ 1978; LIMBIRD et al. 1980a, b). The N protein (which was identified by [^{32}P]-ADP-ribosylation catalyzed by cholera toxin (CASSEL and PFEUFFER 1978; GILL and MEREN 1978) coeluted with [^{3}H]-HBI if [^{3}H]-HBI was incubated with membranes prior to solubilization. [^{32}P] and [^{3}H]-HBI binding material did not coelute if preincubations were carried out with receptor antagonists or GTP prior to solubilization. These results have been interpreted as a direct demonstration of the agonist-induced association of H, R, and N in a ternary complex.

Enzyme activity measured in the presence of Mg^{++} has been shown to be dependent on the presence of both N and C (ROSS et al. 1978). Similarly, the presence of N is necessary for the expression of NaF-stimulated enzyme activity, although NaF may require a specific (though as yet unspecified) component of N (ROSS et al. 1978) and does not require the presence of GTP per se for full activity (DOWNS et al. 1980). It has been demonstrated that the activity of C can be measured directly if Mn^{++} is substituted for Mg^{++} in the adenylate cyclase assay (Ross et al. 1978; Limbird et al. 1979). A mutant of S 49 lymphoma cells (cyc^-) which lacks N but possesses C exhibits activity in the presence of MnATP but not MgATP (Ross et al. 1977).

It has been proposed that the activity expressed by the hormone-responsive adenylate cyclase is determined by the steady state level of a GTP-N-C complex (Cassel and Selinger 1977; Cassel et al. 1977). Since the GTP-N-C complex exhibits a short lifetime in the face of a GTPase activity that appears to be inherent in N (Cassel and Pfeuffer 1978), any situation that decreases the rate of hydrolysis of GTP in the active complex serves to maintain an elevated steady state level of this activated species. For example, a maximally activated system can be generated by the substitution of the stable analogs, Gpp(NH)p or GTP-γ-S, for GTP (Schramm and Rodbell 1975; Ross et al. 1977). These analogs in effect block the inactivation reaction since they are resistant to hydrolysis by the GTPase. Catecholamines have no effect on the maximally activated enzyme in the presence of the stable analogs (Ross et al. 1977; Schramm and Rodbell 1975). They do, however, increase the rate of activation of the system by stable analogs since they accelerate the dissociation of GDP from N and/or accelerate the rate of association of GTP analogs with N (Schramm and Rodbell 1975; Ross et al. 1977; Cassel and Selinger 1978).

In the presence of NAD^+, cholera toxin catalyzes the ADP-ribosylation of a protein which by several criteria appears to be N (Cassel and Pfeuffer 1978; Gill and Meren 1978; Johnson et al. 1978a; Kaslow et al. 1979). Subsequent to ADP-ribosylation, the GTPase activity of N is greatly reduced (Cassel and Selinger 1977) and the kinetics of activation of the system in the presence of GTP become similar to those observed with nonhydrolyzable GTP analogs (Johnson and Bourne 1977).

Although a number of important aspects of hormone-induced activation of adenylate cyclase are still poorly understood or are the subjects of controversy, it is now possible to describe a model for this system based on the proposal of Cassel and Selinger (1977). The model as stated (Fig. 1) incorporates the ideas of a number of workers in this field and describes the interaction of at least three specifiable proteins: The β-adrenergic receptor (R), a guanine nucleotide binding protein (N), and a catalytic protein (C). In the absence of hormone (H), the system is maintained in a basal state with GDP associated with N. The combination of H with R results in the formation of HRN and the concomitant release of GDP from N. HRN is rapidly converted to HR+N-GTP in the presence of GTP. Once formed, N-GTP rapidly interacts with C to form the active enzyme complex, GTP-N-C. The lifetime of GTP-N-C is regulated by the activity of the GTPase which is associated with N. The GTPase converts GTP-N-C back to the ground state, liberating C and P_i. The resulting GDP-N is then free to traverse the cycle again in the presence of hormone and GTP. Thus, it follows that the hormone-occupied receptor acts somewhat as a catalyst of the "on" reaction of the cycle. Under steady state conditions the extent of activation of adenylate cyclase depends on the rate of the "on" reaction which in turn depends on the amount of HR. A more detailed statement of this basic model has been reported by Stadel et al. (1980) and Swillens and Dumont (1980) (also see Chap. 3 of this volume).

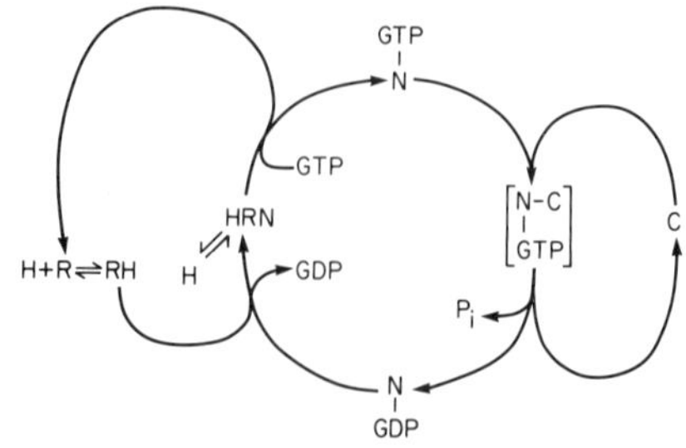

Fig. 1. This model indicates the hypothetical interactions of a catecholamine (H), the β-adrenergic receptor (R), a guanine nucleotide binding protein (N), the catalytic protein (C), and guanine nucleotides (GDP and GTP). The model indicates that RH is able to bind to N, resulting in the release of free GDP and the formation of HRN. The RN complex has higher affinity for H than does free R. The formation of HRN is rate-limiting in the activation process and, in the presence of GTP, HRN is rapidly converted to N-GTP and HR; thus, the role of the hormone-receptor system is to effect the conversion of N-GDP to N-GTP. Once formed, N-GTP interacts with C to form the enzymically active complex C-N-GTP. The lifetime of the active complex is determined by the activity of a GTPase (probably an integral part of N) which hydrolyzes the bound GTP to release P_i, with the subsequent regeneration of C and N-GDP. In the absence of GTP, addition of H leads to the formation of HRN in amounts sufficient to change the apparent K_a of the system for H. Thus, in the absence of GTP, agonists (H) exhibit binding characteristic of interaction at two sites (R and RN). In the intact cells, or upon addition of GTP to membranes, the amount of HRN would be small because its rate of formation is postulated to be the limiting step in the intact system. Under these conditions, agonists (H) would exhibit binding characteristic of the reaction $R+H \rightleftharpoons RH$, namely low affinity binding to a single type of site. Modified from SU et al. (1980)

B. Agonist-Induced Decreases in the Response of Cells to Hormones

I. Early Observations

As mentioned previously, KAKIUCHI and RALL (1968) were first to demonstrate directly that exposure of a cell or tissue to a 1st messenger would cause, with time, a reduction in the response (accumulation of cyclic AMP) of the cell to the 1st messenger. Rabbit brain slices were exposed to either epinephrine or histamine for a period of time after which the slices were washed and rechallenged with the same effector or the alternate effector. This work indicated that the tachyphylaxis, refractoriness, or desensitization [1] was limited to the agent used in the first exposure.

In the wake of the development of simplified assays for cyclic AMP (KUO and DE RENZO 1969; SHIMIZU et al. 1969; GILMAN 1971) and for adenylate cyclase activity (KRISHNA et al. 1968) a number of studies appeared documenting the phenomenon of agonist-induced decreases in the effects of hormones on cyclic AMP

1 In this article we will use the terms tachyphylaxis, desensitization, and refractoriness interchangeably. Their use implies no mechanism and the terminology is used solely to indicate, in an operational sense, that an initial exposure of a cell to some agonist results in a reduced capacity of the cell to respond subsequently

production or on adenylate cyclase activity (for references to these early studies see TERASAKI et al. 1978; PERKINS et al. 1979; LEFKOWITZ et al. 1980). It soon became apparent that this phenomenon was a general property of most cells and that the overt expression of loss of responsiveness to hormones was mediated by more than a single mechanism. Most studies have described variations on the theme of one or the other of two mechanisms. One mechanism, perhaps the major mechanism, involves a reduction in responsiveness that is selective for the inducing hormone. This process of receptor-specific desensitization results in a reduction in the efficacy of the hormone but little or no change in its K_{act} for stimulation of cyclic AMP synthesis. Usually, little or no change in cellular responsiveness to other hormones occurs and when various parameters of adenylate cyclase are examined in homogenates of such desensitized cells no change, other than a selective loss of response to the inducing hormone, is observed. The observation of this mechanism in intact cell assays can be partially obscured if the cell exhibits additional mechanisms for reducing its responsiveness to the hormone. Such is the case with human astrocytoma cells (1321 N 1) which appear to exhibit both receptor-specific and non-specific decreases in hormone responsiveness (SU et al. 1976a, b). These mechanisms can be differentiated using intact cells by taking advantage of the difference in kinetics of the involved reactions in whole cells or by examining the phenomena in broken cell preparations as described below. The receptor-specific mechanism of desensitization has been shown to involve changes in both the properties of receptors and a reduction in the number of receptors per cell.

A second process by which reduction in hormonal responsiveness occurs appears to involve inhibition of the adenylate cyclase system at a site or sites that reduce in a non-specific manner not only the effects of all hormones on adenylate cyclase activity, but reduces basal, Gpp(NH)p- and NaF-stimulated enzyme activities as well. In some cases the effect of cholera toxin also is reduced (TERASAKI et al. 1978).

We will proceed by examining in greater depth both of these general mechanisms for desensitization. Our review of the literature will not be comprehensive but rather will focus on several cell systems that have been extensively examined.

II. Agonist-Induced Desensitization of Intact Cells

In this section we will focus on work with human astrocytoma cells as an example of the overall approach that has been used by many workers in this field. At the same time reference will be made to work with other cell systems in an attempt to establish certain features of the process of desensitization that may be shared by the adenylate cyclase systems of a variety of tissues.

Human astrocytoma cells respond to catecholamines, prostaglandins, and adenosine with an increased rate of synthesis of cyclic AMP (CLARK et al. 1975). Separate structure-specific receptors appear to mediate the actions of each class of agonists through a common pool of catalytic components. The effects of catecholamines occur as a result of interaction with a β_2-adrenergic receptor-linked adenylate cyclase system which exhibits characteristics similar to those observed for this system in a variety of tissues.

1. General Characteristics of Agonist-Induced Desensitization

Catecholamines cause an initial rapid rise in intracellular cyclic AMP content in human astrocytoma cells which is followed by a slower decline back toward basal levels. The characteristic biphasic response of these cells (and cells in general) to norepinephrine is illustrated in Fig. 2. Replacing the medium with medium containing freshly prepared norepinephrine does not reduce the rate of decline of cellular cyclic AMP. Furthermore, exposure of naive cells to medium conditioned by a prior 2-h exposure to cells, elicits a normal rise and fall in cyclic AMP levels. Similar experiments have been used by other investigators to provide evidence that the observed desensitization phenomena are not artifacts of hormone destruction (e.g. Franklin and Foster 1973; Lauzon et al. 1976). Such experiments do not exclude the time-dependent formation of an inhibitor that is retained within the cell, but do exclude the possibility of the formation of an inhibitor of hormone action that is secreted into the medium. However, it should be pointed out that, at least in fat cells, an inhibitory factor that is produced in concert with hormone-stimulated elevations of cyclic AMP levels has been described (Ho and Sutherland 1971). The time course of appearance of the factor corresponded with the time course of desensitization. This observation coupled with the fact that the factor caused naive fat cells to become desensitized to the effects of hormones suggested to Ho and Sutherland that the factor was a "feedback regulator" of hormone action. Although rather extensively studied (Ho et al. 1975; Ho and Sutherland 1975), neither the structure of the feedback regulator nor the mechanism of its inhibition have been elucidated. The factor does not appear to be of general importance in tachyphylaxis phenomena since it does not affect cells other than adipocytes nor does it appear to be produced by other cells. No other "desensitization factors" have been described in any detail although other inhibitors of hormone activation of adenylate cyclase have been discovered (Bitensky et al. 1971; Levey et al. 1975; Fain and Shepherd 1975). To date no compelling evidence exists supporting the involvement in the desensitization process of a factor that is released by cells upon exposure to hormones.

The cyclic AMP content of the medium also increases when many cell types are exposed to hormones. In human astrocytoma cells, changes in the rate of excretion of cyclic AMP are not sufficient to account for the decline in intracellular cyclic AMP. Furthermore, blockade of the egress of cyclic AMP from the cells with dipyridamole does not prevent the decline in cellular content (Clark et al. 1975). An agonist-induced change in the rate of secretion of cyclic AMP has not been implicated as a major mechanism of desensitization in any cell type to date; however, it has not often been analyzed.

When astrocytoma cells are exposed to norepinephrine and then washed free of the agonist, there is a rapid fall in cellular cyclic AMP content (Su et al. 1976a and Fig. 2). When cells so treated are exposed to the same agonists in a second incubation, the elicited rise in cyclic AMP is progressively smaller as the time of the first exposure is increased. It is clear from such studies that the magnitude of the loss of responsiveness to the second challenge of a hormone is related to the time of exposure and to the concentration of hormone in the first exposure (see Fig. 2).

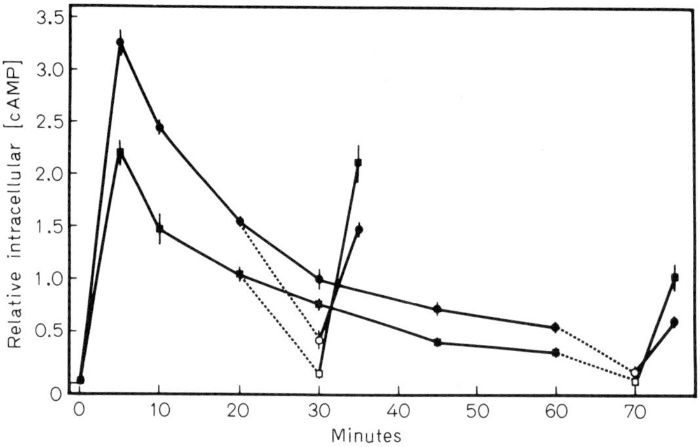

Fig. 2. The effect of time and agonist concentration on desensitization. Cells were incubated with 100 μ*M* norepinephrine (●) or 10 μ*M* norepinephrine (■), and cyclic AMP accumulation measured. At the times indicated, norepinephrine was removed by washing. The decline of cyclic AMP content is shown by the dashed lines. The cells were subsequently challenged with 100 μ*M* norepinephrine for 5 min. Taken from Su et al. (1976a)

Time course relationships and dose response relationships have been examined in a number of other systems. For intact cells, the $t_{1/2}$ for desensitization has been observed to vary from a few minutes (ANDERSON and JAWORSKI 1979) to a few hours (DE VELLIS and BROOKER 1974). Dose-response relationships for desensitization usually have been similar to dose-response relationships for stimulation of cyclic AMP accumulation in the various cells studied; however, notable exceptions exist. Studies with 132INI astrocytoma cells have shown that the dose-response relationship can be related to the time of exposure to agonist. Short term incubation (1 h) of 132INI cells with various concentrations of isoproterenol results in a concentration-effect curve for the induction of desensitization that is essentially the same as that for stimulation of cyclic AMP formation in naive cells. However, long term exposure (24–36 h) of 132INI cells to either PGE_1 or isoproterenol, at concentrations that have little or no effect on cyclic AMP content, result in a greater than 90% desensitization to the inducing agonist with no significant effect on the response of the cells to other 1st messengers. For 132INI cells, the time dependency of the dose-response relationship for desensitization appears to be the result of an agonist-induced, irreversible loss of receptors (see below and SU et al. 1980).

The observation that very low concentrations of either a catecholamine or PGE_1 could cause complete receptor specific desensitization (LEICHTLING et al. 1976) while higher concentrations of either agonist caused a non-specific desensitization (SU et al. 1976a) suggested that more than one process might mediate desensitization in 132INI cells. In order to determine the specificity of the desensitization reaction, cells were first exposed to either isoproterenol or PGE_1 and then responsiveness to the same agonist (homologous) or the alternative agonist (heterologous) was tested in a second incubation (Fig. 3). It is clear from such ex-

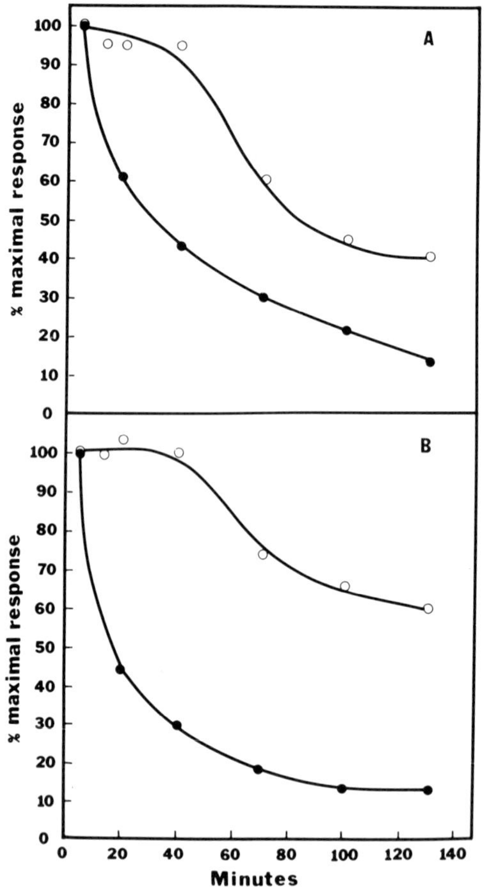

Fig. 3 A, B. Agonist specificity during desensitization to isoproterenol or PGE_1. Cells were first exposed to isoproterenol **A** or PGE_1 **B** for the times indicated. The agonists were then removed by washing and the cells incubated an additional 10 min to allow cyclic AMP content to decline. The cells initially exposed to isoproterenol (**A**) were then challenged with either isoproterenol (●) or PGE_1 (○) for 5 min. The cells initially exposed to PGE_1 **B** were then challenged with either 10 μM PGE_1 (●) or 10 μM isoproterenol (○) for 5 min. Taken from JOHNSON et al. (1978 b)

periments that for 132 INI cells both the rate and extent of homologous[2] desensitization are greater than for heterologous[2] desensitization. Whereas an essentially

2 The term "homologous" desensitization has not been used in a consistent manner by the workers in this field. We will use it as an operational term i.e. to denote the desensitization observed when cells are exposed to an agonist and then rechallenged with the same agonist. Such desensitization will not necessarily only reflect *agonist-specific* desensitization but could include a component of agonist-induced, nonspecific desensitization. Heterologous desensitization also will be used as an operational term to indicate the desensitization observed when cells are exposed to one class of agonists and then challenged with a second class of agonist. Such desensitization is equivalent to "agonist nonspecific" desensitization but could be the result of more than one process. We will use the term agonist-specific desensitization only when it can be shown that no parameter of adenylate cyclase activity is changed except the capacity to respond to the single class of hormones

complete loss of responsiveness to homologous agonists occurs by 3 h, loss of heterologous responsiveness seldom exceeds 50% for PGE_1 or 70% for the catecholamines even with extended exposure of the cells to the inducing, alternative agonist. Similar results have been reported recently by TOUGUI et al. (1980) in an analysis of the effects of PGE_2 and isoproterenol on cyclic AMP accumulation in rat myometrium. To our knowledge, a comparison of the kinetics of these two processes in a single cell type has not been reported by others. In cells that respond to more than one hormone it is important to carry out such studies since the two types of analyses may detect different mechanisms of desensitization. For example, the results of JOHNSON et al. (1978 b) suggest that the loss of response of intact cells to homologous agonists is a composite of a receptor-specific process and a cyclic AMP mediated, non-specific process.

Sotalol, a β-receptor antagonist, causes essentially complete blockade of the effect of isoproterenol on cyclic AMP accumulation but has no effect on the stimulatory action of PGE_1. Exposure of astrocytoma cells to sotalol, at a concentration sufficient to occupy 95% of the β-receptors, produces no desensitization to a subsequent challenge with isoproterenol. However, sotalol blocks isoproterenol-induced desensitization to the effects of either isoproterenol or PGE_1. The action of sotalol is selective in that PGE_1-induced desensitization to isoproterenol occurs to the same extent in the presence of sotalol as in its absence. Experiments such as these (MICKEY et al. 1976; FRANKLIN et al. 1975) have indicated clearly that even though a hormone can induce non-specific desensitization it does so by acting through the same receptor involved in the activation of adenylate cyclase. In every case tested, receptor antagonists have not been capable of causing desensitization but have blocked agonist-induced desensitization. When the properties of partial agonists have been examined, a good correlation between agonist efficacy in terms of stimulation of cyclic AMP formation and efficacy for induction of desensitization has been observed (FRANKLIN et al. 1975; SU et al. 1980). Thus, whereas true antagonists are not able to cause desensitization, even weak partial agonists can be effective desensitizing agents (FRANKLIN and TWOSE 1979; SU et al. 1980). Again, a full analysis of the effects of partial agonists must take into consideration the irreversible nature of the loss of β-adrenergic receptors and, thus, must consider time of exposure (see below).

2. Effects of Cyclic AMP and Cyclic AMP Analogs

Since in many instances the dose-response relationship for hormone-induced cyclic AMP accumulation and hormone-induced desensitization are similar, it is reasonable to speculate that cyclic AMP might mediate desensitization by some feedback mechanism. Furthermore, such a mechanism would be expected to produce a non-specific (heterologous) desensitization. In a few cell systems the effects of cyclic AMP or cyclic AMP analogs have been tested for their effect on hormonal responsiveness. NEWCOMBE et al. (1975) demonstrated that human synoviocytes exhibit agonist-selective desensitization to either catecholamines or prostaglandins of the E type. Conversely, cyclic AMP caused a complete loss of responsiveness of the cells to both types of agonists. Incubation of human astrocytoma cells with dibutycyl-cyclic AMP (Bt_2cyclic AMP) or 8-methylthio-cyclic AMP caused a partial desensitization of human astrocytoma cells to both catecholamines and PGE_1 (SU et al. 1976a). The time course and extent of desensitization were similar to those of agonist-induced heterologous desensitization. TERASAKI et al. (1978) have reported that Bt_2cyclic AMP causes a loss of response to catecholamines in C6-2B rat glioma cells. Furthermore, they reported that treatment of C6-2B cells with cholera toxin led to desensitization to catecholamines (NICKOLS and BROOKER 1979). The effects of both Bt_2 cyclic AMP and cholera toxin and a major portion of the desensitization by catecholamines are blocked by acetoxycycloheximide in C6-2B cells. Thus, it seems reasonable to surmise that at least a portion of the catecholamine-induced desensitization process in C6-2B cells is mediated by cyclic AMP. Cholera toxin also induces a heterologous desensitization in human astrocytoma cells (JOHNSON 1977). In both synoviocytes and astrocytes, cyclic AMP appears to induce a state of hormone *nonspecific* desensitization. C6-2B cells respond only to catecholamines so the typical test for heterologous desensitization cannot be conducted in this cell line. However, based on an analysis of this enzyme activity in membranes prepared from desensitized cells, it appears that catecholamines induce nonspecific inhibition of adenylate cyclase in C6-2B cells (see discussion below). Thus, results to date are consistent with the conclusion that cyclic AMP may mediate hormone-induced, heterologous desensitization by effecting an inhibition of the adenylate cyclase reaction at a site distal to hormone/receptor interaction.

3. Whole Cell Analysis of Rates of Cyclic AMP Synthesis and Degradation

One logical alternative mechanism of nonspecific desensitization of whole cells to hormones would be the cyclic AMP-mediated induction of phosphodiesterase activity that is known to occur in a variety of cells (D'ARMIENTO et al. 1972; MANGANIELLO and VAUGHAN 1972). However, it is difficult using whole cells to unambiguously determine the precise roles of cyclic AMP *synthesis* and cyclic AMP *degradation* when change in cyclic AMP *content* is the parameter being measured. Furthermore, it may not be possible to accurately estimate rates of synthesis and degradation in whole cells based on an analysis of adenylate cyclase and phosphodiesterase activities in homogenates. In view of such concerns, a number of laboratories have developed methods for the analysis of rates of cyclic AMP synthesis and degradation in whole cells (BROWNING et al. 1976; SU et al. 1976b; CLARK and BUTCHER 1979). A pulse-labeling technique was used to determine the relative rates of synthesis of cyclic AMP during a series of 3 min periods throughout a 2-h exposure of human astrocytoma cells to catecholamines (SU et al. 1976b). The results provided convincing evidence that catecholamines induce, within 3 min, a reduction in the rate of cyclic AMP synthesis that further declined over the 2-h period. In related experiments the initial velocity of accumulation of cyclic AMP (5–10 s intervals for 40 s) was measured as an indication of changes in the rate of synthesis of cyclic AMP during desensitization. The results of these experiments indicated that a rapid loss in the capacity of the cells to synthesize cyclic AMP occurs within several minutes of exposure to catecholamines. CLARK and BUTCHER (1979) have used a related approach to detect, continuously, changes in the rate of cyclic AMP synthesis during exposure of WI-38 fibroblasts to catecholamines or prostaglandins. From the results of such studies it has been concluded that changes in the rate of synthesis of cyclic AMP is the primary basis for the loss of response of intact human astrocytoma and WI-38 cells to catecholamines. However, when BARBER et al. (1978) attempted to formulate a mathematical model to describe the kinetics of cyclic AMP accumulation in WI-38 cells continuously exposed to PGE, they observed complex behavior. They concluded that although an agonist-induced decline in cyclic AMP synthetic rate occurred, this change was not sufficient to account completely for the observed reduction in cyclic AMP content.

Degradation of cyclic AMP has been measured directly in whole cells to confirm that a change in phosphodiesterase activity is not significantly involved in agonist-induced desensitization. In human astrocytoma cells, no significant change in degradative rate was observed during a 60 min preincubation with catecholamines or PGE_1 even though, by 60 min, responsiveness to the agonist was reduced to 20%–30% of control. In general, we have not observed agonist induced increases in cyclic AMP degradation rates with less than 90 min of exposure to either PGE_1 or catecholamines. The observed increase in degradative activity (1.5- to 2.0-fold after 2 h) could play only a minor role in the overall desensitization process in 132 INI cells.

Increases in phosphodiesterase activity have been implicated in the loss of response of S 49 lymphoma cells (BOURNE et al. 1973) and C 6-glioma cells (BROWNING et al. 1976) to hormones. However, in both cases it has been shown that increases in phosphodiesterase activity comprise only one component of agonist-induced loss of responsiveness to hormones in these cell lines (SHEAR et al. 1976; TERASAKI et al. 1978). Certainly, in the case of S 49 cells, agonist-specific modification in the response of adenylate cyclase to catecholamines has been demonstrated as has agonist-induced loss of β-adrenergic receptors (SHEAR et al. 1976; SU et al. 1980).

4. Conclusions Based on Studies with Whole Cells

The results obtained regarding agonist-induced changes in the response (cyclic AMP accumulation) of intact cells to hormones can be summarized as follows.

a) In general, agonist-induced desensitization is not the result of destruction of the agonist, increased rates of cyclic AMP secretion, or the secretion of an inhibitory substance into the medium. The agonist-induced secretion by fat cells of an inhibitor of hormone action on fat cells stands as the single exception to the latter point.

b) At least 3 distinct processes have been documented to result in loss of agonist effects on cells, (1) an agonist- or receptor-specific desensitization; (2) an agonist-or receptor-nonspecific desensitization process that involves changes in the rate of synthesis of cyclic AMP and may be mediated by cyclic AMP, and (3) an agonist-induced, cyclic AMP-mediated induction of phosphodiesterase activity which can result in a 1.5- to 3.0-fold increase in degradative capacity after 1.5–3.0 h of exposure to hormone. The latter process is judged to be of minor importance since it occurs late in the overall process of agonist-induced desensitization.

c) Based on studies with antagonists and partial and full agonists of the β-adrenergic receptor, it is apparent that both agonist-specific and agonist-nonspecific desensitization result from the interaction of catecholamines with β-receptors. Conversely, PGE_1-mediated reduction in the response of cells to catecholamines does not result from an interaction of PGE_1 at β-receptors.

III. Agonist-Induced Changes in Adenylate Cyclase and Hormone Binding

Based on current perceptions of the adenylate cyclase system, it is reasonable to expect that agonist-specific desensitization would involve a change in the number or the functional state of the receptor protein specific for the agonist. With the development of radioligand binding assays it became possible to more directly assess the role of receptor modification as a basis for desensitization of cells to hormones. MUKHERJEE et al. (1975, 1976) were the first to apply this technology to a study of catecholamine-induced desensitization of the β-receptor-linked adenylate cyclase system. They demonstrated that administration of catecholamines to frogs led not only to a marked loss in responsiveness of frog erythrocyte adenylate cyclase to catecholamines, but also to a significant loss in binding sites for ^{3}H-dihydroalprenolol, a β-receptor antagonist. Similar changes could be induced in vitro simply by incubating erythrocytes with catecholamines (MICKEY et al. 1975, 1976). Although these original studies emphasized the similarity in time course and extent of loss of β-receptors and catecholamine-responsive adenylate cyclase, it subsequently became apparent that a discrepancy between these two parameters usually existed (SHEAR et al. 1976; JOHNSON et al. 1978b; WESSELS et al. 1978). For example, using EH 118 MG astrocytoma cells, it was shown that incubation of cells with saturating concentrations of isoproterenol resulted in an 80%–90% loss of catecholamine responsiveness in intact cells but only a 20% decrease in the number of β-adrenergic receptors (JOHNSON et al. 1978b). Isoproterenol-stimulated adenylate cyclase activity of homogenates of catecholamine-treated cells was 50%–65% of control. The effects on receptor number and catecholamine-stimulated adenylate cyclase activity were agonist specific since, although incubation of cells for four hours with PGE_1 caused a 60% decrease in catecholamine responsiveness of intact cells, there was no PGE_1-induced change in catecholamine-stimulated adenylate cyclase activity of homogenates and no decrease in the number of β-adrenergic receptors.

SHEAR et al. (1976) also carried out similar experiments using wild type S49 lymphoma cells. Incubation of cells for two hours with isoproterenol resulted in a 60 to 80% reduction in isoproterenol-stimulated adenylate cyclase activity. In contrast, the number of β-adrenergic receptors, as measured by [^{125}I]-hydroxybenzylpindolol ([^{125}I]-HYP) binding decreased only by 25%–45%. In experiments with frog erythrocytes, LEFKOWITZ and co-workers also documented that the catecholamine-induced loss of isoproterenol-stimulated adenylate cyclase activity is substantially greater than the loss of β-adrenergic receptors as measured with radioligands that are β-receptor antagonists (WESSELS et al. 1978, 1979). The observed differences between agonist-induced changes in receptor number and loss of response of adenylate cyclase to agonists called into question the role of receptor loss as the sole basis of desensitization.

More recent studies of the adenylate cyclase system during catecholamine-induced desensitization have focused on an analysis of the kinetics of alterations in receptors and catecholamine-stimulated adenylate cyclase activity and on attempts to define the molecular basis of these alterations. SU et al. (1980) demonstrated that incubation of 132 1N1 astrocytoma cells with isoproterenol results in a rapid ($t_{1/2}$ =

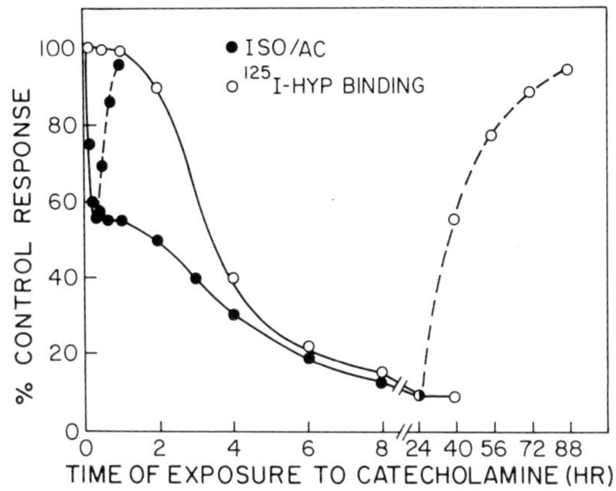

Fig. 4. Time courses of desensitization and reversal of desensitization for isoproterenol-stimulated adenylate cyclase activity and ^{125}I-HYP binding to β-adrenergic receptors. Cells were incubated with 1 μ*M* isoproterenol. At the times indicated, isoproterenol-stimulated adenylate cyclase activity (●) and β-adrenergic receptor density (○) were determined in membrane preparations. The symbols connected by dashed lines indicate the recovery of enzyme activity or β-receptor number after washing the cells free of isoproterenol at the indicated times. The responses illustrated are representative of numerous such experiments

3 min) decrease of isoproterenol-stimulated adenylate cyclase activity to a new essentially steady state level (40%–50% of control) (Fig. 4). During this time there is no alteration in the number of β-adrenergic receptors or in basal or PGE_1-, NaF-, or Gpp(NH)p-stimulated adenylate cyclase activity. Since the amount of adenylate cyclase or β-adrenergic receptors per se do not appear to be altered during this condition, it was suggested that an uncoupling occurs between β-adrenergic receptors and adenylate cyclase during this phase of desensitization (Su et al. 1979). Recovery from this uncoupled state is rapid ($t_{1/2}=7$ min) upon removal of isoproterenol from the medium.

The conclusion that short-term desensitization of 132INI astrocytoma cells involves an uncoupling of β-receptors and adenylate cyclase also is indicated by receptor binding studies (Harden et al. 1979b; Su et al. 1980). As discussed earlier, work from a number of laboratories suggests that in well coupled β-receptor/adenylate cyclase systems, agonists bind with high affinity to the receptors in the absence of GTP. In the presence of GTP, this high affinity complex does not accumulate and only low affinity binding of agonists is observed. In systems that have been uncoupled by chemical means (Lefkowitz et al. 1976; Howlett et al. 1979; Stadel and Lefkowitz 1979) or by genetic selection (Maguire et al. 1977; Ross et al. 1978), agonists bind to β-adrenergic receptors with only low affinity and there is no effect of guanine nucleotides on agonist binding. The agonist binding properties of β-receptors of astrocytoma cells that have been incubated with isoproterenol for 15 min are similar to those of uncoupled receptors. That is, although there is no change in the number of receptors, the apparent affinity of isoproterenol for inhibition of [^{125}I]-HYP binding is reduced by 10-fold; similarly, there is little effect of GTP on the apparent affinity of isoproterenol (Harden et al. 1979b; Su et al. 1980). The time course of the uncoupling reaction as detected by changes in the affinity of agonist binding is similar to the time course of loss of isoproterenol stimulated adenylate cyclase activity. Furthermore, the alteration in agonist binding affinity and the reduction of isoproterenol-stimulated adenylate cyclase activity appear to return to control levels with similar time courses upon removal of isoproterenol.

The uncoupling reaction exhibits the same concentration-effect relationship for isoproterenol (SU et al. 1979) as does stimulation of cyclic AMP accumulation in intact cells ($K_{0.5}$ values approximately 0.03 μ*M*). Also, partial agonists have roughly equivalent partial effects on the degree of uncoupling and the degree of activation of adenylate cyclase (SU et al. 1980).

Although no change in the density of β-adrenergic receptors occurs during a 30–60 min incubation of astrocytoma cells with isoproterenol, continued exposure to agonist results in a precipitous loss of receptors after 2–4 h (SU et al. 1979, 1980). By 24 h the loss of receptors and isoproterenol-stimulated enzyme activity are usually similar (0%–15% of control) (Fig. 4). After long-term incubation with catecholamines, recovery of hormone responsiveness or receptor number occurs slowly ($t_{1/2}$ = 12–16 h) upon removal of isoproterenol from the medium (DOSS, HARDEN, PERKINS, unpublished observations). At intermediate times (2–8 h), the degree of recovery from desensitization is rapid but occurs to an extent dictated by the proportion of receptors remaining (SU et al. 1980).

Since an essentially irreversible reaction appears to be involved in the catecholamine-induced changes of the β-receptor system of astrocytoma cells, the concentration-effect relationship for induction of these alterations changes as a function of time. As stated above, at short times (i.e., during the uncoupling reaction) the concentration-effect curve for desensitization is essentially the same as that for activation of adenylate cyclase. With time, however, the concentration effect curve is shifted to the left. For example, the $K_{0.5}$ for loss of responsiveness is approximately 300 n*M* at 30 min compared to 2.0 n*M* at 24 h (SU et al. 1980). Similarly, although partial agonists cause only a partial uncoupling (compared to isoproterenol) at short times, continued incubation can lead to a virtually complete loss of receptors from the cell.

The results of experiments presented thus far suggest the following kinetic model for agonist-specific desensitization.

$$R \rightleftharpoons R' \rightarrow R''.$$

The initial process can be envisioned to involve the formation of an altered form of the receptor, R′. Although [^{125}I]-HYP still binds to this form of the receptor, the interaction of agonists with R′ does not elicit activation of adenylate cyclase. During a short term incubation with catecholamine, the receptor exists in both the R and R′ forms and the steady state levels of these two forms are determined by the balance of two reactions. The extent of formation of R′ appears to depend on the concentration of isoproterenol. With time, the receptor is converted to a form (R″) that is not detected by [^{125}I]-HYP.

Using genetic mutants of the S 49 lymphoma cell line, SHEAR et al. (1976) were first to show that occupancy alone of β-adrenergic receptors by agonists was insufficient to cause a decline in β-receptor number. Incubation of wild type S 49 cells (which contain a fully responsive β-adrenergic receptor linked adenylate cyclase) with isoproterenol resulted in a decrease in receptor number and catecholamine responsiveness. Also, experiments with an S 49 cell variant that is deficient in cyclic AMP dependent protein kinase yielded similar results. In contrast, incubation with isoproterenol of an S 49 cell variant (cyc^-) that lacks hormone responsiveness resulted in no change in receptor number. Combinations of isoproterenol and Bt_2cyclic AMP also had no effect on β-receptor number in cyc^- cells. The authors concluded that receptor occupancy alone or in combination with elevated cyclic AMP levels was insufficient to cause refractoriness; it was thus proposed that the development of refractoriness required an intact adenylate cyclase (SHEAR et al. 1976). Since it is now known that cyc^- cells lack the guanine nucleotide binding protein while retaining the catalytic unit (ROSS et al. 1978), these studies suggest that coupling per se may be involved in agonist-induced changes of the β-receptor system. Our recent studies support this contention. For example, incubation of HC-1 hepatoma cells with isoproterenol results in an initial uncoupling reaction as measured by agonist competition binding curves and in an eventual large decrease in the number of β-adrenergic receptors (SU et al. 1980). Since this cell line lacks the catalytic protein but possesses both β-receptors and the guanine nucleotide binding protein (i.e., it will couple), the catalytic unit per se would not appear to be involved in the desensitization process.

STADEL and LEFKOWITZ (1979) have presented results in support of an alternate conclusion. They observed that treatment of frog erythrocytes with the sulfhydryl reagent N-ethylmaleimide, which inactivates adenylate cyclase catalytic activity but has no effect on high-affinity nucleotide-sensitive agonist binding, also prevents the receptor desensitization process. They concluded that agonist occupancy of a nucleotide-sensitive receptor complex is insufficient by itself to initiate down-regulation of receptors in the whole cell. They further suggested that productive coupling of the high-affinity agonist receptor complex to an N-ethylmaleimide-sensitive component, possibly adenylate cyclase, is also required for desensitization of the receptors in frog erythrocytes. However, since this experiment involved exposure of intact cells to a non-specific sulfhydryl reagent, N-ethylmaleimide, it is possible that enzymatic steps involved in the desensitization reaction were inhibited in addition to any inhibition of adenylate cyclase.

BOURNE and co-workers (personal communication) have isolated a new variant of the S 49 lymphoma. This cell exhibits β-receptors and an adenylate cyclase that can utilize Mg ATP as substrate, but is not stimulated by catecholamines; high affinity agonist binding is expressed and is inhibited by GTP. In this regard, the adenylate cyclase of these cells exhibits properties similar to that of adenylate cyclase of reticulocytes or frog erythrocytes in the presence of high Mn^{++} (LIMBIRD et al. 1979). The cells appear to be capable of forming the high affinity HRN complex (see Fig. 1) but not of coupling of the GTP-N complex to C. Of pertinence to the above controversy, these S 49 cells do exhibit agonist-induced loss of β-receptors.

Our experiments with cyc^- S 49 lymphoma cells (SU et al. 1980) confirm the results of SHEAR et al. (1976). That is, even long term incubation (24 h) of cells with isoproterenol had no effect on β-receptor number. Surprisingly, a 24-h incubation of the UNC clone of S-49 lymphoma cells with isoproterenol resulted in a 20%–30% loss of β-adrenergic receptors (SU et al. 1980). Since these cells appear to be uncoupled in terms of the interaction of hormone receptors and the guanine nucleotide binding protein, an agonist-induced change in receptor number was somewhat surprising. However, as has been recently pointed out by ROSS and GILMAN (1980), UNC cells appear to undergo a small degree of hormone induced coupling of components of the adenylate cyclase. Thus, a small loss of β-adrenergic receptors during long term incubations with catecholamine might be expected.

Recent work by LEFKOWITZ and co-workers (WESSELS et al. 1979; KENT et al. 1980; LEFKOWITZ et al. 1980) suggest that the mechanisms involved in agonist-induced desensitization of the adenylate cyclase system of frog erythrocytes is in general similar to that of astrocytoma cells. Incubation of erythrocytes with isoproterenol results in a rapid and parallel loss of both isoproterenol-stimulated adenylate cyclase activity and receptor binding of the agonist [^{3}H]-HBI. After a 5-h incubation these activities were both approximately 30% of control. In contrast, β-adrenergic receptor number as measured by the binding of the antagonist [^{3}H]-dihydroalprenolol ([^{3}H]-DHA) was decreased by only 35% over the same time period. The authors concluded that an alteration occurred during desensitization which preferentially interferred with agonist versus antagonist binding. Since the affinity of isoproterenol for inhibition of [^{3}H]-DHA binding was reduced in desensitized membranes, the effects of incubation of cells with isoproterenol appeared to affect the affinity of [^{3}H]-HBI binding. Thus, the results observed in erythrocytes appear to be qualitatively similar to those observed in other cell lines. Recently, computer analysis of ligand binding to β-adrenergic receptors has been carried out with the frog erythrocyte system (KENT et al. 1980). The capacity of agonists to form a high affinity complex in membrane preparations was found to correlate closely with the efficacy of these agonists for stimulation of adenylate cyclase activity. Analysis of ligand binding with membranes from desensitized frog erythrocytes has led to the suggestion that during agonist-induced refractoriness there is an impairment of the capacity of agonists to form a high affinity binding state. This conclusion is similar to that based on studies of agonist-induced desensitization of astrocytoma cells and supports the general conclusion that the process of β-receptor-specific desensitization involves a change in the receptor that prevents its association with the nucleotide binding protein.

In light of the properties of agonist-induced desensitization described above, it is apparent that catecholamine specific desensitization of a number of cell lines occurs through a multistep process. Indeed, although the relative rates of the involved reactions may be considerably different, kinetic data would suggest that at least in two astrocytoma cell lines, in HC-1 hepatoma cells, in S-49 lymphoma cells, in frog erythrocytes, and in fibroblasts (KURZ and PERKINS unpublished observations), a common mechanism of catecholamine-induced desensitization is involved. That is, catecholamine agonists induce a rapid uncoupling reaction which is followed by another reaction(s) which leads to a loss of β-adrenergic receptors from the cell. The molecular basis of these changes is not known. However, a number of reports have appeared which may shed light on the bases of these phenomena.

IV. Separation of Native and Desensitized Beta-Adrenergic Receptors

Using sucrose density gradients, it has recently been demonstrated that short-term incubation of astrocytoma cells with isoproterenol results in an alteration in the

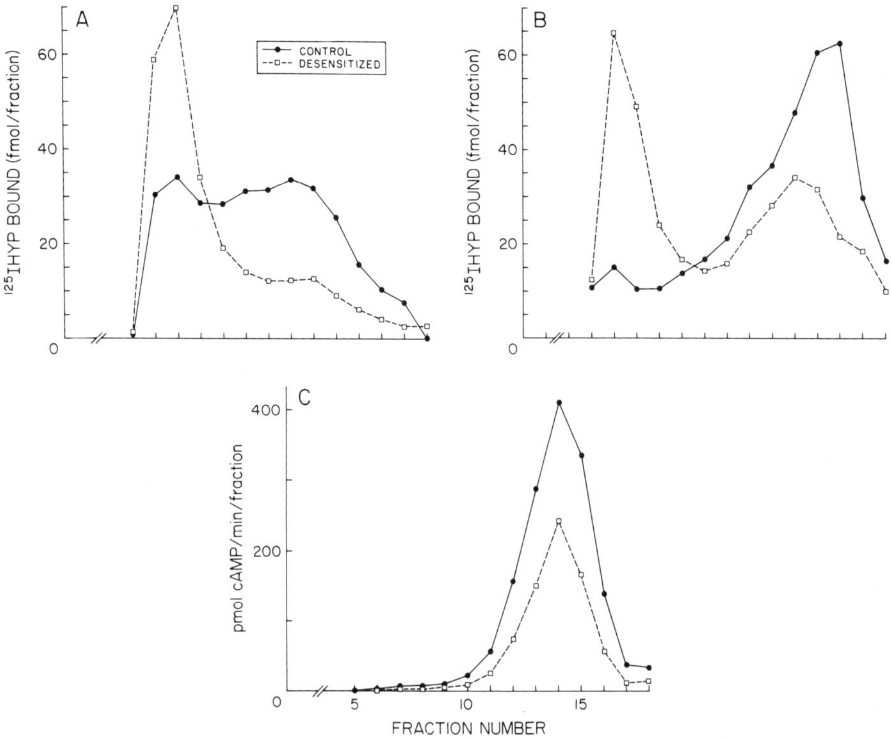

Fig. 5 A–C. Sucrose density gradient distribution of β-adrenergic receptors and adenylate cyclase after short-term incubation of cells with isoproterenol. **A** Cells were incubated with 1 m*M* sodium ascorbate (control) or 1 m*M* sodium ascorbate plus 1 μ*M* isoproterenol (desensitized) for 15 min. Cells were then lysed and centrifuged in a sucrose density gradient. **B** and **C** Cells were treated as described above except that after incubation with isoproterenol, the cells were treated with Concanavalin A before lysis (LUTTON et al. 1979). [^{125}I]-HYP was used to determine β-receptor density **B** in gradient fractions. Isoproterenol-stimulated adenylate cyclase activity **C** was determined in gradient fractions. Taken from HARDEN et al. (1980)

membrane form of β-adrenergic receptors (HARDEN et al. 1980). Isoproterenol or NaF-stimulated adenylate cyclase activities are recovered from gradients as a single peak in preparations from both control and desensitized cells. β-receptors from control cells are also predominately recovered in this fraction. In contrast, receptors from desensitized cells migrate as two peaks (30%–35% and 45%–50% sucrose) (Fig. 5). The "light" peak of β-receptors that is generated during short term incubation with catecholamines exhibits binding properties similar to those of uncoupled receptors. That is, only low affinity agonist binding is observed and there is no effect of GTP on agonist affinity. The time course of appearance and disappearance of the new receptor form upon addition or removal of isoproterenol from the medium correlates with the time courses of decrease or recovery of isoproterenol-stimulated adenylate cyclase activity. At present the significance of the physical form of receptors that appears during desensitization is not known. Clearly, its appearance coincides on a temporal basis with the agonist-induced uncoupling reaction. Whether an alteration in physical form of the receptor is causal in the uncoupling process or occurs as a result of such an event is not known. Also, we can only speculate concerning the cellular localization of the altered receptor form. These receptors do not appear to be associated with the plasma membrane per se since markers of this membrane migrate to heavier equilibrium densities. Possibly, the altered receptors could be associated with the surface membrane initially but upon cell lysis are altered in such a way as to not associate with the plasma membrane. Such an idea would require the further qualification that a domain of receptors exists in the plasma membrane which, due to an agonist-induced aggregation or some other alteration, are affected differently by cell lysis than is the rest of the plasma membrane. A more plausible explanation is that these uncoupled receptors exist in cytoplasmic vesicles that have resulted from an agonist-induced selective endocytosis of receptor protein. Such a phenomenon has been demonstrated with receptors for a number of polypeptide hormones (PASTAN and WILLINGHAM 1981). Unpublished studies (LUTTON and PERKINS) indicate that the "light" receptor peak sediments at an equilibrium density similar to that for Golgi apparatus, lysosomes, and endoplasmic reticulum. Unlike the case with certain polypeptide hormones, there is no agonist ligand available with which β-adrenergic receptors can be visualized; thus the fate of hormone-receptor complexes cannot be followed visually. A number of compounds, methylamine, quinacrine, and dansylcadavarine have been used to block the processes of endocytosis or degradation of polypeptide hormone/receptor complexes (PASTAN and WILLINGHAM 1981). To date we have been unsuccessful in using these compounds to block the isoproterenol-induced loss of β-receptors from 132INI cells. Although receptor loss can be at least partially antagonized by agents such as methylamine or quinacrine, the lack of specificity of these drugs limits the conclusions that can be drawn from such results. In contrast to our experience, REGGIANI et al. (1980) have shown that dansylcadavarine can block the catecholamine-induced loss of β-receptors in BHK cells.

Recently, CHUANG and COSTA (1979) reported that incubation of bullfrog erythrocytes with isoproterenol results in the appearance of a "soluble" fraction of β-adrenergic receptors in cell lysates. Although only 20% of the lost receptors could be accounted for in this fraction, these results suggest that, as discussed above, catecholamines may induce a change in the physical form of β-adrenergic receptors. It should be pointed out that although the new form of receptors appearing in erythrocytes was proposed to be soluble, only a 40,000 g supernatant was used as a source of "soluble" receptors. Such a preparation would likely contain small membrane vesicles. Thus, it is possible that the receptor form that is generated in erythrocytes during incubation with agonists may be similar to that which has been observed in astrocytoma cells (HARDEN et al. 1980).

MALLORGA et al. (1980) have utilized a different approach to assess the mechanisms of agonist-induced desensitization in C6 astrocytoma cells. Incubation of cells with isoproterenol for 2 h resulted in an 80% decrease in responsiveness of intact cells to catecholamines and a 20%–30% decrease in the number of β-adrenergic receptors. These changes were also paralleled by increased activity of phospholipase A_2 and turnover of methylated phospholipids as measured by an increase in the release of arachidonic acid. Quinacrine, which inhibits phospholipase

A_2 (YORIO and BENTLY 1978), blocked the decrease in responsiveness and the reduction in receptor number. In light of these results, MALLORGA et al. proposed that activation of phospholipase A_2 may be intimately involved in the process of desensitization. However, the type of desensitization that was studied in these experiments is not clear. TERASAKI et al. (1978) have shown that catecholamine agonists induce a nonspecific type of desensitization in C 6 cells that, at least during the first several hours, does not involve loss of β-adrenergic receptors. Thus, it is unlikely that only agonist-specific desensitization was studied under the conditions reported by MALLORGA et al. (1980).

V. Catecholamine-Induced Desensitization in C 6–2 B Glioma Cells

BROOKER, DE VELLIS, and co-workers have reported a number of studies of catecholamine-induced refractoriness of the C 6–2 B clone of the rat C 6 glioma cell line. Incubation of this cell line with catecholamines results in a transient elevation of intracellular cyclic AMP levels (DE VELLIS and BROOKER 1974). Rechallenge of catecholamine-preincubated cells with the agonist fails to elicit a full cyclic AMP response. The development of refractoriness appears to occur through decreases in synthetic capacity of the cells since, at least in the 2 B-clone of C 6 cells, there is no evidence for an induction of phosphodiesterase activity (DE VELLIS and BROOKER 1974). Indeed, the refractoriness that develops in the 2 B clone appears to be, at least in part, dependent upon the synthesis of a protein which affects the rate of synthesis of cyclic AMP. For example, if cells are incubated with cycloheximide or actinomycin D, as well as norepinephrine, during the desensitization reaction, the degree of loss of responsiveness is substantially reduced (DE VELLIS and BROOKER 1974; TERASAKI et al. 1978). The protein involved in refractoriness appears to exhibit a high rate of turnover. That is, little recovery of catecholamine responsiveness is observed if cells are desensitized and then placed in catecholamine-free medium; however, if C 6–2 B cells are desensitized and placed in agonist-free medium *plus* cycloheximide the degree of loss of responsiveness is rapidly reduced.

The "refractoriness protein" appears to be synthesized as a result of transcription of a messenger RNA that is relatively long-lived. For example, if cells are incubated for a long period of time with catecholamine plus cycloheximide and then washed and incubated for a short period of time in their absence, the degree of refractoriness upon subsequent challenge is almost identical to that of cells that were preincubated with catecholamine alone. Similarly, actinomycin D does not effect a reversal of refractoriness, as does cycloheximide, during recovery experiments and actinomycin D is much more effective in preventing desensitization during the first hour of catecholamine incubation than at later times.

The induction of refractoriness in C 6–2 B cells appears to involve cyclic AMP as an obligatory mediator, as was discussed previously. The concentration-effect relationship for induction of refractoriness is similar to the concentration-effect relationship for elevation of intracellular cyclic AMP levels. Moreover, preincubation of cells with a phosphodiesterase inhibitor results in the development of refractoriness to catecholamine challenge, and phosphodiesterase inhibitors potentiate the effects of submaximal concentrations of isoproterenol on induction of desensitization. Analogs of cyclic AMP also induce refractoriness. Interestingly, the refractoriness that occurs under these conditions is completely blocked by protein synthesis inhibition whereas norepinephrine-induced refractoriness is only partially blocked by cycloheximide (TERASAKI et al. 1978).

The β-receptor/adenylate cyclase system of C 6–2 B cells also has been examined in membrane fractions prepared from desensitized cells. Although data were not presented, TERASAKI et al. (1978) reported that there was no change in β-receptor number in membranes prepared from desensitized cells. It was unclear, however, as to the time of preincubation of cells with catecholamine prior to assay of receptors. Basal, isoproterenol- and GTP-stimulated enzyme activities decreased in membranes from desensitized cells with a time course similar to that of loss of whole cell responsiveness (TERASAKI et al. 1978). Interestingly, fluoride-stimulated activity did not appear to be significantly changed in desensitized cells.

Since C 6–2 B cells contain only one known hormone receptor that is linked in a stimulatory manner to adenylate cyclase, it is not possible to define the catecholamine-induced changes that occur in C 6–2 B cells as agonist-specific or agonist-nonspecific. However, the fact that cyclic AMP induces refractoriness would suggest that an agonist-nonspecific change occurs in these cells that may be analogous to that reported for cells that respond to more than one type of hormone agonist (NEWCOMBE et al. 1975; SU et al. 1976 a; TOUGUI et al. 1980). The hormone nonspecific nature of the refractoriness induced by catecholamines is supported by the fact that the cyclic AMP response to cholera toxin in C 6–2 B cells is also potentiated by protein synthesis inhibition (NICKOLS and BROOKER, 1980).

Several lines of evidence suggest that the refractoriness that develops in C 6–2 B cells may be of more than one type. For example, although protein synthesis inhibition is effective in blocking refractoriness induced by dibutyryl cyclic AMP alone, this blockade is not complete if refractoriness is induced by norepinephrine. Also, in the one time course experiment that has been reported, there is a trend which suggests that hormone responsiveness may be lost faster than basal activity during incubation of cells with catecholamine (TERASAKI et al., 1978). Finally, our own observations indicate that at least during long-term (2–12 h) incubation of C 6–2 B cells with isoproterenol there is a decrease (down to 20% of control) in the number of β-adrenergic receptors (FREDERICH et al. unpublished). Thus, as with several other cell lines, it appears that multiple mechanisms also may be responsible for the regulation of catecholamine responsiveness C 6–2 B cells. As was pointed out above, one or another of these mechanisms may predominate in a particular cell line and the data of BROOKER and coworkers would suggest that in C 6–2 B cells a hormone-nonspecific type of regulation is of major importance.

VI. Agonist-Induced Desensitization in Cell-Free Preparations

BOCKAERT et al. (1976) first established that desensitization to the effects of hormones on adenylate cyclase could be induced in vitro using broken cell preparations of pig graafian follicles. Desensitization to the effects of luteinizing hormone (LH) in this in vitro system required ATP and Mg^{++}; the process was stimulated by, but did not require, LH. Hormone-stimulated activity was selectively affected as shown by the fact that basal activity did not change. Neither GTP nor adenylyl imidodiphosphate App(NH)p would substitute for ATP in the desensitization reaction, suggesting to the authors an involvement of phosphorylation of one or more of the components of the adenylate cyclase system. This idea was indirectly supported by a subsequent study (HUNZICKER-DUNN et al. 1979) demonstrating reversal of in vitro desensitization to LH upon activation of endogenous phosphatase activity with Mn^{++} or dithiothreitol or by the addition of a partially purified phosphatase preparation. Both studies provided evidence for a lack of involvement of cyclic AMP in the desensitization process observed in vitro.

Recently EZRA and SALOMON (1980) have studied a similar reaction in rat ovarian plasma membrane preparations. In this study, LH caused desensitization of the response of adenylate cyclase to LH but not to NaF. The enzyme system would reconvert to its fully responsive state if the desensitized preparations were incubated in GTP-free medium. Alterations in LH receptor number or binding properties were not detected during in vitro desensitization. The authors proposed that GTP serves as a substrate in a phosphotransferase reaction that leads to phosphorylation of adenylate cyclase in its inactive (GDP-bound) state (see Fig. 1).

Lysine vasopressin-induced desensitization of adenylate cyclase of pig kidney membranes was studied by ROY et al. (1976). As above, ATP and Mg^{++} were required. Unfortunately, other parameters of adenylate cyclase activity were not studied so conclusions regarding the specificity of the observed loss of vasopressin-stimulated enzyme activity could not be made.

To date, only one study of catecholamine-induced desensitization in cell-free systems has appeared. When crude membranes from normal rat kidney cells (NRK-S) were incubated with isoproterenol, a selective loss in the response of adenylate cyclase to isoproterenol occurred; basal and PGE_1-stimulated activities were not altered (ANDERSON and JAWORSKI 1979). An absolute requirement for ATP, GTP, and Mg^{++} was established. Also, increasing concentrations of Mg^{++}, in the presence of ATP and GTP led to desensitization even in the absence of isoproterenol. It may be relevant to successful demonstration of desensitization in cell-free preparations that NRK-S cells exhibit a very rapid rise and fall in intracellular cyclic AMP when exposed to isoproterenol. It is possible that in these cells the capacity of the desensitization reaction is exceptionally great, and thus the probability of observing this reaction in diluted cell free preparations is favorable. IYENGAR (pers. comm.) has shown that incubation of membranes from S 49 lymphoma cells with isoproterenol also leads to agonist-specific desensitization. In this study, ATP was required and its nonphosphorylating analog App(NH)p would not substitute. Neither basal activity nor NaF-, Gpp(NH)p-, or PGE_1-stimulated activities were altered by a prior 45 min incubation of S 49 membranes with isoproterenol. Changes in the number of β-receptors or in their binding properties were not determined in either study.

Although only these few studies of agonist-induced desensitization in cell-free systems have appeared (also see IYENGAR et al. 1980) they appear to exhibit the common feature of requiring phosphorylating conditions for the observance of desensitization. In addition, the reaction does not appear to involve cyclic AMP, eliminating the obvious possibility that cyclic AMP-dependent protein kinases could be functioning in a feed-back regulatory mechanism. Although the data are too limited to determine if these cell-free systems are valid models of the agonist-specific desensitization observed with whole cells, there are at least two positive correlations. First, at least in some of the studies the desensitization appears agonist-specific. Second, the "phosphorylation" is not cyclic AMP-dependent, consistent with the lack of involvement of cyclic AMP in agonist-specific desensitization in whole cells (SHEAR et al. 1976; SU et al. 1980). Careful comparative studies are needed to firmly establish any relation between whole cell and cell-free agonist-induced desensitization reactions.

VII. Desensitization of Gonadotropin Receptor-Linked Adenylate Cyclase

Agonist-induced desensitization of the gonadotropin receptor-linked adenylate cyclase system of rat testis and ovary has been studied in detail. Specific receptors for LH have been identified and shown to be identical with binding sites for human chorionic gonadotropin (hCG). Administration of hCG to rats reduces the response of adenylate cyclase to LH, the number of LH receptors, and the capacity of LH to stimulate steroidogenesis (CONTI et al. 1977; CIGORRAGA et al. 1978; HARWOOD et al. 1978). The decrease in hormone binding to LH receptors appears to involve two processes (CONTI et al. 1977). Early after administration of hCG to rats, LH receptors are tightly occupied by hCG and cannot be detected in standard binding assays. Subsequently, receptors appear to be actually lost from the plasma membrane. The loss in responsiveness of adenylate cyclase occurs early in this process prior to the net loss of LH receptors (CONTI et al. 1977; LAMPRECHT et al. 1977). Low concentrations of hCG appear to lead to limited loss of LH receptors and a more rapid return of the effects of LH on cyclic AMP production and steroidogenesis than occurs after administration of sufficient hCG to cause the marked, net loss of LH receptors. At high concentrations, hCG causes the loss of response of ovarian adenylate cyclase not only to LH, but to epinephrine and NaF as well (HARWOOD et al. 1980). Conversely, epinephrine causes only agonist-specific desensitization (HARWOOD et al. 1979). Interestingly, the recovery of NaF and epinephrine responsiveness occurred within 1–2 days whereas hCG injection into rats caused total loss of response of adenylate cyclase to LH that had not recovered *at all* after 3 days. The hCG-induced loss of response to epinephrine of rat ovarian cell adenylate cyclase is not accompanied by a loss of β-receptors (HARWOOD et al. 1980).

These observations suggest the following series of events following administration to rats of an agonist of the LH receptor. First, the agonist (hCG) binds to the receptor and stimulates adenylate cyclase; the resultant increase in cyclic AMP formation results in increased steroidogenesis. A non-specific (heterologous) desensitization ensues and is completely expressed by 24 h (HSUEH et al. 1977). An LH receptor-specific process may be obscured by the nonspecific desensitization reaction. By 24–48 h, LH receptors undergo a net decline to an extent determined by the concentration of hCG previously administered. The more extensive the loss of LH receptors, the longer the time of recovery of responsiveness to LH. Responsiveness to other hormones returns independently and at a more rapid pace. After a large dose of hCG the recovery of response to epinephrine was complete at least 48 h before the response to LH was regained (HARWOOD et al. 1980).

In addition to the hCG-induced non-specific desensitization of the adenylate cyclase system and the specific loss of LH receptors, there appears to be an hCG-induced alteration in the effect of LH on steroidogenesis distal to the production of cyclic AMP (CIGORRAGA et al. 1978; HARWOOD et al. 1980). Thus, after treatment with hCG, not only is steroidogenesis not stimulated by hCG, LH or epinephrine, but cholera toxin and Bt_2AMP are without effect as well (HARWOOD et al. 1980). CIGORRAGA et al. (1978) have postulated that hCG causes a block of testosterone production at the 17,20-desmolase step. HARWOOD et al. (1980) suggest that such effects of hCG reflect secondary changes in steroid biosynthesis due to feedback inhibition as a result of a local accumulation of steroid produced by desensitizing doses of hCG.

A final interesting observation is that complete responsiveness of adenylate cyclase and of steroidogenesis to high concentrations of LH are recovered, after desensitization, before the full complement of LH receptors is recovered. The results of CONTI et al. (1977) in this regard (see their Fig. 8), suggest the existence of "spare" LH receptors.

The LH receptor-linked adenylate cyclase system is not desensitized by catecolamines whereas the catecholamine-linked adenylate cyclase (apparently in the same cells, HARWOOD et al. 1980) is non-specifically desensitized by the LH receptor agonist, hCG. It is not clear why the LH response would not exhibit heterologous desensitization after injection of epinephrine if, in fact, heterologous desensitization is mediated by cyclic AMP. Clearly, there is much left to be learned about this complex regulatory response. It remains a possibility that hCG-induced loss of LH receptors proceeds by an initial "uncoupling" of receptors from adenylate cyclase followed by a net loss of LH receptors, perhaps by a process of receptor internalization (CONN et al. 1978; AMSTERDAM et al. 1980). The uncoupling event may be difficult to detect if a simultaneous but non-specific desensitization reaction is occurring (HARWOOD et al. 1980).

C. Receptor-Mediated Inhibition of Adenylate Cyclase Activity

To this point we have only considered those mechanisms involved in the regulation of the *activation* of adenylate cyclase by hormone agonists. However, it is now apparent that the interaction of agonists with a number of receptors results initially in *inhibition* rather than activation of adenylate cyclase. The inhibition of adenylate cyclase by a receptor agonist was first reported in 1962 by MURAD and co-workers. Both basal and epinephrine stimulated adenylate cyclase activities were inhibited approximately 30% by carbachol in dog heart homogenates. This carbachol-induced inhibition, which was blocked by atropine, suggested that muscarinic receptor mediated-regulation of cardiac activity might involve the cyclic AMP system. Surprisingly, few studies were reported over the next decade whose purpose was either to confirm or extend this early work of MURAD et al. (1962). However, as early as 1967, ROBISON, BUTCHER, and SUTHERLAND proposed that α-adrenergic receptors might be linked in an inhibitory manner to adenylate cyclase. Subsequently, these and a number of other investigators have demonstrated in a variety of intact cell systems that stimulation of α-adrenergic (ROBISON et al. 1969; MCCARTHY and DE VELLIS 1978), muscarinic cholinergic (GEORGE et al. 1973; LICHTSHTEIN et al. 1979) and opiate (SHARMA et al. 1975a, b) receptors results in decreases in intracellular cyclic AMP levels. The mechanisms responsible for such a reduction in cellular cyclic nucleotide content is not known in detail. However, the demonstration in the last several years of an agonist-induced inhibition of adenylate cyclase in cell free preparations suggests that it is now possible to define the mechanisms through which such a phenomenon might occur. We will review in this section results that have been obtained in cell free preparations. Since these inhibitory effects of, for example, muscarinic receptor agonists have been reported for only a few tissues, it has not yet been unequivocally demonstrated that such an effect is a common property of muscarinic receptors in all tissues. The possibility of multiple muscarinic receptor subtypes (BIRDSALL et al. 1978) also confuses extrapolation of these results. Indeed, results from studies of α-adrenergic receptors suggest that only stimulation of the α_2-subtype leads to an inhibition of adenylate cyclase (TSAI and LEFKOWITZ 1978). Although it seems likely that different receptor types would inhibit adenylate cyclase through a common mechanism, there are enough discrepant results in the studies thus far reported to prevent unequivocal statements concerning commonality of mechanisms.

In general, it has been demonstrated that inhibitory agonists reduce all aspects of adenylate cyclase activity; i.e. basal, and hormone- or NaF-stimulated activities are usually reduced. In several studies the inhibition of basal enzyme activity has occured over the same concentrations of inhibitory agonist and to about the same extent (on a percentage basis) as did the inhibition of hormone stimulated activities. However, opiate inhibition of basal adenylate cyclase activity in NG 108–15 neuroblastoma × glioma cells occurred at lower concentrations of opiate agonists than did inhibition of PGE_1-stimulated activity (BLUME et al. 1979). Although basal and hormone stimulated activities appear to be inhibited to a similar degree in most systems, NaF-stimulated activity is affected differently in a number of systems. For example, muscarinic receptor stimulation in heart membranes reduced NaF-stimulated activity to approximately the same extent as basal or hormone-stimulated activity (JAKOBS et al. 1979). Similarly, NaF-stimulated activity was reduced in platelet membranes by α-adrenergic receptor agonists, although on a percentage basis this effect was not as great as that on basal activity (JAKOBS et al. 1976). In contrast, in NG 108–15 cells α-receptor stimulation reduced basal and hormone-stimulated activities without affecting NaF-stimulated activity (SABOL and NIRENBERG 1979). Also, using dog heart membranes, WATANABE et al. (1978) failed to observe an effect of methacholine on NaF-stimulated activity.

Discrepant results also have been reported on the effects of inhibitory agonists on Gpp(NH)p-stimulated enzyme activity. In NG 108–15 cells, α-receptor agonists had no effect on the apparent rate or extent of activation of adenylate cyclase by Gpp(NH)p (SABOL and NIRENBERG 1979). The effect of an α-agonist on the rate of activation in the presence of a stimulatory hormone was not examined. Similarly, carbachol had no effect on Gpp(NH)p-stimulated activity (maximally activated) in heart membranes (JAKOBS et al. 1979). An examination of the effect of carbachol on the rate of Gpp(NH)p-induced activation of adenylate cyclase was not reported. In contrast to these results, STEER and WOOD (1979) have reported that α-receptor agonists decrease maximally activated Gpp(NH)p-stimulated activity in platelet membranes. It was not clear from this study, however, that an exchange of GTP for Gpp(NH)p did not occur; i.e., it was not shown that the system remained maximally or irreversibly activated once epinephrine and GTP were removed.

It has been reported that guanine nucleotides are necessary for the inhibition of adenylate cyclase that is mediated through muscarinic (JAKOBS et al. 1979; LICHTSHTEIN et al. 1979), α-adrenergic (JAKOBS et al. 1978) and opiate receptors (BLUME et al. 1979). It is by no means clear whether this GTP-dependence directly involves the guanine nucleotide binding protein that is necessary for hormone-stimulated enzyme activity. BLUME et al. (1979) have shown that opiate inhibition of adenylate cyclase in NG 108–15 cells occurs in the presence of either GTP or ITP; ATP, CTP, and GDP would not effectively substitute for GTP or ITP. The order of potency of nucleotides for inhibition of platelet adenylate cyclase by epinephrine was GTP > deoxy-GTP > ITP > UTP > CTP > XTP. An extensive comparison between the nucleotide dependence for stimulation versus inhibition of adenylate cyclase has not been reported. However, JAKOBS et al. (1979) have demonstrated that the half maximal concentration of GTP needed for isoproterenol-stimulated activity in heart membranes was about five-fold less than that required for observation of inhibition by carbachol. In this regard, STEER and WOOD (1979) have proposed that agonist stimulation and inhibition of platelet adenylate cyclase involve two different guanine nucleotide bindings sites. This conclusion was primarily based on the observation that low concentrations of GTP enhanced PGE_1-stimulation while higher concentrations of GTP produced inhibition. The general applicability of these results should be questioned, however, since in contrast to most hormone-stimulated adenylate cyclases, PGE_1-stimulated enzyme activity in platelet membranes exhibited little GTP-dependence.

As was described above, inhibitory agonists interact with the adenylate cyclase system in a number of tissues with properties that suggest interaction of excitatory and inhibitory receptors with a common pool of guanine nucleotide binding sites. However, evidence is not yet available which allow such a conclusion to be made unequivocally. Indeed, the often disparate results obtained with different receptor systems and different tissues do not allow conclusion concerning the commonality of the mechanism of adenylate cyclase inhibition in different systems. However, in a tissue source (NG 108–15 cells) in which several different receptors (α-adrenergic, muscarinic, and opiate) interact with adenylate cyclase in an inhibitory manner, agonist stimulation of each of these receptors appears to inhibit enzyme activity through a common mechanism (SABOL and NIRENBERG, 1979). That is, inhibition of adenylate cyclase activity by α-adrenergic, muscarinic, or opiate receptor agonists alone was not different from that produced by all of the agonists in combination.

To date, the molecular nature of the guanine nucleotide binding site that is involved in the inhibition of adenylate cyclase is not known. However, it is tempting to speculate than this site is the same as that involved in the activation of adenylate cyclase. In such a situation agonist stimulation of inhibitory receptors might sequester or alter guanine nucleotide binding sites in such a way as to reduce their availability for activation of adenylate cyclase through a stimulatory receptor. Such a mechanism, of course, would require an appropriate stoichiometry of the system whereby a reduction of available N would become rate limiting for the activation process.

Recently, RODBELL (1980) has proposed that receptors may inhibit adenylate cyclase through a different guanine nucleotide binding protein than that involved in enzyme activation. This contention is mainly based on results with fat cell membrane preparations in which low concentrations of GTP activate adenylate cyclase while higher concentrations of guanine nucleotide are inhibitory (YAMAMURA et al. 1977). COOPER et al. (1979) have shown that the GTP-stimulatory versus inhibitory processes are differentially affected by a variety of conditions suggesting that these two effects involve separable components. Also, adenosine inhibits the fat cell adenylate cyclase only in the presence of high (inhibitory) concentrations of GTP (LONDOS and WOLFE 1977). These results have led to the proposal that separate nucleotide binding sites are involved in activation and inhibition of adenylate cyclase in fat cells as well as other tissues (RODBELL 1980). Evidence has also been presented for two guanine nucleotide binding sites in the glucagon receptor linked-adenylate cyclase of rat liver (IYENGAR et al. 1979; IYENGAR and BIRNBAUMER 1979).

As discussed earlier, agonists bind to β-adrenergic receptors in a high affinity, slowly dissociable complex. This complex apparently involves a ternary association of agonist, receptor, and nucleotide binding protein, and serves as an obligatory intermediate through which adenylate cyclase is activated by agonists and guanine nucleotides. Interestingly, it also has been demonstrated that guanine nucleotides decrease agonist affinity at receptors (muscarinic, α-adrenergic, and opiate) which have been shown to mediate a GTP-dependent inhibition of adenylate cyclase (U'PRICHARD and SNYDER 1978; TSAI and LEFKOWITZ 1978; BLUME 1978; CHILDERS and SNYDER 1978; ROSENBERGER et al. 1980). The relationship between GTP-dependent inhibition of adenylate cyclase mediated through these receptors and the effects of guanine nucleotides on agonist binding is not yet clear. However, in light of the well documented relationship of effects on β-adrenergic receptor agonist binding and the activation of adenylate cyclase, it is possible that a similar relationship obtains for nucleotide effects on agonist binding at muscarinic, α-adrenergic, and opiate receptors. Thus, binding studies of nucleotide and agonist interactions in these receptor systems may be reflective of a fundamental interaction in the process through which agonist-receptor interactions lead to an inhibition of adenylate cyclase.

The inhibitory effects of receptor agonists on adenylate cyclase activity are dependent in several tissues on the presence of monovalent cations. This dependence was essentially absolute for opiate (BLUME et al. 1979) and muscarinic receptor (LICHTSHTEIN et al. 1979) mediated inhibition in NG 108–15 cells and for α-receptor mediated inhibition with hamster fat cells (AKTORIES et al. 1979). Muscarinic receptor-mediated inhibition of adenylate cyclase in cardiac membranes was at least partially dependent on the presence of Na^+ (JAKOBS et al. 1979). This effect appears to be relatively ion specific; that is, Li^+ is approximately equally active in substituting for Na^+ (BLUME et al. 1979; LICHTSHTEIN et al. 1979), while K^+, choline$^+$, rubidium, and cesium are much less active (BLUME et al. 1979; LICHTSHTEIN et al. 1979; AKTORIES et al. 1979). As was the case with guanine nucleotides, monovalent cations affect the apparent affinity of agonists but not antagonists at muscarinic (ROSENBERGER et al. 1980), opiate (BLUME 1978; CHILDERS and SNYDER 1978; MILLER et al. 1978) and α-adrenergic (U'PRICHARD and SNYDER 1978; TSAI and LEFKOWITZ 1978) receptors. At least in the case of α-adrenergic receptors, the degree of shift in agonist binding curves induced by NaCl appears to be directly related to the intrinsic activity of α-receptor agonists for inhibition of PGE_1-stimulated adenylate cyclase activity in platelet membranes (TSAI and LEFKOWITZ 1978).

The physiological significance of the regulation of adenylate cyclase by agonists acting through inhibitory muscarinic, α-adrenergic, opiate or adenosine receptors is not yet clear. Studies of this level of regulation are rapidly evolving and the elucidation of the molecular mechanisms involved should be forthcoming in the near future.

D. Physiological Significance of the Regulation of Responsiveness to Hormones

Although significant progress has been made in understanding the mechanisms of catecholamine-induced desensitization in homogeneous cell preparations in vitro, the physiological significance of these mechanisms is still not clear. Most in vivo systems are not amenable to the types of analyses that have been carried out with cells maintained in vitro. The instability of catecholamines, cellular heterogeneity, compensatory alterations in the release of other neurotransmitters, hormones or substances, and the difficulty of kinetic analyses all have made difficult the accomplishment of well controlled studies using intact animals. Nevertheless, there is widespread evidence for the occurrence of agonist-induced adaptive phenomena in vivo. Probably the most thoroughly studied system in this regard involves the regulation of cholinergic transmission at the neuromuscular junction (see FAMBROUGH 1979). However, excellent examples of in vivo adaptation at the level of the β-receptor/adenylate cyclase system also have been described. Probably the first and one of the best characterized systems in this context involves the light sensitive alterations that occur in the β-receptor system of the pineal. Indole metabolism of the pineal is maintained under a circadian rhythm that is driven by light-induced changes in the extent of norepinephrine release from the sympathetic innervation of the gland (AXELROD 1974). Not only does sympathetic denervation of the pineal or injection of isoproterenol result in compensatory changes in both β-receptor number and catecholamine stimulated cyclic AMP accumulation (DEGUCHI and AXELROD 1973a, b; STRADA and WEISS 1974; ROMERO and AXELROD 1975), but a diurnal variation in postsynaptic responsiveness is apparent. During a 12-h light cycle, sympathetic activity is reduced and there is a progressive increase in the number of β-adrenergic receptors and response of adenylate cyclase to catecholamines (ROMERO and AXELROD 1975; ROMERO et al. 1975). At the beginning of the subsequent 12 h dark cycle sympathetic activity is greatly increased and, because of a high level of β-receptors and catecholamine responsiveness, large increases in cyclic AMP levels occur with resulting large increases in melatonin synthesis (ROMERO et al. 1975). During this dark phase a desensitization process is initiated. By the end of the twelve hour dark cycle the β-receptor/adenylate cyclase system is at a low level of responsiveness to catecholamines and the number of β-receptors is low. Thus, an argument can be made for an important physiological role of adaptive responses at the level of the β-receptor-linked adenylate cyclase in pineal function.

Other examples of changes in responsiveness of the adenylate cyclase system to catecholamines occurring as a function of alterations in the availability of norepinephrine can be pointed out. For example, intraventricular administration of 6-hydroxydopamine to adult rats or subcutaneous administration to newborn rats results in marked depletion of endogenous norepinephrine stores in the cerebral cortex. This depletion is followed by large increases in the efficacy of catecholamines for stimulation of cyclic AMP synthesis in cortical slices (KALISKER et al. 1973; PALMER et al. 1973) and by elevations in the number of β-adrenergic receptors (SPORN et al. 1976; HARDEN et al. 1977b). In contrast, chronic administration of tricyclic antidepressants, which elevate norepinephrine concentrations at the level of postsynaptic receptors by inhibiting the neuronal reuptake of neurotransmitter, results in a progressive decrease in catecholamine responsiveness of the adenylate cyclase system (VETULANI and SULSER 1975) and a decreased number of β-adrenergic receptors (WOLFE et al. 1978). Indeed, the time course of the adaptive changes in the β-receptor/adenylate cyclase system correspond more closely to the positive psychiatric effects of such drugs than does the initial blockade of the neurotransmitter uptake mechanism. Thus, it has been proposed that the adaptive response that occurs in postsynaptic mechanisms during chronic treatment with antidepressants may be responsible for their therapeutically useful effects (see WOLFE and MOLINOFF 1980).

Adaptive mechanisms in postsynaptic catecholamine receptor systems may also play a role in several other clinical situations. For example, the tachyphylaxis that occurs to the clinical effects of catecholamines during the treatment of asthma may be related to an agonist-induced desensitization of the β-receptor system at the appropriate target cells (GALANT et al. 1978; GREENACRE et al. 1978). Similarly, the withdrawal syndrome that occurs following cessation of chronic propranolol treatment is likely related to increases in responsiveness of postsynaptic β-receptor systems. For example, it has been shown that chronic blockade of β-receptors with propranolol results in an increase in the number of β-receptors in rat heart (GLAUBIGER and LEFKOWITZ 1977).

Whether changes in the β-receptor-linked adenylate cyclase system can be considered of etiological importance in disease states remains to be determined. To various degrees such alterations have been implicated in a number of conditions, including hypertension (see AMIR 1975; WOODCOCK et al. 1979) and asthma (SZENTIVANYI 1968). Also, the hyperthyroidism syndrome is accompanied by alterations in adrenergic receptor mechanisms (HOFFMAN and LEFKOWITZ 1980). Indeed, chronic treatment of experimental animals with high doses of thyroxine results in increases in β-adrenergic receptor number in the heart (WILLIAMS et al. 1977; TSAI and CHEN 1977). Although the implication of changes in β-receptor functions as being causal in disease states is still speculative, there is strong evidence for such a relationship in other receptor systems. For example, in myasthenia gravis (LINDSTROM and DAU 1980) and insulin resistant diabetes (KAHN et al. 1973; SOLL et al. 1975) receptor alterations appear to be of primary importance in the production of the disease state.

Figures 6 and 7 illustrate hypothetical changes in dose-response relationships for a hormone-sensitive adenylate cyclase system upon adaptation to marked changes in hormone concentration. Such formulations may be useful in explaining

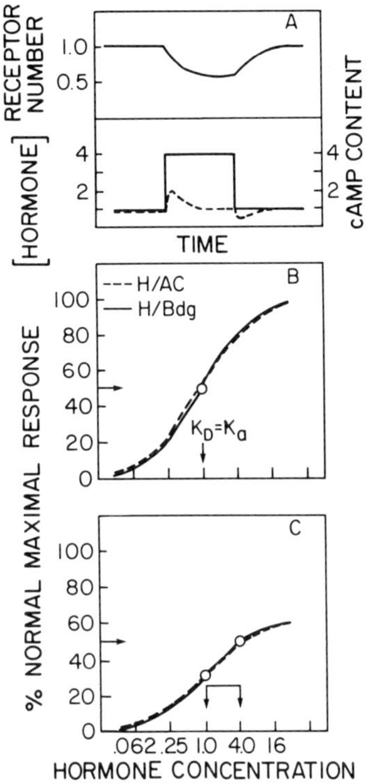

Fig. 6

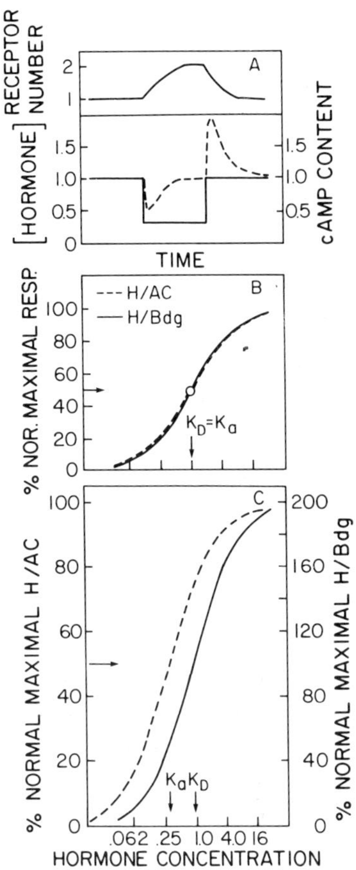

Fig. 7

Fig. 6 A–C. The effect of an adaptive increase in the number of functional receptors on the relation between hormone concentration and receptor occupancy or cellular response. The lower panel of **A** indicates a sequence of step changes in hormone (signal) concentration with the resultant changes in cAMP content. The upper panel indicates changes in the number of functional receptors resulting from changes in hormone concentration. **B** Illustration of the relation of hormone concentration to receptor occupancy (H/Bdg) and activation of adenylate cyclase (H/AC) under normal conditions, i.e. hormone concentration equals one. **C** Illustration of the relation of hormone concentration to receptor occupancy and activation of adenylate cyclase after the adaptation to the increased hormone concentration. The arrows indicate that the normal level of activation of adenylate cyclase is now attained at 4 times the normal hormone concentration due to the reduction in functional receptors. K_D denotes the dissociation constant for binding of receptor to hormone; K_a indicates the concentration of hormone causing half-maximal activation of adenylate cyclase

Fig. 7 A–C. The effect of an adaptive decrease in the number of functional receptors on the relation between hormone concentration and receptor occupancy or cellular response. **A** and **B** see description for 6 A and B. **C** Relation of hormone concentration to receptor occupancy and activation of adenylate cyclase after adaptation to the decreased hormone concentration. The two-fold increase in functional receptors permits the same degree of activation of adenylate cyclase at the reduced hormone concentration. See text for further explanation certain of the examples of adaptive changes shown to occur in vivo as discussed above.

The responses illustrated are for the situation where an abrupt change in the level of a hormone or neurotransmitter occurs, is sustained for a time, and then abruptly returns to normal (Fig. 6 A). When the signal increases, cyclic AMP content abruptly rises but this response is blunted by the loss of functionality of receptors which continues until the cyclic AMP content returns to normal. When the signal abruptly returns to normal, there is a transient fall in cyclic AMP content to below normal levels; this trend is then reversed as the receptors return to the normal functional state. The rate at which the system can respond relative to the rate and extent of change in signal is a measure of the effectiveness of the compensatory mechanism. For example, the time constants involved in the uncoupling and recoupling of β-receptors in human astrocytoma cells are of a magnitude to prevent cyclic AMP excursions from the set point for more than 10–30 min as a result of even very abrupt changes in agonist concentration. If the compensatory response included net reduction in the number of β-receptors, then an abrupt fall in signal intensity would result in a prolonged fall of cyclic AMP below normal because of the time required for synthesis of new receptors.

If under normal conditions the relationship between percent receptor occupancy and percent of maximal cyclic AMP content was strictly related (Fig. 6 B), the changes in signal/response and signal/binding relationships after the adaptive response shown in Fig. 6 A would be as shown in 6 C. The assumption is made that under normal conditions the signal is such that half of the receptors are occupied and thus half the potential response is generated.

After the adaptation, the cell can still generate the normal response, but a greater percentage of the receptors must now be occupied to effect the response (the same absolute number of receptors are occupied), and the maximal response is reduced. In this situation the number of occupied receptors and the level of response remain strictly related (i.e., there is no change in potency).

If on the other hand, as shown in Fig. 7 A, the signal falls and the cell responds by increasing the number of functional receptors, the signal/response and signal/occupancy relationships will change both in relation to each other and in absolute terms. The assumption is made that the previously observed maximal response (see Fig. 7 B) of the cell is retained due to limitations of the system distal to the receptor. It remains the case that 100% of the increased number of receptors will become saturated over the same range of signal intensity as before; however, in this circumstance the "normal" response is achieved by the occupancy of the same number of receptors (a smaller percentage) at a lower signal level. In this case there are "spare" receptors in the sense that the maximal cellular response does not require occupancy of all of the receptors (Fig. 7 C).

In a situation where under *normal* conditions the signal/response and signal/occupancy relationships were as shown in Fig. 7 C, a rise in the signal intensity and the resultant loss of functional receptors still would lead to desensitization even though spare receptors had existed. The result of such a transition would be the case shown in Fig. 7 B. The increased signal would now occupy a higher percentage of the reduced number of receptors (but the same absolute number) and this would achieve the normal response. This adaptation has the effect of shifting the signal/

response curve to the right. Higher signal intensity would then be required to achieve any given level of response but the same maximal response could be attained.

Another physiologically significant circumstance that could involve compensatory regulation would be the chronic use, then acute withdrawal from use, of a drug that either directly or indirectly mimicked the signal. At the moment of withdrawal of the mimicking drug, the system would exist in a down-regulated state but would now be exposed to a normal signal input; this would result in a functional deficit until new receptors could be synthesized or otherwise regain function. The converse situation would obtain upon chronic use then acute withdrawal of a drug that was a direct signal antagonist. A cellular hyper response would be expressed until signal-induced desensitization occurred. The latter process has been proposed as the basis for changes in adrenergic function upon acute withdrawal of the β-adrenergic receptor antagonist propranolol (Aarons et al. 1980).

In summary, there would appear to be a reasonable theoretical basis for alterations in the number of functional receptors resulting in alteration in cellular response to hormones and neurohumors. In addition, a growing body of experimental observations support the existence of cellular mechanisms for adaptation to fluctuating levels of signal molecules. The ability of hormone-sensitive cells to regulate their response to first messengers of the adenylate cyclase system may prove to be only the first level of a series of adaptive responses geared to maintenance of cellular homeostasis in the face of wide variations in signal input. It is clear that such variations frequently occur in response to changes in the environment and as a result of disease. The study of the regulation of cyclic AMP formation in hormone responsive cells is illuminating an elaborate set of regulatory processes that modulate hormone-mediated intercellular communication.

References

Aarons RD, Nies AS, Gal J, Hegstrand LR, Molinoff PB (1980) Elevation in B-adrenergic receptor density in human lymphocytes after propranolol administration. J Clin Invest 65:949–957

Aktories K, Schultz G, Jakobs KH (1979) Inhibition of hamster fat cell adenylate cyclase by prostaglandin E_1 and epinephrine: requirement for GTP and sodium ions. FEBS Lett 107:100–104

Amir MS, Gomell AW, Perhack JL, Ferguson HC, McKinney GR (1975) Observations of cyclic nucleotide metabolism in the hearts and vessels of hypertensive rats. Proc Natl Acad Sci USA 71:4930–4934

Amsterdam A, Berkowitz A, Nimrod A, Kohen F (1980) Aggregation of luteinizing hormone receptors in granulosa cells: A possible mechanism of desensitization to the hormone. Proc Natl Acad Sci USA 77:3440–3444

Anderson WB, Jaworski CJ (1979) Isoproterenol-induced desensitization of adenylate cyclase responsiveness in a cell-free system. J Biol Chem 254:4596–4601

Axelrod J (1974) The pineal gland: A neurochemical transducer. Science 184:1341–1348

Barber R, Clark RB, Kelly LA, Butcher RW (1978) A model of desensitization in intact cells. Adv Cyclic Nucleotide Res 9:507–516

Birdsall NJM, Burgen ASV, Hulme EC (1978) The binding of agonists to brain muscarinic receptors. Mol Pharmacol 14:723–736

Bitensky MW, Russell V, Blanco M (1970) Independent variation of glucagon and epinephrine responsive components of hepatic adenyl cyclase as a function of age, sex, and steroid hormones. Endocrinology 86:154–159

Bitensky MW, Gorman RE, Neufeld AH, King R (1971) A specific, reversible, macromolecular inhibitor of hepatic glucagon responsive adenyl cyclase. Endocrinology 89:1242–1249
Blume AJ (1978) Interaction of ligands with the opiate receptors of brain membranes: Regulation by ions and nucleotides. Proc Natl Acad Sci USA 75:1713–1717
Blume AJ, Lichtshtein D, Boone G (1979) Coupling of opiate receptors to adenylate cyclase: Requirement for Na^+ and GTP. Proc Natl Acad Sci USA 76:5626–5630
Bockaert J, Hunzicker-Dunn M, Birnbaumer L (1976) Hormone-stimulated desensitization of hormone-dependent adenylyl cyclase. J Biol Chem 251:2653–2663
Bourne HR, Tomkins GM, Dion S (1973) Regulation of phosphodiesterase synthesis. Requirement for cyclic adenosine monophosphate-dependent protein kinase. Science 181:952–954
Browning ET, Brostrom CO, Groppi VE, Jr (1976) Altered adenosine cyclic 3′,5′-monophosphate synthesis and degradation by C-6 astrocytoma cells following prolonged exposure to norepinephrine. Mol Pharmacol 12:32–40
Butcher RW, Ho RJ, Meng HC, Sutherland EW (1965) Adenosine 3′,5′-monophosphate in biological materials. J Biol Chem 240:4515–4523
Cassel D, Levkovitz H, Selinger Z (1977) The regulatory GTPase cycle of turkey erythrocyte adenylate cyclase. J Cyclic Nucleotide Res 3:393–406
Cassel D, Pfeuffer T (1978) Mechanism of cholera toxin action: Covalent modification of the guanyl nucleotide-binding protein of the adenylate cyclase system. Proc Natl Acad Sci USA 75:2669–2673
Cassel D, Selinger Z (1977) Mechanism of adenylate cyclase activation by cholera toxin: Inhibition of GTP hydrolysis at the regulatory site. Proc Natl Acad Sci USA 74:3307–3311
Cassel D, Selinger Z (1978) Mechanism of adenylate cyclase activation through the β-adrenergic receptor: Catecholamine-induced displacement of bound GDP by GTP. Proc Natl Acad Sci USA 75:4155–4159
Childers SR, Snyder SH (1978) Guanine nucleotides differentiate agonist and antagonist interactions with opiate receptors. Life Sci 23:759–762
Chuang DM, Costa E (1979) Evidence for internalization of the recognition site of β-adrenergic receptors during receptor sub-sensitivity induced by (–)-isoproterenol. Proc Natl Acad Sci USA 76:3024–3028
Cigorraga SB, Dufau ML, Catt KJ (1978) Regulation of luteinizing hormone receptors and steroidogenesis in gonadotropin-desensitized leydig cells. J Biol Chem 253:4297–4304
Clark RB, Butcher RW (1979) Desensitization of adenylate cyclase in cultured fibroblasts with prostaglandin E_1 and epinephrine. J Biol Chem 254:9373–9378
Clark RB, Su Y-F, Ortmann R, Cubeddu-Ximenez L, Johnson GL, Perkins JP (1975) Factors influencing the effect of hormones on the accumulation of cyclic AMP in cultured human astrocytoma cells. Metabolism 24:343–358
Conn PM, Conti M, Harwood JP, DuFau ML, Catt KJ (1978) Internalisation of gonadotropin-receptor complex in ovarian luteal cells. Nature 274:598–600
Conti M, Harwood JP, Dufau ML, Catt KJ (1977) Effect of gonadotropin-induced receptor regulation on biological responses of isolated rat luteal cells. J Biol Chem 252:8869–8874
Cooper MF, Schlegel W, Lin MC, Rodbell M (1979) The fat cell adenylate cyclase system: Characterization and manipulation of its bimodal regulation by GTP. J Biol Chem 8927–8931
D'Armiento M, Johnson GS, Pastan I (1972) Regulation of adenosine 3′,5′-cyclic monophosphate phosphodiesterase activity in fibroblasts by intracellular concentrations of cyclic adenosine monophosphate. Proc Natl Acad Sci USA 69:459–462
Deguchi T, Axelrod J (1973a) Superinduction of serotonin N-acetyl-transferase and supersensitivity of adenyl cyclase to catecholamines in denervated pineal gland. Mol Pharmacol 9:612–618
Deguchi T, Axelrod J (1973b) Supersensitivity and subsensitivity of the beta-adrenergic receptor in pineal gland regulated by catecholamine transmitter. Proc Natl Acad Sci USA 70:2411–2414
DeLean A, Stadel JM, Lefkowitz RJ (1980) A ternary complex model explains the agonist-specific binding properties of the adenylate cyclase-coupled β-adrenergic receptor. J Biol Chem 255:7108–7117

deVellis J, Brooker G (1974) Reversal of catecholamine refractoriness by inhibitors of RNA and protein synthesis. Science 186:1221–1223
Downs RW, Jr, Spiegel AM, Singer M, Reen S, Aurbach GD (1980) Fluoride stimulation of adenylate cyclase is dependent on the guanine nucleotide regulatory protein. J Biol Chem 255:949–954
Ezra E, Salomon Y (1980) Mechanism of desensitization of adenylate cyclase by lutropin. J Biol Chem 255:653–658
Fain JN, Shepherd RE (1975) Free fatty acids as feedback regulators of adenylate cyclase and cyclic 3′:5′-AMP accumulation in rat fat cells. J Biol Chem 250:6586–6592
Fambrough DM (1979) Control of acetylcholine receptors in skeletal muscle. Physiol Rev 59:165–227
Foster SJ, Perkins JP (1977) Glucocorticoids increase the responsiveness of cells in culture to prostaglandin E_1. Proc Natl Acad Sci USA 74:4816–4820
Franklin TJ, Foster SJ (1973) Hormone-induced desensitization of hormonal control of cyclic AMP levels in human diploid fibroblasts. Nature New Biol 246:146–148
Franklin TJ, Morris WP, Twose PA (1975) Desensitization of beta adrenergic receptors in human fibroblasts in tissue culture. Mol Pharmacol 11:485–491
Franklin TJ, Twose PA (1979) Desensitization of beta receptor mediated cyclic AMP response of cultured fibroblasts by partial agonists. J Cyclic Nucleotide Res 5:19–30
Galant SP, Duriseti L, Underwood S, Insel PA (1978) Decreased beta-adrenergic receptors on polymorphonuclear leukocytes after adrenergic therapy. New Engl J Med 299:933–936
George WJ, Wilkerson RD, Kadowitz PJ (1973) Influence of acetylcholine on contractile force and cyclic nucleotide levels in the isolated perfused rat heart. J Pharmacol Exp Ther 184:228–235
Gill DM, Meren R (1978) ADP-ribosylation of membrane proteins catalyzed by cholera toxin: Basis of the activation of adenylate cyclase. Proc Natl Acad Sci USA 75:3050–3054
Gilman AG (1971) A protein binding assay for adenosine 3′,5′-cyclic monophosphate. Proc Natl Acad Sci 67:305–312
Glaubiger G, Lefkowitz RJ (1977) Elevated beta-adrenergic receptor number after chronic propranolol treatment. Biochem Biophys Res Commun 78:720–725
Greenacre JK, Schofield P, Conolly ME (1978) Desensitization of the beta-adrenoreceptor of lymphocytes from normal subjects and asthmatic patients in vitro. Br J Clin Pharmacol 5:199–206
Haga T, Haga K, Gilman AG (1977a) Hydrodynamic properties of the β-adrenergic receptor and adenylate cyclase from wild type and variant S 49 lymphoma cells. J Biol Chem 252:5776–5782
Haga T, Ross EM, Anderson HJ, Gilman AG (1977b) Adenylate cyclase permanently uncoupled from hormone receptors in a novel variant of S 49 mouse lymphoma cells. Proc Natl Acad Sci USA 84:2016–2020
Harden TK, Cotton CU, Waldo GL, Lutton JK, Perkins JP (1980) Catecholamine-induced alteration in the sedimentation behavior of membrane-bound β-adrenergic receptors. Science 210:441–443
Harden TK, Foster SJ, Perkins JP (1979a) Differential expression of components of the adenylate cyclase system during growth of astrocytoma cells in culture. J Biol Chem 254:4416–4422
Harden TK, Su Y-F, Perkins JP (1979b) Catecholamine-induced desensitization involves an uncoupling of beta-adrenergic receptors and adenylate cyclase. J Cyclic Nucleotide Res 5:99–106
Harden TK, Wolfe BB, Sporn JR, Perkins JP, Molinoff PB (1977a) Ontogeny of beta-adrenergic receptors in rat cerebral cortex. Brain Res 125:99–108
Harden TK, Wolfe BB, Sporn JR, Poulous BK, Molinoff PB (1977b) Effects of 6-hydroxydopamine on the development of the beta adrenergic receptor/adenylate cyclase system in rat cerebral cortex. J Pharmacol Exp Ther 203:132–143
Harwood JP, Conti M, Conn PM, Dufau ML, Catt KJ (1978) Receptors, regulation and target cell responses: Studies in the ovarian luteal cell. Mol Cell Endocrinol 11:121–135

Harwood JP, Dufau ML, Catt KJ (1979) Differing specificities in the desensitization of ovarian adenylate cyclase by epinephrine and human chorionic gonadotropin. Mol Pharmacol 15:439–444

Harwood JP, Richert ND, Dufau ML, Catt KJ (1980) Gonadotropin-induced desensitization of epinephrine action in the luteinized rat ovary. Endocrinology 107:280–288

Haslam RJ, Goldstein S (1974) Adenosine 3′:5′-cyclic monophosphate in young and senescent human fibroblasts during growth and stationary phase in vitro. Biochem J 144:253–263

Ho R-J, Russell TR, Asakawa T, Sutherland EW (1975) Cellular levels of feedback regulator of adenylate cyclase and the effect of epinephrine and insulin. Proc Natl Acad Sci USA 72:4739–4743

Ho R-J, Sutherland EW (1971) Formation and release of a hormone antagonist by rat adipocytes. J Biol Chem 246:6822–6827

Ho R-J, Sutherland EW (1975) Action of feedback regulator on adenylate cyclase. Proc Natl Acad Sci USA 72:1773–1777

Hoffman BB, Lefkowitz RJ (1980) Radioligand binding studies of adrenergic receptors: New insights into molecular and physiological regulation. Ann Rev Pharmacol Toxicol 20:581–608

Howlett AC, Sternweis PC, Macik BA, Van Arsdale PM, Gilman AG (1979) Reconstitution of catecholamine-sensitive adenylate cyclase. J Biol Chem 254:2287–2895

Hsueh AJW, Dufau ML, Catt KJ (1977) Gonadotropin-induced regulation of luteinizing hormone receptors and desensitization of testicular 3′:5′-cyclic AMP testosterone responses. Proc Natl Acad Sci USA 74:592–595

Hunzicker-Dunn M, Derda D, Jungmann RA, Birnbaumer L (1979) Resensitization of the desensitized follicular adenylyl cyclase system to luteinizing hormone. Endocrinology 104:1785–1793

Iyengar R, Birnbaumer (1979) Coupling of the glucagon receptor to adenylyl cyclase by GDP: Evidence for two levels of regulation of adenylyl cyclase. Proc Natl Acad Sci USA 76:3189–3193

Iyengar R, Mintz PW, Swartz TL, Birnbaumer L (1980) Divalent cation induced desensitization of glucagon stimulable adenylyl cyclase in rat liver plasma membranes: GTP dependent stimulation by glucagon. J Biol Chem 225:11875–11882

Iyengar R, Swartz TL, Birnbaumer L (1979) Coupling of glucagon receptor to adenylyl cyclase. J Biol Chem 254:1119–1923

Jakobs KH, Aktories K, Schultz G (1979) GTP-dependent inhibition of cardiac adenylate cyclase by muscarinic cholinergic agonists. Naunyn-Schmiedeburgs Archs Pharmacol 310:113–119

Jakobs KH, Saur W, Schultz G (1976) Reduction of adenylate cyclase activity in lysates of human platelets by the alpha-adrenergic component of epinephrine. J Cyclic Nucleotide Res 2:381–392

Jakobs KH, Saur W, Schultz G (1978) Inhibition of platelet adenylate cyclase by epinephrine requires GTP. FEBS Letters 85:167–170

Johnson GL (1977) Thesis: Studies of the changes in responsiveness of hormone-sensitive adenylate cyclase in cultured cells. University of Colorado School of Medicine

Johnson GL, Bourne HR (1977) Influence of cholera toxin on the regulation of adenylate cyclase by GTP. Biochem Biophys Res Commun 78:792–799

Johnson GL, Kaslow HR, Bourne HR (1978a) Reconstitution of cholera toxin-activated adenylate cyclase. Proc Natl Acad Sci USA 75:3113–3117

Johnson GL, Wolfe BB, Harden TK, Molinoff PB, Perkins JP (1978b) Role of β-adrenergic receptors in catecholamine-induced desensitization of adenylate cyclase in human astrocytoma cells. J Biol Chem 253:1472–1480

Kakiuchi S, Rall TW (1968) The influence of chemical agents on the accumulation of adenosine 3′,5′-phosphate in slices of rabbit cerebellum. Mol Pharmacol 4:367–378

Kalisker A, Rutledge CO, Perkins JP (1973) Effect of nerve degeneration by 6-hydroxydopamine on catecholamine-stimulated adenosine 3′,5′-monophosphate formation in rat cerebral cortex. Mol Pharmacol 9:619–629

Kahn CR, Neville DM, Roth J (1973) Insulin-receptor interaction in the obese-hyperglycemic mouse. J Biol Chem 248:244–250

Kaslow HR, Farfel Z, Johnson GL, Bourne HR (1979) Adenylate cyclase assembled in vitro: Cholera toxin substrates determine different patterns of regulation by isoproterenol and guanosine 5′-triphosphate. Mol Pharmacol 15:472–483

Kent RS, DeLean A, Lefkowitz RJ (1980) A quantitative analysis of beta-adrenergic receptor interactions: Resolution of high and low affinity states of the receptor by computer modeling of ligand binding data. Mol Pharmacol 17:14–23

Krishna G, Weiss B, Brodie BB (1968) A simple, sensitive method for the assay of adenyl cyclase. J Pharmacol Exp Therap 163:379–385

Kuo JF, DeRenzo EC (1969) A comparison of the effects of lipolytic and antilipolytic agents on adenosine 3′,5′-monophosphate levels in adipose cells as determined by prior labeling with adenine – 9– ^{14}C. J Biol Chem 244:2252–2260

Lamprecht SA, Zor U, Salomon Y, Koch Y, Ahren K, Lindner HR (1977) Mechanism of hormonally induced refractoriness of ovarian adenylate cyclase to luteinizing hormone and prostagladin E_2. J Cyclic Nucleotide Res 3:69–83

Lauzon GJ, Kulshrestha S, Starr L, Bar H-P (1976) On the accumulation of adenosine 3′,5′-monophosphate in Ehrlich cells and adenylate cyclase desensitization in response to epinephrine. J Cyclic Nucleotide Res 2:99–114

Lefkowitz RJ, Mullikin D, Caron MG (1976) Regulation of beta-adrenergic receptors by guanyl-5′-yl imidophosphate and other purine nucleotides. J Biol Chem 251:4686–4692

Lefkowitz RJ, Wessels MR, Stadel JM (1980) Hormones, receptors, and cyclic AMP: Their role in target cell refractoriness. Current Topics Cell Reg 17:205–230

Lefkowitz RJ, Williams LT (1978) Molecular mechanisms of activation and desensitization of adenylate cyclase coupled beta-adrenergic receptors. Adv Cyclic Nucleotide Res 9:1–16

Leichtling BJ, Drotar AM, Ortmann R, Perkins JP (1976) Growth of astrocytoma cells in the presence of prostaglandin E_1: Effect on the regulation of cyclic AMP metabolism. J Cyclic Nucleotide Res 2:89–98

Levey GS, Lehotay DC, Canterbury JM, Bricker LA, Meltz GJ (1975) Isolation of a unique peptide inhibitor of hormone-responsive adenylate cyclase. J Biol Chem 250:5730–5733

Lichtshtein D, Boone G, Blume A (1979) Muscarinic receptor regulation of NG 108–15 adenylate cyclase: Requirement for Na^+ and GTP. J Cyclic Nucleotide Res 5:367–375

Limbird LE, Gill DM, Lefkowitz RJ (1980 a) Agonist-promoted coupling of the β-adrenergic receptor with the guanine nucleotide regulatory protein of the adenylate cyclase system. Proc Natl Acad Sci USA 77:775–779

Limbird LE, Gill DM, Stadel JM, Hickey AR, Lefkowitz RJ (1980 b) Loss of β-adrenergic receptor-guanine nucleotide regulatory protein interactions accompanies decline in catecholamine responsiveness of adenylate cyclase in maturing rat erythrocytes. J Biol Chem 255:1854–1861

Limbird LE, Hickey AR, Lefkowitz RJ (1979) Unique coupling of the frog erythrocyte adenylate cyclase system by manganese. J Biol Chem 254:2677–2683

Limbird LE, Lefkowitz RJ (1977) Resolution of β-adrenergic receptor binding and adenylate cyclase activity by gel exclusion chromatography. J Biol Chem 252:799–802

Limbird LE, Lefkowitz RJ (1978) Agonist-induced increase in apparent β-adrenergic receptor size. Proc Natl Acad Sci USA 75:228–232

Lindstrom J, Dau P (1980) Biology of myasthenia gravis. Ann Rev Pharmacol Toxicol 20:337–362

Londos C, Wolff J (1977) Two distinct adenosine-sensitive sites on adenylate cyclase. Proc Natl Acad Sci USA 74:5482–5486

Lutton JK, Frederich RC, Perkins JP (1979) Isolation of adenylate cyclase-enriched membranes from mammalian cells using concanavalin A. J Biol Chem 254:11181–11184

Maguire ME, Ross EM, Gilman AG (1977) β-adrenergic receptor: Ligand binding properties and the interaction with adenylyl cyclase. Adv Cyclic Nucleotide Res 8:1–83

Maguire ME, Van Arsdale PM, Gilman AG (1976) An agonist-specific effect of guanine nucleotides on binding to the beta adrenergic receptor. Mol Pharmacol 12:335–339

Mallorga P, Tallman JF, Henneberry RC, Hirata F, Strittmatter WT, Axelrod J (1980) Mepacrine blocks β-adrenergic agonist-induced desensitization in astrocytoma cells. Proc Natl Acad Sci USA 77:1341–1345

Manganiello V, Vaughan M (1972) Prostaglandin E_1 effects on adenosine 3′:5′-cyclic monophosphate concentration and phosphodiesterase activity in fibroblasts. Proc Natl Acad Sci USA 69:269–273
McCarthy KD, DeVellis (1978) Alpha adrenergic receptor modulation of beta adrenergic, adenosine, and prostaglandin E_1 increased adenosine 3′:5′-cyclic monophosphate levels in primary cultures of glia. J Cyclic Nucleotide Res 4:15–26
Mickey J, Tate R, Lefkowitz RJ (1975) Subsensitivity of adenylate cyclase and decreased β-adrenergic receptor binding after chronic exposure to (–)-isoproterenol in vitro. J Biol Chem 250:5727–5729
Mickey JV, Tate R, Mullikin D, Lefkowitz RJ (1976) Regulation of adenylate cyclase-coupled beta adrenergic receptor binding sites by beta adrenergic catecholamines in vitro. Mol Pharmacol 12:409–419
Miller RJ, Chang K, Leighton J, Cuatrecasas P (1978) Interaction of iodinated enkephalin analogues with opiate receptors. Life Sci 22:379–388
Mukherjee C, Caron MG, Lefkowitz RJ (1975) Catecholamine-induced subsensitivity of adenylate cyclase associated with loss of β-adrenergic receptor binding sites. Proc Natl Acad Sci USA 72:1945–1949
Mukherjee C, Caron MG, Lefkowitz RJ (1976) Regulation of adenylate cyclase coupled β-adrenergic receptors by β-adrenergic catecholamines. Endocrinology 99:347–357
Murad F, Chi YM, Rall TW, Sutherland EW (1962) The effect of catecholamines and choline esters on the formation of adenosine 3′,5′-phosphate by preparations from cardiac muscle and liver. J Biol Chem 237:1233–1238
Newcombe DS, Ciosek CP, Jr, Ishikawa Y, Fahey JV (1975) Human synoviocytes: activation and desensitization by prostaglandins and l-epinephrine. Proc Natl Acad Sci USA 72:3124–3128
Nickols GA, Brooker G (1979) Induction of refractoriness to isoproterenol by prior treatment of C 6–2 B rat astrocytoma cells with cholera toxin. J Cyclic Nucleotide Res 5:435–447
Northup JK, Sternweis PC, Smigel MD, Schleifer LS, Gilman AG (1980) Purification of the regulatory component of adenylate cyclase. Proc Natl Acad Sci 77:6516–6520
Palmer GC, Sulser F, Robison GA (1973) Effects of neurohumoral and adrenergic agents on cyclic AMP levels in various areas of the rat brain in vitro. Neuropharmacol 12:327–337
Pastan IH, Willingham MC (1981) Receptor-mediated endocytosis of hormones in cultured cells. Ann Rev Physiol 43:239–250
Perkins JP, Moore MM (1973) Regulation of the adenosine cyclic 3′,5′-monophosphate content of rat cerebral cortex. Ontogenetic development of the responsiveness to catecholamines and adenosine. Mol Pharmacol 9:774–782
Perkins JP, Su Y-F, Harden TK (1979) Adaptive changes in the responsiveness of adenylate cyclase to catecholamines. Drug Alcohol Depend 4:279–294
Reggiani A, Vernaleone F, Robinson GA (1980) Loss and restoration of sensitivity to catecholamines in cultured BHK cells: Influence of dansylcadavarine. Soc Neurosc Abstr 6:183–187
Robison GA, Butcher RW, Sutherland EW (1967) Adenyl cyclase as an adrenergic receptor. Ann NY Acad Sci 139:703–723
Robison GA, Butcher RW, Sutherland EW (1971) Cyclic AMP. Academic Press, New York, pp 224–228
Robison GA, Arnold A, Hartmann RC (1969) Divergent effects of epinephrine and prostaglandin E_1 on the level of cyclic AMP in human blood platelets. Pharmacol Res Commun 1:325–332
Rodbell M (1980) The role of hormone receptors and GTP-regulatory proteins in membrane transduction. Nature 284:17–21
Romero J, Axelrod J (1975) Regulation of sensitivity to β-adrenergic stimulation in the induction of pineal N-acetyltransferase. Proc Natl Acad Sci USA 72:1661–1665
Romero JA, Zatz M, Kebabian JW, Axelrod J (1975) Circadian cycles in binding of ^3H-alprenolol to β-adrenergic receptor sites in rat pineal. Nature 258:435–436
Rosenberger LB, Yamamura HI, Roeske WR (1980) Cardiac muscarinic cholinergic receptor binding is regulated by Na^+ and guanyl nucleotides. J Biol Chem 255:820–822

Ross EM, Gilman AG (1977) Reconstitution of catecholamine-sensitive adenylate cyclase activity: Interaction of solubilized components with receptor-replete membranes. Proc Natl Acad Sci USA 74:3715–3719
Ross EM, Gilman AG (1980) Biochemical properties of hormone-sensitive adenylate cyclase. Ann Rev Biochem 49:533–564
Ross EM, Howlett AC, Ferguson KM, Gilman AG (1978) Reconstitution of hormone-sensitive adenylate cyclase activity with resolved components of the enzyme. J Biol Chem 253:6401–6412
Ross EM, Maguire ME, Sturgill TW, Biltonen RL, Gilman AG (1977) Relationship between the β-adrenergic receptor and adenylate cyclase. J Biol Chem 252:5761–5775
Roy C, Gullion G, Jard S (1976) Hormone-dependent desensitization of vasopressin-sensitive adenylate cyclase. Biochem Biophys Res Commun 72:1265–1269
Sabol SL, Nirenberg M (1979) Regulation of adenylate cyclase of neuroblastoma X glioma hybrid cells by alpha-adrenergic receptors: 1. Inhibition of adenylate cyclase mediated by alpha-receptors. J Biol Chem 254:1913–1920
Schmidt MJ, Palmer GC, Dettbarn WD, Robison GA (1970) Cyclic AMP and adenyl cyclase in the developing rat brain. Dev Psychobiol 3:53–67
Schramm M, Rodbell M (1975) Persistent active state of the adenylate cyclase system produced by the combined actions of isoproterenol and guanylyl imidodiphosphate in frog erythrocyte membranes. J Biol Chem 250:2232–2237
Schwartz JP, Passoneau JV (1974) Cyclic AMP-mediated induction of the cyclic AMP phosphodiesterase of C-6 glioma cells. Proc Natl Acad Sci USA 71:3844–3848
Sharma SK, Klee WA, Nirenberg M (1975 a) Dual regulation of adenylate cyclase accounts for narcotic dependence and tolerance. Proc Natl Acad Sci USA 72:3092–3096
Sharma SK, Nirenberg M, Klee WA (1975 b) Morphine receptors as regulators of adenylate cyclase activity. Proc Natl Acad Sci USA 72:590–594
Shear M, Insel PA, Melmon KL, Coffino P (1976) Agonist-specific refractoriness induced by isoproterenol. J Biol Chem 251:7572–7576
Shimizu H, Daly JW, Creveling CR (1969) A radioisotopic method for measuring the formation of adenosine 3′,5′-cyclic monophosphate in incubated slices of brain. J Neurochem 18:1157–1161
Soll AH, Kahn CR, Neville DJ Jr, Roth J (1975) Insulin receptor deficiency in genetic and acquired obesity. Clin Invest 56:769–780
Sporn JR, Harden TK, Wolfe BB, Molinoff PB (1976) Beta-adrenergic receptor involvement in 6-hydroxydopamine-induced supersensitivity in rat cerebral cortex. Science 194:624–626
Stadel J, DeLean A, Lefkowitz RJ (1981) Molecular mechanism of coupling in hormone-responsive adenylate cyclase systems. Adv Enzymol, vol 53 (in press)
Stadel JM, Lefkowitz RJ (1979) Multiple reactive sulfhydryl groups modulate the function of adenylate cyclase coupled beta-adrenergic receptors. Mol Pharmacol 16:709–718
Steer ML, Wood AJ (1979) Regulation of human platelet adenylate cyclase by epinephrine, prostaglandin E_1, and guanine nucleotides. J Biol Chem 254:10791–10797
Sternweis PC, Gilman AG (1979) Reconstitution of catecholamine-sensitive adenylate cyclase. J Biol Chem 254:3333–3340
Strada SJ, Weiss B (1974) Increased response to catecholamines of the cyclic AMP system of rat pineal gland induced by decreased sympathetic activity. Arch Biochem Biophys 160:197–204
Su YF, Cubeddu-Ximenez L, Perkins JP (1976 a) Regulation of adenosine 3′:5′-monophosphate content of human astrocytoma cells: desensitization to catecholamines and prostaglandins. J Cyclic Nucleotide Res 2:257–270
Su YF, Johnson GL, Cubeddu-Ximenez L, Leichtling BH, Ortmann R, Perkins JP (1976 b) Regulation of adenosine 3′:5′-monophosphate content of human astrocytoma cells: Mechanism of agonist-specific desensitization. J Cyclic Nucleotide Res 2:271–285
Su YF, Harden TK, Perkins JP (1979) Isoproterenol-induced desensitization of adenylate cyclase in human astrocytoma cells. J Biol Chem 254:38–51
Su YF, Harden TK, Perkins JP (1980) Catecholamine-specific desensitization of adenylate cyclase: Evidence for a multistep process. J Biol Chem 255:7410–7419

Sutherland EW, Oye I, Butcher RW (1965) The action of epinephrine and the role of the adenyl cyclase system in hormone action. Rec Prog Horm Res 21:623–646
Sventivanyi A (1968) The beta adrenergic theory of the atopic abnormality in bronchial asthma. J Allergy 42:203–232
Swillens S, Dumont JE (1980) A unifying model of current concepts and data on adenylate cyclase activation by β-adrenergic agonists. Life Sci 27:1013–1028
Terasaki WL, Brooker G, de Vellis J, Inglish D, Hsu C-Y, Moylan RD (1978) Involvement of cyclic AMP and protein synthesis in catecholamine refractoriness. Adv Cyc Nucleotide Res 9:33–52
Tougui Z, Khac LD, Harbon S (1980) Modulation of cyclic AMP content of the rat myometrium: Desensitization to isoproterenol, PGE_2, and prostacyclin. Mol Cell Endocr 20:17–34
Tsai JS, Chen A (1977) L-triiodothyronine increases the level of beta-adrenergic receptors in rat heart: Effects of thyroidectomy. Eur J Pharmacol 43:207–208
Tsai BS, Lefkowitz RJ (1978) Agonist-specific effects of monovalent and divalent cations on adenylate cyclase-coupled alpha-adrenergic receptors in rabbit platelets. Mol Pharmacol 14:540–548
U'Prichard DC, Snyder SH (1978) Guanyl nucleotide influence on [^{3}H]-ligand binding to alpha-noradrenergic receptors in calf brain membranes. J Biol Chem 253:3444–3452
Vauquelin G, Geynet P, Hanoune J, Strosberg AS (1977) Isolation of adenylate cyclase-free, β-adrenergic receptor from turkey erythrocyte membranes by affinity chromatography. Proc Natl Acad Sci USA 74:3710–3714
Vetulani J, Sulser F (1975) Action of various antidepressant treatments reduces reactivity of noradrenergic cyclic AMP-generating system in limbic forebrain. Nature 257:495–496
Watanabe AM, McConnaughey MM, Strawbridge RA, Fleming JW, Jones LR, Besch RR Jr (1978) Muscarinic cholinergic receptor modulation of β-adrenergic receptor affinity for catecholamines. J Biol Chem 253:4833–4835
Wessels MR, Mullikin D, Lefkowitz RJ (1978) Differences between agonist and antagonist binding following beta-adrenergic receptor desensitization. J Biol Chem 253:3371–3373
Wessels MR, Mullikin D, Lefkowitz RJ (1979) Selective alteration in high affinity agonist binding: A mechanism of beta-adrenergic receptor desensitization. Mol Pharmacol 16:10–20
Williams LT, Lefkowitz RJ (1977) Slowly reversible binding of catecholamine to a nucleotide-sensitive state of the β-adrenergic receptor. J Biol Chem 252:7207–7213
Williams LT, Lefkowitz RJ, Watanabe AM, Hathaway DR, Besch HR Jr (1977) Thyroid hormone regulation of beta-adrenergic receptor number. J Biol Chem 252:2787–2789
Wolfe BB, Harden TK, Molinoff PB (1976) Beta-adrenergic receptors in rat liver: Effects of adrenalectomy. Proc Natl Acad Sci USA 73:1343–1347
Wolfe BB, Harden TK, Sporn JR, Molinoff PB (1978) Presynaptic modulation of beta-adrenergic receptors in rat cerebral cortex after treatment with antidepressants. J Pharmacol Exp Ther 207:446–457
Wolfe BB, Molinoff PB (1980) Postsynaptic effects of antidepressants on adrenergic receptor number and responsiveness. In: Mann, Gershon, Friedman (eds) Depression and antidepressants: Implications for cause and treatment. Raven Press, New York
Woodcock EA, Funder JW, Johnston CI (1979) Decreased cardiac β-adrenergic receptors in deoxycorticosterone-salt and renal hypertensive rats. Circ Res 45:560–565
Yamamura H, Lad PM, Rodbell MJ (1977) GTP stimulates and inhibits adenylate cyclase in fat cell membranes through distinct regulatory processes. J Biol Chem 252:7964–7966
Yorio T, Bentley PJ (1978) Phospholipase A and the mechanism of action of aldosterone. Nature 271:79–82

CHAPTER 5

Guanylate Cyclase: Regulation of Cyclic GMP Metabolism

C. K. MITTAL and F. MURAD

Overview

Guanylate cyclase that is present in soluble and particulate fractions of homogenates from most tissues has different properties, indicating the presence of multiple forms. While the subcellular distribution of the enzyme can be altered, it is unknown whether or not the soluble and particulate activities represent similar proteins. The soluble enzyme has been purified from several tissues, including rat liver and lung, and the holoenzyme is about 150,000 Daltons. Crude and purified preparations of the enzyme can be activated with a variety of agents including azide, nitroprusside, and other nitro compounds. The common feature of this class of activators is their ability to form nitric oxide which can activate purified enzyme. Another free radical, probably hydroxyl radical, can activate the enzyme. Guanylate cyclase is also activated by fatty acids, detergents and other materials. Guanylate cyclase and cyclic GMP levels are sensitive to oxidizing and reducing agents in incubations, and hormone effects on this system may be mediated through alterations in the redox state of preparations. Increases in cyclic GMP levels in intact cells with nitro compounds are not dependent upon calcium in incubations, in contrast to the effects of acetylcholine and other hormones. Activation of guanylate cyclase results in alterations in the properties of the enzyme. Activated enzyme can utilize Mg^{2+} as effectively as Mn^{2+} for cofactor while native enzyme prefers Mn^{2+}. Enzyme activated with various agents can also catalyze the formation of cyclic AMP from ATP. The significance of this alternate pathway for cyclic AMP synthesis is not known. While many hormones can increase cyclic GMP levels in various tissues, these agents have failed to activate guanylate cyclase in cell-free preparations. However, a heat-stable enterotoxin from *Escherichia coli* increases cyclic GMP levels in intestinal mucosa and can activate particulate guanylate cyclase from intestinal mucosal homogenates from several species. The increase in cyclic GMP in this tissue may mediate the effects of the enterotoxin on fluid secretion in the intestine. While cyclic GMP may function as a second messenger in some systems, it is proposed that cyclic GMP may also act as a feedback regulator to alter the redox state of preparations.

A. Introduction

Guanosine 3′,5′-monophosphate (cyclic GMP) was discovered in 1963 in rat urine during studies which searched for the presence of other cyclic nucleotides in a biological system (ASHMAN et al. 1963). In contrast to the discovery of cyclic AMP (RALL et al. 1957), cyclic GMP was not found as an integral component of a pre-existing and thoroughly examined regulatory process. The structural similarity of cyclic GMP and cyclic AMP stimulated many laboratories to accumulate in various biological systems experimental data that relate to the metabolism and function of cyclic GMP. Investigators have examined numerous biological systems in search of a role for cyclic GMP. Despite tremendous efforts most of the approaches have failed to define a regulatory role of this cyclic nucleotide. Thus, the apparent parallelisms between the cyclic GMP and cyclic AMP system each with their respective cyclases, phosphodiesterases, and kinases, have probably distracted investigators and, in retrospect, have impeded some of the expected progress in the field of cyclic GMP research.

The intracellular levels of cyclic GMP are regulated by a) guanylate cyclase that catalyses its formation from GTP and, b) phosphodiesterase that degrades it to 5′-GMP and, thus, terminates the effect of cyclic GMP on some function(s). During the past decade numerous papers have been published dealing with cyclic GMP synthesis, metabolism, and possible functions. A comprehensive and encyclopedic review of the work in these areas would not be feasible or constructive. The main object of this review is to summarize some of the existing experimental data dealing with cyclic GMP synthesis and to propose some unifying concepts and some new hypotheses. Cyclic nucleotide hydrolysis will be discussed in another chapter of this book.

The sensitivity of the cyclic GMP system to cholinergic drugs, adrenergic agonists, nitrocompounds, oxygen radicals, reductants, unsaturated fatty acids and their metabolites, prostaglandins, etc. will permit investigators an opportunity to examine the cyclic GMP system with some new approaches and perspectives. Hopefully, some of the concepts and generalities learned in this system will be applicable to other enzymes and mechanisms of information transfer. While various neurotransmitters and autacoids alter cyclic GMP levels in intact tissues in the presence of extracellular calcium, they usually do not reproduceably alter the activity of guanylate cyclase in broken cell preparations in a physiologically relevant way. Thus, hormone-receptor interactions are apparently less tightly coupled to guanylate cyclase activation and cyclic GMP synthesis than the adenylate cyclase-cyclic AMP system. Presumably an understanding of the intermediate reactions and processes should permit reconstitution of cell-free systems and hormonal effects on cyclic GMP synthesis. The other categories of substances such as nitrocompounds, oxidizing and reducing substances, and *E. coli* heat-stable enterotoxin modulate both the activity of guanylate cyclase in broken cells as well as cyclic GMP levels in intact-cell systems. The effects of some of these latter materials may be traced to alterations in cyclic GMP metabolism in tissues and have provided some ideas about functions of cyclic GMP. Furthermore, an understanding of the mechanisms of these effects could also provide valuable clues about hormonal regulation of cyclic GMP synthesis.

B. Guanylate Cyclase

I. Activities in Various Tissues and Subcellular Distribution

Guanylate cyclase has been detected in virtually all tissues and organisms throughout the phylogenetic kingdom. These include plants (GOLDBERG et al. 1973), insects (FILBURN and WYATT 1976; CATALAN et al. 1976), bacteria (SUN et al. 1974; MACCHIA et al. 1975), birds, fish, and mammals (see KIMURA and MURAD 1975a and references therein). The subcellular distribution of guanylate cyclase has been a subject of considerable interest to investigators. It has been assumed that knowledge of the subcellular location of guanylate cyclase would provide some information relating to mechanisms of hormonal regulation of the enzyme and functions of cyclic GMP. In most tissues, enzyme activity is associated with both the soluble and particulate fractions of homogenates. However, the activity ratio between soluble and particulate fractions varies in different tissues and species. For example, in liver, lung, kidney (KIMURA and MURAD 1974; CHRISMAN et al. 1975) and platelets (BÖHME et al. 1974; GLASS et al. 1977a), the activity is predominately soluble, whereas with intestinal mucosa (ISHIKAWA et al. 1969) and fibroblast cultures (WALLACH and PASTAN 1976b) the activity is predominately particulate, and with sea urchin sperm (GRAY et al. 1976) it is exclusively particulate. A large proportion of the particulate activity is latent and is expressed only when organelles are treated with detergents such as Triton X-100, Lubrol PX, and bile salts. Treatment of guanylate cyclase preparations with detergents has not only solubilized guanylate cyclase but also has caused an activation of the enzyme (HARDMAN et al. 1973; KIMURA and MURAD 1975a, b). In general, detergents increase activities of particulate preparations from liver, parotid, gland, uterus, cerebellum, heart, sperm, etc. 3- to 12-fold (HARDMAN et al. 1973; KIMURA and MURAD 1974, 1975a, b, c; WHITE 1975; FREY et al. 1977; NAKAZAWA et al. 1976; NEER and SUKIENNIK 1975; DURHAM 1976; SIEGEL et al. 1976; GARBERS et al. 1975b; GRAY et al. 1976; SULAKHE et al. 1976a). Detergents have also been shown to increase the cytosolic enzyme activity 30%–100%. While the mechanism of detergent activation of soluble enzyme is not known, few experiments have addressed this question. Treatment of particulate or soluble guanylate cyclase with detergents may alter some of its properties. In the case of rat renal medulla (NEER and SUKIENNIK 1975) guanylate cyclase undergoes changes in both its physical and kinetic properties following detergent treatment. GARBERS (1976) also observed a change in the sea urchin particulate enzyme when it was solubilized with Lubrol PX and purified. This enzyme preparation displayed a change in the kinetic behavior from one of positive cooperativity with respect to Mn-GTP to one of classical Michaelis-Menten kinetics. However, investigations from this laboratory (KIMURA and MURAD 1974) have indicated no change in the kinetic behavior of particulate enzyme from rat heart with respect to GTP following detergent treatment in that the positive cooperativity with respect to GTP persisted after Triton X-100 treatment.

In liver, particulate guanylate cyclase activity has been associated with purified plasma membranes, endoplasmic reticulum, Golgi (KIMURA and MURAD 1975a, c) and nuclear membranes (EARP et al. 1977). Activity has also been reported in sarcoplasmic reticulum from heart (WHITE 1975), nuclear membranes from uterus

(SIEGEL et al. 1976), rod outer segments from retina (ZIMMERMAN et al. 1976; VIRMAUX et al. 1976), mitochondria and synaptosomes from cerebellum (NAKAZAWA et al. 1976), plasma membranes from parotid (DURHAM 1976), skeletal muscle, heart, and cerebral cortex (SULAKHE et al. 1976a, b), plasma membranes, endoplasmic reticulum, mitochondria, and lysosomes from spleen lymphoid cells (WATSON et al. 1976) and sea urchin sperm flagella (GRAY and DRUMMOND 1976). Thus, guanylate cyclase is associated with virtually every organelle in one tissue or another.

The significance of the soluble and particulate forms of guanylate cyclase is not understood. Several distinguishing physical and kinetic characteristics have been noted in the two enzyme forms (KIMURA and MURAD 1974, 1975a, b, c; CHRISMAN et al. 1975) which may be of regulatory significance. However, it is not known if the two forms of the enzyme are similar proteins or contain similar subunits or if they have different primary structures. The differences noted in the soluble and particulate forms of the enzyme in crude preparations may be attributed to a number of factors such as the dissimilarity of the proteins or the inherent differences in the enzyme's environment and assay conditions (KIMURA and MURAD 1975b). Until the soluble and particulate forms of the enzyme are purified and their structures compared, this question can not be answered definitively. An immunological approach to this question has been performed. Antibody to purified sea urchin particulate enzyme cross reacts with mammalian particulate enzyme but not soluble enzyme (GARBERS 1978a). This study suggests that soluble and particulate forms of guanylate cyclase are antigenically different. However, additional studies are needed to confirm the immunological characterization utilizing purified soluble and particulate mammalian enzymes as the antigens to obtain antibody.

Other reports have also indicated that there are specific areas of distribution of guanylate cyclase within a tissue or cell. For example, in normal rat kidney (CRAVEN and DERUBERTIS 1976; TSAI et al. 1978) particulate cyclase activity increases progressively from the renal cortex to the medulla. Retina represents an excellent example of a discrete localization of guanylate cyclase activity. In this tissue the enzyme is greatest in the rod outer segments compared to other organelles of the photoreceptor cell (GORIDIS et al. 1973). With intestine the relative specific activity of guanylate cyclase is much greater in the brush border of the mucosal villus cells than in the proliferating crypt cells (QUILL and WEISER 1975; DEJONGE 1975). In the case of sea urchin sperm the head region has very low activity compared to the flagella where it is considerably higher (GRAY and DRUMMOND 1976). Much of the guanylate cyclase in brain is soluble and found in synaptosomal preparations (GORIDIS and MORGAN 1973).

II. Altered Enzyme Activity and Subcellular Distribution

There have been reports in the literature that indicate that changes in the subcellular distribution of guanylate cyclase activity occur during cell growth. Regenerating rat liver, fetal tissues, transplantable hepatomas, and renal tumors have been used as model systems for proliferation in these studies. In regenerating rat liver after partial hepatectomy the homogenate activity was unaltered (KIMURA and MURAD 1975d). However, the particulate activity increased and the soluble activity decreased within 12–18 h of partial hepatectomy (KIMURA and MURAD 1975d). Using the same model system other investigators have shown as much as an 11-fold increase in the plasma membrane associated activity (GORIDIS et al. 1977). Although little is known about the mechanisms involved in these changes in soluble and particulate enzyme activities, the shift in enzyme distribution was prevented by cycloheximide indicating that *de novo* synthesis of protein(s) may be involved in this process (KIMURA and MURAD 1975d). Using immunocytochemical techniques in regenerating liver it has been shown that cyclic GMP associated with nuclear membranes is elevated and this correlates with increased nuclear guanylate cyclase activity (EARP et al. 1977).

In fetal rat liver, particulate guanylate cyclase activity is increased while soluble activity is decreased compared to adult liver (KIMURA and MURAD 1975d). Within one to two weeks of birth, however, the enzyme distribution is reversed and the ratio of activities is similar to adult liver. In similar studies with fetal rat heart (VESELY et al. 1976), enzyme activity in the $37,000 \times g$ supernatant fraction is increased, remains elevated for several weeks after birth, and then declines to approach adult levels. Guanylate cyclase activity increases in the contralateral kidney within several hours of unilateral nephrectomy and the increase in activity preceeds the development of renal hypertrophy (SCHOLONDORFF and WEBER 1976).

Processes associated with abnormal growth and neoplasia have also been shown to be associated with alterations in guanylate cyclase activity and subcellular distribution. These have included both spontaneous as well as experimentally induced neoplasms. Transplantable renal tumors and hepatomas propagated in rats have been studied in this context. In several types of transplantable solid tumors the distribution of guanylate cyclase activity is altered and most of the homogenate activity is associated with the particulate fractions (CRISS et al. 1976a, b). Compared to normal tissues, these tumors contained similar amounts of guanylate cyclase activity when assayed in homogenates. The shift in the distribution of guanylate cyclase in favor of the particulate fraction may result from the translocation of enzyme from the cytosolic compartment or may represent independent turnover and/or regulation of the soluble and particulate enzymes. The increased levels of cyclic GMP in these tumors correlated with the ratio of particulate to soluble guanylate cyclase activities in these preparations. The increases in particulate guanylate cyclase activity could be due to increases in free radical formation (COMMONER et al. 1970) as a result of catalase deficiency in tumors such as hepatomas (DOUNCE and SHANEWISE 1950). A lack of catalase would lead to the accumulation of hydrogen peroxide which can enhance the formation of free radicals as discussed later in this review. The ratio of particulate to soluble activity is also increased in other chemically induced tumors. For example, in rats, ethionine-induced hepatomas have a similar shift in patterns of enzyme activities (DERUBERTIS and CRAVEN 1977d). However, the opposite is the case with human renal adenocarcinomas where soluble activity is increased (KIM et al. 1977). The alterations in total guanylate cyclase activity and its subcellular distribution have suggested that cyclic GMP may be related to cellular growth and differentiation. However, a clear association of cyclic GMP with these processes has not been established. Obviously more studies are required in these complex areas to support or refute the role of guanylate cyclase and cyclic GMP in proliferation.

Because of changes in the urinary excretion of cyclic GMP in rats after hypophysectomy (HARDMAN et al. 1966) or thyroparathyroidectomy (HARDMAN et al. 1969), many laboratories have examined the status of the guanylate cyclase-cyclic GMP system in animals and patients with various alterations in their endocrine and metabolic state. VESELY and co-workers reported that decreases in cytosolic guanylate cyclase activity occurred in a number of tissues with diabetes mellitus induced with streptozotocin (VESELY et al. 1977a). The decreases in guanylate cyclase activity in diabetic rats were restored to normal levels with insulin administration. We have also observed a decrease in soluble guanylate cyclase activity in several rat tissues with streptozotocin-induced diabetes (ICHIHARA and MURAD 1979). However, we were not successful in reversing the decrease in soluble guanylate cyclase activity with insulin administration. The role of the guanylate cyclase-cyclic GMP system in diabetes mellitus and insulin action is not known and additional studies are required. As discussed later in this review other effects of insulin on cyclic GMP have been reported.

What has become apparent from these metabolic studies and a number of models of proliferation are that the activity of guanylate cyclase and the subcellular distribution of the enzyme can be modified. The mechanisms for these effects, the role of guanylate cyclase and cyclic GMP in these processes, and the possible translocation or independent regulation of particulate and soluble enzyme are important issues that remain unanswered.

III. Kinetic and Physical Properties of the Multiple Forms

1. Molecular Size and Subunits

The molecular size of guanylate cyclase has varied considerably in different tissues and reports. Reports of the size have ranged from 30,000 Daltons with the soluble enzyme from *E. coli* (MACCHIA et al. 1975) to as large as 900,000 Daltons with particulate guanylate cyclase from rat lung (CHRISMAN et al. 1975). The molecular weight of cytosolic guanylate cyclase has been reported to about 150,000–200,000 Daltons from most mammalian tissues (NEER and SUKIENNIK 1975) whereas the particulate enzyme is generally about 300,000–400,000 Daltons (MURAD et al. 1979). The physical characteristics of the soluble enzyme from different mammalian tissues are quite similar. Soluble enzyme from rat renal medulla has been shown to possess a sedimentation co-efficient (S_{20}) of 6.3 S, a Stokes' radius of 4 Å, a partial specific volume of 1.4 and an axial ratio of 7 (NEER and SUKIENNIK 1975). Similar physical constants have been observed with soluble enzyme for rat testis (NEER and MURAD 1979). The electrophoretic mobilities of soluble enzyme from numerous rat tissues have also been compared in our laboratory. We have found that soluble enzymes from various tissues co-migrate on non-denaturing polyacrylamide gels which suggests that the soluble enzymes from different tissues have similarities in size and charge. The properties of the particulate kidney enzyme from rat are quite different. This enzyme has a molecular weight of about 300,000 Daltons a S_{20w} of 10 S and a Strokes' radius of about 105 Å (NEER and SUKIENNIK 1975). A partially purified preparation of guanylate cyclase from rat synaptosomal fractions has been recently reported to be about

480,000 Daltons (NAKENE and DEGUCHI 1978). Most of the studies dealing with the physical and kinetic properties of guanylate cyclase have been done with crude or partially purified preparations. Sea urchin sperm particulate guanylate cyclase has been purified and is 182,000 Daltons (GARBERS 1976). Purified soluble guanylate cyclase from rat liver homogenates has a molecular weight of 150,000 Daltons (BRAUGHLER et al. 1979 a). Purified rat lung soluble enzyme has a molecular weight of 151,000 Daltons (GARBERS 1979).

Guanylate cyclase from several sources has been shown to have two nonidentical subunits. The purified particulate enzyme from sea urchin has two subunits of 118,000 and 75,000 Daltons on sodium dodecylsulfate gels (GARBERS 1976). Soluble liver and lung guanylate cyclases purified to apparent homogeneity also contain two dissimilar subunits on sodium dodecylsulfate gels. However, some differences in the size of the subunits from these tissues was observed. For example, with liver the subunits are about 100,000 and 45,000 Daltons (MURAD et al. 1979) whereas with the soluble lung enzyme they are 79,000 and 74,000 Daltons (GARBERS 1979). The two dissimilar subunits may represent the regulatory and catalytic subunits of guanylate cyclase and the regulatory subunit could obviously influence the activity of the catalytic subunit in a variety of ways. The specific activity of partially purified or purified soluble enzyme can be increased when the protein concentration is increased in assays (BRAUGHLER et al. 1979 a; MITTAL et al. 1979 b). These studies indicate that a mechanism of protein association and dissociation could influence enzyme activity. More detailed studies are required to ascertain the exact role of these subunits and protein interactions in guanylate cyclase activity.

In more recent studies (MURAD et al. 1980), we have found that purified soluble guanylate cyclase from rat lung contains two similar subunits of about 72,000 Daltons and the holoenzyme is about 150,000 Daltons.

2. Effect of GTP and Other Nucleotides

Guanylate cyclase catalyzes the conversion of GTP to cyclic GMP and pyrophosphate (GARBERS et al. 1975 a) and the enzyme has an absolute requirement for divalent cation as cofactor (see below). The metal-GTP complex is believed to serve as substrate for guanylate cyclase reaction (HARDMAN et al. 1969). Mn^{2+}, Mg^{2+} or Ca^{2+} may serve as the sole cation cofactor to satisfy the metal requirement in cyclase reactions. As discussed in the next section, Mn^{2+} is the preferred cation with native enzyme and the activity with Mn^{2+} is 10- to 20-fold greater than the activity with Mg^{2+} or Ca^{2+}. However, activated enzyme can utilize Mg^{2+} as effectively as Mn^{2+} (KIMURA et al. 1976; KATSUKI et al. 1977 a).

Kinetic analyses of guanylate cyclase activities in a variety of tissues and species exhibit either classical or atypical behavior with respect to GTP. With cytosolic enzyme, double-reciprocal plots are linear and classical Michaelis-Menten kinetics are observed. In contrast double-reciprocal plots with particulate enzyme or solubilized particulate enzyme are curvilinear (KIMURA and MURAD 1974; CHRISMAN et al. 1975; GARBERS 1976). The curved nature of double-reciprocal plots with particulate guanylate cyclase is consistent with positive cooperative behavior with respect to GTP and Hill coefficients greater than 1.0. The apparent Km for GTP using different metals as sole cation cofactors have varied over a considerable range depending on the source of the enzyme, the divalent cation used, and the presence of different modulators (activators and inhibitors). With soluble enzyme preparations the apparent Km for GTP in the presence of Mn^{2+} as divalent cation has been: rat lung 12 μM (Chrisman et al. 1975), bovine lung 300 μM (White 1975), rat liver 35 μM (Kimura and Murad 1975c; Kimura et al. 1975a), rat kidney cortex and inner medulla 110 μM (Craven and DeRubertis 1976), rat heart 65 μM (Kimura and Murad 1974), mouse parotid 140 μM (Durham 1976), and calf uterus 70 μM (Siegel et al. 1976). The particulate guanylate cyclase from various tissues requires higher GTP concentrations for expression of half-maximal activity and shows a positive cooperative behavior. The apparent $S_{0.5}$ values for particulate guanylate cyclase from rat lung (Chrisman et al. 1975), rat liver (Kimura and Murad 1975c), rat kidney (Craven and DeRubertis 1976), rat intestine (De Jonge 1975), rat cerebral cortex (Sulakhe et al. 1976a), calf uterus (Siegel et al. 1976), rabbit skeletal muscle (Sulakhe et al. 1976a), sea urchin sperm (Garbers et al. 1975b), mouse brain (Troyer and Ferrendelli 1976), and bovine retine (Goridis et al. 1973) have varied from 10 to 300 μM with Hill coefficients of 1.4–1.7. The anomalous behavior of crude particulate guanylate cyclase from many mammalian tissues with respect to GTP may be attributed to the cellular heterogeneity of the tissues used or the mixture of organelles in the preparations that contain different forms of the enzyme. However, the nonlinear double-reciprocal plots and Hill coefficients greater than 1.0 have been interpreted to indicate that the particulate enzyme is cooperative and has two or more interactive sites for GTP. The kinetic characteristics of guanylate cyclase from particulate fractions of rat intestine vary with anatomical location (De Jonge 1975). The enzyme in particulate fractions of brush border exhibited a positive cooperative behavior with a Hill coefficient of 1.6 using Mn-GTP. In contrast, guanylate cyclase in microsomes from the upper villous cells showed classical Michaelis-Menten behavior with a Hill coefficient of 1.2. In some instances the use of detergents to solubilize particulate enyzme may also alter the kinetic characteristics of crude particulate guanylate cyclase. For example, detergent treatment altered the properties of the enzyme from rat lung (White 1975) and not from rat heart (Kimura and Murad 1974). The crude enzyme from sea urchin sperm exhibited positive cooperative behavior with GTP; whereas after solubilization with Lubrol PX and purification, the preparation gave classical kinetics without cooperativity (Garbers 1976). The loss of cooperativity with solubilization and purification was not associated with a significant change in molecular size.

Other nucleotides have also been examined as substrates for guanylate cyclase. Guanyl-5′-yl imidodiphosphate (Gpp(NH)p), a synthetic analogue of GTP that is resistent to GTPases and hydrolysis, is utilized about 5% to 20% as well as GTP with soluble liver (Mittal et al. 1977) or particulate fibroblast enzyme (Nesbitt et al. 1976). This is in contrast to the adenylate cyclase system where App(NH)p is about one-half as effective as ATP as substrate. Garbers and co-workers reported that particulate enzyme from sea urchin sperm or soluble enzyme from rat lung can catalyze the formation of cyclic IMP and cyclic deoxy GMP from ITP and deoxy GTP, respectively (Garbers et al. 1975b). With both enzyme preparations deoxy GTP was much more effective as substrate than ITP which was about 10% as effective as GTP.

Addition of ATP to guanylate cyclase reaction mixtures has usually been found to be inhibitory (HARDMAN et al. 1969; WHITE and AURBACH 1969). Inhibition has also been observed with ADP and AMP (MITTAL et al. 1977). The tissues in which ATP inhibition has been described include lung (HARDMAN et al. 1969), heart (KIMURA and MURAD 1974), hepatomas (CRISS et al. 1976b), kidney and renal tumors (CRISS et al. 1976a), small intestine (ISHIKAWA et al. 1969), lymphocytes (DEVILLER et al. 1975), parotid gland (DURHAM 1976), and uterus (SIEGEL et al. 1976). Inhibition of guanylate cyclase activity be adenine nucleotides could be of physiological significance and may represent a potential regulatory mechanism. Generally the soluble enzyme is more sensitive to ATP inhibition than the particulate activity (KIMURA and MURAD 1974; CRISS et al. 1976b). The concentrations of ATP required for half-maximal inhibition are 0.4 and 1.0 mM with soluble and particulate fractions, respectively. Recent studies from this laboratory have indicated that soluble guanylate cyclase activity may be changed with some metabolic conditions such as fasting and streptozotocin-induced diabetes mellitus (ICHIHARA and MURAD 1979). Soluble guanylate cyclase activity in several tissue preparations was increased as a result of 72 h of fasting. These changes in enzyme activity with feeding and fasting may be associated with alterations in the levels of inhibitory nucleotides. In these studies the particulate activity in most tissues remained unaltered during fasting and refeeding which is consistent with less sensitivity of the particulate enzyme to nucleotide inhibition. Differences in the sensitivity of soluble and particulate forms of the enzyme to nucleotide inhibition may provide mechanisms for the apparent independent regulation of each enzyme form with some metabolic conditions.

Cyclic GMP formation and accumulation in hepatocytes are sensitive to alterations in GTP and ATP levels as one would expect from the studies with guanylate cyclase in cell-free systems (DOHI and MURAD, unpublished observations).

As discussed below, native guanylate cyclase does not catalyze the formation of cyclic AMP from ATP or does so at a very low rate that is not measurable with most assay methods (MITTAL and MURAD 1977; MITTAL et al. 1979a). However, activation of guanylate cyclase permits the enzyme to catalyze the formation of cyclic AMP from ATP and the rate of cyclic AMP formation is about 1%–15% of the rate of cyclic GMP formation (MITTAL and MURAD 1977; MITTAL et al. 1979a; ARNOLD et al. 1977).

3. Effects of Cations

As discussed earlier, native guanylate cyclase from most tissue sources prefers Mn^{2+} as the sole cation cofactor (HARDMAN and SUTHERLAND 1969; ISHIKAWA et al. 1969; WHITE and AURBACH 1969). Generally Mg^{2+} is about 5%–20% as effective as Mn^{2+} and Ca^{2+} is somewhat less effective. A number of other cations are minimally effective or ineffective. The apparent Ka for Mn^{2+} is about 0.2 mM which is considerably higher than the concentrations of Mn^{2+} found in tissues (about 1–10 μM). Most enzyme preparations have two apparent sites for cation interaction; one is the metal-nucleotide catalytic site and a second site is for free metal (KIMURA and MURAD 1974; CHRISMAN et al. 1975; KIMURA et al. 1976). With some preparations Mg^{2+} or Ca^{2+} may be used effectively when the Mn^{2+} concentrations are suboptimal. GARBERS and co-workers have suggested that Ca-GTP is an effective substrate for the enzyme (GARBERS et al. 1975b, 1978). Generally high concentrations of Ca^{2+} (millimolar concentrations) increase the activity of the soluble enzyme and decrease the activity of the particulate enzyme (KIMURA and MURAD 1974; CHRISMAN et al. 1975; CRISS et al. 1976a, b). However, several exceptions to this general rule do exist and may be related to the concentration of Mn^{2+} or Mg^{2+} in incubations. It is quite unlikely that the effects of high Ca^{2+} concentrations serve a physiologically relevant control mechanism. However, it has been reported that Ca^{2+} at micromolar concentrations can activate particulate enzyme from fibroblast cultures (WALLACH and PASTAN 1976a). The effects of Ca^{2+} on cyclic GMP levels in intact cells probably occur through other mechanisms as discussed below.

Activation of guanylate cyclase with azide or various nitroso compounds changes the cation requirements of the enzyme (KIMURA et al. 1976; KATSUKI et al. 1977a). Activated enzyme can use Mg^{2+} as effectively as Mn^{2+} as sole cation cofactor. These studies suggest that Mg^{2+} may, in fact, be the physiological cofactor. The effects of millimolar concentrations of Ca^{2+} on guanylate cyclase activity are also altered after activation of the enzyme with azide and other agents (KIMURA et al. 1976).

4. Antibodies Against Guanylate Cyclase

Recently it has demonstrated that soluble and particulate forms of guanylate cyclase are antigentically different (GARBERS 1978a). In this study antibodies generated against highly purified sea urchin sperm particulate guanylate cyclase cross-reacted with particulate guanylate cyclase from various rat tissues and not with the soluble form of the enzyme. The antibody preparation used also did not cross-react with sea urchin adenylate cyclase or cyclic GMP phosphodiesterase. Thus, adenylate cyclase and the soluble form of guanylate cyclase are antigentically different from particulate guanylate cyclase. These observations have not been confirmed with other antibody preparations, and reservations must be made about the similarities or dissimilarities of the soluble and particulate forms of the enzyme. The availability of antibodies to guanylate cyclase should also provide a very important tool in understanding the physiological role and regulation of this enzyme.

IV. Activation

1. Detergents

Triton X-100, Lubrol PX, and other detergents have been shown to increase both particulate and soluble guanylate cyclase activity from most mammalian tissues (HARDMAN et al. 1973; KIMURA and MURAD 1974, 1975b). Bile salts such as cholate and deoxycholate have been reported to have similar effects (ICHIHARA et al. 1977). In fact, several reports in the literature pertaining to the activation of guanylate cyclase by hormones have been attributed to contaminating bile salts in the hormone preparations (ICHIHARA et al. 1977). The degree of activation of different enzyme preparations by detergents varies considerably. Soluble enzyme preparations are usually activated about 50%–100% whereas the particulate preparations have shown between 200%–1,100% activation with a variety of detergents (KIMURA and MURAD 1975a, b). Activation of guanylate cyclase in particulate fractions is in large part attributed to the solubilization phenomenon and expression of latent enzyme activity (KIMURA and MURAD 1974). The mechanism of detergent activation of cytosolic enzyme is obscure at the present time but it appears to differ from the activation caused spontaneously in air or by nitrocompounds as described in later sections. As discussed earlier the kinetic and physical properties of the solubilized particulate enzyme may or may not be altered utilizing different tissue preparations. As would be predicted the maximal solubilization and/or activation of the preparation occurs with concentrations of detergents near their critical micel concentrations. Concentrations of detergents less than or greater than the critical micel concentrations are less effective (WALDMAN et al., unpublished observations).

2. Spontaneous Activation

BÖHME and associates described the phenomena of spontaneous activation of platelet guanylate cyclase (BÖHME et al. 1974). This activation was observed with both crude homogenates and cell-free supernatant fractions. Addition of dithiothreitol or dithioerythritol reduced or prevented the apparent activation of platelet guanylate cyclase that occurred when enzyme was incubated in air. Similar examples of autoactivation of the enzyme have been described with preparations from rat lung and other tissues (CHRISMAN et al. 1975). Activation of guanylate cyclase in a post-mitochondrial supernatant fraction of rat lung homogenates was found to be dependent on temperature, Cu^{2+}, and the presence of oxygen (WHITE et al. 1976). The activation was inhibited by thiols, bovine serum albumin, KCN and sodium diethyldithiocarbamate, a Cu^{2+} chelator. The mechanism of activation in these crude preparations was thought to be due to the non-enzymatic production of H_2O_2 (WHITE et al. 1976). As discussed below, activation of crude or purified guanylate cyclase can occur with a reactive oxygen species such as hydroxyl radical which is dependent on H_2O_2 production (MITTAL and MURAD 1977a, b). Spontaneous activation of guanylate cyclase has also been described with crude soluble enzyme from rat uterus and this activation exhibited all of the characteristics described for lung preparations (KRASKA et al. 1977). While the mechanism of spontaneous activation is not completely understood, it is very likely that the formation of oxygen containing free radicals are involved which leads to the oxidation and activation of guanylate cyclase as described below.

3. Effects of Azide and Nitro Compounds

Sodium azide (NaN_3) is a nucleophilic agent and at high concentrations is a potent inhibitor of cellular metabolism (DIXON and WEBB 1958). It is also employed to inhibit phosphatase reactions, and sodium azide was introduced in guanylate cyclase incubations in order to inhibit GTPase in crude preparations when reaction mixtures were devoid of a GTP-regenerating system. Using rat liver crude cytosolic or particulate guanylate cyclase preparations, sodium azide markedly increased the enzyme activity (KIMURA et al. 1975a). This provided investigators with one of the earliest experimental agents to manipulate guanylate cyclase activity in broken cell preparations. The activation of guanylate cyclase by azide was concentration dependent, required oxygen, exhibited a lag phase that was abolished by preincubation, was tissue specific, was reversible, and its effect was enhanced with thiols. Other nucleophilic agents that also activated guanylate cyclase in some crude preparations included hydroxylamine, phenylhydrazine, and sodium nitrite (Kimura et al. 1975a, b). Since our early reports of these effects, guanylate cyclase activation by azide, hydroxylamine, and sodium nitrite has been demonstrated in numerous tissues including rat adipocytes (Levilliers et al. 1976), kidney (Kimura et al. 1975a; DeRubertis and Graven 1976c; Criss et al. 1976a), guinea pig gastric mucosa (Greissen 1978), lung (Mittal et al. 1978), cerebellum (Kimura et al. 1975a, b), and cerebral cortex (Deguchi 1977). The effect of sodium azide on guanylate cyclase activity can be modified by a number of compounds. For example, dithiothreitol, cysteine, glutathione, and $FeCl_2$ enhanced azide activation of guanylate cyclase in crude liver supernatants whereas detergents, cyanide, hydrogen peroxide, and other oxidizing agents were inhibitory (Kimura et al. 1975a; Katsuki et al. 1977a; Mittal et al. 1977). Modulation of the azide effect on enzyme activity led to the hypothesis that an oxidative-reductive process was involved in this activation mechanism (Mittal et al. 1977; Katsuki et al. 1977a; Arnold et al. 1977; Murad et al. 1978). Subsequent studies have proven this to be the case (Braughler et al. 1979a, b) (see below).

Sodium azide activation of crude guanylate cyclase preparations was not a universal effect but was tissue specific (Kimura et al. 1975a; Mittal et al. 1975). Of several crude tissue preparations examined, the activation occurred with supernatant fractions of liver and kidney but not with other tissues including heart, lung, cerebral cortex, spleen, etc. Azide activation was also observed with particulate fractions from liver, kidney, cerebral cortex, cerebellum but not with lung, heart, and spleen (Kimura et al. 1975a). Classical mixing experiments with crude supernatant fractions from different tissues indicated that a protein(s) was required for azide activation and that some protein(s) inhibited the azide effect which could explain the apparent tissue specificity of the phenomenon (Mittal et al. 1975). The activator factor from rat liver supernatant fractions that is required for the azide effect has been purified and characterized (Mittal et al. 1977). It is a heat labile macromolecular of about 200,000 Daltons that is sensitive to protease and chymotrypsin but not to trypsin, and the protein factor copurified with catalase. The addition of either the activator factor or commercial bovine catalase to azide insensitive tissue preparations such as partially purified liver guanylate cyclase or to the supernatant fraction from cerebral cortex (Mittal et al. 1977), kidney tumors (Criss et al. 1976a), hepatomas (Criss et al. 1976b), and bovine tracheal smooth muscle (Katsuki et al. 1977a) resulted in activation of the enzyme. Horse radish peroxidase, bacterial cytochrome b_2 or cytochrome C reductase were also able to substitute for the activator factor (Mittal et al. 1977). Heart and lung supernatant fractions contained inhibitory materials (Kimura et al. 1975a; Mittal et al. 1975, 1977) as described in a later section.

Sodium azide inhibition of catalase has been described (Keilin and Hartree 1954). This work demonstrated that sodium azide is peroxidatively oxidized by catalase to form a catalase-nitric oxide complex which explains the inhibition. We thought that the formation of a catalase-NO complex might be involved in the mechanism of azide activation of guanylate cyclase, and we tested a variety of nitro compounds and amines in crude and partially purified guanylate cyclase systems. Agents tested and found to activate guanylate cyclase included antihypertensives such as sodium nitrite, nitroglycerine, and sodium nitroprusside (KATSUKI et al. 1977a). Other activators have included potent chemical carcinogens such as nitrosamines (DERUBERTIS and CRAVEN 1976a), phenylhydrazine (KIMURA et al. 1975a), hydrazine (VESLEY and LEVY 1977), and 4-nitro quinoline 1-oxide (DERUBERTIS and CRAVEN 1977c). Streptozotocin a nitrosourea that damages beta cells in pancreatic islets and causes experimental diabetes mellitus has also been shown to activate guanylate cyclase (VESLEY et al. 1977b). Activation of enzyme by many of these compounds occurs with most tissue preparations in contrast to the effects of azide and hydroxylamine which are tissue specific. There are, however, some preparations such as particulate fractions from bovine tracheal smooth muscle (KATSUKI et al. 1977a), Balb 3T3 fibroblasts (ICHIHARA et al. 1979), and several other cell cultures (MURAD and GILMAN, unpublished observations) that are not activated by nitro compounds. The reason for a lack of activation with these preparations is unknown but may be attributed to some inhibitory material(s) in these crude particulate preparations.

A striking similarity among many of the effective compounds that activate guanylate cyclase is the presence of a nitro or nitroso group in their structures (KATSUKI et al. 1977a). Generally low concentrations of reducing agents such as ascorbic acid, dithiothreitol and cysteine promote the activation of guanylate cyclase by these compounds whereas oxidizing agents are inhibitory (KIMURA et al. 1975a; MITTAL et al. 1977; ARNOLD et al. 1977; KATSUKI et al. 1977a). The effects of some oxidizing and reducing agents on guanylate cyclase activation have varied from one tissue to another. 25 m*M* dithiothreitol or cysteine completely inhibited guanylate cyclase activation by N-methyl-N-nitro-N-nitrosoguanidine (MNNG) in liver (DERUBERTIS and CRAVEN 1977a) and by nitrosourea in renal cortex (DERUBERTIS and CRAVEN 1977b) while 10 m*M* dithiothreitol was effective in inhibiting sodium nitroprusside activation in the same tissues. In contrast, 5 m*M* dithiothreitol enhanced nitroprusside activation of platelet and smooth muscle guanylate cyclase (BÖHME et al. 1978). Similar observations have been reported with soluble guanylate cyclase from rat lung (LIEBEL et al. 1978). We suspected that activation of guanylate cyclase by various nitro compounds could be due to the removal of their nitro or nitroso groups with the formation of some active nitrogen species. The introduction of nitric oxide gas in guanylate cyclase incubations caused an instantaneous activation of the enzyme from a variety of tissues (KATSUKI et al. 1977a; ARNOLD et al. 1977). Activation of guanylate cyclase by nitric oxide was also altered by redox agents in that reducing agents potentiated the activation by nitric oxide while oxidizing agents were inhibitory (KATSUKI et al. 1977a; ARNOLD et al. 1977; BRAUGHLER et al. 1979b). The proposal that NO is the active species for azide, hydroxylamine, and nitro compound activation of the enzyme (KATSUKI et al. 1977a; ARNOLD et al. 1977) is further substantiated by the observations that decomposition of 4-nitroquinoline-1-oxide and MNNG are required prior to the activation of enzyme (DERUBERTIS and CRAVEN 1977a, c).

In recent studies it has been postulated that heme is required for nitric oxide activation of guanylate cyclase with the formation of a nitroxide-heme intermediate (CRAVEN et al. 1979). However, we have found that in addition to heme a variety of other materials such as albumin, thiols and sugars will also enhance activation of homogeneous enzyme and that nitric oxide at sufficiently low concentrations can activate purified enzyme in the absence of these compounds (BRAUGHLER et al. 1979b). Thus, nitric oxide can interact directly with guanylate cyclase, but the nature of this interaction is not presently known. As discussed below, an attractive hypothesis is the oxidation of guanylate cyclase by nitric oxide and other free radicals. Activation of guanylate cyclase by nitric oxide and other reactive free radicals is at present a unique mechanism for the reversible regulation of an enzyme's activity.

4. Oxidation and Hydroxyl Radical

Nitric oxide as well as various nitro compounds are potent oxidizing agents. Blockade of the nitric oxide effect by methylene blue, $K_3Fe(CN)_6$ and antioxidants such as butylated hydroxyanisole is consistent with an oxidative mechanism (KATSUKI et al. 1977a; ARNOLD et al. 1979; MITTAL and MURAD 1977b). Although the formation and/or release of nitric oxide may explain the effects of various nitro compounds on guanylate cyclase and cyclic GMP levels and may be of pharmacological importance, the physiological significance of these effects is unknown. Since nitric oxide is required in micromolar concentrations to increase cyclic GMP synthesis (ARNOLD et al. 1977; BRAUGHLER et al. 1979b) and levels of nitrites, nitrates, and amines in animal tissues exceed these levels (WHELAN 1935), it is possible that under appropriate redox conditions sufficient quantities of nitric oxide may be formed in tissues and regulate cyclic GMP synthesis. This hypothesis requires additional investigation.

The availability of H_2O_2 or another oxidant in tissues could be rate-limiting in generating nitric oxide from some precursor. Thus, we examined the effect of superoxide dismutase on guanylate cyclase activity (MITTAL and MURAD 1977a) since this enzyme is a major source of H_2O_2 in tissues (CHANCE et al. 1979). Bovine erythrocyte superoxide dismutase activated partially purified rat liver guanylate cyclase (MITTAL and MURAD 1977a). The activation by superoxide dismutase was increased with the addition of nitrate reductase which was added to increase the formation of superoxide anion, the substrate for superoxide dismutase. These effects were prevented by inhibitors of superoxide dismutase (thiols and KCN), catalase, and antioxidants such as butylated hydroxyanisole, hydroquinone, and catecholamines (MITTAL and MURAD 1977a). Because of the apparent involvement of both superoxide anion and hydrogen peroxide in this activation, it was proposed that hydroxyl radicals were formed in incubations by the interaction of superoxide anions and H_2O_2 and that the hydroxyl radicals or a similar reactive species caused the activation of guanylate cyclase (MITTAL and MURAD 1977a, b). Recently, catecholamines (LIANG and SACKTOR 1978; MITTAL and MURAD, unpublished observations) and paraquat (VESELY et al. 1979) have been found to activate guanylate cyclase and these effects probably involve the formation of superoxide anions and hydroxyl radicals. However, the possible involvement of other reactive free radicals has not been excluded and another viable candidate should also include peroxy radicals. Hydroxyl radicals are ubiquitous in any aerobic system and are formed under a variety of conditions that are associated with increases in cyclic GMP levels. Thus, oxygen free radicals represent a category of naturally occurring substances that are excellent candidates as physiological regulators of guanylate cyclase activity and cyclic GMP synthesis (MITTAL and MURAD 1977 a, b). The formation of oxygen radicals are associated with processes such as phagocytosis (TAUBER and BABIOR 1977), prostaglandin synthesis (EGAN et al. 1976) and, platelet aggregation (WHITE et al. 1977). Some of these processes are accompanied by increased cyclic GMP levels.

The mechanism of guanylate cyclase activation by hydroxyl radical or another reactive oxygen species is not understood. Hydroxyl radical is much more reactive as an oxidant than is nitric oxide. All of these reactive radicals have extremely short half-lives in the order of milliseconds or seconds which prevent investigators from conducting a number of important experiments. Furthermore, the formation of free radicals in a complex biological system leads to the formation of other free radicals as electrons are transferred from one material to another. Thus, it becomes quite difficult if not impossible to establish which free radical is the "activating species" for guanylate cyclase regulation.

Guanylate cyclase appears to be a very unique enzyme that it is regulated by oxidizing agents (MITTAL and MURAD 1977 a, b). Undoubtedly other enzymes will also be found to be activated by reactive free radicals and the chemistry of guanylate cyclase regulation may have broad implications in biological systems. In any event, very subtle alterations in the redox state of incubations can dramatically influence the formation and/or scavenging of various free radicals and the activation of guanylate cyclase. With pencil, paper and schemes of metabolic pathways, numerous models can be proposed to couple membrane perturbations by hormones and drugs to altered redox state, free radical formation, and cyclic GMP synthesis. Some models of coupling these processes are proposed in a later section.

5. Fatty Acids and Their Metabolites

In recent years many investigators have described activation of guanylate cyclase by unsaturated fatty acids and their metabolites using preparations from a variety of tissues. Tissues examined have included human platelets (GLASS et al. 1977 a; HIDAKA and ASANO 1977), Balb 3 T 3 fibroblasts (WALLACH and PASTAN 1976 b; ICHIHARA et al. 1979), rat cerebral cortex (ASAKAWA et al. 1978), fat cells (ASAKAWA et al. 1976), liver (ICHIHARA et al. 1979; BRAUGHLER et al. 1979 a), and lung (GARBERS et al. 1978). In most of these studies activation has required unsaturated fatty acids containing two or more double bonds (GLASS et al. 1977 a). There have, however, been some exceptions in that guanylate cyclase from fibroblasts can be activated by saturated fatty acids (WALLACH and PASTAN 1976 b). The activation of enzyme by fatty acids is generally greater when Mg^{2+} is employed as the sole cation cofactor rather than Mn^{2+}.

The mechanism of guanylate cyclase activation by unsaturated fatty acids is unknown. There are several conflicting reports in the literature that deal with this issue. While some reports have suggested that metabolites of fatty acids activate the enzyme, there are other reports that emphasize the direct interaction of fatty acids with the enzyme. A kinetic study of fatty acid activation using crude soluble preparations of platelet guanylate cyclase suggested that the enzyme contains a single binding site for fatty acids (GLASS et al. 1977a). In this study a structural requirement for unsaturated fatty acid activation was found in which a 1,4,7-octatriene group was needed with the first double bond in the $\omega 6$ position. In contrast, in another study (WALLACH and PASTAN 1976b) with particulate Balb 3T3 guanylate cyclase it was concluded that there were no apparent structural requirements for fatty acid activation. GARBERS has demonstrated with hydrophobic chromatography that rat lung soluble guanylate cyclase absorbs to alkyl functions on a sepharose matrix which indicates that guanylate cyclase has a hydrophobic site (GARBERS 1978b). Hydrophobic chromatography has also been used to purify soluble guanylate cyclase from rat liver (BRAUGHLER et al. 1979a).

Some investigators have speculated that unsaturated fatty acids must be converted to hydroperoxides and endoperoxides by the action of lipoxygenase or cyclooxygenase, respectively, in order to be effective activators (HIDAKA and ASANO 1977). Partially purified soluble guanylate cyclase is activated by hydroperoxides of either arachidonic, oleic, or linoleic acid (HIDAKA and ASANO 1977). GOLDBERG and co-workers in contrast to their earlier report with platelet enzyme (GLASS et al. 1977a) suggested that guanylate cyclase from splenic cells is activated by hydroperoxides and prostaglandin endoperoxides and the effect was reversed by thiols such as dithiothreitol (GRAFF et al. 1978). These observations suggest that guanylate cyclase activation by some fatty acids is also a result of oxidation as may be the case with nitro compounds and hydroxyl radical as discussed above (MITTAL and MURAD 1979b). Cyclo-oxygenase inhibitors such as aspirin and indomethacin which block the formation of prostaglandin endoperoxide did not alter unsaturated fatty acid activation of particulate guanylate cyclase from Balb 3T3 fibroblasts (ICHIHARA et al. 1979). The addition of soybean lipoxygenase to unsaturated fatty (arachidonic, oleic or linoleic) to form the hydroperoxides did not enhance their effect on enzyme activity. In fact, lipoxygenase decreased the effects of fatty acids and shifted their dose response curve to the right (ICHIHARA et al. 1979). Thus, with particulate guanylate cyclase from Balb 3T3 fibroblasts, unsaturated fatty acids and not their peroxides are able to activate the enzyme. In contrast to the report of GLASS and co-workers (GLASS et al. 1977a), we have found that arachidonate and oleate are about equally effective in activating either crude or purified guanylate cyclase from rat liver (ICHIHARA et al. 1979; BRAUGHLER et al. 1979a). These results are consistent with the observations of WALLACH and PASTAN (WALLACH and PASTAN 1976b). At present the effects of fatty acids appear complex and the active species and mechanism of activation have not been resolved. It appears that with some tissues and species the unmodified unsaturated fatty acid is effective while in other preparations the fatty must be converted to a peroxide or another metabolite in order to observe activation. The effects may also be related to the purity of the enzyme preparations and fatty acids used. EGAN and co-workers have recently reported that the conversion of arachidonate by cyclo-oxygenase to the endoperoxides and prostaglandins leads to the formation of a reactive free radical which may be hydroxyl radical (EGAN et al. 1976). Thus, some of the effects of fatty acids may be mediated through the formation of a reactive oxygen species and oxidation of guanylate cyclase as discussed earlier and other effects may be due to hydrophobic interactions with the enzyme.

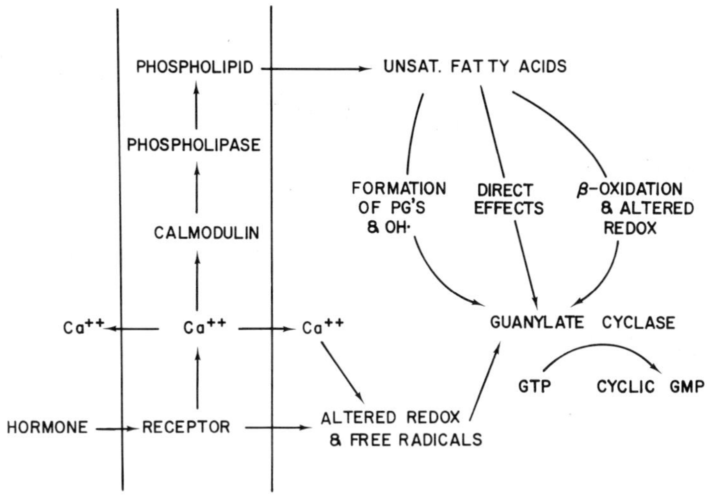

Fig. 1. Proposed scheme for hormone, fatty acid, and redox regulation of cyclic GMP synthesis

6. Phospholipase and Lipoxygenase

The effects of phospholipases on guanylate cyclase activity have also been examined. Of the several lipases tested, only phospholipase A_2 has been shown to activate guanylate cyclase (MITTAL and MURAD 1978). The tissue preparations examined have included skeletal muscle (SULAKHE et al. 1976a), cerebral cortex (MITTAL and MURAD 1978) and neuroblastoma cells (ZWILLER et al. 1976). Phospholipase A_2 releases fatty acids from the beta position of phospholipids and this is a major source of arachidonic acid for prostaglandin synthesis (DERKSEN and COHEN 1975). However, the effects of phospholipase A_2 on guanylate cyclase activity are not altered with cyclo-oxygenase inhibitors such as aspirin or indomethacin or with oxidizing and reducing agents (MITTAL and MURAD 1978). The mechanism of phospholipase A_2 activation of guanylate cyclase is different from that of nitro compounds or hydroxyl radical and resembles the activation by unsaturated fatty acids (ICHIHARA et al. 1979). Lysolecithin has also been shown to enhance guanylate cyclase activity from fibroblast (SHIER et al. 1976) and neuroblastoma cultures (ZWILLER et al. 1976). It is attractive to speculate that various agents which perturb the plasma membrane and increase cyclic GMP levels in cells produce their effects by the stimulation of phospholipase A_2 (Fig. 1). The enzyme is sensitive to Ca^{2+} ions (DERKSEN and COHEN 1975), and basal as well as hormonally-induced changes in intracellular levels of cyclic GMP are frequently dependent upon Ca^{2+} concentration (SCHULTZ et al. 1973). It has also been reported recently that phospholipase A_2 is influenced by calmodulin and phenothiazines (WONG and CHEUNG 1979; SCHREY and RUBIN 1979). Thus, hormonal effects on cyclic GMP levels in intact cells may be mediated through Ca^{2+}, calmodulin, phospholipase A_2 activation, and release of unsaturated fatty acids which may increase guanylate cyclase activity through one of several possible mechanism as discussed earlier and summarized in Fig. 1. This hypothesis and the very indirect effect of hormones on guanylate cyclase activity may explain the difficulty that numerous laboratories have encountered with obtaining reproduceable and relevant hormone effects on guanylate cyclase in cell-free systems.

Soybean lipoxygenase has also been examined for its effects on guanylate cyclase activity. Preliminary observations indicate that lipoxygenase activates guanylate cyclase in a variety of rat tissues (MITTAL and MURAD 1978). Lipoxygenase is also activated by Ca^{2+} (KOCH 1968) and may be substituted for phospholipase A_2 in the hypothetical scheme in Fig. 1. As with phospholipase A_2 effects, lipoxygenase activation of guanylate cyclase is not altered by oxidizing or reducing agents (MITTAL and MURAD 1978). Thus, lipoxygenase activates guanylate cyclase by a mechanism that appears to be different from that of nitro compounds.

7. Effects of Proteins

Bovine serum albumin and other proteins are usually employed in assays of partially purified or purified guanylate cyclase in order to maintain an effective concentration of total protein and prevent enzyme denaturation or adherence to glassware. The activity of highly purified soluble guanylate cyclase from rat liver or lung is increased by bovine serum albumin (MITTAL et al. 1979 b; BRAUGHLER et al. 1979 a). The rates of cyclic GMP formation are linear in the absence or presence of albumin implying that albumin increases the activity of the enzyme. Similar observations have been reported using highly purified soluble guanylate cyclase from rat lung (GARBERS 1979). The effects of bovine serum albumin on guanylate cyclase are not altered in the presence of oxidizing or reducing agents or free radical scavengers such as butylated hydroxyanisole. Thus, the mechanism of activation by albumin does not involve an oxidative mechanism and appears different from that of nitro compounds or hydroxyl radical. Heat treatment of albumin at 90 °C for 2 min totally abolishes the activation phenomenon (MITTAL et al. 1979 b). Activation is also observed with delipidated preparations of albumin and a variety of other proteins (MITTAL et al. 1979 b). These include human serum albumin, bovine γ-globulin, human γ-globulin, and ovalbumin. Similar effects have also been observed with heat denatured preparations of purified guanylate cyclase. The precise mechanism of modulating guanylate cyclase activity with proteins is not understood. However, this phenomenon may explain the increases in specific activity of crude or purified preparations of enzyme when assayed at greater protein concentrations (BRAUGHLER et al. 1979 a).

V. Altered Kinetic Properties with Activation

1. Cation Dependence

Significant changes occur in the properties of guanylate cyclase following its activation with sodium azide, nitric oxide, and nitroso compounds. While the molecular size of the enzyme as determined by gel filtration is not altered (KIMURA et al. 1975 a), the effects of cations on guanylate cyclase activity are changed significantly (KIMURA et al. 1976). The native soluble and particulate guanylate cyclase from a variety of tissues in rat and other animals prefers Mn^{2+} to Mg^{2+} as the sole cation cofactor (HARDMAN and SUTHERLAND 1969; WHITE and AURBACH 1969; ISHIKAWA et al. 1969). With 4 m*M* cation the activity of the enzyme with Mn^{2+} is generally 5- to 20-fold greater than with Mg^{2+} as cofactor. After activation of guanylate cyclase with sodium azide, similar activities are obtained with either cation as sole cofactor (KIMURA et al. 1976). These alterations in cation preference are also observed when the enzyme is activated with nitric oxide (ARNOLD et al. 1977), sodium nitroprusside (KATSUKI et al. 1977a), and nitrosoureas (DERUBERTIS and CRAVEN 1977b). Following activation with azide, guanylate cyclase loses its dependence for free Mn^{2+} but retains its dependence for free Mg^{2+} (KIMURA et al. 1976). Native soluble guanylate cyclase from rat liver is increased with millimolar concentrations of Ca^{2+} in the presence of Mn^{2+} or Mg^{2+}, whereas following activation Ca^{2+} either does not stimulate the enzyme or inhibits guanylate cyclase (KIMURA et al. 1976). Thus, high concentrations of Ca^{2+} can stimulate, inhibit, or have no effect on guanylate cyclase depending upon the activation state of the enzyme.

2. Effects of Nucleotides

Generally the increases in guanylate cyclase activity that have been described are associated with increases in maximal velocity utilizing substrate concentrations well above the K_m for GTP. The apparent K_m for GTP may or may not be changed depending upon the preparation used. For example, with crude liver supernatant fractions the K_m for GTP increased from 35 μ*M* to 120 μ*M* following activation with azide (KIMURA et al. 1975a). With partially purified (MITTAL et al. 1977) or highly purified (BRAUGHLER et al., unpublished observations) guanylate cyclase, the K_m for GTP may not change with activation. Similar observations have been reported with rat kidney guanylate cyclase (DERUBERTIS and CRAVEN 1976c). As discussed earlier, ATP inhibition of guanylate cyclase is competitive and the degree of inhibition by ATP is not altered when guanylate cyclase is activated (MITTAL et al. 1977). The concentration of ATP that results in a 50% inhibition of partially purified cerebral cortex guanylate cyclase is about 0.25 m*M* with or without activation (MITTAL et al. 1977). ATP at 0.5 m*M* also inhibited basal and azide-activated crude soluble guanylate cyclase from rat liver 60%–80% (KIMURA et al. 1975a). Addition of cyclic AMP and Mg^{2+} with and without cyclic AMP dependent protein kinase to incubations did not alter the ATP inhibition of guanylate cyclase. Nor was basal activity or the activation of guanylate cyclase with azide altered with 10 m*M* NaF (KIMURA et al. 1975a; MITTAL et al. 1977). These observations suggest that ATP inhibition is probably due to the competition of ATP and GTP for the catalytic site of the enzyme, and ATP inhibition is unaltered in the native or activated enzyme. However, as discussed below active but not inactive enzyme can utilize either nucleotide as substrate to form the corresponding cyclic nucleotide (MITTAL and MURAD 1977a; MITTAL et al. 1979b).

3. Formation of Cyclic AMP

Adenylate cyclase is generally localized to membranous fractions of mammalian tissues (SUTHERLAND et al. 1962) with the exception of rat testis where it is also found in the cytosolic fraction (BRAUN and DODS 1975). Studies of the molecular properties of mammalian adenylate cyclase and guanylate cyclase and their sensitivity towards various hormones and effectors have clearly revealed that the two enzymes are different. Soluble adenylate cyclase and guanylate cyclase activities in rat testis are separable and have different physical properties (NEER and MURAD 1979). Furthermore, the S49 lymphoma cell variant (AC^-) with deficient adenylate cyclase activity when assayed with Mg^{2+} has normal levels of guanylate cyclase activity (MURAD and GILMAN, unpublished observations).

Guanylate cyclase preparations activated with compounds such as sodium azide, sodium nitroprusside, sodium nitrite, nitrosamines and nitric oxide can catalyze the conversion of ATP to cyclic AMP (MITTAL and MURAD 1977a; MITTAL et al. 1979b; ARNOLD et al. 1977; BRAUGHLER et al. 1979a). Although the rates of cyclic AMP formation by activated guanylate cyclase are about 1%–15% of the rates of cyclic GMP formation, the rate of production of cyclic AMP by these preparations is comparable to crude adenylate cyclase from rat liver. Heat denaturation of liver supernatants preparations resulted in parallel losses in the formation of both cyclic nucleotides. Activation of guanylate cyclase by hydroxyl radical and unsaturated fatty acids also led to similar alterations in the substrate specificity of purified guanylate cyclase (MITTAL et al. 1979b). The formation of cyclic AMP and cyclic GMP in the presence of nitro compounds occurred with similar protein peaks after chromatography on gel filtration or ion exchange columns. Highly purified preparations of soluble guanylate cyclase from rat liver catalyzed the formation of cyclic GMP and cyclic AMP indicating that both catalytic activities resided in the same protein (BRAUGHLER et al. 1979a). The significance of cyclic AMP formation by activated guanylate cyclase is not presently known. However, another pathway for cyclic AMP formation may exist in tissues in addition to the classical adenylate cyclase pathway. Perhaps some hormone and drug effects in the literature on cyclic AMP accumulation in tissues and adenylate cyclase activity in intact cell and cell-free systems may be associated with guanylate cyclase activation and cyclic AMP formation. These studies also suggest that the catalytic regions of guanylate cyclase and adenylate cyclase are quite similar. Since most adenylate cyclase preparations are contaminated with guanylate cyclase, it is not known if the reciprocal relationship exists i.e. that adenylate cyclase can form cyclic GMP. Other nucleotides have not been examined as potential substrates for activated guanylate cyclase.

VI. Inactivation

1. Hemeproteins

As discussed earlier, crude guanylate cyclase preparations could be activated with azide only from certain tissues. The lack of an azide response in certain tissues was attributed to either the absence of a required macromolecular activator factor or the presence of some inhibitory material that prevented the activation of guanylate cyclase (KIMURA et al. 1975a). The inhibitors in supernatant fractions of rat lung and heart and bovine lung have been purified and characterized (MITTAL et al. 1978). The inhibitory substances from these tissues was nondialyzable, heat labile, and trypsin resistant. Gel filtration chromatography of heart and lung supernatant fractions separated guanylate cyclase from the inhibitory materials, and the partially purified enzyme could be activated with azide. The inhibitory materials cochromatographed with hemoglobin and myoglobin which inhibited activation of guanylate cyclase by sodium azide, sodium nitrite, sodium nitroprusside, nitric oxide, hydroxyl radical (MITTAL et al. 1978; MITTAL and MURAD 1977a, b), and fatty acid peroxides (HIDAKA and ASANO 1977). Human hemoglobin and horse myoglobin also inhibited the activation of guanylate cyclase by various nitro compounds (MITTAL et al. 1978). The effects of heme proteins were dependent upon their state of oxidation in that reduced hemoglobin (Fe^{2+}) was effective in causing the inactivation of activated enzyme whereas the oxidized form (methemoglobin, Fe^{3+}) was ineffective. In some preparations basal guanylate cyclase activities have been suppressed by hemoglobin. Thus, basal guanylate cyclase activity may represent some preactivated enzyme in preparations. This may be attributed to endogenous activation or autoactivation of the enzyme as discussed earlier.

2. Other Inhibitors

Other inhibitors of guanylate cyclase have also been described and the effects of nucleotides have been discussed earlier. LEHOTAY and co-workers reported on an inhibitor of guanylate cyclase obtained from the perfusion media of rat liver (LEHOTAY et al. 1977). This inhibitor was effective against both guanylate cyclase and adenylate cyclase and the chemical nature of this substance is not known. The same laboratory has also described another substance from balsam pear that inhibits guanylate cyclase (CLAFLIN et al. 1978). This material inhibits guanylate cyclase activity in several preparations including normal prostate and prostatic adenocarcinoma (CLAFLIN et al. 1978). The structure of the latter inhibitor is also unknown.

C. Cyclic GMP Levels in Tissues

I. Effects of Hormones and Autacoids

1. Effects of Pituitary Hormones and Thyroid Function

Following the discovery of cyclic GMP in rat urine in 1963 (ASHMAN et al. 1963), attempts were made to obtain some correlation between several hormonal processes and cyclic GMP levels in tissues and extracellular fluids such as urine and plasma. The first in this series of experiments examined the urinary excretion of cyclic GMP in animals with endocrine organ ablation. It was observed that hypophysectomy, adrenalectomy and thyroparathyroidectomy of rats led to a decreased urinary excretion of cyclic GMP (HARDMAN et al. 1966, 1969). These alterations in urinary cyclic GMP could be reversed by treating the animals with some individual hormones or a mixture of several hormones. For example, the effects of hypophysectomy were reversed with thyroxine and hydrocortisone or a mixture of anterior pituitary hormones. Treatment with hydrocortisone or thyroxine reversed the effects of adrenalectomy and thyroparathyroidectomy, respectively, on cyclic GMP excretion. Similar studies were performed with guanylate cyclase activities in various tissues after adrenalectomy (STEINER et al. 1972; THOMPSON and WILLIAMS 1974). These observations supported the view that hormones were involved in the regulation of cyclic GMP metabolism, perhaps analogous to the numerous studies with cyclic AMP system. Some clinical studies have also suggested an association of cyclic GMP metabolism with endocrine function. Urinary excretion of cyclic GMP is altered in some patients with certain diseases. Patients with hyperparathyroidism have significant increases in urinary cyclic GMP as well as cyclic AMP and the levels decrease to normal following surgical removal of adenomatous or hyperplastic parathyroid glands (MURAD and PAK 1972; MURAD 1973). Patients with cystic fibrosis (MURAD et al. 1975) or Cushing's syndrome (WRAY et al. 1975) also have elevated levels of cyclic GMP in their urine.

There have been several problems in these areas of investigation that have prevented the formulation of some unifying concepts about hormones and cyclic GMP synthesis. First, the effects of hormonal deprivation on intracellular levels of cyclic GMP have been inconsistent with different tissues, and second the alterations induced in the activity of guanylate cyclase have not paralleled the cyclic GMP levels in a particular tissue. For example, adrenalectomy of rats increased the cyclic GMP content of skeletal muscle (STEINER et al. 1972) and decreased it in liver while the guanylate cyclase activity increased in liver following adrenalectomy (THOMPSON and WILLIAMS 1974). Estrogen treatment has also been shown to increase the levels of cyclic GMP in rat uterus (KUEHL et al. 1974; FLANDROY and GALAND 1975). However, none of the hormones which alter cyclic GMP levels in in vivo or in intact cell systems has had convincing, reproduceable, or physiologically relevant effects on guanylate cyclase activity in cell-free systems. This has obviously been a significant obstacle to overcome in the field and may be related to indirect effects of hormones on cyclic GMP synthesis as discussed earlier (Fig. 1).

2. Choline Esters

Choline esters were the first humoral agents to be shown to alter cyclic GMP levels in an in vitro system. GEORGE and co-workers demonstrated an elevation of cyclic GMP levels in perfused rat heart after treatment with acetylcholine (GEORGE et al. 1970). Many reports have since emerged describing effects of acetylcholine and the cholinesterase resistant analogue, carbachol, on cyclic GMP levels in a variety of tissues (see GOLDBERG and HADDOX 1977; MURAD et al. 1978, 1979 and reference therein). These included preparations of smooth muscle (LEE et al. 1972; SCHULTZ et al. 1973; MURAD and KIMURA 1974; KATSUKI and MURAD 1977), lung (STONER et al. 1974; KIMURA et al. 1974), umbilical artery (CLYMAN et al. 1975b), thyroid (YAMASHITA and FIELD 1972; SANDE et al. 1975; CHAMPION and JACQUEMIN 1978; SANDE et al. 1979), pancreas (CHRISTOPHE et al. 1976; KAPOOR and KRISHNA 1978), liver (POINTER et al. 1976), adipose tissue (ILLIANO et al. 1973), kidney (DERUBERTIS and CRAVEN 1978), neuroblastoma (BARTFAI et al. 1978; STRANGE 1978; RICHELSON et al. 1978), and leukocytes (IGNARRO and CECH 1975; SANDLER et al. 1975). The effects of cholinergic agents on cyclic GMP in most tissues have been shown to be mediated by muscarinic receptors because of their sensitivity to atropine. However, in skeletal muscle the acetylcholine-induced elevations in cyclic GMP levels were blocked by nicotine rather than atropine (NESTLER et al. 1978; GARBER et al. 1978). Generally, choline esters increase cyclic GMP levels in a tissue if any effect on cyclic nucleotides is to be observed. However, with rat glioma C6 cell cultures carbachol decreases cyclic GMP levels and this effect is probably mediated through nicotinic cholinergic receptors (BOTTENSTEIN and VELLIS 1978). Choline esters also decrease cyclic AMP levels and adenylate cyclase activity in some preparations and these effects are prevented with atropine (MURAD et al. 1962).

The mechanism(s) of choline ester-induced alterations in cyclic GMP levels are obscure. While muscarinic-cholinergic activation of cardiac guanylate cyclase in a cell free system has been described (GEORGE et al. 1975), these observations have not been reproduced by many other investigators working with preparations from heart or other tissues. The effects of cholinergic agents and most other hormones on cyclic GMP accumulation in intact tissues are dependent upon the presence of calcium ions in the incubation media (SCHULTZ et al. 1973). However, pancreatic acinar cell preparations have been an exception to this general requirement of extracellular calcium (CHRISTOPHE et al. 1976). This report describes an increse in cyclic GMP with choline esters in the absence of added calcium. While high concentrations of calcium can alter guanylate cyclase activity as discussed earlier, it seems quite unlikely that the effects of choline esters or other hormones on guanylate cyclase activity in intact cells are mediated by direct effects of calcium on the enzyme. The effects of hormones and calcium are probably mediated through other indirect processes as proposed earlier.

3. Histamine

Histamine also elevates cyclic GMP levels in human umbilical artery (CLYMAN et al. 1975b), tracheal smooth muscle (KATSUKI and MURAD 1977), lung (STONER et al. 1973), and mouse neuroblastoma (RICHELSON 1978; TAYLOR and RICHELSON 1979). As with choline esters, the increase in cyclic GMP levels is dependent upon extracellular calcium ions. These effects are probably mediated through histamine H_1 receptors since the effects are prevented with pyrilamine. In contrast histamine-H_2 receptors have been associated with changes in cyclic AMP levels (KAKIUCHI and RALL 1968).

4. Alpha-Adrenergic Agonists

There is ample evidence to demonstrate that adrenergic agents interact with tissues through either alpha- or beta-receptor sites. Agonists and antagonists that alter the beta-receptor have been shown to influence the adenylate cyclase-cyclic AMP system (MURAD et al. 1962; ROBISON et al. 1967, 1971). While alpha-adrenergic agonists may decrease cyclic AMP synthesis and cyclic AMP levels in some tissues, these compounds have increased cyclic GMP levels in several tissues such as ductus deferens (SCHULTZ et al. 1973, 1975), parotid (BUTCHER 1975), pineal (O'DEA and ZATZ 1976), and hepatocytes (POINTER et al. 1976). Increases in cyclic GMP levels in response to alpha-adrenergic agonists are dependent upon the presence of Ca^{2+} in the extracellular medium and are prevented with alpha-adrenergic blocking agents such as phentolamine and phenoxybenzamine (SCHULTZ et al. 1973; BUTCHER 1975). However, several reports have described beta-adrenergic blockade of norepinephrine-induced increases in cyclic GMP levels with cultured C_6 glioma cells (BOTTENSTEIN and VELLIS 1978; SCHWARTZ 1976).

5. Insulin

While insulin can alter cyclic AMP levels in several tissues (BUTCHER et al. 1968; CRAIG et al. 1969), several reports have also described effects of insulin on cyclic GMP metabolism. Insulin increases cyclic GMP levels in adipose tissue (ILLIANO et al. 1973; FAIN and BUTCHER 1976), hepatocytes (POINTER et al. 1976), and intestinal mucosa (BRASITUS et al. 1976). Recently nonsuppressible insulin-like protein (NSILP) has been shown to alter cyclic AMP and cyclic GMP levels in a reciprocal manner using rat liver preparations (LEICHTER and POFFENBARGER 1978). NSILP decreased glucagon or norepinephrine stimulated cyclic AMP levels and increased the levels of cyclic GMP. The effects of insulin on cyclic GMP levels in adipose tissue (FAIN and BUTCHER 1976) or liver cells (POINTER et al. 1976) have required calcium in the incubation medium.

6. Ionophore A-23187

Ionophore A-23187, an antibiotic that facilitates the movement of divalent cations across biological membranes, increases cyclic GMP levels in a variety of tissues. The ionophore increases cyclic GMP in tracheal smooth muscle (KATSUKI and MURAD 1977), leukocytes (SMITH and IGNARRO 1975), thyroid slices (SANDE et al. 1979), adipose tissue (FAIN and BUTCHER 1976), parotid (BUTCHER 1975), umbilical artery (CLYMAN et al. 1975a), liver cells (POINTER et al. 1976), urinary bladder (SAHIB et al. 1978), cerebellum (FERRENDELLI et al. 1976), renal cortex (DERUBERTIS and CRAVEN 1976b), pancreatic acinar cells (CHRISTOPHE et al. 1976), etc. The increases in cyclic GMP levels by ionophore are usually dependent upon extracellular calcium ions and are abolished by calcium chelating agents.

7. Other Agents

Several other agents have also been reported to alter intracellular levels of cyclic GMP in a variety of tissues. Serotonin elevates cyclic GMP levels in leukocytes (SANDLER et al. 1975), platelets (DAVIES et al. 1976; AGARWAL and STEINER 1976), and human umbilical artery (CLYMAN et al. 1975a, b). The effects of serotonin cyclic GMP levels with umbilical artery are observed in the absence of extracellular calcium in contrast to the effects of other hormones and autacoids. In platelets, the serotonin-induced increases in cyclic GMP levels are blocked by methysergide (DAVIES et al. 1976).

Cyclic GMP levels are also increased by bradykinin (CLYMAN et al. 1975a, b), cholecystokinin (CHRISTOPHE et al. 1976), erythropoeitin (RODGERS et al. 1976), angiotensin (BUONASSI and VENTER 1976), arachidonic acid, prostaglandin endoperoxide (GLASS et al. 1977b) and ascorbic acid (SANDLER et al. 1975; GOLDBERG et al. 1978; SCHOEPEFLIN et al. 1978). The mechanism of cyclic GMP elevation in various tissues by each of these agents is not completely understood. Some of these materials increase guanylate cyclase activity in broken-cell preparations. Arachidonate (WALLACH and PASTAN 1976b; GLASS et al. 1977a), prostaglandin endoperoxide and hydroperoxide (HIDAKA and ASANO 1977; GRAFF et al. 1978) increase the activity of guanylate cyclase from platelets, spleen cells, and other tissues as discussed earlier. The effects of secretin and probably pancreozymin on guanylate cyclase (THOMPSON et al. 1973) can be attributed to bile salt contaminants in crude preparations. Bile salts have been shown to activate guanylate cyclase in a manner similar to detergents and pure or synthetic preparations of these peptides are inactive on the enzyme (ICHIHARA et al. 1977). Although the effects of ascorbate or dehydroascorbate (SANDLER et al. 1975; GOLDBERG et al. 1978) could be due to redox effects or free radical formation as discussed earlier, it has recently been reported that the effect of ascorbate on guanylate cyclase may be due to alterations in the pH of incubations (ATKINSON et al. 1979). Recently pyruvate and lactate have also been shown to increase cyclic GMP synthesis in rat hepatocytes (DOHI and MURAD 1979). These effects are observed only in the presence of extracellular calcium ions and are prevented with a variety of metabolic and electron transport inhibitors. The increases in cyclic GMP in adrenal cells and adrenal carcinoma with ACTH (SHARMA et al. 1974; PERCHELLET et al. 1978) have not been confirmed in several other laboratories. An explanation for tis discrepancy is not apparent.

II. Effects of Nitro Compounds and Nitric Oxide

This category of effectors is comprised of a variety of nitro compounds that have proved to be useful in increasing cyclic GMP synthesis in both intact- and broken-cell systems. The effects of sodium azide, hydroxylamine, nitroglycerin, sodium nitroprusside, nitrosamines, nitrosoureas, and nitric oxide on guanylate cyclase have already been discussed. These compounds have also been tested for their effects on cyclic GMP levels in intact-cell systems and all of these agents elevate cyclic GMP in a variety of tissues (KIMURA et al. 1975b; KATSUKI et al. 1977; KATSUKI and MURAD 1977; ARNOLD et al. 1977; DERUBERTIS and CRAVEN 1976c; SCHULTZ et al. 1977; DIAMOND and JANIS 1978). The increase in cyclic GMP levels in intact cell systems by these compounds is not dependent upon extracellular calcium ions (KIMURA et al. 1975b; DERUBERTIS and CRAVEN 1976c; KATSUKI and MURAD 1977). Many of these compounds are smooth muscle relaxants and vasodilators or carcinogenic materials. While it has been suggested that the pharmacological effects of these compounds are mediated through increased cyclic GMP synthesis (see MURAD et al. 1978, 1979 and references therein), additional studies are needed. One characteristic that is common with these agents is their ability to form nitric oxide as discussed in an earlier section. Nitric oxide can activate guanylate cyclase and increase cyclic GMP levels in most tissues examined (ARNOLD et al. 1977). These agents may provide investigator probes to increase cyclic GMP and evaluate the effects of the nucleotide in various systems and processes in the presence and absence of extracellular calcium.

III. Effect of E. Coli Heat-Stable Enterotoxin

Enterotoxigenic strains of *Escherichia coli* may elaborate one of two enterotoxins, heat-labile toxin (LT) or a heat-stable toxin (ST), which can cause diarrheal disease in humans and animals (SACK 1975). While LT acts in a manner similar to cholera toxin by activating adenylate cyclase and increasing cyclic AMP levels (GILL et al. 1976), ST has no effect on cyclic AMP levels (HUGHES et al. 1978a). In contrast to cholera toxin and LT, ST has no time lag in the induction of fluid secretion in the intestine. Filtrates from ST-producing strains of *E. coli* purified ST increase cyclic GMP levels in rabbit and mouse intestine (Hughes et al. 1978a; Field et al. 1978). ST also activates guanylate cyclase from intestinal mucosa preparations (Field et al. 1978; Hughes et al. 1978b). The effects of ST on guanylate cyclase activation are specific for particulate enzyme from intestinal mucosa preparations, and effects are not observed with soluble enzyme from intestine or soluble or particulate enzyme from a number of other tissues (Guerrant et al. 1980). A role for cyclic GMP in the action of ST on intestinal mucosa is further suggested by the observations that 8-Bromo cyclic GMP causes a concentration and time dependent increase in intestinal fluid secretion (Hughes et al. 1978a). The effects of ST on intestinal guanylate cyclase activation or intestinal fluid secretion are prevented with indomethacin, or antioxidants which suggest that prostaglandin metabolism and/or redox events are involved in the mechanism (Guerrant et al. 1980).

D. Conclusion and Hypotheses

In spite of the considerable literature that currently exists dealing with guanylate cyclase and cyclic GMP, two major questions still remain unresolved. These include: a) the precise mechanisms and sequence of events whereby hormones increase guanylate cyclase activity and cyclic GMP accumulation in tissues and b) the physiological function(s) of cyclic GMP. Numerous agents and perturbations can increase cyclic GMP accumulation in intact cells or guanylate cyclase activity in cell-free systems as discussed earlier. At present, two mechanisms continue to emerge as likely candidates for the physiological regulation of guanylate cyclase. These include free radicals such as hydroxyl radical or perhaps other reactive species of oxygen and unsaturated fatty acids. As pointed out earlier, these mechanisms may not be distinct and separate since the effects of unsaturated fatty acids and their peroxides may also result from the formation of free radicals and oxidation of guanylate cyclase. Unfortunately the methods required to answer these important questions are quite difficult, generally insensitive, and will not permit investigators to assay the activating species particularly when working with crude or impure systems. As mentioned earlier, guanylate cyclase appears to be a rather unique enzyme which appears to be oxidized and activated by free radicals. While the chemistry of this process should prove interesting, intuitively one would expect that other enzymes may be regulated in a similar fashion and that the information gained in this currently complex process may be applicable to other proteins and biological problems. Thus, information obtained with the guanylate cyclase-cyclic GMP system may have greater significance and broad application to other enzymes and processes.

The function(s) of cyclic GMP has been as evasive as the mechanisms of its hormonal regulation. Many attempts have been made to associate and correlate cyclic GMP with various processes and some of these studies have been presented here. Unfortunately, on closer inspection, usually the temporal correlation of cyclic GMP accumulation with some process or the precise dose-response comparisons of a hormone effect on a process and cyclic GMP accumulation have failed to support possible cause and effect associations of cyclic GMP and some physiologic event. Two processes that presently look quite promising for cyclic GMP regulation include smooth muscle relaxation induced by nitro compounds and intestinal secretion resulting from *E. coli* heat-stable enterotoxin. The data generated in these areas have suggested roles for cyclic GMP in smooth relaxation and intestinal secretion as discussed earlier. Obviously additional studies are required to support these hypotheses.

Another hypothesis that we have formulated with respect to another function of cyclic GMP has been developed as a result of our work with free radical and redox modulation of guanylate cyclase. If, indeed, guanylate cyclase is a monitor or sensor of altered redox state and free radical formation in tissues, perhaps the product of this process (i.e. cyclic GMP) is formed to permit a feedback control of these events. The formation of cyclic GMP may result from a distal and yet highly coupled and complex redox process occurring as a result of hormone interaction or membrane perturbation. The activation of guanylate cyclase and the production of cyclic GMP may then lead to a negative or even positive feedback signal to influence metabolism and redox events that led to guanylate cyclase activation. Redox events and free radical formation and/or scavenging could obviously be influenced in a variety of ways. This feedback hypothesis is testable and should stimulate new approaches to examine possible functions of cyclic GMP. The activation of cyclic GMP-dependent protein kinase and the phosphorylation of some protein or enzyme may or may not be involved in this feedback mechanism. In some processes such as smooth muscle relaxation, intestinal secretion, and undoubtedly other processes, cyclic GMP may behave as a traditional messenger analogous to the cyclic AMP system. However, in other processes, cyclic GMP may not be a messenger as we have come to know cyclic AMP's functions. In this feedback model there is no compelling reason to expect that the effects of cyclic GMP are antagonistic or synergestic to those of cyclic AMP.

References

Agarwal KC, Steiner M (1976) Effect of serotonin on cyclic nucleotides of human platelets. Biochem Biophys Res Commun 69:962–969

Arnold WP, Mittal CK, Katsuki S, Murad F (1977) Nitric oxide activates guanylate cyclase and increases guanosine 3′,5′-monophosphate levels in various tissue preparations. Proc Natl Acad Sci USA 74:3203–3207

Asakawa T, Scheinbaum I, Ho R (1976) Stimulation of guanylate cyclase activity by several fatty acids. Biochem Biophys Res Commun 73:131–148

Asakawa T, Takenoshita M, Uchida S, Tanaka S (1978) Activation of guanylate cyclase in synaptic plasma membranes of cerebral cortex by free fatty acids. J Neurochem 30:161–166

Ashman DF, Lipton R, Melicow MM, Price TD (1963) Isolation of adenosine 3′,5′-monophosphate and guanosine 3′,5′-monophosphate from rat urine. Biochem Biophys Res Commun 11:330–334

Atkinson JP, Weiss A, Ito M, Kelly J, Parker CW (1979) Effects of ascorbic acid and sodium ascorbate on cyclic nucleotide metabolism in human lymphocytes. J Cyclic Nucleotide Res 5:107–123

Bartfai T, Breakefield XO, Greengard P (1978) Regulation of synthesis of guanosine 3′,5′-monophosphate in neuroblastoma cells. Biochem J 176:119–127

Böhme E, Graff H, Schultz G (1978) Effects of sodium nitroprusside and smooth muscle relaxants on cyclic GMP formation in smooth muscle and platelets. Adv Cyclic Nucleotide Res 9:131–143

Böhme E, Jung R, Mechler I (1974) Guanylate cyclase in human platelets. In: Hardman JG, O'Malley BW (eds) Methods in enzymology, vol 38. Academic Press, New York, pp 199–202
Bottenstein JE, Vellis JD (1978) Regulation of cyclic GMP, cyclic AMP and lactate dehydrogenase by putative neurotransmitters in the C 6 rat glioma cell line. Life Sci 23:821–834
Brasitus TA, Field MA, Kimberg DV (1976) Intestinal mucosal cyclic GMP: Regulation and relation to ion transport. Am J Physiol 231:275–282
Braughler JM, Mittal CK, Murad F (1979 a) Purification of soluble guanylate cyclase from rat liver. Proc Natl Acad Sci USA 76:219–222
Braughler JM, Mittal CK, Murad F (1979 b) Effects of thiols, sugars, and proteins on nitric oxide activation of guanylate cyclase. J Biol Chem 254:12450–12454
Braughler JM, Mittal CK, Murad F (unpublished observations)
Braun T, Dods RF (1975) Development of a Mn^{2+} sensitive, soluble adenylate cyclase in rat testis. Proc Natl Acad Sci US 72:1097–1101
Buonassi V, Venter JC (1976) Hormone and neurotransmitter receptors in an established vascular endothelial cell line. Proc Natl Acad Sci USA 73:1612–1616
Butcher FR (1975) The role of calcium and cyclic nucleotides in α-amylase release from slices of rat parotid: Studies with the divalent cation ionophore A-23187. Metabolism 24:409–418
Butcher RW, Baird CE, Sutherland EW (1968) Effects of lipolytic and antilipolytic substances on adenosine 3′,5′-monophosphate levels in isolated fat cells. J Biol Chem 243:1705–1712
Catalan RE, Castillon MP, Municio (1976) Cyclic nucleotide cyclase variation during development of the insect *ceratitis capitata.* Biochem Biophy Res Commun 69:914–919
Chance B, Sie H, Boveris A (1979) Hydroperoxide metabolism in mammalian organs. Physiol Rev 59:527–605
Champion S, Jacquemin C (1978) The role of cyclic GMP in the action of carbamylcholine on the cyclic AMP levels in the dog thyroid. FEBS Letters 92:156–158
Chrisman TD, Garbers DL, Parks MA, Hardman JG (1975) Characterization of particulate and soluble guanylate cyclases from rat lung. J Biol Chem 250:374–381
Christophe JP, Frandsen EK, Conton TP, Krishma G, Gardner JD (1976) Action of cholecystokinin, cholinergic agents, and A-23187 on accumulation of guanosine 3′,5′-monophosphate in dispersed guinea pig pancreatic acinar cells. J Biol Chem 251:4640–4645
Claflin AJ, Vesely D, Hudson JL, Bagwell CB, Lehotay DC, Lo TM, Fletcher MA, Block NL, Levey GS (1978) Inhibition of growth and guanylate cyclase activity of an undifferentiated prostate carcinoma by an extract of the balsam pear (momordica charantia abbreviata). Proc Natl Acad Sci USA 75:989–993
Clyman RI, Blacksin AS, Sandler JA, Manganiello VC, Vaughan M (1975 a) The role of calcium in regulation of cyclic nucleotide content in human umbilical artery. J Biol Chem 250:4718–4721
Clyman RI, Sandler JA, Manganiello VC, Vaughan M (1975 b) Gaunosine 3′,5′-monophosphate and adenosine 3′,5′-monophosphate content of human umbilical artery: Possible role in perinatal arterial patency and closure. J Clin Invest 55:1020–1025
Commoner B, Woolum JC, Senturia BH, Ternberg JL (1970) The effects of 2-acetylaminefluorene and nitrite on free radicals and carcinogenesis in rat liver. Cancer Res 30:2091–2097
Craig JW, Rall TW, Larner J (1969) The influence of insulin and epinephrine on adenosine 3′,5′-monophoshate and glycogen transferase in muscle. Biochim Biophys Acta 177:213–219
Craven PA, DeRubertis FR (1976) Properties and subcellular distribution of guanylate cyclase activity in rat renal medulli: Correlation, with tissue content of guanosine 3′,5′-monophosphate. Biochemistry 15:5131–5137
Craven PA, DeRubertis FR, Pratt DW (1979) Electron spin resonance study of the role of NO. catalase in the activation of guanylate cyclase by NaN_3 and NH_2OH. Modulation of enzyme responses by heme proteins and their nitrosyl derivatives. J Biol Chem 254:8213–8222

Criss WE, Murad F, Kimura H (1976a) Properties of guanylate cyclase from rat kidney cortex and transplantable kidney tumors. J Cyclic Nucleotide Res 2:11–19
Criss WE, Murad F, Kimura H, Morris HP (1976b) Properties of guanylate cyclase in adult rat liver and several Morris hepatomas. Biochim Biophys Acta 445:500–508
Davies T, Davidson MML, McClenaghan MD, Say A, Haslam R (1976) Factors affecting platelets cyclic GMP levels during aggregation induced by collagen and by arachidonic acid. Thromb Res 9:387–405
Deguchi T (1977) Activation of guanylate cyclase in cerebral cortex of rat by hydroxylamine. J Biol Chem 252:596–601
DeJonge HR (1975) Properties of guanylate cyclase and levels of cyclic GMP in rat intestine villous and crypt cells. FEBS Letters 55:143–152
Derksen A, Cohen P (1975) Pattern of fatty acid release from endogenous substrates by human platelet homogenates and membranes. J Biol Chem 250:9342–9347
DeRubertis FR, Craven PA (1976a) Calcium-independent modulation of cyclic GMP and activation of guanylate cyclase by nitrosamines. Science 193:897–899
DeRubertis FR, Craven PA (1976b) Calcium-dependent regulation of guanosine 3′,5′-monophosphate in renal cortex: Effects of ionophore A-23187 and tetracaine and evidence for independent control of adenosine 3′,5′-monophosphate. Metabolism 25:1113–1127
DeRubertis FR, Craven PA (1976-gc) Properties of the guanylate cyclase-guanosine 3′,5′-monophosphate system of rat renal cortex. Activation of guanylate cyclase and calcium-independent modulation of tissue guanosine 3′,5′-monophosphate by sodium azide. J Biol Chem 251:4651–4658
DeRubertis FR, Craven PA (1977a) Activation of hepatic guanylate cyclase by N-methyl-N′-nitro-N-nitrosoguanidine, effects of thiols, N-ethyl maleimide and divalent cations. J Biol Chem 252:5804–5814
DeRubertis FR, Craven PA (1977b) Activation of renal cortical and hepatic guanylate cyclase-guanosine 3′,5′-monophosphate by nitroso ureas, divalent cation requirements and relations to thiol reactivity. Biochim Biophys Acta 499:337–351
DeRubertis FR, Craven PA (1977c) Stimulation of soluble guanylate cyclase activity and cellular accumulation of cyclic guanosine 3′,5′-monophosphate by the carcinogen 4-nitro-quinoline 1-oxide. J Natl Cancer Inst 59:1741–1745
DeRubertis FR, Craven P (1977d) Increased guanylate cyclase activity and guanosine 3′,5′-monophosphate content in etionine-induced hepatomas. Cancer Res 37:15–21
DeRubertis FR, Craven PA (1978) Alterations in rat renal cortical and medullary guanosine 3′,5′-monophosphate accumulation by oxygen and calcium-dependent and independent mechanisms: Evidence for a calcium-independent action of oxygen in renal inner medulla. Metabolism 27:855–868
Deviller P, Cille Y, Betuel H (1975) Guanylate cyclase activity of human blood lymphocytes. Enzyme 19:300–313
Diamond J, Janis RA (1978) Increases in cyclic GMP levels may not mediate relaxant effects on sodium nitroprusside, verapamil, and hydralazine in rat vas deferens. Nature 271:472–473
Dixon M, Webb EC (eds) (1958) Enzymes. Academic Press, New York
Dohi T, Murad F (1979) Pyruvate and other metabolites increase cyclic GMP levels in rat hepatocytes and kidney cortex. Fed Proc 38:432
Dohi T, Murad F (unpublished observations)
Dounce AL, Shanewise RP (1950) Liver catalase of tumor bearing and leprous rats. Cancer Res 10:103–107
Durham JP (1976) Guanylate cyclase: Assay and properties of the particulate and supernatant enzymes in mouse parotid. Eur J Biochem 61:535–544
Earp HS, Smith P, Huang-Ong SH, Steiner AL (1977) Regulation of hepatic nuclear guanylate cyclase. Proc Natl Acad Sci USA 74:946–950
Egan RW, Paxton J, Kuehl FA (1976) Mechanism of irreversible self-deactivation of prostaglandin synthetase. J Biol Chem 251:7329–7335
Fain JN, Butcher FR (1976) Cyclic guanosine 3′,5′-monophosphate and the regulation of lipolysis in rat fat cells. J Cyclic Nucleotide Res 2:71–78

Ferrendelli JA, Rubin EH, Kinscherf DA (1976) Influence of divalent cations on regulation of cyclic GMP and cyclic AMP levels in brain tissue. J Neurochem 26:740–748
Field M, Graf LH, Laird WJ, Smith PL (1978) Heat-stable enterotoxin of Escherichia Coli: In vitro effects on guanylate cyclase activity, cyclic GMP concentration, and ion transport in small intestine. Proc Natl Acad Sci USA 75:2800–2804
Filburn CR, Wyatt GR (1976) Adenylate and guanylate cyclases of cecropia silkmoth fat body. J Insect Physiol 22:1635–1640
Flandroy L, Galand P (1975) Oestrogen-related changes in uterine and vaginal cyclic AMP and cyclic GMP. Arch Int Physiol Biochim 83:965
Frey WH 2d, Boman BM, Newman D, Goldberg ND (1977) Manganese and magnesium-dependent activities of soluble and particulate guanylate cyclase from sheep kidney outer medulla. J Biol Chem 252:4298–4304
Garber AJ, Entman ML, Birnbaumer L (1978) Cholinergic stimulation of skeletal muscle alanine and glutamine formation and release: Evidence for mediation by a nicotinic cholinergic receptor and guanosine 3′,5′-monophosphate. J Biol Chem 253:7924–7930
Garbers DL (1976) Sea urchin sperm guanylate cyclase; purification and los of cooperativity. J Biol Chem 251:4071–4077
Garbers DL (1978a) Sea urchin sperm guanylate cyclase antibody cross reactivity with various rat tissue guanylate cyclases. J Biol Chem 253:1898–1901
Garbers DL (1978b) Immobilization of rat lung soluble guanylate cyclase on alkyl-agarose gels. J Cyclic Nucleotide Res 4:271–279
Garbers DL (1979) Purification of soluble guanylate cyclase from rat lung. J Biol Chem 254:240–243
Garbers DL, Chrisman TD, Hardman JG (1978) Guanylate cyclase: Purification of particulate and soluble forms of the enzyme and response to the soluble form to cations and fatty acids. In: Folco G, Paoletti R (eds) Molecular biology and pharmacology of cyclic nucleotides. Elsevier, Amsterdam, pp 43–55
Garbers DL, Chrisman TD, Suddath JL, Hardman JG (1975a) Formation of pyrophosphate by soluble guanylate cyclase from rat lung. Arch Biochem Biophys 166:135–138
Garbers DL, Dyer EL, Hardman JG (1975b) Effects of cations on guanylate cyclase of sea urchin sperm. J Biol Chem 250:382–387
Garbers DL, Suddath JL, Hardman JG (1975c) Enzymatic formation of inosine 3′,5′-monophosphate and of 2′-deoxyguanosine 3′,5′-monophosphate. Inosinate and deoxyguanylate cyclase activity. Biochim Biophys Acta 377:174–185
George WJ, Ignarro LJ, White LE (1975) Muscarinic stimulation of cardiac guanylate cyclase. Recent Adv Stud Cardiac Struct Metab 7:381–390
George WJ, Polson JB, O'Toole AG, Goldberg ND (1970) Elevation of guanosine 3′,5′-monophosphate in rat heart after perfusion with acetylcholine. Proc Natl Acad Sci USA 66:398–403
Gill DM, Evans DJ, Evans DG (1976) Mechanism of activation of adenylate cyclase in vitro by polymixin-released, heat-labile enterotoxin of Escherichia Coli. J Infect Dis 133:S 103–S 107
Glass DB, Frey WH 2d, Carr DW, Goldberg ND (1977a) Stimulation of human platelet guanylate cyclase by fatty acids. J Biol Chem 252:1279–1285
Glass DB, Gerrard JM, Townsend D, Carr DW, White JG, Goldberg ND (1977b) The involvement of prostaglandin endoperoxide formation in the elevation of cyclic GMP levels during platelet aggregation. J Cyclic Nucleotide Res 3:37–44
Goldberg ND, Graff G, Haddox MK, Stephenson JH, Glass DB, Moser ME (1978) Redox modulation of splenic cell soluble guanylate cyclase activity: activation by hydrophilic and hydrophobic oxidants represented by ascorbic and dehydroascorbic acids, fatty acid hydroperoxides and prostaglandin endoperoxides.
Goldberg ND, Haddox MK (1977) Cyclic GMP metabolism and involvement in biological regulation. Ann Rev Biochem 46:823–896
Goldberg ND, O'Dea RF, Haddox MK (1973) Cyclic GMP. Adv Cyclic Nucleotide Res 3:155–223
Goridis C, Morgan IG (1973) Guanylate cyclase in rat brain subcellular fractions. FEBS Letters 34:71–73

Goridis C, Virmaux N, Urban PF, Mandel P (1973) Guanylate cyclase in mammalian photoreceptor. FEBS Letters 30:163–166

Goridis C, Zwiller J, Reutter W (1977) Guanylate cyclase and cyclic nucleotide concentrations during liver regeneration after experimental injury. Biochem J 164:33–39

Graff G, Stephenson JH, Glass DB, Haddox MK, Goldberg ND (1978) Activation of soluble splenic cell guanylate cyclase by prostaglandin endoperoxide and fatty acid hydroperoxides. J Biol Chem 253:7662–7676

Gray JP, Drummond GI (1976) Guanylate cyclase of sea urchin sperm: subcellular localization. Arch Biochem Biophys 172:31–38

Gray JP, Drummond GI, Luk DW, Hardman JG, Sutherland EW (1976) Enzymes of cyclic nucleotide metabolism in invertebrate and vertebrate sperm. Arch Biochem Biophy 172:20–30

Greissen M (1978) Effects of Ca^{2+}, Mg^{2+}, NaN_3, cholinergic agents and gastrointestinal hormones on guanylate cyclase from guinea pig gastric mucosa. Gastroenterology 74:263–270

Guerrant RL, Hughes JM, Chang B, Robertson DC, Murad F (1980) Activation of rat and rabbit intestinal guanylate cyclase by the heat-stable enterotoxin of Escherichia coli: Studies of potential receptors and intermediates. J Infect Dis (in press)

Hardman JG, Chrisman TD, Gray JP, Suddath JL, Sutherland EW (1973) Guanylate cyclase: Alteration of apparent subcellular distribution and activity by detergents and cations. In: Pharmacology and the future of man, vol 5. Karger, Basel, pp 143–145

Hardman JG, Davis JW, Sutherland EW (1966) Measurement of guanosine 3′,5′-monophosphate and other cyclic nucleotides: Variations in urinary excretion with hormonal state of the rat. J Biol Chem 241:4812–4815

Hardman JG, Davis JW, Sutherland EW (1969) Effect of some hormonal and other factors on the excretion of guanosine 3′,5′-monophosphate and adenosine 3′,5′-monophosphate in rat urine. J Biol Chem 244:6354–6362

Hardman JG, Sutherland EW (1969) Guanylate cyclase, an enzyme catalizing the formation of guanosine 3′,5′-monophosphate from guanosine triphosphate. J Biol Chem 244:6363–6370

Hidaka H, Asano T (1977) Stimulation of human platelet guanylate cyclase by unsaturated fatty acid peroxide. Proc Natl Acad Sci USA 74:3657–3661

Hughes JM, Murad F, Chang B, Guerrant RL (1978a) Role of cyclic GMP in the action of heat-stable enterotoxin of Escherichia Coli. Nature 271:755–756

Hughes JM, Murad F, Guerrant RL (1978b) Studies of guanylate cyclase (G-cyclase) activation by heat-stable enterotoxin (ST) of Escherichia Coli. Clin Res 26:524A

Ichihara K, El-Zayat M, Mittal CK, Murad F (1979) Fatty acid activation of guanylate cyclase from fibroblasts and liver. Arch Biochem Biophys 197:44–51

Ichihara K, Larner J, Kimura H, Murad F (1977) Activation of liver guanylate cyclase by bile salts and contaminants in crude secretin and pancreozymin preparations. Biochim Biophys Acta 481:734–740

Ichihara K, Murad F (1979) Guanylate cyclase activity in rat liver and other tissues with starvation and streptozotocin-induced diabetes mellitus. Arch Biochem Biophys 194:292–298

Ignarro LJ, Cech SY (1975) Lysosomal enzyme secretion from human neutrophils mediated cyclic GMP: Inhibition of cyclic GMP accumulation and neutrophil function by glucocorticoids. J Cyclic Nucleotide Res 1:283–292

Illiano G, Tell GP, Siegel MI, Cuatrecasas P (1973) Guanosine 3′,5′-cyclic monophosphate and the action of insulin and acetylcholine. Proc Natl Acad Sci USA 70:2443–2447

Ishikawa E, Ishikawa S, Davis JW, Sutherland E (1969) Determination of guanosine 3′,5′-monophosphate in tissues and of guanylate cyclase in rat intestine. J Biol Chem 244:6371–6376

Kakiuchi S, Rall TW (1968) The influence of chemical agents on the accumulation of adenosine 3′,5′-phosphate in slices of rabbit cerebellum. Mol Pharmacol 4:367–378

Kapoor CL, Krishma G (1978) A possible role for guanosine 3′,5′-monophosphate in the stimulus-secretion coupling in exocrine pancreas. Biochim Biophys Acta 544:102–122

Katsuki S, Arnold WP, Mittal CK, Murad F (1977a) Stimulation of guanylate cyclase by sodium nitroprusside, nitroglycerine, and nitric oxide in various tissue preparations and comparison to the effects of sodium azide and hydroxylamine. J Cyclic Nucleotide Res 3:1829–1833

Katsuki S, Arnold WP, Murad F (1977b) Effect of sodium nitroprusside, nitroglycerin, and sodium azide on levels of cyclic nucleotides and mechanical activity of various tissues. J Cyclic Nucleotide Res 3:239–247

Katsuki S, Murad F (1977) Regulation of adenosine 3′,5′-monophosphate and guanosine cyclic 3′,5′-monophosphate levels and contractility in bovine tracheal smooth muscle. Mol Pharmacol 13:330–341

Keilin D, Hartree EF (1954) Reactions of methemoglobin and catalase with peroxide and hydrogen donors. Nature 173:720–723

Kim JK, Frohnert PP, Hoi YS, Barnes LD, Farrow GM, Dousa TP (1977) Enzymes of cyclic 3′,5′-nucleotide metabolism in human renal cortex and renal adenocarcinoma. Kidney Int. 12:172–183

Kimura H, Mittal CK, Murad F (1975a) Activation of guanylate cyclase from rat liver and other tissues by sodium azide. J Biol Chem 250:8016–8022

Kimura H, Mittal CK, Murad F (1975b) Increases in cyclic GMP levels in brain and liver with sodium azide an activator of guanylate cyclase. Nature 257:700–702

Kimura H, Mittal CK, Murad F (1976) Appearance of magnesium guanylate cyclase activity in rat liver with sodium azide activation. J Biol Chem 251:7769–7773

Kimura H, Murad F (1974) Evidence for two different forms of guanylate cyclase in rat heart. J Biol Chem 249:6910–6916

Kimura H, Murad F (1975a) Subcellular localization of guanylate cyclase. Life Sci 17:837–844

Kimura H, Murad F (1975b) Two forms of guanylate cyclase in mammalian tissues and possible mechanisms for their regulations. Metabolism 24:439–445

Kimura H, Murad F (1975c) Localization of particulate guanylate cyclase in plasma membranes and microsomes of rat liver. J Biol Chem 250:4810–4817

Kimura H, Murad F (1975d) Increased particulate and decreased soluble guanylate cyclase activity in regenerating liver, fetal liver, and hepatoma. Proc Natl Acad Sci USA 72:1965–1969

Kimura H, Thomas E, Murad F (1974) Effects of decapitation, ether and pentobarbital on guanosine 3′,5′-phosphate and adenosine 3′,5′-phosphate levels in rat tissues. Biochim Biophys Acta 343:519–528

Koch RB (1968) Calcium ion activation of lipoxidase. Arch Biochem Biophys 125:303–307

Kraska RC, Stephenson JH, Goldberg ND (1977) Spontaneous oxidative activation and inactivation of rat uterine guanylate cyclase. Fed Proc 36:686

Kuehl FA, Ham EA, Zanetti ME, Sanford CH, Nieol SE, Goldberg ND (1974) Estrogen-related increases in uterine guanosine 3′,5′-cyclic monophosphate levels. Proc Natl Acad Sci USA 71:1866–1870

Lee TP, Kuo JF, Greengard P (1972) Role of muscarinic cholinergic receptors in regulation of guanosine 3′,5′-monophosphate content in mammalian brain, heart muscle, and intestinal smooth muscle. Proc Natl Acad Sci USA 69:3287–3291

Lehotay DC, Levey GS, Vesely DL, Bornet EP, Ray MV, Entman ML, Schwartz A (1977) The effect of adenylate cyclase inhibitor (ACI) on guanylate cyclase, phosphodiesterase and other enzymes in heart. J Cyclic Nucleotide Res 3:55–65

Leichter SB, Poffenbarger PL (1978) The effects of nonsuppressible insulin-like protein (NSILP) on cyclic nucleotide metabolism in rat liver. Biochem Biophys Res Commun 84:403–410

Levilliers J, Pairault J, Lecot F, Laudat MH (1976) Activation of guanylate cyclase by sodium azide in rat adipocytes. FEBS Letters 63:323–327

Liang CT, Sacktor B (1978) The stimulation by catecholamines of guanylate cyclase activity in a cell-free system. J Cyclic Nucleotide Res 4:97–111

Liebel MA, Lad PJ, White AA (1978) Reversible stimulation of soluble rat lung guanylate cyclase by sodium nitroprusside in the presence of thiols. Fed Proc 37:1537

Macchia V, Varrone S, Weissbach H, Miller DL, Pastan I (1975) Guanylate cyclase in Escherichia Coli purification and properties. J Biol Chem 250:6214–6217

Mittal CK, Arnold WP, Murad F (1978) Characterization of protein inhibitors of guanylate cyclase activation from rat heart and bovine lung. J Biol Chem 253:1266–1271
Mittal CK, Braughler JM, Ichihara K, Murad F (1979a) Synthesis of adenosine 3′,5′-monophosphate by guanylate cyclase, a new pathway for its synthesis. Biochim Biophys Acta 585:333–342
Mittal CK, Braughler JM, Murad F (1979b) Modulation of purified guanylate cyclase activity. Fed Proc Fed Am Soc Expt Biol 38:318
Mittal CK, Kimura H, Murad F (1975) Requirement for a macromolecular factor for sodium azide activation of guanylate cyclase. J Cyclic Nucleotide Res 1:261–269
Mittal CK, Kimura H, Murad F (1977) Purification and properties of a protein required for sodium azide activation of guanylate cyclase. J Biol Chem 252:4384–4390
Mittal CK, Murad F (1977a) Formation of adenosine 3′,5′-monophosphate by preparations of guanylate cyclase from rat liver and other tissues. J Biol Chem 252:3136–3140
Mittal CK, Murad F (1977b) Activation of guanylate cyclase by superoxide dismutase and hydroxyl radical: A physiological regulator of guanosine 3′,5′-monophosphate formation. Proc Natl Acad Sci USA 74:4360–4364
Mittal CK, Murad F (1977c) Properties and oxidative regulation of guanylate cyclase. J Cyclic Nucleotide Res 3:381–391
Mittal CK, Murad F (1978) Effects of phospholipase A_2 and lipoxygenase on rat cerebral cortex guanylate cyclase activity. Fed Proc 37:1689
Mittal CK, Murad F (unpublished observations)
Murad F (1973) Clinical studies and applications of cyclic nucleotides. Adv Cyclic Nucleotide Res 3:355–383
Murad F, Arnold WP, Mittal CK, Braughler JM (1979) Properties and regulation of guanylate cyclase and some proposed functions for cyclic GMP. Adv Cyclic Nucleotide Res 11:175–205
Murad F, Chi YM, Rall TW, Sutherland EW (1962) Adenyl cyclase III. The effects of catecholamines and choline esters on the formation of adenosine 3′,5′-phosphate by preparations from cardiac muscle and liver. J Biol Chem 237:1233–1238
Murad F, Gilman AG (unpublished observations)
Murad F, Kimura H (1974) Cyclic nucleotide levels in incubations of guinea pig trachea. Biochim Biophys Acta 343:275–286
Murad F, Lewicki JA, Brandwein HJ, Mittal CK, Waldman SA (1980) Guanylate cyclase: Purification, properties, free radical activation, radiolabeling and preparation of hybridoma antibodies. Adv Cyclic Nucleotide Res 14 (in press)
Murad F, Mittal CK, Arnold WP, Ichihara K, Braughler JM, El-Zayat M (1978) Properties and regulation of guanylate cyclase: Activation by azide, nitrocompounds and hydroxyl radical and effects of heme containing proteins. In: Folco G, Paoletti R (eds) Molecular biology and pharmacology of cyclic nucleotides. Elsevier Press, North Holland, pp 33–42
Murad F, Moss WM, Johanson AJ, Seldon RF (1975) Urinary excretion of adenosine 3′,5′-monophosphate and guanosine 3′,5′-monophosphate in normal children and those with cystic fibrosis. J Clin Endocr Metab 40:552–559
Murad F, Pak CY (1972) Urinary excretion of adenosine 3′,5′-monophosphate and guanosine 3′,5′-monophosphate. New Engl J Med 286:1382–1387
Nakazawa K, Sano M, Saito TC (1976) Subcellular distribution and properties of guanylate cyclase in rat cerebellum. Biochem Biophys Acta 444:563–570
Nakene M, Deguchi T (1978) Purification and properties of guanylate cyclase from the synaptosomal soluble fraction of rat brain. Biochim Biophys Acta 525:275–285
Neer EJ, Murad F (1979) Separation of soluble adenylate cyclase and guanylate cyclase from the mature rat testis. Biochim Biophys Acta 583:531–534
Neer EJ, Sukiennik EA (1975) Guanylate cyclase from the rat renal medulla, physical properties and comparison with adenylate cyclase. J Biol Chem 250:7905–7909
Nesbitt JA, Anderson WB, Miller Z, Pastan I, Russell TR, Gaspodorowicz, D (1976) Guanylate cyclase and cyclic guanosine 3′,5′-monophosphate phosphodiesterase activities and cyclic guanosine 3′,5′-monophosphate levels in normal and transformed fibroblasts in culture. J Biol Chem 251:2344–2352

Nestler EJ, Beam KG, Greengard P (1978) Nicotinic cholinergic stimulation increases cyclic GMP levels in vertebrate skeletal muscle. Nature 275:451
O'Dea RF, Zatz M (1976) Catecholamine-stimulated cyclic GMP accumulation in the rat pineal: Apparent presynaptic site of action. Proc Natl Acad Sci USA 73:3398–3402
Perchellet JP, Shankar G, Sharma RK (1978) Regulatory role of guanosine 3′,5′-monophosphate in adrenocorticotropic hormone-induced steroidogenesis. Science 199:311–312
Pointer RH, Butcher FR, Fain JN (1976) Studies on the role of cyclic guanosine 3′,5′-monophosphate and extracellular Ca^{2+} in the regulation of glycogenolysis in rat liver cells. J Biol Chem 251:2987–2992
Quill H, Weiser MW (1975) Adenylate and guanylate cyclase activities and cellular differentiation in rat small intestine. Gastroenterology 69:470–478
Rall TW, Sutherland EW, Berthet J (1957) The relationship of epinephrine and glucagon to liver phosphorylase. IV. Effect of epinephrine and glucagon on the reactivation of phosphorylase in liver homogenates. J Biol Chem 224:463–475
Richelson E (1978) Histamine H_1 receptor-mediated guanosine 3′,5′-monophosphate formation by cultured mouse neuroblastoma cells. Science 201:69–71
Richelson E, Prendergast FG, Romero SD (1978) Muscarinic receptor-mediated cyclic GMP formation by cultured nerve cells-ionic dependence and effects of local anesthetics. Biochem Pharmacol 27:2039–2048
Robison GA, Butcher RW, Sutherland EW (1967) Adenyl cyclase as an adrenergic receptor. Ann NY Acad Sci 139:703–723
Robison GA, Butcher RW, Sutherland EW (eds) (1971) Cyclic AMP. Academic Press, New York
Rodgers GM, Fisher JW, George WJ (1976) Elevated cyclic GMP concentrations in rabbit bone marrow culture and mouse spleen following erythropoietic stimulation. Biochem Biophys Res Commun 70:287–294
Sack RB (1975) Human diarrheal disease caused by enterotoxigenic Escherichia Coli. Ann Rev Microbiol 29:333–353
Sahib MK, Schwartz JH, Handler JS (1978) Inhibition of toad urinary bladder sodium transport by carbamylcholine: possible role of cyclic GMP. Am J Physiol 235:F 586–F 591
Sande JV, Decoster C, Dumont JE (1975) Control and role of cyclic 3′,5′-guanosine monophosphate in the thyroid. Biochim Biophys Res Commun 62:168–175
Sande JV, Decoster C, Dumont JE (1979) Effects of carbamylcholine and ionophore A-23187 on cyclic 3′,5′-AMP and cyclic 3′,5′-GMP accumulation in dog thyroid slices. Mol Cell Endocrine 14:45–57
Sandler JA, Gallin JI, Vaughan M (1975) Effects of serotonin, carbamylcholine, and ascorbic acid on leukocyte cyclic GMP and chemotaxis. J Cell Biol 67:480–484
Schoepeflin GS, Goetzl EJ, Austen KF (1978) The predominant contribution of platelets to baseline and ascorbate-stimulated increments in cyclic GMP in human mononuclear leukocyte preparations. Cell Immunol 35:330–339
Scholondorff D, Weber H (1976) Cyclic nucleotide metabolism in compensatory renal hypertrophy and neonatal kidney growth. Proc Natl Acad Sci USA 73:524–528
Schrey MP, Rubin RP (1979) Characterization of a calcium-mediated activation of arachidonic acid turnover in adrenal phospholipids by corticotropin. J Biol Chem 254:11234–11241
Schultz G, Hardman JG, Schultz K, Baird CE, Sutherland EW (1973) The importance of calcium ions for the regulation of guanosine 3′,5′-cyclic monophosphate levels. Proc Natl Acad Sci USA 70:3889–3893
Schultz G, Schultz K, Hardman JG (1975) Effect of norepinephrine on cyclic nucleotide levels in the ductus deferens of the rat. Metabolism 24:429–437
Schultz KD, Schultz K, Schultz G (1977) Sodium nitroprusside and other smooth muscle relaxants increase cyclic GMP levels in rat ductus deferens. Nature 265:750–751
Schwartz JP (1976) Catecholamine-mediated elevation of cyclic GMP in the rat C-6 glioma cell lines. J Cyclic Nucleotide Res 2:287–296
Sharma RK, Ahmed NK, Sutliff LS, Brush JS (1974) Metabolic regulation of steroidogenesis in isolated adrenal cells of the rat. ACTH regulation of cyclic GMP and cyclic AMP levels and steroidogenesis. FEBS Letters 45:107–110

Shier WT, Baldwin JH, Nilsen-Hamilton M, Thanassi NM (1976) Regulation of guanylate and adenylate cyclase activities by lysolecithin. Proc Natl Acad Sci USA 73:1586–1590

Siegel MI, Puca GA, Cuatrecasas P (1976) Guanylate cyclase: Existence of different forms and their regulation by nucleotides in calf uterus. Biochim Biophys Acta 438:310–323

Smith RJ, Ignarro LJ (1975) Bioregulation of lysosomal enzyme secretion from human neutrophils: Role of guanosine 3′,5′-monophosphate and calcium in stimulus-secretion coupling. Proc Natl Acad Sci USA 72:108–112

Steiner AL, Pagliara AS, Chase LR, Kipnis DM (1972) Radioimmunoassay for cyclic nucleotides II. adenosine 3′,5′-monophosphate and guanosine 3′,5′-monophosphate in mammalian tissues and body fluids. J Biol Chem 247:1114–1120

Stoner J, Manganiello VC, Vaughan M (1973) Effect of bradykinin and indomethacin on cyclic GMP and cyclic AMP in lung slices. Proc Natl Acad Sci USA 70:3830–3833

Stoner J, Manganiello VC, Vaughan M (1974) Guanosine 3′,5′-monophosphate and guanylate cyclase activity in guinea pig lung: Effect of acetylcholine and cholinesterase inhibitors. Mol Pharm 10:155–161

Strange PG (1978) A method for measuring relative changes in guanosine, 3′,5′-monophosphate in mouse neuroblastoma cells on muscarinic cholinergic stimulation. Biochem J 176:583–590

Sulakhe SJ, Leung NL, Sulakhe PV (1976a) Properties of particulate, membrane-associated and soluble guanylate cyclase from cardiac muscle, skeletal muscle, cerebral cortex and liver. Biochem J 157:713–719

Sulakhe PV, Sulakhe SJ, Leung NL, St. Louis PJ, Hickie RA (1976b) Guanylate cyclase subcellular distribution in cardiac muscle, skeletal muscle, cerebral cortex and liver. Biochem J 157:705–712

Sun IC, Shapiro L, Rosen OM (1974) Purification and characterization of guanylate cyclase from *caulobacter crescentus*. Biochem Biophys Res Commun 61:193–203

Sutherland EW, Rall TW, Menon T (1962) Adenyl cyclase: Distribution, preparation and properties. J Biol Chem 237:1220–1227

Tauber AT, Babior BM (1977) Evidence for hydroxyl radical production by human neutrophils. J Clin Invest 60:374–379

Taylor JE, Richelson E (1979) Desensitization of histamine H_1 receptor mediated cyclic GMP formation in mouse neuroblastoma cells. Mol Pharmacol 15:462–471

Thompson WJ, Williams RH (1974) Effects of adrenalectomy on 3′,5′-guanosine monophosphate metabolism of rat liver and other tissues. Arch Biochem Biophys 165:468–477

Thompson WJ, Williams RH, Little SA (1973) Activation of guanyl cyclase and adenyl cyclase by secretin. Biochim Biophys Acta 302:329–337

Troyer EW, Ferrendelli JA (1976) Two forms of guanylate cyclase in cerebellum. Fed Proc 35:456

Tsai SC, Manganiello VC, Vaughan M (1978) Activation of kidney guanylate cyclase by cobalt. Arch Biochem Biophys 189:155–160

Vesely DL, Castro A, Levey GS (1977a) Decreased rat hepatic guanylate cyclase activity in streptozotocin-induced diabetes mellitus. Diabetes 26:308–313

Vesely DL, Chown J, Levey GS (1976) Guanylate cyclase activity in fetal, neonatal and adult rat heart. J Mol Cell Cardiol 8:909–913

Vesely DL, Levey GS (1977) Hydrazine activation of guanylate cyclase: potential application to tobacco carcinogenesis. Biochem Biophys Res Commun 74:780–784

Vesely DL, Rovere LE, Levey GS (1977b) Activation of guanylate cyclase by streptozotocin and l-methyl-l-nitrosourea. Cancer Res 37:28–31

Vesely DL, Watson B, Levey GS (1979) Activation of liver guanylate cyclase by paraquat: Possible role of superoxide anion. J Pharm Exp Ther 209:162–164

Virmaux N, Nullans G, Goridis C (1976) Guanylate cyclase in vertebrate retina: Evidence for specific association with rod outer segments. J Neurochem 26:233–235

Waldman S, Gibbs J, Mittal CK, Murad F (unpublished observations)

Wallach D, Pastan I (1976a) Stimulation of membranous guanylate cyclase by concentrations of calcium that are in the physiological range. Biochem Biophys Res Commun 72:859–865

Wallach D, Pastan I (1976b) Stimulation of guanylate cyclase of fibroblasts by free fatty acids. J Biol Chem 251:5802–5809
Watson J, Nilsen-Hamilton M, Hamilton RT (1976) The subcellular distribution of adenylate and guanylate cyclase in murine lymphoid cells. Biochemistry 15:1527–1534
Whelan M (1935) Determination of nitrate in animal tissues. J Lab Clin Med 70:755–757
White AA (1975) Guanylate cyclase activity in heart and lung. Adv Cyclic Nucleotide 5:353–373
White AA, Aurbach GD (1969) Detection of guanyl cyclase in mammalian tissues. Biochim Biophys Acta 191:686–697
White AA, Crawford KM, Patt CS, Lad PJ (1976) Activation of soluble guanylate cyclase from rat lung by incubation or by hydrogen peroxide. J Biol Chem 251:7304–7312
White JG, Rao GHR, Gerrard JM (1977) Effects of nitroblue tetrazolium and vitamin-E on platelet ultrastructure and aggregation and secretion. Am J Pathol 88:387–402
Wong PYK, Cheung WY (1979) Calmodulin stimulates human platelet phospholipase A_2. Biochem Biophys Res Commun 90:473–480
Wray HL, Corrigan DF, Burton J, Schaaf M, Earle JM (1975) Elevated urinary cyclic GMP in patients with Cushing's syndrome. In: Proc 57th Annual Meeting of the Endocrine Society p 362
Yamashita K, Field JB (1972) Elevation of cyclic guanosine 3′,5′-monophosphate levels in dog thyroid sliced caused by acetylcholine and sodium fluoride. J Biol Chem 247:7062–7066
Zimmerman WF, Daemen FS, Bonting SL (1976) Distribution of enzyme activities in subcellular fractions of bovine retina. J Biol Chem 251:4700–4705
Zwiller J, Ciesielski-Treska J, Mandel P (1976) Effect of lysolecithin of guanylate cyclase and adenylate cyclase activities in neuroblastoma cells in culture. FEBS Letter 69:286–290

CHAPTER 6

Cyclic Nucleotide Phosphodiesterases

M. M. Appleman, M. A. Ariano, D. J. Takemoto, and R. H. Whitson

Overview

Cyclic nucleotide phosphodiesterase activity plays an important role in cellular function. A variety of cellular activities are regulated through mechanisms controlling the level of cyclic AMP or cyclic GMP. These mechanisms include synthesis, degradation, efflux, and sequestration of cyclic nucleotides within the cell. Hydrolysis by phosphodiesterase is the unique mechanism for the degradation of the cyclic nucleotides. Despite their ubiquitous occurrence in biological systems, the cyclic nucleotides are not known to be substrates for any other enzymatic reaction.

The term "cyclic nucleotide phosphodiesterase" is meant to refer to a class of enzymes rather than slight variations in a single catalytic function. Several distinct hydrolytic activities have in common the same basic chemical reaction: the substrate is a nucleotide with a 3′,5′-phosphodiester and the product is a 5′-nucleotide. The phosphodiesterases isolated from various tissues differ in their substrate specificities, in their substrate affinities, and in their sensitivity to a variety of natural and artificial effectors. This variability has been observed even among the enzymes isolated from a single tissue. In fact, the presence of multiple forms of phosphodiesterase within a single tissue or cell type has been a consistent finding. In order to discuss this topic intelligently, a measure of order must be introduced. Accordingly, for the purposes of this review, we divide the commonly occurring phosphodiesterases into three broad categories: cyclic AMP phosphodiesterases, cyclic GMP phosphodiesterases, and cyclic nucleotide phosphodiesterases. This system of classification is based solely on the relative rates of hydrolysis of cyclic AMP and cyclic GMP at presumed physiological substrate conditions. The classification of a given enzyme refers only to its reported in vitro properties, and does not imply a judgment as to its true physiological role. This system of classification may not agree with that of all workers; almost all forms of phosphodiesterase will hydrolyze both cyclic AMP and cyclic GMP to some extent, and thus assays carried out at very high substrate concentrations will lead to different characterizations. Also, in much of the literature, the phosphodiesterases in a given tissue are described only in terms of a particular method of resolution. Such classifications do not lend themselves very readily to comparisons with work in other tissues using other methods. This is especially true of resolutions based on molecular size, which may give confusing results due to aggregation of similar enzymes to form enzymes with identical catalytic properties. Finally, there are certainly phosphodiesterases which do not fit into any of the above three categories and probably serve highly specific cell functions, for example, light-activated retinal cyclic GMP phosphodiesterase and cyclic CMP phosphodiesterase.

Each of the three categories of cyclic nucleotide phosphodiesterase, as defined above, is found in most mammalian tissues upon thorough examination, i.e., extracted by a number of methods, assayed in both soluble and particulate fractions, resolved by ion exchange chromatography, electrophoresis, or density gradient centrifugation, and assayed using both cyclic AMP and cyclic GMP over a range of substrate concentrations. The relative proportions of the three forms, however, vary greatly from organ to organ: in most brain tissues the cyclic GMP phosphodiesterase predominates; liver has a higher content of the low affinity, non-selective cyclic nucleotide phosphodiesterase, a lesser amount of the cyclic AMP phosphodiesterase, and very little of the cyclic GMP hydrolyzing enzyme; fat contains a relatively high proportion of the high affinity cyclic AMP phosphodiesterase. Even within cells, the distribution of the different enzyme forms is not uniform. While cytological localization is just beginning on mammalian phosphodiesterase, it appears, from work on homogenates, that the low affinity form is soluble and cytoplasmic, while the other two forms are particulate; cyclic GMP phosphodiesterase has been cytochemically visualized on the plasma membrane (at the synaptic junction of neurons), and cyclic AMP phosphodiesterase is found in microsomal preparations. The findings that the different forms of phosphodiesterase exhibit distinct catalytic and regulatory properties and different subcellular localizations suggest that they are indeed different enzymes, synthesized under separate genetic control and serving distinct physiological roles in regulating cellular activities. This is not a concept which is totally accepted; a number of investigators have obtained data which can be interpreted as indicating the interconvertability of the different cyclic nucleotide phosphodiesterases or the presence of a common subunit.

The question of the functions of the different forms of cyclic nucleotide phosphodiesterase must now be answered. A straight-forward pharmacological approach to this problem would be the first choice: introduce selective inhibitors of the different forms into the cells or organs of interest and observe which biological activities have been perturbed. This has not been successful. The best inhibitors of phosphodiesterase activity are methylxanthines and they are not selective to a great degree. Also cyclic nucleotides are so critical to cell function that the use of inhibitors leads to a great variety of direct and indirect changes. Another approach has been to analyze the phosphodiesterase activities extracted from cells which are functionally perturbed. Thus, insulin enhances the low K_m cyclic AMP phosphodiesterase of certain tissues; one role of this insulin-regulated enzyme may be in the regulation of carbohydrate and lipid metabolism. Similarly, the low K_m cyclic AMP phosphodiesterase is elevated in malignant cells. The enzyme might therefore be involved in regulating cell proliferation. Further evidence of discrete enzyme-function relationships comes from the localization of phosphodiesterase. The finding of high levels of cyclic GMP phosphodiesterase in neural tissues and the subcellular localization of this enzyme at postsynaptic membranes implicate this enzyme in some aspect of neurotransmission. Finally, the possibility exists that an enzyme considered to be hydrolytic is, in reality, a transferase utilizing the energy of the phosphate bond to carry out as yet undiscovered reactions.

To present the subject of cyclic nucleotide phosphodiesterase from a more historical point of view, one could cite the publications which mark significant milestones in research on the enzyme activity. First must come the original report of cyclic AMP hydrolysis (SUTHERLAND and RALL 1958); then, after a gap of some years, the report of resolvable forms of phosphodiesterase (THOMPSON and APPLEMAN 1971). Next, the three important findings of the regulation of the different forms of the enzyme by insulin (SENFT et al. 1968; LOTEN and SNEYD 1970), by cyclic GMP (BEAVO et al. 1970, 1971), and by calcium and a protein factor (CHEUNG 1970; KAKIUCHI and YAMAZAKI 1970). Further developments to watch for might include immunological evidence for or against interconversion of various forms, cytochemical and immunocytochemical localization of different forms within the cell, development of highly specific inhibitors and, ultimately, the association of the various cyclic nucleotides with specific cellular activities.

In this review we will deal with a number of the subjects mentioned above. We will not discuss in any depth topics which have been recently reviewed: phosphodiesterase inhibitors (CHASIN and HARRIS 1976), the methods of phosphodiesterase assay (THOMPSON et al. 1979 b), the calcium dependent regulator protein (Calmodulin) (CHEUNG and STORM, this volume). Recent general reviews dealing with these enzymes should also be mentioned (WELLS and HARDMAN 1977; AMER and KRIEGBAUM 1975).

A. The Phosphodiesterase Reaction and Assay

The term "cyclic nucleotide phosphodiesterase" covers a variety of enzymatic activities which have in common the catalysis of the following reaction (E.C.3.1.4.17):

3′,5′-cyclic nucleotide monophosphate $+H_2O \rightarrow$ 5′-nucleotide monophosphate.

These enzymes are specific for this cyclic diester; they will not hydrolyse the phosphodiester bonds of either the 2′,3′-cyclic nucleotides which occur as intermediates of ribonuclease (E.C.3.1.27.5) or of nucleic acids, which are hydrolysed by a different group of enzymes also referred to as phosphodiesterases (E.C.3.1.4.1).

Substrate specificity towards the base portion of the cyclic nucleotide is a major criterion for differentiating the various forms of phosphodiesterase. In general these enzymes will hydrolyse the two naturally occurring purine cyclic nucleotides, cyclic AMP and cyclic GMP, with some activity also towards cyclic IMP. Early reports of phosphodiesterase activity against pyrimidine nucleotides were not clearly defined as either alternative substrates of the above enzymes or as independent enzyme activities. Recently the occurrence of a cyclic cytidine monophosphate phosphodiesterase has been reported (KUO et al. 1978; CHEUNG and BLOCH 1978); the existence of this enzyme seems to be less controversial than that of its substrate (GAION and KRISHNA 1979).

The phosphodiesterase assay has been the subject of some controversy (WELLS and HARDMAN 1977). The most frequently used methods depend on the hydrolysis of radiolabeled substrates and the conversion of the 5′-nucleotide primary product to a more readily separated nucleoside in a second reaction catalyzed by 5′-nucleotidase or alkaline phosphatase. Early work, in which the inorganic phosphate produced in the second reaction was measured, gave misleading results; the high substrate concentrations necessary to detect activity interfered with the identification of the different forms of the enzymes. The major difficulty in current assay methods is optimizing recovery of the radioactive nucleoside products with the maximum separation from the cyclic nucleotide substrates. Ion exchange media are used either in batch treatment, small columns, or thin layer chromatographic plates to carry out this separation. With proper caution, most of the assay variations will give satisfactory results for phosphodiesterase preparations from most tissues. A physiologically meaningful substrate concentration is an important consideration; at millimolar concentrations of cyclic nucleotide, kinetic differentiation between cyclic AMP and cyclic GMP phosphodiesterases is impossible. All cyclic nucleotide phosphodiesterases require a divalent metal for catalysis, and magnesium at concentrations from 2 to 10 m*M* is used in most assays. The pH optimum of cyclic nucleotide phosphodiesterases from most tissues is near pH 8′, a somewhat lower pH (pH 7.4) is recommended for studies of complex kinetics and other regulatory phenomena.

B. Multiple Forms of Phosphodiesterase

I. Identification and Separation

The discovery that multiple enzyme forms catalyzing the hydrolysis of cyclic nucleotides occur in most mammalian tissues (THOMPSON and APPLEMAN 1971) meant that a great deal of the published research on the enzyme had to be re-evaluated. It was necessary to determine whether these early reports dealt with a single form dominant in the tissue or with a mixture of enzymes. The subsequent demonstration that each of the forms of phosphodiesterase is subject to unique regulatory controls suggested that these enzymes serve an important role in the regulation of cellular activities. Although there has been a surge of information regarding the various types of phosphodiesterases in many tissues, a clear-cut cohesive picture of their importance in physiological processes has not yet emerged. Some tentative conclusions can, however, be made: (1) there are three general categories of phosphodiesterase activity into which most of the enzymes that have been described will fit; (2) the catalytic properties of the enzymes in each category do not vary greatly from tissue to tissue; (3) many tissues have all three types of activity; (4) the relative proportion of the three types varies among different cell types.

The criteria used to differentiate the cyclic nucleotide phosphodiesterases are primarily catalytic: substrate specificity, substrate affinity (apparent K_m and regulatory properties such as cooperative kinetic behavior or sensitivity to activation by specific factors. These criteria have proven to be more consistent than association with certain subcellular fractions, which, as discussed below, can be a function of laboratory techniques or physical properties of the enzyme molecule. Molecular weight or charge characteristics of the phosphodiesterases seem to be sensitive to proteolysis, disulfide linkage and other influences during purification so that a number of isozymes may be derived from a single enzyme. For these reasons, resolution of active fractions on gel electrophoresis or related methods is not in itself an acceptable indication of unique enzymatic activities. The demonstration that the separated forms have distinguishable catalytic activities is required to substantiate such an interpretation.

The three categories of generally occurring cyclic nucleotide phosphodiesterases do not have generally accepted names. In much of the literature, they are referred to by the order in which they appear during a separation procedure, by their substrate affinity, or by their method of regulation in vitro. In this review we will use a nomenclature based on the naturally occurring substrate for which the enzyme shows the greatest activity under presumed physiological concentrations:

Cyclic AMP phosphodiesterase is a high affinity enzyme with comparatively little activity against cyclic GMP. It frequently exhibits negative cooperative kinetic behavior with an apparent low K_m of 1–5 μ*M* for cyclic AMP.

Cyclic GMP phosphodiesterase is a relatively high affinity enzyme (K_m of 3–8 μ*M*) which is frequently activated by Ca^{++} and calmodulin. This enzyme has a lower affinity for cyclic AMP (K_m of 200 μ*M*) but its maximum velocity with this substrate is greater than that for cyclic GMP.

Cyclic nucleotide phosphodiesterase is an enzyme with a very low affinity for either cyclic AMP or cyclic GMP (K_m 50–200 μ*M*). Preparations of this enzyme from some tissues show positive cooperativity and activation of cyclic AMP hydrolysis by low concentrations of cyclic GMP.

This last designation is unfortunately somewhat confusing, inasmuch as "cyclic nucleotide phosphodiesterase" is also the generic term applied to enzymes of all three of these categories.

The resolution of these forms of phosphodiesterase in mammalian tissue preparations has been carried out by almost every available technique of protein separation. In many tissues, cyclic GMP phosphodiesterase and cyclic AMP phosphodiesterase, are relatively enriched in the particulate fraction of homogenates. The reliability of this last type of separation is limited, however, in that in some tissues these forms may have a non-specific or reversible association with the particulate fraction (this point is discussed more extensively in Sect. C). The particulate phosphodiesterases must be solubilized before they can be resolved by conventional techniques and adequately characterized. Solubilization by sonication (TERASAKI and APPLEMAN 1975), detergent (LOVELL-SMITH et al. 1977), hypotonic salts (LOTEN et al. 1978) and by freezing and thawing (WHITSON and APPLEMAN in press) have been reported. Soluble preparations may be resolved by ion exchange chromatography (RUSSELL et al. 1973; WELLS et al. 1975), gel filtration (THOMPSON and APPLEMAN 1971), density gradient centrifugation (THOMPSON et al. 1973) and polyacrylamide gel electrophoresis (WEISS et al. 1974). Not all methods will resolve all forms of phosphodiesterase, and nearly all of these methods are susceptible to enzyme instability and artifacts.

The choice of appropriate assay conditions is critical for correctly identifying the different phosphodiesterase forms derived from a separation procedure. A pre-

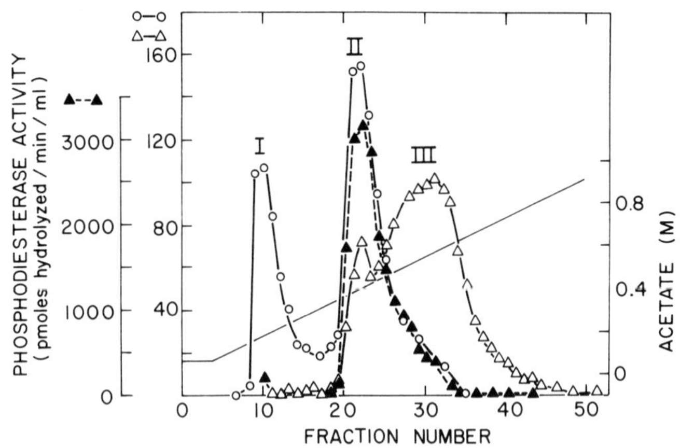

Fig. 1. DEAE-Cellulose Chromatography of Rat Liver Phosphodiesterase. The supernatant fraction of a fresh sonicated homogenate of rat liver was applied to DEAE cellulose and eluted with a sodium acetate gradient. The phosphodiesterase activity was measured under the following conditions: 1 μ*M* cyclic AMP *(open triangles)*, 100 μ*M* cyclic AMP *(closed triangles)*, and 1 μ*M* cyclic GMP (open circles). Under these conditions the peaks labelled I, II, and III are shown to be cyclic GMP phosphodiesterase, cyclic nucleotide phosphodiesterase, and cyclic AMP phosphodiesterase, respectively

liminary scan, using two or more of the following assay conditions, will give a general impression of the types of activity present: (1) cyclic AMP as substrate at a concentration of 1 μ*M* or less; (2) cyclic GMP as substrate at a concentration of 1 μ*M*; (3) cyclic AMP as substrate at a concentration of 100 μ*M*; (4) cyclic AMP as substrate at a concentration of 1 μ*M* in the presence of 1 μ*M* cyclic GMP as activator; (5) cyclic GMP as substrate at 1 μ*M* or cyclic AMP as substrate at 100 μ*M* in the presence of Ca^{++} and calmodulin. The choice of assay conditions depends on the interests of the investigator; each method tends to emphasize the activity of a different form: (1) for cyclic AMP phosphodiesterase, (2) and (5) for cyclic GMP phosphodiesterase, and (3) and (4) for cyclic nucleotide phosphodiesterase. None of these methods alone will yield unambiguous identification of any form of phosphodiesterase.

Figure 1 is illustrative of the principles outlined above. It depicts a DEAE chromatogram of the supernatant fraction from a sonicated rat liver homogenate. The fractions were assayed under conditions (1), (2), and (3) as described. The three major peaks of activity are clearly distinguished by varying assay conditions. After such a preliminary screening, the various activities can now be examined by more extensive kinetic analysis.

The methods discussed in this section can only be considered as guidelines and will not yield definitive results for all preparations. The point which should be stressed, however, is that determination of the regulatory properties of the phosphodiesterases in a given tissue absolutely requires a clear physical separation of the forms and a rigorous analysis under a variety of assay conditions.

II. Properties

As outlined above, the phosphodiesterases within each of the three major categories tend to have similar kinetic and regulatory properties. Some of these properties will be discussed in detail in later sections. In this section we present a series of tables (Tables 1–3) which give a survey of the phosphodiesterases in each category from a variety of mammalian tissues. In compiling these tables, we have limited our considerations to studies in which the phosphodiesterases were clearly separated and examined under a variety of assay conditions. This is by no means an exhaustive list, but it does give a fair impression of the findings with respect to these enzymes.

The tables list the similarities among the enzymes of a given category. Other properties, such as the subcellular distribution of the enzymes and their chromatographic behavior were somewhat less consistent and are not listed. Although the enzymes listed as either cyclic AMP phosphodiesterases or cyclic GMP phosphodiesterases clearly showed greater hydrolytic activity for their nominal substrate, there was a substantial variation in the degree of substrate specificity. For example, the cyclic AMP phosphodiesterases varied from absolute specificity for cyclic AMP (dog kidney, THOMPSON et al. 1979a) to enzymes with substantial cyclic GMP hydrolytic activity (rat liver, WHITSON and APPLEMAN, in press; rat adipose tissue, LOVELL-SMITH et al. 1977).

The properties listed in the tables provide a framework for a systematic examination of the regulatory properties of phosphodiesterases in addition tissues. If, for example, one were to isolate a low K_m cyclic GMP phosphodiesterase in a tissue, it would be prudent to test the preparation for Ca^{++} and calmodulin activation. It is readily apparent, however, that this pattern is not invariate, and one should not discount the possibility of a given preparation responding to a given effector without adequate testing. For example, the tables show examples of cal-

Table 1. Cyclic AMP phosphodiesterase in mammalian tissues

Tissue	Apparent K_m	Kinetics or regulation	Reference
Liver			
Rat	1 μM	Negative cooperativity	RUSSELL et al. (1972)
Rat	1 μM	Negative cooperativity, cyclic GMP inhibited	WHITSON and APPLEMAN (in press)
Rat	3 μM	Negative cooperativity, cyclic GMP insensitive	WHITSON and APPLEMAN (in press)
Rat (Hepatocytes)	1 μM	Negative cooperativity, insulin stimulated	LOTEN et al. (1978)
Adipose			
Rat	1 μM	Negative cooperativity	RUSSELL et al. (1972)
Rat (Adipocytes)	1 μM	Negative cooperativity, insulin stimulated	LOVELL-SMITH et al. (1977)
Heart			
Rat (Sinoatrial node)	0.8 μM	Negative cooperativity	TANIGUCHI et al. (1978)

Table 1 (Continued)

Tissue	Apparent K_m	Kinetics or regulation	Reference
Rat	1 μ*M*	Negative cooperativity	GAIN and APPLEMAN (1978)
Bovine	N.D.	Ca^{++}-sensitive	CARPENDO et al. (1975)
Kidney			
Rat	1 μ*M*	Negative cooperativity	RUSSELL et al. (1972)
Dog	2.2 μ*M*	Linear kinetics	THOMPSON et al. (1979)
Rat	4 μ*M*	Linear kinetics	GULYASSY et al. (1975)
Brain			
Rat	1 μ*M*	Negative cooperativity	RUSSELL et al. (1972)
Rat „P 1"	1 μ*M*	Ca^{++}, calmodulin activated	SAKAI et al. (1977)
Brain (cont'd)			
Rat „P3"	1 μ*M*	Ca^{++}, calmodulin insensitive	SAKAI et al. (1977)
Rat (Striatum)	1 μ*M*	Negative cooperativity	FREDHOLM (1977)
Superior cervical ganglion			
Bovine	N.D.	Ca^{++}-inhibited	BODREAU and DRUMMOND (1975)
Skeletal muscles			
Rat	1 μ*M*	Negative cooperativity	RUSSELL et al. (1972)
Rat	4 μ*M*	Negative cooperativity, insulin stimulated	GAIN and APPLEMAN (1976)
Rabbit	1 μ*M*	Linear kinetics	HUANG and KEMP (1971)
Smooth muscle			
Bovine (Aorta)	3 μ*M*	Negative cooperativity	ILIEN et al. (1978)
Porcine coronary artery	1 μ*M*	Negative cooperativity	WELLS et al. (1975)
Lung			
Human	1 μ*M*	Negative cooperativity	BERGSTRAND and LUNDQUIST (1978)
Rat (Mast cells)	0.4 μ*M*	Linear kinetics	BERGSTRAND et al. (1978)
Mammary			
Human	1 μ*M*	Negative cooperativity	SINGER et al. (1976)
Lymphocyte			
Human	2 μ*M*	Linear kinetics	WEDNER et al. (1979)
Human	1 μ*M*	Negative cooperativity	TAKEMOTO et al. (1978)
Human	1 μ*M*	Linear kinetics	THOMPSON et al. (1976)
Lymphocyte (cont'd)			
Mouse	10 μ*M*	Linear kinetics	HAIT and WEISS (1977)
Monocytes			
Human	1.4 μ*M*	Linear kinetics	THOMPSON et al. (1976)
Platelets			
Human	1.0 μ*M*	Linear kinetics	HIDAKA and ASANO (1976)
Epidermis			
Porcine	1 μ*M*	Negative cooperativity	ADACHI et al. (1976)

Table 2. Cyclic GMP phosphodiesterase in mammalian Tissues

Tissue	Apparent K_m	Kinetics or regulation	Reference
Liver			
Rat	6 μM	Ca^{++}, calmodulin activated	RUSSELL et al. (1973)
Rat	N.D.	Ca^{++}, calmodulin insensitive	WHITSON and APPLEMAN (in press)
Heart			
Rabbit (Sinoatrial node)	2 μM	Ca, calmodulin activated Cyclic GMP activates cyclic AMP hydrolysis	TANIGUCHI et al. (1978)
Brain			
Rat	1.3 μM	Ca^{++}, calmodulin, α Tocopherol activated	SAKAI et al. (1977)
Superior cervical ganglion			
Bovine	N.D.	Ca^{++} inhibited, calmodulin activated	BODREAU and DRUMMOND (1975)
Skeletal muscle			
Rat	0.3 μM	Non-linear kinetics, Ca^{++} calmodulin activated	GAIN and APPLEMAN (1978)
Smooth muscle			
Human aorta	3 μM	Ca^{++}, calmodulin activated	HIDAKA et al. (1978)
Bovine aorta	7 μM	Non-linear kinetics, calmodulin insensitive	ILIEN et al. (1978)
Porcine coronary artery	2 μM	Ca^{++}, calmodulin activated	WELLS et al. (1975)
Lung			
Human	0.4 μM	Linear kinetics	BERGSTRAND and LUNDQUIST (1978)
Rat (Mast cells)	0.2 μM	Linear kinetics	BERGSTRAND et al. (1978)
Mammary			
Human	1 μM	Ca^{++}, calmodulin activated	SINGER et al. (1976)
Anterior pituitary			
Rat	10 μM	Ca^{++}, calmodulin activated	AZHAR and MENON (1977)
Lymphocytes			
Human	6.7 μM	Linear kinetics, calmodulin N.D.	WEDNER et al. (1979)
Monocytes			
Human	40 μM	Linear kinetics, calmodulin, N.D.	WEDNER et al. (1979)
Platelets			
Human	0.5 μM	Negative cooperativity, cyclic GMP activates cyclic AMP hydrolysis	HIDAKA and ASANO (1976)
Kidney			
Rat	1 μM	Not characterized	GULYASSY et al. (1975)
Epidermis	6.7 μM	Linear kinetics	ADACHI et al. (1976)

Table 3. Cyclic nucleotide phosphodiesterase in mammalian tissues

Tissue	Apparent K_m		Kinetics and Regulation	Reference
	Cyclic AMP	Cyclic GMP		
Liver				
Rat	20 μ*M*	40 μ*M*	Positive cooperativity, cyclic GMP activates cyclic AMP hydrolysis	RUSSELL et al. (1973) TAKEMOTO and APPLEMAN (in press)
Rat	87 μ*M*	30 μ*M*	Cyclic GMP activates cyclic AMP hydrolysis, positive cooperativity	MOSS et al. (1977)
Rat	>20 μ*M*	>20 μ*M*	N.D.	LOTEN et al. (1978)
Brain				
Bovine		N.D.	Cyclic GMP activates cyclic AMP hydrolysis, Ca^{++} inhibits	BODREAU and DRUMMOND (1975)
Skeletal muscle				
Rat	80 μ*M*	24 μ*M*	Cyclic AMP activates cyclic AMP hydrolysis	GAIN and APPLEMAN (1978)
Kidney				
Rat	40 μ*M*		Not characterized	GULYASSY et al. (1975)
Platelet				
Human	40 μ*M*	40 μ*M*	N.D.	HIDAKA and ASANO (1976)
Smooth muscle				
Bovine (Aorta)	15 μ*M*	15 μ*M*	Non-linear kinetics Ca, calmodulin activated	ILIEN et al. (1978)
Heart				
Rabbit (Sinoatrial mode)	2 μ*M*	1.2 μ*M*	Cyclic GMP activates cyclic AMP hydrolysis	TANIGUCHI et al. (1978)
Epidermis				
Porcine	25 μ*M*	33 μ*M*	Cyclic GMP activates cyclic AMP hydrolysis	ADACHI et al. (1976)
Lung				
Rat	25 μ*M*	30 μ*M*	No activation of cyclic AMP hydrolysis by cyclic GMP	BERGSTRAND et al. (1978)

modulin activation of both cyclic AMP and cyclic nucleotide phosphodiesterase. Similarly, although negative cooperativity is most generally associated with cyclic AMP phosphodiesterase, there are enzymes from all three categories which appear to exhibit this kinetic anomaly. Finally, cyclic GMP activation of cyclic AMP hydrolysis, usually a property of cyclic nucleotide phosphodiesterases, is also found in some cyclic GMP phosphodiesterases.

The enzymes listed in a given category are more remarkable for their similarities, however, than for their differences. Because enzymes from such a wide range of specialized tissues have such similar properties, the control of cyclic nucleotide hydrolysis may be relatively uniform from tissue to tissue. The adaptation to specialized cell function seems to be reflected more in the varied ratios of the three major forms of phosphodiesterase than in differences in their subcellular distribution.

III. In vitro Alterations and Possible Interconversions

The existence of multiple forms of cyclic nucleotide phosphodiesterases in preparations from a single tissue must be distinguished from enzyme alterations taking place during laboratory operations. Such artificial, in vitro alterations of phosphodiesterase structure and activity may be reported together with the genuine occurrence of distinct forms of phosphodiesterase that are not derived from each other or from a common precursor. The resulting confusion can best be settled by testing the various forms for immunological cross reactivity, but this procedure requires purified enzymes and has been performed in only two cases (THOMPSON et al. in press; WHITSON and APPLEMAN, in press). Until such techniques are more generally available, investigators working on the cyclic nucleotide phosphodiesterases must be aware that the molecular, kinetic, and regulatory properties of these enzymes are very susceptible to factors which are not easily controlled.

The concentrations-dependent interconversion of apparently different forms of cyclic nucleotide phosphodiesterase from rat liver and human platelets was first reported by PICHARD and CHEUNG (1976). These workers employed sucrose density gradients to separate three soluble activities with apparent molecular weights of 22,000, 75,000, and 140,000 Daltons. They noted that each peak individually gave rise to the other two forms upon re-centrifugation. Each peak had distinct kinetic characteristics and substrate preferences. The relative amounts of the three forms were dependent on the concentration of enzyme in the preparation, but independent of overall protein concentration. The interconversion could also be affected by the substrate analogue, dibutyryl cyclic AMP, but was insensitive to added calmodulin. The enzyme concentration-dependence and the comparatively low molecular weights of the enzymes described by PICHARD and CHEUNG suggest that they observed the aggregation and disaggregation of the subunits of a single enzyme form. Such disaggregation has been reported for the low K_m cyclic AMP phosphodiesterase of rat lung mast cells upon dilution (BERGSTRAND et al. 1978).

Freezing and thawing is another treatment frequently affects phosphodiesterases. FREDHOLM (1977) has shown that in rat striatum, a soluble cyclic AMP phosphodiesterase activity was reduced when the tissue was frozen and thawed before homogenization. A specific reduction in this form of phosphodiesterase also occurred when the rats were subjected to chronic haloperidol treatments. A similar form of the enzyme, present in the particulate fraction of the homogenates of this tissue, was not affected by haloperidol treatments. The reduction of the activity of the soluble cyclic AMP phosphodiesterase by haloperidol could be prevented by the addition of calcium to the homogenate, although the calmodulin concentration of the tissue was apparently unaffected by the drug.

A low K_m high molecular weight phosphodiesterase in human peripheral blood lymphocytes has been shown to undergo a transition to a high K_m, low molecular weight enzyme when the cell extract is frozen and thawed (THOMPSON et al. 1976). This change was also observed when the extract was kept at 4 °C in tris-glycerol buffer. Incubation of rat renal cortex homogenates at 4 °C increased the cyclic AMP phosphodiesterase activity (STREWLER et al. 1978). This activation was more efficient in hypotonic buffers and did not occur if the supernatant fraction was incubated alone. The activation was accompanied by a decreased sedimentation velocity for the enzyme. The activation of cyclic AMP phosphodiesterase from rat erythrocytes by slow freezing and thawing or hypertonic salt treatments (2 *M* NaCl) has also been reported (SHEPPARD and TSIEN 1974). SAKAI et al. (1978) have observed similar changes in rat liver phosphodiesterase when that tissue was frozen and thawed before homogenization. There was increased activity in the supernatant and lower sensitivity of the soluble enzymes to alpha tocopherol phosphate and trypsin. These changes occurred in other tissues with a high lysosomal content and were prevented by inhibitors of lysosomal proteases.

Proteolysis activates and alters phosphodiesterases. CHEUNG (1970) has shown trypsin activation of brain cyclic GMP phosphodiesterase. This treatment reduced the sensitivity of the enzyme to calmodulin, EGTA and alpha tocopherol and changed the molecular weight of the enzyme from 150,000 to 80,000 Daltons (SAKAI et al. 1977). Trypsin has also been shown to affect the activity of a membrane-bound low K_m cyclic AMP phosphodiesterase in intact chick embryo fibroblasts (RUSSELL and PASTAN 1973), but this treatment caused a loss, rather than a gain, in activity in this cell type. EPSTEIN et al. (1978) have investigated the proteolytic activation of BHK fibroblast and rat uterus phosphodiesterase by trypsin. Trypsin treatment caused the activation of a large (6.7 S) calmodulin-sensitive, high affinity phosphodiesterase which hydrolyzed both cyclic AMP and cyclic GMP. Concomitant with trypsin activation there was a loss of calmodulin sensitivity, a conversion to a lower molecular weight form (3.6 S), and a loss of the ability to hydrolyze cyclic GMP. Storage of this enzyme at 4 °C also brought about these changes. Calmodulin activation of the enzyme did not lead to a reduction in the apparent molecular weight. Both high and low molecular weight forms were present in extracts of sonically disrupted cells even in the presence of benzamidine and other serine protease inhibitors. The higher molecular weight form had non-linear kinetic behavior for cyclic AMP hydrolysis; the low molecular weight enzyme gave linear kinetics with a lower K_m.

BERGSTRAND et al. (1978) have shown that a form of phosphodiesterase from human lung tissue is subject to considerable variation. This enzyme has a high affinity for cyclic GMP and cyclic AMP, appears to be soluble, and is extremely thermo-labile. The activity exists in multiple forms of molecular weight 200,000, 120,000, 105,000, and 78,000 Daltons and has non-linear kinetics at substrate concentrations which are substantially above its K_m (0.4 μM). The elution of the enzyme from DEAE cellulose columns is affected by Ca^{++} or EGTA. The peak isolated in the presence of EGTA is calmodulin-sensitive; trypsin also activates this form of the enzyme.

The rat liver particulate cyclic AMP phosphodiesterase is also subject to proteolytic alterations upon solubilization (WHITSON and APPLEMAN, in press). Detergent solubilization released a high molecular weight form of phosphodiesterase with a high affinity and negatively cooperative kinetics for both cyclic AMP and cyclic GMP. When the enzyme was solubilized by freezing and thawing, two forms of low K_m cyclic AMP phosphodiesterase, separable on DEAE cellulose, were released. One form (B) was identical to the detergent-soluble enzyme with respect to its elution of DEAE, its substrate specificity, its kinetic properties, and its sensitivity to inhibition by cyclic GMP, although it had a lower molecular weight. Cyclic AMP hydrolysis by both this form and the detergent-solubilized form was strongly inhibited by cyclic GMP ($K_i = 1\ \mu M$). The other freeze-thaw solubilized form (A) eluted from DEAE cellulose at lower salt concentrations, had negatively cooperative kinetics and a high affinity for cyclic AMP, but cyclic GMP would not inhibit the hydrolysis of cyclic AMP by this enzyme. Prolonged storage of the rat liver particulate fraction at $-20\ °C$ reduced the amount of the B form relative to the A form. Freezing and thawing also released a cyclic GMP phosphodiesterase from the rat liver particulate fraction. This enzyme had kinetic properties similar to the calmodulin-sensitive cyclic GMP phosphodiesterase, and eluted from DEAE columns at the same salt concentrations, but it was not calmodulin sensitive. Antibodies raised against the B form of cyclic AMP phosphodiesterase cross reacted against both the A form of cyclic AMP phosphodiesterase and the freeze thaw solubilized cyclic GMP phosphodiesterase, but not against the soluble cyclic nucleotide phosphodiesterase or calmodulin-sensitive phosphodiesterase.

LOTEN et al. (in press) indicate that many of the methods used to solubilize phosphodiesterase from rat liver expose the enzyme to limited proteolysis. A cyclic AMP phosphodiesterase solubilized by simple suspension of the particulate fraction in hypotonic buffer was not released in the presence of benzamidine, a protease inhibitor. This enzyme could also be solubilized by detergent and remained soluble after the removal of detergent. When the detergent solubilization was performed in the presence of benzamidine, however, the enzyme was much larger and was insoluble after the removal of detergent.

These examples show that phosphodiesterases are highly susceptible to modification by both exogenous and endogenous proteases. These proteolytic alterations often leave the enzymes with hydrolytic activities which are equal to or greater than their unmodified activities. Many of the endogenous proteases may be lysosomal in origin, and nearly all of the methods used to solubilize particulate phosphodiesterases could also be expected to burst lysosomes. The solubilization process may involve cleavage of the soluble active portion of those molecules from hydrophobic anchors. Thus, unless precautions are taken to avoid proteolysis it is unlikely that these enzymes are released in their native forms.

The ability to alter the properties of phosphodiesterases by these various treatments has led some investigators (PICHARD and CHEUNG 1976) to suggest that the major forms of phosphodiesterase are interconvertable or that the major forms possess a common subunit (EPSTEIN et al. 1978). The immunological studies in this laboratory indicate that while the particulate phosphodiesterases appear to be related, they are distinct from the soluble cyclic nucleotide phosphodiesterase and the calmodulin activated cyclic GMP phosphodiesterase.

The alteration of phosphodiesterases by endogenous proteases has a significance beyond the production of artifacts in in vitro manipulations. Some of the forms of phosphodiesterase produced by these modifications also appear in fresh tissues prepared in the presence of protease inhibitors (EPSTEIN et al. 1978; VAUGHAN et al. in press). Proteolytic modification of the phosphodiesterases could alter both the regulatory properties and the subcellular localization of the enzymes; such modifications could profoundly alter intracellular cyclic nucleotide metabolism.

C. Localization

The subcellular localization of the different forms of cyclic nucleotide phosphodiesterase very likely determines their biological functions. There is remarkably little data in the literature, however, with respect to the association of specific forms of phosphodiesterase with isolated subcellular fractions. Some work has been done with differential and sucrose density gradient centrifugation, but the enzymes from these particulate sub-fractions have seldom been solubilized, isolated or rigorously characterized. As discussed above, the susceptibility of particulate phosphodiesterases to proteolysis creates confusion as to the subcellular localization of the various forms of phosphodiesterase.

Both soluble and particulate forms of enzymes from all three categories have been reported. MANGANIELLO and VAUGHAN (1973) found that cyclic AMP phosphodiesterase from fat cells was mainly particulate. Most of this activity was in a fraction which sedimented at centrifugal forces indicating that it is associated with plasma membrane, mitochondria or microsomes. A number of subfractions containing cyclic AMP phosphodiesterase were demonstrable in liver (THOMPSON et al. 1973) and adipose tissue (SAKAI et al. 1974; KONO et al. 1975) using sucrose density gradient centrifugation. These studies indicate that the cyclic AMP phosphodiesterase, unlike adenylate cyclase, is not uniquely associated with the plasma membrane and may even be more concentrated in microsomal fractions in these cells. RUSSELL and PASTAN (1973), reported that trypsin externally applied to chick embryo fibroblasts affects the low K_m cyclic AMP phosphodiesterase. Although this enzyme must therefore be associated in some manner with the plasma membrane, it was also found in the soluble portion of the cell. Low affinity, cyclic nucleotide phosphodiesterase has been reported in a particulate fraction of rat liver homogenates (MOSS et al. 1977; WHITSON and APPLEMAN in press). This enzyme also binds tightly to hexyl sepharose, and the association with both the resin and the particulate fraction may represent nonspecific adsorption. The binding of the cyclic nucleotide phosphodiesterase to the particulate fraction could be disrupted by detergents, but not by freezing and thawing (MOSS et al. 1977; WHITSON and APPLEMAN in press). This form of phosphodiesterase is probably a soluble form, although one report (IGNARRO et al. 1973) has indicated that it is associated with a lysosomal fraction from liver. No form of the phosphodiesterase is restricted to, or even enriched, in mitochondrial or nuclear fractions of those cells analysed to date.

Cytochemical methods allow localization of phosphodiesterase hydrolytic activity in intact tissue, and consequently provide for identification of specific cell types exhibiting enzymatic reactivity as well as designation of specific sub-cellular sites for this activity. Adequate preservation of tissue fine structure and of phosphodiesterase hydrolytic activity are critical requirements. Equally important is the ability to differentiate the various forms of the enzyme after modification by the fixation procedure.

Cyclic nucleotide phosphodiesterase activity has been cytochemically localized in a number of different tissues using an electron microscopic modification of the classical GOMORI lead salt method (1939). An electron-dense, lead phosphate precipitate is visualized following the sequential actions of endogenous phosphodiesterase and exogenous 5′-nucleotidase on the cyclic nucleotide in the incubation medium. It is necessary to use high substrate concentrations in these studies because of the stoichiometry of the detection and the diffusion-penetration problems incurred when manipulating aldehyde-fixed tissue slices. Unfortunately, under these conditions, biochemically assessed hydrolysis rates for both cyclic AMP and cyclic GMP are similar for most forms of the phosphodiesterase and enzyme characterization is complex (BEAVO et al. 1970; MINNEMAN 1976; AZHAR and MENON 1977).

Following incubation with 3 m*M* cyclic AMP, phosphodiesterase activity has been visualized at the newt neuromuscular junction (PARDOS and LENTZ 1976). Enzymatic activity was localized on membranes of synaptic vesicles of the axon terminal as well as in the sarcoplasmic reticulum. The reaction product was restricted to the longitudinal tubules of the reticulum, while the junctional folds of the motor end plate were not reactive. These authors suggest that this localization pattern implicates a role for cyclic AMP in the functions of synaptic vesicles and sarcoplasmic reticulum in skeletal muscles. Renal medullary interstitial cells exhibited phosphodiesterase activity on the cytoplasmic border of cisternal membranes (FLORENDO et al. 1978). Since the membranes of the dilated cisternae are related to prostaglandin synthesis, and possibly to their antihypertensive function as well, these authors propose that cyclic AMP, and perhaps phosphodiesterase directly, share a role in these two processes. Cells of the collecting tubule and loop of Henle showed diffuse deposition of the lead phosphate precipitate representing cyclic AMP hydrolytic sites. Incubation of mouse retinas with either cyclic AMP or cyclic GMP (3 m*M*) showed localization of phosphodiesterase activity on the photoreceptor outer segment lamellae (ROBB 1974). Analysis by other methods indicates that this enzyme is a cyclic GMP phosphodiesterase. Cyclic AMP phosphodiesterase activity has been localized on the apical and lateral surfaces of plasma membranes and the outer surface of the endoplasmic reticulum in thyroid tissue slices (KALDERON and RAVANSHENAS 1974). Phosphodiesterase activity has also been visualized in isolated human lymphocytes on nuclear membranes (COULSON and KENNEDY 1971) and in mouse lung alveoli where cyclic AMP hydrolytic activity was restricted to the membranes of the Type I pneumocytes and endothelial cells (SINGER, BENNETT, and SHERWIN, personal communication). PDE activity has been cytochemically visualized in liver slices at the microvilli border of hepatocytes (ARIANO, unpublished observations).

INDIVERI et al. (1974) have reported phosphodiesterase localization on the perimeter of quiescent thyroidocytes. One hour following TSH treatment, staining is no longer uniformly distributed around the cell, but has become concentrated at the apex of the thyroidocytes so that the rims of sectioned follicles appear to be heavily stained. After two hours, the lead precipitate representing phosphodiesterase activity is widely spread throughout the cytoplasm of the thyroidocyte. The cell membrane is once again uniformly stained, with no detectable difference between the apical portion of the cell and other cell parts. These authors conclude that phosphodiesterase distribution within the thyroidocyte changes under the influence of TSH and that a different enzyme localization appears to coincide with cyclic AMP action as the cell carries out synthesis of triiodothyronine. The alterations of phosphodiesterase distribution are consistent with the modular role of hormone action that has been attributed to cyclic AMP in this tissue.

It is felt that aldehyde fixation may destroy the activity of the soluble forms of cyclic nucleotide phosphodiesterase. As a consequence, the enzyme cytochemistry method is probably detecting only membrane-associated phosphodiesterase forms in the different experimental tissues thus far examined. Nevertheless, each tissue appears to have a characteristic distribution of these hydrolytic activities.

D. Activators and Inhibitors

The activities of the cyclic nucleotide phosphodiesterases are affected by a wide variety of agents, endogenous, naturally occurring or synthetic in nature. Activators and inhibitors have been shown to act at the catalytic sites, at allosteric effector sites or on the process of enzyme synthesis. Unfortunately, many of the compounds reported cannot be discussed in a systematic fashion; their actions have been observed either on tissue preparations which include a number of different forms of phosphodiesterase or under assay conditions which are not definitive. In many cases, the modulation of phosphodiesterase activity is suggested only on the indirect evidence of a substance which modifies cyclic, nucleotides levels in vivo. In this chapter, we will consider only those agents acting in a specific manner on one or another form of phosphodiesterase. As there are recent or concurrent reviews dealing with the actions on the cyclic nucleotide phosphodiesterases of the calcium dependent regulatory protein (CDR or Calmodulin – CHEUNG – this volume) and the methylxanthines or related inhibitory drugs (CHASIN and HARRIS 1976) these will only be touched upon in this chapter.

I. Cyclic Nucleotides as Effectors of Enzyme Activity

Among the most interesting and most thoroughly investigated effectors of the phosphodiesterases are the cyclic nucleotides themselves. The occurrance of two purine cyclic nucleotides, cyclic CMP, and three or more forms of phosphodiesterase provides many opportunities for interactions. These interactions are of special interest in view of the hypothesis of GOLDBERG et al. (1973) that there is a reciprocal relationship between the cyclic nucleotides. The simplest case, inhibition of the hydrolysis of either cyclic nucleotide by the other, can be demonstrated in any form of the phosphodiesterase. The relative concentrations of substrate and inhibitor required to show this effect, however, vary tremendously with the form of the enzyme and even for the analogous form, from one tissue to another. Some of these effects, which seem especially important, will be discussed here; their significance as in vitro phenomena may be greater, however, than their actual role in the more complex environment within the intact cell. The interactions to be discussed are: (1) the cyclic GMP activation of cyclic AMP hydrolysis by the low affinity cyclic nucleotide phosphodiesterase, (2) the inhibition by cyclic GMP of the high affinity cyclic AMP phosphodiesterase, and (3) the induction of this latter enzyme by its substrate cyclic AMP.

The activation of cyclic AMP hydrolysis by cyclic GMP is, perhaps, the first reported direct interaction of these cyclic nucleotides. FRANKS and MACMANUS (1971) and BEAVO et al. (1970, 1971) first noted that in rat thymic lymphocytes and liver, respectively, the addition of micromolar cyclic GMP activated cyclic AMP hydrolysis at 10^{-7} to 10^{-5} M substrate concentrations. Cyclic AMP did not activate cyclic GMP hydrolysis. In lymphocytes this activation occurs through an increase in the V_{max} of a high K_m, particulate enzyme; cyclic GMP was a competitive inhibitor of the low K_m cyclic AMP phosphodiesterase. BEAVO et al. reported this same stimulation process in particulate fractions from rat liver, brain, kidney, heart, and thymus and in soluble fractions from liver and thymus. Cyclic IMP was a less effective activator than cyclic GMP. Ammonium sulfate fractionation studies suggested that this activation phenomenon resided in a single enzyme.

A similar activation has been reported in other tissues and cell types. Fractionation of rat adipose tissue extracts by Sepharose gel filtration resulted in three peaks of cyclic AMP phosphodiesterase activity, one peak being activated by 0.1–10 μM cyclic GMP (KLOTZ and STOCK 1972). This effect was most dramatic at pH 6.0 and began to disappear at pH 7.5. The enzyme exhibited kinetics which were positively cooperative, becoming linear upon activation by cyclic GMP. This type of enzyme fits and allosteric model in which substrate analogues should activate at low concentrations and inhibit at high concentration (MONOD et al. 1965).

This allosteric mechanism was confirmed by TERASAKI and APPLEMAN (1975) in an enzyme from rat heart partially purified on DEAE-cellulose. The cyclic GMP-activated enzyme had the same V_{max} and K_m for both nucleotides. The activation, which caused a change to linear kinetics, could be abolished by sulfhydryl modifying reagents without affecting basal catalytic activity. This and other kinetic data suggested a single enzyme with separate catalytic and regulatory sites. The catalytic site could accomodate both cyclic nucleotides while the regulatory site greatly favored binding of cyclic GMP ($K_a = 3 \times 10^{-6}$ M). The catalytic site would have an equal but low affinity for both nucleotides in its activated state.

This two-site hypothesis has since been studied using a highly purified enzyme from rat liver (TAKEMOTO and APPLEMAN in preparation). This enzyme exhibited positively cooperative kinetics, was activated by cyclic GMP and required the presence of free sulfhydryl groups. The different pH optima for activation (pH 7.5) and basal activity (pH 7.0) support the model of a two-site enzyme. Immobilization of the enzyme on a Sepharose-guanine affinity column resulted in a reversible loss of activation without a loss of basal enzyme activity.

The physiological role of this enzyme is hard to assess since its high K_m(10^{-5} M) would allow the regulatory process to be effective only at very high nucleotide cleotide concentrations. The higher cyclic AMP levels in cells, together with low cyclic GMP activation, suggest that this enzyme may function as a cyclic GMP-activated cyclic AMP phosphodiesterase. This enzyme may be responsible, in part, for some of the observed reciprocal changes in cyclic AMP and cyclic GMP levels. As will be discussed later, however, this activity is often very low or absent in malignant cells (SINGER et al. 1976) and human lymphocytes (TAKEMOTO et al. 1978);

in both of these types of cells reciprocal changes in cyclic nucleotides occur during mitogenic stimulation of cell growth. Some evidence regarding the role of cyclic GMP activation may come from the fact that it does not always occur. The high K_m cyclic nucleotide phosphodiesterase is *not* so stimulated in, for example, lung (BERGSTRAND et al. 1978).

The inhibition by cyclic GMP of the low K_m cyclic AMP phosphodiesterase would not be of potential significance except for two facts: (1) the concentration of the guanine nucleotide required for inhibition is relatively low and (2) the degree of inhibition varies between tissues and between cells of the same type at different stages of growth. THOMPSON and APPLEMAN (1971) and RUSSELL et al. (1972, 1973) reported that the cyclic AMP phosphodiesterase isolated from sonicated extracts of a number of rat tissues was inhibited by cyclic GMP at concentrations which the cyclic GMP itself was only very slowly hydrolyzed. The same enzyme, when isolated from rat liver by detergent solubilization and ion exchange chromatography, was found to have quite similar affinities for both cyclic AMP and cyclic GMP, whether they were substrate or inhibitor, and in both roles they exhibited negative cooperative kinetics (WHITSON and APPLEMAN in press). In view of the substantially greater hydrolytic activity of this enzyme for cyclic AMP, cyclic GMP is possibly more important as an inhibitor than as a substrate. Reciprocal inhibition of cyclic AMP and cyclic GMP hydrolysis has also been reported for the high affinity phosphodiesterase of human lung tissue (BERGSTRAND et al. 1978), but a similar low K_m enzyme isolated from dog kidney does not hydrolyze cyclic GMP and is only slightly inhibited by cyclic GMP with K_i of 300 μM (THOMPSON et al. 1979a).

Two additional experimental results might provide some evidence regarding the reality of cyclic GMP inhibition of cyclic AMP phosphodiesterase. In rat liver, where the detergent solubilized enzyme is sensitive to inhibition, preparations involving freezing and thawing for solubilization yield two active fractions on chromatography over DEAE cellulose columns (WHITSON and APPLEMAN in press). One of these fractions contains a cyclic AMP phosphodiesterase which, like the single fraction obtained by detergent solubilization, is inhibited by cyclic GMP. The other peak contains an enzyme with similar kinetic properties for cyclic AMP hydrolysis but without susceptibility to cyclic GMP inhibition; this property may be associated with a subunit which can be lost during processing. On the other hand "resting" human peripheral blood lymphocytes possess a low K_m cyclic AMP phosphodiesterase which is inhibited by cyclic GMP. Transformation of these cells by the mitogen concanavalin A, leads to the appearance over a period of hours of increased activity and decreased sensitivity to cyclic GMP inhibition (TAKEMOTO et al. 1979). It is not yet clear whether this is *de novo* enzyme synthesis or enzyme modification but it suggests a biological role for this cyclic GMP effect.

The induction of a cyclic AMP phosphodiesterase by elevated cyclic AMP levels has been noted in a number of rapidly dividing and malignant cell systems. SENFT et al. (1968) noted an increase in a liver enzyme following adrenalectomy. DUTTAGUPTA et al. (1978) reported an increase in cyclic AMP phosphodiesterase of fibroblasts after adding theophylline, MIX, adenosine, or 2-chloradenosine, to the media. Kinetically, the enzyme was identical to the non-induced species. In 3T3 and SV3T3 fibroblasts (D'ARMIENTO et al. 1972), chick embryo fibroblasts

(RUSSELL and PASTAN 1974), L929 cells (MANGANIELLO and VAUGHAN 1972) and rat astrocytoma cells (UZUNOV et al. 1973), exposure to high cyclic AMP levels also results in an increase in the activity of a low K_m cyclic AMP phosphodiesterase. Cyclic GMP or high K_m cyclic AMP hydrolytic activities are not affected. ROSS et al. (1977) also reported increases in the low K_m cyclic AMP phosphodiesterase from cultured hepatoma cells incubated with dibutyryl cyclic AMP and MIX. These cyclic AMP-induced increases require both protein and RNA synthesis; inhibition of induction occurs through addition of actinomycin D or cycloheximide. The half-life of the induced enzyme varies from 1 to 2 h in the cultured cells.

The mechanism by which cyclic AMP induces the phosphodiesterase was partly elucidated through the use of S49 mutants insensitive to the lethal effects of high cyclic AMP (INSEL et al. 1975; COFFINO et al. 1976). The insensitivity probably is due to the lack of the regulatory subunit of a cyclic AMP-dependent protein kinase. These cells will accumulate cyclic AMP when treated with isoproterenol, PGE_1, cholera toxin or other agents which activate adenylate cyclase; however induction of a low K_m cyclic AMP phosphodiesterase does not occur. This would suggest that the induction of the enzyme requires the action of a kinase, perhaps at the nuclear level. Indeed, a transcriptional mechanism is suggested by the work of MURRAY and RUSSELL (1978). 3T3-L fibroblasts which are simultaneously treated with MIX and cycloheximide for 12 h, rinsed, then treated with actinomycin D show a twofold increase in the enzyme within 90 min. Exposure of these cells to 1.0 μM bromodeoxyuridine inhibits phosphodiesterase induction; addition of thymidine removes the block by the analog, suggesting that the induction of the cyclic AMP phosphodiesterase occurs at the level of DNA transcription.

The function, in normal cells, of this specific enzyme induction would be to control high cellular levels of cyclic AMP which may affect cell growth. As discussed below, however, a permanent and irreversible induction occurs in many malignant cells. The resulting lowered cyclic AMP levels may be a factor in uncontrolled malignant cell growth.

II. Other Effectors

In addition to the cyclic nucleotides themselves, a wide array of substances appear to interact with the binding sites of the phosphodiesterases. These include cyclic nucleotide derivatives, such as N^6, O^2, dibutyryl-, N^6, monobutyryl-, and O^2, monobutyryl- cyclic AMP (JARRETT and SMITH 1974), and 8-substituted cyclic nucleotides (HARRIS et al. 1973; SASAKI et al. 1978). Pyrimidine cyclic nucleotides are generally poor inhibitors of the purine cyclic nucleotide phosphodiesterase (KUO et al. 1978). Purine nucleotide mono-, di-, and triphosphates inhibit some phosphodiesterase preparations, but they require high concentrations; these compounds may actually function as chelators (TESHIMA et al. 1974). NADH appears to interact with the catalytic sites of both high and low K_m phosphodiesterases (see Sect. E.3). In addition, a wide variety of pharmacological agents appear to compete with cyclic nucleotides for phosphodiesterase binding sites (see Sect. E.4).

All the phosphodiesterases isolated to date require divalent metal cations for activity; therefore metal chelators will act as inhibitors. As mentioned above, cyclic nucleotide phosphodiesterase requires an active sulfhydryl for activation by cyclic

GMP (TAKEMOTO and APPLEMAN in preparation), and other phosphodiesterases appear to require a sulfhydryl for activity (DRUMMOND and PERROT-YEE 1961). This may explain the reported effects of oxidizing agents (KONO et al. 1975) on phosphodiesterase activities.

Inhibitors specific to the individual forms of phosphodiesterase could be quite useful in elucidating the roles of these enzymes in the control of cell function. The major forms of phosphodiesterase exhibit distinct substrate preferences, substrate affinities and subcellular distributions; therefore discrete intracellular pools of cyclic nucleotides may mediate various cell functions. Specific inhibitors could be useful in exploring this possibility. There is one rather debilitating factor in the quest for such inhibitors, however, and it relates to the properties of the enzymes themselves. In view of the wide variety of inhibitors interacting with the cyclic nucleotide binding sites, these sites appear to be somewhat non-selective. This seems especially true for the low K_m cyclic AMP phosphodiesterase (WHITSON and APPLEMAN, unpublished observations). The introduction into a cell of agents in sufficient quantity to affect the lower affinity phosphodiesterases would almost certainly inhibit the low K_m enzyme as well. Therefore, attempts to produce specific inhibitors by modifying analogues of cyclic nucleotides may not succeed. A more valid approach may be to take advantage of specific regulating properties of the enzymes. For example, the calmodulin-activated enzyme may be selectively inhibited by agents which compete with the enzyme for calmodulin binding. In light of the pleiotropic effects of calmodulin, however, it could be difficult to distinguish the effects of such agents on phosphodiesterase from effects on other components of the Ca^{++} and cyclic nucleotide regulatory systems. In any case, it is clear that fruitful investigations of phosphodiesterase inhibitors must be performed with isolated, well characterized enzymes.

E. Physiological Roles of the Cyclic Nucleotide Phosphodiesterases

All of the information presented above on the properties of the various forms of phosphodiesterase in mammalian tissues is of little importance if the activity of these enzymes cannot be associated with specific cellular activities. This association could be demonstrated experimentally by relating the phosphodiesterase and the physiological activity either cytologically (a physical relationship), chronologically (a temporal change in one paralleled by a change in the other), or pharmacologically (parallel responses to a drug or other effector). As mentioned previously, research in this area has been hindered by the necessity to show effects on one, out of perhaps three similar enzymes. This section will discuss 4 biological activities involving cyclic nucleotide phosphodiesterases: development, cell growth and malignancy, responses to insulin and other metabolic hormones, and neurotransmission.

I. Development

Development and growth-related events are thought to be triggered by changes in the circulating levels of certain hormones whose actions may be mediated by cyclic nucleotides. Hormone-sensitive adenylate cyclase systems appear in various organs at specific times during the course of mammalian fetal development. Ontogenic changes in cyclic nucleotide phosphodiesterases have also been reported in the silkworm, *Bombyx mori*, in which at least three forms of phosphodiesterase, similar to those found in other tissues, change dramatically during development (MORISHIMA 1975a, b).

Dramatic or consistent changes have not been found in studies of the development of phosphodiesterases in most mammalian systems. Early studies dealt primarily with growth related changes in the high K_m cyclic AMP phosphodiesterases (WEISS 1971; GENGLER and FORTE 1972; SMOAKE et al. 1974; STRADA et al. 1974). DAVIS and KUO (1976) reported ontogenetic changes in the hydrolysis of cyclic AMP and cyclic GMP in guinea pig lung, liver, brain, and heart. Developmental patterns varied; for example, the specific activity of a soluble enzyme from lung was higher in the fetus than neonate or adult but a reverse situation was observed in the enzyme from brain.

TANIGAWA et al. (1976) have reported on the developmental changes in chick embryo liver phosphodiesterases resolved by DEAE-cellulose column chromatography. In the early stages of embryonic development, a low K_m, calmodulin-sensitive, cyclic GMP phosphodiesterase predominated; this enzyme was substantially reduced in adult chick liver. In contrast, cyclic GMP hydrolysis was not selectively decreased in rat liver during development (SICARD and APRILLE 1977), rather, the specific activities of all forms decreased.

SINGER et al. (1978) determined that all three major forms of phosphodiesterases, as well as calmodulin, were present in early fetal rabbit tissues but sensitivity of cyclic GMP hydrolysis to calmodulin appeared only during the later stages of development. In general, the changes in rabbit liver closely resembled those reported for rat liver. The early appearance of all forms of this enzymatic activity in most tissues indicates an essential, but at present unknown role in biological regulation.

Developmental changes in cyclic CMP phosphodiesterase have also been reported. The activity of this enzyme has been shown to decrease with development in a variety of guinea pig tissues (HELFMAN et al. 1975) and in rat liver (SHOJI et al. 1978).

II. Transformation and Malignancy

Normal and malignantly transformed cells contain different quantities of cyclic nucleotides. Frequently, malignantly transformed cells contain less cyclic AMP and more cyclic GMP than do normal cells. The trends for malignancy-associated changes in phosphodiesterase are, as might be expected, a diminished capacity to hydrolyze cyclic GMP and an increased capacity to hydrolyze cyclic AMP. An increase in low K_m cyclic AMP phosphodiesterase activity was reported to occur in malignant, as well as cyclic AMP-"differentiated," neuroblastoma cells (KUMAR et al. 1975). A decrease in K_m for cyclic AMP and an increased total catalytic activity have been observed in the phosphodiesterase activity from rat hepatoma (RHOADS et al. 1972; SCHRODER and PLAGEMAN 1972) and from 3T3 and SV-3T3 mouse fibroblast cultures (D'ARMIENTO et al. 1972). The 3T3 cells contain three forms of phosphodiesterase (identified by DEAE-cellulose chromatography) but the SV transformed cells contain predominantly the low K_m cyclic AMP hydrolyzing enzyme which is also higher in specific activity (TAKEMOTO and APPLEMAN, unpublished).

The lymphocyte has been the object of considerable growth-related research because of its availability in various growth conditions: normal peripheral cells, mitogen-transformed, leukemic, and transformed cell lines in culture. An additional advantage is that normal lymphocytes present a more limited array of phosphodiesterase activities than liver and other tissues. Murine leukemic L5178Y and L1210 cells, grown in ascites fluid, have more low K_m cyclic AMP and cyclic GMP phosphodiesterase than normal lymphocytes (HAIT and WEISS 1976). The observed increase is mainly due to an increase in the V_{max} of the enzyme. The enzymes from normal and leukemic lymphocytes have similar pH optima, activities and sensitivities to various pharmacological inhibitors (HAIT and WEISS 1977). Both normal and leukemic lymphocytes have calmodulin but lack calcium-sensitive phosphodiesterase and cyclic GMP-activated cyclic AMP phosphodiesterase. WEISS and WINCHURCH (1978), using the AKR mouse as a model system, have found that older leukemic mice have a higher percent of large blast-like cells than their normal young congeners. The specific activity of cyclic AMP phosphodiesterase is higher in extracts from these large, blast-like cells than in extracts from normal cells. EPSTEIN et al. (1977) have also reported an elevation of low K_m cyclic AMP and cyclic GMP hydrolytic activities in murine leukemic lymphocytes. The enzyme involved differs physically from the enzyme derived from human leukemic cells.

The human lymphocyte has a single dominant form of cyclic AMP phosphodiesterase which shows dramatic changes in activity during periods of accelerated cell proliferation. THOMPSON et al. (1976), EPSTEIN et al. (1977), and TAKEMOTO et al. (1978) have shown that leukemic lymphocytes and virally-transformed lymphocytes in culture contain significantly elevated levels of cyclic AMP phosphodiesterase when compared to normal peripheral cells. The induced enzyme has the same high affinity and substrate specificity but differs in being almost completely insensitive to inhibition by cyclic GMP (TAKEMOTO et al. 1978). SHER et al. (1976) have reported a decrease rather than an increase in cyclic AMP phosphodiesterase from leukemic as opposed to normal lymphocytes. These workers noted, however, that the lymphocyte preparation included a significant number of platelets, cells which contain up to 10,000 times as much phosphodiesterase activity as lymphocytes (PICHARD et al. 1973; THOMPSON et al. 1976). It is possible that the reduction in phosphodiesterase activity may reflect the higher degree of contamination of the leukemic cells by platelets. An in vitro demonstration of the proliferation-related changes in phosphodiesterase has been carried out by TAKEMOTO et al. (1979) using mitogen transformation of human peripheral lymphocytes. Exposure of these cells to concanavalin A induces the appearance of elevated levels of cyclic AMP phosphodiesterase; the "new" enzyme is inhibited only slightly by cyclic GMP. The induction process seems to require an elevation in cyclic GMP, but cyclic GMP probably acts at an early step in the protein synthetic process rather than through any direct interaction with the enzyme molecule.

III. Insulin and Other Metabolic Hormone Actions

BUTCHER, PARK, and SUTHERLAND (BUTCHER et al. 1966) first suggested that some of the actions of insulin might be mediated via a decrease in intracellular cyclic AMP levels and demonstrated such a decrease in rat livers perfused with the hormone. Cyclic AMP levels in liver were elevated in rats made diabetic with anti-insulin serum (JEFFERSON et al. 1968). This biochemical response could conceivably be brought about either by a decrease in the activity of adenylate cyclase or by an increase in the activity of phosphodiesterase. HEPP and RENNER (1972) demonstrated an insulin-mediated decrease in the adenylate cyclase activity of rat liver, but this was dependent upon prior stimulation of the enzyme by epinephrine.

The stimulation of phosphodiesterase by insulin was first reported by LOTEN and SNEYD (1970). This effect was demonstrated in the particulate fraction of homogenates of rat adipocytes which were treated with insulin. These results in adipocytes have been repeated by many other investigators (MANGANIELLO and VAUGHAN 1973; SAKAI et al. 1974; KONO et al. 1975, 1977; LOVELL-SMITH et al. 1977; SOLOMON 1975; SOLOMON et al. 1977). Similar results have also been shown in isolated hepatocytes (ALLAN and SNEYD 1975; LOTEN et al. 1978), and in skeletal muscle (GAIN and APPLEMAN 1978). These effects of insulin were shown to be dose- and time-dependent (MANGANIELLO and VAUGHAN 1973; LOTEN et al. 1978) and reversible (KONO et al. 1975). The levels of insulin which induce the effect on phosphodiesterase are in the same range as those which are required for stimulation of gluconeogenesis and the reduction of intracellular cyclic AMP levels (KONO et al. 1975).

Supporting evidence for the involvement of phosphodiesterase in insulin action has come from studies of tissues of diabetic animals. THOMPSON et al. (1973) demonstrated a decline in the activity of a high affinity cyclic AMP phosphodiesterase in the particulate fraction of streptozotocin-treated diabetic rats. This decline could be reversed by administration of insulin or growth hormone to the treated animals. Phosphodiesterase declined, also, in the plasma membrane fraction of liver from alloxan-treated rats; this activity could be restored by insulin, but only when the hormone was administered to the intact animal (PILKIS et al. 1974). A decrease in enzyme activity in diabetic tissues and its restoration by insulin was also shown in the phosphodiesterase of rat heart (DAS 1973) and rat plasma (HEMMINGTON et al. 1973).

A relatively consistent picture of the insulin-sensitive phosphodiesterase has emerged in these studies. The enzyme is a high affinity, cyclic AMP-preferring enzyme with negatively-cooperative kinetics. This enzyme is not activated by Ca^{++} or calmodulin. It corresponds to the cyclic AMP phosphodiesterase described above. The primary consequence of the stimulation by insulin is an increase in the V_{max} of the enzyme. Only two reports (TRIA et al. 1976; HOUSE et al. 1972) suggest that insulin could be effective in less than intact cells[1].

1 In a recent report, MARCHMONT and HOUSLAY (1980a) have characterized the cyclic AMP phosphodiesterase activity of rat liver plasma membranes as including an intrinsic high Km form and a peripheral low Km, negative cooperative form. The latter enzyme is activated and phosphorylated through the incubation of a cell-free membrane preparation with ATP, high concentrations of cyclic AMP, and physiological concentrations of insulin (MARCHMONT and HOUSLAY, 1980b). The insulin-stimulated cyclic AMP phosphodiesterase has been purified and characterized (MARCHMONT et al., 1981).

Although the effects of insulin seem to be largely on cyclic AMP hydrolysis, an insulin-mediated effect on cyclic GMP hydrolysis has also been reported (LOVELL-SMITH et al. 1977). The insulin-sensitive phosphodiesterase is classified as a cyclic AMP phosphodiesterase yet it can hydrolyze cyclic GMP with equal affinity and very much lower velocity. As discussed above, this and other properties of this enzyme are affected by the method of cell disruption and enzyme preparation. The failure of many investigators to report an effect on cyclic GMP hydrolysis may be attributed to this fact.

Insulin has been widely demonstrated to bind to receptors on the external membranes of its target cells and this event is believed to initiate many, if not all, of the effects of the hormone. If phosphodiesterase activation were an early step in the transmission of the insulin binding signal to the intracellular machinery, the activation process could be closely linked, both temporally and spatially, with hormone binding. The stimulation of phosphodiesterase does occur with a rapid time-course; demonstrable activation is seen after a two-minute incubation (MANGANIELLO and VAUGHAN 1973). However, the subcellular location of this enzyme is probably not in direct proximity to that of insulin binding. MANGANIELLO and VAUGHAN (1973) noted that insulin stimulated phosphodiesterase was in the particulate fraction of rat adipocyte homogenates. THOMPSON et al. (1973) also showed this in liver homogenates from streptozotocin-treated rats using discontinuous sucrose-density gradients. They further fractionated this preparation and found low K_m cyclic AMP phosphodiesterase in three distinct fractions; activity in only one fraction was affected by insulin treatment. The same methods were used to demonstrate a single peak of insulin-stimulated activity in isolated adipocytes from normal rats (SAKAI et al. 1974). KONO et al. (1975) compared the distribution of the enzyme with the distribution of various markers for specific subcellular fractions and found that only a small portion of the insulin-stimulated phosphodiesterase co-migrated with insulin-binding activity and adenylate cyclase. The majority of the activated phosphodiesterase was associated with a "light microsomal" fraction bearing enzymatic markers characteristic of endoplasmic reticulum. In a later paper, however, KONO et al. (1977) found that [^{125}I]-labelled insulin also accumulated in a non-plasma membrane fraction.

Although these results could be interpreted as indicating that both the insulin receptor and the insulin-sensitive phosphodiesterase reside in a portion of the membrane which is internalized when insulin binds, these events appear to occur independently. Insulin binding and insulin stimulated phosphodiesterase in particulate fractions are readily dissociable by detergents (APPLEMAN and TERASAKI 1975). Thus, the insulin-stimulated phosphodiesterase, although to a large extent particulate, does not appear to be plasma membrane bound. Significant amounts of soluble, insulin-sensitive cyclic AMP phosphodiesterase have been found in rat skeletal muscle preparations (GAIN and APPLEMAN 1978).

The fact that the insulin-stimulation of phosphodiesterase generally requires intact cells, and the finding that the receptor and the enzyme do not reside in the same membrane fraction, suggest that some intermediate messenger must transmit the hormonal signal. Several investigators have sought to identify this messenger by examining the conditions which affect the insulin stimulation of phosphodiesterase. Several mechanisms have been ruled out: the stimulation process does not seem to involve calmodulin or cyclic GMP, as the insulin-sensitive enzyme is not activated by these agents (LOVELL-SMITH et al. 1977; SAKAI et al. 1974). The stimulation of the enzyme does not involve protein synthesis (LOTEN and SNEYD 1970), but the possibility of protein phosphorylation has not been ruled out. Recent reports of mediators of insulin action (JARRETT and SEALS 1979; LARNER et al. 1979) may soon provide new evidence regarding the molecular mechanism of the activation.

Phosphodiesterase does not seem to be made more resistant to degradative agents by insulin treatment. No difference in the susceptibility of insulin-treated enzyme versus control enzyme to heat, ethanol, urea or trypsin, has been observed by APPLEMAN and TERASAKI (1975). A number of agents have been tested for their effects on the insulin stimulation of phosphodiesterase. KONO et al. (1975) reported that glutathione, dithiothreitol, EDTA, and EGTA prevent or reverse the stimulation of the enzyme when added to the cell homogenate after insulin treatment. The EGTA inhibition was quite complex; "aged" homogenates were unaffected, and the effect of EGTA on fresh homogenate could be overcome by Ni^{++}, Co^{++}, and sulfhydryl reagents. Homogenization in a nitrogen atmosphere also abolished the insulin activation of phosphodiesterase. These results were interpreted as indicating that the enzyme was stimulated via an oxidation process. LOTEN et al. (1978), however, found no effects of either EGTA or mercaptoethanol on insulinstimulation of phosphodiesterase; MANGANIELLO and VAUGHAN (1973) reported an increase in the activity of both basal and insulin-stimulated phosphodiesterase, when assayed in the presence of EGTA and dithiothreitol. There is also a report of the potential involvement of the enzyme glutathione-insulin transhydrogenase in the insulin-stimulation of phosphodiesterase: antiserum against this enzyme, when added to the adipocyte preparation, prevented the stimulation of phosphodiesterase (PHELPS and VARANDINE 1977). The effects of the oxidizing agents may be related to this finding but these agents are quite generalized in their actions and the effects on phosphodiesterase may be secondary to many other factors.

If the details of the mechanism of insulin-stimulation of phosphodiesterase are to be known, it will be necessary to separate the enzyme from its membrane environment in the activated state as well as to resolve the insulin-stimulated enzyme from other forms of the enzyme. The insulin-sensitive phosphodiesterase of rat adipocytes has been solubilized from the particulate fraction via non-ionic detergents (LOVELL-SMITH et al. 1977). The enzyme retained its activation when solubilized, but more than one fraction of activity was seen when the solubilized material was applied to ECTEOLA cellulose, and the stimulated fraction could not be identified. The insulin-stimulated phosphodiesterase of rat hepatocytes could also be solubilized in an activated state by suspension of the particulate fraction in hypotonic buffers (LOTEN et al. 1978). Solubilization methods for the cyclic AMP phosphodiesterase probably allow some proteolysis to occur (see Sect. C.3). Proteolysis would be especially debilitating in the studies which attempt to isolate the insulin-stimulated enzyme in its native state. It is quite possible that the portion of the enzyme which regulates the stimulation by insulin is lost upon solubilization. LOTEN et al. (1980) have suggested that the fragment of the enzyme containing the hydrophobic residues which attach it to the membrane may be phosphorylated. Thus, studies which attempt to demonstrate [^{32}P]-labelling of phosphodiesterase in the presence of insulin may fail because the phosphorylated residue is lost upon solubilization. In this regard it is noteworthy that in adipocytes, which characteristically have a low lysosomal content, the detergent-solubilized enzyme is large enough to be excluded from a 1.5×10^6 Dalton exclusion limit gel (LOVELL-SMITH et al. 1977). While this may mean that in adipocytes, proteolytic degradation of the phosphodiesterase does not occur, numerous attempts to demonstrate phosphorylation of this enzyme have failed.

A phosphodiesterase with properties similar to the insulin-sensitive enzyme is also activated by numerous other agents. It is interesting that many of these agents are concerned with the regulation of lipid metabolism, although many of them have physiological effects opposite to those of insulin.

There is good evidence for the involvement of thyroid hormones in the control of cyclic AMP metabolism through the regulation of phosphodiesterase. ARMSTRONG and STOUFFER (1974) found that the adipocytes of hypothyroid rats failed to give a lipolytic response to epinephrine or glucagon but could respond to dibutyryl cyclic AMP. Epinephrine did increase lypolysis but only in the presence of theophylline. The soluble, low affinity cyclic nucleotide phosphodiesterase was unaltered in adipocytes from hypothyroid animals, but the particulate low K_m phosphodiesterase was markedly increased. VAN INWEGEN et al. (1975) showed that thyroxine (T 3) could restore the lipolytic response to epinephrine and glucagon, and could return the phosphodiesterase activity to the control level in these animals.

Altered lipid metabolism has also been shown to involve changes in phosphodiesterase activity in the genetically obese mouse. In this case the low affinity cyclic nucleotide phosphodiesterase showed a marked increase in activity. The enzyme was altered in both its kinetic and electrophoretic properties (KAPLAN et al. 1973). The low affinity enzyme is also increased in lactating rat mammary gland in correlation with a sharp rise in lipogenesis (SAPAG-HAGAR and GREENBAUM 1973).

Growth hormone has been reported to affect phosphodiesterase activity in two biological systems. THOMPSON et al. (1973) showed that growth hormones, as well as insulin, could reverse the negative effects of streptozotocin on rat liver and kidney phosphodiesterase. The stimulating effects of growth hormone on glucose transport in diaphragms from hypophysectomized rats is prevented by phosphodiesterase inhibitors (RILLEMA et al. 1973). The insulin stimulation of glucose transport was not prevented by these drugs.

Glucagon has been shown to stimulate phosphodiesterase directly in isolated hepatocytes (LOTEN et al. 1978). The responsive enzyme was apparently identical to that stimulated by insulin. This activation of a single enzyme by hormones which tend to drive metabolism in opposite directions may seem paradoxical. The stimulation of the enzyme by glucagon may represent an "overshoot" control mediated by cyclic AMP derived from the stimulation of adenylate cyclase. The induction of cyclic AMP phosphodiesterase in this case could be similar to that produced by the nucleotide and its derivatives in fibroblasts, etc. (D'ARMIENTO et al. 1972).

A less direct relationship between glucagon and phosphodiesterase was reported by CLARK and JARRETT (1978) who found that the redox state of hepatocytes exerted a controlling influence on the response to glucagon. They also found that NADH, but not NAD^+, inhibited the low K_m cyclic AMP phosphodiesterase.

They speculated that phosphodiesterase provided the link between glucagon action and the cytosolic redox states. It is not surprising that NADH inhibits phosphodiesterase; Blue Dextran Sepharose and Cibacron blue dyes which bind enzymes using NADH as cofactors can also bind all three forms of phosphodiesterase (MORILL et al. 1979; ASHTON and POLYA 1978; TAKEMOTO and WHITSON unpublished). This probably involves purine binding sites since nicotinic acid itself does not affect phosphodiesterase (SKIDMORE et al. 1971).

Epinephrine stimulates the phosphodiesterase of isolated adipocytes (ZINMAN and HOLLENBERG 1974; KONO et al. 1977; SOLOMON 1975). This effect on the particulate low K_m enzyme is probably secondary to the increase in cyclic AMP. Propanolol, a β-adrenergic antagonist, and ACTH also stimulates this enzyme (ZINMAN and HOLLENBERG 1974; LOVELL-SMITH et al. 1977). The gut peptide referred to as "non-supressable insulin-like protein" has been shown to stimulate low K_m cyclic AMP phosphodiesterase in rat liver slices (LEICHTER and POFFENBARGER 1978). All of these agents act through plasma-membrane receptors. In fact, a general perturbation of the plasma membrane such as by the binding of concanavalin A, may be sufficient for stimulation of the low K_m cyclic AMP phosphodiesterase (DESAI et al. 1976).

The prostaglandins have been implicated in the actions of cyclic nucleotides, but most of their effects seem to be via the modulation of adenylate cyclase activation by other agents. L-cell fibroblasts do exhibit an apparent change in subcellular distribution of cyclic AMP phosphodiesterase in the presence of prostaglandin E_1 (BACALAO and RIEBER 1972; MANGANIELLO and VAUGHAN 1972). The absence of an effect of Prostaglandin E_1 on hepatic phosphodiesterase was reported by DERUBERTIS et al. (1974) even though this hormone inhibited the glucagon-mediated increase in cyclic AMP.

Adrenal steroids have also been shown to affect cyclic AMP phosphodiesterase activity. Adrenalectomized rats had increased adipocyte phosphodiesterase activity which could be restored by administration of cortisone (SCHONHÖFER et al. 1972). MANGANIELLO and VAUGHAN (1973) reported that the cyclic AMP phosphodiesterase of rat adipocytes was decreased 20 h after the rats were injected with dexamethasone; SENFT et al. (1968) reported that the enzyme was decreased 20 h after the administration of 6-methyl-prednisolone. Cortisol inhibits phosphodiesterase (SCHMIDTKE et al. 1976).

The fact that the high affinity particulate cyclic AMP phosphodiesterase of adipocytes is affected by so many diverse agents indicates that this enzyme is acutely in tune with the metabolic state of the cell. It is not always clear to what extent the effects on this enzyme follow directly in the metabolic "cascade" of the effects of the hormone. The role of the enzyme may vary from one of primary control of the cyclic AMP signal to acting as an off-switch for the hormones which stimulate adenylate cyclase. The development and integration of a great deal of data on cyclic nucleotide metabolism will be needed before the relative role of phosphodiesterase activity can be truly assessed.

IV. Neurotransmission

Cyclic nucleotides have a role in neural tissues that is equal in complexity to the role they play in the regulation of energy metabolism in other tissues. Neural tissues are therefore very rich in the enzymes of cyclic nucleotide metabolism. Phosphodiesterase activity is highest in tissue of the central nervous system (BUTCHER and SUTHERLAND 1962; CHEUNG and SALGANICOFF 1967; WEISS and COSTA 1968; BRECKENRIDGE and JOHNSTON 1969; CHEUNG 1967) the level of this enzymatic activity varies in different brain regions. A variety of phosphodiesterases with different affinities for cyclic nucleotides and varying degrees of sensitivity to calcium and calmodulin are present in brain homogenates. It is therefore very difficult to interpret the results from studies of enzyme activity in homogenates in terms of the functioning, intact central nervous system.

Three primary experimental approaches are used to elucidate the functional role of phosphodiesterase in the nervous system. First, biochemical measurements are made of changes in the enzyme activity when various neuropharmacological agents are administered. Second, electrophysiological analyses are used to ascertain biophysical and pharmacological responses of neurons, or other innervated target cells, to drugs which have been shown to alter the dynamics of the cyclic nucleotide system. Third, the anatomical approach, applying the techniques of light microscopic histochemistry (SHANTA et al. 1966), immunocytochemistry, and electron microscope cytochemistry (FLORENDO et al. 1971), is used to localize phosphodiesterase in the areas associated with specific cell functions.

A variety of cyclic nucleotide phosphodiesterase exist in the nervous system. The brain phosphodiesterase form which predominates in most preparations is specifically activated by calmodulin, an endogenous protein activator (CHEUNG 1970; KAKIUCHI and YAMAZAKI 1970), in a calcium-dependent mechanism (KAKIUCHI et al. 1975; TEO and WANG 1973). Calmodulin first interacts with calcium and then binds to the phosphodiesterase, forming an active holoenzyme (KAKIUCHI et al. 1973; WOLF and BROSTROM 1974; LIN et al. 1974). This enzyme will hydrolyze cyclic AMP at very high concentrations (GNEGY et al. 1976a, b; UZUNOV et al. 1976) but cyclic GMP is the preferred substrate at physiological levels (ARIANO and APPLEMAN 1979; FILBURN et al. 1978; GREENBERG et al. 1978). Nevertheless, many studies of phosphodiesterases from nervous tissue have been performed using only cyclic AMP as a substrate and consequently this enzyme is referred to as a "high K_m cyclic AMP phosphodiesterase" in some of the literature. It should not be confused with the high K_m cyclic nucleotide phosphodiesterase which has equal affinities for cyclic AMP and cyclic GMP and is activatable by the latter nucleotide.

The multiple forms of phosphodiesterase have been resolved in brain preparations by THOMPSON and APPLEMAN (1971), UZUNOV and WEISS (1972) and PLEDGER et al. (1974). Attempts to purify the major brain enzyme using affinity chromatography on calmodulin-coupled to Sepharose initially yielded only partial success (WATTERSON and VANAMAN 1976; MIYAKE et al. 1977) although a recent report (MORILL et al. 1979) describes a complete purification combining this method with Blue Dextran Sepharose affinity chromatography. Copurification of phosphodiesterase with other proteins on calmodulin-Sepharose (KLEE 1977; KLEE and KRINKS 1978) is not surprising in view of recent evidence of the ability of the protein activator to regulate a number of different enzymes (CHEUNG and STORM, this volume).

The levels of the phosphodiesterases vary considerably in different regions of mammalian brain (GREENBERG et al. 1978; HOWLETT and BRECKENRIDGE 1978;

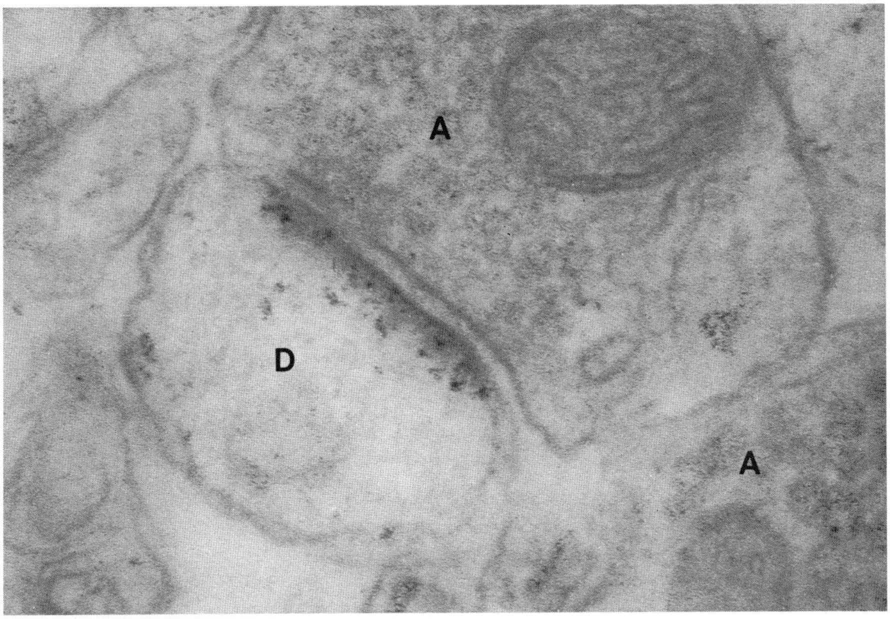

Fig. 2. Electron Microscope Cytochemical Localization of Phosphodiesterase. Unstained rat corpus striatum showing localization of phosphodiesterase activity following incubation with 1 m*M* cyclic GMP, 2 m*M* calcium chloride, and excess calmodulin. Deposition of the electron-dense, lead phosphate precipitate, representing calcium and CDR-sensitive cyclic GMP PDE activity can be seen on synaptic vesicles in the axon terminal bouton *A*, and juxtaposed to the subsynaptic density of the post synaptic region *D*. The amount of reaction product associated with the neurotransmitter-containing synaptic vesicles is considerably less than the precipitate localized at the subsynaptic thickening. Magnification, × 195,000. (Micrograph taken by Dr. M. Ariano)

Egrie et al. 1977). The highest specific activities appear to be associated with soluble or particulate fractions from lysed synaptosomes (DeRobertis et al. 1967; Gaballah and Popoff 1971; Miki and Yoshida 1972). These investigations report a low K_m (1–5 μM) cyclic GMP hydrolytic activity, and both low K_m (10–20 μM) and high K_m (about 350 μM) cyclic AMP hydrolytic activity. All forms of synaptosomal phosphodiesterases are predominantly in the soluble phase with the fraction in the particulate phase dependent on the means of tissue homogenization.

A number of investigators have studied the subcellular localization of phosphodiesterase in the central nervous system (Florendo et al. 1971; Adinolfi and Schmidt 1974; Ariano and Adinolfi 1977). The enzyme activity is found juxtaposed to the postsynaptic density of a large number of asymmetrical, axodendritic synapses (Fig. 2). Correlation of the anatomical localization with the different enzymatic properties known to exist for the multiple forms of phosphodiesterase activity showed that the postsynaptically visualized form in the rat neocortex and corpus striatum is a cyclic GMP-preferring enzyme (Ariano and Appleman 1979). Visualization of this hydrolytic activity requires the presence of calcium and calmodulin. Anatomical differentiation of the other phosphodiesterase enzymes demonstrated biochemically in the nervous system awaits their purification and the production of specific antibodies.

During the development of rat cerebral cortex there is a 5-fold increase in total phosphodiesterase activity during the first 20 postnatal days (STRADA et al. 1974). This increase correlates with the onset of synaptogenesis occurring in this central nervous system region (ADINOLFI 1972), and the phosphodiesterase can be cytochemically localized at developing synaptic densities (ADINOLFI and SCHMIDT 1974). Changes in enzyme activity during development reflect the appearance of specific cell types and physiological maturation of afferent inputs.

Calmodulin may be of special importance in neural tissues. In rat cerebral cortex, the protein activator increased the activity of the resolved cyclic GMP phosphodiesterase fraction by 5- to 10-fold (ARIANO and APPLEMAN 1979; UZUNOV and WEISS 1972) but had little or no effect on the other enzyme forms. Immunohistochemical demonstration of calmodulin shows this compound to be localized at postsynaptic sites which, in preliminary work, appear morphologically equivalent to the active phosphodiesterase sites (GRAB et al. 1979; WOOD et al. 1979). Physiological investigations (GNEGY et al. 1976a, b) have shown that the activator is released from membrane sites following increased transsynaptic activity. Together these findings suggest the hypothesis that calmodulin activated phosphodiesterase has a very important role in synaptic transmission. Indulging in speculation, in the quiescent state, membrane-associated phosphodiesterase would be bound as an apoenzyme; with the initiation of synaptic activity and subsequent release of calcium, the activator would be displaced from its membrane-binding protein and could then migrate to form active enzyme complexes of the adenyl cyclase and cyclic GMP phosphodiesterase. As the levels of cyclic nucleotide decreased due to the increased rate of hydrolysis and the sequestration of calcium ions, the calmodulin would dissociate from the phosphodiesterase and be recaptured by the membrane-binding protein.

The calmodulin-activated phosphodiesterase is uniquely sensitive to inhibition by the antipsychotic drug, trifluoperazine (WEISS 1975). It has been demonstrated that this selective inhibition is due to the interaction of trifluoperazine with the activator protein (LEVIN and WEISS 1978a, b, 1979). This may prove to be the mechanism by which these antipsychotics function at the molecular level.

Other inhibitors of the phosphodiesterase enzymes in brain have proven to be less specific with regard to the particular form of the enzyme which they inhibit in vitro. Despite this, many are quite active pharmacologically. In addition to the commonly used methyl xanthines, caffeine, theophylline, aminophylline etc., there are the 3,4-dialkoxybenzyl-2-imidazolidinones, such as RO 20-1724 (SHEPPARD et al. 1972; SCHULTZ 1974); a 3-4-dialkoxyphenyl-2-pyrrolidone, ZK 62711 (SCHWABE et al. 1976); 1-H-pyrazolo[3,4b] pyridines such as SQ 20009 (BEER et al. 1972); benzodiazepines such as diazepam, medezepam, and chlordiazepoxide (BEER et al. 1972; DALTON et al. 1974); tricyclic antidepressants (HONDA and INAMURA 1968); flavonoids (RUCKSTUHL et al. 1979); cannabinoids (DOLBY and KLEINSMITH 1977) and others. The interested investigator is referred to the monograph of AMER and KREIGBAUM (1975) and the excellent reviews of DALY (1977) and CHASIN and HARRIS (1976). Naturally occurring effectors of phosphodiesterases of one form or another have also been demonstrated to act on the brain enzymes: calcium and other divalent cations (FERRENDELLI et al. 1976; BOUDREAU and DRUMMOND 1975; FILBURN et al. 1978; MIYAKE et al. 1977); cyclic GMP (FILBURN et al. 1978); calmodulin (UZUNOV et al. 1976; SAKAI et al. 1977); and inhibitory factors involved in the calmodulin system (KANAMORI et al. 1979; RICHMAN and KLEE 1978; DAVIS and DALY 1978).

A development of very great potential in the elucidation of the role of phosphodiesterases in neural processes is the isolation by KIGER and his colleagues (KIGER and GOLANTY 1977; KIGER 1977; DAVIS and KIGER 1978; BYERS et al. 1980) of *Drosophila* mutants with various deletions of the phosphodiesterase gene producing major behavioral changes in this organism. The extensive knowledge of the genome for *D. melanogaster* has allowed identification of two regions of the X chromosome (3 D 3, 3 D 4, and 90 E-91 B) which increase cyclic AMP phosphodiesterase activity when present as duplications. Two different regions of the X chromosome (5 D-9 C and 88 C-91 B) elevate cyclic GMP phosphodiesterase activity when present as duplications; however, a single locus is not responsible. These studies substantiate biochemical inferences of the presence of at least two distinct and separate phosphodiesterase enzymes by demonstrating the existence of two genes controlling hydrolytic activity of cyclic AMP and cyclic GMP in the fruit fly. The occurrence of deficiencies for the cyclic AMP phosphodiesterase locus and its presence in insertional translocations makes it possible to obtain sibs with different gene dosages and enzyme activities by segregation within a single stock. A genetic study of adult *Drosophila* nullosomic for chromomere 3 D 4 demonstrated that this region of the genome is also necessary for male fertility, normal female fertility, and normal oogenesis. In addition, BENZER and his co-workers (personal communication) have shown that *Drosophila* with this defect are unable to learn an avoidance task and the strain has been named "Dunce." This system provides the opportunity to study the phenotypic result of a deletion of the gene controlling phosphodiesterase activity.

F. Prognosis

Future progress in the field of phosphodiesterase research will undoubtedly benefit from the powerful tools of immunology, biochemistry and molecular biology which have already begun to show great promise in other areas. Although the isolation and characterization of the forms of phosphodiesterase in various tissues has made an important contribution to this field, future efforts should tend away from the simple cataloguing of these activities.

Reports of the complete purification of the major forms of phosphodiesterase are just now beginning to emerge. Analysis of the purified enzymes should facilitate a more complete understanding of the regulation of their activities. The availability of purified enzymes will also allow the production of specific antibodies which will provide a whole new arsenal of techniques for the study of these enzymes: analysis of the immunological cross-reactivities of the various forms of phosphodiesterase may help to settle the question of whether the enzymes have similar regions or common subunits; fluorescent-labelled antibodies will allow the localization of specific forms of phosphodiesterase without the uncertainties which encumber other methods. The production of monoclonal antibodies to specific regions the enzyme molecules may allow a "molecular dissection" of their catalytic and allosteric sites.

The application of genetic analysis to the phosphodiesterase field is still in its early infancy. Studies on mutants with deletions or modifications of the genes for specific forms of phosphodiesterase hold great promise for unravelling the complex role of these enzymes in cellular metabolism. To date, relatively little use has been made of the S-49 system in examining phosphodiesterases. In the distant future, the production of cDNA probes to phosphodiesterase genes may provide the keys to the understanding of the induction of these enzymes in transformed or malignant cells.

Some problems relating to the elucidation of the true role of the phosphodiesterases seem less tractable. The determination of the in vivo availability of the potential substrates and inhibitors of these enzymes is such a problem. The ultimate solution to this problem will require a detailed knowledge of the subcellular distribution of the cyclic nucleotides and many of the other components of this regulatory system.

References

Adachi K, Levine V, Halprin KM, Iizuka H, Yoshikawa K (1976) Multiple forms of cyclic nucleotide phosphodiesterase in pig epidermis. Biochim Biophys Acta 429:498–507

Adinolfi AM (1972) The organization of the paramembranous densities during postnatal maturation of synaptic junctions in the cerebral cortex. Exp Neurol 34:383–393

Adinolfi AM, Schmidt SY (1974) Cytochemical localization of cyclic nucleotide phosphodiesterase activity at developing synapses. Brain Res 76:21–31

Allan EA, Sneyd JGT (1975) Activation of cyclic AMP phosphodiesterase of isolated rat hepatocytes by insulin. Biochem Biophys Res Commun 62:594–601

Amer MS, Kriegbaum WE (1975) Cyclic nucleotide phosphodiesterase: properties activators, inhibitors, structure-activity relationships, and possible role in drug development. J Pharmaceut Sci 64:1–37

Appleman MM, Terasaki WL (1975) Regulation of cyclic nucleotide phosphodiesterases. Adv Cyclic Nucleotide Res 5:153–168

Armstrong KJ, Stouffer JE (1974) Effects of thyroid hormone deficiency on cyclic adenosine 3′,5′-monophosphate and control of lipolysis in fat cells. J Biol Chem 249:4226–4231

Ariano MA, Adinolfi AM (1977) Subcellular localization of cyclic nucleotide phosphodiesterase in the caudate nucleus. Exp Neurol 55:84–94

Ariano MA, Appleman MM (1979) Biochemical characterization of postsynaptically localized cyclic nucleotide phosphodiesterase. Brain Res 177:301–309

Ashton AR, Polya GM (1978) The specific interaction of cibacron and related dyes with cyclic nucleotide phosphodiesterase and lactate dehydrogenase. Biochem J 175:501–506

Azhar S, Menon KMJ (1977) Cyclic nucleotide phosphodiesterases from rat anterior pituitary: characterization of multiple forms and regulation by protein activator and Ca^{+2}. Eur J Biochem 73:73–82

Bacalao J, Rieber M (1972) Modified distribution of some cyclic AMP PDE activities following growth of L cell fibroblasts in presence of prostaglandin E_1. Exp Cell Res 75:518–521

Beavo JA, Hardman JG, Sutherland EW (1970) Hydrolysis of cyclic guanosine and adenosine 3′,5′-monophosphate by rat and bovine tissues. J Biol Chem 245:5649–5655

Beavo JA, Hardman JE, Sutherland EW (1971) Stimulation of adenosine 3′,5′-monophosphate hydrolysis by guanosine 3′,5′-monophosphate. J Biol Chem 246:3841–3846

Beer B, Chasin M, Clody DE, Vogel JR, Horovitz ZP (1972) Cyclic adenosine monophosphate phosphodiesterase in brain: effect on anxiety. Science 176:428–430

Bergstrand H, Lundquist B, Schurmann A (1978) Rat mast cell high affinity cyclic nucleotide phosphodiesterase; separation and inhibitory effects of two antiallergic agents. Mol Pharmacol 14:848–855
Bergstrand H, Lundquist B (1978) Partial purification and characterization of cyclic nucleotide phosphodiesterases from human bronchial tissue. Mol Cell Biochem 21:9–21
Boudreau RJ, Drummond GI (1975) The effect of Ca^{+2} on cyclic nucleotide phosphodiesterases of superior cervical ganglion. J Cyclic Nucleotide Res 1:219–228
Breckenridge B McL, Johnston RE (1969) Cyclic 3′,5′-nucleotide phosphodiesterase in brain. J Histochem Cytochem 17:505–511
Butcher RW, Sneyd JGT, Park CR, Sutherland EW (1966) Effect of insulin on adenosine 3′,5′-monophosphate in the rat epididymal fat pad. J Biol Chem 241:1651–1653
Butcher RW, Sutherland E (1962) Adenosine 3′,5′-phosphate in biological materials. J Biol Chem 237:1244–1250
Byers D, Davis RL, Kiger JA Jr. (1981) Defect in cyclic AMP phosphodiesterase due to the dunce mutation of learning in Drosophila melanogaster. Nature 289:79–81
Carpendo F, Gaion RM, Fassima G (1975) Calcium and papaverine interaction with soluble cardiac phosphodiesterase. Biochem Pharm 24:2069–2073
Chasin M, Harris DN (1976) Inhibitors and activators of cyclic nucleotide phosphodiesterase. Adv Cyclic Nucleotide Res 7:225–264
Cheng Y, Bloch A (1978) Demonstration, in leukemia L-1210 cells, of a phosphodiesterase acting on 3′,5′-cyclic CMP but not on 3′,5′-cyclic GMP. J Biol Chem 253:2522–2524
Cheung WY (1970) Cyclic 3′,5′-nucleotide phosphodiesterase, demonstration of an activator. Biochem Biophys Res Commun 38:533–538
Cheung WY (1967) Properties of cyclic 3′,5′-nucleotide phosphodiesterase from rat brain. Biochemistry 6:1079–1087
Cheung WY, Salganicoff L (1967) Cyclic 3′,5′-nucleotide phosphodiesterases: localization and latent activity in rat brain. Nature 214:90–91
Clark MG, Jarrett IG (1978) Responsiveness to glucagon by isolated rat hepatocytes controlled by the redox state of the cytosolic nicotinamide adenine dinucleotide couple acting on adenosine 3′,5′-cyclic monophosphate phosphodiesterase. Biochem J 176:805–816
Coffino P, Bourne HR, Friedrich U, Hochman J, Insel PA, Lemarie I, Melmon ML, Tomkins GM (1976) Molecular mechanisms of cyclic AMP action: a genetic approach. Recent Prog Horm Res 32:669–685
Coulson AS, Kennedy LA (1971) Lymphocyte membrane enzymes. II. Cyclic 3′,5′-adenosine monophosphatase located on unstimulated human small lymphocyte nuclear membranes. Blood 38:485–490
Dalton C, Crowley HJ, Sheppard H, Schallek W (1974) Regional cyclic nucleotide phosphodiesterase activity in cat central nervous system: effects of benzodiazepines. Proc Soc Exp Biol Med 145:407–410
Daly JW (1977) Cyclic nucleotides in the nervous system. Plenum Press, New York
Das I (1973) Effects of diabetes and insulin on the rat heart adenyl cyclase, cyclic AMP phosphodiesterase and cyclic AMP. Horm Metab Res 5:330–339
D'Armiento M, Johnson GS, Pastan I (1972) Regulation of adenosine 3′,5′-cyclic monophosphate phosphodiesterase activity in fibroblasts by intracellular concentrations of cAMP. Proc Natl Acad Sci USA 69:459–462
Davis CW, Daly JW (1978) Calcium-dependent 3′,5′-cyclic nucleotide phosphodiesterase. Inhibition of basal activity at physiological levels of potassium ions. J Biol Chem 253:8683–8686
Davis CW, Kuo JK (1976) Ontogenetic changes in levels of phosphodiesterase for adenosine 3′,5′-monophosphate in the lung, liver, brain, and heart from guinea pigs. Biochim Biophys Acta 444:554–562
Davis RL, Kiger JA (1978) Genetic manipulation of cyclic AMP levels in *Drosophila melanogaster*. Biochem Biophys Res Commun 81:1180–1186
DeRobertis E, DeLorez-Arnaiz GR, Butcher RW, Sutherland EW (1967) Subcellular distribution of adenyl cyclase and cyclic phosphodiesterase in rat brain cortex. J Biol Chem 242:3487–3493

DeRubertis FR, Zensen TV, Curnow RT (1974) Inhibition of glucagon-mediated increases in hepatic cyclic AMP by prostaglandin E_1 and E_2. Endocrinology 95:93–101
Desai K, Marion M, Appleman MM (1976) Conconavalin A stimulation of low Km cyclic AMP phosphodiesterase of rat adipocytes – an insulin-like effect. Fed Proc 35:1438
Dolby TW, Kleinsmith LJ (1977) Cannabinoid effects on adenylate cyclase and phosphodiesterase activities of mouse brain. Can J Physiol Pharm 55:934–942
Donnelly JE (1978) Effects of zinc chloride on the hydrolysis of cyclic GMP and cyclic AMP by the activator-dependent cyclic nucleotide phosphodiesterase from bovine heart. Biochim Biophys Acta 522:151–160
Drummond GI, Perrott-Yee S (1961) Enzymatic hydrolysis of adenosine 3′,5′-phosphoric acid. J Biol Chem 236:1126–1129
Duttagupta C, Rifas L, Makman MH (1978) Regulation of cyclic nucleotide phosphodiesterase activity in human lung fibroblasts. Biochim Biophys Acta 523:385–394
Egrie JC, Campbell JA, Flangas AL, Siegel FL (1977) Regional, cellular, and subcellular distribution of calcium-activated cyclic nucleotide phosphodiesterase and calcium-dependent regulator in porcine brain. J Neurochem 28:1207–1213
Epstein PM, Mills JS, Ross CP, Strada SJ, Hersh EM, Thompson WJ (1977) Increased cyclic nucleotide phosphodiesterase activity associated with proliferation and cancer in human and murine lymphoid cells. Cancer Res 37:4016–4023
Epstein PM, Pledger WJ, Gardner EA, Stancel GM, Thompson WJ, Strada SJ (1978) Activation of mammalian cyclic AMP phosphodiesterases by trypsin. Biochim Biophys Acta 527:442–455
Ferrendelli JA, Rubin EH, Kinscherf DA (1976) Influence of divalent cations on regulation of cyclic GMP and cyclic AMP levels in brain tissue. J Neurochem 26:741–748
Filburn CR, Colpo F, Sacktor B (1978) Regulation of cyclic nucleotide phosphodiesterases of cerebral cortex by Ca^{+2} and cGMP. J Neurochem 30:337–346
Florendo NT, Barrnett RJ, Greengard P (1971) Cyclic 3′,5′-nucleotide phosphodiesterase: cytochemical localization in cerebral cortex. Science 171:745–747
Florendo NT, Pitcock JA, Muirhead EE (1978) Cyclic 3′,5′-nucleotide phosphodiesterase; cytochemical localization in rat renomedullary interstitial cells. Phosphodiesterase in rat renal medulla. J Histochem Cytochem 26:441–451
Franks DJ, MacManus JP (1971) Cyclic GMP stimulation and inhibition of cyclic AMP phosphodiesterase from thymic lymphocytes. Biochem Biophys Res Commun 42:844–849
Fredholm B (1977) Decreased adenosine 3′,5′-monophosphate phosphodiesterase activity in rat striatum following chromic haloperidol. Med Biol 55:61–69
Gaballah S, Popoff C (1971) Cyclic 3′,5′-nucleotide phosphodiesterase in nerve endings of developing rat brain. Brain Res 25:220–222
Gain KR, Appleman MM (1978) Regulation of the phosphodiesterases of muscle tissue. Adv Cyclic Nucleotide Res 9:221–232
Gaion RM, Krishna G (1979) Cytidylate cyclase: the product isolated by the method of Cech and Ignarro is not cytidine 3′,5′-monophosphate. Biochem Biophys Res Commun 86:105–111
Gengler WR, Forte LR (1972) Neonatal development of rat kidney adenyl cyclase and phosphodiesterase. Biochim Biophys Acta 279:367–372
Gnegy ME, Costa E, Uzunov P (1976a) Regulation of transsynaptically elicited increase of 3′,5′-cyclic AMP by endogenous phosphodiesterase activator. Proc Natl Acad Sci USA 73:352–355
Gnegy ME, Costa E, Uzunov P (1976b) Release of phosphodiesterase activator (PDEA) from subcellular fractions of rat brain. Fed Proc 35:609
Goldberg ND, Haddox MK, Dunham E, Lopez E, Hadden W (1973) The Yin Yang hypothesis of biological control: opposing influences of cyclic GMP and cyclic AMP in the regulation of cell proliferation and other processes. In: Clarkson B, Baserga R (eds) Control of proliferation in animal cells. Cold Spring Harbor Laboratory, pp 609–625
Gomori G (1939) Microtechnical demonstration of phosphatase in tissue sections. Proc Soc Exp Biol Med 42:23–26
Grab DJ, Berzins K, Cohen RS, Siekevitz P (1979) Presence of calmodulin in postsynaptic densities isolated from canine cerebral cortex. J Biol Chem 254:8690–8696

Greenberg LH, Troyer E, Ferrendelli JA, Weiss B (1978) Enzymatic regulation of the concentration of cyclic GMP in mouse brain. Neuropharmacology 17:737–745
Gulyassy PF, Farrand JR, Gugler HD (1975) Multiple cyclic nucleotide phosphodiesterases in rat kidney. Kidney Int 8:284–291
Hait WN, Weiss B (1976) Increased cyclic nucleotide phosphodiesterase activity in leukaemic lymphocytes. Nature 259:321–323
Hait WN, Weiss B (1977) Characteristics of the cyclic nucleotide phosphodiesterases of normal and leukemic lymphocytes. Biochim Biophys Acta 497:86–100
Harris DN, Chasin M, Phillips MB, Goldenberg H, Samaniego S, Hess SM (1973) Effects of cyclic nucleotides on activity of cyclic 3′,5′-adenosine monophosphate phosphodiesterase. Biochem Pharmacol 22:221–228
Helfman DM, Brackett NL, Kuo JF (1975) Depression of cytidine 3′,5′-cyclic monophosphate phosphodiesterase activity in development tissues of guinea pigs. Proc Natl Acad Sci USA 75:4422–4425
Hemmington J, Dunn AS, Chenoweth ME (1973) Cyclic nucleotide phosphodiesterase activity in the plasma and erythrocytes of normal and diabetic rats. Biochim Biophys Acta 304:552–559
Hepp KD, Renner R (1972) Insulin action on the adenyl cyclase system: antagonism to activation by lipolytic hormones. Fed Eur Biochem Soc Letters 20:191–194
Hidaka H, Asano T (1976) Human blood platelet 3′,5′-cyclic nucleotide phosphodiesterase. Biochim Biophys Acta 429:485–497
Hidaka H, Yamaki T, Yamabe H (1978) Two forms of Ca^{++} dependent cyclic 3′,5′-nucleotide phosphodiesterase from human aorta and effect of free fatty acids. Arch Biochem Biophys 187:315–321
Honda R, Imamura H (1968) Inhibition of cyclic 3′,5′-nucleotide phosphodiesterase by phenothiazine and reserpine derivatives. Biochim Biophys Acta 161:267–269
House PDR, Poulis P, Weideman J (1972) Isolation of a plasma-membrane subfraction from rat liver containing an insulin sensitive cyclic AMP phosphodiesterase. Eur J Biochim 24:429–437
Howlett AC, Breckenridge B McL (1978) The distribution of cyclic nucleotide phosphodiesterase in the hypothalamus. J Neurochem 31:321–327
Huang X-C, Kemp RG (1971) Properties of a phosphodiesterase with high affinity for adenosine 3′,5′-cyclic phosphate. Biochemistry 10:2278–2283
Ignarro LJ, Kossikoff N, Slywka J (1973) Release of enzymes from a rat liver lysosome fraction. J Pharm Exp Ther 186:86–99
Ilien B, Stierle A, Lughier C, Stoclet JG, Landry Y (1978) Separation of three cyclic nucleotide phosphodiesterases from bovine aorta. Biochem Biophys Res Commun 83:486–492
Indiveri F, Barabino A, Fontana R, Polleri A (1974) Thyrotropin effect on cyclic 3′,5′-nucleotide phosphodiesterase intracellular distribution. Horm Res 5:65–75
Insel PA, Bourne HR, Coffino P, Tomkins GM (1975) Cyclic AMP-dependent protein kinases: pivotal role in regulation of enzyme induction and growth. Science 190:896–898
Jarrett L, Seals JR (1979) Pyruvate dehydrogenase activation in adipocyte mitochondria by an insulingenerated mediator from muscle. Science 206:1407–1408
Jarrett L, Smith RM (1974) Mode of action of N^6-dibutyryl cyclic 3′,5′-monophosphate on fat cell metabolism. Diabetes 23:29–40
Jefferson LS, Exton JH, Butcher RW, Sutherland EW, Park CR (1968) Role of adenosine 3′,5′-monophosphate in the effects of insulin and anti-insulin serum on liver metabolism. J Biol Chem 243:1031–1038
Kakiuchi S, Yamazaki R (1970) Calcium-dependent phosphodiesterase activity and its activating factor (PAF) from brain: studies on cyclic 3′,5′-nucleotide phosphodiesterase. Biochem Biophys Res Commun 41:1104–1110
Kakiuchi S, Yamazaki R, Teshima Y, Uenishi K (1973) Regulation of nucleotide cyclic 3′,5′-monophosphate phosphodiesterase activity from rat brain by a modulator and Ca^{+2}. Proc Natl Acad Sci USA 70:3526–3535
Kakiuchi S, Yamazaki R, Teshima Y, Uenishi K, Miyamoto Ca^{+2}/Mg^{+2}-dependent cyclic nucleotide phosphodiesterase and its activator protein. Adv Cyclic Nucleotide Res 5:163–178

Kalderon AE, Ravanshenas SF (1974) Localization of 3′,5′-cyclic nucleotide phosphodiesterase activity in isolated thyroid cells and intact thyroid tissue. Histochemistry 39:229–242

Kanamori R, Creveling CR, Daly JW (1979) Calcium-dependent cyclic nucleotide phosphodiesterase. Inhibition of basal activity by heat-stable factors from rat cerebrum. Biochim Biophys Acta 582:434–447

Kaplan JC, Pachard AL, Laudat MH, Laudat P (1973) Kinetic and electrophoretic abnormality of cyclic AMP phosphodiesterase in genetically obese mouse adipocytes. Biochem Biophys Res Commun 51:1009–1014

Kiger JA, jr (1977) The consequences of nullosomy for a chromosomal region affecting cyclic AMP phosphodiesterase activity in *Drosophila*. Genetics 85:623–628

Kiger JA, jr, Golanty E (1977) A cytogenetic analysis of cyclic nucleotide phosphodiesterase activities in *Drosophila*. Genetics 85:609–622

Klee CB (1977) Conformational transition accompanying the binding of Ca^{+2} to the protein activator of 3′,5′-cyclic adenosine monophosphate phosphodiesterase. Biochemistry 16:1017–1024

Klee CB, Krinks MH (1978) Purification of cyclic 3′,5′-nucleotide phosphodiesterase inhibitory protein by affinity chromatography in activator protein coupled to sepharose. Biochemistry 17:120–126

Klotz U, Stock K (1972) Influence of cyclic guanosine-3′,5′-monophosphate on the enzymatic hydrolysis of adenosine-3′,5′-monophosphate. Naunyn-Schmiedebergs Arch Pharmacol 274:54–62

Kono T, Robinson FW, Sarver JA (1975) Insulin sensitive phosphodiesterase. J Biol Chem 250:7826–7835

Kono T, Robinson FW, Sarver JA, Vega EV, Pointer RH (1977) Actions of insulin in fat cells. J Biol Chem 252:2226–2233

Kumar S, Becker G, Prasad KN (1975) Cyclic adenosine 3′,5′-monophosphate phosphodiesterase activity in malignant and cyclic adenosine 3′,5′-monophosphate induced "differentiated" neuro blastoma cells. Cancer Res 35:82–87

Kuo JF, Brackett NL, Shoji M, Tse J (1978) Cytidine 3′:5′-monophosphate phosphodiesterase in mammalian tissues. J Biol Chem 253:2518–2521

Larner J, Galasko G, Cheng K, DePaoli-Roach AH, Huang L, Daggy P, Kellogg J (1979) Generation by insulin of a chemical mediator that controls protein phosphorylation and dephosphorylation. Science 206:1408–1410

Leichter SB, Poffenbarger PL (1978) The effects of NSILP on cyclic nucleotide metabolism in liver. Biochem Biophys Res Commun 84:403–410

Levin RM, Weiss B (1978a) Characteristics of the cyclic nucleotide phosphodiesterases in a transplantable pheochromocytoma and adrenal medulla of the rat. Cancer Res 38:915–920

Levin RM, Weiss B (1978b) Specificity of the binding of trifluoperazine to the calcium-dependent activator of phosphodiesterase and to a series of other calcium-binding proteins. Biochim Biophys Acta 540:197–204

Levin RM, Weiss B (1979) Selective binding of antipsychotics and psychoactive agents to the calcium-dependent activator of cyclic nucleotide phosphodiesterase. J Pharm Exp Ther 208:454–459

Lin S, Riggs AD (1972) LAC operator analogues: bromo-deoxyuridine substitution in the LAC operator affects the rate of dissociation of the LAC repressor. Proc Natl Acad Sci USA 69:2574–2576

Lin YM, Liu YP, Cheung WY (1974) Cyclic 3′,5′-nucleotide phosphodiesterase. Purification, characterization, and active form of the protein activator from bovine brain. J Biol Chem 249:4943–4954

Londesborough J (1977) Characterization of an adenosine, 3′:5′-cyclic monophosphate phosphodiesterase from baker's yeast. Biochemical J 163:467–476

Loten EG, Sneyd JGT (1970) Activation of adipose tissue cyclic 3′:5′-nucleotide phosphodiesterase by insulin. Proc Univ Otago Med School 48:44–45

Loten EG, Assimacopoulis-Jeannet FD, Exton JH, Park CR (1978) Stimulation of a low Km phosphodiesterase from liver by insulin and glucagon. J Biol Chem 253:746–757

Loten EG, Francis S, Corbin JD (1980) Proteolytic solubilization and modification of hormone sensitive cyclic nucleotide phosphodiesterase. J Biol Chem 255:7838–7849
Lovell-Smith CJ, Manganiello V, Vaughan M (1977) Solubilization and characterization of hormone responsive phosphodiesterase activity of rat fat cells. Biochim Biophys Acta 497:447–458
Manganiello V, Vaughan M (1972) Prostaglandin E_1 effects on adenosine 3′:5′-cyclic monophosphate concentration and phosphodiesterase activity in fibroblasts. Proc Natl Acad Sci USA 69:269–273
Manganiello V, Vaughan M (1973) An effect of insulin cyclic adenosine 3′:5′-monophosphate phosphodiesterase activity in fat cells. J Biol Chem 248:7164–7170
Marchmont R, Houslay M (1980a) A peripheral and an intrinsic enzyme constitute the cyclic AMP phosphodiesterase activity of rat liver plasma membranes. Biochem J 187:381–392
Marchmont R, Houslay M (1980b) Insulin triggers cyclic AMP-dependent activation and phosphorylation of a plasma membrane cyclic AMP phosphodiesterase. Nature 286:904–907
Marchmont R, Ayad S, Houslay M (1981) Purification and properties of the insulin-stimulated cyclic AMP phosphodiesterase from rat liver plasma membranes. Biochem J 195:645–652
Miki N, Yoshida H (1972) Purification and properties of cyclic AMP phosphodiesterase from rat brain. Biochim Biophys Acta 268:166–174
Minneman KP (1976) Cyclic nucleotide phosphodiesterase in rat neostriatum: properties of the enzyme in crude homogenates. J Neurochem 27:1173–1180
Miyake M, Daly JW, Creveling CR (1977) Purification of calcium-dependent phosphodiesterases from rat cerebrum by affinity chromatography on activator-protein sepharose. Arch Biochem Biophys 181:39–45
Monod J, Wyman J, Changeux JP (1965) On the nature of allosteric transitions: a plausible model. J Mol Biol 12:88–118
Morill ME, Thompson ST, Stellwagen E (1979) Purification of a cyclic nucleotide phosphodiesterase from bovine brain using blue dextran sepharose chromatography. J Biol Chem 254:4371–4374
Morishima I (1975a) Cyclic nucleotide phosphodiesterase in silkworm. Developmental changes of cyclic AMP and cyclic GMP phosphodiesterase. Biochim Biophys Acta 403:106–112
Morishima I (1975b) Cyclic nucleotide phosphodiesterase in silkworm. Characterization of cyclic GMP phosphodiesterase. Biochim Biophys Acta 410:310–317
Moss J, Manganiello VC, Vaughan M (1977) Substrate and effector specificity of a guanosine 3′:5′-monophosphate phosphodiesterase from rat liver. J Biol Chem 252:5211–5215
Murray T, Russell TR (1978) Effect of 5-bromodeoxyuridine on the induction of adenosine 3′,5′-monophosphate phosphodiesterase in 3T3-L fibroblasts. Arch Biochem Biophys 190:705–711
Pardos GJ, Lentz TL (1976) Cytochemical localization of cyclic 3′,5′-nucleotide phosphodiesterase activity in the neuromuscular junction and skeletal muscle of the newt. Brain Res 107:355–364
Phelps BH, Varandini PI (1977) Stimulation of insulin of cyclic AMP phosphodiesterase: role of glutathione insulin transhydrogenase. Biochem Biophys Res Commun 75:302–310
Pichard AL, Cheung WY (1976) Cyclic 3′:5′-nucleotide phosphodiesterase interconversion of multiple forms and their effects on enzyme activity and kinetics. J Biol Chem 251:5726–5737
Pichard AL, Hanoune J, Kaplan JC (1973) Multiple forms of cyclic adenosine 3′,5′-monophosphate phosphodiesterase of human blood platelets. Biochim Biophys Acta 315:370–377
Pilkis SJ, Exton JH, Johnson RA, Park CR (1974) Effects of glucagon on cyclic AMP and carbohydrate metabolism in livers from diabetic rats. Biochim Biophys Acta 343:250–267

Pledger WJ, Stancel GM, Thompson WJ, Strada SJ (1974) Separation of multiple forms of cyclic nucleotide phosphodiesterases from rat brain by isoelectric focusing. Biochim Biophys Acta 370:242–248

Rhoads AR, Morris HP, West WL (1972) Cyclic 3′,5′-nucleotide monophosphate PDE activity in hepatoma of different growth rates. Cancer Res 32:2651–2655

Richman PG, Klee CB (1978) Interaction of ^{125}I-labelled Ca^{+2}-dependent regulator protein with cyclic nucleotide phosphodiesterase and its inhibitory protein. J Biol Chem 253:6323–6326

Rillema JD, Kostyo JL, Gimpel LP (1973) Inhibition of metabolic effects of growth hormone by various inhibitors of cyclic nucleotide phosphodiesterase. Biochim Biophys Acta 297:527–539

Robb RM (1974) Histochemical evidence of cyclic nucleotide phosphodiesterase in photoreceptor outer segments. Invest Ophthal 13:740–747

Ross PS, Manganiello VC, Vaughan M (1977) Regulation of cyclic nucleotide phosphodiesterase in cultured hepatoma cells by dexamethasone and N^6,O^2-dibutyryl adenosine 3′:5′-monophosphate. J Biol Chem 252:1448–1452

Ruckstuhl M, Beretz A, Anton R, Landry Y (1979) Flavonoids are selective cyclic GMP phosphodiesterase inhibitors. Biochem Pharm 28:535–538

Russell TR, Pastan I (1973) Plasma membrane cyclic AMP phosphodiesterase of cultured cells and its modification after trypsin treatment. J Biol Chem 248:5835–5840

Russell TR, Pastan I (1974) Cyclic adenosine 3′:5′-monophosphate and cyclic guanosine 3′:5′-monophosphate phosphodiesterase activites are under separate genetic control. J Biol Chem 249:7764–7769

Russell TR, Terasaki WJ, Appleman MM (1973) Separate phosphodiesterases for the hydrolysis of 3′,5′-cyclic adenosine monophosphate and 3′,5′-cyclic guanosine monophosphate in rat liver. J Biol Chem 248:1334–1340

Russell TR, Thompson WJ, Schneider RW, Appleman MM (1972) 3′:5′-cyclic adenosine monophosphate phosphodiesterase: negative cooperativity. Proc Natl Acad Sci USA 9:1791–1795

Sapag-Hagar M, Greenbaum AL (1973) Changes in the activity of adenyl cyclase and cyclic AMP phosphodiesterase and in the levels of 3′-5′ cyclic AMP in rat mammary gland during pregnancy and lactation. Biochem Biophys Res Commun 53:982–987

Sakai T, Thompson WJ, Lavis VR, Williams RH (1974) Cyclic nucleotide phosphodiesterase activities from isolated fat cells: correlation of subcellular distribution with effects of nucleotides and insulin. Arch Biochem Biophys 162:331–339

Sakai T, Yamanaka H, Tanaka R, Makino H, Kasai H (1977) Stimulation of cyclic nucleotide phosphodiesterases from rat brain by activator protein, proteolytic enzymes and a vitamin E derivative. Biochim Biophys Acta 483:121–134

Sakai T, Makino H, Tanaka R (1978) Increased activity of cyclic AMP phosphodiesterase from frozen and thawed rat liver. Biochim Biophys Acta 522:477–484

Sasaki Y, Kodaira R, Nozawa R, Yokota T (1978) Inhibitory effect of 8-substituted adenosine derivaties on Ca^{++} and modulator protein-dependent phosphodiesterase activity. Biochem Biophys Res Commun 84:277–284

Schmidtke J, Wienker T, Flugel M, Engel W (1976) In vitro inhibition of cyclic AMP phosphodiesterases by cortisol. Nature 262:593–594

Schonhofer PA, Skidmore IF, Paul MI, Ditzion BR, Pauk GL, Krishna G (1972) Effects of glucocorticoids on adenyl cyclase and phosphodiesterase activity in fat cell homogenates. Naunyn-Schmiedebergs Arch Pharm 273:257–282

Schroder J, Plagemann PGW (1972) Cyclic 3′,5′-nucleotide phosphodiesterase of Novikoff rat hepatoma, mouse L, and HeLa cells growing in suspension culture. Cancer Res 32:1082–1087

Schultz J (1974) Inhibition of 3′,5′-nucleotide phosphodiesterase in guinea pig cerebral cortex slices. Archives Biochem Biophys 163:15–20

Schwabe U, Miyaki M, Ohga Y, Daly JW (1976) 4-(3-cytopentyloxy-4-methoxyphenyl)-2-pyrrolidone (ZK 62711): a potent inhibitor of cyclic AMP-phosphodiesterases in homogenates and tissue slices from rat brain. Mol Pharm 12:900–910

Senft G, Schultz G, Munske G, Hoffman M (1968) Effects of glucocorticoids and insulin of 3′,5′ AMP phosphodiesterase activity in adrenalectomized rats. Diabetologia 4:330

Shanta TR, Woods WD, Waitzman MB, Bourne GH (1966) Histochemical method for localization of cyclic 3′,5′-nucleotide phosphodiesterase. Histochemie 7:177–190

Sheppard H, Tsien WH (1974) The effects of slow freezing and thawing and hypertonic NaCl on the hydrolytic activity of rat erythrocyte cyclic AMP phosphodiesterase and its sensitivity to the inhibitor D,L 4-(3-butoxy-4-methoxybenzyl)-2-imidizolidinone. Biochim Biophys Acta 341:489–495

Sheppard H, Wiggan G, Tsun WH (1972) Structure-activity relationships for inhibitors of phosphodiesterase from erythrocytes and other tissues. Adv Cyclic Nucleotide Res 1:103–112

Sher NS, Quagliata F, Malathi VG, Faig D, Melton FA, Silber R (1976) Cyclic adenosine 3′:5′-monophosphate phosphodiesterase activity in normal and chronic lymphocytic leukemia lymphocytes. Cancer Res 36:3958–3962

Shoji M, Brackett NL, Kuo JF (1978) Cyclic CMP phosphodiesterase decreased activity in regenerating and developing liver. Science 202:826–828

Sicard RE, Aprille JR (1977) Adenylate cyclase and cyclic nucleotide phosphodiesterases in the developing rat liver. Biochim Biophys Acta 500:235–245

Singer AL, Sherwin RP, Dunn AS, Appleman MM (1976) Cyclic nucleotide phosphodiesterases in nonneoplastic human mammary tissues. Cancer Res 36:60–66

Singer AL, Dunn A, Appleman MM (1978) Cyclic nucleotide phosphodiesterase and protein activator in fetal rabbit tissues. Arch Biochem Biophys 187:406–413

Skidmore IF, Schönhöfer PS, Kritchevsky D (1971) Effects of nicotinic acid and some of its homologues on lipolysis, adenyl cyclase, phosphodiesterase, and cyclic AMP accumulation in isolated fat cells. Pharmacology 6:330–342

Smoake JA, Song S-Y, Cheung WY (1974) Cyclic 3′,5′-nucleotide phosphodiesterase. Distribution and developmental changes of the enzyme and its protein activator in mammalian tissues and cells. Biochim Biophys Acta 34:402–411

Solomon SS (1975) Effect of insulin and lipolytic hormones on cyclic AMP phosphodiesterase in normal and diabetic rat adipose. Endocrinology 96:1336–1373

Solomon SS, Polazzolo M, King LE (1977) Cyclic nucleotide phosphodiesterase: insulin activation detected in adipose by gel electrophoresis. Diabetes 26:967–979

Strada SJ, Uzunov P, Weiss B (1974) Ontogenetic development of a phosphodiesterase activator and the multiple forms of cyclic AMP phosphodiesterase of rat brain. J Neurochem 23:1097–1104

Strewler GJ, Manganiello VC, Vaughan M (1978) Phosphodiesterase activators from rat kidney cortex. J Biol Chem 253:390–394

Sutherland EW, Rall TW (1958) Fractionation and characterization of a cyclic adenine ribonucleotide formed by tissue particles. J Biol Chem 232:1077–1091

Takemoto DJ, Lee W-NP, Kaplan SA, Appleman MM (1978) Cyclic AMP phosphodiesterase in human lymphocytes and lymphoblasts. J Cyclic Nucleotide Res 4:123–132

Takemoto DJ, Kaplan SA, Appleman MM (1979) Cyclic guanosine 3′,5′-monophosphate and phosphodiesterase in mitogen stimulated human lymphocytes. Biochem Biophys Res Commun 90:491–497

Tanigawa Y, Shimoyama M, Tai J, Fujii K, Ueda I (1976) Reciprocal changes in Ca^{+}/protein activator-dependent and -independent cyclic AMP phosphodiesterase during the development of chick embryos. Biochem Biophys Res Commun 73:19–24

Taniguchi TS, Fujiwara M, Lee J, Hidaka H (1978) Cyclic 3′:5′-nucleotide phosphodiesterase of rabbit sino-atrial node. Biochim Biophys Acta 552:465–473

Teo TS, Wang JH (1973) Mechanism of activation of a cyclic adenosine 3′,5′-monophosphate phosphodiesterase from bovine heart by calcium ions. Identification of the protein activator as a Ca^{+2} binding protein. J Biol Chem 248:5950–5966

Terasaki WL, Appleman MM (1975) The role of cyclic GMP in the regulation of cyclic AMP hydrolysis. Metabolism 24:311–319

Teshima Y, Yamazaki R, Kakiuchi S (1974) Effects of ATP on the activity of nucleoside 3′,5′-cyclic monophosphate phosphodiesterase from brain. J Neurochem 22:789–791

Thompson WJ, Appleman MM (1971) Multiple cyclic nucleotide phosphodiesterase activities from rat brain. Biochemistry 10:311–316
Thompson WJ, Epstein PM, Strada SJ (1979a) Purification and characterization of high affinity cyclic adenosine monophosphate phosphodiesterase from dog kindey. Biochemistry 14:5228–5237
Thompson WJ, Little SA, Williams RH (1973) Effect of insulin and growth hormone on rat liver cyclic nucleotide phosphodiesterase. Biochemistry 12:1889–1894PThompson WJ, Ross CP, Pledger WH, Strada SJ, Banner RL, Hersh EM (1976) Cyclic adenosine 3′:5′-monophosphate phosphodiesterase: distinct forms in human lymphocytes and monocytes. J Biol Chem 251:4922–4929
Thompson WJ, Terasaki WL, Epstein PM, Strada SJ (1979b) Assay of cylcic nucleotide phosphodiesterase and resolution of multiple forms of the enzyme. Adv Cyclic Nucleotide Res 10:69–92
Tria E, Scapin S, Cocco C (1976) Insulin-sensitive adenosine 3′,5′-cyclic monophosphate phosphodiesterase of hepatocyte plasma membranes. Biochim Biophys Acta 496:77–87
Uzunov P, Lehne R, Revuelta AV, Gnegy ME, Costa E (1976) A kinetic analysis of the cyclic nucleotide phosphodiesterase regulation by the endogenous protein activator. A study of rat brain and frog sympathetic ganglia. Biochim Biophys Acta 422:326–334
Uzunov P, Shein HM, Weiss B (1973) Cyclic AMP phosphodiesterase in cloned astrocytoma cells. Norepinephrine induces a specific enzyme form. Science 180:304–306
Uzunov P, Weiss B (1972) Separation of multiple molecular forms of cyclic adenosine 3′,5′-monophosphate phosphodiesterase in rat cerebellum by polyacrylamide gel electrophoresis. Biochim Biophys Acta 284:220–226
Van Inwegen RG, Robinson GA, Thompson WJ (1975) Cyclic nucleotide phosphodiesterases and thyroid hormones. J Biol Chem 250:2452–2456
Watterson DM, Vanaman TC (1976) Affinity chromatography purification of a cyclic nucleotide phosphodiesterase using immobilized modulator protein, a troponin-C like protein from brain. Biochem Biophys Res Commun 73:40–46
Wedner HJ, Chan BY, Parker CS, Parker CW (1979) Cyclic nucleotide phosphodiesterase activity in human peripheral blood lymphocytes. J Immunol 123:725–732
Weiss B (1975) Differential activation and inhibition of the multiple forms of cyclic nucleotide phosphodiesterase. Adv Cyclic Nucleotide Res 5:195–211
Weiss B (1971) Ontogenic development of adenyl cyclase and phosphodiesterases in rat brain. J Neurochem 18:469–477
Weiss B, Costa E (1968) Regional and subcellular distribution of adenyl cyclase and 3′,5′-cyclic nucleotide phosphodiesterase in brain and pineal. Biochem Pharmacol 17:2107–2116
Weiss B, Fertel R, Figlin R, Uzunov P (1974) Selective alteration of the activity of the multiple forms of adenosine 3′,5′-monophosphate phosphodiesterase of rat cerebrum. Mol Pharm 10:615–625
Weiss B, Winchurch RA (1978) Analysis of cyclic nucleotide phosphodiesterases in lymphocytes from normal and aged leukemic mice. Cancer Res 38:1274–1280
Wells JN, Hardman JG (1977) Cyclic nucleotide phosphodiesterases. Adv Cyclic Nucleotide Res 8:119–143
Wells JN, Wu XJ, Baird CE, Hardman JG (1975) Phosphodiesterases from porcine coronary arteries: inhibition of separated forms by xanthines papaverine and cyclic nucleotides. Mol Pharmacol 11:775–783
Wolff DJ, Brostrom CO (1974) Calcium-binding phosphoprotein from pig brain: identification as a calcium dependent regulator of brain cyclic nucleotide phosphodiesterase. Arch Biochem Biophys 163:349–358
Wood JG, Wallace R, Whitaker J, Cheung WY (1979) Studies of the localization of calmodulin and its binding protein in brain. Soc Neurosci 5:310–319
Zinman B, Hollenberg CH (1974) Effect of insulin and lipolytic agents on rat adipocyte low Km cyclic adenosine 3′:5′-monophosphate phosphodiesterase. J Biol Chem 249:2182–2187

CHAPTER 7

Calmodulin Regulation of Cyclic AMP Metabolism

W. Y. CHEUNG and D. R. STORM

Overview

Calmodulin is a relatively small regulatory protein which mediates many Ca^{2+} effects, including the stimulation of brain adenylate cyclase and Ca^{2+}-sensitive phosphodiesterase. In both cases, calmodulin forms a complex with the apoenzyme in the presence of Ca^{2+}. The Ca^{2+}-sensitive phosphodiesterase is a tetramer comprised of two catalytic and two calmodulin subunits. Brain may contain distinct forms of adenylate cyclase, one calmodulin-sensitive and another calmodulin-insensitive. Neither form has been vigorously characterized, and the relationship between the two forms remains to be established.

The stimulation of phosphodiesterase by calmodulin does not require any other cofactor. In contrast, stimulation of adenylate cyclase by calmodulin appears to require a regulatory subunit in addition to the catalytic subunit. Preliminary evidence suggests that this subunit is the guanyl nucleotide regulatory subunit. There are a number of interesting similarities between Ca^{2+} and hormone stimulation of adenylate cyclase. Both types of regulation are mediated by specific receptors that are distinct entities, and both apparently require a guanyl nucleotide regulatory subunit.

Although our understanding of the regulation of adenylate cyclase and phosphodiesterase by calmodulin has advanced significantly, it is still not possible to predict the net effect of Ca^{2+} fluxes on the intracellular concentrations of cyclic AMP or cyclic GMP. To do so would require detailed information on the absolute Ca^{2+} dependencies of the two enzymes, as well as the relative affinities of calmodulin for its other receptors.

A. Introduction

Since its discovery as an activator of cyclic 3′,5′-nucleotide phosphodiesterase in the late 1960's, calmodulin has been found to regulate a wide variety of specific enzymes in the presence of Ca^{2+}. Binding of Ca^{2+} induces a conformational change in calmodulin, allowing it to interact with a variety of enzyme systems. Calmodulin as a protein is unique; it has no known intrinsic enzyme activity, but has the ability to regulate the activity of numerous enzymes. The regulation of adenylate cyclase and cyclic nucleotide phosphodiesterase by calmodulin strongly suggests that the protein plays a pivotal role in coordinating cyclic nucleotide mediated events in response to fluctuations of intracellular Ca^{2+}. The purpose of this chapter is to review the general properties of calmodulin and its role in the regulation of adenylate cyclase and phosphodiesterase activities.

1 10 20 30
* * * * * *
Ac-A-D-Q-L-T-E-E-Q-I-A-E-F-K-E-A-F-S-L-F-D-K-D-G-N-G-T-I-T-T-K-E-L-G-T-V-M-R-

40 50 60 70
* * * * * *
S-L-G-Q-N-P-T-E-A-E-L-Q-D-M-I-N-E-V-D-A-D-G-N-G-T-I-D-F-P-E-F-L-T-M-M-A-R-

80 90 100 110
* * * * * *
K-M-K-D-T-D-S-E-E-E-I-R-E-A-F-R-V-F-D-K-D-G-N-G-Y-I-S-A-A-E-L-R-H-V-M-T-N

120 130 140
* * * * * *
L-G-E-K-L-T-D-F-F-V-D-F-M-I-R-E-A-N-I-D-G-D-G-E-V-N-Y-E-E-F-V-Q-M-M-T-A-K

Fig. 1. Complete amino acid sequence of bovine brain calmodulin. The molecule consists of 148 amino acids with an acetylated alanine at one end and lysine at the other. The asterisk (*) denotes putative Ca^{2+}-binding groups, which are arranged into 4 domains, each furnishing 1 Ca^{2+}-binding site. (From WATTERSON et al. 1980)

The abbreviations: *A*, alanine; *D*, aspartate; *E*, glutamate; *F*, phenylalanine; *G*, glycine; *H*, histidine; *I*, isoleucine, *K*, lysine; *L*, leucine; *M*, methionine; *N*, asparagine; *P*, proline; *Q*, glycine; *R*, arginine; *S*, serine, *T*, threonine; *V*, valine; *Y*, tyrosine. The lysine at position 115 is trimethylated.

B. Calmodulin

I. Properties of Calmodulin

Calmodulin is a calcium-binding protein with multiple functions, found in all eukaryotes yet examined (CHEUNG 1980a). It was discovered serendipitously as an activator of phosphodiesterase (CHEUNG 1967, 1969a, 1970a, b; for a detailed account, see CHEUNG 1980b), and was subsequently found to be a regulator of numerous key enzymes and cellular processes. Mounting evidence now indicates that the protein serves as a major mediator of Ca^{2+} actions (WOLFF and BROSTROM 1979; KLEE et al. 1980; WANG and WAISMAN 1979; CHEUNG 1980a, b).

Prominent features of calmodulin are its thermal stability and acidic nature (CHEUNG 1971). Calmodulin is a single polypeptide with 148 amino acids and a molecular weight of 16,700. The complete amino acid sequence of bovine brain calmodulin as elucidated by WATTERSON et al. (1980) is presented in Fig. 1. Some 30% of its amino acids consist of aspartic and glutamic acid residues, accounting for a pI of about 4.3. The protein contains no cysteine, hydroxyproline or tryptophan, but it does contain a trimethylated lysine at position 115. The lack of cysteine and hydroxyproline allows calmodulin to assume a highly flexible tertiary structure permitting it to interact with its receptor proteins – a feature highly desirable in view of its multiple functions. Because of its high ratio of phenylalanine (8 residues) to tyrosine (2 residues), calmodulin exhibits an unusual ultraviolet absorption spectrum with 5 peaks; 253, 259, 365, 269, and 272, and a shoulder at 282 – a spectrum characteristic of the fine structure of phenylalanine (LIU and CHEUNG 1976).

Additional striking features of calmodulin are its lack of tissue or species specificity (CHEUNG 1971) and its similar physical chemical properties to those of troponin c (STEVENS et al. 1976; WATTERSON et al. 1976), a calcium-binding protein involved in the regulation of skeletal and cardiac muscle contraction. More than 70% of the amino acid sequence in calmodulin appears homologous with that of troponin c. Indeed, bovine brain calmodulin substitutes effectively for troponin c in the activation of muscle actomyosin adenosine triphosphatase. However, troponin c does not substitute for calmodulin (AMPHLETT et al. 1976).

The primary structure of calmodulin appears to be well conserved throughout the phylogenetic scale. Calmodulin from different sources shows biological as well as immunological cross-reactivity. A comparison of the amino acid sequence of calmodulin from bovine brain with that from rat testes (DEDMAN et al. 1978) and bovine uterus (GRAND and PERRY 1978) indicates that the primary structure of the three proteins is virtually identical. As pointed out by VANAMAN and co-workers, calmodulin can be subdivided into four domains, each having a homologous sequence. The homology is greatest when the first domain, residues 8–40, is aligned with the third, residues 81–113, and the second domain, residues 44–76, is aligned with the fourth, residues 117–148 (WATTERSON et al. 1980).

On the basis of the crystal structure of parvalbumin and its amino acid sequence homology to troponin c, KRETSINGER and BARRY (1975) have predicted the three-dimensional structure of rabbit skeletal muscle troponin c and have identified in each domain the prospective calcium-binding residues; the corresponding residues in calmodulin are indicated in Fig. 1. The sequence immediately adjacent to these calcium-binding residues contains appropriate hydrophobic groups necessary for the helices that have been predicted to be essential for calcium-binding.

Attempts have been made to relate the structure and function of calmodulin. One approach has been chemical modification of certain selected groups. Modification of the single histidine residue or both tyrosine residues does not affect the biological properties of calmodulin, whereas derivatization of methione, lysine or arginine residue, or of carboxyl groups impaired the activity of calmodulin to various extents (WALSH and STEVENS 1977, 1978; RICHMAN and KLEE 1978; LIU and CHEUNG 1976). Another approach has been controlled tryptic digestion of calmodulin to peptides of various sizes (WALSH and STEVENS 1977). The results of these studies suggest that the biological activity of calmodulin depends heavily on the integrity of the entire polypeptide. Conceivably, the role of calmodulin in cellular function is of such fundamental importance that a mutation resulting in a gross change of its primary sequence may be lethal to the cell.

Calmodulins isolated from widely divergent species have similar, if not identical, amino acid sequences and share several basic characteristics. Each calmodulin contains a trimethyllysine[1] and four calcium-binding sites with high affinities; in the presence of calcium, the protein undergoes a conformational change, assuming a more helical structure which is stable during electrophoresis in alkaline polyacrylamide gel containing 8 M urea. Further, each calmodulin forms a complex with troponin I and the complex is also stable under similar conditions. In addition, each calmodulin stimulates Ca^{2+}-dependent phosphodiesterase and the stimulation is blocked by trifluoperazine, a phenylthiazine derivative often used as an antipsychotic agent (CHEUNG 1980a). Calmodulin binds 2 moles of trifluoperazine with a dissociation constant of 10^{-6} M. Upon binding trifluoperazine, calmodulin becomes biologically inactive (LEVIN and WEISS 1976, 1977).

The high affinity of trifluoperazine for calmodulin has been cleverly exploited to devise a highly specific affinity column for the isolation of calmodulin from tis-

[1] Please see *note added in proof* (p 323).

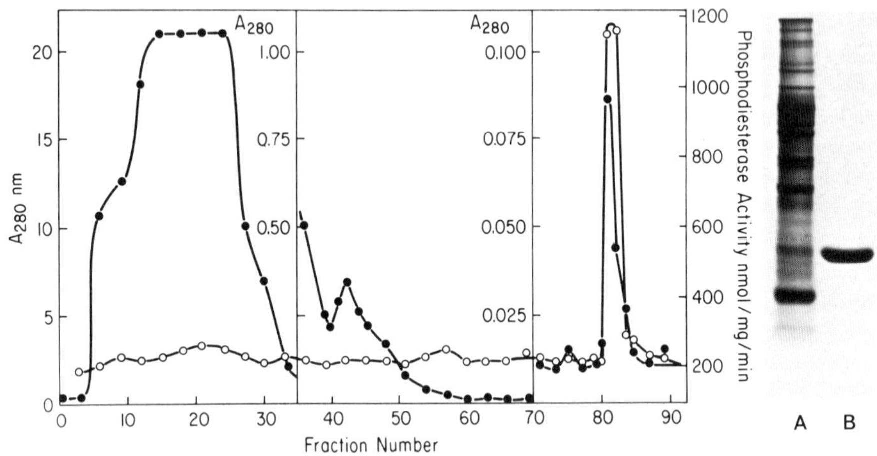

Fig. 2. Preparation of calmodulin by a fluphenazine-Sepharose affinity column. A high speed supernatant fluid of a bovine brain extract (prepared from 100 g of brain cerebrum) was applied to the fluphenazine-Sepharose column (2.6 × 29 cm) at a flow rate of 1 ml/min. The column was washed with 100 ml of PIPES (pH 7.0) containing 1 m*M* $CaCl_2$ (fractions 1–35), and was eluted first with 350 ml of 10 m*M* PIPES (pH 7.0) containing 1 m*M* $CaCl_2$ and 0.5 *M* NaCl (fractions 36–70), and then with 250 ml of 50 m*M* Tris-HCl (pH 8.0) containing 0.5 *M* NaCl and 10 m*M* EGTA (fractions 71–95). Ten millimeter fractions were collected and aliquots were assayed for calmodulin with the phosphodiesterase system. A_{280}, ○–○–○; phosphodiesterase activity, ●–●–●. The affinity column can be used repeatedly over a period of at least 6 months. The column was regenerated by washing with 1 liter of 50 m*M* Tris-HCl (pH 8.0) containing 1.0 *M* NaCl and 10 m*M* EGTA followed by 1 liter of 10 m*M* PIPES (pH 7.0) containing 1 m*M* $CaCl_2$. The SDS gel on the right compares the protein pattern of calmodulin before (lane **A**) and after (lane **B**) the fluphenazine-Sepharose affinity column. (From WALLACE et al. 1980)

sue extracts (JAMIESON and VANAMAN 1979; CHARBONNEAU and CORMIER 1979). In bovine brain, the only protein bound to a fluophenazine-agarose affinity column appears to be calmodulin (WALLACE et al. 1980). WEISS and his colleagues speculated that the action of the phenothiazines is through the inactivation of calmodulin (WEISS and HIRT 1977).

II. Preparation and Assay of Calmodulin

As noted, calmodulin is heat-stable and acidic, and these properties were exploited in early studies to prepare large quantities of calmodulin from bovine brain, a tissue highly rich in calmodulin (LIN et al. 1974). More recently, the high affinity of calmodulin for certain phenothiazines (see above) has permitted the isolation of pure calmodulin from bovine brain in essentially one affinity chromatography step. Figure 2 depicts the elution profile of calmodulin in a fluphenazine-agarose affinity column.

Calmodulin has been measured by its ability to stimulate a Ca^{2+}-dependent phosphodiesterase. This assay has been widely used because Ca^{2+}-dependent phosphodiesterase was the first known calmodulin-response system and the en-

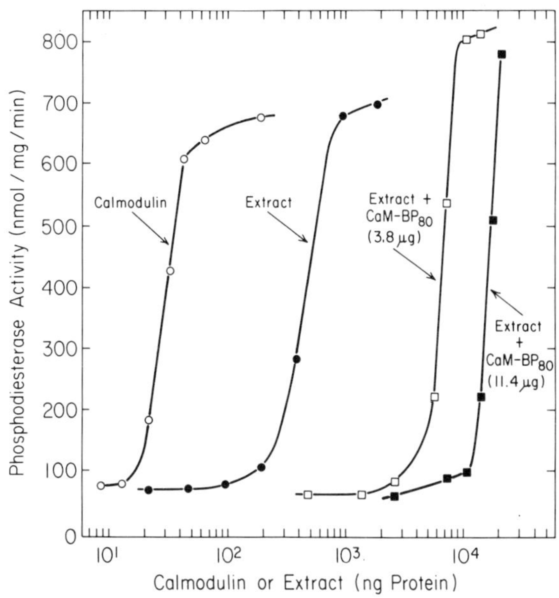

Fig. 3. Assay of calmodulin by Ca^{2+}-dependent phosphodiesterase in the presence or absence of CaM-BP_{80}. Note that the presence of CaM-BP_{80} causes a shift in the response curve, causing an underestimation of tissue level of calmodulin. In the absence of CaM-BP_{80}, the amount of calmodulin in the brain extract was 74 mg calmodulin/kg tissue. In the presence of 3.8 and 11.4 µg CaM-BP_{80} per reaction mixture, the level of calmodulin became 5.8 and 2.2 mg/kg, respectively. The sensitivity of the assay is approximately 20 ng or 0.1 pmol of calmodulin. (From WALLACE and CHEUNG 1979)

zyme is easily prepared in bulk. The assay is simple and is sensitive enough (0.1 pmole) to quantify the calmodulin content of most tissues. (The level of calmodulin in many tissues is in the micromolar range.) However, some tissues contain calmodulin-binding proteins as well as various enzymes that depend on calmodulin for activity (WANG et al. 1980; SOBUE and KAKIUCHI 1980). All these proteins are potential competitors for calmodulin and can interfere with the assay of phosphodiesterase, leading to an underestimation of the tissue content of calmodulin (Fig. 3). Although brief heat treatment removes the majority of these proteins, there is at least one calmodulin-binding protein (CaM-BP_{70}) which is heat stable (SHARMA et al. 1978). A recently developed radioimmunoassay for calmodulin allows the measurement of calmodulin in the presence of calmodulin-binding proteins without apparent interference (Fig. 4) (WALLACE and CHEUNG 1979; CHAFOUDEAS et al. 1979). The sensitivity of the enzyme assay mentioned above is in the range of 0.1 pmol of calmodulin, and that of the radioimmunoassay is comparable (WALLACE and CHEUNG 1969) or about 10-fold higher (CHAFOUDEAS et al. 1979).

III. Calmodulin-Regulated Cellular Processes

The enzymes which are regulated by calmodulin are fundamental for coordinating cellular activity. Calmodulin controls the metabolism of cyclic nucleotides and

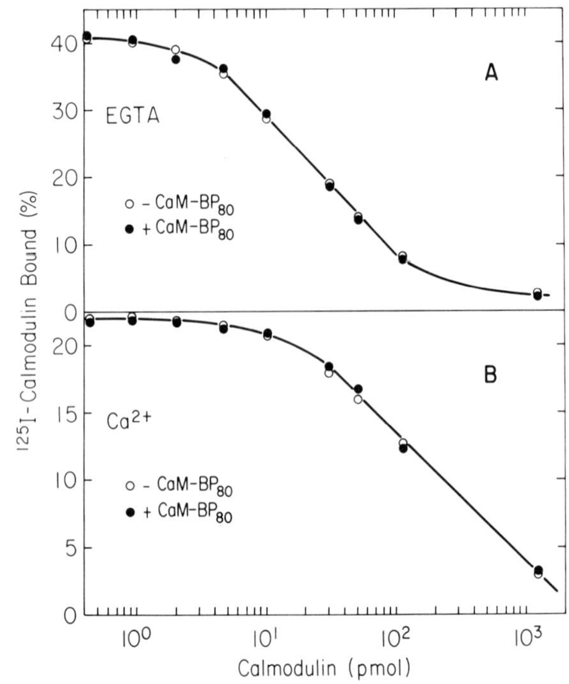

Fig. 4 A, B. Radioimmunoassay of calmodulin in the presence or absence of CaM-BP_{80}. Note that CaM-BP_{80} in the presence of Ca^{2+} or EGTA does not affect the measurement of calmodulin. The sensitivity of the assay is approximately 0.1 pmol of calmodulin. (From WALLACE and CHEUNG 1979)

other key enzymes and cellular processes as enumerated in Table 1. Adenylate cyclase and phosphodiesterase are the only known enzymes involved in the metabolism of cyclic AMP. In the brain (BROSTROM et al. 1975, 1976, 1977; CHEUNG et al. 1975; LYNCH et al. 1977) and adrenal (LE DONNE and COFFEE 1979) calmodulin regulates the activities of both enzymes. In other tissues, including heart (TEO et al. 1973; KAKIUCHI et al. 1975), liver, kidney (KAKIUCHI et al. 1975) and pituitary (AZHAR and MENON 1977), at least one form of phosphodiesterase is regulated by calmodulin.

Recently, NAGAO et al. (1979) and SUZUKI et al. (1981) demonstrated that calmodulin activates guanylate cyclase in a protozoan *(Tetrahymena pyriformis)*. Mammalian guanylate cyclase requires Ca^{2+} for activity; it remains to be shown whether the effect of Ca^{++} in the mammalian enzyme is mediated through calmodulin.

Muscle contraction results from the force generated by an actin myosin complex in response to the cellular level of Ca^{2+}. In skeletal and cardiac muscle the receptor of Ca^{2+} is troponin C, whereas in smooth muscle and non-muscle cells the receptor is calmodulin. Calmodulin is absolutely required for the activity of myosin light chain kinase of smooth muscle or non-muscle cells (WAISMAN et al. 1978; DABROWSKA and HARTSHORNE 1978; HATHAWAY and ADELSTEIN 1979; YAGI et al. 1978). The enzyme catalyzes the phosphorylation of a 20,000-Dalton light

Table 1. Calmodul in regulated enzymes and cellular processes

Enzyme or cellular process
Cyclic nucleotide phosphodiesterase
Adenylate cyclase
Guanylate cyclase
Phosphorylase kinase
Ca^{2+}-dependent protein kinase
NAD kinase
Phospholipase A_2
Ca^{2+}-ATPase
Ca^{2+} transport in erythrocytes
Ca^{2+} transport in sarcoplasmic reticulum
Phosphorylation of membranes
Neurotransmitter release
Disassembly of microtubules

The enzymes listed above govern the metabolism of cyclic nucleotides, prostaglandins, glycogen, pyridine nucleotides, and diverse cellular activities such as muscle contraction, cell movement, neurotransmitter release and disassembly of microtubules. For details and references, see the text

chain of myosin. Phosphorylation of the light chain is obligatory for actomyosin ATPase activity.

Phosphorylase kinase controls the hydrolysis of glycogen. The enzyme is complex, consisting of four tetramers $(\alpha\beta\gamma\delta)_4$; the δ-subunit is calmodulin, tightly bound to the ternary structure. Phosphorylase activity is further increased by an exogenous calmodulin (COHEN et al. 1978; DEPAOLI-ROACH et al. 1979).

Calmodulin regulates a Ca^{2+}-dependent protein kinase found in a membranous fraction of various tissues (SCHULMAN and GREENGARD 1978 a, b). In synaptosomes, the influx of Ca^{2+} following depolarization leads to the phosphorylation of two membrane proteins with molecular weights of 80,000 and 86,000 (SCHULMAN and GREENGARD 1978 a, b). These polypeptides also serve as substrates for cyclic AMP-dependent protein kinase (SIEGHART et al. 1979). DELORENZO et al. (1979) noted that calmodulin-regulated protein phosphorylation caused the release of neurotransmitters from synaptic vesicles. In addition, YAMAUCHI and FUJISAWA (1979) showed that tryptophan 5-monooxygenase, a key enzyme in the biosynthesis of serotonin, is activated by a calmodulin-induced phosphorylation.

NAD kinase activity in both pea seedling (ANDERSON and CORMIER 1978) and sea urchin oocyte (EPEL et al. 1981) is governed by calmodulin. The kinase controls the ratio of NAD/NADP which determines the direction of certain metabolic pathways. One of the early events following fertilization of sea urchin oocytes is an increase of free Ca^{2+}, and a rapid conversion of NAD to NADP (EPEL 1964). Calmodulin is present at high levels in sea urchin oocyte (HEAD et al. 1979) and the rapid increase of Ca^{2+} following fertilization of the egg may be related to the conversion of NAD to NADP.

Phospholipase A_2 is a key enzyme that governs the availability of arachidonic acid in many tissues. The level of free arachidonic acid is usually low, a rate-limiting step in the synthesis of endoperoxides, thromboxanes, prostaglandins, and prostacyclin. The enzyme depends on Ca^{2+} for activity; in human platelets, the effect of Ca^{2+} appears to be mediated through calmodulin (WONG and CHEUNG 1979). It will be important to determine if this finding can be substantiated with a purified enzyme and if the enzyme activity in other tissues is regulated through a similar mechanism.

Calmodulin regulates the activity of Ca^{2+}-ATPase of plasma membrane and endoplasmic reticulum (GOPINATH and VINCENZI 1977; JARRETT and PENNISTON 1977, 1978; HANAHAN et al. 1978; KATZ and REMTULLA 1978; LYNCH and CHEUNG 1979; SOBUE et al. 1979). As intracellular Ca^{2+} increases, calmodulin activates the membrane Ca^{2+}-ATPase, which extrudes the Ca^{2+} into the extracellular space, lowering the Ca^{2+} concentrations to a steady state level, an excellent example of feedback control. Thus, calmodulin acts not only as a mediator of Ca^{2+} actions but also as a modulator of its intracellular level.

Using indirect immunofluorescence, WELSH et al. (1978) found that calmodulin-specific antibodies decorated the mitotic spindle during the division of various cultured cells. The fluorescence appeared most intense at the poles of the spindle. Agents such as colcemid and N_2O which disrupt microtubule structure altered the distribution of fluorescence (WELSH et al. 1979). Subsequently, MARCUM et al. (1978) showed that calmodulin accelerated the disassembly of microtubules.

WOOD et al. (1980) have localized calmodulin in mouse basal ganglia. They found that the protein is associated primarily with the postsynaptic densities (PSD) and the dendritic microtubules. This observation is consistent with the work of GRAB et al. (1979) which shows that calmodulin is associated with the PSD isolated by subcellular fractionation. The PSD also contains a Ca^{2+}-dependent protein kinase whose activity may be regulated by calmodulin (GRAB and SIEKVIETZ 1979).

HARPER et al. (1980) studied the distribution of calmodulin in rat liver, skeletal muscle, and adrenal. In liver, calmodulin was associated with the nucleus, plasma membrane and glycogen granules. Skeletal muscle calmodulin was located at the I band and at the intermyofibrillar region, probably in association with glycogen granules and the sarcoplasmic reticulum, respectively. In the adrenal, the endocrine state of the animal affected the distribution of calmodulin within the cell. Administration of ACTH to the animal greatly increased calmodulin in the nuclei, implying a role of calmodulin with certain nuclear functions.

In view of the extensive involvement of Ca^{2+} in cell functions, it would not be surprising that future studies will extend the role of calmodulin to other cellular processes.

IV. Mechanism of Action

The mode of action of calmodulin was first established with phosphodiesterase (TEO and WANG 1973; TESHIMA and KAKIUCHI 1974; LIN et al. 1974, 1975; LIU and CHEUNG 1976), and subsequently with adenylate cyclase (BROSTROM et al. 1975; CHEUNG et al. 1975; LYNCH et al. 1976) and other enzyme systems. Although present in high concentrations in various tissues, calmodulin itself is not active; the active form is the Ca^{2+}·calmodulin complex. Once bound to Ca^{2+}, calmodulin assumes a more helical conformation and becomes the active species, which binds reversibly to the apoenzyme of phosphodiesterase, or adenylate cyclase, resulting in the formation of an active holoenzyme. The sequence of events associated with the stimulation of phosphodiesterase or adenylate cyclase by calmodulin may be depicted as follows:

$$\underset{}{Ca^{2+}} + \underset{\text{inactive}}{(CaM)} \rightleftarrows \underset{\text{active}}{(CaM^{*} \cdot Ca^{2+})}$$

$$(E) + \underset{\text{inactive}}{CaM^{*} \cdot Ca^{2+}} \rightleftarrows \underset{\text{active}}{(E^{*} \cdot CaM^{*} \cdot Ca^{2+})}\,.$$

According to this scheme, Ca^{2+}, and not calmodulin, is the rate-limiting factor, and the cellular flux of calcium plays a crucial role in regulating the enzyme activities. Subsequently, erythrocyte ATPase, myosin light chain kinase, and adenosine dinucleotide (NAD) kinase have been found to be stimulated by calmodulin, and the mechanism of stimulation appears similar (for reviews, see CHEUNG 1980a; WANG and WAISMAN 1979; KLEE et al. 1980; WOLFF and BROSTROM 1979).

Calmodulin also regulates the activities of phosphorylase kinase, Ca^{2+}-dependent protein kinase and a protozoan guanylate cyclase. It remains to be shown whether the mechanism of action of calmodulin on these enzyme systems is similar to the one depicted above. In the case of phosphorylase kinase, calmodulin appears to be bound intrinsically to the holoenzyme, which has the subunit structure of $(\alpha\beta\gamma\delta)_4$, δ being calmodulin; exogenous calmodulin further stimulates phosphorylase kinase activity.

C. Adenylate Cyclase

I. Ca^{2+} Sensitivity of Adenylate Cyclase

Regulation of adenylate cyclase by Ca^{2+} has been reported for a number of tissues and it has been proposed that cyclic AMP levels may be controlled by fluctuations in the amount of free intracellular Ca^{2+} (DRUMMOND and DUNCAN 1970; VON HUNGEN and ROBERTS 1973; TADA et al. 1975; BROSTROM et al. 1975; CHEUNG et al. 1975). Ca^{2+} inhibition of adenylate cyclase has been observed with many tissues examined: adipose (BIRNBAUMER et al. 1969), liver (HEPP et al. 1970; POHL et al. 1971), heart (TADA et al. 1975), leukocytes (STOLC 1977), brain (JOHNSON and SUTHERLAND 1973; VON HUNGEN and ROBERTS 1973), and turkey erythrocytes (HANSKI et al. 1977). Although the mechanism for Ca^{2+} inhibition has not been elucidated, HANSKI et al. (1977) have proposed that turkey erythrocyte adenylate cyclase is inhibited by Ca^{2+} through interaction with a specific inhibitory site. Kinetic studies revealed that Ca^{2+} had no effect on the binding of α-adrenergic agonists or antagonists, or of Mg-ATP, or Mg^{2+}. The only effect of Ca^{2+} was to reduce the V_{max} of the reaction in a cooperative manner. In contrast, Ca^{2+} inhibition of the heart enzyme has been reported to be competitive with Mg^{2+} (DRUMMOND and DUNCAN 1970). There is no evidence that Ca^{2+} inhibition of cardiac adenylate cyclase is mediated through calmodulin.

Although Ca^{2+} inhibits adenylate cyclase in many tissues, it also stimulates the enzyme in others, these include brain (BRADHAM et al. 1970; VON HUNGEN and ROBERTS 1973; MACDONALD 1975), heart (AVDONIN and TKACHUK 1978), glial tumor cells (BROSTROM et al. 1976), pancreatic islets (VALVERDE et al. 1979), silkworm pupal fat body (MORISHIMA 1979), and chicken embryo fibroblasts (LAPORTE et al. 1979b). BRADHAM et al. (1970) originally observed that EGTA inhibited adenylate cyclase in homogenates of calf cerebral cortex and that Ca^{2+} and Sr^{2+} but not Mg^{2+} or Ba^{2+} reversed this inhibition. It was proposed that Ca^{2+} was required for the accumulation of cyclic AMP in brain. Subsequently, VON HUNGEN and ROBERTS (1973) noted that brain adenylate cyclase exhibited a biphasic response to Ca^{2+}, with stimulation occurring at low Ca^{2+} (micromolar or below) and inhibition at higher Ca^{2+} concentrations (greater than 50 μM). These investigators suggested that alterations in the movement of extracellular Ca^{2+} into the cell (e.g., Ca^{2+} influx accompanying membrane depolarization) may stimulate cyclic AMP synthesis. Similar Ca^{2+} dependencies have been reported for porcine (BROSTROM et al. 1975) and rat (CHEUNG et al. 1975) brain adenylate cyclase. Ca^{2+} stimulation of adenylate cyclase is not unique to brain cerebral cortex; a similar effect on the cyclase from pancreatic islets (VALVERDE et al. 1979) and adrenal medulla (LEDONNE and COFFEE 1979) has been reported. Since stimulation of the enzyme occurs at low Ca^{2+} concentrations, detection of Ca^{2+} stimulation requires careful manipulation of free Ca^{2+} by EDTA or EGTA buffers. It is relatively easy to overlook Ca^{2+} stimulation and it seems likely that Ca^{2+}-stimulated adenylate cyclase activities in other tissues will be discovered in the future. It should be emphasized that the inhibition of adenylate cyclase by EGTA and its reversal by Ca^{2+} is not in itself sufficient evidence for a stimulatory effect of Ca^{2+}; the chelator may inhibit the enzyme directly.

II. Regulation of Adenylate Cyclase by Calmodulin

In 1975, it was discovered that calmodulin mediates the Ca^{2+} stimulation of brain adenylate cyclase (BROSTROM et al. 1975; CHEUNG et al. 1975; BROSTROM 1978b). Exogenous calmodulin had no appreciable effect on the enzyme activity of a detergent-solubilized adenylate cyclase; the preparation already contained high levels of endogenous calmodulin. However, calmodulin was separated from the enzyme when a Lubrol-PX solubilized preparation was submitted to ECTEOLA-cellulose chromatography. This treatment resulted in a significant loss of adenylate cyclase activity. Exogenous calmodulin, in the presence of Ca^{2+}, markedly stimulated the cyclase activity of this enzyme. The stimulation by calmodulin was dependent upon the presence of free Ca^{2+}; the addition of a molar excess of EGTA rapidly reversed the stimulation. Ca^{2+} did not stimulate adenylate cyclase in the absence of calmodulin. Furthermore, calmodulin from bovine brain cross-activated rat brain adenylate cyclase and vice versa. These studies indicate that calmodulin mediates Ca^{2+} stimulation of brain adenylate cyclase. Ca^{2+} stimulation of adenylate cyclase in glial tumor cells (BROSTROM et al. 1976), adrenal medulla (LEDONNE and COFFEE 1979), and pancreatic islets (VALVERDE et al. 1979) is also mediated by calmodulin.

The observation that the phenothiazines inhibit brain adenylate cyclase activities (UZUNOV and WEISS 1971) supports the hypothesis that calmodulin regulates brain adenylate cyclase. The phenothiazines bind to calmodulin in the presence of Ca^{2+} (LEVIN and WEISS 1979), and antagonize the interaction between calmodulin and adenylate cyclase. In addition, calcineurin (KLEE et al. 1979), which is a calmodulin-binding protein (CaM-BP_{80}), inhibits Ca^{2+} stimulation of brain adenylate cyclase (WALLACE et al. 1978). Neither the phenothiazines nor CaM-BP_{80} affect the basal activity of adenylate cyclase.

The mechanism for the stimulation by calmodulin of adenylate cyclase has not been fully elucidated. Calmodulin appears to form a complex with adenylate cyclase in the presence of Ca^{2+}; this possibility was originally suggested by the differential behavior of the two proteins on a Sephadex G-200 column in the presence of EGTA or Ca^{2+} (LYNCH et al. 1976). In the presence of EGTA, adenylate cyclase and calmodulin were well resolved, whereas in the presence of free Ca^{2+}, some of the calmodulin eluted with the peak of adenylate cyclase activity. This result suggested that calmodulin and adenylate cyclase form a complex in the presence, but not in the absence of Ca^{2+}, as described in Sect. B.IV.

Direct evidence for complex formation between brain adenylate cyclase and calmodulin was obtained using calmodulin-Sepharose chromatography (WESTCOTT et al. 1979). Calmodulin-Sepharose was evaluated as a tool for the purification of brain adenylate cyclase since there was already evidence for complex formation between the two proteins; moreover, calmodulin-Sepharose had been successfully used for the purification of CaM-BP_{80} (KLEE and KRINKS 1978; WALLACE et al. 1979) and the Ca^{2+}-sensitive phosphodiesterase (LAPORTE et al. 1979a). Partially purified adenylate cyclase from bovine brain cortex was fractionated into two distinct forms using calmodulin-Sepharose affinity chromatography. The major form of the enzyme (designated F-1), comprising 80% of the applied activity, did not bind to the affinity column in the presence of Ca^{2+} and was insensitive to calmodulin. The remaining 20% of the activity (designated F-2) was bound to the column and was stimulated by calmodulin.

The existence of two forms of adenylate cyclase resolvable by a calmodulin-Sepharose column reflects one of two possibilities. BROSTROM et al. (1977) proposed that brain cortex contains separate calmodulin-sensitive and calmodulin-insensitive forms of adenylate cyclase. Alternatively, adenylate cyclase is a multi-subunit enzyme complex (ORLY and SCHRAMM 1976; PFEUFFER 1977; WELTON et al. 1977; ROSS et al. 1978) and purification procedures may have removed one or more subunits required for the sensitivity to calmodulin. Further examination revealed that F-1 was insensitive to both Gpp(NH)p and NaF (TOSCANO et al. 1979). Sensitivity to these effectors was restored by incubating F-1 with a detergent-solubilized factor from the brain extract. The factor was sensitive to heat, trypsin digestion, and N-ethylmaleimide. Restoration of adenylate cyclase to an activation by calmodulin further requires the presence of Gpp(NH)p. These observations suggested that brain adenylate cyclase requires a guanyl nucleotide-binding subunit for calmodulin sensitivity. It is conceivable that F-2 as isolated contained a functional guanyl nucleotide regulatory subunit. In contrast, Ca^{2+}-sensitive phosphodiesterase, a tetramer comprised of two catalytic subunits and two calmodulins, does not require additional factors for calmodulin sensitivity (LAPORTE et al. 1979a).

The relationship between the calmodulin-sensitive adenylate cyclase and other adenylate cyclase activities has not been established. Cholera toxin stimulation of adenylate cyclase requires several factors including the toxin A subunit, NAD, a nucleotide triphosphate and a macromolecule from the tissue extract (GILL and KING 1975). The identity of the macromolecular factor has not been determined; however, MOSS and VAUGHAN (1977) reported that the activation of a detergent-solubilized brain adenylate cyclase by cholera toxin required calmodulin. ENOMOTO and GILL (1979) found a similar factor in pigeon erythrocytes, the molecular weight being 15,000–20,000. Calmodulin did not substitute for this factor. Brain adenylate cyclase may be unique, and these two studies are not necessarily contradictory.

BROSTROM et al. (1977, 1978a) proposed that adenylate cyclase of rat cerebral cortex consisted of two contributing components: a calmodulin-independent activity seen in the absence of calmodulin, and calmodulin-dependent activity which is the increment obtained in the presence of calmodulin. The different kinetic properties and stability of the two activities suggested two distinct forms of adenylate cyclase. Although this is an attractive hypothesis, it is also possible that only one form of the enzyme exists which is regulated by calmodulin. A partially purified calmodulin-sensitive adenylate cyclase from brain expresses a basal activity in the absence of calmodulin (WESTCOTT and STORM unpublished results) or in the presence of high levels of CaM-BP_{80} (WALLACE et al. 1978).

Adenylate cyclase in rat striatal homogenates is responsive to adenosine, Ca^{2+} and dopamine. PREMONT et al. (1977) proposed that the receptors for these agonists are distinct, since the effect of the three effectors was additive. Furthermore, the accumulation of cyclic AMP in rat slices via stimulation of alpha-adrenergic receptors is dependent on the presence of extracellular Ca^{2+} whereas that occurring via stimulation of β-receptors is less dependent (SCHWABE and DALY 1977). The relationship between the calmodulin-sensitive adenylate cyclase and other adenylate cyclase activities can not be defined until further characterization of the enzyme components.

D. Cyclic Nucleotide Phosphodiesterase

I. Multiple Forms of Phosphodiesterase

Early studies of phosphodiesterase from bovine brain (CHEUNG 1969b) and frog erythrocytes (ROSEN 1970) indicated the presence of two forms of phosphodiesterase. Using a sensitive radioactive assay, THOMPSON and APPLEMAN (1971) detected three forms of phosphodiesterase in rat brain as well as several other tissues. The existence of multiple forms of phosphodiesterase has since been extensively studied in numerous tissues. These forms exhibit distinct kinetic properties, substrate specificities and cellular localizations. In general, there are three basic forms of mammalian phosphodiesterase. One is membrane bound with a high affinity for cyclic AMP. The other two forms are cytosolic and are capable of hydrolyzing both cyclic AMP and cyclic GMP. One of them has approximately the same affinity for cyclic AMP and cyclic GMP, and cyclic GMP stimulates the hydrolysis of cyclic AMP, but not vice versa. The other form displays a higher affinity for cyclic GMP than for cyclic AMP. This is the form that is dependent on calmodulin for activity (for reviews, see WELLS and HARDMAN 1977; THOMPSON et al. 1979; LIN and CHEUNG 1980; APPLEMAN et al. this volume). Other studies on multiple phosphodiesterase forms employed electrophoresis (MOON and CHRISTIANSEN 1971) and isoelectric focusing (STRADA and PLEDGER 1975) and identified as many as seven different forms of phosphodiesterase in rat and rabbit tissues. Brain contained four forms, whereas the other tissues contained from one to three forms of phosphodiesterase. All seven forms of phosphodiesterase catalyze the hydrolysis of both cyclic AMP and cyclic GMP.

Many studies on the multiple forms of phosphodiesterase used DEAE-cellulose column chromatography, which resolved the enzyme into three forms (HIDAKA et al. 1977; DONNELLY 1977; TANIGUCHI et al. 1978; SAKAI et al. 1978), designated as Forms I, II, and III according to the order of emergence from the column. Forms I and II generally have a low K_m value for cyclic GMP than for cyclic AMP, and the reverse is true for Form III. Form II may be distinguished from Form I in that cyclic GMP stimulates the hydrolysis of cyclic AMP (HIDAKA et al. 1977). Form I in most cases is Ca^{2+}-dependent and the extent of activation varies with the tissue.

Brain generally has a more complex phosphodiesterase pattern. HIDAKA et al. (1977) resolved human brain phosphodiesterase into four activity peaks by a DEAE column. Peaks II, III, and IV appear comparable to Forms I, II, and III of other mammalian phosphodiesterases, respectively.

It is difficult to correlate with certainty the individual forms of phosphodiesterase obtained by different separation methods, since some methods are based on differences in charge, and others in size. The designation of different forms of phosphodiesterase is primarily based on the physical behavior by a particular separation method. Similarity in phyiscal and kinetic properties may be found for the same form of phosphodiesterase from different species, but significant differences may exist in different tissues obtained from the same species; for instance, the extent of Ca^{2+}-dependent activation of phosphodiesterase may be different. One possible explanation for the apparent discrepancies is that the enzyme is contaminated by other forms of phosphodiesterase; another is that tissue protease decreases the sensitivity of phosphodiesterase to calmodulin (CHEUNG 1971). SAKAI (1978) showed that the molecular forms of phosphodiesterase may be altered by lysosomal protease during the course of purification, especially when the tissue has been frozen and thawed. Ca^{2+}-dependent phosphodiesterase invariably was found less responsive to Ca^{2+} after purification (CHEUNG and LIN 1974; PLEDGER et al. 1974). Moreover, conditions of enzyme purification affect the properties of the different enzyme forms (VAN INWEGEN et al. 1976), giving rise to differences between the same form of phosphodiesterase from different tissues.

The conversion of one form of phosphodiesterase into another has been noted. Using Sepharose 4 B chromatography, CHEUNG and LIN (1974) resolved phosphodiesterase partially purified from bovine brain into two activity peaks. When an aged rather than a fresh sample was used, the activity corresponding to a higher molecular weight increased at the expense of that corresponding to a lower molecular weight. The conversion was prevented by prior incubation of the aged sample with β-mercaptoethanol. Aggregation and deaggregation caused by a change of the ionic strength of the medium (TISDAL 1975; SCHRODER and RICHENBERG 1973) or of the enzyme concentration (PICHARD and CHEUNG 1976) have also been reported.

In summary, although it is easy to demonstrate the presence of multiple forms, it is difficult to fully characterize these forms. Caution should be exercised in the interpretation of the apparent multiple forms. Artifacts arising during the course of purification through tissue protease could well generate some of the molecular forms reported in the literature.

II. Ca^{2+}-Dependent Phosphodiesterase

KAKIUCHI and YAMAZAKI (1970) first demonstrated the existence of a Ca^{2+} + Mg^{2+}-dependent phosphodiesterase in rat brain; this is the form stimulated by calmodulin. Subsequently, the enzyme was detected in various mammalian tissues. The Ca^{2+}-dependent phosphodiesterase from mammalian brain and heart has been extensively characterized. Although the kinetic constants reported from different laboratories differ, some generalizations may be made. The enzyme catalyzes the hydrolysis of both cyclic AMP and cyclic GMP. At millimolar concentration, the rate of cyclic AMP hydrolysis is higher than that of cyclic GMP. The reverse is true at micromolar substrate concentrations. Since the activated enzyme appears to have a higher affinity for cyclic GMP than for cyclic AMP, the enzyme has been referred to as a cyclic GMP phosphodiesterase by some investigators. Tissue level of cyclic AMP is generally 10 times or higher than cyclic GMP, and the enzyme probably acts on both cyclic AMP and cyclic GMP in vivo (KAKIUCHI et al. 1975; WANG et al. 1975; CHEUNG et al. 1978).

The Ca^{2+}-dependent phosphodiesterase has been purified to apparent homogeneity from bovine heart (LAPORTE et al. 1979a) and brain (SHARMA et al. 1980; KRINKS and KLEE 1980; MORRILL et al. 1979). The enzyme consists of two identical subunits, each with a molecular weight of approximately 59,000. In the presence of Ca^{2+}, each catalytic unit binds one mole of calmodulin. Calmodulin increases the V_{max} of the enzyme without affecting its K_m for the substrate. The enzyme displays normal Michaelis-Menten kinetics. Ca^{2+}-dependent phosphodiesterase is activated by a brief treatment with trypsin or other endoproteases. A comparison of the stimulation of phosphodiesterase by calmodulin with that by trypsin may shed light on the molecular structure of phosphodiesterase. Activation of phosphodiesterase by calmodulin proceeds stoichiometrically and reversibly, whereas that by trypsin proceeds catalytically and irreversibly (CHEUNG 1971). Activation by trypsin reduces the molecular weight of phosphodiesterase and prevents the enzyme from interacting with, and responding to calmodulin. Stimulation by either agent produces an activated enzyme with comparable final activities (CHEUNG and LIN 1974). Figure 5 depicts a speculative scheme on the molecular events associated with the stimulation of phosphodiesterase by both agents. Binding of calmodulin to phosphodiesterase is believed to cause the enzyme to assume a new conformation, with the catalytic site more accessible to the substrate. On the other hand, limited trypsinization of phosphodiesterase decreases its molecular size with a similar consequence – an enzyme with a more exposed substrate site. Although the scheme in Fig. 5 assumes that the polypeptide released from phosphodiesterase contains the calmodulin-binding domain, this is not a necessary condition; other alternatives are possible.

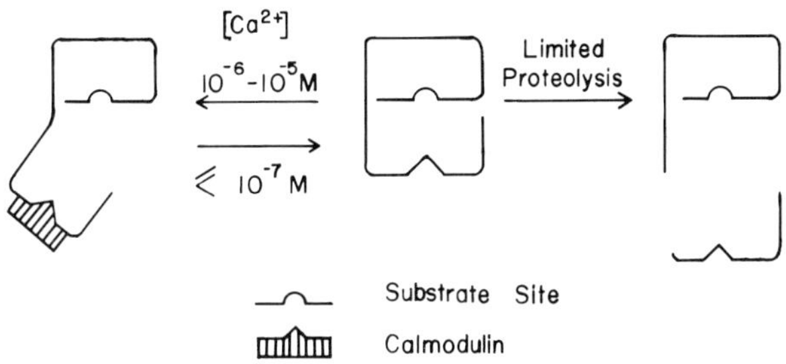

Fig. 5. A hypothetical scheme depicting the activation of phosphodiesterase by calmodulin or by trypsin. Binding of calmodulin renders phosphodiesterase to assume a new conformation with the catalytic site more accessible to the substrate. Dissociation of the ternary complex returns the enzyme to its prestimulated conformation. The process is reversible, and the equilibrium is controlled by the cellular flux of Ca^{2+}. Limited trypsinization releases a polypeptide from phosphodiesterase. Removal of the polypeptide exposes the substrate site, and the result is phenomenally equivalent to the effect of calmodulin – an increase in V_{max} with little or no change in the K_m for cAMP. The cleavage of phosphodiesterase by trypsin is not reversible, and is probably not physiologically. (From LIN and CHEUNG 1980)

E. Biological Significance

The biological significance of calmodulin stimulation of adenylate cyclase and phosphodiesterase activities has not been fully defined. However, the existence of calmodulin-sensitive adenylate cyclase and phosphodiesterase activities in a single cell type (BROSTROM et al. 1979) suggests that calmodulin may play a key role in mediating Ca^{2+} regulation of cyclic nucleotide levels. Ca^{2+} influx is a critical component of the flow of ions comprising nerve depolarizations (BAKER et al. 1971). The observation that calmodulin stimulates the activities of both adenylate cyclase and phosphodiesterase may appear paradoxical. At micromolar concentrations of substrate, the Ca^{2+}-dependent phosphodiesterase catalyzes the hydrolysis of cyclic GMP faster than cyclic AMP, and the Ca^{2+} influx may result in an increase of cyclic AMP and a decrease of cyclic GMP (BROSTROM et al. 1975). On the other hand, PIASCIK et al. (1980) noted that the sensitivity of adenylate cyclase and phosphodiesterases to Ca^{2+} is different. Brain adenylate cyclase is activated by a lower concentration of Ca^{2+} and, at a concentration of Ca^{2+} which activates phosphodiesterase, adenylate cyclase is already inhibited. The influx of Ca^{2+} across the plasma membrane into the cell may cause a sequential activation of adenylate cyclase and the phosphodiesterase, resulting in a transient accumulation of cyclic AMP, as shown in Fig. 6. A third postulate has been suggested by GNEGY et al. (1976a), who observed that a purified cyclic AMP-dependent protein kinase, in the presence of cyclic AMP and ATP stimulated the release of calmodulin from a particulate fraction of rat brain and adrenal medulla. The released calmodulin appeared to decrease the K_m of Ca^{2+}-dependent phosphodiesterase in the cytosol for cyclic AMP (GNEGY et al. 1976b; HANBAUER et al. 1979). This notion appears difficult to reconcile with the general finding that in the cytosolic compartment the concentration of calmodulin is considerably higher than that of phosphodiesterase.

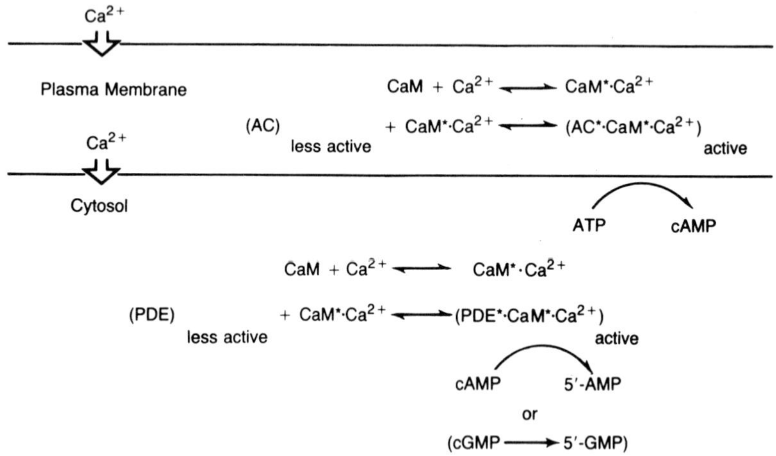

Fig. 6. Speculative scheme depicting the regulation of brain cAMP metabolism by the cellular flux of Ca^{2+}. CaM, calmodulin; AC, adenylate cyclase; PDE, phosphodiesterase; the asterisk (*) indicates the active conformation. See text for details. (Modified from CHEUNG et al. 1978)

Because calmodulin plays an important role in the regulation of cyclic AMP metabolism, fluctuations in the levels of calmodulin could be physiologically significant. Adenylate cyclase activities are decreased in many malignant cell types including chicken embryo fibroblasts transformed by Rous sarcoma virus (ANDERSON and PASTAN 1975). ANDERSON and PASTAN reported that normal chicken embryo fibroblasts contain an adenylate cyclase activator that is lacking in transformed fibroblasts. WATTERSON et al. (1976) noted that calmodulin levels are higher in transformed than in normal fibroblasts.

LAPORTE et al. (1979b) examined a temperature-dependent mutant of Rous sarcoma virus and found that a decrease in adenylate cyclase and an increase in calmodulin accompanied transformation. Moreover, the half-life of calmodulin was more than twice that of the enzyme, suggesting that the decline in adenylate cyclase activity in the transformed cells did not result from a change in the amount of calmodulin. In addition, adenylate cyclase associated with normal cells was Ca^{2+}-dependent, but that of the transformed cell was not. It thus appears that calmodulin may not modulate adenylate cyclase of transformed cells; either the adenylate cyclase system or the membrane environment in these cells may have been altered.

Acknowledgement. The work from our laboratories has been supported by grants NS 08059 (W.Y.C.). PCM 78-03188 (D.R.S.) the American Heart Association of Washington (D.R.S.) and by ALSAC (W.Y.C.). D.R.S. is a recipient of NIH Research Career Development Award AI 00310. W.Y.C. is grateful to Dr. E.G. Krebs for his generous support and kind hospitality while he was on sabbatical leave as a Faculty Scholar, Josiah Macy, Jr., Foundation, in his Laboratory of Molecular Pharmacology, Howard Hughes Medical Institute, University of Washington. We thanks Ms. Marion Turner for excellent assistance in the preparation of the manuscript.

References

Amphlett GW, Vanaman TC, Perry SV (1976) Effect of the troponin C-like protein from bovine brain (brain modulator protein) on the Mg^{2+}-stimulated ATPase of skeletal muscle actomyosin. FEBS Lett 72:163–168

Anderson JM, Cormier MJ (1978) Calcium-dependent regulator of NAD kinase in higher plants. Biochem Biophys Res Commun 84:595–602

Avdonin PV, Tkachuk VA (1978) Participation of the Ca^{2+}-dependent activator in the regulation of the adenylate cyclase activity of the heart by calcium ions. Dokl Akad Nauk SSSR 238:726–729

Azhar S, Menon KMJ (1977) Cyclic nucleotide phosphodiesterase from rat anterior pituitary. Characterization of multiple forms and regulation by protein activation and Ca^{2+}. Eur J Biochem 73:73–82

Baker PF, Hodgkin AL, Ridgway EB (1971) Depolarization and calcium entry in squid giant axons. J Physiol 218:709–755

Birnbaumer L, Pohl SL, Rodbell M (1969) Adenyl cyclase in fat cells. J Biol Chem 244:3468–3476

Bradham LS, Holt DA, Sims M (1970) The effect of Ca^{2+} on the adenyl cyclase of calf brain. Biochim Biophys Acta 201:250–260

Brostrom CO, Brostrom MA, Wolff DJ (1977) Calcium-dependent adenylate cyclase from rat cerebral cortex. J Biol Chem 252:5677–5685

Brostrom CO, Huang YC, Breckenridge BM, Wolff DJ (1975) Identification of a calcium-binding protein as a calcium-dependent regulator of brain adenylate cyclase. Proc Natl Acad Sci USA 72:64–68

Brostrom MA, Brostrom CO, Breckenridge BM, Wolff DJ (1976) Regulation of adenylate cyclase from glial tumor cells by calcium and a calcium-binding protein. J Biol Chem 251:4744–4750

Brostrom MA, Brostrom CO, Breckenridge BM, Wolff DJ (1978 b) Calcium-dependent regulation of brain adenylate cyclase. Adv Cyclic Nucleotide Res 9:85–99

Brostrom MA, Brostrom CO, Wolff DJ (1978 a) Calcium-dependent adenylate cyclase from rat cerebral cortex: activation by guanine nucleotides. Arch Biochem Biophys 191:341–350

Brostrom MA, Brostrom CO, Wolff DJ (1979) Calcium dependence of hormone-stimulated cAMP accumulation in intact glial tumor cells. J Biol Chem 254:7548–7557

Chafoudeas JG, Dedman JR, Munjoal RP, Means AR (1979) Calmodulin. Development and application of a sensitive radioimmunoassay. J Biol Chem 254:10262–10267

Charbonneau H Cormier MJ (1979) Purification of plant calmodulin by fluphenazine-sepharose affinity chromatography. Biochem Biophys Res Commun 90:1039–1047

Cheung WY (1967) Cyclic 3′,5′-nucleotide phosphodiesterase: pronounced stimulation by snake venom. Biochem Biophys Res Commun 29:478–482

Cheung WY (1969 a) Cyclic 3′,5′-nucleotide phosphodiesterase preparation of a partially inactive enzyme and its subsequent stimulation by snake venom. Biochem Biophys Acta 191:303–315

Cheung WY (1969 b) Cyclic 3′,5′-nucleotide phosphodiesterase. A continuous titrimetric assay. Anal Biochem 28:182–191

Cheung WY (1970 a) Cyclic 3′,5′-nucleotide phosphodiesterase. Demonstration of an activator. Biochem Biophys Res Commun 38:533–538

Cheung WY (1970 b) Cyclic nucleotide phosphodiesterase. Adv Biochem Pharmacol 3:51–56

Cheung WY, Bradham LS, Lynch TJ, Lin YM, Tallant EA (1975) Protein activator of cyclic 3′,5′-nucleotide phosphodiesterase of bovine or rat brain also activates its adenylate cyclase. Biochem Biophys Res Commun 66:1055–1062

Cheung WY (1971) Cyclic nucleotide phosphodiesterase: evidence for and properties of a protein activator. J Biol Chem 246:2859–2869

Cheung WY, Lin YM (1974) Purification and characterization of cyclic 3′,5′-nucleotide phosphodiesterase from bovine brain. Methods Enzymol 38:223–239 (Part C)

Cheung WY, Lynch TJ, Wallace RW (1978) An endogenous Ca^{2+}-dependent activator protein of brain adenylate cyclase and cyclic nucleotide phosphodiesterase. Adv Cyclic Nucleotide Res 9:233–251
Cheung WY (1980a) Calmodulin plays a pivotal role in cellular regulation. Science 207:19–27
Cheung WY (1980b) Calmodulin – an introduction. In: Calcium and Cell Function. Vol 1. Cheung WY (ed) Academic Press, New York, p 1–12
Cohen P, Burchell A, Foulkes JG, Cohen TN, Vanaman TC, Nairn AL (1978) Identification of the Ca^{2+}-dependent modulator protein as the fourth subunit of rabbit skeletal muscle phosphorylase kinase. FEBS Lett 92:287–293
Dabrowska R, Hartshorne DJ (1978) A Ca^{2+}- and modulator-dependent myosin light chain kinase from non-muscle cells. Biochem Biophys Res Commun 85:1352–1359
Dedman JR, Jackson RL, Schneiber WE, Means AR (1978) Sequence homology of the Ca^{2+}-dependent regulator of cyclic nucleotide phosphodiesterase from rat testis with other Ca^{2+}-binding proteins. J Biol Chem 253:343–346
DeLorenzo RJ, Freedman SD, Yobe WB, Maurer SC (1979) Stimulation of Ca^{2+}-dependent neurotransmitter release and presynaptic nerve-terminal protein phosphorylation by calmodulin and a calmodulin-like protein isolated from synaptic vesicles. Proc Natl Acad Sci USA 76:1838–1842
Depaoli-Roach AA, Gibbs JB, Roach PJ (1979) Calcium and calmodulin activation of muscle phosphorylase kinase: effect of tryptic proteolysis. FEBS Lett 105:321–324
Donnelly TE (1977) Properties of the activator-dependent cyclic nucleotide phosphodiesterase from bovine heart. Biochim Biophys Acta 480:193–203
Drummond GI, Duncan L (1970) Adenyl cyclase in cardiac tissues. J Biol Chem 245:976–983
Enomoto K, Gill MD (1979) Requirement for guanine triphosphate in the activation of adenylate cyclase by cholera toxin. J Supramol Struc 10:51–60
Epel D (1964) A primary metabolic change of fertilization: interconversion of pyridine nucleotide. Biochem Biophys Res Commun 17:62–68
Epel D, Patton C, Wallace RW, Cheung WY (1981) Calmodulin activates NAD kinase of sea urchin eggs: an early event following fertilization. Cell 23:543–549
Garret HW, Penniston JT (1977) Partial purification of the Ca^{2+}-Mg^{2+} ATPase activator from human erythrocytes: its similarity to the activator of 3′,5′-cyclic nucleotide phosphodiesterase. Biochem Biophys Res Commun 77:1210–1216
Gill MD, King CA (1975) The mechanism of action of cholera toxin in pigeon erythrocyte lysates. J Biol Chem 250:6424–6432
Gnegy ME, Uzunov P, Costa E (1976a) Regulation of dopamine stimulation of striatal adenylate cyclase by an endogenous Ca^{2+}-binding protein. Proc Natl Acad Sci USA 73:3887–3890
Gnegy ME, Costa E, Uzunov P (1976b) Regulation of transsynaptically elicited increase of 3′,5′-cyclic AMP by endogenous phosphodiesterase activator. Proc Natl Acad Sci USA 73:352–355
Gopinath RM, Vincenzi FF (1977) Phosphodiesterase protein activator mimics red blood cell cytoplasmic activator of (Ca^{2+}-Mg^{2+})-ATPase. Biochem Biophys Res Commun 77:1203–1209
Grab DL, Siekevitz P (1979) Calmodulin-dependent protein kinase activity in postsynaptic densities. J Cell Biol 83:603 (Abstract)
Grab DJ, Berzins K, Cohen P, Siekevitz (1979) Presence of calmodulin in postsynaptic densities isolated from canine cerebral cortex. J Biol Chem 254:8690–8696
Grand RJA, Perry SV (1978) The amino acid sequence of the troponin C-like protein (modulator protein) from bovine uterus. FEBS Lett 92:137–142
Hanahan DJ, Taverna RD, Glynn DD, Ekholm JE (1978) The interaction of Ca^{2+}/Mg^{2+} ATPase activator protein and Ca^{2+} with human erythrocyte membranes. Biochem Biophys Res Commun 84:1009–1015
Hanbauer I, Gimble J, Sankaron K, Sherard R (1979) Modification of striatal cyclic nucleotide phosphodiesterase by calmodulin: regulation by opiate dopamine receptor activation. Neuropharmacology 18:859–864

Hanski E, Sevilla N, Levitzki A (1977) The allosteric inhibition by calcium of soluble and partially purified adenylate cyclase from turkey erythrocytes. Eur J Biochem 76:513–520
Harper JF, Cheung WY, Wallace RW, Huang HL, Levine SN, Steiner AL (1980) Localization of calmodulin in rat tissues. Proc Natl Acad Sci USA 77:366–370
Hathaway DR, Adelstein RS (1979) Human platelet myosin light chain kinase requires the calcium-binding protein calmodulin for activity. Proc Natl Acad Sci USA 76:1653–1657
Head JF, Mader S, Kaminer B (1979) Calcium-binding modulator protein from the unfertilized egg of the sea-urchin egg Arbacia punctulata. J Cell Biol 80:211–218
Hepp KD, Edel R, Wieland O (1970) Hormone action on liver adenyl cyclase. J Biochem 17:171–177
Hidaka H, Yamak T, Ochiai Y, Asano T, Yamabe H (1977) Cyclic 3′,5′-nucleotide phosphodiesterase determined in various human tissues by DEAE-cellulose chromatography. Biochem Biophys Acta 484:398–407
Jamieson GA, Vanaman TC (1979) Calcium-dependent affinity chromatography of calmodulin on an immobilized phenothiazine. Biochem Biophys Res Commun 90:1048–1056
Johnson RA, Sutherland EW (1973) Detergent-dispersed adenylate cyclase from rat brain. J Biol Chem 248:5114–5121
Kakiuchi S, Yamazaki R (1970) Calcium dependent phosphodiesterase activity and its activating factor (PAF) from brain studies on cyclic 3′,5′-nucleotide phosphodiesterase. Biochem Biophys Res Commun A 1:1104–1110
Kakiuchi S, Yamazaki R, Teshima Y, Uenishi K, Miyamoto E (1975) Ca^{2+}/Mg^{2+}-dependent cyclic nucleotide phosphodiesterase and its activator protein. Adv Cyclic Nucl Res 5:163–178
Kakiuchi S, Yamazaki R, Teshima Y, Uenishi K, Miyamoto E (1975) Multiple cyclic nucleotide phosphodiesterase activities from rat tissues and occurrence of a calcium-plus-magnesium-ion-dependent phosphodiesterase and its protein activator. Biochem J 146:109–120
Katz S, Remtulla MA (1978) Phosphodiesterase protein activator stimulates calcium transport in cardiac microsomal preparations enriched in sarcoplasmic reticulum. Biochem Biophys Res Commun 83:1373–1379
Klee CB, Krinks MH (1978) Purification of cyclic 3′,5′-nucleotide phosphodiesterase inhibitory protein by affinity chromatography on activator protein coupled to Sepharose. Biochem 17:120–126
Klee CB, Crouch TH, Krinks MH (1979) Calcineurin: A calcium- and calmodulin-binding protein of the nervous system. Proc Natl Acad Sci USA 76:6270–6273
Klee CB, Crouch TH, Richman P (1980) Calmodulin. Ann Rev Biochem 49:489–515
Kretsinger RH, Barry CD (1975) The predicted structure of the calcium-binding component of troponin. Biochem Biophys Acta 405:40–52
Krinks MH, Klee CB (1980) Proteolytic activation of Ca^{2+}-dependent cyclic nucleotide phosphodiesterase. Fed Proc 39:1624 (Abstract)
LeDonne NC, Coffee CJ (1979) Properties of the bovine adrenal medulla adenyl cyclase. Fed Proc 38:317 (Abstract)
LaPorte DC, Toscano WA, Storm DR (1979a) Cross-linking of iodine ^{125}labelled calmodulin to the Ca^{2+}-sensitive phosphodiesterase purified from bovine heart. Biochemistry 18:2820–2825
LaPorte DC, Gidwitz S, Weber MJ, Storm DR (1979b) Relationship between changes in calmodulin and adenylate cyclase during viral transformation. Biochem Biophys Res Commun 86:1169–1177
Levin RM, Weiss B (1976) Mechanism by which psychotropic drugs inhibit adenosine cyclic 3′,5′-monophosphate phosphodiesterase of brain. Mol Pharmacol 12:581–589
Levin RM, Weiss B (1977) Binding of trifluoperazine to the calcium-dependent activator of cyclic nucleotide phosphodiesterase. Mol Pharmacol 13:690–697
Levin RM, Weiss B (1979) Selective binding of antipsychotics and other psychoactive agents to the calcium-dependent activator of cyclic nucleotide phosphodiesterase. J Pharm Exp Ther 208:454–459
Lin YM, Liu YP, Cheung WY (1974) Cyclic 3′,5′-nucleotide phosphodiesterase. Purification, characterization and active form of the protein activator from bovine brain. J Biol Chem 249:4943–4954

Lin YM, Liu YP, Cheung WY (1975) Cyclic 3′,5′-nucleotide phosphodiesterase. Ca^{2+}-dependent formation of bovine brain enzyme activator complex. FEBS Lett 49:356–360
Lin YM, Cheung WY (1980) Ca^{2+}-dependent cyclic nucleotide phosphodiesterase. In: Calcium and Cell Function. Vol 1. Cheung WY (ed) Academic Press, New York p 79–107
Liu YP, Cheung WY (1976) Cyclic 3′,5′-nucleotide phosphodiesterase. Ca^{2+} confers more helical conformation to the protein activator. 251:4193–4198
Lynch TJ, Cheung WY (1979) Human erythrocytes Ca^{2+}-ATPase: mechanism of stimulation by Ca^{2+}. Arch Biochem Biophys 194:165–170
Lynch TJ, Tallant EA, Cheung WY (1976) Ca^{2+} dependent formation of brain adenylate cyclase-protein activator complex. Biochem Biophys Res Commun 68:616–625
Lynch TJ, Tallant EA, Cheung WY (1977) Rat brain adenylate cyclase. Further studies on its stimulation by a Ca^{2+}-binding protein. Arch Biochem Biophys 182:124–133
MacDonald I (1975) Differentiation of fluoride-stimulated and non-fluoride stimulated components of beef brain cortex adenylate cyclase by calcium ions, EGTA, and triton X-100. Biochim Biophys Acta 397:244–253
Marcum JM, Dedman JR, Brinkley BR, Mean AR (1978) Control of microtubule assembly-dissambly by calcium-dependent regulator protein. Proc Natl Acad Sci USA 75:3771–3775
Moon E, Christiansen RO (1971) Adenosine 3′,5′-monophosphate phosphodiesterase: multiple molecular forms. Science 173:540–541
Morishima I (1979) Adenylate cyclase in silkworm: regulation by calcium and a calcium-binding protein. Agric Biol Chem 43:1127–1131
Morrill ME, Thompson ST, Stillwagen E (1979) Purification of a cyclic nucleotide phosphodiesterase from bovine brain using blue dextran-sepharose chromatography. J Biol Chem 254:4371–4374
Moss J, Vaughan M (1977) Choleragen activation of solubilized adenylate cyclase: requirement for GTP and protein activator for demonstration of enzymatic activity. Proc Natl Acad Sci USA 74:4396–4400
Nagao S, Suzuki Y, Watanabe Y (1979) Activation by a calcium-binding protein of guanylate cyclase in Tetrahymena pyriformis. Biochem Biophys Res Commun 90:261–268
Orly J, Schramm M (1976) Coupling of catecholamine receptor from one cell with adenylate cyclase from another cell by cell fusion. Proc Natl Acad Sci USA 73:4410–4414
Pfeuffer T (1977) GTP-bind proteins in membranes and the control of adenylate cyclase activity. J Biol Chem 252:7224–7234
Piascik MT, Wisler PL, Johnson CL, Potter JD (1980) Ca^{2+}-dependent regulation of guinea pig brain adenylate cyclase. J Biol Chem 255:4176–4181
Pichard AL, Cheung WY (1976) Cyclic nucleotide phosphodiesterase. Interconvertible multiple forms and their effects on enzyme activity and kinetics. J Biol Chem 251:5726–5737
Pledger WJ, Stancel GM, Thompson WJ, Strada JJ (1974) Separation of multiple forms of cyclic nucleotide phosphodiesterase from rat brain by isoelectro focusing. Biochem Biophys Acta 370:242–248
Pohl SL, Birnbaumer L, Rodbell M (1971) Glucagon-sensitive adenyl cyclase system in plasma membranes of rat liver. J Biol Chem 246:1849–1856
Premont J, Perez M, Bockaert J (1977) Adenosine-sensitive adenylate cyclase in rat striatal homogenates and its relationship to dopamine- and Ca^{2+}-sensitive adenylate cyclases. Mol Pharm 13:662–670
Richman PG, Klee CB (1978) Conformation-dependent nitration of the protein activator of cyclic adenosine 3′,5′-monophosphate phosphodiesterase. Biochemistry 17:928–935
Rosen OM (1970) Preparation and properties of a cyclic 3′,5′-nucleotide phosphodiesterase isolated from frog erythrocytes. Arch Biochem Biophys 137:435–441
Ross EM, Haga T, Howlett AC, Schwarzmeier J, Schleifer LS, Gilaman AG (1978) Hormone-sensitive adenylate cyclase; resolution and reconstitution of some components necessary for regulation of the enzyme. Adv Cyclic Nucleotide Res 9:53–68
Sakai T, Makimo H, Tanaka R (1978) Increase activity of cyclic AMP phosphodiesterase from frozen-thawed rat liver. Biochem Biophys Acta 522:477–490
Schroder J, Richenberg HV (1973) Partial purification and properties of the cyclic AMP and the cyclic GMP phosphodiesterase of bovine liver. Biochem Biophys Acta 302:50–63

Schulman H, Greengard P (1978 a) Stimulation of brain membrane protein phosphorylation by calcium and an endogenous heat-stable protein. Nature 271:478–479

Schulman H, Greengard P (1978 b) Ca^{2+}-dependent protein phosphorylation system in membranes from various tissues, and its activation by calcium dependent regulator. Proc Natl Acad Sci USA 75:5432–5436

Schwabe U, Daly JW (1977) The role of calcium ions in accumulations of cAMP elicited by alpha and beta adrenergic agonists in rat brain slices. J Pharm and Exp Therap 202:134–142

Sharma RK, Wirch E, Wang JH (1978) Inhibition of Ca^{2+}-activated cyclic nucleotide phosphodiesterase reaction by a heat-stable inhibitor protein from bovine brain. J Biol Chem 253:3575–3580

Sharma RK, Wang T, Wirch E, Wang JH (1980) Purification and properties of bovine brain calmodulin-dependent cyclic nucleotide phosphodiesterase. Fed Proc 39:1624 (Abstract)

Sieghart W, Forn J, Greengard P (1979) Ca^{2+} and cyclic AMP regulate phosphorylation of same two membrane-associated proteins specific to nerve tissue. Proc Natl Acad Sci 76:2475–2479

Sobue K, Ichida S, Yoshida H, Yamazaki R, Kakuichi S (1979) Occurrence of a Ca^{2+}- and modulator protein activatable ATPase in the synaptic plasma membrane of brain. FEBS Lett 99:199–202

Sobue K, Kakuichi S (1980) Modulator binding proteins of testis. Biochem Biophys Res Commun 93:850–856
USA 2475–2479

Stevens FC, Walsh J, Ho HC, Teo TS, Wang JH (1976) Comparison of calcium-binding proteins. J Biol Chem 251:4495–4500

Stole V (1977) Mechanism of regulation of adenylate cyclase activity in human polymorphonuclear leukocytes by calcium, guanosyl nucleotides and positive effectors. J Biol Chem 252:1901–1907

Strada SJ, Pledger WJ (1975) The role of cyclic nucleotides in cell growth and development: regulation and characterization of cyclic nucleotide phosphodiesterases in mammalian cells. In: Weiss B (ed) Cyclic nucleotide disease. University Park Press, Baltimore, pp 3–34

Suzuki Y, Nagao S, Abe K, Hirabayashi T, Watanabe Y (1981) Tetrahymena calcium-binding protein is indeed calmodulin. J Biochem 89:333–336

Taniguchi T, Fujisawa M, Lee JJ, Hidaka H (1978) Cyclic 3′,5′-nucleotide phosphodiesterase of rabbit sinoatrial node. Biochem Biophys Acta 522:465–476

Teo TS, Wang JH (1973) Mechanism of activation of a cyclic adenosine 3′:5′-monophosphate phosphodiesterase from bovine heart by calcium ions. J Biol Chem 248:5950–5955

Teo TS, Wang TH, Wang JH (1973) Purification and the properties of the protein activator of bovine heart cyclic adenosine 3′:5′-monophosphate phosphodiesterase. J Biol Chem 248:588–595

Teshima Y, Kakiuchi S (1974) Mechanism of stimulation of Ca^{2+}-dependent phosphodiesterase from rat cerebral cortex by the modulator protein and Ca^{2+}. Biochem Biophys Res Commun 56:489–495

Thompson WJ, Terrasaki WL, Epstein PM, Strada SJ (1979) Assay of cyclic nucleotide phosphodiesterase and resolution of multiple molecular forms of the enzyme. Adv Cyclic Nucleotide Res 10:69–92

Thompson WJ, Appleman MM (1971) Multiple cyclic nucleotide phosphodiesterase activities from rat brain. Biochemistry 10:311–316

Tisdal MJ (1975) Characterization of cyclic 3′,5′-monophosphate phosphodiesterase from Walker Carcinoma sensitive and resistant to bifunctional alkylating agent. Biochem Biophys Acta 397:134–143

Tada M, Kirchberger A, Iorio JAM, Katz AM (1975) Control of sarcolemmal adenylate cyclase and sodium, potassium-activated ATPase activities. Circulation Res 36:8–17

Toscano WA, Westcott KR, LaPorte DC, Storm DR (1979) Evidence for a dissociable subunit required for calmodulin stimulation of brain adenylate cyclase. Proc Natl Acad Sci USA 76:5582–5586

Uzunov P, Weiss B (1971) Effects of phenothiazine tranquilizers on the cAMP system of rat brain. Neuropharmacology 10:697–708
Valverde I, Vandermeers A, Anjaneyulu R, Malaisse WJ (1979) Calmodulin activation of adenylate cyclase in pancreatic islets. Science 206:225–227
Van Inwegan RG, Pledger WJ, Strada SJ, Thompson WJ (1976) Characterization of cyclic nucleotide phosphodiesterase with multiple separation techniques. Arch Biochem Biophys 175:700–709
Von Hungen K, Roberts S (1973) Catecholamine and Ca^{2+} activation of adenylate cyclase systems in synaptosomal fractions from rat cerebral cortex. Nature New Biol 242:58–60
Waisman DM, Singler TJ, Wang JH (1978) The modulator-dependent protein kinase. A multifinctional protein kinase activatable by the Ca^{2+}-dependent modulator protein on the cyclic nucleotide system. J Biol Chem 253:3387–3390
Wallace RW, Cheung WY (1979) Calmodulin. Production of an antibody in rabbit and development of a radioimmunoassay. J Biol Chem 254:6564–6571
Wallace RW, Lynch TJ, Tallant EA, Cheung WY (1978) An endogenous inhibitor protein of brain adenylate cyclase and cyclic nucleotide phosphodiesterase. Arch Biochem and Biophys 187:328–334
Wallace RW, Tallant EA, Cheung WY (1980) Assay, preparation, and properties of calmodulin. In: Calcium and Cell Function, vol 1. Cheung WY (ed) Academic Press, New York p 13–40
Walsh M, Stevens FC (1977) Chemical modification studies on the Ca^{2+}-dependent protein modulator of cyclic nucleotide phosphodiesterase. Biochemistry 16:2742–2749
Walsh M, Stevens FC (1977) Characterization of tryptic fragments obtained from bovine brain protein modulator of cyclic nucleotide phosphodiesterase. J Biol Chem 252:7440–7443
Walsh M, Stevens FC (1978) Chemical modification studies on the Ca^{2+}-dependent protein modulator. The role of methionine residues in the activation of cyclic nucleotide phosphodiesterase. Biochemistry 17:3924–3930
Wang JH, Teo TS, Ho HC, Stevens FC (1975) Bovine heart protein activator of cyclic nucleotide phosphodiesterase. Adv Cyclic Nucleotide Res 5:179–194
Wang JH, Waisman DM (1979) Calmodulin and its role in the second messenger system. Current Topics in Cellular Regulation 15:47–107
Wang JH, Sharma RK, Tam ST (1980) Calmodulin-binding proteins. In: Calcium and Cell Function, vol 1. Cheung WY (ed) Academic Press, New York p 305–328
Watterson DM, Van Eldik LJ, Smith RE, Vanaman TC (1976) Calcium-dependent regulatory protein of cyclic nucleotide metabolism in normal and transformed chicken embryo fibroblasts. Proc Natl Acad Sci USA 73:2711–2715
Watterson DM, Sharief F, Vanaman TC (1980) The complete amino acid sequence of the Ca^{2+}-dependent modulator protein (calmodulin) of bovine brain. J Biol Chem 255:962–975
Weiss B, Hirt WN (1977) Selective cyclic nucleotide phosphodiesterase inhibitors as potential therapeutic agents. Ann Rev Pharmacol Toxicol 441–447
Wells JN, Hardman JG (1977) Cyclic nucleotide phosphodiesterase. Adv Cyclic Nucleotide Res 8:119–143
Welsh MJ, Dedman JR, Brinkley BR, Means AR (1979) Tubulin and calmodulin: effects of microtubules and microfilament inhibitors on localization in the mitotic apparatus. J Cell Biol 81:624–634
Welsh MJ, Dedman JR, Brinkley BR, Means AR (1978) Calcium-dependent regulator protein: localization in mitotic apparatus of eukaryotic cells. Proc Natl Acad Sci 75:1867–1871
Welton AF, Lad PM, Newby AC, Yamamura H, Nicosia S, Rodbell M (1977) Solubilization and separation of the glucagon receptor and adenylate cyclase in guanine nucleotide-sensitive states. J Biol Chem 252:5947–5956
Westcott KR, LaPorte DC, Storm DR (1979) Resolution of adenylate cyclase sensitive and insensitive to Ca^{2+} and calmodulin by calmodulin-Sepharose affinity chromatography. Proc Natl Acad Sci USA 76:204–208

Wolff DJ, Brostrom CO (1979) Properties and functions of the Ca^{2+}-dependent regulator protein. Adv Cyclic Nucleotide Res 11:28–88
Wong PYK, Cheung WY (1979) Calmodulin stimulates human platelet phospholipase A_2. Biochem Biophys Res Commun 90:473–480
Wood J, Wallace RW, Whitaker JN, Cheung WY (1980) Immunocytochemical localization of calmodulin and a heat-labile calmodulin-binding protein in basal ganglia of mouse brain. J Cell Biol 84:66–76
Yagi K, Yazawa M, Kakiuchi S, Ohshima M, Uenishi K (1978) Identification of an activator protein for myosin light chain kinase as the Ca^{2+}-dependent modulator protein 253:1308–1340
Yamauchi T, Fujisawa H (1979) Activation of tryptophan 5-monooxygenase by calcium-dependent regulator protein. Biochem Biophys Res Commun 90:28–35

Note Added in Proof

Although the lysine at residue 115 of calmodulin from bovine brain as well as many tissues is trimethylated, no such lysine derivative is found in calmodulin from others. Calmodulin not methylated is fully active in stimulating phosphodiesterase and myosin light chain kinase.

CHAPTER 8

Radioimmunoassay Techniques for Cyclic Nucleotides

G. BROOKER

Overview

Analysis of cyclic nucleotides by radioimmunoassay is discussed with respect to the preparation of antisera, [^{125}I]-labeled cyclic nucleotides, preparation of samples for analysis, assay, protocols, methods of data calculation and evaluation of assay validity. The increase in assay sensitivity upon 2′0-acylation is discussed with respect to the advantages of acetylation over succinylation. Details are given of methodology for total automation of the radioimmunoassay by the Gamma-Flo automated assay system.

A. Introduction

The discovery of cyclic AMP (RALL and SUTHERLAND 1958 a) and the introduction of the second messenger concept of hormone action was an attractive concept which explained the mechanism for the action of a variety of hormones. Consequently, a large number of investigators in a wide variety of fields desired to quantify cyclic AMP in order to test its involvement in almost every biological and biochemical system then under study. History tells the story – cyclic AMP and possibly other cyclic nucleotides are important regulators of a multitude of biological processes.

The original assay (RALL and SUTHERLAND 1958 b) for cyclic AMP was difficult to perform. Only a small handful of people could measure cyclic AMP for more than a decade after the isolation and identification of the cyclic nucleotide. The assay involved a cascade of events which required an inactive liver protein kinase (as we now know it); other events could be triggered not only by cyclic AMP, but also by calcium ion. It is now reasonable to see why the assay was hard to control on a daily basis; because the component reactions were unknown, "trouble shooting" was in many cases as logical as distilling well water from Willowick, Ohio (a suburb of Cleveland) for preparation of the assay reagents.

The information gathered by the mid-sixties was so impressive that it stimulated the development of a large number of assay methods for cyclic nucleotides (GREENGARD et al. 1972), many of which were more sensitive and/or easier to perform than the phosphorylase activation method. The first competitive protein binding assay for cyclic AMP and cyclic GMP was based upon our discovery that cyclic AMP phosphodiesterase and cyclic GMP phosphodiesterase had a low K_m form (10^{-6} *M*) (BROOKER et al. 1968) in addition to the previously described enzyme with a K_m of about 10^{-4} *M*. Subsequent discovery of endogenous binding proteins for cyclic nucleotides associated with cyclic nucleotide-dependent protein kinases (K_m 10^{-9} M) became the basis for fairly sensitive radioisotope displacement methods developed by GILMAN (1970) and BROWN et al. (1972). While these methods are still very useful, they do not offer the sensitivity or selectivity that is obtained with the radioimmunoassay method. The radioimmunoassay developed by STEINER et al. (1972) resulted from their discovery that high affinity antibodies to cyclic nucleotides and high specific activity [^{125}I]-labeled 2′-O-succinyl tyrosine methyl ester derivatives of cyclic nucleotides could be produced.

More recent improvements in cyclic nucleotide assays are the result of the important discovery by Cailla and her associates that base catalyzed succinylation by succinic anhydride would quantitatively lead to 2′O-succinyl cyclic nucleotides (CAILLA et al. 1973b). These derivatives have 40–100 times the affinity for the cyclic nucleotide antisera than do the unsuccinylated cyclic nucleotides. This provides a method to markedly increase assay sensitivity. Further improvement of assay sensitivity resulted when acetic anhydride was used (HARPER and BROOKER 1975) yielding 2′O-acetyl derivatives of cyclic nucleotides. Acetylation gave comparable sensitivity to the succinylation method without the interference of the remaining succinate anion and the inconvenience of adding a dry reagent to each tube. Finally, the assay has been reduced to total automation (BROOKER et al. 1976), allowing the *complete* assay of cyclic nucleotides at 60 samples per hour in the absence of any human intervention.

Several review articles and books have been written on cyclic nucleotide assay methods (GREENGARD et al. 1972; BROOKER et al. 1979a; DELAAGE et al. 1978; STEINER 1979). We have recently provided detailed directions for the preparation of reagents and performance of the acetylation radioimmunoassay of cyclic nucleotides (BROOKER et al. 1979b).

B. Preparation and Characterization of Antisera

The successful immunoassay of cyclic nucleotides requires good reagents and, most especially, antisera demonstrating high affinity and selectivity for the cyclic nucleotide to be assayed. If the selectivity of the antisera is high, then samples can contain contaminating substances without altering assay results. Excellent antisera can be raised in rabbits (KIMURA et al. 1974), rats (CAILLA et al. 1973a), and goats (BROOKER et al. 1979b); in our hands (BROOKER et al. 1979b), goats produced antisera of better quality than the antisera raised in rabbits.

Animals are immunized with 2′0-Succinyl cyclic AMP coupled to protein. The quality of the antisera is monitored for titer and selectivity between each antigenic boost. In this way, animals which are not producing can be eliminated and the good ones kept for continued antisera production.

Figure 1 demonstrates the specificity of well-characterized antisera against cyclic AMP (goat) and cyclic GMP (rabbit). The cyclic AMP antisera distinguishes between cyclic AMP and cyclic GMP by greater than 5 orders of magnitude; conversely, the cyclic GMP antisera recognizes cyclic GMP better than cyclic AMP. Both antisera reject other nucleotides by 6–10 orders of magnitude. In most cases, such selectivity eliminates concern about other nucleotide substances interfering with the analysis of cyclic nucleotides.

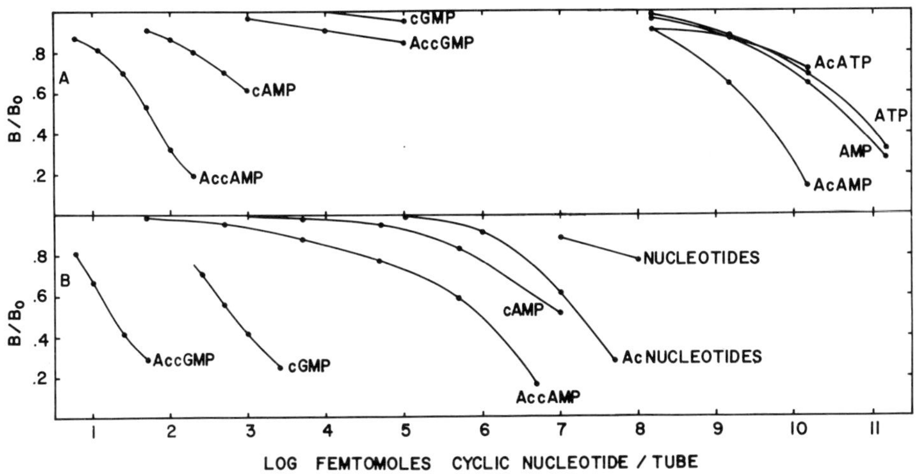

Fig. 1 A, B. Sensitivity and specificity of anti-cyclic AMP and anti-cyclic GMP sera. The effects of various nucleotides on the binding of ScAMP [^{125}I]-TME to their respective antibodies are shown. The amount of nonradioactive nucleotide per assay tube (in fmoles) is plotted on the logarithmic abscissa. Displacement is denoted by B/B_0, the ratio of the amount of radioactivity bound in the presence of competing nucleotide (B) to that bound in the absence of nonradioactive nucleotide (B_0). Cyclic AMP and cyclic GMP assays were performed as described using anti-cyclic AMP serum 122G at a 1:100,000 dilution panel **A** and anti-cyclic GMP serum 358G diluted to 1:10,000 panel **B**. The prefix of "Ac" denotes that the nucleotide(s) had been treated with TEA and acetic anhydride. "Nucleotides" refers to the use of a mixture containing the indicated concentration of each of seven purines. The purines contained in the "nucleotides" solution include adenosine, 5′-AMP, ADP, ATP, guanosine, 5′-GMP, and GTP. In addition, one-third as much GDP as GTP was included in the "nucleotides" solution. Thus the actual total concentration of nucleotides and nucleosides is 7.3 times the amount indicated on the abscissa of panel **B**. Since each nucleotide is capable of competing for antibody binding sites (with low affinity), actual displacement of the binding curve by any one purine would be shifted by approximately 1 log unit to the right. (Reprinted from Brooker et al. 1979b, by permission of Raven Press)

It is important to know the approximate concentration of contaminating nucleosides and bases in samples and to refer to cross reaction curves (as in Fig. 1) for assurance that the antisera being used has sufficient selectivity for the needs of a particular assay. Users of commercial kits should develop such data or obtain it from the kit manufacturer.

C. Preparation and Purification of [^{125}I]-Labeled Cyclic Nucleotides

The preparation of [^{125}I]-labeled succinyl tyrosine methyl ester derivatives of cyclic nucleotides was initially described by Steiner and colleagues (Steiner et al. 1972). Several improvements in the labeling, purification and storage of these derivatives have been described by Cailla and her associates (Delaage et al. 1978) and Brooker (Brooker et al. 1979b). Our laboratory prefers purification by descending paper chromatography because of its ease and the fact that the radioiodinated product is greater than 98% pure. Cailla prefers purification on Sephadex G-25 because it separates small molecules in the order of their iodination state, giving a mono-iodinated derivative with the same specific radioactivity as the initial iodide (DELAAGE et al. 1978).

An important difference in the iodination conditions from those originally reported by STEINER (1972) should be noted. The use of less chloramine T (BROOKER et al. 1979b) prevents the formation of a number of radioactive impurities which are difficult to remove by subsequent purification procedures. We have found that the iodinated derivatives are stable for more than three months when stored in 50% l-propanol.

D. Preparation of Samples for Assay

While the labeled antigen and antisera are very important components of a successful cyclic nucleotide radioimmunoassay, the extraction of cyclic AMP from tissues and the preparation of biological fluids for analysis need be performed with great care. It is important that this step in the assay quantitatively extract the cyclic nucleotide and preserve its integrity once extracted; the extractant should not interfere with the subsequent assay nor lead to non-enzymatic production of cyclic nucleotide. These possibilities should be verified whenever a departure from a well-characterized method is initiated. When possible, cyclic nucleotide standards should be made in media closely resembling the samples being measured.

It is preferable not to purify samples from an initial extract or physiological fluid before analysis; such manipulations inevitably reduce assay reproducibility. Previous purification of samples is necessary whenever the antisera specificity is insufficient to discriminate against the presence of substances other than the specific cyclic nucleotide being measured. The decision to remove interfering substances should be made after an evaluation of assay validity as discussed at the end of this chapter.

Several factors can contribute to assay variability and inaccuracies during the preparation of samples for analysis. The simplest extraction method is the use of 0.1 *N* HCl to inactivate the enzymes involved in cyclic nucleotide metabolism and to remove cyclic nucleotides from cells. This method works well for many tissues and has been used for suspension and monolayer cell cultures. Upon modest dilution of the sample, there is no interference in the assay; furthermore solutions of 0.1 *N* HCl can be directly acetylated after addition of triethylamine and acetic anhydride. It is important to use a sufficient volume of 0.1 *N* HCl (or other extractant) so that the tissue components do not significantly raise the pH of the extract above that of the extracting solution. Trichloroacetic acid (usually 5%) is also a popular extractant. However, the trichloroacetic acid must be neutralized or extracted before reliable analysis can be performed. Water saturated ethyl ether (3 extractions with 3–5 volumes) removes a sufficient amount of the acid to allow an adequate assay of the aqueous extract. In some cases, it can be neutralized with trishydroxymethylaminoethane or with solid calcium carbonate (TIHON et al. 1977). However, neutralization can leave the sample basic which is not desirable, because cyclic nucleotides (especially 2′0-acetylated or -succinylated derivatives) are unstable when kept in alkaline solutions. In addition, the salt concentration is about 0.3 *M* when 5% trichloroacetic acid is neutralized. Perchloric acid (1 *N*) can also be used to extract cyclic nucleotides from tissues. The perchlorate anion is removed as insoluble potassium perchlorate after neutralization of the sample with potassium hydroxide or potassium carbonate (BAGNARA and FINCH 1972).

When purification of samples by column chromatography of samples is necessary, it is important to determine purification recovery. In somc cases, inclusion of trace [^{3}H]-cyclic nucleotide is possible, but the amount which can be added is limited by the specific activity of the tritiated cyclic nucleotide. The "tracer" should not contribute sufficient cyclic nucleotide to the sample, thereby altering assay results. For example, inclusion of 22,200 dpm of [^{3}H]-cyclic AMP (specific activity 10 Ci/mmole) to a sample introduces 1,000 femtomoles of cyclic AMP. Thus, when femtomole amounts of cyclic nucleotides are being measured, it may not be possible to monitor recovery for each sample; prior determination of the range of recovery for the purification scheme may be desirable. A wide variety of purification methods have been devised (GREENGARD et al. 1972); in general, they involve anion or cation exchange chromatography.

E. Acetylation or Succinylation

STEINER et al. (1972), in their initial description of the radioimmunoassay of cyclic nucleotides, demonstrated that 2′0-succinyl, 2′0-succinyl tyrosine methyl ester, 2′0-monobutyryl, and 2′0-monooctanoyl derivatives of cyclic AMP interacted with cyclic nucleotide antisera about two orders of magnitude better than cyclic AMP itself. This is not surprising since the antigen used to immunize animals is a 2′0-succinyl cyclic nucleotide-protein conjugate. Thus, the discovery that cyclic nucleotides could be readily and quantitatively converted to 2′0-succinyl cyclic nucleotides in basic aqueous solution (CAILLA et al. 1973b) made possible the assay of cyclic nucleotides in small samples of biological material.

The succinylation procedure has certain disadvantages. The most serious of these is that succinate ion, a by-product of the hydrolysis of succinic anhydride, remains in high concentrations and, unfortunately, interferes either by cross-reacting with the antibody or by altering the ionic environment. Dilution of succinylated samples or separation of succinate ion by column chromatography eliminates the interference but either reduces assay sensitivity or introduces assay variability.

The acetylation method involving acetic anhydride was developed to capitalize upon the increased sensitivity possible by 2′0-substitution. The acetate ion, the product of the hydrolysis of excess acetic anhydride, does not interfere with the assay. Acetylation provides the basis for a rather simple, highly sensitive assay method (HARPER and BROOKER 1975). The complete assay is set up in 50 m*M* sodium acetate, pH 4.75, so that acetate ion, remaining after the acetylation, can not alter the pH of the assay mixture. Sufficient triethylamine is provided to initially make the solution basic to catalyze the acetylation, and then to neutralize the excess acetic acid which results from the hydrolysis of acetic anhydride. Thus, the final solution is acidic, a condition in which the acetyl cyclic nucleotide is relatively stable. There has been the suggestion that multiple sites of acetylation can occur with cyclic GMP (FRANDSEN and KRISHNA 1976); however, using the acetylation as initially described (HARPER and BROOKER 1975), only one site of acetylation occurs (BROOKER et al. 1979b). Cailla (DELAAGE et al. 1978) has suggested that potassium hydroxide can be substituted for triethylamine and that this reduces secondary reactions on cyclic GMP.

F. Assay Procedure

We have recently described the step by step method for the acetylation immunoassay of cyclic AMP and cyclic GMP (BROOKER et al. 1979b). The typical assay is conducted in a volume of 300 μl and contains the appropriate amounts of 2′0-[^{125}I]-succinyl tyrosine methyl ester cyclic nucleotide and antisera sufficient to bind 30%–40% of the radioligand in the absence of 2′0-acetyl cyclic nucleotide. Standards are prepared and acetylated on the day of the assay. It is preferable to prepare standards in the same media as the unknown samples. Once sufficient time has elapsed, antibody-bound radioligand is separated from free radioligand by the introduction of a slurry of charcoal in a solution of bovine albumin. The albumin is important because it occupies sites upon charcoal which absorb antibody and would lead to underestimation of the amount of bound radioactivity. The bound radioactivity should be determined with a gamma counter. The time for counting should be sufficient to yield counting accuracy of 1%–2%.

G. Data Analysis

It is now well established how to calculate radioimmunoassay data (RODBARD et al. 1968). The "C-fit" technique for data reduction (BROOKER et al. 1979b), a method involving a log-log iterative regression analysis, has worked well for a wide variety of immunoassays including the cyclic nucleotide immunoassays. Samples with B/B_0 values either less than 0.15 or greater than 0.9 cannot accurately be determined; samples yielding such observations should either be diluted or concentrated (as required) and reassayed.

H. Evaluation of Assay Validity

Displacement of radioligand binding to specific cyclic nucleotide antibodies is suggestive that cyclic nucleotide is present. The better the selectivity of the antisera, the higher the probability that cyclic nucleotide is present. Several procedures can be used to confirm that a sample actually contains a cyclic nucleotide. Evaluation of "parallelism" over a wide range of sample dilutions is a useful technique to help confirm the authenticity of the cyclic nucleotide. Thus, dilutions of the sample should follow the standard curve, and "spiking" the sample with authentic cyclic nucleotide should lead to assayable recovery of the added cyclic nucleotide. Finally, one could subject the sample to purification before assay and confirm that no assayable sample is lost.

I. Total Automation of the Cyclic AMP and Cyclic GMP Assay with the Gamma-Flo Assay System

We have developed an integrated and totally automated radioimmunoassay system (BROOKER et al. 1976) which can perform a wide variety of radioimmunoassays including those for cyclic AMP and cyclic GMP. We have recently described the performance of the cyclic AMP method (BROOKER and MURAD 1980). The system, in a continuous flow mode, picks up sample (25 μl), combines it with radioligand and antisera, allows a precise incubation time of 5–6 min (sufficient time for ligand binding to the antisera), separates free from bound radioligand, counts the bound radioligand in a static mode, calculates the standard curve and compares unknown samples to that curve. Sixty samples per hour are measured in this manner. This system is now commercially available from E.R. SQUIBB.

Acknowledgement. The achievement of the currently sensitive, specific, and simple analytical methods for cyclic nucleotides are the result of many investigators' unswerving and relentless efforts.

Celinda Johnson provided excellent assistance in the preparation of this manuscript. Contributions to the assay from my laboratory were supported by NIH grants HL 15985, HL 19242, and AM 22125.

References

Bagnara AS, Finch LR (1972) Quantitative extraction and estimation of intracellular nucleoside triphosphates. Anal Biochem 45:24–34

Brooker G, Greengard P, Robison GA (1979a) Advances in cyclic nucleotide research, vol 10. Raven Press, New York

Brooker G, Harper JF, Terasaki WL, Moylan RD (1979b) Radioimmunoassay of cyclic AMP and cyclic GMP. In: Brooker G, Greengard P, Robison GA (eds) Advances in cyclic nucleotide research, vol 10. Raven Press, New York, pp 1–33

Brooker G, Murad F (1980) Automated Gamma-Flo™ immunoassay of urinary cyclic AMP. Clin Chem (in press)

Brooker G, Terasaki WL, Price MG (1976) Gammaflow: A completely automated radioimmunoassay system. Science 194:270–276

Brooker G, Thomas LJ, Appleman MM (1968) The assay of adenosine 3′,5′-cyclic monophosphate in biological materials by enzymatic radioisotopic displacement. Biochemistry 7:4177–4181

Brown BL, Ekins RP, Albano JDM (1972) Saturation assay for cyclic AMP using endogenous binding protein. In: Greengard P, Robison GA (eds) Advances in cyclic nucleotide research, vol 2. Raven Press, New York, pp 25–40

Cailla HL, Cros GS, Jolu EJP, Delaage MA, Depied RC (1973a) Comparison between rat and rabbit anticyclic AMP antibodies – specificity toward acyl derivatives of cyclic AMP. Anal Biochem 56:383–393

Cailla HL, Racine-Weisbuch MS, Delaage MA (1973b) Adenosine 3′,5′-cyclic monophosphate assay at 10^{-15} mole level. Anal Biochem 56:394–407

Delaage MA, Roux D, Cailla HL (1978) Recent advances in cyclic nucleotide radioimmunoassay. In: Folco G, Paoletti R (eds) Molecular biology and pharmacology of cyclic nucleotides. Elsevier, Holland, pp 155–171

Frandsen EK, Krishna G (1976) A simple ultrasensitive method for the assay of cyclic AMP and cyclic GMP in tissues. Life Sci 18:529–542

Gilman AG (1970) A protein binding assay for adenosine 3′:5′-cyclic monophosphate. Proc Natl Acad Sci USA 67:305–312

Greengard P, Robison GA, Paoletti R (1972) Advances in cyclic nucleotide research, vol 2. Raven Press, New York

Harper JF, Brooker G (1975) Femtomole sensitive radioimmunoassay for cyclic AMP and cyclic GMP after 2′0 acetylation by acetic anhydride in aqueous solution. J Cyclic Nucleotide Res 1:207–218
Kimura H, Thomas E, Murad F (1974) Effects of decapitation, ether and pentobarbital on guanosine 3′,5′-phosphate and adenosine 3′,5′-phosphate levels in rat tissues. Biochim Biophys Acta 343:519–528
Rall TW, Sutherland EW (1958a) Formation of a cyclic adenine ribonucleotide by tissue particles. J Biol Chem 232:1065–1076
Rall TW, Sutherland EW (1958b) Fractionation and characterization of a cyclic adenine ribonucleotide formed by tissue particles. J Biol Chem 232:1077–1091
Rodbard D, Rayford PO, Cooper JA, Ross GT (1968) Statistical quality control of radioimmunoassays. J Clin Endocrinol Metal 28:1412–1418
Steiner AL (1979) Cyclic AMP and cyclic GMP. In: Jaffe BM, Behrman HR (eds) Methods of hormone radioimmunoassay. Academic Press, New York, pp 3–17
Steiner AL, Parker CW, Kipnis DM (1972) Radioimmunoassay for cyclic nucleotides. I. Preparation of antibodies and iodinated cyclic nucleotides. J Biol Chem 247:1106–1113
Tihon C, Goren MB, Spitz E, Rickenberg HV (1977) Convenient elimination of trichloroacetic acid prior to radioimmunoassay of cyclic nucleotides. Anal Biochem 80:652–653

CHAPTER 9

Immunocytochemistry of Cyclic Nucleotides and Their Kinases

C. L. Kapoor and A. L. Steiner

Overview

The immunocytochemical localization of cyclic nucleotides and cyclic nucleotide-dependent protein kinases is a useful tool to identify the site of action of cyclic nucleotides in cells. This approach utilizes specific antibodies which detect antigenic molecules in tissue sections and cells. The detection of these antigens is done either by fluorescence or by bright field microscopy. Cyclic nucleotide immunocytochemistry utilizes cAMP and cGMP antisera which are routinely screened by their radioimmunoassays for crossreactivity with structurally related nucleotides. Cyclic nucleotide-dependent protein kinase immunocytochemistry utilizes antisera to type I cAMP-dependent protein kinase regulatory subunit (R^I), type II cAMP-dependent protein kinase regulatory subunit (R^{II}), catalytic subunit (C), type I cAMP-dependent holoenzyme ($R_2^IC_2$) and cGMP dependent protein kinase (G-kinase); all these antisera are screened for their specificity by immunodiffusion and immunoprecipitation techniques.

Antisera toward R^I, R^{II}, and C subunit have shown complete identity for their respective subunit or holoenzyme containing the subunit; crossreactivity against other subunits and holoenzyme has not been observed. G-kinase antisera have shown complete identity for G-kinase; crossreactivity with cAMP-dependent protein kinases and their subunits has not been observed. The formation of a fused precipitin band for the pure antigen and a crude antigen from tissue homogenates is taken as a criteria for the selection of antiserum for protein kinases immunocytochemistry. Thus far, it has been possible to produce antisera that recognize determinants present on both the holoenzyme and isolated subunits of type I cAMP dependent protein kinase; however, immunocytochemical data of tissue staining patterns suggest that antisera recognize determinants unique to the subunits versus holoenzyme and vice-versa. In the future it will be important to prepare subunit versus holoenzyme specific antibodies for immunocytochemistry by either selective immunoabsorption or by the hybridoma monolocal technique.

In order to quantitate the tissue and species crossreactivity of various R^I antisera for immunocytochemistry, a sensitive and specific radioimmunoassay for R^I was developed. This technique is found useful as an initial screen for selecting various R^I antisera for immunocytochemistry. The radioimmunoassay has also detected a site specific R^I antiserum which differentiated immunoreactivity of R^I, R^I·cAMP and $R_2^IC_2$. It seems possible that development of similar radioimmunoassays for C, R^{II}, and G-kinase may be useful for probing the structure, function, and antigenic determinants of cyclic nucleotide dependent protein kinases.

A. Introduction

By making possible the visualization of organelles in situ, cytochemistry has aided the interpretation of biochemical analysis of subcellular fractions from tissue homogenates. It has added a new dimension to the understanding of structural and regulatory functions of various cellular constituents. Immunocytochemical techniques can be more sensitive than standard biochemical techniques, because the antigen-antibody reaction visualizes the antigen only in limited compartments of cells or tissues which are not detectable by biochemical techniques (BLOOM et al. 1978).

It has previously been shown that specific antibodies raised against cyclic nucleotides can be used to localize cyclic nucleotides in subcellular compartments of cells (WEDNER et al. 1972; FALLON et al. 1974; STEINER et al. 1976; SPRUILL and STEINER 1979). Similar techniques are currently being extended to gain information about the immunocytochemical localization of cyclic nucleotide dependent protein kinases. This chapter, summarizes the current aspects of cytochemical approaches that are applicable for localization of cyclic nucleotides and their protein kinases. These approaches attempt to identify the site(s) of action of cyclic nucleotides and their kinases in tissue or cultured cells (STEINER et al. 1976), thus providing a means for assessing potential sites of action of cyclic nucleotides and their kinases under various physiological and fundamental processes. This review will focus on methodological and theoretical consideration of this immunocytochemical approach and will not concentrate on the data generated by this methodology.

B. Principles

Antibodies tagged covalently with a fluorescent dye offer a means for the precise identification and microscopic localization of antigens. This technique was introduced originally by COONS et al. (1941). Fluorescein isothiocynate and tetramethyl rhodamine isothiocynate are the most commonly utilized fluorochromes which fluoresce brightly in the visible spectrum when excited at the appropriate wave length. The fluorochrome molecules can bind covalently to antibodies without impairing their ability to fluoresce and without affecting the binding of antibody to the antigen in the tissue section. These techniques have recently been reviewed by STERNBERGER (1974) and KAWAMURA (1977). In fluorescence immunocytochemistry, the two most commonly employed approaches are the direct and indirect methods. The direct technique utilizes labelled antibody to localize, in a single step, the antigen in the tissue section. By this method, two or more antibodies may be labelled with different spectrum fluorochromes and used to localize different antigens in the same tissue section (e.g. fluorescein-isothiocynate coupled to a specific cGMP antiserum and tetramethyl rhodamine isothiocynate coupled to a cAMP antiserum localize cAMP and cGMP respectively in the same section, PRYZVANSKY et al. 1980). Alternatively, the indirect technique, which is the most commonly used method, employs an unlabelled antiserum as the first step, followed by a commercially available, fluorochrome-labelled second antiserum. The indirect technique offers the advantage of increased sensitivity because several fluorescently-labelled secondary antibodies react with the primary immunoglobulin of the species in which the primary antiserum was made.

Another commonly employed approach for the localization of cellular antigens is bright field immunocytochemistry. This technique involves conjugation of antibodies to the enzyme horse radish peroxidase which can develop a histochemically demonstrable reaction product by catalyzing a reaction with its substrates at the site of the antigens in the tissue (GRAHAM and KARNOFSKY 1966). Peroxidase covalently linked to specific antibodies forms an electron dense precipitate at the site of the antigen when exposed to hydrogen peroxide and diaminobenzidine (DAB), an electron donor. Because the insoluble oxidation product is osmiophillic, the antigen-antibody reaction site can be observed by light as well as electron microscopy. Other substrates can be used with peroxidase to give various colored products (NAKANE 1968, 1970).

Although direct conjugation of horse radish peroxidase may give higher resolution in localizing the antigen, the unlabelled antibody enzyme-bridge technique is often the method of choice (STERNBERGER et al. 1970). The latter method was developed to enhance the sensitivity and overcome problems of specificity which can arise during conjugation of an enzyme to primary antibodies. In the bridge technique, a multivalent unlabelled secondary immunoglobulin forms a bridge between the primary immunoglobulin, which is bound to the cellular antigen, and antibody against the enzyme marker peroxidase, which is made in the same species as the primary antibody. Using this technique, only the commercially available, bridge antiserum and peroxidase-antiperoxidase (PAP) reagent are required to visualize any number of antigens which have had antibodies raised to them in a particular species. The PAP reagent represents an immunologically rather than chemically, labelled gammaglobulin and is used for the detection of antigen-antibody reactions through the "bridge antiserum."

C. Techniques

The immunocytochemical localization of cyclic nucleotides and their kinases routinely utilizes unfixed frozen sections. Alternatively, the tissue sections can be fixed in 2% paraformaldehyde prior to the staining procedure. The staining pattern of various antigens does not appear to differ from unfixed frozen or fixed frozen tissue sections. Generally, the freshly removed pieces of tissue or the incubated tissue slices (after hormonal stimulation) are gently placed at the center of an aluminum foil "boat" which contains optimal cutting temperature (OCT) compound. The samples are then cooled in the dry ice-acetone bath. Moreover, when experimental conditions require freezing the tissue instantaneously after hormonal stimulation, the pieces of tissue are frozen immediately on a cardboard by immersion in prechilled 2-methyl butane in liquid nitrogen (KAPOOR and KRISHNA

1977). Frozen sample blocks are placed in a cryostat chamber and allowed to equilibrate at −20 °C, and sections of 4–6 μ are cut with a cryostat microtome. These sections are placed on slides or on a coverslip and air-dried for 30–60 min. Sections either can be processed for immunocytochemistry staining or can be stored at −20 °C in a sealed box. For optimal staining, it is advisable to use the freshly cut sections from a tissue.

The air-dried sections are equilibrated with phosphate buffered saline (PBS) for 10 min by 2–3 changes of PBS. For cytochemistry utilizing fluorochrome conjugated second antibody, slides or coverslips are removed from the PBS, and excess PBS is drained and blotted. The sections are then covered with appropriately diluted cyclic nucleotide or protein kinase antiserum for 30 min in a moist chamber. Incubation of tissue sections can be increased to overnight by utilizing higher dilutions of first antibody (CUMMING et al. 1980). The section on the slides or coverslips are then rinsed three times with PBS for 3 min and excess PBS is drained and blotted. Slides or coverslips are then covered with appropriately diluted fluorochrome or rhodamine conjugated (Miles or Cappel Labs) second antibody (anti-primary antisera IgG) for 30 min. The tissue sections are then rinsed 3 times with PBS for 3 min and excess PBS is drained and blotted. Slides are mounted with 50% glycerin in PBS and are examined by using a dark field fluorescence microscope equipped with an epi-illumination system through an appropriate filter.

For bright field immunocytochemistry utilizing the peroxidase-antiperoxidase unlabelled antibody enzyme method, sections are equilibrated with PBS, and excess PBS is drained and blotted. The sections on the slide or coverslips are incubated with 1% (v/v) of normal serum from the animal species of the "bridging antiserum" or with 0.5%–5% of gelatin solution in PBS for 30 min at room temperature. This procedure greatly reduces the nonspecific staining due to residual antibodies. The slides or coverslips are drained and blotted and then covered with a primary antiserum generated against a cyclic nucleotide or a protein kinase subunit at optimal dilution for overnight. The slides or coverslips are again rinsed with PBS 3 times for 10 min; excess PBS is drained and blotted. Specimens are incubated with peroxidase-antiperoxidase complex (PAP), washed thoroughly with PBS for 30 min and then incubated with a mixture of diaminobenzidine (0.025%) and H_2O_2 (0.01%) for 5–8 min at room temperature. Specimens are then washed several times with PBS and mounted with 50% glycerin in PBS. Slides are examined by using a bright field microscope.

For staining of cultured or isolated cells, one of the following procedures is utilized: a) cells are placed on gelatin coated slides and dried using a stream of air (this procedure is necessary to ensure penetration of antiserum into the cells); or b) for cells grown on the coverslips, the coverslips are removed from the media and rinsed with PBS. The cells are fixed with 1%–10% formaldehyde in PBS for 10–30 min and then made permeabile by immersing in methanol (−20 °C) for 4 min and acetone (−20 °C) for 10 min (WEBER et al. 1975).

D. Cyclic Nucleotide Immunocytochemistry

The potential role(s) of cAMP and cGMP in cellular metabolism have made it increasingly desirable to measure these nucleotides in heterogenous tissues accumulating cyclic nucleotide during various physiological processes (WEDNER et al. 1972; STEINER et al. 1976). Moreover, it is equally important to identify in a cell the subcellular localization of a cyclic nucleotide. For detailed reviews of this technique, the reader is referred to STEINER et al. (1976) and SPRUILL and STEINER (1979).

I. Preparation of Antisera

Rabbits, guinea pigs, and goats are generally utilized for the production of cAMP and cGMP antisera. However, the choice of each animal has some practical advantages and disadvantages. For example, the use of rabbit and guinea pig is handy, easily accommodated in general laboratory facilities and is less expensive in upkeep. The goat, on the other hand, provides a much larger volume of serum with each bleeding.

The cyclic nucleotides are rendered immunogenic by coupling them to human serum albumin (HSA) or keyhole limpet hemocyanin (KLH) through a 2′-0 succinyl derivative of the cyclic nucleotide (STEINER et al. 1972) by using the water soluble ethyl-3-(3-dimethyl-aminopropyl)-carbo-diimide (EDC). The 2′-0 monosuccinyl derivative of cAMP or cGMP (15 mg) is reacted with HSA or KLH (30 mg) in 3 ml of distilled water in the presence of EDC-HCl (15.5 mg). The pH of the reaction mixture is maintained at 5.5, and the reaction is allowed to proceed for 18 h at room temperature in the dark. The mixture is dialyzed extensively against PBS after the reaction is completed. The conjugates exhibit a maximum absorbance at 260 nm compared with unconjugated protein which has a maximum absorbance of 280 nm. Freshly conjugated cyclic nucleotides are thoroughly emulsified in an equal volume of Freund's complete adjuvant. Injections of 1 mg conjugate per rabbit are made in the foot pad of the animal or in the back of the animal. Booster injections of 0.15 mg are made at 4–6 week intervals at multiple sites into the back of the animal, and the animals are bled via a vein seven days after the booster. The sera are fractionated with 40% ultrapure saturated ammonium sulfate; the resulting pellets are resuspended, dialyzed extensively against phosphate buffer saline (pH 7.4), clarified by centrifugation, and stored at −20 °C.

The injection of 2′-0-succinylated-cyclic nucleotide conjugates results in the production of a heterogenous population of antibodies of various avidity and specificity. Each individual antiserum is checked for crossreactivity against ATP, 5′-AMP, GTP, and other cyclic nucleotides by radioimmunoassay (STEINER et al. 1972), so that antiserum can be utilized witout purification for immunocytochemistry. Antisera with virtually no crossreactivity with ATP, 5′AMP, and GTP are routinely found during the course of immunization, and these antisera also usually show less than 0.1% crossreactivity with other cyclic nucleotides.

II. Testing of Antisera for Specificity of Staining

After characterization, various cyclic nucleotide antisera are checked for specificity of staining, relative to background, in each tissue under investigation. Various useful controls are employed. It is important to obtain a preimmune serum from an animal prior to the injection of cyclic nucleotide protein conjugate. This preimmune serum should be used to stain the tissue at the same IgG concentration and in the same manner as the immunized serum. This control will reveal any natural antibodies to the antigen that were present in the serum prior to immunization. Alternately, a cyclic nucleotide depleted antisera (ORTEZ et al. 1980) prepared by passage of a cyclic nucleotide antisera over cyclic nucleotide-sepharose could be utilized as a control.

One of the better criteria to check the specificity of staining is the absorption of the specific cyclic nucleotide antiserum with the respective cyclic nucleotide antigen. Both liquid and solid phase absorption procedures are commonly employed for the absorption of staining patterns (STEINER et al. 1976; SPRUILL and STEINER 1979). Liquid phase absorption involves blocking antibody binding sites by preincubation of primary or directly fluorochrome labelled antibody with an excess of the respective cyclic nucleotide. In this procedure, specific staning is substantially eliminated or diminished when an antiserum is incubated with the appropriate concentrations of the respective cyclic nucleotide for 16–20 h at 4 °C. A final check of the specificity of staining is carried out by incubation of the cyclic nucleotide antiserum with other nucleotides such as ATP, 5′AMP, GTP or other cyclic nucleotides (STEINER et al. 1976) which should not modify the cyclic nucleotide staining pattern. Because of the heterogeneity of antibodies (BAUVOIS and LENG 1976) present in varius cyclic nucleotide antisera obtained from different animals, the concentration required to diminish specific staining with a particular cyclic nucleotide varies from one antiserum to another. It is our general experience that with certain antisera, staining is completely blocked by 1 μ*M* or less of the respective cyclic nucleotide, while with other antisera, a much higher concentration of cyclic nucleotide (1 m*M*) is required to eliminate the staining. With still other antisera, the staining pattern is not blocked with 1 m*M* concentrations of the respective cyclic nucleotide even though it is eliminated by solid phase absorption. Thus, although the blocking of antibody binding by excess of antigen is helpful in assessing the specificity of an antigen, it is not a definite answer of the question of specificity. A final check of specificity by solid phase absorption of the antiserum should be carried out.

A confirmatory test of staining specificity is carried out by utilizing affinity chromatography of the specific cyclic nucleotide antiserum. In this procedure, a cyclic nucleotide is covalently coupled to cyanogen bromide activated sepharose (FALLON et al. 1974). The sepharose column is washed with a 1% solution of gelatin to reduce nonspecific serum protein adhesion to the columns. Antibodies are selectively removed by passage of the antiserum over the respective cyclic nucleotide-immobilized sepharose column. The selective removal of antibodies is checked by radioimmunoassay (STEINER et al. 1972). Antibodies eluted from the column under appropriate conditions (FALLON et al. 1974) should reveal the staining pattern of the respective cyclic nucleotide. The specificity of the antiserum is determined by selective removal of cAMP antibodies by cAMP immobilized sepharose column, but not by cGMP-immobilized column; conversely, cGMP antibodies are selectively removed by cGMP-immobilized sepharose column but not by cAMP sepharose columns. In this way both positive and negative controls are obtained for each antiserum.

In the enzyme conjugated antibody method, specific enzyme (usually horse radish peroxidase) labelled antibodies are utilized (NAKANE and KAWOI 1974). These enzyme conjugated antibodies provide the highest possible antigenic staining specificity because the antisera are utilized at higher dilution and this reduces nonspecificity problems associated with the fluorochrome labelled antibodies. These enzyme conjugated antibodies can be utilized directly or indirectly similar to the way fluorochrome labelled antibodies have been used. The reaction product of the conjugated antibodies can be visualized at both the light and electron microscopic levels.

In the unlabelled antibody enzyme method, the controls for specificity are similar to that of fluorescence microscopy. In addition, to check the specificity of the bridging antiserum and peroxidase-antiperoxidase reagents: a) a dark background should be observed if the bridging antiserum, applied in the second step, is not added; b) nonspecific staining of the bridging antiserum PAP complex and reagents should be checked by excluding the primary antiserum from the staining procedure and specific staining should be substantially reduced by this procedure.

III. Methodological Considerations

There are several questions that are still unanswered in evaluating cyclic nucleotide immunocytochemistry. These include: (1) What moiety of cAMP binding protein is being detected by this methodology? and (2) How great are the losses of cyclic nucleotides during the staining procedure? There is evidence to show that in different tissues, 10%–40% of cyclic nucleotides are retained after the staining procedures (CUMMING et al. 1980; ORTEZ et al. 1980). Moreover, we believe that the cytochemical technique detects tightly bound and, most probably, the physiologically active moiety of cyclic nucleotides, because free and loosely bound cyclic nucleotides are washed out during the staining procedure. However, the question of what moiety of cyclic nucleotide is detected by this methodology is still not resolved. Our recent data of a striking co-localization of cGMP and cGMP-dependent protein kinase (STEINER et al. 1978a) in a variety of different tissues would tend to support the idea that cyclic nucleotide immunocytochemistry is detecting cyclic nucleotides bound to their receptor proteins, most likely the receptor subunits of cyclic nucleotide dependent protein kinases. Further support of this hypothesis is obtained from the similar pattern of localization of cAMP and type I regulatory subunit of cAMP-dependent protein kinase in several tissues. Taken together, it appears that binding of cyclic nucleotides to receptor proteins still allows recognition of cyclic nucleotide determinants by antibodies. Proof of this hypothesis, however, remains to be demonstrated.

E. Protein Kinase Immunocytochemistry

In recent years, evidence has accumulated to show that cAMP is the mediator of many hormonally-mediated processes. Most, if not all, of the effects of cAMP on these processes are mediated through phosphorylation events carried out by cAMP-dependent protein kinases (RUBIN and ROSEN 1975; NIMMO and COHEN

1977; GREENGARD 1978; EARP and STEINER 1978; KREBS and BEAVO 1979; see also this volume). At least two forms of cAMP-dependent protein kinases have been identified in various mammalian tissues (WALSH et al. 1968; KUO and GREENGARD 1969; KREBS 1972; RUBIN and ROSEN 1975; CORBIN et al. 1975). The two forms are commonly referred to as type I and type II; the physiochemical (RUBIN and ROSEN 1975; CORBIN et al. 1975; HOFMANN et al. 1975; ROSEN and ERLICHMAN 1975; POTTER and TAYLOR 1979; ZOLLER et al. 1979) and immunochemical (FLEISHER et al. 1976; KAPOOR et al. 1979 a, b) properties of their regulatory subunits differ, while the catalytic subunits appear similar (BEAVO et al. 1975; SUGDEN et al. 1976; ZOLLER et al. 1979). Both of these enzymes are tetramers consisting of a cAMP dimeric regulatory subunit and two catalytic subunits. Activation of the enzyme occurs by a cAMP promoted dissociation of the holoenzyme to yield two catalytic subunits and a dimeric regulatory subunit-cAMP complex (RUBIN and ROSEN 1975; HOFMANN et al. 1975; BEAVO et al. 1975). cGMP-dependent protein kinases are also known to be present in various biological systems (KUO and GREENGARD 1970; KUO 1974; KUO et al. 1978). While the cGMP-dependent protein kinases phosphorylate in vitro many of the same substrates of cAMP-dependent protein kinases, generally the former enzymes have low affinities for these substrates (LINCOLN and CORBIN 1977; HASHIMOTO et al. 1976; LINCOLN and CORBIN 1978 a, b). The cGMP-dependent protein kinases are different from cAMP-dependent protein kinases because the regulatory subunits and catalytic subunits are not dissociated upon activation by cGMP (GILL et al. 1976; LINCOLN et al. 1977). The enzyme appears to be similar to cAMP-dependent protein kinase in certain physical properties, amino acid composition and substrate specificity (LINCOLN and CORBIN 1977; LINCOLN et al. 1978 a, b).

The mechanism(s) by which the cyclic nucleotide-dependent protein kinases and their components are translated into various physiological processes are, in many cases, not known (EARP and STEINER 1978; GREENGARD 1978; KREBS and BEAVO 1979). One metod to explore the regulatory function of an enzyme is to determine its cellular and subcellular localization. For this reason, we have (STEINER et al. 1978) utilized an immunocytochemical approach for examining and monitoring the subcellular distribution of cyclic nucleotide-dependent protein kinases and their subunits in a variety of tissues and cultured cells. In addition, several independent studies in the literature indicate that one method of regulating the function of a protein kinase may involve the translocation of a regulatory and/or catalytic subunit of protein kinase to various sites within the cell (JUNGMANN et al. 1974; CASTAGNA et al. 1975; COSTA et al. 1976; SPEILVOLGEL et al. 1977; KUROSAWA et al. 1979; CHO CHUNG 1980). An immunocytochemical approach might be helpful in supporting the concept of translocation of these subunits after hormonal stimulation.

I. Purification of Immunogens

The most common method of purification of cyclic nucleotide receptor proteins is by cyclic nucleotide affinity chromatography. The detailed methodology and potential problems of purification have been reviewed recently (DILLS et al. 1979). For preparations of bulk amounts of immunogens for immunocytochemical work, the cyclic nucleotide-dependent protein kinase and their subunits are purified from bovine or rabbit sources. The regulatory subunits of type I and type II cAMP-dependent protein kinases are purified from bovine skeletal muscle and bovine heart muscle by utilizing DEAE fractionation and N^6-H_2N $(CH_2)_2$-cAMP immobilized Sepharose affinity chromatography (DILLS et al. 1979). Bovine heart muscle type II regulatory R^{II} subunit (R^{II}) often requires further purification on a Sephacryl S-200 column to remove trace contaminants and is isolated as a mixture of phospho and dephospho forms. The catalytic subunit of cAMP-dependent protein kinase (C) is purified from bovine skeletal muscle by utilizing "tricolumn" modification (BECHTEL et al. 1977). Cyclic nucleotide-free type I regulatory subunit (R^I_u) and R^{II}_u prepared by urea treatment as described by BUILDER et al. (1980). Recombined type I cAMP-dependent protein kinase holoenzyme ($R^I_2C_2$) is prepared by combining cyclic nucleotide free R^I subunit with a slight molar excess of C subunit in the presence of 10 mM 2-(N-morpholine) ethanesulfonic acid (MES) buffer (pH 6.5), 15 mM β-mercaptoethanol, 0.1 mM EGTA, containing 0.1 mM ATP and 1 mM $MgCl_2$. Excess C subunit is then removed by passage over a Sephadex CM-50 column. The recombined enzyme is sedimented as a single species with a s_{20W} value of approximately 7.0 as described by SCHWECHLEIMER and HOFMANN (1977). Bovine lung cGMP-dependent protein kinase (G-kinase) is prepared by utilizing 8-(2 amino-ethyl) amino cAMP-Sepharose affinity chromatography and gel filtration on Sephacryl S-200 as described (DILLS et al. 1979). It is advisable to check the purity of all immunogens routinely by SDS polyacrylamide disc gel electrophoresis (WEBER and OSBORN 1969).

II. Preparation and Selection of Antisera

Initial attempts to generate antisera toward the regulatory subunits of cAMP-dependent protein kinases from bovine heart muscle, and bovine skeletal muscle were reported independently from several laboratories (FLEISHER et al. 1976; HOFMANN et al. 1977; UNO et al. 1977; STEINER et al. 1978; KAPOOR et al. 1979a, b). The immunocytochemical procedure for localization of various cyclic nucleotide-dependent protein kinases and their subunits utilizes antibodies generated toward R^I, C, R^{II}, reconstituted type I protein kinase (STEINER et al. 1978; KOIDE et al. 1981) and cGMP-dependent protein kinase (SPRUILL et al. 1981) produced in guinea pig, rabbit or goat. The choices of each animal has some practical advantages and disadvantages. The methods of the immunization of various polypeptide hormones and enzymes have been recently reviewed (FELBER 1974).

Pure preparations of R^I, C, R^{II}, and G-kinase diluted (200 µg/ml) in PBS, thoroughly emulsified in equal volumes of Freund's complete adjuvant, are injected at multiple sites in the back of male albino New Zealand rabbits weighing 1–2 kg. Because the bovine skeletal muscle type I protein kinase holoenzyme can dissociate under certain conditions (BEAVO et al. 1974; HOFMANN et al. 1975), immunization of the reconstituted enzyme is performed with 10 mM ES pH 6.5, 0.1 mM EGTA, 15 mM β-mercaptoethanol, 50 mM NaCl containing 0.1 mM ATP, and 1 mM Mg^{++} which favors the association of the asymmetric holoenzyme structure. Multiple site subcutaneous injections of immunogen (100–200 µg) are usually made after shaving the fur of rabbits on the backs and lateral surfaces of the fore and hind limbs. In a few instances, animals are sensitized by injecting in the foot pad and intramuscularly (Sect. D). Booster injections are made at 3–4 week intervals and the animals are bled from the ear vein one week after each booster injection. The antisera toward bovine heart muscle R^{II} was generated in a goat by injecting 1.2 mg of pure immunogen every two weeks for six weeks (STEINER et al. 1978a). The animals were bled two days after the last injection. Each individual antisera is checked, as described in the cyclic nucleotide immunocytochemistry section, for avidity and specificity. The production of antibodies against various immunogens is routinely checked after each booster by Ouchterlony immunodiffusion technique (OUCHTERLONY 1966). Some antisera obtained after injection of C or $R_2^IC_2$ of cAMP-dependent protein kinase contained nonprecipitating antibodies to C that could be detected by double antibody techniques. The avidity of R^I and type I protein kinase antisera during the course of immunization are checked by $[^{125}I]$ R^I binding in the radioimmunoassay developed recently in our laboratory (KAPOOR et al. 1979a, b). The development of similar radioimmunoassays for catalytic subunit and cGMP-dependent protein kinase should facilitate screening of these antisera. For immunocytochemical purposes, antisera are fractionated with 40% ammonium sulfate and resuspended pellets are dialysed against PBS (pH 7.2) or 10 mM MES buffer pH 7.0 as described (Sect. D of this chapter).

III. Characterization of Antisera for Immunocytochemistry

The assessment of specificity of antisera is a major task for the immunocytochemical localization of cyclic nucleotide-dependent protein kinases. Two general approaches are commonly employed to demonstrate the immunological specificity of the antiserum: a) several types of immunodiffusion experiments are performed to define the specificity of the major precipitating antibodies in the antisera; b) immunoprecipitation techniques are employed to confirm and quantitate the specificity of the antiserum. Usually, the precipitin band formed as a result of the reaction of a partially purified fraction from crude tissue obtained during the early steps in antigen purification fuses completely with the precipitin band formed by the pure subunit indicating that the immunizing proteins are homogenous within the limit of detection. Figure 1 depicts an example of this experiment with an R^I antiserum.

1. Cyclic AMP-Dependent Protein Kinases and Their Subunit Antisera

The antisera raised toward beef skeletal muscle regulatory subunit, catalytic subunit, reconstituted type I holoenzyme, and beef heart muscle type II regulatory subunit are characterized by Ouchterlony immunodiffusion and immunoprecipitation techniques as described (KOIDE et al. 1981). R^I, R^{II}, and C subunit have shown crossreactivity by yielding a single precipitin band when tested either with their respective free subunit or the holoenzyme complex containing the subunit. The precipitin line formed to the isolated subunits fused completely with those formed to the holoenzyme complex indicating that the determinants present in the isolated

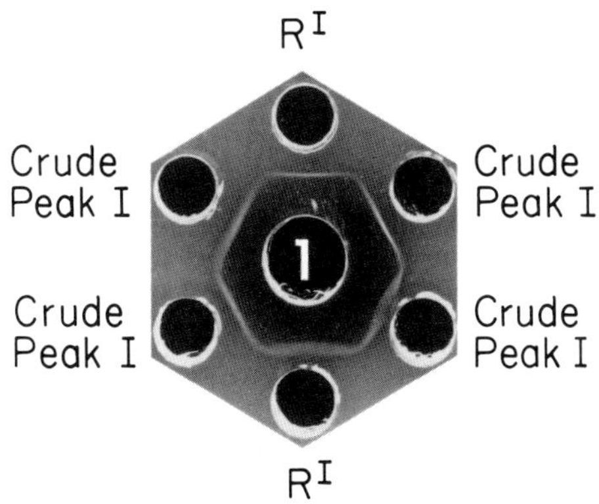

Fig. 1. Immunodiffusion of R^{I} antiserum. Anti-R^{I} (180 μg) R^{I}-cA (63 μg), crude peak I (148 μg of crude enzyme prepared on DE 52 column (20 μl aliquot). Note fusion of precipitin lines between partially purified peak I enzyme and purified R^{I} subunit. (Taken from KOIDE et al. 1981)

subunits are also present in the holoenzyme complex. Crossreactivity against other subunits and cGMP-dependent protein kinase was not observed. Antisera to type I protein kinase catalytic subunit have crossreacted equally well with the type II beef heart muscle catalytic subunit providing further evidence for their structural identity. HOFMANN et al. (1977) have shown that an R^{I} antiserum does not crossreact with the R^{II} subunit, nor does R^{II} antiserum crossreact with the R^{I} subunit as evidenced by immunoprecipitation titration. These results help to confirm that these regulatory subunits differ in the physiochemical properties (ROSEN and ERLICHMAN 1975; HOFMANN et al. 1975) and point out the sensitivity of these immunological techniques. Antisera from animals injected with reconstituted type I holoenzyme have shown a reaction of complete identity against type I holoenzyme, R^{I} and C. The antibodies produced against C subunit after immunization with the type I holoenzyme were not of precipitating type as judged by the double antibody precipitation technique. The presence of such nonprecipitating antibodies have been described in other systems (DEDMAN et al. 1978). Furthermore, immunoprecipitation data indicate that both cAMP binding and cAMP-dependent phosphotransferase activities are recovered in the immunoprecipitates formed from the reaction of holoenzyme with any of the antibodies (KOIDE et al. 1981). These results suggest that none of the antibodies completely blocked either the cAMP binding site or the catalytic site of the enzyme. This is in agreement with other systems in which antibodies raised against various enzymes did not eliminate enzyme activity (BOYD 1966; SMITH et al. 1973; KYTE 1974).

2. Cyclic GMP-Dependent Protein Kinase Antisera

The production of specific antisera toward bovine lung cGMP dependent protein kinase has been recently described (SPRUILL et al. 1981). Antiserum has been shown to crossreact by yielding a single precipitin band against the kinase. Crossreactivity toward R^{I}, C, and reconstituted type I holoenzyme was not observed. The antisera did exhibit crossreactivity with the structurally related R^{II} subunit (LINCOLN and CORBIN 1977) of cAMP-dependent protein kinase at several fold (5- to 6-fold) higher protein concentrations of R^{II} after a variety of immunochemical tests (SPRUILL et al. 1981). We believe that these reactions between cGMP kinase antisera and R^{II} subunit most likely represent cGMP-kinase contamination by R^{II} subunit in some of the preparations of G-kinase. Immunoreactivity of cGMP-dependent protein kinase antisera can be completely blocked by immunoprecipitation of antisera with pure cGMP-kinase but not by cAMP-dependent protein kinase components. SPRUILL et al. (1981) have further suggested that even though type II cAMP-dependent protein kinase appears to be structurally related, there are sufficient differences in their sequence and/or conformation which provide unique immunological determinants.

IV. Determination of Specificity of Protein Kinase Antisera by Radioimmunoassay

The use of cyclic nucleotide radioimmunoassays in determining the competitive crossreactivity of various nucleoside and nucleotides for immunocytochemistry is outlined in Sect. D of this chapter. One of the better approaches toward answering the question of tissue and species specificity of various protein kinase antisera will be to use radioimmunoassay to measure the crossreactivity among different tissues and species. FLEISCHER et al. (1976) have first reported the development of a radioimmunoassay for the type II cAMP-dependent protein kinase. In that study, antisera were prepared in guinea pigs by injecting purified bovine heart muscle type II protein kinase. They demonstrated that both the type II holoenzyme and the type II cAMP-regulatory subunit (R^{II}) competed in an identical manner for $[^{125}I]$-R^{I} or $[^{125}I]$-holoenzyme. However, C and partially purified type I protein kinase eluted from DEAE cellulose did not crossreact in the immunoassay. These observations suggested that the type II protein kinase holoenzyme antisera was recognizing determinants on the R^{II} subunit, but not on the catalytic subunit. They further demonstrated that intertissue and interspecies similarities and differences exist in the structure of these protein kinases. RUBIN et al. (1979) extended these studies to soluble and particulate fractions of bovine cerebral cortex by utilizing a type II protein kinase immunoassay. They reported that the bovine cerebral cortex cAMP-dependent protein kinases possess many of the characteristics of the bovine heart soluble type II protein kinase and possess antigenic determinants that are related to, but readily distinguishable from, determinants of type II heart soluble protein kinase. In order to determine the specificity of various R^{I} antisera for immunocytochemical localization, we have recently developed a radioimmunoassay for the type I bovine skeletal muscle regulatory subunit which can detect picograms quantities of R^{I} or type I protein kinase. Various R^{I} antisera were tested for their titer, avidity and specificity. Antisera which do not exhibit crossreactivity with R^{II}, C, and G-kinase isolated from the same species are utilized for immunocytochemical purposes. Figure 2 depicts an example of the screening of an R^{I} antiserum. Protein kinases isolated from bovine tissues and rat tissue have shown crossreactivity in the immunoassay. However, the degree of parallel and nonparallel crossreactivity varies from tissue to tissue (KAPOOR et al. 1979b).

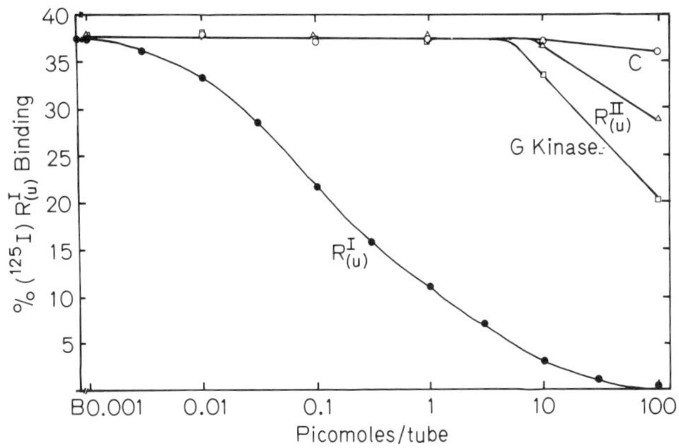

Fig. 2. Competitive crossreactivity of: *A* bovine skeletal muscle C; *B* bovine heart muscle R^{II}, and *C* bovine lung G-kinase. Various increasing concentrations of competitive agents were incubated and inhibition of $3R^{I}$ antiserum binding of $^{125}I\text{–}R_{(u)}$ was determined by radioimmunoassay. Values are the mean of duplicate determinations. (Taken from KAPOOR et al. 1979b, with permission)

V. Testing of Antisera for Specificity of Staining

After immunochemical-characterization, the protein kinase antisera are routinely checked for their working dilution with each tissue for immunofluorescence and bright field immunocytochemistry as outlined in the Sect. C of this chapter. The specificity of the staining patterns for the protein kinases and their subunits is confirmed by showing loss of specific staining after absorption of the antisera with their respective subunit or holoenzyme, but no change after absorption by heterologous antigens. For absorption, undiluted antisera are incubated with the appropriate concentrations of homologous or heterologous antigens at 37 °C for 30 min and then at 4 °C for 18–24 h. Antigen-antibody complexes are removed by centrifugation at 9,000 × *g* for 5 min in a Beckman microfuge. Supernatants are collected and diluted appropriately for immunocytochemical staining. However, with a few antisera, the staining pattern completely diminishes by liquid phase absorption technique. These antisera are absorbed by utilizing solid phase affinity chromatography. We have recently utilized R^{I} and R^{II} subunit bound to $N^6\text{-}CH_2(CH_2)2$-cAMP-immobilized Sepharose gel (DILLS et al. 1979) for solid phase absorption of specific antibodies against R^{I} and R^{II} for several antisera. The antibodies are incubated as described for liquid phase absorption techniques. Specificity of absorption can be confirmed by running the appropriate controls. For example, R^{I} bound to $N^6\text{-}CH_2\,(CH_2)_2$-cAMP Sepharose resin should not absorb R^{II} antibodies and vicevesa. Moreover, removal of specific antibodies can also be checked by radioimmunoassay of the specific antigen (KAPOOR et al. 1979a, b).

VI. Methodological Considerations

1. Crossreactivity of Antibodies for Holoenzyme and Their Subunits

In our experience with antisera to the type I cAMP-dependent protein kinase, crossreactivity is observed between the undissociated holoenzyme antisera with its regulatory and catalytic subunit as well as between subunit antisera and undissociated holoenzyme. This means that antisera utilized for cAMP-dependent protein kinase immunocytochemistry may not be either subunit specific or holoenzyme specific, but rather have cross specificities. However, differences are observed in the staining patterns of antisera obtained by injection of type I protein kinase as compared to the patterns of R^I and C subunits in control and hormone stimulated tissues (KOIDE et al. 1981). This suggests that some of the antibodies in these antisera are recognizing determinants unique to the subunit versus the holoenzyme and vice-versa (for further discussion see Sect. F, Future Perspectives).

2. Avidity of an Antiserum for Available Determinants in Tissue

The avidity of an antisera for available antigenic determinants in tissue is an important aspect of protein kinase immunocytochemistry. We have recently utilized several R^I antisera obtained by injection of type I protein kinase for protein kinase immunocytochemistry. The specificity of each antisera was checked by several immunochemical parameters as discussed above. In rat liver, two different antisera for R^I gave similar but clearly different staining patterns and two holoenzyme antisera also differed in their localization. In every instance the staining pattern of an antiserum could be blocked by absorption of the antiserum by its immunogen prior to the tissue staining procedure. Figure 3 depicts an example of the different staining patterns of the R^I antisera in rat liver. These findings suggest that the detection of an antigen in a tissue section depends partly upon the avidity of the population of antibodies for the antigen and partly upon the antigenic determinants that are recognized by the antibodies for the antigen while it is bound in the tissue. These findings emphasize the need to compare a large number of antisera generated to a protein kinase or subunit. The absence of staining at a particular site in a cell does not necessarily indicate that the antigen is not present at that site. It is possible that the antibodies generated to the antigen will not recognize it in tissue because the specific determinants are not available to the antibodies when the antigen is embedded in the cell.

3. Losses and Redistribution of Protein Kinases

The present studies utilize unfixed or paraformaldehyde fixed tissue sections in which there is a repeated washing of the sections with PBS. The high concentration of saline has been shown to activate cAMP-dependent protein kinases (CORBIN et al. 1973) by dissociation of the subunits. If dissociation of protein kinase occurs during the staining procedure, or if soluble complexes are present in the tissue, soluble protein kinase subunits could be lost during the washing procedure or possibly translocate into the different site(s) of the cells or tissue. One approach which has been utilized to assess the possibility of such changes involves an analysis of various fixation procedures prior to the application of an antiserum. The site of localization of various antigens does not appear to differ from sections which are fixed with paraformaldehyde and other fixatives prior to the application of the primary antiserum. However, relatively high concentrations of glutaraldehyde tend to elim-

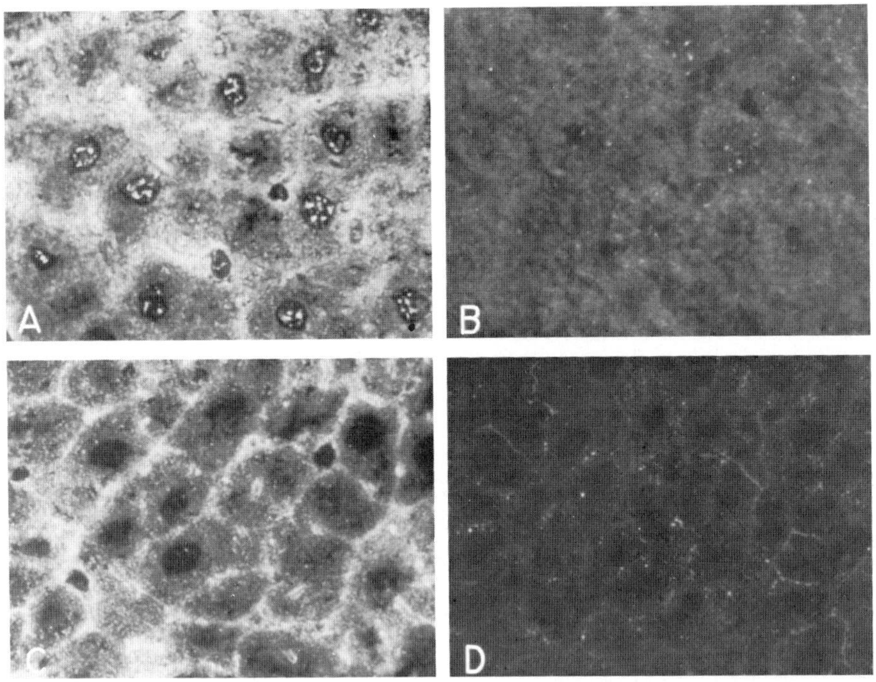

Fig. 3 A–D. Immunocytochemical localization of R^I subunit in rat liver by two different R^I antisera. **A** R^I subunit is localized in the sinusoid areas, in hepatocyte cytoplasm and within nuclei by antiserum, No 1. **B** Fluorescence for R^I subunit was eliminated following absorption of the R^I subunit antiserum with R^I subunit. **C** R^I subunit is observed in plasma membrane area in cell cytoplasm but not in hepatocyte nuclei by antiserum, No 2. **D** R^I subunit fluorescence was eliminated following absorption of R^I subunit antiserum with R^I prior to staining ($\times 600$).

inate fluorescence (unpublished observation). Further insight into these issues can be gained by quantitating the extent of losses of protein kinases and their subunit from tissue sections in both fixed and unfixed tissue.

4. Species Specificity of Protein Kinase Immunocytochemistry

Protein kinases isolated from various rat tissues exhibit nonparallel crossreactivities in the bovine skeletal muscle type I and bovine heart muscle type II radioimmunoassays (Fleischer et al. 1976; Kapoor et al. 1979a, b). These observations suggest that both similarities and differences in the structural properties of protein kinases exist between species. In order to explore this possibility immunocytochemically, a preliminary survey of a variety of tissues and isolated cells from rat, guinea pig, mouse and hamster was carried out; a specific staining pattern for protein kinases and their subunits occurred in all tissues examined. The localization patterns were quite similar to those in bovine tissues (unpublished observations). Although protein kinases prepared from rat tissues crude extracts exhibit nonparallel crossreactivity in the R^I immunoassay, they are precipitated by the combined action of antisera specific to R^{I} and R^{II} (KAPOOR et al. 1979b). It should be pointed out that this immunocytochemical procedure is not run at equilibrium conditions as in the radioimmunoassays, and thus, a large proportion of the antibody might contribute to immunocytochemical staining pattern.

F. Future Perspectives

In recent years, evidence has accumulated suggesting that the mechanism of action of cAMP and cGMP is mediated by specific cyclic nucleotide-dependent protein kinases. Specific antibodies to cyclic nucleotides, their protein kinases and various subunits of protein kinases have been utilized to localize these molecules in a number of different tissues whose regulation are affected by cyclic nucleotides. In many tissues cAMP is localized in the cytoplasm and on the plasma membrane of cells (STEINER et al. 1976). Nuclear membrane and nuclear element localization is also observed in some tissues (STEINER et al. 1976; SPRUILL and STEINER 1976). Immunocytochemical techniques have been particularly useful in identifying the cell type(s) generating increased amounts of cAMP in heterogenous tissues. cGMP has been localized on plasma membrane and in nuclear elements of cells suggesting that this nucleotide may serve as a regulator of nuclear directed events and in membrane function (STEINER et al. 1976). The immunocytochemical localization of both cAMP and cGMP has been useful in determining differences in the relative subcellular distribution of cyclic nucleotides and in showing, after hormonal stimulation, the apparent intracellular redistribution of cyclic nucleotides when there is no change in the total tissue concentration of the cyclic nucleotide (KOIDE et al. 1978).

We have recently characterized several antisera to both cAMP and cGMP-dependent protein kinases and various protein kinase subunits so that immunocytochemical studies on the cellular and subcellular distribution of these cyclic nucleotide dependent protein kinases could be initiated. Immunocytochemical localization of various components of cAMP-dependent protein kinase has provided evidence for the possible translocation of cAMP-dependent protein kinase subunits from cytoplasm to nucleus (STEINER et al. 1978; KOIDE et al. 1981). These findings should be helpful in exploring the site of action of cyclic nucleotides during hormonal action. Moreover, it is becoming evident that the action of cyclic nucleotides is closely associated with another second messenger, calcium, which binds in many tissues to a ubiquitous calcium binding protein, calmodulin (CHEUNG 1971; see also this volume). We have recently studied the localization of calmodulin in a variety of rat tissues by immunocytochemical procedures (HARPER et al. 1980) and found that it is present in various parts of the cell, including the nucleus. We believe that in the future it will be important to study the localization of both calmodulin as an effector of the action of calcium and protein kinases as an effector of cyclic nucleotides in the same tissue sections.

It seems likely that immunocytochemical techniques predominantly detect cyclic nucleotides that are tightly bound to tissue receptors, because the procedures involve washing of tissue sections. We believe that, in the future, experiments in various tissues with a variety of fixation conditions will help to determine optimal conditions for the retention of cyclic nucleotides and their kinases.

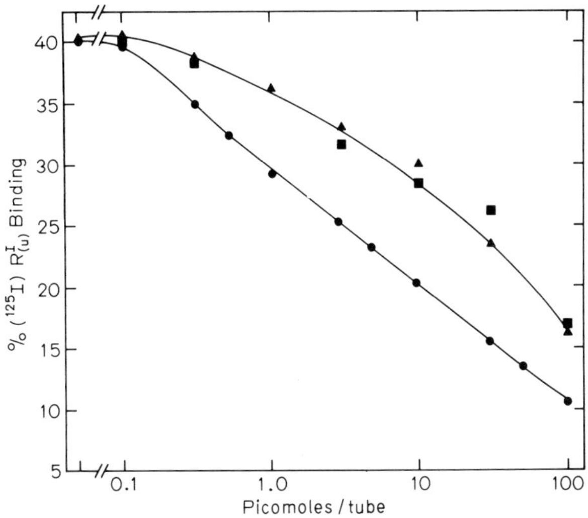

Fig. 4. Comparison of inhibition of site specific 3 R^I, No 3 antiserum binding of ^{125}I–$R^I_{(u)}$ by increasing concentrations of $R^I_{(u)}$, $R^I_2C_2$, and $R^I_2C_2$ dissociated with cyclic AMP in the immunoassay. Tubes containing various amounts of $R^I_2C_2$ were preincubated at room temperature for 30 min with 10 μM cyclic AMP and inhibition of antiserum binding of ^{125}I–$R^I_{(u)}$ was determined by radioimmunoassay. ●—●, $R^I_{(u)}$, ▲—▲, $R^I_2C_2$ ■—■, $R_2C_2 + 10\ \mu M$ cyclic AMP. Each value is the mean of triplicate determinations. (Taken from KAPOOR et al. 1979b, with permission)

The specificity of these immunocytochemical procedures lies primarily with the antiserum to the respective immunogen. Antisera contain a heterogenous population of antibody molecules that vary from preparation to preparation, making the standardization of immunocytochemical techniques a difficult task. We believe that the use of monoclonal antibodies produced by the hybridoma technique (MELCHER et al. 1978) may help to improve cyclic nucleotides and protein immunocytochemistry because the technique permits the production of permanent cell lines capable of continuous production of monoclonal antibodies. Preliminary information gained from utilizing cAMP antibodies generated by the hybridoma method suggest that this approach may be valuable in producing highly specific cAMP antibodies. The application of these antibodies should reduce the nonspecific background of immunocytochemistry.

As discussed in the Sect. D of this chapter, antibodies utilized for cyclic nucleotide immunocytochemistry are characterized by radioimmunoassay. Recently this approach has been applied for determining the specificity of R^I and R^{II} antibodies. It will be important to screen a number of antisera for protein kinase immunocytochemistry by utilizing radioimmunoassays for these antigens. In general, various type I regulatory subunit antisera characterized by radioimmunoassay do not differentiate between R^I, R^I·cAMP or $R^I_2C_2$ in the immunoassay. However, a site specific antiserum was found which differentiated immunoreactivity of R^I, R^I·cAMP or R_2C_2 in the immunoassay (Fig. 4). These findings suggested that specific binding of cAMP and catalytic subunit to R^I cause blocking of R^I antigenic determinants and/or causes a change in the conformation of the molecule (Kapoor et al. 1979a, b). In the future, the availability of site specific antibodies should facilitate studies designed to probe the antigenic determinants of proteolytic fragments of R^I and R^{II} (Potter and Taylor 1979; Srivastava and Stellwagon 1978). The antibodies generated to each fragment could be utilized as a molecular probe for determining multiple, different domain specific antigenic sites of the regulatory subunits (Benjamin and Teale 1978) in tissue sections by immunocytochemical techniques.

Crossreactivity is observed between one of the type I protein kinase holoenzyme antisera and its regulatory and catalytic subunit (Sect. E). The antisera utilized do not appear by immunodiffusion to be either subunit specific or holoenzyme specific, but rather have crossreactivity. However, differences are observed in the localization patterns of type I protein kinases versus R^I and C, both in control and hormonally stimulated tissues. These findings suggest that some of the antibodies in these antisera recognize determinants in tissue sections unique to the subunit, relative to the holoenzyme, and vice-versa. In future experiments, selective absorption of R^I and C subunit antisera with type I holoenzyme may yield a population of antibodies which have only subunit-specific determinants; and selective absorption of type I holoenzyme antiserum with R^I and C subunit might yield a population of antibodies which only recognize specific conformational determinants formed by the recombination of R^I and C subunits. Alternatively, it may be possible to produce monoclonal hybridoma antibodies (Melchers et al. 1978) which have immunological recognition specificity of the subunit versus holoenzyme. The potential utility of such techniques for immunocytochemical localization of various antigens was recently demonstrated in various systems (Cuello et al. 1979).

Acknowledgements. The original research work of our collaborators, especially Drs. Y. Koide, W.A. Spruill, and J.A. Beavo, is gratefully acknowledged. Dr. R.D. Cumming provided valuable criticism of the manuscript. The work reviewed here was supported by United States Public Health Service Grant AM 19438. We would also like to express our appreciation to Ms. Celeste Layton for her care in the preparation of this manuscript.

References

Bauvois B, Leng M (1976) Heterogenity of antibodies to adenosine 5′monophosphate. Biochem Biophys Acta 454:1–8

Beavo JA, Bechtel PJ, Krebs EG (1974) Criteria for the classification of protein kinase. Methods Enzymol 38C:299–308

Beavo JA, Bechtel PJ, Krebs EG (1975) Mechanism of control for cAMP-dependent protein kinase from skeletal muscle. Adv Cyclic Nucleotide Res 5:241–251

Bechtel PJ, Beavo JA, Krebs EG (1977) Purification and characterization of catalytic subunit of skeletal muscle adenosine 3′:5′ monophosphate-dependent protein kinase. J Biol Chem 252:2691–2697

Benjamin DC, Teale JM (1978) The antigenic structure of bovine serum albumin. Evidence for multiple different domain specific antigenic determinants. J Biol Chem 253:8087–8092

Bloom FE, Rossier J, Battenberg EL, Bayon A, French E, Henricksen SJ, Siggen GR, Segal D, Browne R, Ling N, Guillemin R (1978) β-endorphin: cellular localization, electrophysiological and behavioral aspects. Adv Biochem Psychopharm 18:89–108

Boyd WC (1966) Fundamentals of immunology, 4th edn. Interscience Publ, New York, p 338

Builder S, Beavo JA, Krebs EG (1980) The mechanism of activation of bovine skeletal muscle protein kinase by adenosine 3′:5′ monophosphate. J Biol Chem 255:3514–3519
Castagna M, Palmar WK, Krebs DA (1975) Nuclear protein kinase activity in perfused rat liver stimulated with dibutyryl adenosine cyclic 3′:5′-monophosphate. Eur J Biochem 55:193–199
Cheung WY (1971) Cyclic 3′:5′ nucleotide phosphodiesterase: Evidence and properties of protein activator. J Biol Chem 246:2859–2869
Cho-Chung YS (1980) Cyclic AMP and its receptor protein in tumor growth regulation in vivo. J Cyclic Nucleotide Res 6:163–177
Coons AH, Creech JH, Jones RN (1941) Immunological properties of antibody containing a fluorescent group. Proc Soc Exp Biol Med 47:200–202
Corbin JD, Reiman EM, Walsh DA, Krebs EG (1973) Regulation of adenosine 3′5′-monophosphate-dependent protein kinase. J Biol Chem 248:1813–1821
Corbin JD, Kelly SL, Park CR (1975) The distribution and dissociation of cyclic adenosine 3′:5′-monophosphate dependent protein kinase in adipose, cardiac, and other tissues. J Biol Chem 250:218–225
Costa E, Kurosawa A, Guidotti (1976) Activation and nuclear translocation of protein kinase during transsynaptic induction of tysosine 3-monooxygenase. Proc Natl Acad Sci USA 73:1058–1062
Cuello AC, Galfre G, Milstein C (1979) Detection of substance P in the central nervous system by a monoclonal antibody. Proc Natl Acad Sci USA 763:3532–3536
Cumming R, Arbuthnott G, Steiner AL (1979) Characterization of immunofluorescent cyclic GMP positive dimers in the central nervous system. J Cyclic Nucleo Res 5:463–467
Cumming R, Dickinson S, Arbuthnott A (1980) Cyclic nucleotide losses during tissue processing for immunocytochemistry. J Histocytochem 28:54–55
Dedman JR, Welsh MJ, Means AR (1978) Ca^{++} dependent regulator production and characterization of a monospecific antibody. J Biol Chem 253:7517–7521
Dills WL, Goodwin CD, Lincoln TM, Beavo JA, Bechtel PJ, Corbin JD, Krebs EG (1979) Purification of cyclic nucleotide receptor proteins by cyclic nucleotide affinity chromatography. Adv Cyclic Nucleo Res 10:199–217
Earp HS, Steiner AL (1978) Compartmentalization of cyclic nucleotide-mediated hormone action. Ann Rev Pharmacol Toxicol 18:431–459
Fallon EJ, Agrawal R, Furth E, Steiner AL, Cowden R (1974) Cyclic guanosine and adenosine 3′,5′-monophosphates in canine thyroid. Localization by immunofluorescence. Science 184:1089–1091
Felber JB (1974) Radioimmunoassay of polypeptide hormones and enzymes. In: Glick D (ed) Methods of biochemical analysis, vol 22. John Wiley and Sons, New York, pp 1–194
Fleischer N, Rosen OM, Reichlin M (1976) Radioimmunoassay of bovine heart protein kinase. Proc Natl Acad Sci USA 73:54–58
Gill GN, Holdy KE, Walton GM, Kanstein CB (1976) Purification and characterization of 3′:5′-cyclic GMP dependent protein kinase. Proc Natl Acad Sci USA 73:3918–3922
Graham RC, Jr, Karnovsky MJ (1966) The early stages of absorption of injected horse radish peroxidase in the proximal tubules of mouse kidney: ultrastructural cytochemistry by a new technique. J Histochem Cytochem 14:291–302
Greengard P (1978) Phosphorylated proteins as physiological effectors. Protein phosphorylation may be a final common pathway for many biological regulatory agents. Science 13:146–152
Harper JF, Cheung WY, Wallace RW, Huang HL, Levine SN, Steiner AL (1980) Localization of calmodulin in rat tissues. Proc Natl Acad Sci USA 77: 366–370
Hashimoto E, Takeda M, Nishizuka Y, Hamana K, Iwai K (1976) Studies on the sites in histone phosphorylated by adenosine 3′:5′-monophosphate-dependent and guanosine 3′:5′-monophosphate dependent protein kinase. J Biol Chem 251:6287–6293
Hofmann F, Beavo JA, Bechtel PJ, Krebs EG (1975) Comparison of adenosine 3′:5′ monophosphate dependent protein kinase from rabbit skeletal muscle and bovine heart muscle. J Biol Chem 250:7795–7800

Hofmann F, Bechtel PJ, Krebs EG (1977) Concentrations of cyclic AMP-dependent protein kinase subunits in various tissues. J Biol Chem 252:1441–1447

Jungmann RA, Hiestand PC, Schweppe JS (1974) Mechanism of action of gonadotropins IV. Cyclic adenosine monophosphate dependent translocation of ovarian cytoplasmic cyclic adenosine monophosphate binding protein and protein kinase to nuclear receptor site. Endocrinology 94:168–183

Kapoor CL, Krishna G (1977) Hormone induced cyclic guanosine monophosphate secretion from guinea pig pancreatic lobules. Science 196:1003–1005

Kapoor CL, Beavo JA, Steiner AL (1979a) Effect of cyclic AMP and bovine skeletal muscle protein kinase catalytic subunit (C) on the immunoreactivity of bovine skeletal muscle regulatory subunit (R^I). Fed Proc 38:263

Kapoor CL, Beavo JA, Steiner AL (1979b) Radioimmunoassay of the regulatory subunit of type I cAMP-dependent protein kinase. J Biol Chem 254:12427–12432

Kawamura A, Jr (1977) In: Fluorescent antibody techniques and their applications. University Park Press, Baltimore

Koide Y, Earp HS, Ong SH, Steiner AL (1978) Alteration in the intracellular distribution of cGMP and guanylate cyclase activity during rat liver regeneration. J Biol Chem 253:4431–4445

Koide Y, Beavo JA, Kapoor CL, Spruill WA, Huang H, Levine SL, Ong S, Bechtel PJ, Yönt WJ, Steiner AL (1981) Hormonal effect on the immunolyts. Chemical location of cAMP-dependent protein kinase in rat tissues. Endocrinology. In Press.

Krebs EG (1972) Protein kinase. Curr Top Cell Regul 5:99–133

Krebs EG, Beavo JA (1979) Phosphorylation dephosphorylation of enzyme. Ann Rev Biochem 48:923–959

Kuo JF, Greengard P (1969) Cyclic nucleotide dependent protein kinases IV. Widespread occurrence of adenosine 3′:5′-monophosphate dependent protein kinase in various tissue and phyla of animal kingdom. Proc Natl Acad Sci USA 64:1349–1355

Kuo JF, Greengard P (1970) Cyclic nucleotide-dependent protein kinases. VI Isolation and partial purification of a protein kinase activated by guanosine 3′:5′-monophosphate. J Biol Chem 245:2493–2498

Kuo JF (1974) Guanosine 3′:5′-monophosphate-dependent protein kinase in mammalian tissues. Proc Natl Acad Sci USA 71:4037–4041

Kuo JF, Shoji M, Kuo W (1978) Molecular and physiological aspects of mammalian cyclic GMP-dependent protein kinases. Ann Rev Pharmacol Toxicol 18:341–355

Kurosawa A, Guidotti A, Costa E (1979) Nuclear translocation of cyclic AMP-dependent protein kinase subunit during the transsynaptic activation of gene expression in rat adrenal medulla. Mol Pharm 15:115–130

Kyte J (1974) The reaction of sodium and potassium adenosine diphosphate with specific antibodies. Implication for the mechanisms of transport. J Biol Chem 249:3652–3660

Lincoln TM, Dills WL, Corbin JD (1977) Purification and subunit composition of guanosine 3′:5′-monophosphate-dependent protein kinase from bovine lung. J Biol Chem 252:4269–4275

Lincoln TM, Corbin JD (1977) Adenosine 3′:5′ dependent protein kinase. Possible homologous proteins. Proc Natl Acad Sci USA 74:3239–3243

Lincoln TM, Flockhart DA, Corbin JD (1978a) Studies on the structure and mechanism of the guanosine 3′:5′ monophosphate-dependent protein kinase. J Biol Chem 253:6002–6009

Lincoln TM, Corbin JD (1978b) On the role of cAMP and cGMP-dependent protein kinase in cell function. J Nucleotide Res 4:3–14

Melchers F, Potter M, Warner NH (1978) Lymphocyte hybridomas. In: Current topics microbiology-immunology. Springer, Berlin Heidelberg New York, pp 1–24

Nakane PK (1968) Simultaneous localization of multiple tissue antigens using the peroxidase-labelled antibody method: A study of pituitary glands of rat. J Histochem Cytochem 16:557–571

Nakane PK (1970) Classification of anterior pituitary cell types with immunoenzyme histochemistry. J Histochem Cytochem 18:9–20

Nakane PK, Kawoi A (1974) Peroxidase-labelled antibody: A new method of conjugation. J Histochem Cytochem 22:1084–1104
Nimmo HG, Cohen P (1977) Hormonal control of protein phosphorylation. Adv Cyclic Nucleotide Res 8:145–226
Ortez RA, Sikes RW, Sperling HG (1980) Immunocytochemical localization of cyclic GMP in goldfish retinae. J Histochem Cytochem 28:263–270
Ouchterlony O (1966) In: Ackroyd JF (ed) Immunological methods. FA Davis Company, Philadelphia, p 55
Potter RL, Taylor SS (1979) Relationship between structural domain and function in the regulatory subunit of cyclic AMP-dependent protein kinase I and II from bovine skeletal muscle. J Biol Chem 254:2413–2418
Pryzwansky KB, Steiner AL, Spitznagel JK, Kapoor CL (1980) Compartmentalization of cyclic AMP during Phagocytosis by Human Neutrophillic Granulocytes. Science 211:407–410
Rosen OM, Erlichman J (1975) Reversible autophosphorylation of cyclic 3′:5′ monophosphate dependent protein kinase from bovine cardiac muscle. J Biol Chem 250:7788–7794
Rubin CS, Rosen OM (1975) Protein phosphorylation. Ann Rev Biochem 44:831–887
Rubin CS, Rangel-Aldao Sarkar D, Erlichman J, Fleischer N (1979) Characterization of membrane associated and cytosolic cAMP-dependent protein kinase. Physiochemical and immunological studies on bovine cerebral cortex protein kinase. J Biol Chem 254:3797–3805
Schechheimer K, Hofmann K (1977) Properties of regulatory subunit of cyclic AMP-dependent protein kinase (peak I) from rabbit skeletal muscle prepared by urea treatment of the holoenzyme. J Biol Chem 252:7690–7696
Smith L, Davies HC, Reichlin M, Margoliash E (1973) Separate oxidase and reductase reaction sites on cytochrome C demonstrated with purified site-specific antibodies. J Biol Chem 248:237–243
Spielvolgel AM, Medneik MI, Eppenberger V, Jungmann RA (1977) Evidence for the identity of nuclear and cytoplasmic adenosine 3′:5′ monophosphate dependent protein kinase from porcine ovaries and nuclear translocation of cytoplasmic enzyme. Eur J Biochem 73:199–212
Spruill WA, Steiner AL (1976) Immunocytochemical localization of cyclic nucleotide during testicular development. J Cyclic Nucleotide Res 2:225–239
Spruill WA, Steiner AL (1979) Cyclic nucleotide and protein kinase immunocytochemistry. Adv Cyclic Nucleotide Res 10:169–186
Spruill WA, Koide Y, Huang H, Levine SL, Ong S, Steiner AL, Beavo JA (1981 in press) Immunocytochemical localization of cGMP-dependent protein kinase in endocrine tissues. Endocrinology
Srivastava AK, Stellwagon RH (1978) Presence of the sites for interacting with cAMP and catalytic subunit on small fragments of protein kinase regulatory subunit. J Biol Chem 253:1752–1755
Steiner AL, Parker CW, Kipnis DM (1972) Radioimmunoassay for cyclic nucleotide I. Preparation of antibodies and iodinated cyclic nucleotides. J Biol Chem 247:1106–1113
Steiner AL, Ong SH, Wedner HJ (1976) Cyclic nucleotide immunocytochemistry. Adv Cyclic Nucleotide Res 116:115–155
Steiner AL, Koide Y, Earp HS, Bechtel PJ, Beavo JA (1978) Compartmentalization of cyclic nucleotide and cylcic AMP dependent protein kinase in rat liver. Immunocytochemical demonstration. Adv Cyclic Nucleotide Res 9:691–705
Steiner AL, Koide Y, Spruill WA, Beavo JA (1978a) The distribution of cyclic nucleotide and their protein kinases in tissues: An immunocytochemical approach. In: Stoclet JC (ed) Advances in pharmacology and therapeutics, vol III. Pergamon Press, Oxford, New York, p 207
Sternberger LA, Hardy PH, Jr, Cueulis JJ, Mayer HG (1970) The unlabelled antibody enzyme method of immunocytochemistry: Preparation and properties of soluble antigen-antibody complex (horse radish peroxidase-anti-horse radish peroxidase and its use in identification of spirochetes. J Histochem Cytochem 18:315–333

Sternberger LA (1974) Immunofluorescence. In: Osler A, Weiss (eds) Immunocharacterization, Foundation in Immunology Series. Prentice Hall Englewood Cliff, NJ, p 18
Sugden PH, Holladay LA, Reiman EM, Corbin JD (1976) Purification and characterization of the catalytic subunit of adenosine 3′:5′ monophosphate dependent protein kinase from bovine liver. Biochem J 159:409–422
Uno I, Ueda T, Greengard P (1977) Adenosine 3′:5′-monophosphate regulated phosphoprotein system of neural membrane II. Solubilization, purification, and some properties of an endogenous adenosine 3′,5′-monophosphate dependent protein kinase. J Biol Chem 252:5164–5174
Walsh DA, Perkins JP, Krebs EG (1968) An adenosine 3′:5′ monophosphate-dependent protein kinase from rabbit skeletal muscle. J Biol Chem 243:3763 3765
Weber K, Osborn M (1969) The reliability of molecular weight determination by dodicylsulfate-polyacrylamide gel electrophoresis. J Biol Chem 244:3580–3583
Weber K, Bibrung T, Osborn M (1975) Specific visualization of tubulin containing structure in tissue culture cell by immunofluorescence. Exp Cell Res 95:111–120
Wedner JH, Hoffer BJ, Battenberg E, Steiner AL, Parker CW, Bloom FE (1972) A method detecting intracellular cyclic adenosine monophosphate by immunofluorescence. J Histochem Cytochem 20:293–295
Zoller MJ, Kerlarage AR, Taylor SS (1979) Structural comparison of cAMP-dependent protein kinase I and II from porcine skeletal muscle. J Biol Chem 254:2408–2412

Section II:
Biochemistry of Protein Phosphorylation

CHAPTER 10

Protein Phosphorylation: An Overview

P. GREENGARD

The work of Edwin Krebs and his colleagues over a number of years culminated in the discovery of a cyclic AMP-dependent protein kinase in skeletal muscle (WALSH et al. 1968). Moreover, Krebs and his group provided strong evidence to suggest that this cyclic AMP-dependent protein kinase was involved in mediating the effects of epinephrine in the regulation of carbohydrate metabolism in skeletal muscle. Upon the appearance of this important paper by Krebs and his colleagues, Drs. Miyamoto, Kuo and I looked for and found a cyclic AMP-dependent protein kinase in mammalian brain (MIYAMOTO et al. 1969). Of particular interest, our results indicated that this enzyme was present in high concentration in brain and that it was greatly enriched in those subcellular fractions containing synaptic material. These results suggested that cyclic AMP-dependent protein kinase in the brain might be involved not only in regulation of carbohydrate metabolism, but also in regulation of some of the molecular processes underlying synaptic transmission, and possibly of other physiological processes occurring in the nervous system as well. For these reasons we proposed the hypothesis (KUO and GREENGARD 1969), shown in Fig. 1, that in various tissues the diverse effects of cyclic AMP, both metabolic and physiological, were achieved through regulating the activity of this one class of enzyme, cyclic AMP-dependent protein kinase. According to this hypothesis, the hormone in the endocrine system or the neurotransmitter in the nervous system activates an adenylate cyclase; the activated adenylate cyclase causes an increased conversion of ATP to cyclic AMP; the newly formed cyclic AMP activates a cyclic AMP-dependent protein kinase; the activated protein kinase causes the phosphorylation of a substrate protein, converting it from the dephospho-form to the phospho-form; the phosphorylated substrate, through one or more steps, leads to the metabolic or physiological response characteristic of the hormone or neurotransmitter in question.

Direct evidence for the validity of this hypothesis has recently been obtained in the case of neuronal function. In collaboration with two groups of workers, Leonard Kaczmarek and Felix Strumwasser at the California Institute of Technology (KACZMAREK et al. 1980) and Vincent Castellucci, Eric Kandel, and James Schwartz at Columbia University (CASTELLUCCI et al. 1980), it has been possible to demonstrate that the injection of homogeneous catalytic subunit of cyclic AMP-dependent protein kinase into appropriate neurons and neurosecretory cells mimics the physiological response observed upon synaptic activation of these cells. Those studies have provided direct evidence for a causal relationship between protein phosphorylation and physiological response in such cells.

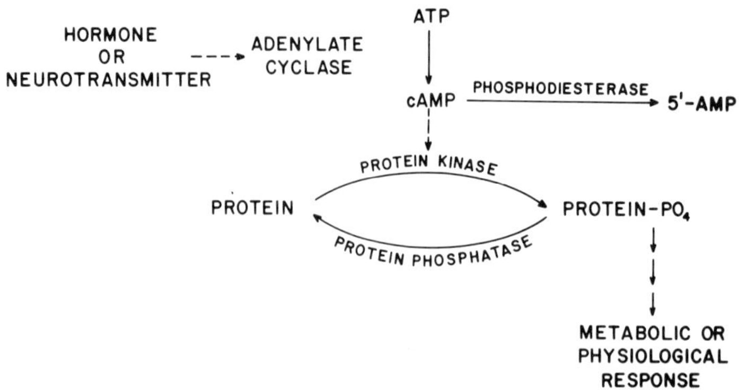

Fig. 1. Schematic diagram of the apparent role played by protein phosphorylation in mediating the biological effects of those hormones and neurotransmitters that act through cyclic AMP. The appropriate hormone or neurotransmitter stimulates a specific adenylate cyclase present in the target tissue, causing an increased conversion of ATP to cyclic AMP; the newly formed cyclic AMP is either hydrolyzed by a phosphodiesterase to 5′-AMP or else it activates a protein kinase, leading to the phosphorylation of a substrate protein; the substrate protein, through one or more steps, controls the rate of some metabolic or physiological response; the phosphorylation of this substrate protein leads to an increase or decrease in the rate of this response. (Taken from GREENGARD 1978)

Work carried out in many laboratories now makes it seem likely that most, and perhaps all, of the diverse effects of cyclic AMP in eukaryotic organisms are indeed achieved through regulation of protein kinases (for reviews, see KREBS 1972; RUBIN and ROSEN 1975; NIMMO and COHEN 1977; GREENGARD 1978). In fact, a large body of work carried out in our own and other laboratories has suggested that the scheme shown in Fig. 1 concerning the importance of protein phosphorylation should be extended.

Thus, we have been led to hypothesize (GREENGARD 1978) that protein phosphorylation is involved in mediating not only the actions of cyclic AMP and those hormones and neurotransmitters whose effects are achieved through altering cyclic AMP levels, but the actions of a large number of other regulatory agents as well (Fig. 2). For instance, it was shown that each of the major classes of steroid hormones listed in Fig. 2 affected the state of phosphorylation of a specific protein present in its target tissue (LIU and GREENGARD 1974, 1976), and that this effect of each of the steroid hormones was achieved only in its target tissue. Moreover, the effects were obtained with physiological doses of the steroid hormones and required *de novo* protein synthesis. A class of protein kinases exists that is activated specifically by low concentrations of cyclic GMP rather than of cyclic AMP (KUO and GREENGARD 1969; KUO et al. 1971). This class of enzymes was termed cyclic GMP-dependent protein kinases. It seems likely, at the present time, that many, and possibly all, of the physiological effects of cyclic GMP, and of those regulatory agents that act through altering cyclic GMP levels, are achieved through regulation of these cyclic GMP-dependent protein kinases. Work in numerous laboratories, including the studies of Javier Forn in our own research group (FORN and GREENGARD 1976), has indicated that insulin affects the state of phosphorylation

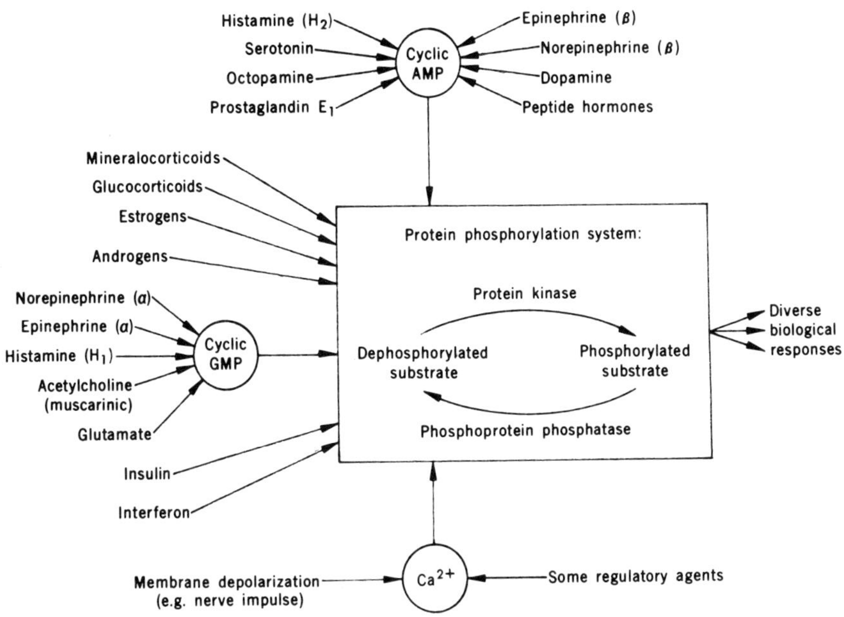

Fig. 2. Schematic diagram of postulated role played by protein phosphorylation in mediating some of the biological effects of a variety of regulatory agents. The diagram gives examples of regulatory agents some of whose effects may be mediated through regulation of the phosphorylation of specific proteins, and is not intended to be complete. In addition to cyclic AMP and a variety of neurotransmitters and hormones whose effects are mediated through cyclic AMP, these regulatory agents include cyclic GMP, a variety of neurotransmitters and hormones whose effects are mediated through cyclic GMP, Ca^{2+} and agents whose effects are mediated through Ca^{2+}, several classes of steroid hormones, insulin, and interferon. For brevity, the numerous peptide hormones whose effects are known to be mediated through cyclic AMP and the various regulatory agents believed to act through translocation of Ca^{2+} are not listed individually. It seems likely that some, but not necessarily all, of the biological responses elicited by any given one of the regulatory agents shown are mediated through the protein phosphorylation system; for simplicity, pathways from regulatory agent to biological response that do not involve protein phosphorylation are not shown. (Taken from GREENGARD 1978)

of specific proteins present in its target tissues (for reviews, see NIMMO and COHEN 1977; GREENGARD 1978). Studies in several laboratories (LEBLEU et al. 1976; ROBERTS et al. 1976; ZILBERSTEIN et al. 1976) have indicated that certain of the actions of interferon are achieved through the induction of a double-stranded RNA-dependent protein kinase in the host tissue. Studies in our own (SCHULMAN and GREENGARD 1978 a, b; KENNEDY and GREENGARD 1981) and other (DABROWSKA et al. 1978; YAGI et al. 1978; PERRY et al. 1978; COHEN et al. 1978) laboratories led to the demonstration of the existence of calcium/calmodulin-dependent protein kinases in nervous and other tissues which appear to mediate certain of the actions of calcium acting as a second messenger.

Since the publication in 1978 of the scheme shown in Fig. 2, evidence has been obtained to suggest that other biologically active agents may achieve at least certain of their effects through activation of protein kinases. For example, Stanley Cohen and his colleagues at Vanderbilt University School of Medicine have been able to demonstrate an epidermal growth factor-dependent protein kinase which may be responsible for mediating certain of the biological effects of epidermal growth factor (CARPENTER et al. 1978, 1979). Finally, the transforming actions of a number of oncogenic viruses would appear to be achieved in part through the induction of specific protein kinases in the affected tissues (e.g. COLLETT and ERIKSON 1978; LEVINSON et al. 1978); the results suggest that the induced protein kinases are obligatorily involved in the sequence of events leading to cell transformation by the viruses (LANGAN 1980; PURCHIO et al. 1981).

In conclusion, there seems little doubt that protein phosphorylation represents a biological regulatory mechanism of paramount importance. Moreover, by detecting, purifying, and characterizing the specific endogenous substrate proteins whose state of phosphorylation is altered by any given regulatory agent, a path is opened to understanding the detailed molecular mechanism by which that regulatory agent achieves its biological effects (GREENGARD 1981).

References

Carpenter G, King C, Jr, Cohen S (1978) Epidermal growth factor stimulates phosphorylation in membrane preparations *in vitro*. Nature 276:409–410

Carpenter G, King C, Jr, Cohen S (1979) Rapid enhancement of protein phosphorylation in A-431 cell membrane preparations by epidermal growth factor. J Biol Chem 254:4884–4891

Castellucci VF, Kandel ER, Schwartz JH, Wilson F, Nairn AC, Greengard P (1980) Intracellular injection of catalytic subunit of cyclic AMP-dependent protein kinase simulates facilitation of transmitter release underlying behavioral sensitization in *Aplysia*. Proc Natl Acad Sci USA 77:7492–7496

Cohen P, Burchell A, Foulkes JF, Cohen PTW, Vanaman TC, Nairn AC (1978) Identification of the Ca^{2+}-dependent modulator protein as the fourth subunit of rabbit skeletal muscle phosphorylase kinase. FEBS Lett 92:287–293

Collett MS, Erikson RL (1978) Protein kinase activity associated with the avian sarcoma virus src gene product. Proc Natl Acad Sci USA 75:2021–2024

Dabrowska R, Sherry JMF, Aromatorio DK, Hartshorne DJ (1978) Modulator protein as a component of the myosin light chain kinase from chicken gizzard. Biochemistry 17:253–258

Forn J, Greengard P (1976) Regulation by lipolytic and antilipolytic compounds of the phosphorylation of specific proteins in isolated intact fat cells. Arch Biochem Biophys 176:721–733

Greengard P (1978) Phosphorylated proteins as physiological effectors. Science 199:146–152

Greengard P (1981) Intracellular signals in the brain. In: The Harvey Lecture Series. Academic Press, New York, in press

Kaczmarek LK, Jennings KR, Strumwasser F, Nairn AC, Walter U, Wilson FD, Greengard P (1980) Microinjection of catalytic subunit of cAMP-dependent protein kinase enhances calcium action potentials of bag cell neurons in cell culture. Proc Natl Acad Sci USA 77:7487–7491

Kennedy MB, Greengard P (1981) Two calcium/calmodulin-dependent protein kinases, which are highly concentrated in brain, phosphorylate Protein I at distinct sites. Proc Natl Acad Sci USA 78:1293–1297

Krebs EG (1972) Protein kinases. Curr Top Cell Regul 5:99–133

Kuo JF, Greengard P (1969) Cyclic nucleotide-dependent protein kinases. IV. Widespread occurrence of adenosine 3′,5′-monophosphate-dependent protein kinase in various tissues and phyla of the animal kingdom. Proc Natl Acad Sci USA 64:1349–1355

Kuo JF, Wyatt GR, Greengard P (1971) Cyclic nucleotide-dependent protein kinases. IX. Partial purification and some properties of guanosine 3′,5′-monophosphate-dependent and adenosine 3′,5′-monophosphate-dependent protein kinases from various tissues and species of Arthropoda. J Biol Chem 246:7159–7167

Langan T (1980) Malignant transformation and protein phosphorylation. Nature 286:329–330

Lebleu B, Sen GC, Shaila S, Cabrer B, Lengyel P (1976) Interferon, double-stranded RNA, and protein phosphorylation. Proc Natl Acad Sci USA 73:3107–3111

Levinson AD, Oppermann H, Levintow L, Varmus HE, Bishop JM (1978) Evidence that the transforming gene of avian sarcoma virus encodes a protein kinase associated with a phosphoprotein. Cell 15:561–572

Liu AY-C, Greengard P (1974) Aldosterone-induced increase in protein phosphatase activity in toad bladder. Proc Natl Acad Sci USA 71:3869–3873

Liu AY-C, Greengard P (1976) Regulation by steroid hormones of phosphorylation of specific protein common to several target organs. Proc Natl Acad Sci USA 73:568–572

Miyamoto E, Kuo JF, Greengard P (1969) Cyclic nucleotide-dependent protein kinases. III. Purification and properties of adenosine 3′,5′-monophosphate-dependent protein kinase from bovine brain. J Biol Chem 244:6395–6402

Nimmo HG, Cohen P (1977) Hormonal control of protein phosphorylation. Adv Cyclic Nucleotide Res 8:145–266

Perry SV, Cole HA, Frearson N, Moir AJG, Nairn AC, Solaro RJ (1978) Phosphorylation of the myofibrillar proteins. Proc 12th FEBS Meeting 54:147–159

Purchio AF, Erikson E, Collett MS, Erikson RL (1981) Comparison of the Rous sarcoma virus transforming gene product, pp 60^{src}, and its homologue, pp 60^{sarc}, from normal cells. Cold Spring Harbor Symp 44:in press

Roberts WK, Hovanessian A, Brown RE, Clemens MJ, Keer IM (1976) Interferon-mediated protein kinase and low-molecular-weight inhibitor of protein synthesis. Nature 264:477–480

Rubin CS, Rosen OM (1975) Protein phosphorylation. Annu Rev Biochem 44:831–887

Schulman H, Greengard P (1978a) Stimulation of brain membrane protein phosphorylation by calcium and an endogenous heat-stable protein. Nature 271:478–479

Schulman H, Greengard P (1978b) Ca^{2+}-dependent protein phosphorylation system in membranes from various tissues, and its activation by "calcium-dependent regulator." Proc Natl Acad Sci USA 75:5432–5436

Walsh DA, Perkins JP, Krebs EG (1968) An adenosine 3′,5′-monophosphate dependent protein kinase from rabbit skeletal muscle. J Biol Chem 243:3763–3765

Yagi K, Yazawa M, Kakiuchi W, Ohshima M, Uenishi K (1978) Identification of an activator protein for myosin light chain kinase as the Ca^{2+}-dependent modulator protein. J Biol Chem 253:1338–1340

Zilberstein A, Federman P, Schulman L, Revel M (1976) Specific phosphorylation *in vitro* of a protein associated with ribosomes of interferon-treated mouse L cells. FEBS Lett 68:119–124

CHAPTER 11

Cyclic AMP-Dependent Protein Phosphorylation

J. A. BEAVO and M. C. MUMBY

Overview

Since the early findings that most if not all effects of cAMP are mediated via its binding to the regulatory subunit of cAMP-dependent protein kinase, a considerable amount of effort has been expended delineating the mechanism(s) by which cAMP activates protein kinase and what the functions of the active kinase are in the cell. Some of the more important general observations can be summarized as follows. Two major types of cAMP-dependent protein kinase are known; both types exist as asymetric tetramers composed of two catalytic subunits (C) and one dimeric regulatory subunit (R_2). Upon exposure to a saturating dose of cAMP a three species complex of cAMP, regulatory and catalytic subunit is formed. At what are probably physiological concentrations of enzyme, this complex dissociates into two free, active catalytic subunits and one R dimer having 4 cyclic AMP molecules bound. The K_d of the R:C interaction for either major isozyme type appears to be about 0.5 n*M*. The association and dissociation rates are probably fast enough at physiological enzyme concentrations to allow protein kinase to mediate rapidly oscillating processes in vivo.

Once activated, the enzyme is capable of promoting the phosphorylation of a large number of proteins in vitro. However, current data indicates that many of these substrates are probably not physiologically important. This chapter presents a number of criteria which are useful in determining if cyclic nucleotide-dependent phosphorylations are meaningful in vivo events. It also discusses the types of changes in enzyme function usually caused by phosphorylation. A number of relatively new physiologically important substrates for cAMP-dependent protein kinase are also described. These include myosin light chain kinase, the src gene product, calmodulin-dependent cyclic nucleotide phosphodiesterase and ATP citrate lyase. Of these, only the light chain kinase has as yet been shown to be regulated both in vitro and in vivo by cAMP-dependent phosphorylation.

In addition to research on the substrates of cAMP-dependent protein kinase, this chapter describes current research on the regulation, physiology and possible differential function of the two major isozymes of this enzyme. Numerous studies have shown that protein kinase is compartmentalized within the cell and that the function of an isozyme may be regulated by its location. Protein kinase also appears to be translocated between cellular compartments during treatment of certain cells with agents that raise intracellular cAMP levels. The use of mutants with altered protein kinase has led to an understanding of the role of the enzyme in regulation of processes, such as cell growth, which do not lend themselves to simple biochemical analysis. Studies on the regulation of the amounts of regulatory and catalytic subunits have indicated that while in most tissues both subunits are present in stoichiometric amounts, certain cell types appear to differentially regulate R and C.

A. Introduction

Considerable interest has been shown in recent years concerning the mechanisms by which cAMP exerts its many effects in tissues. Much of this interest has been channeled into studies of how cAMP activates protein kinase and into what proteins are physiological substrates for the activated kinase. In this chapter we would like to discuss these two major aspects of regulation by cAMP of cellular function. The first section addresses a number of aspects of the basic mechanism(s) thought to be operative in the control of protein kinase activity in the cell. The second part describes some of the specific enzymes known to be regulated by cyclic nucleotide-dependent phosphorylation/dephosphorylation and discusses in both general and specific terms how phosphorylation modifies the function or activity of the enzymes. Since many of these topics have been the subject of several other recent reviews, or are treated in more detail in other chapters of this volume, emphasis will be placed largely on properties of the enzymes and enzyme substrates that illustrate specific points and on those which have only very recently been described. For similar reasons many of the citations will refer only to one or two pertinent early articles and then to more comprehensive recent articles. When they exist, recent review articles will be cited.

B. History of Discovery of cAMP-Dependent Protein Kinase

cAMP-dependent protein kinase was originally discovered as a trace contaminant in highly purified preparations of muscle phosphorylase kinase. In the early 1960's KREBS and co-workers were working to elucidate the mechanism by which cAMP promoted the conversion of inactive dephospho phosphorylase to the active phosphorylated form. During the course of these studies, highly purified preparations of phosphorylase kinase, the only enzyme known to catalyze this phosphorylation, were isolated. Further studies indicated that cAMP could activate phosphorylase kinase. However, these investigators were never able to demonstrate stoichiometric binding of cAMP to the phosphorylase kinase enzyme, and, perhaps more importantly, the cAMP-dependent activation was found upon closer investigation to be a time and ATP-dependent catalytic phenomenon. These observations then led to a search for and the discovery of another kinase, originally termed phosphorylase kinase kinase, which conferred cAMP sensitivity to the system. It was soon discovered that this enzyme catalyzed the phosphorylation of a rather wide range of substrates. In particular it became evident that part of the phosphorylation of histone described by LANGAN (1968) and of glycogen synthase described by FRIEDMAN and LARNER (1963) could be catalyzed by cAMP-dependent protein kinase (SODERLING et al. 1970). Since that time a large number of other physiological, and non-physiological, substrates have been described. These include many of the key regulatory enzymes of energy metabolism, a number of structural proteins, and a large number of proteins of as yet unknown function (for review see KREBS and BEAVO 1979).

C. Nomenclature and General Activation Mechanism of Protein Kinase

Until the discovery of cAMP-dependent protein kinase, most protein kinases were named with respect to the protein substrates whose phosphorylation they catalyzed. Hence, the early names phosvitin kinase, histone kinase, casein kinase, phosphorylase kinase kinase, glycogen synthese kinase, etc. However, given the multiplicity of substrates of the cAMP-dependent protein kinase, this system was no longer practical and therefore it was named with respect to its regulatory properties. Even the term cAMP-dependent protein kinase has turned out to be less than satisfactory with the discovery that isozymes of protein kinase exist. The total number of such isozymes present in most tissues is still not clear. However, at least two general types are present in the soluble fraction of most tissues. These are usually designated as type I or II on the basis of their elution profile from DEAE cellulose. Recently, two subclasses of type II enzyme have been described based largely on differential immunoreactivity (ERLICHMAN et al. 1980).

The general overall mechanisms by which cAMP activates protein kinase were originally postulated from the results of dilution experiments in which the apparent specific activity in the absence of cAMP increased with dilution. On the basis of these observations it was speculated that cAMP activated protein kinase by binding to the enzyme and causing its dissociation to the active form (BROSTROM et al. 1970). This speculation was soon demonstrated experimentally (GILL and GAREN et al. 1971; TAO et al. 1970; YAMAMURA et al. 1970; REIMAN et al. 1971) by direct measurement in several systems. The basic activation mechanism defined by these studies is given in Eq. (1) below:

$$\underset{\text{inactive}}{\mathrm{RC+cAMP}} \rightleftharpoons \underset{\text{active.}}{\mathrm{RcAMP+C}} \tag{1}$$

The inactive holoenzyme (RC) was shown to be composed of two dissimilar types of subunit: one was catalytically active (C) and the other was the regulatory subunit (R) which in the absence of cAMP bound to and inhibited the C subunit. The R subunit was also shown to contain the binding sites for cAMP.

More recently it has been found that for both isozyme types the inactive holoenzyme consists of two 40,000 Dalton catalytic subunits and one regulatory subunit dimer (49,500–55,000 Daltons) (see BEAVO et al. 1975 and ROSEN et al. 1975 for summaries of this data). The current shorthand nomenclature for cAMP-dependent protein kinase takes both the isozyme type and subunit structure into account by using subscripts to designate the number of molecules and superscripts to designate the isozyme type (e.g., the type I holoenzyme $= R^{I}_{2}C_{2}$ and the type II holoenzyme $= R^{II}_{2}C_{2}$).

D. Isolation of cAMP-Dependent Protein Kinase

Both of the major soluble isozymes of muscle protein kinase have been purified in large quantity to homogeneity even though the enzymes represent less than 0.05% of the total protein in these tissues. The early methods for purifying the holoenzyme form of protein kinase (BEAVO et al. 1974a; RUBIN et al. 1974) are tedious and suffer from relatively poor yields. However, the more recent use of hydrophobic chromatography (HOPPE and WAGNER 1977) seems to be an improvement in this regard. For the R subunits large scale isolation has been possible due primarily to the availability of highly specific and effective cAMP affinity resins (DILLS et al. 1978; RAMSEYER et al. 1974). In the case of the catalytic subunits, advantage has been taken of the fact that its interaction with CM or DEAE cellulose resins is greatly altered by binding to the acidic R subunit. Therefore, specific absorption (BEAVO et al. 1975) or elution (SUGDEN et al. 1976) of the C subunit is possible. The inactive holoenzyme form of protein kinase has in recent years usually been prepared by in vitro recombination of the C subunit with R that has had bound cyclic nucleotide removed (BUILDER et al. 1980).

E. Physical Properties of Protein Kinase

The major physical properties of both isozyme types are shown in Table 1. To date, no large chemical or physical differences have been noted for the C subunit isolated from either isozyme type. However, one report does indicate different proteolytic fragments can be obtained from C isolated from the two isozyme types and therefore indirectly indicates that some difference may exist (ZOLLER et al. 1979). Several isoelectric forms of catalytic subunit have been found (CHEN and WALSH 1971) but each is present in both isozyme types.

On the other hand, the R subunits can be distinguished easily on the basis of their apparent molecular weight and the amount of phosphate they contain. The R^{I} monomer is about 49,000–50,000 Daltons and usually contains 0.2–0.5 mol phosphate/monomer while the R^{II} is about 54,000 Daltons with 2–3 moles of phosphate. All of the subunits have blocked N terminal amino acids. One rather remarkable physical property of the regulatory subunits is the ease by which they can be renatured from denatured states. In early studies, the cAMP binding capacity could be restored after guanidine HCl treatment (DILLS et al. 1975). Currently, one of the most effective means of removing cAMP from the R·cAMP complex is by

Table 1. Physical properties of cAMP dependent protein kinases

Subunit	R^{I}	R^{II}	C
Molecular weight of monomer	49,500	53,000	41,300
Endogenous phosphate content of monomer	0.2–0.5 mol	2–3 mol	2–3 mol
S_{20w}	5.3	5.4	3.6
Axial ratio	–	12.0	3.0

The data are taken largely from BEAVO et al. (1975) and ROSEN et al. (1975)

treating the subunit with high concentrations of urea which allows separation of the nucleotide from the protein (BUILDER et al. 1980). Upon removal of the urea the protein seems fully active as measured by its ability to bind cAMP and recombine with C subunit (SCHWECHEIMER and HOFMANN 1977; BUILDER et al. 1980). Recently, it has been observed in the authors laboratory (BUILDER, S.E., unpublished observations) that complete restoration of cAMP binding activity is also possible even after boiling the protein. A striking property of both isozyme types is the marked asymmetry in solution. Axial ratios of about 12 have been determined for both R_2cAMP_4 and R_2C_2 complexes. However, the free C subunits have an axial ratio of about 3. Given the fact that R is a very asymmetric dimer and R_2C_2 is also very asymmetric, it seems probable that the C subunits must bind to opposite sides of the R dimer well away from the R:R interaction site.

All of the subunits (R^I, R^{II}, and C) contain endogenous phosphate, some of which is serine phosphate inserted by cAMP or cGMP-dependent protein kinase. The functions and effects of the cyclic nucleotide-dependent phosphorylation of R subunit are discussed below. However, little is known about the nature or functions of the other phosphorylations, especially on the C subunit. At least one of the endogenous phosphates on the C is very resistant to base hydrolysis and, therefore, unlikely to be serine phosphate (BECHTEL et al. 1977).

F. Naturally Occurring Mechanisms Modulating Protein Kinase Activity

I. Heat Stable Protein Inhibitors

1. Muscle Inhibitor

In addition to the R subunit, at least two relatively low molecular weight protein inhibitors of cAMP-dependent protein kinase have been described. However, unlike R both are heat stabile. The first one identified is found in many tissues and has been most extensively studied in skeletal muscle (WALSH et al. 1970). Recently, this form has been purified to homogeneity and characterized (DE MAILLE et al. 1977; MCPHERSON et al. 1979). The native inhibitor is 22,000 MW and may exist in vivo as a mixture of several charge variants. To date, this inhibitor (I) has been found to be specific for inhibiting the C subunit of cAMP-dependent protein kinase. It does not affect the activity of even as similar an enzyme as cGMP-dependent protein kinase. The K_d for the I-C interaction has been estimated to be on the order of 5 nM or about 10-fold higher than the K_d for the R:C interactions (see below). Little is known about the physiological role, if any, of inhibitor in the tissues. WALSH and ASHBY (1973) reported that "I-like activity" in rat tissue extracts could be modulated by the hormone status of the animal; this observation may turn out to be physiologically important. In most tissues I is present in 20% or less than the concentration of C. Because of this, unless it turns out that the inhibitor and part of the C subunit are functionally compartmentalized in the same subcellular area, it seems unlikely that this inhibitor could be modulating protein kinase activity at saturating cAMP levels. However, fluctuations in inhibitor levels may affect the concentration of cAMP necessary to produce meaningful activation of the kinase by low levels of cAMP (Beavo et al. 1974). The amount of inhibitor present is sufficient to affect the basal activity and therefore indirectly the "set point" for cAMP stumulation.

2. Testis Inhibitor

A different heat-stable protein which inhibits cAMP-dependent protein kinase has been isolated from rat testis. Multiple forms of this inhibitor were found in whole testcs, while a single major form was present in Sertoli cell-enriched testes (Beale et al. 1977). The purified protein had a molecular weight of 20,000 and contained a high percentage of acidic residues and proline as distinguished from the skeletal muscle inhibitor. The testis inhibitor also acted as an inhibitor of Ca^{2+}-activated cyclic nucleotide phosphodiesterase, an effect that seemingly would be antagonistic to its effect on protein kinase except, perhaps, in rapidly oscillating systems. Inhibitor activity from Sertoli cell-enriched rat testis was shown to be regulated in vivo by exogenous follicle-stimulating hormone as well as by the age of the animals (Tash et al. 1979). At an age when activity was highest, it was estimated that this inhibitor could inactivate as much as 41% of the total catalytic subunit present in Sertoli cell enriched testes. These results suggested that stimulation of inhibitor activity by FSH may play a significant role in modulating cAMP-dependent protein kinase activity in this tissue.

II. Nucleotide Effects

1. cAMP

Probably the most important modulator of cAMP-dependent protein kinase activity is, of course, cAMP. The early proposals to explain how cAMP activates protein kinase were discussed above and current ideas about the stoichiometry and mechanism of activation are discussed in a later section of this chapter. Nevertheless, there are several other aspects of the effects of cAMP on protein kinase activity that merit discussion here. The first point is whether cAMP concentrations in vivo are such that meaningful regulation is possible. Early studies indicated that cAMP levels in most tissues were at least an order of magnitude higher than the reported K_d's for cAMP binding or K_a's for activation. However, more recent evidence indicated that the concentration of protein kinase in most cell types was considerably higher than originally anticipated and that when these higher concentrations of enzyme were taken into account activation occured at physiological levels of cAMP (Beavo et al. 1975). The kinetic reasons for this decrease in apparent affinity at higher enzyme concentration are due in part to the fact that activation requires a dissociation reaction and therefore the equilibrium position is dependent on the concentration of all the reactants [see Eq. (1)]. The kinetic and molecular mechanisms responsible for these observations have been discussed extensively by a number of investigators (Ogez and Segal 1976; Swillens et al. 1974).

2. ATP

Although cAMP is the most dramatic effector of protein kinase activity, other nucleotides also have some effect. In particular, ATP may play an important role.

Two mechanisms or effects of ATP on protein kinase activity are known. The first is a direct binding of ATP to the type I holoenzyme that causes an increase in the apparent K_a for cAMP activation of the enzyme (HADDOX et al. 1972; BEAVO et al. 1975). The second mechanism is due to the enzymatic transfer of the phosphate from ATP to the regulatory subunit (ROSEN et al. 1975; GEAHLEN and KREBS 1980a, b).

High affinity binding of ATP to a crude preparation of protein kinase was originally reported by HADDOX et al. (1972) and later confirmed for the pure enzyme (BEAVO et al. 1975). When measured by millipore filtration assays, two mol of ATP are bound per mol holoenzyme with an apparent affinity of about 50 nM. The location of the high affinity ATP binding site is not yet clear. Experiments with analogs of ATP imply that the site is likely to be on the C subunit and may be the same as the active site (HOPPE et al. 1977). However, experiments with ATP photoaffinity analogs imply that either the C or R subunit can be covalently modified depending on the enzyme preparation used (OWENS and HALEY 1978; HOPPE and FRIEST 1979). If the high affinity site is the active site on the C subunit, the affinity for ATP is apparently increased by the presence of R. One reason for this conclusion is that the high affinity ATP binding site cannot be detected by filter assays on either the isolated R or C subunits but is easily seen on the holoenzyme. Although most easily detected on the type I isozyme, a high affinity binding site also can be seen on the type II isozyme if AMPPNP, the nonhydrolyzable derivative of ATP, is used (BEAVO, BECHTEL, and KREBS, unpublished observations). The effect of ATP binding to the type I isozyme is to increase the concentration of cAMP required for activation of the enzyme (BEAVO et al. 1975) and possibly also to increase the rate and/or extent of recombination of the subunits to yield the holoenzyme. The exact mechanism by which these effects occur is not clear although at least part of the effect may be due to a simple mass action on the multiple equilibria present in the activation process. It is also possible that ATP directly or allosterically affects the on or off rate of cAMP binding. To date no form of the protein kinase has been shown to bind both ATP and cAMP simultaneously.

III. ATP-Dependent Phosphorylations

1. Autophosphorylation of R^{II}

ATP has a complex effect on type II enzymes. As noted above, a high affinity site can be detected with AMPPNP as substrate. However, this is not seen with ATP, for several reasons. First, the type II enzyme has an appreciable endogenous ATPase activity which makes such studies difficult. Secondly, and probably more importantly, ATP is used as a substrate for auto-phosphorylation of the R^{II} subunit by C subunit even in the absence of cAMP (ROSEN and ERLICHMAN 1975). Finally, the phosphorylated enzyme apparently does not bind ATP with high affinity as evidenced by decreased AMPPNP binding to the phospo-form of the enzyme (BEAVO, BECHTEL, and KREBS, unpublished). To date, two effects of phosphorylation have been described. The first is to increase the apparent affinity for cAMP binding (BEAVO et al. 1975). The second is to decrease the observed rate of recombination of the subunits (RANGEL-ALDAO and ROSEN 1976).

2. Phosphorylation of R^I

Very recently it has been reported that the type I regulatory subunit also can be phosphorylated (GEAHLEN and KREBS 1980a). The phosphorylation is stoichiometric but not very rapid. However, in this case, the phosphorylation is catalyzed by cGMP-dependent protein kinase. Other studies (HOFMANN et al. 1975) have shown that essentially no phosphate is incorporated into R^I by the C subunit of cAMP-dependent protein kinase. Little is known about the physiological significance, if any, of the phosphorylation. Preliminary evidence indicates that it does occur in vivo, that the binding of cAMP to R may be altered by the phosphorylation, and that the phosphorylated R^I is much less inhibitory toward C (GEAHLEN and KREBS 1980b; GEAHLEN et al. 1981). Early studies also indicated that less than stoichiometric amounts of endogenous phosphate could be measured in purified preparations of R^I (BEAVO et al. 1975). Similarly, the charge variants of R^I noted in two dimensional electrophoresis by several workers studying various transformed cell lines (COFFINO et al. 1976; PRASHAD et al. 1979) may be due to the state of phosphorylation of the R^I. It is not yet known if this phosphorylation affects ATP binding.

G. Substrates of cAMP-Dependent Protein Kinase

As mentioned in the introductory paragraphs, cAMP-dependent protein kinase will catalyze the transfer of phosphate to serine or threonine residues of a large number of proteins in vitro. However, in most studies carried out with intact cells, many fewer proteins are found to have their state of phosphorylation specifically increased by cAMP. For this reason, considerable effort has been directed toward understanding what mechanism(s) control the "phosphorylatability" of a protein by cAMP-dependent protein kinase.

I. Primary Sequence Determinants

Most of the definitive research has centered around determination of the primary sequence of the substrates near the phosphorylated serine. These results can be summarized as follows. Nearly all of the natural substrates of the kinase have a pair of positively charged, amino acids (lysine and/or arginine) two or three residues toward the N-terminal side of the phosphoserine. Studies with synthetic peptides indicate that these basic groups affect both the K_m and V_{max} of the reaction (KEMP et al. 1975, 1977; HUANG et al. 1979). In addition, work with peptides derived from phosphorylase indicate that negative determinants can also exist. For example, removal of the first arginine residue on the carboxyl terminal side of the phosphoserine in the phosphorylase phosphopeptide makes this peptide a good substrate for cAMP-dependent protein kinase whereas the complete peptide is a very poor substrate (KEMP et al. 1977).

II. Higher Order of Structure Effects

The primary sequence of a protein is obviously not the only determinant of substrate specificity for cAMP-dependent protein kinase since many proteins have basic residues on the N-terminal side of one or more of their serine residues. This fact was illustrated very graphically in the studies of BYLUND and KREBS (1975) where they showed that denaturation of a number of proteins caused them to become reasonably good substrates for cAMP-dependent protein kinase whereas the same proteins in their natural conformations were not phosphorylated. Taken together these results indicate that the appropriate primary sequence is essential and contains sufficient information for a peptide to be a good substrate but that higher orders of structure can modify the local conformation or availability of the site such that it will not be phosphorylated.

H. Physiological Effects of Phosphorylation

I. Criteria for Evaluating Physiological Roles of Cyclic Nucleotide-Dependent Phosphorylation Reactions

Since the phosphorylation of a large number of proteins can be stimulated by cAMP in vitro and especially since an even larger number can be phosphorylated when partially denaturated, the physiological relevance of many of these phosphorylations can and should be questioned. In addition, it is becoming increasingly clear that many proteins are regulated by multiple phosphorylation/dephosphorylation reactions often catalyzed by distinct kinases or phosphatases. For these reasons a series of criteria based on properties of known substrates have been suggested (KREBS 1973; NIMMO and COHEN 1977; KREBS and BEAVO 1979) for determining whether a given phosphorylation is likely to be physiologically relevant. These properties are summarized below for cAMP-dependent phosphorylation. Most of them are also relevant, however, to phosphorylation catalyzed by other kinases.

The first criterion that has proven useful for evaluating the relevance of phosphorylation is that the stoichiometry of physiologically important phosphorylations is usually at least 1:1 (e.g., 1 phosphate per monomer of protein substrate). Experimentally, if the observed phosphorylation is stoichiometric then it is less likely to be due to partial denaturation of a small percentage of the proteins during isolation and assay. Ideally, stoichiometric phosphorylation should be demonstrated in both broken and intact cell preparations.

The second general criterion is that the rate of phosphorylation be rapid enough to correspond to alteration of the proposed physiological function associated with the protein substrate. For example, a very slow rate of phosphorylation of one of the contractile proteins of a fast twitch muscle is unlikely to be directly associated with immediate control of tension development in the filament. That is not to say that it could not mediate some other slower regulatory component. Again, ultimately this should be demonstrated with intact cell systems.

The third general criterion usually associated with physiologically important phosphorylations is that the functional properties of the substrate be altered in a manner proportionate to the extent of phosphorylation. This does not necessarily mean that there must be a 1:1 correlation since cooperative interactions or effects of opposing processes may intervene. However, examples of maximal functional effects seen at minimal degrees of phosphorylation are immediately suspect.

The fourth general criterion usually demonstrable for physiologically meaningful cAMP-dependent phosphorylation reactions is that the degree of phosphorylation of the substrate should correlate with the level of cAMP in the compartment of the cell containing the kinase and substrate. Given the increasing evidence for compartmentalization, this criterion has become increasingly difficult to demonstrate experimentally.

In the last few years the isolation of somatic cell lines having mutations in protein kinase (see Sect. I.4) has suggested another criterion for establishing physiological relevance of a cAMP-dependent phosphorylation. For those cases where the phosphorylation of interest occurs in the mutant cell lines, then one would expect to see a diminished response in both degree of phosphorylation and function in the mutant cell lines as opposed to the wild type. To date few putative cAMP-mediated processes have been tested in this way but it is expected that many will be, particularly as new cell lines are developed. The studies of cyclic nucleotide effects on steroidogenesis in mutant cell lines of mouse adrenal cortex may provide the best example to date of this approach (for review see SCHIMMER et al. 1979).

II. Methods of Demonstrating Criteria, Problems Encountered, and Interpretation of Results

Demonstration of the first two criteria for physiological relevance (i.e., rapid and stoichiometric phosphorylation) often have been relatively easy especially in broken cell preparations. In these cases, exogenous protein substrates and ^{32}P labeled ATP of known specific activity can be used to determine directly the amount and rate of phosphorylation. However, an increasing number of substrates have been shown to be phosphophorylated at multiple sites and therefore some method for determining the actual site of phosphorylation has to be used. These methods usually involve cleavage of the phospho proteins by cyanogen bromide or proteases followed by separation and quantitation of specific phosphopeptides (e.g. SODERLING 1976).

Demonstration of rapid stoichiometric phosphorylation in intact cell preparations nearly always has been more difficult since neither the protein nor nucleotide substrates can enter most cells. In this case, specific methods of isolating a particular phosphoprotein product from the many which are phosphorylated must be devised. The most generally useful procedure has been specific affinity chromatography or specific antibody precipitation. Quantitation of the amount of phosphate in a specific site of the protein often must still be made. Alternatively, one must make the assumption that the observed specific activity of the total cellular [^{32}P]ATP (usually labeled by preincubation with [^{32}P]Pi) is the same as that pool used by the kinase and that the amount of ^{32}P in the protein is proportional to the degree of total phosphorylation. This assumption often may not be true in many whole cell preparations. Finally, all of the manipulations usually have to be carried out under conditions where the degree of phosphorylation has not been changed during isolation. Usually, high concentrations of metal ion chelator and fluoride are included in the buffers to minimize kinase and phosphatase activity.

More recently, high resolution two-dimensional gel electrophoretic techniques (O'FARRELL 1975) have been employed as a method of determining the time course and degree of cyclic nucleotide-dependent phosphorylation of specific proteins (e.g. STEINBERG and COFFINO 1979). Since the cellular proteins can be labeled to high specific activity with radioactive amino acids, this method should allow studies on phosphorylation of trace proteins to be carried out which would be very difficult using other techniques. The recent application of computer assisted image analysis to these two-dimensional separation techniques should greatly accelerate progress in this area.

Demonstration of the third criterion (i.e., correlation of phosphorylation with function) obviously has all of the same potential problems as described for the first two since the degree of phosphorylation must be measured. In addition, conditions must be devised that not only allow the state of phosphorylation to be maintained, but also allow the state of activity or function of the phosphorylated substrate to be maintained and measured. In the case of those enzyme substrates activated by proteases or those requiring divalent metal ion, this is not always a trivial problem. A more detailed description of the methodology and problems of determining rates and degrees of phosphorylation of several muscle phosphoproteins has recently been presented (MANNING et al. 1980); interested readers are referred to that review for specific examples.

The final criterion, correlation of phosphate content and function with cAMP levels, also has proven difficult with many systems. Reasons for this include the probability that several "pools" of cAMP, protein kinase, and protein substrates are present in many cells and the certainty that they are present among different cell types within tissues. As a result, negative correlations between cAMP content and degrees of phosphorylation or activity are not convincing demonstrations that cAMP-dependent phosphorylation is an important regulatory mechanism for a given physiological process. Conversely, positive correlations by themselves do not prove such cause and effect relationships. In addition, this criterion assumes that other regulatory processes have not occurred (e.g., activation of a protein phosphatase) which would mask changes in phosphorylation or activity. A number of attempts have been made to bypass some of these possible problems by trying to correlate the degree of phosphorylation and activity with changes in the state of activation (activity ratio) of protein kinases (CORBIN and KEELY 1976). Some of these studies have been very useful, especially those indicating an increase in activity ratio where little or no change in cAMP content can be measured. However, many recent studies have not critically evaluated the necessary controls described in the original papers. A discussion of some of the potential pitfalls in this approach has appeared recently (PALMER et al. 1980). Few, if any, of the studies are really free from the potential pitfalls present if differential modulation of cellular and subcellular pools of cAMP occurs in the in vivo system being tested. Therefore, the test for correlation of phosphate content, function, protein kinase activation and cAMP levels should be used in conjunction with the other criteria listed above in order to be useful. In many cases the use of mutant cell lines having defects in cAMP generation or function may offer an extremely useful way of indirectly correlating cAMP levels with function.

III. Effects of Phosphorylation on the Activity or Function of the Substrate

When one wants to demonstrate a change in the activity or function of a protein substrate upon phosphorylation in order to be convinced of its physiological importance, the question arises as to what changes are likely to be encountered. Of course, each enzyme or structural protein has a different function(s) in the cell and therefore, the details of any change will depend on the function altered. However, to date only a few different parameters of function have been shown to be altered by phosphorylation and therefore are the prime initial candidates for study. Far and away, the most common change is an alteration in the apparent affinity of a substrate or an effector for the enzyme. For example, phosphorylation of phosphorylase kinase by protein kinase results in a decrease in the apparent K_m for its substrate phosphorylase (FISCHER et al. 1971). Similarly, the phosphorylation of glycogen synthase by protein kinase results in a decrease in the affinity for the allosteric activator, glucose-6- phosphate (ROSELL-PEREZ and LARNER 1964). Affinity changes have also been seen upon phosphorylation with pyruvate kinase (ENGSTROM 1978), hormone sensitive lipase (KHOO et al. 1976), tyrosine hydroxylase (AMES et al. 1978), and, most recently, myosin light chain kinase (ADELSTEIN and EISENBERG 1980). Experimentally, therefore, it is obvious that putative functional differences in a protein caused by phosphorylation/dephosphorylation are not likely to be evident unless measurements are made under conditions which would detect activity differences due to shifts in affinity (e.g. subsaturating levels of substrates and effectors).

IV. Proteins of Known Function as Substrates

As discussed previously, an enormous number of polypeptides have been shown to be phosphorylated in vitro by cAMP-dependent protein kinase. However, the number that can be shown to be phosphorylated in vivo in a cAMP-dependent manner is much smaller. When one considers only those well defined proteins involved in known functions (usually enzymes or enzyme modulators) that are rapidly and stoichiometrically phosphorylated, the list narrows predominately to that shown in Table 2. The details of most of these posphorylation reactions and the effects they have on function have been reviewed recently (KREBS and BEAVO 1979; ADELSTEIN and EISENBERG 1980; STULL 1980) and therefore will not be discussed in detail. Interested readers are referred to these previous reviews and the references in Table 2. A few of the reactions have been described only recently and are discussed in a later section on new substrates. In many cases where it is known that multiple phosphorylation sites exist on the protein substrates, it is indicated in the Table. It is probable that additional sites will be found on some of these known phosphoproteins as they are studied more extensively and that completely new substrates of known function will be identified. Many other proteins of less defined function have been shown experimentally to have increased degrees of phosphorylation in response to cAMP.

Table 2. Proteins having known function which are phosphorylated by cAMP-dependent protein kinase

Protein substrate	Multiple Phosphorylation Sites	Reference
1. Phosphorylase kinase	(+)	KREBS et al. (1959) CARLSON et al. (1980)
2. Glycogen synthase	(+)	FRIEDMAN and LARNER (1963) SODERLING (1979)
3. Hormone-sensitive triglyceride lipase		RIZACK (1964) STEINBERG (1976)
4. Fructose 1, 6 biphosphatase		MENDICINO, J. (1966) RIOU et al. (1977)
5. Ribosomal protein S-6	(+)	BLAT and LOEB (1971) GRESSNER and WOOL (1976)
6. Hydroxymethyl glutayl CoA reductase		BEG et al. (1973) GIBSON and INGEBRITSEN (1978)
7. Acetyl CoA carboxylase	(+)	CARLSON and KIM (1973) HARDIE and COHEN (1978)
8. Liver pyruvate kinase	(+)	LJUNGSTROM et al. (1974) ENGSTROM (1978)
9. Cholesterol ester hydrolase		TRZECIAK and BOYD (1974) KHOO et al. (1976)
10. cAMP-dependent protein kinase (R^{II})	(+)	ERLICHMAN et al. (1974)
11. Cardiac troponin	(+)	COLE and PERRY (1975) STULL (1980)
12. Phosphofructokinase		BRAND and SOLING (1975)
13. Tyrosine hydroxylase		MORGENROTH et al. (1975) AMES et al. (1978)
14. Phosphoprotein phosphatase inhibitor		HUANG and GLINSMAN (1975) COHEN (1978)
15. Phenylalanine hydroxylase		MILSTEIN et al. (1976) DONLON and KAUFMAN (1978)
16. Tryptophan hydroxylase		HAMON et al. (1977)
17. Glycerophosphate acyltransferase		NIMMO and HOUSTON (1978)
18. Myosin light chain kinase		ADELSTEIN et al (1978, 1980)
19. ATP citrate Lyase		JANSKI et al. (1979)
20. $pp60^{src}$	(+)	COLLETT et al. (1979)
21. Calmodulin sensitive cyclic nucleotide phosphodiesterase	(+)	SHARMA et al. (1980)

I. Current Areas of Research

I. Current Research on cAMP-Dependent Protein Kinase

The following sections describe some of the more recent areas of research regarding the regulation of protein kinase and its substrates. The large interest in this enzyme has led to generation of such a voluminous literature that only a few selected aspects relating directly to regulation in vivo are described. This is not to imply that other areas have not progressed rapidly. In particular, studies on the catalytic mechanisms of the reaction, protein structure and chemistry, and effects of synthetic modulators of activity have all been very interesting. Similarly, only those relatively recently discovered substrates of the kinase, for which considerable evidence exists for physiological regulation by phosporylation, are mentioned.

1. Stoichiometry and Properties of cAMP Binding

The original studies on subunit structure and cyclic nucleotide binding strongly indicated that for either protein kinase isozyme type a single molecule of cAMP could be bound per R monomer (BEAVO et al. 1975; ROSEN et al. 1975). More recent studies, however (CORBIN et al. 1978; BUILDER et al. 1980) indicate that 2 mol of cAMP are bound per R^{II} monomer at equilibrium and saturating cAMP concentration. Furthermore, these studies imply that one of the sites is greatly affected by the conditions of measurement. For example, only one mol/mol is bound when tested by most Millipore filter binding assays. However, when measured by equilibrium dialysis, optical density, or fluorescence quenching methods, the second mol of bound cAMP is seen. Interestingly, the affinities for both sites are similar, having an apparent K_d in the n*M* range. However, the sites can be distinguished by measurements of on and off rates (SMITH et al. 1980; RANNELS and CORBIN 1980).

Current research is being directed towards the questions of the function of the sites as related to cAMP activation of the kinase. Studies are also being carried out in several laboratories on the nature of the cAMP binding site. With regard to this latter point, the studies with fluorescent analogs of cAMP are of interest. In the case of beef skeletal muscle (Type I) R, nearly all of the fluorescence from both cAMP molecules is quenched. However, with the R^{II} from beef heart, it appears that binding to one of the sites causes quenching and binding to the other may actually enhance fluorescence (LAPORTE et al. 1980). Finally, with R isolated from brain, it is reported that only enhancement of fluorescence is seen (GABIBOV et al. 1979). It will be of interest to see what difference in structural features are responsible for these observations and to see if such behavior is a general feature diagnostic of the particular isozyme type. The studies of ERLICHMAN et al. (ERLICHMAN et al. 1980) indicating immunologically distinct subclasses of type II kinase may be significant in this regard.

From a theoretical viewpoint, the finding of multiple cAMP binding sites opens up the possibility of differential modulation of protein kinase activity by factors which may selectively affect one site. The state of phosphorylation of R seems a prime candidate for such putative effects. From a practical viewpoint, the multiplicity of binding sites considerably complicates the interpretation of many of those studies using cAMP affinity labels. It will be interesting to see which site(s) is being labeled in such studies.

2. Mechanism of Activation and Inactivation

a) Activation

The generalized equation for the activation of protein kinase has been known for a number of years and was discussed earlier [see Eq. (1)]. As more details about the stoichiometry of subunit structure and nucleotide binding are elucidated, the possible number of intermediates in this general expression have become greater. The original expression [e.g., Eq. (1)] really depicts only the end points of a complex multiple equilibrium. For example, we now know that in vitro, at least, the protein kinase holoenzyme can exist in a complex with ATP. Furthermore, a small but potentially variable percentage of free C subunit can be complexed with one or more heat stable inhibitor proteins. Finally, the enzyme can exist as phospho or dephospho forms at sites on the R subunit and possibly also the C subunit. All of these observations increase the number of potential forms that may be present in vivo during the activation reaction by cAMP. When one takes into account the observations that the holoenzyme is a heterologous tetramer having possibilities of cooperative subunit-subunit and subunit-nucleotide interactions, the complexity of the activation process becomes unmanageably complex.

Currently, there is little direct physical or kinetic data that indicates which one of the many possible intermediate forms of protein kinase is quantitatively most important in vivo. However, a number of generalizations probably can be made based on data derived from experiments carried out under conditions that put limits to the possible number of forms. Some of these are discussed below.

The most stable subunit species appear to be R_2C_2, R_2-$cAMP_4$, C, CI-complex, and $R_2C_2{:}ATP_2$ since only these forms have been isolated. Little definitive evidence for anything more than the transient existence of R_2C with or without nucleotide bound has been shown. However, as described below, some sort of RC cAMP "ternary" complex does exist transiently and under some conditions may even make up an appreciable percentage of the total enzyme. R_2C_2 probably exists as the phosphorylated form for the type II kinase and as the ATP complex for the type I kinase; however, these points have not been demonstrated experimentally in vivo.

Although we still do not know all of the details of the activation process, two recent observations have provided added information. The first is the observation (CORBIN et al. 1978, BUILDER et al. 1980) that both isozymes of protein kinase can bind two mol cAMP per monomer. Although it is not clear whether one or both cAMP sites must be occupied in order for the enzyme to be activated, at saturation, both sites are occupied. The second observation (CHAU et al. 1980; BUILDER et al. 1980) is that cAMP must first bind to the holoenzyme to form a "ternary complex" which under physiological conditions goes on to dissociate into free active C subunit and R_2 $cAMP_4$ complex. The term "ternary complex" in this case refers to the fact that three different species are involved but does not specify the subunit structure or stoichiometry of binding to the complex.

In some ways these observations raise more questions than they answer, especially with regard to the number of possible molecular species potentially involved in the activation process. Under physiological conditions, the limiting activation equation for the type I enzyme is therefore likely to be:

$$R^{I}_2C_2{:}\text{Mg ATP}_2 + 4\,\text{cAMP} \rightleftharpoons R^{I}_2{:}\text{cAMP}_4 + 2\,\text{C} + 2\,\text{MgATP}. \qquad (2)$$

Obviously, transient multiple molecular species of "ternary" complex or even "quaternary complex" are possible. At saturating cAMP, 4 mol of cAMP are probably bound to R_2^I, and ATP is probably always tightly bound to $R_2^IC_2$ and less tightly to C. It is not clear whether both nucleotides can be bound to protein kinase at the same time. Neither is it yet entirely clear what the phosphorylation state of R^I or C is in vivo and whether it changes or affects the activation of the enzyme.

For the type II enzyme, it is likely that the predominant species is phospho R^{II} (p-R^{II}) since ATP levels are much higher than the K_m for the phosphorylation and since the reaction can be catalyzed intramolecularly. Some direct confirmation of this hypothesis has been presented (RANGEL ALDAO et al. 1979). In these experiments with fresh extracts, the predominant species of R^{II} labeled by photo affinity probes was p-R^{II}. The most likely in vivo activation equation for the type II isozyme then becomes:

$$\text{p-}R_2^{II}C_2 + 4\ \text{cAMP} \rightleftharpoons \text{p-}R_2^{II}\text{:cAMP}_4 + 2\ \text{C}. \tag{3}$$

In this case MgATP does not bind with high affinity to P-$R_2^{II}C_2$.

Since ATP is a substrate for the kinase reaction it obviously does bind with some affinity to C. Equations 2 and 3 are interesting since they would imply that ATP may have a lesser effect on favoring the reassociated form of the enzyme with the type II than with the type I enzyme (e.g., ATP does not bind with high affinity to phospho $R_2^{II}C_2$ and therefore does not pull the reaction to the left). Experimental data indicate that the phosphorylated $R_2^{II}C_2$ requires less cAMP for half maximal activation and that the rate of reassociation of p-R^{II} with C is slower than dephospho R^{II} (BEAVO et al. 1975; ROSEN et al. 1975). However, under the conditions used to date, no dissociation of type II holoenzyme by MgATP has been demonstrated, implying that it may bind with similar affinity to p-$R_2^{II}C_2$ and to C.

b) Inactivation

There have been an enormous number of studies on the activation of protein kinase and on the phosphorylation reactions catalyzed by the active form of the enzyme. Relatively speaking, there have been fewer reports on the reversal of the phosphorylation event and very few studies on the reversal of the activation process of the kinase. Since cAMP-dependent phosphorylation events are postulated to regulate many rapidly oscillating physiological processes, it is somewhat surprising that more attention has not been paid to the rates of reversal (i.e., dephosphorylation and recombination of R and C).

The normal inactivation of protein kinase is thought to be mediated primarily by one of two mechanisms in vivo. The first is a reversal of the activation process observed in Eq. 2 above, the so called "recombination" reaction. The other inactivation reaction in vivo is due to association with one of the heat stable inhibitors of protein kinase discussed above. Surprisingly, very few studies have been carried out to describe and define the mechanism and possible regulation of R and C recombination. Early studies (TAO 1972; MIYAMOTO et al. 1973; CORBIN et al. 1975; BEAVO et al. 1975) indicated that a number of factors including salt concentration, MgATP, and the nature of the protein substrate can all affect the equilibrium position of the association/dissociation reaction [Eq. (1)].

Recently, preliminary evidence (BUILDER et al. 1980, and unpublished) indicates that other agents may effect the rates of subunit recombination. For example, mixed histones decrease the rate of recombination by at least 2 orders of magnitude. Contrary to the implications of early equilibrium studies (MIYAMOTO et al. 1971; TAO 1972), the effect seems greater with type II enzyme than with type I. Whether these observations are in vitro artifacts or if they relate to the control of in vivo phosphorylation, particularly in the nucleus, remains to be seen.

The data from measurements of the rates of activation and inactivation (BUILDER et al. 1980) taken together with the early data for measurements of protein kinase concentration (BEAVO et al. 1975; HOFMANN et al. 1975) have several interesting implications. For example, they imply that activation/deactivation can occur rapidly enough for posphorylation (and, presumably dephosphorylation) to be mechanistically involved in rapidly oscillating processes. Such conclusions are made more tenable given the increasing evidence that high local concentrations of the enzyme may exist in the cell (see below) and, therefore, the bimolecular activation and inactivation processes could occur very rapidly. Unfortunately, to date, relatively little data on in vivo rates of specific dephosphorylation of endogenous substrates or their subcellular localization are available.

3. Compartmentalization

a) Localization of Protein Kinase

A number of observations over the past several years have indicated that cAMP-dependent protein kinase may exist at least in part in vivo in areas of local concentration. Much of the evidence for this statement has been reviewed recently (SPRUILL and STEINER 1979; KAPOOR and STEINER, this volume) and will, therefore, be only briefly mentioned here. In early experiments it was shown that, although most of the enzyme appears to be easily solubilized, some of it remains firmly attached to particulate fractions. More recently, these particulate forms have been more extensively investigated (RUBIN et al. 1979; UNO et al. 1977). With the realization that cyclic nucleotide-dependent phosphorylation may be altering gene expression (JOHNSON, this volume) considerable effort has been put forward towards studying the nature of nuclear protein kinase.

b) Translocation of cAMP-Dependent Protein Kinase

Cyclic AMP-dependent protein kinase has been implicated in increased gene activity during the induction of a number of enzymes by cAMP. It has been postulated that activated nuclear cAMP-dependent protein kinase phosphorylates chromatin associated proteins resulting in increased synthesis of specific messenger RNA. This hypothesis would require the presence of cAMP-dependent protein kinase in the nucleus; however, a number of studies indicated that nuclear protein kinase activity was insensitive to cAMP. This has led to the further postulation that cytoplasmic cAMP-dependent protein kinase is translocated to the nucleus where it is then active in phosphorylation of nuclear proteins. This topic has been reviewed by JUNGMANN and RUSSELL (JUNGMANN and RUSSELL 1977) and only a brief description of systems where translocation has been described will be presented here.

Translocation of cAMP-dependent protein kinase has been reported to occur as the result of treating tissues with analogs of cAMP or hormones which elevate intracellular cAMP. These studies have typically been done by isolating nuclei and cytosol from cells after treatment and comparing protein kinase and/or [^{3}H]-cAMP binding in the nuclear and cytoplasmic fractions. The accuracy of these determinations thus depends on the ability to isolate an intact nuclear fraction not contaminated with cytoplasmic elements. The importance of proper isolation conditions was pointed out by KEELY et al. (KEELY et al. 1975) who demonstrated that high ionic strength (150 mM KCl) was necessary to prevent non-specific binding of free catalytic subunit to the particulate fraction of rat heart. A number of studies describing translocation of protein kinase have been done under conditions which should prevent non-specific binding of catalytic subunit to nuclei.

Translocation of protein kinase has been reported in a number of different tissue and cell types. Nuclei isolated from rat heart perfused with glucagon or dibutyryl-cAMP were found to contain 3–5 times as much free catalytic subunit as control nuclei, suggesting a cAMP-mediated redistribution of cytoplasmic catalytic subunit (PALMER et al. 1974; CASTAGNA et al. 1975). These nuclei were isolated at low ionic strength and possible contamination of nuclei with free cytoplasmic catalytic subunit cannot be ruled out. Jungmann and co-workers described the translocation of both regulatory and catalytic subunits to the nucleus upon gonadotropin treatment of porcine ovaries (JUNGMANN et al. 1974; JUNGMANN et al. 1975). This translocation was also shown to occur when nuclei were isolated in nonaqueous medium to prevent cytoplasmic contamination of nuclei (JUNGMANN et al. 1976). The induction of tyrosine hydroxylase in the adrenal medulla has been reported to involve the translocation of cytoplasmic protein kinase activity to the nucleus (KUROSAWA et al. 1976; COSTA et al. 1976). Dibutyryl cAMP treatment of responsive Walker 256 mammary carcinoma resulted in a 3-fold increase in nuclear cAMP binding (CHO-CHUNG et al. 1977). An increase in nuclear cAMP binding and protein kinase activity has also been observed in mouse neuroblastoma cells (PRASAD et al. 1979; HOLLENBECK et al. 1979).

4. Genetic Studies, Mutant Cell Lines

Cyclic-AMP-dependent protein kinase has been implicated as the mediator of most, if not all, processes regulated by cAMP. One of the more difficult postulated roles to establish for this enzyme is the regulation of cell growth and morphology. In most instances, biochemical demonstration of an obligatory role of protein kinase has been impossible due to the complexity and uncharacterized nature of the processes involved. One extremely useful approach has been the use of somatic cell genetics to produce mutants with altered cAMP-dependent protein kinase activity. These mutant cells can be used to examine the role of protein kinase in any cellular process which can be conveniently measured. This section will briefly describe recent studies on protein kinase from mutant cell lines.

Selection of cell lines having mutations affecting protein kinase has been facilitated because high intracellular concentrations of cAMP cause growth inhibition and/or death of a number of cultured cell lines. Agents commonly used to raise intracellular cAMP concentrations have included: the nucleotide itself; analogs of cAMP; or compounds such as isoproterenol or cholera toxin which activate adenylate cyclase. Mutants deficient in some aspect of the cAMP system have been obtained by isolating cells which are able to grow in high concentrations of cAMP or hormone. In most cases, the mutants have altered cAMP-dependent protein kinase or adenylate cyclase activities. Cells with mutations in protein kinase have been used to show that functional cAMP-dependent protein kinase is required to observe a number of the physiological effects of cAMP.

Studies with protein kinase mutants of S49 lymphoma cells have demonstrated three basic types of mutations (for review, see COFFINO et al. 1976). The first type has normal levels of protein kinase but requires higher levels of cAMP for activation. The second type contains reduced amounts of apparently normal kinase while the third type lacks detectable cAMP-dependent protein kinase activity. The protein kinase deficient mutants have been shown to be resistant to the effects of cAMP on several processes including growth regulation, protein modification, protein repression and induction of a number of proteins including cyclic nucleotide phosphodiesterase and ornithine decarboxylase (INSEL et al. 1975; INSEL and FENNO 1978; STEINBERG and COFFINO 1979). This provides strong evidence that functional protein kinase is required to mediate all of the observed effects of cAMP on these systems. Mutants basically similar to those described in mouse lymphoma cells have also been observed in a number of other cell lines.

The most common mutations affecting cAMP-dependent protein kinase have caused an increase in the apparent activation constant for cAMP and are often due to altered regulatory subunits. S49 lymphoma mutants of this type were shown to require a ten fold higher concentration of cAMP for half maximal activation of protein kinase in cell extracts. This ten-fold increase in cAMP concentration was also observed for half maximal binding of cAMP to the enzyme (INSEL et al. 1975). Evidence that this alteration in cAMP activation was due to a structural mutation was obtained from experiments showing that the mutant kinase was much less sensitive to heat inactivation than wild type enzyme. Furthermore, reconstitution experiments showed that alterations in mutant regulatory subunits and not catalytic subunits were responsible for the observed differences in protein kinase (HOCHMAN et al. 1975). Two-dimensional gel analysis of mutant regulatory subunits showed that these subunits contained charge differences consistent with a single amino acid substitution (STEINBERG et al. 1977). These observations suggested that many of the cells with altered affinity for cAMP contained mutations in the structural gene for regulatory subunit. A similar approach has been utilized to obtain mutants with apparently altered regulatory subunits in a number of other cell lines. Mutants resistant to the growth inhibitory effects of 8-bromo cAMP were obtained from Y-1 cells, an ACTH-sensitive mouse adrenal cortical tumor cell line (GUTMANN et al. 1978). The mutants required a 7-fold higher cAMP concentration for half maximal activation of protein kinase activity compared to wild type, although the total amount of kinase activity was similar for the two cell types. Based on separation by DEAE-cellulose chromatography, 70% of total kinase activity in these cells is type I and this isozyme required 4-fold higher concentrations of cAMP. The type II isozymes from the mutant and wild type cells were indistinguishable. The apparent decrease in affinity of the mutant type I enzyme for cAMP could suggest an alteration in the regulatory subunit of the type I isozyme. However, these experiments cannot rule out an effect on catalytic subunit which would require a higher cAMP concentration for activation. It was subsequently shown (RAE et al. 1979) that resistant mutants also had decreased ACTH-stimulated steroidogenesis suggesting that the steroidogenic effects of ACTH were mediated through cAMP-dependent protein kinase.

Mutants resistant to the cytotoxic effects of dibutyryl cAMP were isolated from mouse neuroblastoma cells (SIMANTOV and SACHS 1975). The resistant cells had a 50% decrease in the total amount of cAMP-dependent protein kinase and there was a slight decrease in the apparent association constant of the cAMP binding proteins. Additionally, the cAMP binding proteins from the mutant cells were much more temperature sensitive than the wild type. The resistant mutants did not show cAMP mediated induction of acetylcholinesterase activity, increased acetylcholine receptors or axon formation as was observed in the non-resistant wild type cells. This suggests, as in the case of other cell types, that a functional cAMP-dependent protein kinase is required for the observed biochemical and morphological changes induced by cAMP.

In addition to mutations affecting the regulatory subunit of protein kinase, several reports have described mutations affecting the catalytic subunit. One rare type of mutant obtained by selecting S49 lymphoma mutants resistant to dibutyryl cAMP had no detectable cAMP-dependent protein kinase activity and 15%–20% of the cAMP binding activity of wild type S49 cells (STEINBERG et al. 1978). Biochemical analysis of these mutants indicated that this phenotype was not due to: production of an inhibitor of protein kinase; production of a mutant regulatory subunit unable to dissociate from catalytic subunit; or excess regulatory subunit that was not associated with catalytic subunit. These authors suggested that the kinase negative phenotype of these mutants was due to alteration of a gene regulating expression of the protein kinase structural genes, since fusion of mutant and wild type cells resulted in hybrids which were kinase negative.

Evidence for a mutation affecting the catalytic subunit of cAMP-dependent protein kinase also has been obtained in Chinese hamster ovary (CHO) cells. The mutant cells required 100-fold higher concentrations of cAMP for activation of protein kinase than the wild type (EVAIN et al. 1979). The mutant also had increased amounts of type I holoenzyme and little or no type II kinase. It was shown that the mutant catalytic subunit had a decreased affinity for ATP and a distinct protein substrate specificity compared with the wild type enzyme suggesting an altered catalytic subunit in the mutant cells.

5. Differential Regulation of R and C

In general, it has been shown that the regulatory and catalytic subunits of cAMP-dependent protein kinase are present at the same concentration in most tissues. HOFMAN et al. (1977) showed that R and C were present in a 1:1 molar ratio in rabbit skeletal muscle, heart, kidney, liver, and brain. The amounts of R and C were found to range from a low of 0.23 micromoles/kg wet weight in liver to 0.78 micromoles/kg wet weight in brain. A similar result was obtained with heart extracts from several species where the ratio of R to C ranged from 0.80 to 0.94

(CORBIN et al. 1977). The R/C ratio in rat brain was found to be unity and remained constant during development of newborn rats (LOHMANN et al. 1978). These findings suggest that the synthesis and degradation of regulatory and catalytic subunits is coupled in a manner that maintains equimolar amounts of these subunits in most tissues.

In contrast to the results described above where R and C were present in equimolar amounts in normal tissues, mouse neuroblastoma cells have been shown to differentially regulate the amounts of R and C. Using a hybrid cell line derived by fusion of rat glioma cells with mouse neuroblastoma cells, WALTER et al. (WALTER et al. 1979) showed that addition of dibutryl cAMP to the growth medium caused a three-fold increase in the amount of the type I cAMP binding protein. There was no effect of this treatment on either R^{II} or cAMP-dependent protein kinase activity. The induced cAMP binding protein was shown to be type I regulatory subunit of protein kinase by 8-N_3-(^{32}P)cAMP labeling and by its ability to inhibit free catalytic subunit. The ability of the hybrid cells to induce R^{I} was shared with the neuroblastoma but not the glioma parent cell types. Thus, the levels of R^{I} in this cell type appear to be regulated independently of catalytic subunits or R^{II}. These authors also found that R^{I} in this cell type is present as the free subunit and not complexed with C. Induction of cAMP binding proteins has also been described in a different line of mouse neuroblastoma cells. Dibutyryl cAMP treatment caused a 3-fold increase in cAMP binding activity in both the soluble and particulate fractions of these cells (PRASAD and ROSENBERG 1978). This protein was later identified as R^{I} by H_3-cAMP labeling and two-dimensional gel analysis (PRASAD et al. 1979). These authors did not examine the effects of dibutryl-cAMP treatment on catalytic subunits or R^{II} so it is not known if induction by cAMP of these cells is specific for R^{I} as described by WALTER et al. (WALTER et al. 1979). These results suggest the possibility that in some cells R may serve a function not directly related to regulation of protein kinase activity. It will be interesting to see if this is a general phenomenon or is limited to special cell types, perhaps only transformed ones.

II. Current Research on Substrates of Protein Kinase

1. Multiple Phosphorylations

Recently, considerable interest has been generated by the observations that many of the substrates for cAMP-dependent protein kinase can be phosphorylated at multiple sites. Among the more important reasons for this interest is the fact that multiple phosphorylation events allow possible multiple input into the control of the enzyme and therefore increase the ability of the enzyme to respond appropriately to many divergent signals. A list of the substrates known to be phosphorylated rapidly at multiple sites is indicated in Table 2. In some cases it is known that the phosphorylation of different sites is preferentially catalyzed by different kinases while in others it appears that C subunit is probably the catalyst for all of the sites (SODERLING 1979; STULL 1980; ENGSTROM 1978). In the latter case, however, the rate of reaction at each site is often different. For example, the most studied multisite system at present is glycogen synthase (for recent review, see SODERLING

1979). With this enzyme, as many as 6 phosphoserines per 90,000 Daltons have been reported (SMITH et al. 1971). Furthermore, a minimum of 3 different kinases (cAMP-dependent protein kinase, phosphorylase kinase and 1 and 2 and other Ca^{2+}-sensitive kinases) and two phosphatases are thought to catalyze the phosphorylation/dephosphorylation events. To make the situation even more complex, several kinases and phosphatases have overlapping substrate specificities.

The effects of the multisite phosphorylations are probably even more complex than the reactions themselves. Taken together, however, at least two generalizations relevant to the present discussion can be made. The first is that phosphorylation of one site may affect either the phosphorylation or dephosphorylation of other sites; this then allows a mechanism for one signal (hormonal or otherwise) to modify the effect of another signal. Examples exist for both additive and antagonistic effects. The second major generalization is that the effects of each separate phosphorylation or function are not necessarily equal or even in the same direction. For example, phosphorlyation of one of the sites of synthase causes partial inactivation whereas phosphorylation of another site causes essentially complete inactivation under physiological conditions. Finally, in several instances, it is known that allosteric effectors of the enzyme substrate can effect not only the activity of the enzyme but also the rate or extent of phosphorylation at specific sites. For example, several allosteric effectors of pyruvate kinase have effects on the state of phosphorylation of the enzyme (PILKUS et al. 1978).

2. Recently Identified Substrates

a) Myosin Light Chain Kinase

One of the more interesting and exciting new substrates for cAMP-dependent protein kinase is myosin light chain kinase (see SCHULMAN, this volume). The reasons for this excitement lie in the potential importance of this particular phosphorylation event in the hormonal control of smooth muscle and other contractile functions (for review see ADELSTEIN and EISENBERG 1980; STULL 1980). Briefly, it has been known for a number of years now that the mechanism for control of the myosin ATPase activity (and therefore presumably contraction) in smooth muscle is mediated by calcium. More recently it has been shown that a major effect of calcium is due to its interaction with calmodulin which in turn is necessary for myosin light chain kinase activity. It is this calcium mediated phosphorylation of the myosin light chains that is required for increases in ATPase activity. As a corollary to this, the activity state of the myosin light chain kinase is likely to be a control point for muscle contraction. The effect of the cAMP-dependent phosphorylation of the light chain kinase is to decrease ability of the light chain kinase to be activated by the calcium-calmodulin complex. Kinetically, the affinity of light chain kinase for calmodulin is decreased by the phosphorylation (ADELSTEIN and EISENBERG 1980). This inhibition of the ATPase activity by phosphorylation presumably explains the smooth muscle relaxant effect of many agents known to increase cAMP in the tissue. For example, the vasodilatory and bronchodilatory effects of most β-receptor agonists are likely to be mediated, at least in part, by this mechanism.

b) Calcium Calmodulin-Dependent Cyclic Nucleotide Phosphodiesterase (Ca PDE)

Another recently discovered substrate of cAMP-dependent protein kinase is the Ca^{2+} calmodulin-stimulated cyclic nucleotide phospodiesterase (SHARMA et al. 1980). Phosphorylation by C subunit is relatively rapid, occuring at approximately the same rate as with mixed histones. At least 1 mol of phosphate per 60,000 Dalton monomer can be incorporated. The amount of endogenous phosphorylation has not yet been reported. All of these data are consistant with possible physiological relevance of the phosphorylation. However to date, in vivo phosphorylation has not been demonstrated nor have any changes in activity or function of the purified enzyme been shown to be altered by the phosphorylation. However, ELDIK and WATTERSON (1979) have reported a decrease in activity of this enzyme in a crude system which may be due to phosphorylation by cAMP-dependent protein kinase. As a result, any conclusions about possible regulatory roles for the phosphorylation must await further experimentation.

c) ATP Citrate Lyase

Several reports have been published recently that indicate that ATP citrate lyase is a phosphoprotein (LINN and SRERE 1979) and that the degree of phosphorylation may vary with the hormonal state of the tissue (JANSKI et al. 1979; RAMAKRISHNA and BENJAMIN 1979). For the liver enzyme at least two moles of acid stable, alkali labile phosphate can be detected per lyase tetramer. Treatment of hepatocytes with glucagon causes an increase in the incorporation of ^{32}P into specific immunoprecipitates of the enzyme. To date, no differences in activity or function of the enzyme have been found due to the phosphorylation. These studies therefore suggest that cAMP-dependent phosphorylation may play a role in the control of this enzyme. However, the physiological significance of the phosphorylation will be difficult to ascertain until functional changes are established. Very recently GUY et al. (1980) have reported that homogenous preparations of the lyase can be rapidly and stoichiometrically phosphorylated by cAMP dependent protein kinase.

d) src Gene Product, pp 60^{src}

Another recently discovered substrate of the cAMP-dependent protein kinase is the protein product of the avian sarcoma virus src gene, pp 60^{src} (COLLETT et al. 1979). The potential implications of regulation of this protein by cAMP-dependent phosphorylation are enormous since the src protein is required for virally induced transformation of many cells. However, the details of how and even if this phosphorylation is physiologically important are not yet clear. A major reason for this lack of definitive information is that the natural function(s) for pp 60^{src} in the cell is not known. It is known that pp 60^{src} can act as a cAMP-independent protein kinase that slowly catalyzes the transfer of phosphate from ATP to tyrosine residues on several proteins (HUNTER and SEFTON 1980). However, the natural substrates for this kinase are not yet known.

The evidence that pp 60^{src} is a substrate for cAMP dependent protein kinase is much stronger (COLLETT et al. 1979). These investigations have shown that a serine on the amino terminal $^2/_3$ of the protein is phosphorylated in vivo and in vitro. In vitro this phosphorylation is stimulated by cAMP and inhibited by the skeletal muscle heat stable inhibitor of the kinase. However, considerable information on the rates, function, and stoichiometry of the phosphorylation still have to be obtained. Recently it has been shown that small amounts of a similar or identical protein, pp 60^{sarc}, is present in non-infected cells (Sefton et al. 1980). These observations raise the possibility that this protein may play a role in normal growth and proliferation in the cell. A tyrosine kinase immunologically related to pp 60^{src}, and which probably corresponds to pp 60^{sarc}, has been shown to be part of a protein kinase cascade which regulates the efficiency of the Na^+K^+-ATPase in Ehrlich ascites tumor cells (Spector et al. 1981). The kinase was shown to be the initial enzyme in a cascade of protein kinases which phosphorylate on tyrosine residues and results in the phosphorylation of the β-subunit of Na^+K^+-ATPase. These enzymes were shown to be present at lower levels in normal murine brain which explains the reduced efficiency of the pump in the transformed cell line. Three of the kinases, including the pp 60^{sarc}-like kinase, were phosphorylated on serine residues by cAMP-dependent protein kinase. Future studies on the functions of tyrosine kinases and their regulation by cAMP-dependent phosphorylation should have enormous importance in understanding cell growth and tumorigenesis.

References

Adelstein RS, Eisenberg E (1980) Regulation and kinetics of the actin-myosin-ATP interaction. Ann Rev Biochem 49:921–956

Adelstein RS, Conti MA, Hathaway DR, Klee CR (1978) Phosphorylation of smooth muscle myosin light chain kinase by the catalytic subunit of adenosine 3′:5′-monophosphate-dependent protein kinase. J Biol Chem 253:8347–8350

Ames MM, Lerner P, Lovenberg W (1978) Tyrosine hydroxylase, activation by protein phosphorylation and end product inhibition. J Biol Chem 253:27–31

Beale EG, Dedman JR, Means AR (1977) Isolation and characterization of a protein from rat testis which inhibits cyclic AMP-dependent protein kinase and phosphodiesterase. J Biol Chem 252:6322–6327

Beavo JA, Bechtel PJ, Krebs EG (1974) Activation of protein kinase by physiological concentrations of cyclic AMP. Proc Natl Acad Sci USA 71:3580–3583

Beavo JA, Bechtel PJ, Krebs EG (1974a) Preparation of homogeneous cyclic AMP-dependent protein kinase(s) and its subunits from rabbit skeletal muscle. Methods Enzymol 38:299–308

Beavo JA, Bechtel PJ, Krebs EG (1975) Mechanisms of control for cyclic AMP dependent protein kinase from skeletal muscle. Adv Cyclic Nucleotide Res 5:241–251

Bechtel PJ, Beavo JA, Krebs EG (1977) Purification and characterization of catalytic subunit of skeletal muscle adenosine 3′,5′-monophosphate-dependent protein kinase. J Biol Chem 252:2691–2697

Beg ZH, Allmann DW, Gibson DM (1973) Modulation of 3-hydroxy-3-methylglutaryl coenzyme A reductase activity with cAMP and with protein fractions of rat liver cytosol. Biochem Biophys Res Commun 54:1362–1369

Blat C, Loeb JE (1971) Effect of glucagon on phosphorylation of some rat liver ribosomal proteins in vivo. FEBS Lett 18:124–126

Brand IA, Soling HD (1975) Activation and inactivation of rat liver phosphofructokinase by phosphorylation-dephosphorylation. FEBS Lett 57:163–168

Note added in proof: Data regarding the phosphorylation of the beta-subunit on Na^+K^+-ATPase by a protein kinase cascade, including a pp 60^{SARC}-like enzyme has been retracted (Science Vol. 213, Sept. 18, 1981)

Brostrom MA, Reimann EM, Walsh DA, Krebs EG (1970) Cyclic 3′,5′-AMP stimulated protein kinase from cardiac muscle. Adv Enzyme Reg 8:191–203

Builder SE, Beavo JA, Krebs EG (1980) Stoichiometry of cAMP and 1,N^6-etheno-cAMP binding to protein kinase. J Biol Chem 255:2350–2354

Builder SE, Beavo JA, Krebs EG (1980a) The mechanism of activation of bovine skeletal muscle protein kinase by adenosine 3′:5′ monophosphate. J Biol Chem 255:3514–3519

Bylund DB, Krebs EG (1975) Effect of denaturation on the susceptibility of proteins to enzymic phosphorylation. J Biol Chem 250:6355–6361

Carlson CA, Kim KH (1973) Regulation of hepatic acetyl coenzyme A carboxylase by phosphorylation and dephosporylation. J Biol Chem 248:378–380

Carlson GM, Bechtel PJ, Graves DJ (1980) Chemical and regulatory properties of phosporylase kinase and cAMP dependent protein kinase. Adv Enzymol 50:41–115

Castagna M, Palmer WR, Walsh DA (1975) Nuclear protein kinase activity in perfused rat liver stimulated with dibutyryl-adenosine cyclic 3′,5′-monophosphate. Eur J Biochem 55:193–199

Chau V, Huang LC, Romero G, Biltonen RL, Huana C (1980) Kinetic studies on the dissociation of adenosine cyclic 3′,5′-monophosphate from the regulatory subunit from rabbit skeletal muscle. Biochemistry 19:924–928

Chen LJ, Walsh DA (1971) Multiple forms of hepatic adenosine 3′:5′-monophosphate dependent protein kinase. Biochemistry 10:3614–3620

Cho-Chung YS, Clair T, Porper R (1977) Cyclic AMP-binding proteins and protein kinase during regression of Walker 256 mammary carcinoma. J Biol Chem 252:6342–6348

Coffino P, Bourne HR, Friedrich U, Hochman J, Insel PA, Lemaire I, Melmon KL, Tomkins GM (1976) Molecular mechanisms of cylcic AMP action: A genetic approach. Rec Prog Hormones Res 32:669–684

Cohen P (1978) The role of cyclic AMP-dependent protein kinase in the regulation of glycogen metabolism in mammalian skeletal muscle. Curr Top Cell Reg 14:118–196

Cole HA, Perry SV (1975) The phosphorylation of troponin I from cardiac muscle. Biochem J 149:525–533

Collett MS, Erickson E, Erickson RL (1979) Structural analysis of the avian sarcoma virus transforming protein: sites of phosphorylation. J Virol 29:770–781

Corbin JD, Keely SL (1976) Characterization and regulation of heart adenosine 3′,5′ monophosphate-dependent protein kinase isozymes. J Biol Chem 252:910–918

Corbin JD, Keely SL, Park CR (1975) The distribution and dissociation of cyclic adenosine 3′,5′-monophosphate-dependent protein kinases in adipose, cardiac, and other tissues. J Biol Chem 250:218–225

Corbin JD, Sugden PH, Lincoln TM, Keely SL (1977) Compartmentalization of adenosine 3′,5′-monophosphate and adenosine 3′,5′-monophosphate-dependent protein kinase in heart tissue. J Biol Chem 252:3854–3861

Corbin JD, Sugden PH, West L, Flockhart DA, Lincoln TM, McCarthy D (1978) Studies on the properties and mode of action of the purified regulatory subunit of bovine heart adenosine 3′,5′ monophosphate dependent protein kinase. J Biol Chem 253:3997–4003

Costa E, Kurosawa A, Guidotti A (1976) Activation and nuclear translocation of protein kinase during transsynaptic induction of tyrosine 3-monooxygenase. Proc Natl Acad Sci USA 73:1058–1062

DeMaille JG, Peters KA, Fischer EH (1977) Isolation and properties of the rabbit skeletal muscle inhibitor of adenosine 3′,5′ monophosphate dependent protein kinases. Biochemistry 16:3080–3086

Dills WL, Beavo JA, Bechtel PJ, Krebs EG (1975) Purification of rabbit skeletal muscle protein kinase regulatory subunit using cyclic adenosine 3′,5′ monophosphate affinity chromatography. Biochem Biophys Res Commun 62:70–77

Dills WL, Goodwin CO, Lincoln TM, Beavo JA, Bechtel PJ, Corbin JD, Krebs EG (1978) Purification of cyclic nucleotide receptor proteins by cyclic nucleotide affinity chromatography. Adv Cyclic Nucleotide Res 10:199–217

Donlon J, Kaufman S (1978) Glucagon stimulation of rat hepatic phenyl adenosine hydroxylase through phosphorylation in vivo. J Biol Chem 253:6657–6659

Eldik LJV, Watterson DM (1979) Phosphorylation and calmodulin stimulated activities. Fed Proc 38:789

Engstrom L (1978) The regulation of liver pyruvate kinase by phosphorylation-dephosphorylation. Curr Top Cell Regul 13:29–51

Erlichman J, Rosenfeld R, Rosen OM (1974) Phosphorylation of a cyclic adenosine 3′,5′-monophosphate-dependent protein kinase from bovine cardiac muscle. J Biol Chem 249:5000–5003

Erlichman J, Sarkar D, Fleischer N, Rubin CW (1980) Identification of two subclasses of type II cAMP dependent protein kinases: Neural specific and non-neural protein kinases. J Biol Chem 256:8179–8184

Evain D, Gottesman M, Pastan I, Anderson WB (1979) A mutation affecting the catalytic subunit of cyclic AMP-dependent protein kinase in CHO cells. J Biol Chem 254:6931–6937

Fischer EH, Heilmeyer LMG, Haschke RH (1971) Phosphorylase and the control of glycogen degradation. Curr Top Cell Regul 4:211–251

Friedman DL, Larner J (1963) Studies on UDPG-x-glucan transglycosylase III. Interconversion of two forms of muscle UDPG-x-glucan transglycosylase by a phosphorylation-dephosphorylation reaction sequence. Biochemistry 2:669–675

Gabibov AG, Kochetkov SN, Sashchenko LP, Severin ES (1979) Biochim Biophys Acta 569:445–452

Geahlen RL, Krebs EG (1980a) Regulatory subunit of the type I cAMP dependent protein kinase as an inhibitor and substrate of the cGMP-dependent protein kinase. J Biol Chem 255:1164–1169

Geahlen RL, Krebs EG (1980b) Studies on the phorphorylation of the type I cAMP dependent protein kinase. J Biol Chem 255:9375–9379

Geahlen RL, Allen SM, Krebs EG (1981) Effect of phosphorylation on the regulatory subunit of the type I cAMP-dependent protein kinase. J Biol Chem 256:4536–4540

Gibson DM, Ingebritsen TS (1978) Reversible modulation of liver hydroxymethyl glutaryl CoA reductase. Life Sci 23:2649–2664

Gill GN, Garren ID (1971) Role of the receptor in the mechanism of adenosine 3′,5′-cyclic monophosphate. Proc Natl Acad Sci USA 68:786–790

Greaser ML, Gergely J (1971) Reconstitution of troponin activity from three protein components. J Biol Chem 246:4226–4233

Gressner AM, Wool IG (1976) Influence of glucagon and cyclic adenosine 3′:5′-monophosphate on the phosphorylation of rat liver ribosomal protein S6. J Biol Chem 251:1500–1504

Gutmann MS, Rae PA, Schimmer BP (1978) Altered cyclic AMP-dependent protein kinase activity in a mutant adrenocortical tumor cell line. J Cell Physiol 97:451–460

Guy PS, Cohen P, Hardie DG (1980) Rat mammary gland ATP-citrate lyase is phosphorylated by cyclic AMP-dependent protein kinase. FEBS Lett 109:205–208

Haddox MK, Newton NE, Hartle DK, Goldberg ND (1972) ATP (Mg^{2+}) induced inhibition of cyclic AMP reactivity with a skeletal muscle protein kinase. Biochem Biophys Res Commun 47:653–661

Hamon M, Bourgoin S, Artasud F, Hery F (1977) Rat brain stem tryptophan hydroxylase: mechanism of activation by calcium. J Neurochem 28:811–818

Hardie DG, Cohen P (1978) The regulation of fatty acid biosynthesis: Simple procedure for the purification of acetyl CoA carboxylase from lactating rabbit mammary gland, and its phosphorylation by endogenous cyclic AMP-dependent and -independent protein kinase activities. FEBS Lett 91:1–7

Hochman J, Insel PA, Bourne HR, Coffino P, Tomkins GM (1975) A structural gene mutation affecting the regulatory subunit of cyclic AMP-dependent protein kinase in mouse lymphoma cells. Proc Natl Acad Sci USA 72:5051–5055

Hofmann F, Beavo JA, Bechtel PJ, Krebs EG (1975) Comparison of adenosine 3′,5′-cyclic monophosphate-dependent protein kinases from rabbit skeletal and bovine heart muscle. J Biol Chem 250:7795–7801

Hofmann F, Bechtel PJ, Krebs EG (1977) Concentrations of cyclic AMP-dependent protein kinase subunits in various tissues. J Biol Chem 252:1441–1447

Hollenbeck RA, Chuang DM, Costa E (1979) Translocation of cytosol protein kinase into nuclei and the induction of tyrosine hydroxylase in NBD-2 neuroblastoma cells. Brain Res 171:481–487

Hoppe J, Freist W (1979) Localization of the high-affinity ATP site in adenosine 3′:5′-dependent protein kinase type I. Eur J Biochem 93:141–146

Hoppe J, Marutzky R, Freist W, Wagner KG (1977) Mechanism of activation of the protein kinase I from rabbit skeletal muscle. Eur J Biochem 80:369–372

Hoppe J, Wagner KG (1977) An improved method for the purification of cAMP-dependent protein kinase from rabbit muscle using hydrophobic chromatography. Febs Lett 74:95–98

Huang FL, Glinsman WH (1975) Inactivation of rabbit muscle phosphorylase phosphatase by cyclic AMP-dependent kinase. Proc Natl Acad Sci USA 72:3004–3008

Huang TS, Feramisco JR, Glass DB, Krebs EG (1979) Specificity consideration relevant to protein kinase activation and function. Miami Winter Symposia, vol 16, pp 449–461

Hunter T, Sefton BM (1980) Transforming gene product of rous sarcoma virus phosphorylates tyrosine. Proc Natl Acad Sci USA 77:1311–1315

Insel PA, Bourne HR, Coffino P, Tomkins GM (1975) Cyclic AMP-dependent protein kinase: Pivotal role in regulation of enzyme induction and growth. Science 190:896–898

Insel PA, Fenno J (1978) Cyclic AMP-dependent protein kinase mediates a cyclic AMP-stimulated decrease in ornithine and S-adenosyl methionine decarboxylase activities. Proc Natl Acad Sci USA 75:862–865

Janski AM, Srere PA, Cornell NW, Veech RL (1979) Phosphorylation of ATP citrate lyase in response to glucogen. J Biol Chem 254:9365–9368

Jungmann RA, Hiestand PC, Schweppe JS (1974) Mechanism of action of gonadotropin. IV. Cyclic adenosine monophosphate-dependent translocation of ovarian cytoplasmic cyclic adenosine monophosphate-binding protein and protein kinase to nuclear acceptor sites. Endocrinology 94:168–183

Jungmann RA, Lee SG, DeAngelo AB (1975) Translocation of cytoplasmic protein kinase and cyclic adenosine monophosphate binding protein to intracellular acceptor sites. Adv Cyclic Nucleotide Res 5:281–306

Jungmann RA, Christensen ML, Schweppe JS, Mednicks MI, Spielvogel AM (1976) Cyclic AMP-mediated nuclear translocation of cytoplasmic cAMP-dependent protein kinases: Identity of the nuclear and cytoplasmic enzymes. In: Cyclic nucleotides and the regulation of cell growth. Dowden, Hutchinson and Ross, Stroudsburg, pp 225–252

Jungman RA, Russell DH (1977) Cyclic AMP, cyclic AMP-dependent protein kinase, and the regulation of gene expression. Life Sci 20:1787–1798

Keely SL, Corbin JD, Park CR (1975) On the question of translocation of heart cAMP-dependent protein kinase. Proc Natl Acad Sci USA 72:1501–1504

Kemp BE, Bylund DV, Huang TS, Krebs EG (1975) Substrate specificity of the cyclic AMP-dependent protein kinase. Proc Natl Acad Sci USA 72:3448–3452

Kemp BE, Graves DJ, Benjamini E, Krebs EG (1977) Role of multiple basic residues in determining the substrate specificity of cyclic AMP-dependent protein kinase. J Biol Chem 252:4888–4894

Khoo JC, Steinberg D, Huang JJ, Vagelos PR (1976) Triglyceride, diglyceride, monoglyceride, and cholesterol ester hydrolases in chicken adipose tissue activated by adenosine 3′,5′ monophosphate dependent protein kinase. J Biol Chem 251:2882–2890

Krebs EG (1973) The mechanism of hormonal regulation by cyclic AMP. In: Endocrinology proceedings of the 4th international congress. Excerpta Medica, Amsterdam, pp 17–29

Krebs EG, Graves DJ, Fischer E (1959) Factors affecting the activity of muscle phosphorylase b kinase. J Biol Chem 234:2867–2873

Krebs EG, Beavo JA (1979) Phosphorylation-dephosphorylation of enzymes. Ann Rev Biochem 48:923–959

Kurosawa A, Guidotti A, Costa E (1976) Induction of tyrosine 3-monooxygenase elicited by carbamylcholine in intact and denervated adrenal medulla: Role of protein kinase activation and translocation. Mol Pharmacol 12:420–432

Langan TA (1968) Histone phosphorylation: Stimulation by adenosine 3′,5′-monophosphate. Science 162:579–581

LaPorte DC, Builder SE, Storm DR (1980) Spectroscopic studies of the cAMP binding sites of the regulatory subunits of types I and II protein kinase. J Biol Chem 255:2343–2349

Linn TC, Srere PA (1978) Identification of ATP citrate lyase a phosphoprotein. J Biol Chem 254:1691–1698

Ljungstrom O, Hjelmquist G, Engstrom L (1974) Phosphorylation of purified rat liver pyruvate kinase by cyclic 3′,5′-AMP-stimulated protein kinase. Biochim Biophys Acta 358:289–298

Lohmann SM, Walter U, Greengard P (1978) Protein kinases in developing rat brain. J Cyclic Nucleotide Res 4:445–452

Manning DR, DiSalvo J, Stull JT (1980) Protein phosphorylation: Quantitative analysis in vivo and in intact cell systems. Mol Cell Endocrinol 19:1–19

McPherson JM, Whitehouse S, Walsh DA (1979) Possibility of shape conformers of the protein inhibitor of the cyclic adenosine monophosphate-dependent protein kinase. Biochemistry 18:4835–4845

Mendicino J, Beaudreau C, Bhattacharyya RN (1966) Reversible inactivation of D-fructose 1,6-diphosphatase by adenosine triphosphate and cyclic 3′,5′-adenosine monophosphate. Arch Biochem Biophys 116:436–445

Miyamoto E, Petzold GL, Kuo JF, Greengard P (1973) Dissociation and activation of adenosine 3′,5′ monophosphate dependent and guanosine 3′,5′ monophosphate dependent protein kinases by cyclic nucleotides and by substrate protein. J Biol Chem 248:179–189

Miyamoto E, Petzold GL, Harris JS, Greengard P (1971) Dissociation and concomitant activation of adenosine 3′,5′ monophosphate dependent protein kinase by histone. Biochem Biophys Res Commun 44:305–312

Milstein S, Abita SP, Chang N, Kaufman S (1976) Hepatic phenylalanine 4-monooxygenase is a phosphoprotein. Proc Natl Acad Sci USA 73:1591–1593

Morgenroth VH III, Hegstrand LR, Roth RH, Greengard P (1975) Evidence for involvement of protein kinase in the activation by adenosine 3′,5′-monophosphate of brain tyrosine 3-monooxygenase. J Biol Chem 250:1946–1948

Nimmo HG, Cohen P (1977) Hormonal control of cAMP dependent protein kinase. Adv Cyclic Nucleotide Res 8:145–266

Nimmo HG, Houston B (1978) Rat adipose-tissue glycerol phosphate acyltransferase can be inactivated by cyclic AMP-dependent protein kinase. Biochem J 176:607–610

O'Farrell PH (1975) High resolution two-dimensional electrophoresis of proteins. J Biol Chem 250:4007–4021

Ogez JR, Segel IH (1976) Interaction of cyclic adenosine 3′,5′ monophosphate with protein kinase: Equilibrium models. J Biol Chem 251:4551–4556

Owens JR, Haley BE (1978) Use of photo affinity analogs to determine the mechanism of ATP regulation of a membrane-bound cAMP-activated protein kinase. J Supra Mol Struct 9:57–68

Palmer WK, Castagna M, Walsh DA (1974) Nuclear protein kinase activity in glucagon-stimulated perfused rat livers. Biochem J 142:469–471

Palmer WK, McPherson JM, Walsh DA (1980) Critical controls in the evaluation of cAMP dependent protein kinase activity ratios as indices of hormonal action. J Biol Chem 255:2663–2666

Pilkus SJ, Pilkus J, Claus TH (1978) The effect of fructose diphosphate and phosphoenol pyruvate on cAMP-mediated inactivation of rat hepatic pyruvate kinase. Biochem Biophys Res Commun 81:139–146

Prasad N, Rosenberg RN (1978) Induction of cyclic AMP-binding proteins by ditbutyryl cyclic AMP in mouse neuroblastoma cells. Biochim Biophys Acta 539:459–469

Prasad N, Rosenberg RN, Wischmeyer B, Ulrich C, Sparkman D (1979) Induction of adenosine 3′,5′-monophosphate binding proteins by N^6, O^2-dibutyryladenosine 3′,5′-monophosphate in mouse neuroblastoma cells. Analysis by two-dimensional gel electrophoresis. Biochemistry 18:2717–2725

Rae PA, Gutmann NS, Tsao J, Schimmer BP (1979) Mutations in cyclic AMP-dependent protein kinase and corticopin (ACTH)-sensitive adenylate cyclase affect adrenal steroidogenesis. Proc Natl Acad Sci USA 76:1896–1900

Ramakrishna S, Benjamin WB (1979) Fat cell protein phosphorylation identification of phosphoprotein 2 as ATP-citrate lyase. J Biol Chem 254:9232–9236
Ramseyer J, Kaslow JR, Gill GN (1974) Purification of the cAMP receptor protein by affinity chromatography. Biochem Biophys Res Commun 59:813–821
Rangel-Aldao R, Kupiec JW, Rosen OM (1979) Resolution of the phosphorylated and dephosphorylated cAMP-binding proteins of bovine cardiac muscle by affinity labeling and two-dimensional electrophoresis. J Biol Chem 254:2499–2508
Rangel-Aldao R, Rosen OM (1976) Dissociation and reassociation of the phosphorylated and nonphosphorylated forms of adenosine 3′:5′-monophosphate-dependent protein kinase from bovine cardiac muscle. J Biol Chem 251:3375–3380
Rannels SR, Corbin JD (1980) Two intrachain cAMP binding sites of cAMP-dependent protein kinases. J Biol Chem 255:7085–7088
Reimann EM, Walsh DA, Krebs EG (1971) Purification and properties of rabbit skeletal muscle adenosine 3′,5′-monophosphate-dependent protein kinases. J Biol Chem 246:1986–1995
Riou JP, Claus TH, Flockhard DA, Corbin JD, Pilkis SJ (1977) In vivo and in vitro phosphorylation of rat liver fructose-1,6-bisphosphatase. Proc Natl Acad Sci USA 74:4615–4619
Rizak MA (1964) Activation of an epinephrine-sensitive lipolytic activity from adipose tissue by adenosine 3′,5′-phosphate. J Biol Chem 239:392–395
Rosell-Perez M, Larner J (1964) Studies on UDPG: 21.4 glucan x-4 glucosyltransferase. VI Specificity and structural requirements for the activator of the D form of the dog muscle enzyme. Biochemistry 3:773–778
Rosen OM, Erlichman J, Rubin CS (1975) Molecular structure and characterization of bovine heart protein kinase. Adv Cyclic Nucleotide Res 5:253–263
Rosen OM, Erlichmann J (1975) Reversible autophosphorylation of a cyclic 3′,5′-AMP-dependent protein kinase from bovine cardiac muscle. J Biol Chem 250:7786–7794
Rubin CS, Erlichmann J, Rosen OM (1974) Cyclic AMP-dependent protein kinase from bovine heart muscle. Methods Enzymol 38:299–308
Rubin CS, Rangel-Aldao R, Sarkar D, Erlichman J, Fleischer N (1979) Characterization and comparison of membrane-associated and cytosolic cAMP-dependent protein kinases. J Biol Chem 254:3797–3805
Schimmer BP, Rae PA, Gutmann NS, Watt VM, Tsao J (1979) Genetic dissection of ACTH action in adrenal tumor cells. Cold Spring Harbor Conf Cell Prolif 6:281–297
Schwecheimer K, Hofmann F (1977) Properties of regulatory subunit of cAMP-dependent protein kinase (peak I) prepared by urea treatment of the holoenzyme. J Biol Chem 252:7690–7696
Sefton BM, Hunter T, Beeman K (1980) Relationship of polypeptide products of the transforming gene of Rous sarcoma virus and the homologous gene of vertebrates. Proc Natl Acad Sci USA 77:2059–2063
Sharma RK, Wang TH, Wirch E, Wang JH (1980) Purification and properties of bovine brain calmodulin dependent cyclic nucleotide phosphodiesterase. J Biol Chem 255:5916–5923
Simontov R, Sachs L (1975) Temperature sensitivity of cyclic adenosine 3′,5′-monophosphate-binding proteins and the regulation of growth and differentiation in neuroblastoma cells. J Biol Chem 250:3236–3242
Smith CH, Brown NE, Larner J (1971) Glycogen synthesis II. Some chemical characteristics of the enzyme protein and of its change on interconversion. Biochim Biophys Acta 242:81–88
Smith SB, Siegel JB, Krebs EG (1980) Adenosine 3′,5′ monophosphate binding and tertiary structure changes in the regulatory subunit of protein kinase. Fed Proc 39:2117
Soderling TR (1976) Regulation of glycogen synthetase: Effects of trypsin on the structure, activity, and phosphorylation of the skeletal muscle enzyme. J Biol Chem 251:4359–4364
Soderling TR (1979) Regulatory functions of protein multisite phosphorylation. Mol Cell Endocrinol 16:157–179

Soderling TR, Hickenbottom JP, Reimann EM, Hunkeler FL, Walsh DA, Krebs EG (1970) Inactivation of glycogen synthetase and activation of phosphorylase kinase by muscle adenosine 3′,5′-monophosphate-dependent protein kinases. J Biol Chem 245:6317–6328
Spector M, O'Neal S, Racker E (1981) Regulation of phosphorylation of the β-subunit of the Ehrlich ascites tumor $Na^{+}K^{+}$-ATPase by a protein kinase cascade. J Biol Chem 256:4219–4227
Spruill WA, Steiner AL (1979) Cyclic nucleotide and protein kinase immunocytochemistry. Adv Cyclic Nucleotide Res 10:169–186
Steinberg D (1976) Interconvertible enzymes in adipose tissue regulated by cyclic AMP-dependent protein kinase. Adv Cyclic Nucleotide Res 7:157–198
Steinberg RA, Coffino P (1979) Two-dimensional gel analysis of cAMP effects in cultured S49 mouse lymphoma cells: Protein modifications, inductions, and repressions. Cell 18:719–733
Steinberg RA, Wetters TV, Coffino P (1978) Kinase-negative mutants of S49 mouse lymphoma cells carry a trans-dominant mutation affecting expression of cAMP-dependent protein kinase. Cell 15:1351–1361
Steinberg RA, O'Farrel PH, Frederick U, Coffino P (1977) Mutations causing charge alterations in regulatory subunits of the cAMP-dependent protein kinase of cultured S49 lymphoma cells. Cell 10:381–391
Stull JT (1980) Phosphorylation of contractile proteins in relation to muscle function. Adv Cyclic Nucleotide Res 13:39–93
Sugden PH, Holladay LA, Reimann EM, Corbin JD (1976) Purification and characterization of the catalytic subunit of adenosine 3′,5′-cyclic monophosphate-dependent protein kinase from bovine liver. Biochem J 159:409–422
Swillens W, Van Cauter E, Dumont JE (1974) Protein kinase and cyclic 3′,5′-AMP: Significance of binding and activation constants. Biochim Biophys Acta 364:250–259
Tash JS, Dedman JR, Means AR (1979) Protein kinase inhibitor in sertoli cell-enriched rat testis. J Biol Chem 254:1241–1247
Tao M (1972) Dissociation of rabbit red blood cell cyclic AMP dependent protein kinase I by protamine. Biochem Biophys Res Commun 46:56–61
Tao M, Salas MI, Lipmann F (1970) Mechanisms of activation by adenosine 3′:5′-cyclic monophosphate of a protein phosphokinase from rabbit reticulocytes. Proc Natl Acad Sci USA 67:408–414
Tessmer F, Graves DJ (1973) The phosphorylase kinase reaction on a peptide derived from glycogen phosphorylase. Biochem Biophys Res Commun 50:1–7
Trzeciak WH, Boyd GS (1974) Activation of cholesteryl esterase in bovine adrenal cortex. Eur J Biochem 46:201–207
Uno I, Ueda T, Greengard P (1977) Adenosine 3′,5′-monophosphate-regulated phosphoprotein system of neuronal membranes. J Biol Chem 252:5164–5174
Walsh DA, Ashby CD (1973) Protein kinases: Aspects of their regulation and diversity. Recent Prog Horm Res 29:329–359
Walsh DA, Ashby CD, Gonzalez C, Calkins D, Fischer E, Krebs E (1971) Purification and characterization of a protein inhibitor of adenosine 3′,5′-monophosphate-dependent protein kinases. J Biol Chem 246:1977–1985
Walter U, Costa M, Breakefield XO, Greengard P (1979) Presence of free cyclic AMP receptor protein and regulation of its level by cyclic AMP in neuroblastoma-glioma hybrid cells. Proc Nat Acad Sci USA 76:3251–3255
Yamamura H, Takeda M, Kumon A, Nishizuka Y (1970) Adenosine 3′,5′ cyclic-phosphate-dependent and independent histone kinases from rat liver. Biochem Biophys Res Commun 40:675–682
Zoller MJ, Kerlavage AR, Taylor SS (1979) Structural comparisons of cAMP dependent protein kinases I and II from porcine muscle. J Biol Chem 254:2408–2412.

CHAPTER 12

Cyclic GMP-Dependent Protein Phosphorylation

J. F. Kuo and M. Shoji

Overview

Experimental evidence suggests that most, if not all, of the effects of cyclic GMP may be mediated through activation of cyclic GMP-dependent protein kinase, leading to phosphorylation and subsequent modification in the properties of its endogenous substrate proteins. Cyclic GMP-dependent protein kinase and cyclic AMP-dependent protein kinase appear to be homologous proteins, each having many similar molecular and catalytic properties. However, these two enzymes have distinct biological roles, as suggested by differences in: (1) their specificity for, and mechanism of activation by, cyclic nucleotides; (2) their specificity for substrate proteins; (3) the regulation of their phosphorylating activity by modulator proteins; and (4) their involvement in tissue physio-pathology. Many of the physiologic substrate proteins for cyclic GMP-dependent protein kinase have not as yet been identified, and the exact molecular events leading to cellular response are still unknown.

A. Introduction

Cyclic AMP-dependent protein kinase (A-PK) was originally discovered in rabbit skeletal muscle and identified as phosphorylase kinase kinase (Walsh et al. 1968). This enzyme activity was subsequently shown to occur in every tissue examined from a number of species and phyla of the animal kingdom (Kuo and Greengard 1969). Because A-PK is ubiquitous, it has been proposed, as a unifying hypothesis, that the diverse physiologic and pharmacologic effects of cyclic AMP may be mediated through activation of the phosphotransferase activity of A-PK. During the course of searching for the possible existence of the target enzyme for the then new cyclic nucleotide, cyclic GMP, it was observed that the crude protein kinase preparation from the lobster tail muscle was activated equally well by a low concentration (0.5 μM) of cyclic AMP and cyclic GMP (Kuo and Greengard 1970a). A protein kinase activated preferentially by cyclic GMP was eventually separated from the more conventional A-PK by DEAE-cellulose chromatography, and the presence of cyclic GMP-dependent protein kinase (G-PK) was thus recognized (Kuo and Greengard 1970a). A more exhaustive study revealed a widespread occurrence of G-PK in arthropod tissues (Kuo et al. 1971), and, although at much lower levels, in mammalian tissues, such as the brain, uterus, and bladder (Kuo and Greengard 1970a; Greengard and Kuo 1970). By analogy to the A-PK system, it has been hypothesized that the presumed and diverse effects of cyclic GMP may be mediated through activation of G-PK (KUO and GREENGARD 1970a; GREENGARD and KUO 1970; KUO et al. 1971). Both A-PK and G-PK have since become the subjects of intensive investigation, largely owing to their central positions in many biological processes thought to be involved with these two cyclic nucleotides. The subjects of A-PK (KREBS 1972; WALSH and KREBS 1973; WALSH and ASHBY 1973; RUBIN and ROSEN 1975; NIMMO and COHEN 1977; KREBS and BEAVO 1979; GLASS and KREBS 1980), arthropod G-PK (KUO and GREENGARD 1974), and mammalian G-PK (KUO et al. 1978a; KUO 1980; GLASS and KREBS 1980) have been reviewed.

B. Enzymology

I. Distribution

The level of G-PK in the lobster tail muscle is about the same as that of A-PK (KUO and GREENGARD 1970a). The enzyme level is also comparable to that of A-PK in many insect tissues, suggesting that this cyclic GMP-dependent class of protein kinase is generally abundant in arthropod tissues. It is worth mentioning that the fat body of Cecropia silkmoth pupae or larvae contains almost exclusively G-PK, with only a minute amount of A-PK (KUO et al. 1971). G-PK from the silkworm was also purified and studied by TAKAI et al. (1976). G-PK has been reported to be present in *Tetrahymena*, with its level higher than that of A-PK (MUROFUSHI 1974). In contrast, only very low activity of G-PK, compared to the overwhelming A-PK activity, was detected initially in certain mammalian tissues (KUO and GREENGARD 1970a; GREENGARD and KUO 1970). Subsequent studies, while confirming the general low level of G-PK in mammals, revealed that the lung (KUO 1974), cerebellum (KUO 1974), and artery (KUO et al. 1977) contain G-PK levels that are comparable to those of A-PK also present in the same tissues. In a survey of several rat tissues for the cyclic GMP-binding activity, LINCOLN et al. (1976) also observed that the lung and cerebellum contain the highest level of the cyclic GMP target protein. As a matter of fact, the G-PK level was found to be even higher than A-PK in the fetal (not the adult) lung of guinea pigs (KUO 1975a).

G-PK has been partially purified from several mammalian tissues. These include the rat cerebellum (HOFMANN and SOLD 1972), the rat pancreas (VAN LEEMPUT-COUTREZ et al. 1973), the pig lung (NAKAZAWA and SANO 1975), the bovine cerebellum (TAKAI et al. 1975), the guinea pig fetal lung (KUO et al. 1976a), the calf heart (SHOJI et al. 1977a), and the bovine aorta (SHOJI et al. 1977b). The enzyme has been purified to homogeneity from the bovine lung (GILL et al. 1976; LINCOLN et al. 1977) and the bovine heart (FLOCKERZI et al. 1978). Properties of the enzyme will be discussed later in this chapter (Sect. B.II).

It seems that there is a positive correlation between the tissue levels of G-PK and cyclic GMP, since the lung (MURAD et al. 1971; KUO et al. 1972) and cerebellum (FERRENDELLI et al. 1970; STEINER et al. 1970) have been shown to contain the highest concentration of cyclic GMP among many tissues examined. Skeletal muscle is probably one of the few tissues that does not contain detectable G-PK activity under the above assay conditions (KUO 1974). As to the subcellular distribution of G-PK, a great majority of the enzyme activity is in the soluble (13,000 or 30,000 × *g* supernatant) fractions from a number of tissues. Only a minute amount of the G-PK activity is associated with the particulate (13,000 or 30,000 × *g* pellets) fraction, with or without solubilization by 0.3% Triton X-100 (KUO 1974, 1975a).

II. Molecular Properties and Mechanism of Action of Cyclic GMP

The molecular weight (M_r) of G-PK from a number of arthropod and mammalian tissues has been estimated to be between 120,000–165,000, with the sedimentation coefficient ranging from 6.6 to 7.8 S (MIYAMOTO et al. 1973; KUO et al. 1974; TAKAI et al. 1975; KUO et al. 1976a; GILL et al. 1976; SHOJI et al. 1977a, b; LINCOLN et al. 1977; ROCHETTE-EGLY and CASTAGNA 1978; FLOCKERZI et al. 1978). The enzyme from the lobster tail muscle (MIYAMOTO et al. 1973; KUO et al. 1974), as well as that from the guinea pig fetal lung (KUO et al. 1976b), fetal calf heart (SHOJI et al. 1977a), and bovine aorta (SHOJI et al. 1977b), can be completely dissociated to yield what appears to be the free catalytic (C) subunit when the holoenzyme is incubated with both cyclic GMP (50–100 µ*M*) and histone (400 µg/ml). Only a partial dissociation of the holoenzyme is achieved in the presence of either cyclic GMP or histone alone. Cyclic AMP, in the presence or absence of histone, is less effective than cyclic GMP. In order to obtain the putative C subunit, it is essential that the cyclic GMP and histone be present, whether dissociation be shown either by sucrose density gradient centrifugation or by Sephadex G-200 gel filtration. The isolated C subunit has a molecular size (about 50,000–60,000; 3.8–4.9 S) smaller than the holoenzyme, does not bind cyclic GMP, and is not stimulated by cyclic GMP. The cyclic GMP-binding, or regulatory (R), subunit component of G-PK has not been identified in the studies described above, probably due to its instability under the experimental conditions.

Dissociation of the calf lung G-PK by substrate proteins, such as histone and protamine, was reported by KOBAYASHI and FANG (1976), who showed the presence of not only the C subunit but also the cyclic GMP-binding R subunit. The dissociation of G-PK holoenzyme is further supported by the observation that cyclic GMP, at concentrations as low as 1 µ*M*, caused a complete dissociation of the silkmoth fat body holoenzyme (6.6 S) into the C subunit (3.7 S) and R subunit (2.4 S) (ROCHETTE-EGLY and CASTAGNA 1978). It appears that sulfhydryl groups are involved in the interaction of the subunits, since a) mercaptoethanol is required for complete dissociation of the holoenzyme, b) in the absence of mercaptoethanol, cyclic GMP causes formation of a species of the enzyme having an intermediate molecular size (4.6 S) and possessing both the catalytic and cyclic GMP-binding activities, and c) pretreatment of the holoenzyme with sulfhydryl agents, such as N-ethylmaleimide and p-chloromercuribenzoate, inhibits the cyclic GMP-induced dissociation (ROCHETTE-EGLY and CASTAGNA 1978). The reducing agent is required for the dissociation, but not for the activation, of the holoenzyme by cyclic GMP, suggesting that activation and dissociation of G-PK probably are independent molecular events.

The phosphotransferase activity of the C subunit of G-PK from the bovine aorta and guinea pig fetal lung is inhibited by the R subunit of bovine heart A-PK, as in the case of the C subunit of A-PK, indicating that a cross-interaction of the subunits from the two classes of protein kinases is possible (Shoji et al. 1977b). Similar observations had been made earlier for an interaction between the C subunit of G-PK from the lobster tail muscle and the R subunit of A-PK from the bovine brain (Miyamoto et al. 1973). The resulting "hybrid" holoenzyme (6.7 S) is stimulated preferentially by cyclic AMP instead of by cyclic GMP, similar to authentic A-PK. On the other hand, the hybrid holoenzyme is dependent upon the stimulatory modulator of G-PK for its maximal phosphotransferase activity, similar to authentic G-PK. The subject of the modulator will be dealt with later in this chapter (Sect. D.I). These observations clearly indicate that a) a functional C subunit of G-PK can be obtained, and b) a tissue-, species-, or phylum-specificity for the interaction and combination between the C and R subunit, each from different classes of protein kinases, appears to be lacking.

It has been shown that A-PK consists of R and C subunits and that the mechanism of action of cyclic AMP on the A-PK holoenzyme involves binding of cyclic AMP to the R subunit with a concomitant dissociation of the holoenzyme, yielding the fully active C subunit (for reviews, see Krebs 1972; Rubin and Rosen 1975). The activation and dissociation of A-PK by cyclic AMP have been shown to occur both in vitro and in vivo (Corbin et al. 1975). The observations made for the arthropod and mammalian G-PK suggest that the subunit structure of G-PK and the mechanism of action of cyclic GMP are somewhat similar to those for A-PK and cyclic AMP. This is particularly so in the case of the silkmoth fat body G-PK, which has been shown to be dissociated by a low concentration of cyclic GMP (1 μ*M*) to yield both C and R subunits in the presence of 6 m*M* mercaptoethanol (Rochette-Egly and Castagna 1978). However, these findings remain to be confirmed using either the same arthropod G-PK, or better, the mammalian enzyme. The similarities between G-PK and A-PK in terms of activation by their respective cyclic nucleotides cannot be firmly established, unless the R and C subunits from G-PK are isolated and these two subunits can reassociate to yield the cyclic GMP-dependent holoenzyme. So far, G-PK from the lobster and a number of mammalian sources has been dissociated to yield the C subunit only in the presence of an exceedingly high concentration (50 μ*M* or higher) of cyclic GMP. Little or no dissociation of the holoenzyme is observed in the presence of histone and 0.5 μ*M* cyclic GMP, a concentration that maximally stimulates G-PK. It appears, therefore, that formation of the C subunit, at least from the lobster and mammalian G-PK, may not be a physiological process. It is suggested that, unlike the effect of cyclic AMP on A-PK, G-PK may be activated by cyclic GMP without undergoing physical dissociation of the holoenzyme molecule (Gill et al. 1976, 1977; Shoji et al. 1977b; Lincoln et al. 1977; Kuo et al. 1978a).

More recently, G-PK has been purified to homogeneity from the bovine lung by means of either 8-NH_2-$(CH_2)_2$-NH-cyclic AMP-Sepharose (Gill et al. 1976, 1977) or 8-NH_2-$(CH_2)_2$-S-cyclic GMP-Sepharose (Lincoln et al. 1977) affinity chromatography. The G-PK activity bound to the affinity gels was eluted by a high concentration (0.1 or 20 m*M*) of cyclic GMP. It was noted that the enzyme thus recovered retains the original molecular size (M_r, 150,000–165,000; 6.9–7.8 S) and, after extensive dialysis to remove the bound cyclic GMP used for elution from the affinity gels, exhibits the ability to bind added cyclic GMP. This is in contrast to

A-PK, which is dissociated into subunits upon chromatographing on the cyclic AMP affinity gels (RAMSEYER et al. 1976; DILL et al. 1975). The homogeneous G-PK from the bovine lung displays a single polypeptide of a M_r of 74,000 (GILL et al. 1976, 1977) or 81,000 (LINCOLN et al. 1977) in sodium dodecyl sulfate (SDS) polyacrylamide gel electrophoresis, indicating that the holoenzyme is a dimer composed of identical subunits. It appears that dissociation of G-PK into its monomeric form by SDS requires the presence of a reducing agent, dithiothreitol or mercaptoethanol (GILL et al. 1977). The requirement of a reducing agent has also been shown for the dissociation by cyclic GMP of the silkmoth fat body G-PK into the R and C subunits (ROCHETTE-EGLY and CASTAGNA 1978), as discussed above. The bovine lung G-PK binds 2 moles of cyclic GMP per mole of the holoenzyme, corresponding to 1 mole of cyclic GMP for each monomer (LINCOLN et al. 1977; GILL et al. 1977). A subunit of the bovine lung G-PK can also be obtained by limited proteolysis of the holoenzyme (LINCOLN et al. 1977). The subunit thus generated has a M_r of 75,000, being somewhat smaller than the subunit (M_r= 81,000) produced by the SDS treatment, and possesses both the cyclic GMP-binding and phosphotransferase activity. INOUE et al. (1976) reported earlier that limited proteolysis of the silkmoth G-PK (M_r=140,000) yields a catalytic fragment (M_r=34,000) with a full catalytic activity and a binding fragment (M_r=36,000) with a cyclic GMP-binding activity. It is possible, therefore, that formation of the C subunit from G-PK caused by cyclic GMP and histone (MIYAMOTO et al. 1973; KUO et al. 1976a; SHOJI et al. 1977a, b), by protamine or histone (KOBAYASHI and FANG 1976), and by cyclic GMP in the presence of mercaptoethanol (ROCHETTE-EGLY and CASTAGNA 1978) can occur after limited proteolytic cleavage or modification of the holoenzyme or its monomeric subunit. Because partially purified G-PK preparations have been used in these studies, a proteolytic enzyme contamination is a distinct possibility. Formation of subunits or fragments of G-PK under various incubation conditions reported by a number of investigators is summarized in Scheme 1. MCCUNE and GILL (1979) and BUSS et al. (1979) have shown that binding of cyclic GMP to G-PK displays a positive cooperativity, a phenomenon which is not seen for cyclic AMP binding to A-PK. One explanation of this cooperativity may be the site: site interaction of the two cyclic GMP-binding regulatory domains of the two monomeric subunits of G-PK.

There is no evidence to indicate the occurrence of isozymes of G-PK in either arthropod or mammalian tissues. In this respect, it is intriguing that MUROFUSHI (1974) has detected three G-PK peaks (and one A-PK peak) upon chromatographing the Triton X-100 extract of *Tetrahymena* cilia on DEAE-cellulose. It is not clear whether these multiforms of G-PK are unique to the organism or are due to the effects of the detergent used for extraction.

A-PK and G-PK appear to be homologous enzymes evolved from a common ancestral protein (LINCOLN and CORBIN 1977, 1978a; GILL 1977). This concept is based upon their similarities in physical characteristics (molecular size and shape), cyclic nucleotide binding (2 moles per mole of holoenzyme), affinity for ATP and cyclic nucleotides, amino acid composition (70%–90% homology), self phosphorylation, and protein substrate specificity. Comparison of amino acid composition of G-PK and A-PK is presented in Table 1. General properties of A-PK and G-PK are summarized and compared in Table 2. It is further suggested that G-PK is a

Scheme 1. Formation of subunits or fragments of G-PK under in vitro conditions

"R_2C_2"
(M_r = 140,000–160,000)

By Histone + cGMP → C (M_r = 50,000)
Lobster muscle (Miyamoto et al., 1973; Kuo et al., 1974)
Guinea pig fetal lung (Kuo et al., 1976b)
Calf fetal heart (Shoji et al., 1977a)
Bovine aorta (Shoji et al., 1977b)

By Protamine or histone → $R + C$
Calf lung (Kobayashi and Fang, 1976)

By cGMP + mercaptoethanol → R (2.4 S) + C (3.7 S)
Silkworm fat body (Rochette-Egly and Castagna, 1978)

By SDS + mercaptoethanol → 2 identical fragments (M_r = 80,000)
Bovine lung (Gill et al., 1976, 1977; Lincoln et al., 1977)

By Proteolysis → 2 RC (M_r = 75,000)
Bovine lung (Lincoln et al., 1977)

By Proteolysis → 2 R (M_r = 34,000) + 2 C (M_r = 36,000)
Silkworm (Inoue et al., 1976)

Table 1. Comparison of amino acid compositions of G-PK from the bovine lung and A-PK (Type II) from the bovine heart. The mole-% of the amino acid is given in the parentheses

Amino acid	G-PK mole/148,000 molecular weight[a]	G-PK mole/165,000 molecular weight[b]	A-PK mole/186,000 molecular weight[b]
Aspartic acid	122 (9.1)	178 (11.2)	207 (11.8)
Threonine	74 (5.5)	80 (5.1)	75 (4.3)
Serine	80 (5.8)	108 (6.9)	99 (5.7)
Glutamic acid	160 (12.0)	203 (12.9)	235 (13.4)
Proline	68 (5.1)	73 (4.7)	85 (4.9)
Glycine	98 (7.3)	129 (8.2)	122 (6.9)
Alanine	82 (6.1)	110 (7.0)	120 (6.8)
Half-cystine	18 (1.4)	22 (1.4)	26 (1.5)
Valine	78 (5.8)	80 (5.1)	105 (6.0)
Methionine	34 (2.5)	39 (2.5)	36 (2.1)
Isoleucine	82 (6.1)	60 (3.8)	73 (4.2)
Leucine	122 (9.1)	127 (8.1)	150 (8.6)
Tyrosine	44 (3.3)	39 (2.5)	51 (2.9)
Phenylalanine	60 (4.5)	62 (4.0)	95 (5.4)
Histidine	30 (2.2)	19 (1.2)	34 (2.0)
Lysine	104 (7.8)	127 (8.1)	138 (7.9)
Arginine	68 (5.1)	99 (6.3)	91 (5.2)
Tryptophane	14 (1.1)	19 (1.2)	16 (0.9)

[a] Assuming G-PK is a dimer of a subunit of 74,000 molecular weight. Taken and calculated from Gill et al. (1977)
[b] Taken and calculated from Lincoln and Corbin (1977)

Table 2. Comparison of properties of A-PK and G-PK[a]

Parameter	A-PK	G-PK
Distribution	Ubiquitous (rich in all tissues)	Ubiquitous (rich in arthropod tissues and mammalian lung, cerebellum)
Molecular weight	140,000–190,000	120,000–190,000
Stokes radius	50–58 Å	52–59 Å
Frictional ratio (f/f_0)	1.6	1.5
K_a for cAMP or cGMP	0.01–0.08 μ*M*	0.01–0.10 μ*M*
Positive cooperativity for cAMP or cGMP binding	No	Yes
Self-phosphorylation	Yes (Type II)	Yes
Metal ion activator	$Mg^{2+} > Co^{2+} > Mn^{2+}$	$Mg^{2+} > Co^{2+} > Mn^{2+}$
Substrate protein (on molar basis)	Phosphorylase kinase, glycogen synthase, pyruvate kinase > histones	Histones > phosphorylase kinase, glycogen synthase, pyruvate kinase
Requirements of SH reagents for activity	No	Yes
Inhibitory modulator	Inhibition	No effect
Stimulatory modulator	No effect	Stimulation
Polydeoxyribonucleotides	Inhibition	Stimulation
Subunit composition	2R and 2C	2 Identical RC, possible C
Action of cAMP or cGMP	Activation with dissociation	Activation without dissociation (?)

[a] Data are taken, collectively, from the relevant references cited in the text

dimer composed of two identical protomers in isologous association with the chains arranged in the antiparallel fashion, which prevents dissociation into respective subunits upon activation by cyclic GMP. This is in contrast to A-PK, which has a similar structure with an axis of symmetry and a discontinuity in each chain, which permits dissociation of the holoenzyme into respective subunits upon activation by cyclic AMP (GILL 1977; LINCOLN and CORBIN 1978a).

III. General Catalytic Properties

The K_a values for cyclic GMP of G-PK, either of arthropod or mammalian origin, range from 0.03 to 0.2 μ*M*, which are generally 30 to 200 times lower than the K_a values of the same enzyme for cyclic AMP (for a review see KUO et al. 1978a). This compares with the K_a values for cyclic GMP of A-PK from the same tissues (ranging from 10 to 200 μ*M*), which are generally 30 to 200 times higher than those for cyclic AMP. The most potent activator of G-PK is 8-bromo cyclic GMP, with a K_a value 2 to 5 times lower than that of cyclic GMP (KUO 1974; KUO et al. 1976a; SHOJI et al. 1977a, b). Many other cyclic nucleotides, including cyclic AMP and its derivatives, are also capable of maximally stimulating G-PK activity. Their order of potency in activating G-PK from the lobster tail muscle (Kuo and Greengard 1974; Kuo et al. 1974), the guinea pig fetal lung (Kuo et al. 1976a), or the fetal calf heart (Shoji et al. 1977a) is as follows: 8-bromo cyclic GMP > cyclic GMP > 8-benzylamino cyclic AMP > 8-bromo cyclic AMP > 8-bromo cyclic IMP > 8-benzylamino cyclic GMP > 8-methylthio cyclic AMP > N^2-monobutyryl cyclic GMP > cyclic AMP > cyclic IMP > 8-benzylamino cyclic IMP > N^6-monobutyryl cyclic AMP > 2′-O-monobutyryl cyclic GMP > N^6-monobutyryl cyclic AMP > 2′-O-monobutyryl cyclic GMP > cyclic UMP > cyclic CMP. 2′-O-Monobutyryl cyclic AMP and N^6-2′-O-dibutyryl cyclic AMP are essentially inactive.

As shown for A-PK (Krebs 1972; Walsh and Krebs 1973), ATP is the only known direct phosphoryl donor for protein phosphorylation catalyzed by G-PK of either arthropod (Kuo and Greengard 1970a) or mammalian origin (Kuo et al. 1976a). Cyclic GMP has been shown to lower the K_m for ATP, from 0.6 μ*M* in its absence to 0.2 μ*M* in its presence, for histone phosphorylation catalyzed by the guinea pig fetal lung G-PK (Kuo et al. 1976a). A decrease in the K_m for ATP of the bovine brain A-PK by cyclic AMP has been reported earlier (Miyamoto et al. 1969). It is not clear, however, if this increase in the ATP affinity is common to the two classes of protein kinases from all sources.

Of many divalent ions examined, Mg^{2+} and Co^{2+} are effective in stimulating G-PK from a variety of sources, with optimal concentrations of 20–40 m*M* and 0.5 m*M*, equivalent to K_m values of about 5–10 m*M* and 0.1 m*M*, respectively (reviews see Kuo and Greengard 1974; Kuo et al. 1978a). Some investigators have reported that the optimal concentration of Mg^{2+} is as high as 100 m*M* for G-PK, when the enzyme is assayed in the absence of the added stimulatory modulator (Takai et al. 1976). The effects of stimulatory modulator will be dealt with later in this chapter. The maximal activity stimulated by Co^{2+} is about 40%–80% of that by Mg^{2+} for the mammalian G-PK (Kuo et al. 1976a; Shoji et al. 1977a, b). Co^{2+}, however, stimulates lobster G-PK (Kuo and Greengard 1970a) and mammalian A-PK (Kuo et al. 1970) to a greater extent than does Mg^{2+}. Although without effect on G-PK when present alone, Ca^{2+} antagonizes the stimulatory effect of Mg^{2+} or Co^{2+} on G-PK, with a K_i of about 5–10 m*M* (Kuo et al. 1976a). A similar antagonistic effect of Ca^{2+} was noted earlier for A-PK (Miyamoto et al. 1969). Mn^{2+}, Co^{2+}, or Zn^{2+}, each at 5 m*M*, almost completely inhibits the activity of bovine lung G-PK stimulated by 10 m*M* Mg^{2+} (Gill et al. 1977). It should be mentioned that Takai et al. (1979) have recently reported that Ca^{2+} (with a K_m of about 50 μ*M*) stimulates a protein kinase from the rat cerebellum; this enzyme requires certain phospholipids as a cofactor, and is independent of calmodulin for its activation by Ca^{2+}.

The optimal pH for the G-PK activity is from 6.5 to 8.5 for most of the enzyme preparations. For the purified G-PK, a higher activity is usually obtained in sodium acetate or Tris/Cl buffer than in potassium phosphate buffer (Takai et al. 1975; Kuo et al. 1976a). A higher cyclic GMP-stimulated activity in crude extracts of tissues, however, is seen in phosphate buffer (Hofmann and Sold 1972; Kuo 1974).

C. Substrate Proteins

I. Autophosphorylation of G-PK

The homogeneous preparation of G-PK from the bovine lung has been shown to be autophosphorylated by its own phosphotransferase activity, incorporating 2 moles of phosphate per mole of the holoenzyme through an intramolecular mechanism (DE JONGE and ROSEN 1977). The enzyme is also phosphorylated by the C subunit of A-PK through an intermolecular mechanism (LINCOLN et al. 1978). The nature of the phosphorylation sites in either case has not been determined. The significance of self-phosphorylation in the regulation of G-PK remains obscure. However, it appears to be involved in self-inhibition of G-PK. Cyclic GMP itself inhibits the autophosphorylation reaction, but cyclic AMP or histone enhances it (LINCOLN et al. 1978; DE JONGE and ROSEN 1977).

The phosphotransferase activity of G-PK can be activated by blocking or removal of arginine, presumably located in the regulatory domain of the enzyme, by treating the holoenzyme with either 2,3-butanedione (an arginine blocking agent) or trypsin (hydrolyzing at the arginine site). Interestingly, these treatments result in inhibition of its self-phosphorylation reaction (LINCOLN et al. 1978). Autophosphorylation of the A-PK and the significance of the reaction have been dealt with in a review (GLASS and KREBS 1980).

II. Histone

Earlier studies with the arthropod G-PK indicated that histone is generally a far better phosphate acceptor than any other non-histone protein shown to be phosphorylated effectively by A-PK (KUO and GREENGARD 1970b, 1974). Of three histone subfractions studied, all exhibit a higher affinity toward G-PK than A-PK, both purified from the lobster tail muscle. The K_m values for histone H 1 B (lysine-rich), H 2 B (slightly lysine-rich), and H 4 (arginine-rich) of the lobster muscle G-PK are about 80, 5, and 20 μg/ml, respectively. The corresponding values for the lobster muscle A-PK are >200, >200, and 50 μg/ml, respectively. Results qualitatively similar to the arthropod protein kinases are also seen for the enzymes from various mammalian tissues (KUO et al. 1976a; SHOJI et al. 1977a, b). For example, the K_m values for the above three histone subfractions (obtained from commercial sources and therefore not pure) of the guinea pig fetal lung G-PK are about 100, 150, and 50 μg/ml, respectively, compared to >400, >300, and about 100 μg/ml, respectively, for A-PK from the same tissue (Fig. 1). The substrate specificity for A-PK and G-PK is also evidenced by comparing the radioautograph of an electrophoretogram of trypsin digestion of mixed histone phosphorylated by the two classes of protein kinases either from the bovine cerebellum (TAKAI et al. 1975) or the silkworm (NISHIYAMA et al. 1975). Other investigators have reported that the histone subfractions most readily phosphorylated by G-PK (assayed in the absence of added stimulatory modulator) are H 2 B, H 1, and H 2 A, with the arginine-rich subfractions, H 3 and H 4, being much poorer substrates (FLOCKERZI et al. 1978; YAMAMOTO et al. 1977; HASHIMOTO et al. 1976).

Although G-PK phosphorylates many other proteins, histone probably is the only class of substrate that is phosphorylated almost equally well by both classes

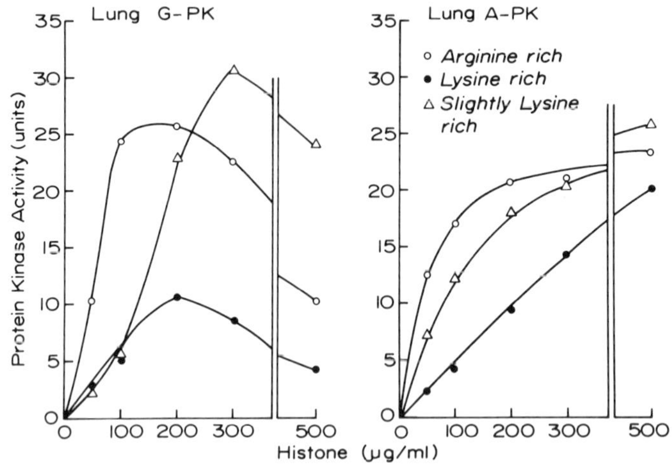

Fig. 1. Comparison of the ability of various histone subfractions to serve as substrates for G-PK and A-PK, both purified from the guinea pig fetal lung. (Taken, with permission, from Kuo et al. 1976a)

of enzymes. The sites of phosphorylation in histone subfractions catalyzed by A-PK have been investigated (Langan 1969; Shlyapnikov et al. 1975; Hashimoto et al. 1975; Yamamoto et al. 1977). G-PK phosphorylates the same serine residues in histones H 1, H 2 B, and H 4 as does A-PK (Hashimoto et al. 1976; Kuroda et al. 1976). Although detailed kinetics on the phosphorylation of the sites have not been carried out, a possibility exists that the rates of phosphorylation of the individual sites by the two enzymes may differ. In this respect, it is of some significance that A-PK phosphorylates equally well Ser-32 and Ser-36 in histone H 2 B, whereas G-PK preferentially phosphorylates Ser-32 (Yamamoto et al. 1977; Hashimoto et al. 1976; Nishizuka et al. 1979). This observation has been confirmed by Glass and Krebs (1979), who studied the kinetics of phosphorylation by A-PK and G-PK of two synthetic oligopeptides, Arg-Lys-Arg- $\underset{32}{\text{Ser}}$ -Arg-Lys-Glu and Arg-Lys-Glu- $\underset{36}{\text{Ser}}$ -Tyr-Ser-Val, corresponding to the amino acid Ser-32 and Ser-36 in histone H 2 B. Lincoln and Corbin (1977) have shown that A-PK phosphorylates a number of synthetic peptides at greater rates than does G-PK, and that the ratios of phosphorylation rates of the individual peptides by the two protein kinases are not the same. An interpretation of these observations may be that requirements of amino acid sequence near the serine residue for A-PK and G-PK are somewhat different, although the presence of arginine near the phosphorylation site is important for both.

III. Non-Histone Proteins

1. Phosphorylase Kinase

Because of many similarities in the molecular and catalytic properties of A-PK and G-PK, it is not unexpected that G-PK also phosphorylates a number of proteins shown to be endogenous, physiological substrates for A-PK. It has been shown that a homogeneous preparation of the lung G-PK phosphorylates (LINCOLN and CORBIN 1977, 1978a) and activates (KHOO et al. 1977; KHOO and GILL 1979) phosphorylase kinase from the rabbit skeletal muscle, as does A-PK (SODERLING et al. 1970). However, such phosphorylation requires a far greater amount of G-PK than A-PK, and this is probably the main reason that little or no phosphorylation of the enzyme was detected in earlier studies in which partially purified preparations of G-PK were used (INOUE et al. 1976; TAKAI et al. 1975; KUO et al. 1976a; NISHIYAMA et al. 1975). The sites of phosphorylation by G-PK, compared to those by A-PK, remain unknown. Because of the extremely low level of G-PK in the skeletal muscle (KUO 1974; LINCOLN et al. 1976) and the failure to demonstrate its phosphorylation by G-PK in vivo, it is unlikely that phosphorylase kinase is a physiological substrate for G-PK.

2. Glycogen Synthase

Although an earlier study showed that glycogen synthase was not phosphorylated by the partially purified G-PK (NISHIYAMA et al. 1975), subsequent studies by LINCOLN and CORBIN (1977, 1978a) indicated that it is phosphorylated and inactivated by homogeneous G-PK as well as A-PK (SODERLING 1970). Again, as in the case of phosphorylase kinase, G-PK is far less effective than A-PK, suggesting that glycogen synthase is an unlikely physiological substrate for G-PK. It is not clear whether the sites phosphorylated by G-PK are the same as by A-PK.

3. Hormone-Sensitive Lipase and Cholesterol Ester Hydrolase

These two lipid metabolizing enzymes are phosphorylated and activated by G-PK (KHOO et al. 1977; KHOO and GILL 1979) as well as by A-PK (CORBIN et al. 1970; HUTTUNEN and STEINBERG 1971; KHOO et al. 1974, 1976). However, they are far better substrates for A-PK than for G-PK. Since only the partially purified preparations of the lipase and hydrolase were used in these studies, it is not clear whether the two classes of protein kinases are phosphorylating the same or different sites. Two observations suggest that neither of the proteins is a likely substrate for G-PK in vivo. First, the G-PK level in the adipose tissue is exceedingly low (KUO 1974), and second, a much higher degree of cyclic nucleotide specificity is exhibited by A-PK and G-PK in phosphorylation of these enzymes, compared to phosphorylation of histone, thus making cross-activation of G-PK by cyclic AMP less likely to occur (KHOO and GILL 1979).

4. Pyruvate Kinase

Hepatic pyruvate kinase is phosphorylated and inactivated by A-PK (for a review, see ENGSTRÖM 1978) as well as by G-PK (LINCOLN and CORBIN 1977). As in the case of other non-histone proteins, phosphorylation by G-PK is much lower than by A-PK. This is also true when a synthetic peptide, "Kemptide" (Leu-Arg-Arg-Ala-Ser-Leu-Gly), corresponding to the amino acid sequence around the phosphorylation site in pyruvate kinase, is used as substrate (SHOJI et al. 1978a).

5. Troponin Inhibitory Subunit (TN-I)

TN-I has been shown to be phosphorylated by G-PK (Blumenthal et al. 1978; Lincoln and Corbin 1978b) as well as by A-PK, as has been reported earlier (Cole and Perry 1975; Stull and Buss 1977). Although it is not clear whether the sites of phosphorylation by G-PK are the same as by A-PK, the two protein kinases appear to phosphorylate common sites, because the phosphorylation by the individual enzymes is not additive. Compared to other non-histone proteins, TN-I is a comparatively good substrate for G-PK, suggesting that additional specificity determinants for G-PK may be present in the TN-I molecule (Lincoln and Corbin 1978b). This hypothesis can be tested by comparing the kinetics of phosphorylation by G-PK of synthetic peptides of TN-I; phosphorylation of these peptides by A-PK has been studied recently by Kemp (1979).

An increased cyclic AMP content and increased contractility in the heart elicited by β-adrenergic stimulation is associated with an increased TN-I phosphorylation (England 1976, 1977). This contrasts with the observations that increased cyclic GMP content and decreased contractility induced by cholinergic stimulation are associated with a decreased TN-I phosphorylation (England 1976). The failure to observe in vivo phosphorylation of TN-I by G-PK (England 1976, 1977) may be in part related to either the low V_{max} for its phosphorylation by G-PK (Blumenthal et al. 1978) or the low level of G-PK in the heart (Kuo 1974; Lincoln et al. 1976), or both. The in vitro phosphorylation of TN-I by the purified G-PK thus seems to represent one example showing an overlapping substrate specificity of the two classes of protein kinases.

6. Other Proteins

Fructose 1,6-biphosphatase (Lincoln and Corbin 1977, 1978a; Riou et al. 1977), myelin basic protein (Shoji et al. 1978a; Carnegie et al. 1974), and liver ribosomal proteins (Chihara-Nakashima et al. 1977) have been shown to be phosphorylated by both A-PK and G-PK. It has not been determined whether G-PK also phosphorylates other proteins shown to be substrates for A-PK, such as denatured lysozyme, protein phosphatase inhibitor-1, phosphofructokinase, RNA polymerase II, myosin light chain kinase, tyrosine hydroxylase, and denatured glycogen phosphorylase (a review see Glass and Krebs 1980).

IV. Membrane and Cytoplasmic Proteins

With the possible exceptions of histone subfractions and TN-I, all the proteins discussed above (Sect. C.III) are phosphorylated to a much lesser extent by G-PK than by A-PK. It appears that these proteins, including histone and TN-I, probably are not endogenous, nor physiological, substrate proteins for G-PK. Recent studies, however, do indicate the presence of certain endogenous proteins whose phosphorylation in a cell free system are stimulated specifically or preferentially by cyclic GMP. Those proteins have not yet been purified and established as the specific substrate proteins for G-PK assayed under the standard incubation conditions. Furthermore, little is known about both the characteristics of the endogenous proteins and possible alterations in their properties after being phosphorylated.

CASNELLIE and GREENGARD (1974) reported two membrane-associated proteins (molecular weights of 130,000 and 100,000) in mammalian smooth muscle preparations, whose phosphorylation is preferentially stimulated by cyclic GMP. Subsequently, phosphorylation of a membrane protein (with a smaller molecular weight of 86,000) from the brush border of the rat intestinal epithelium specifically stimulated by cyclic GMP was reported; interestingly, phosphorylation of cytoplasmic proteins from the same tissue is cyclic AMP-specific (DE JONGE 1976). Similar cyclic GMP-specific phosphorylation of a membrane protein from the intestinal epithelium was reported by SHALTZ et al. (1978). Two similar proteins (with molecular weights of 130,000 and 240,000) were reported to occur also in the membrane from the guinea pig vas deferens (WALLACH et al. 1977).

Another source of endogenous proteins for G-PK is the nervous tissue. A cyclic GMP-specific phosphorylation of an endogenous protein (molecular weight of 23,000) was reported in the soluble fraction from the rabbit cerebellum (SCHLICHTER et al. 1978). The presence of this substrate protein is of interest, particularly in view of the fact that the cerebellum is one of the few tissues that contains a high concentration of cyclic GMP (FERRENDELLI et al. 1970; STEINER et al. 1970) and of its receptor enzyme, G-PK (KUO 1974; KUO et al. 1976c). In addition to the soluble protein mentioned above, cyclic GMP-stimulated phosphorylation of proteins in the membrane has been reported. These include the 60,000 molecular weight protein from the human brain (BOEHME et al. 1978) and a protein of similar molecular weight from the nerve root of *Aplysia californica* (RAM and ERLICH 1978).

D. Factors Influencing Protein Phosphorylation

I. Stimulatory Modulator

The phosphorylation of substrate proteins by protein kinases is also influenced by other factors. A heat-stable protein, which inhibits A-PK, was originally isolated from the skeletal muscle extract (WALSH et al. 1971). This protein inhibitor exerts its effect by interacting with the C subunit of A-PK (ASHBY and WALSH 1972, 1973). During the course of investigating the possible effects of crude preparations of this factor (prepared by precipitating with 5% trichloroacetic acid the protein from the supernatant of the heated extract), it was noted, unexpectedly, that these preparations stimulate G-PK (DONNELLY et al. 1973a, b). Because of the dual effects (i.e. inhibiting A-PK and stimulating G-PK), the factor(s) was named *protein kinase modulator(s)*, instead of protein inhibitor. The crude modulator preparations obtained from many rat and guinea pig tissues, with the possible exception of skeletal muscle, augment phosphorylation of histone (used as a general substrate for both A-PK and G-PK) catalyzed by G-PK from mammalian sources. This phenomenon helped establish the ubiquitous occurrence of G-PK in mammals (KUO 1974, 1975a), and led to its purification from some representative mammalian tissues (KUO et al. 1976a, c; SHOJI et al. 1977a, b). DONNELLY et al. (1973a, b) purified the modulator from the lobster tail muscle to homogeneity, and found it still possessed the ability to inhibit A-PK and stimulate G-PK, as it did in its crude state.

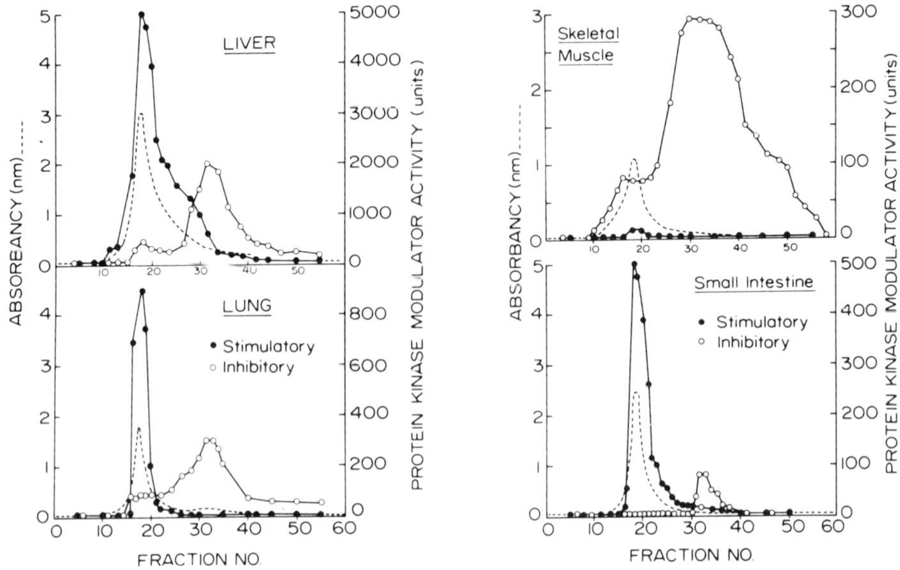

Fig. 2. Separation on Sephadex G-100 of stimulatory modulator of G-PK from inhibitory modulator of A-PK in the crude protein kinase modulator preparations obtained from the rat liver, lung, skeletal muscle, and small intestine. (Taken, with permission, from Kuo et al. 1976d)

These latter experiments raised two additional questions: a) does the purified protein inhibitor of A-PK (Walsh et al. 1971; Ashby and Walsh 1972, 1973) also stimulate G-PK, and b) do the two distinct activities of the modulator seen in its crude form obtained from many mammalian tissues reside in the same protein, as the modulator from the lobster muscle, or in different proteins.

Answers to the above questions have come from the studies of Kuo and Kuo (1976) and Kuo et al. (1976d), who have demonstrated that the inhibitory and stimulatory activity components of the crude protein kinase modulator prepared from a number of rat and guinea pig tissues can be separated from each other by means of Sephadex G-100 gel filtration. The chromatographic patterns of the two modulator activities from the rat liver, lung, skeletal muscle, and small intestine are illustrated in Fig. 2. In contrast to rat and guinea pig tissues, the two modulator activities from the lobster tail muscle were eluted as a single peak when the crude modulator preparation was chromatographed under the same conditions (Kuo et al. 1976d), in agreement with earlier reports of Donnelly et al. (1973a, b) that the dual activities of the factor from the lobster muscle reside in the same protein. The isolated stimulatory modulator from mammalian tissues augments histone phosphorylation by G-PK from either mammalian or arthropod tissues, without affecting the A-PK from the same sources. The isolated inhibitory modulator, presumably the same as the protein inhibitor described earlier (Walsh et al. 1971), depresses the phosphorylating activity of A-PK, without affecting the G-PK activity. These phenomena are illustrated in Fig. 3. It is now clear that the stimulatory and inhibitory modulators of protein kinases from mammalian sources are separate proteins, which are present as a mixture of different ratios in tissues. For example,

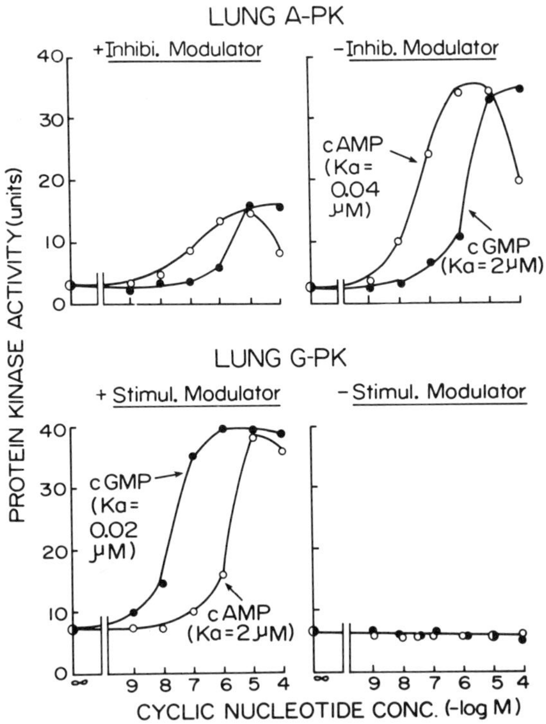

Fig. 3. Differential effects of stimulatory modulator and inhibitory modulator isolated from the rat lung on the phosphorylating activity of G-PK and A-PK, both purified from the guinea pig fetal lung. (Taken, with permission, from Kuo et al. 1976d)

skeletal muscle contains almost exclusively the inhibitory modulator, small intestine contains almost exclusively the stimulatory modulator, and most other tissues, including the liver and lung, contain both modulators at comparable levels (Fig. 2). This contrasts with the lobster muscle modulator which possesses dual activities as a single entity. Furthermore, the studies indicate that the stimulatory modulator, which specifically augments the G-PK activity, is different from inhibitory modulator (protein inhibitor), which specifically inhibits the A-PK activity.

It has been shown that stimulatory modulator, or crude modulator preparations containing the stimulatory modulator component, is required for the maximal phosphorylation of histone by G-PK (Kuo 1974, 1975a; Kuo et al. 1976a, c; Shoji et al. 1977a, b). There is little or no histone phosphorylation stimulated by cyclic GMP in the absence of added stimulatory modulator (Fig. 3). The requirement for the factor is further demonstrated by the observations that it augments histone phosphorylation catalyzed by the C subunit of G-PK, as in the case of the G-PK holoenzyme, in the presence of G-PK (Donnelly et al. 1973b; Kuo et al. 1976b; Shoji et al. 1977a, b, 1978a). Inhibitory modulator, on the other hand, has no effect on this reaction. The requirement of stimulatory modulator for a maximal phosphorylation of histone by G-PK has been confirmed elsewhere (Yamamoto et al. 1977; Mackenzie and Donnelly 1979; Gill and Walton 1979). The stimula-

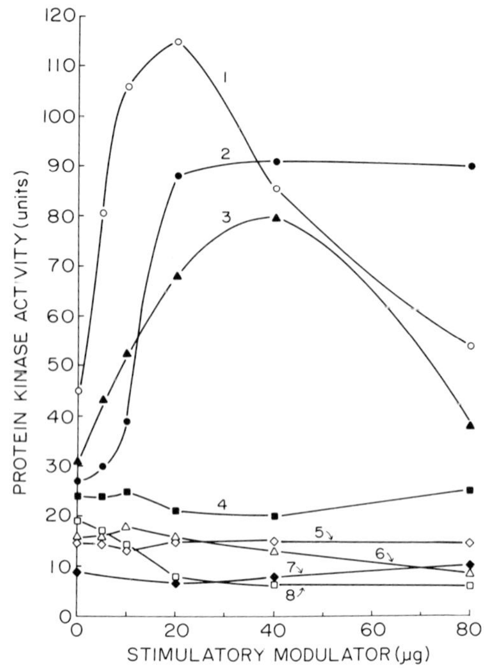

Fig. 4. Effects of stimulatory modulator on phosphorylation of various proteins by G-PK purified from the guinea pig fetal lung. Curves *1*, slightly lysine-rich histone (40 μg); *2*, arginine-rich histone (40 μg); *3*, mixed histone (40 μg); *4*, troponin (10 μg); *5*, Kemptide, Leu-Arg-Arg-Ala-Ser-Leu-Gly (10 μg); *6*, protamine (40 μg); *7*, glycogen synthase (40 μg); *8*, myelin basic protein (40 μg). (Taken, with permission, from SHOJI et al. 1978a)

tory effect of the factor has also been noted for phosphorylation of ribosomal proteins by G-PK (CHIHARA-NAKASHIMA et al. 1977). There appears to be no tissue-, species-, or phylum-specificity for the action of the modulator on histone phosphorylation by G-PK. Lack of specificity had been shown earlier for the action of inhibitory modulator on A-PK (WALSH et al. 1971; ASHBY and WALSH 1972, 1973). It has been speculated that the two modulator activities may be artifacts because of the drastic procedures (i.e. boiling and acid precipitation) employed for their initial isolation. This possibility seems unlikely, however, since the modulators can be prepared using mild methods conventionally used for the preparations of labile proteins (KUO and KUO 1976; KUO et al. 1976d). Several recent observations that the modulator activities alter under certain physio-pathologic conditions also support the contention that the modulators are not only the natural constituents of the cells, but also play certain biological roles regulating the effects of cyclic AMP and cyclic GMP at the level of protein kinases. Physio-pathologic aspects of protein kinases and their modulators will be dealt with later in this chapter (Sect. E).

As discussed earlier (Sect. C), histone is by far the best substrate for G-PK, and its phosphorylation is stimulated in a dose-dependent manner by stimulatory modulator (Fig. 4). Other phosphate acceptors, such as troponin, a synthetic peptide Leu-Arg-Arg-Ala-Ser-Leu-Gly, protamine, glycogen synthase, and myelin

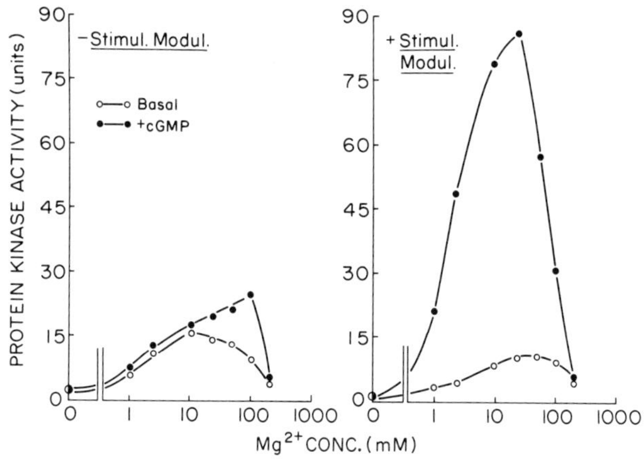

Fig. 5. Comparison of the effects of stimulatory modulator and cyclic GMP on the guinea pig fetal lung G-PK activity as a function of Mg^{2+} concentration. (Taken, with permission, from Kuo et al. 1978b)

basic protein, are also phosphorylated by G-PK, but this reaction is not stimulated by the factor (Fig. 4). In contrast, phosphorylation by A-PK of all phosphate acceptors mentioned above is inhibited by inhibitory modulator. Although histone is a very effective substrate for G-PK in vitro, it is not clear whether it is phosphorylated by G-PK in vivo. Nevertheless, it is possible that the unique property of stimulatory modulator represents a determinant of the substrate specificity for G-PK, and that phosphorylation of endogenous substrate proteins by G-PK may be dependent upon stimulatory modulator or factors with a similar action.

Studies of some workers have shown that the G-PK activity can be demonstrated in the absence of added stimulatory modulator (Nakazawa and Sano 1975; Takai et al. 1975; Gill et al. 1976, 1977; Lincoln et al. 1977; Lincoln and Corbin 1977; de Jonge and Rosen 1977). It has not been determined if greater histone phosphorylation occurs in the presence of added modulator in these studies. The discrepancy appears to be due to differences in assay conditions employed rather than a difference in the enzyme species studied. One likely reason for the discrepancy is the exceedingly high Mg^{2+} concentrations (50 or 100 m*M*) employed in these worker's assay systems. It has been shown recently (Shoji et al. 1978a; Kuo et al. 1978b) that the optimal Mg^{2+} concentration for G-PK activity in the absence of stimulatory modulator is about 100 m*M*, compared to about 10–20 m*M* in its presence, with maximal activity for the former condition being about only one fifth of that for the latter. Moreover, at 100 m*M* Mg^{2+} the G-PK activity is stimulated only up to 2-fold by cyclic GMP, and this stimulation is independent of the factor. At 10 m*M* Mg^{2+}, on the other hand, the phosphorylation is stimulated up to 10-fold by cyclic GMP, and this stimulation is dependent on the modulator. These observations are illustrated in Fig. 5. It appears that stimulatory modulator is indeed required not only for maximal activity of G-PK but also for maximal stimulation of the enzyme by cyclic GMP when it is assayed under a more physiologic condition of 10 m*M* Mg^{2+}.

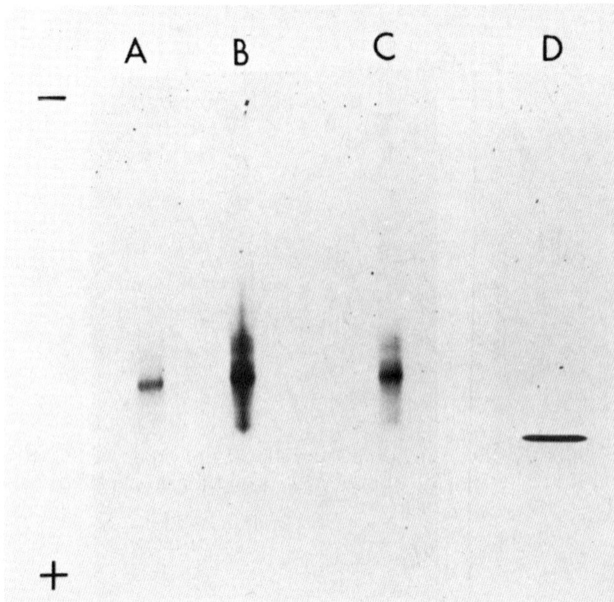

Fig. 6 A–D. Analytical and SDS-polyacrylamide gel electrophoresis of stimulatory modulator purified from the dog heart. **A** the modulator (9 μg) electrophoresed in the analytical gel. **B** the modulator (18 μg) in the analytical gel as in A. **C** the reassociated modulator (12 μg) obtained by dialyzing the SDS-treated modulator followed by electrophoresis in analytical gel as in **A** and **B. D** the same modulator preparation as in **A** and **B**, but treated with 3% SDS and 1% mercaptoethanol and subjected to the SDS gel electrophoresis. (Taken, with permission, from Shoji et al. 1978 a)

Shoji et al. (1978 a) have purified stimulatory modulator from the dog heart to homogeneity. The preparation appears as a single protein band on SDS-polyacrylamide gel electrophoresis, with a minimum M_r of 34,000 (Fig. 6). Analytical polyacrylamide gel electrophoresis reveals several bands. A similar electrophoretic pattern is seen for the reassociated modulator obtained by dialyzing the SDS-treated modulator. The apparent M_r of the native and the reassociated modulator, the latter being as active as the former, is estimated to be about 180,000. It seems that stimulatory modulator of G-PK can exist in multiple forms of molecular aggregates consisting of a common subunit, and the dissociation-reassociation of the modulator subunit is a reversible process. As shown in Table 3, the minimum M_r and amino acid composition of stimulatory modulator from the dog heart are strikingly similar to that of the lobster muscle protein kinase modulator (a factor possessing dual activities), suggesting that they may be homologous proteins evolved from a common ancestral protein. Protein inhibitor (inhibitory modulator) of A-PK has been purified from skeletal muscle to homogeneity (Demaille et al. 1977). Both the M_r and amino acid composition of the factor are different from stimulatory modulator (Table 3), indicating that these two functionally distinct modulators are not homologous proteins. As shown earlier for the lobster muscle protein kinase modulator (Donnelly et al. 1973 a, b), stimulatory modulator augments

Table 3. Comparison of amino acid compositions of protein kinase modulator (with dual activities) from the lobster muscle, stimulatory modulator of G-PK from the dog heart, and inhibitory modulator of A-PK from the rabbit skeletal muscle. The mole-% of the amino acid is given in the parentheses

Amino acid	Protein kinase modulator (mole/34,000 mol. wt. subunit)[a]		Stimulatory modulator (mole/34,000 mol. wt. subunit)[b]		Inhibitory modulator (mole/11,300 mol. wt.)[c]	
Aspartic acid	41	(14.0)	34	(11.5)	12	(12.2)
Threonine	9	(3.1)	9	(3.0)	8	(8.2)
Serine	11	(3.8)	15	(5.1)	11	(11.2)
Glutamic acid	69	(23.5)	72	(24.3)	14	(14.3)
Proline	2	(0.7)	0	(0.0)	1	(1.0)
Glycine	9	(3.1)	3	(1.0)	9	(9.2)
Alanine	33	(11.3)	36	(12.2)	15	(15.3)
Half-cystine	1	(0.3)	1	(0.3)	0	(0.0)
Valine	13	(4.4)	8	(2.7)	3	(3.1)
Methionine	9	(3.1)	5	(1.7)	0	(0.0)
Isoleucine	7	(2.4)	9	(3.0)	5	(5.1)
Leucine	32	(10.9)	35	(11.8)	7	(7.1)
Tyrosine	4	(1.4)	6	(2.0)	1	(1.0)
Phenylalanine	6	(2.0)	2	(0.7)	1	(1.0)
Histidine	1	(0.3)	1	(0.3)	1	(1.0)
Lysine	26	(8.9)	48	(16.2)	4	(4.1)
Arginine	19	(6.5)	12	(4.1)	6	(6.1)
Tryptophane	1	(0.3)	1	(0.3)	0	(0.0)

[a] The modulator from the lobster tail muscle possessing both activities to inhibit A-PK and to stimulate G-PK. Taken and calculated from DONNELLY et al. (1973a)
[b] Taken and calculated from SHOJI et al. (1978a)
[c] Taken and calculated from DEMAILLE et al. (1977)

histone phosphorylation catalyzed by the G-PK holoenzyme or its C subunit via interaction with the substrate protein, but not with the enzyme (SHOJI et al. 1978a). This is in contrast to the action of inhibitory modulator of A-PK, which interacts with the enzyme, but not with the substrate (ASHBY and WALSH 1972, 1973). The stimulatory modulator-histone complex, which is at least partly insoluble, is a far better phosphate acceptor than the free histone for G-PK; on the other hand, the free and complexed histones serve equally well as substrates for A-PK.

II. Polyanions, Polycations, and Basic Polypeptides

Modifications of basic substrate proteins (histone subfractions) through polyanionic interactions render the complexes superior phosphate acceptors specifically for G-PK, and appear not to be limited to stimulatory modulator (a highly acidic protein with a pI of about 4). Recent studies from the group of Nishizuka indicate that other polyanions, such as DNA, poly- or oligodeoxyribonucleotides, poly-glutamate and heparin, are all capable of stimulating the cyclic GMP-dependent phosphorylation of histone by G-PK (Nishizuka et al. 1979; Hashimoto et al. 1979, Kuroda et al. 1979).

In fact, polydeoxyribonucleotide stimulates phosphorylation of histone H2A by G-PK while inhibiting its phosphorylation catalyzed by A-PK, thus changing substrate specificity of the two protein kinases for this protein. That is to say, histone H2A is a better substrate for A-PK in the absence of the substrate modifier, but it becomes a better substrate for G-PK in its presence. Mackenzie and Donnelly (1979) reported the presence of contaminating polydeoxyribonucleotides in many histone preparations, which, they contend, may in part account for the apparent non-requirement of stimulatory modulator for histone phosphorylation by G-PK observed earlier by some investigators. The interactions of these polyanionic, acidic modifiers with the polycationic, basic substrate (histone), which probably make serine residues in histone more accessible to G-PK, may be an important determinant of substrate specificity for G-PK. Compared to histone, phosphorylation of some phosphate acceptors (such as troponin, glycogen synthase and synthetic peptide) by G-PK is not augmented by stimulatory modulator (Shoji et al. 1978a), suggesting that there may be a lack of "proper" interactions between the modulator and these substrates. Cyclic GMP-stimulated phosphorylation of membrane proteins has been reported (Casnellie and Greengard 1974; de Jonge 1976; Schlichter et al. 1978). It is not clear whether the phosphorylation by G-PK of these membrane proteins, if isolated and purified, is dependent upon the presence of the polyanionic substrate modifiers. In view of the specific requirement for G-PK of the complexed substrate, instead of the free substrate, it has been suggested that a topographic relationship between the protein kinase and substrate proteins (and their modifiers) may be crucial for the functional specificity of the two protein kinases. This contention seems to be further supported by a specific requirement of phospholipids for the activity of a calcium-dependent protein kinase described recently (review see Nishizuka et al. 1979).

It has been reported that polycations, such as polyamines (spermine, spermidine, and putrescine) inhibit histone phosphorylation catalyzed by either A-PK or G-PK (Takai et al. 1976; Rochette-Egly and Castagna 1978). The mode of action of the polyamines, however, has not yet been determined. Yamamoto et al. (1979) have reported recently that phosphorylation of various proteins (casein, phosvitin, and cytoplasmic proteins of the liver) by a cyclic nucleotide-independent protein kinase (casein kinase) is stimulated by basic polypeptides, such as polylysine, histone or protamine. The basic polypeptides appear to exert their actions by interacting with substrate proteins rather than with the enzyme.

From the foregoing, it becomes apparent that the phosphotransferase activity of these different classes of protein kinases is specifically regulated by a number of substances with a wide molecular diversity. These observations are summarized and compared in Table 4. The functional specificity of various classes of protein kinases, therefore, is determined at least in part by the following factors: the activator ligands (cyclic AMP, cyclic GMP, and calcium), regulatory substances (such as modulators and phospholipids), substrate proteins, and the topographic relationship among these factors and their respective protein kinases.

Table 4. Regulation by various substances of phosphorylation of protein substrates catalyzed by different classes of protein kinases

Protein kinase	Substrate protein	Regulatory substance	Site of interaction	Effect
cGMP-dependent	Histones	Stimulatory modular	Substrate	Stimulation[a]
	Histones	DNA or polydeoxyribonucleotides, other polyanions	Substrate	Stimulation[b]
	Histones	Polyamines		Inhibition[c]
	Ribosomal proteins	Stimulatory modulator	Substrate	Stimulation[d]
cAMP-dependent	Casein, histone, protamine, phosphorylase kinase and all other substrates	Protein inhibitor (Inhibitory modulator)	Enzyme	Inhibition[e]
Ca-dependent	Histones	Phospholipids	Enzyme	Stimulation[f]
Cyclic nucleotide-independent (casein kinase)	Casein, phosvitin, cytoplasmic proteins	Basic polypeptides	Substrate	Stimulation[g]

[a] DONNELLY et al. (1973a, b); KUO (1974); KUO and KUO (1976); KUO et al. (1976d), SHOJI et al. (1978a); YAMAMOTO et al. (1977); MACKENZIE and DONNELLY (1979)
[b] HASHIMOTO et al. (1979); NISHIZUKA et al. (1979); MACKENZIE and DONNELLY (1979)
[c] TAKAI et al. (1976); ROCHETTE-EGLY and CASTAGNA (1978)
[d] CHIHARA-NAKASHIRMA et al. (1977)
[e] WALSH et al. (1971); ASHBY and WALSH (1972, 1973)
[f] TAKAI et al. (1979) (this enzyme activity being independent of calmodulin)
[g] YAMAMOTO et al. (1979)

E. Physio-Pathologic Aspects of the G-PK System

The effects of cyclic AMP and cyclic GMP have been shown to be mutually opposing (i.e. Yin-Yang) in many instances where the cellular functions are regulated bidirectionally (GOLDBERG et al. 1975). The evidence for the roles of A-PK is quite compelling. However, the studies concerning the functional role of G-PK are largely indirect and sometimes speculative. Intracellular concentrations of cyclic AMP and cyclic GMP, by virtue of their nature as established or presumed second messengers, could fluctuate widely and acutely. This would pose some difficulties in obtaining their true values that would be characteristic of some specific physiologic or pathologic conditions or processes to be investigated. It is conceivable that intracellular levels of A-PK and G-PK, the more stable target enzymes for the more transient cyclic nucleotides, and their modulators, could serve as additional or alternate parameters for characterization.

One example of involvements of protein kinases in physiologic processes is ontogenesis (KUO 1975a). It has been shown that G-PK levels in the guinea pig fetal

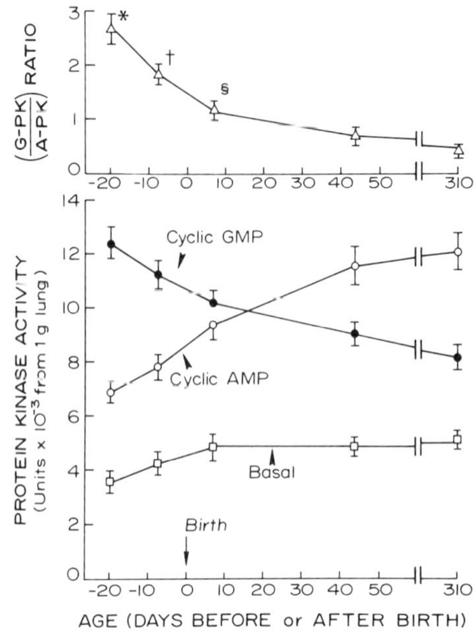

Fig. 7. Comparison of the activity and ratio of G-PK and A-PK in the lung of developing guinea pigs. *Significantly different from the fetus (7 days before birth, $P<0.01$), and all other groups ($P<0.005$). + Significantly different from the neonate and all other groups ($P<0.01$). § Significantly different from the 310-day adult ($P<0.05$). (Taken, with permission, from Kuo 1975a)

lung, the tissue that contains the highest G-PK level (over 3 times higher than that of A-PK) in all mammalian tissues examined, declines progressively during development (Fig. 7). The lung A-PK, on the other hand, is lowest in the fetus, highest in the adult, with an intermediate level in the neonate. Thus the G-PK to A-PK ratio in the lung is highest in the fetus, and declines progressively as the tissue matures, suggesting that the cyclic GMP system may play a more prominent role than the cyclic AMP system in the early stage of lung development or maturation. A similar reciprocal change in the G-PK and A-PK is also seen in the developing heart (Kuo 1975a). Interestingly, patterns of changes in G-PK and A-PK, opposite to those for the lung and heart, are noted for the brain of the developing guinea pig in the same studies. No apparent change either in the levels of G-PK and A-PK or their ratios is seen for the liver during development. These results suggest that a balance between the activity levels of G-PK and A-PK, and hence the effects of cyclic GMP and cyclic AMP, may be crucial in the ontogenesis of certain tissues.

Protein kinases may be involved in some pathology of tissues as well. It has been shown that the cardiac level of G-PK in spontaneously hypertensive (SH/N) rats is lower than that in normotensive control (WKY/N) rats, whereas the level of A-PK is the same in both groups (Kuo et al. 1976e). The levels of G-PK and A-PK in the lung, which serves as a control tissue, are the same in both groups of rats, indicating that decreased G-PK level in the heart is probably related to cardiac hypertrophy secondary to hypertension. It has been suggested that cyclic AMP and cyclic GMP may mediate the cardiac stimulatory and inhibitory effects of catecholamines and acetylcholine, respectively (for a review see GOLDBERG et al. 1975). In a state of sustained elevated blood pressure, the heart adapts to function more efficiently. Physiologically, this manifests itself as a compensatory cardiac hypertrophy. Biochemically, a decreased ratio of cyclic GMP to cyclic AMP effects may be desirable; this could be accomplished by either depressing the G-PK level, or increasing the A-PK level, or both. The observation that G-PK is specifically depressed in the hypertension-induced cardiac hypertrophy is of some physiological significance, because A-PK is present in a far greater amount than G-PK in the heart, and consequently it would require a great deal of the newly synthesized A-PK to lower significantly the G-PK to A-PK ratio. It is not clear whether the depressed G-PK level is a result of cardiac hyperterophy secondary to hypertension, or is an intrinsic characteristic of the genetically hypertensive rat. In this respect, it will be interesting to see if the G-PK level is depressed in the young SH/N rat prior to the onset of hypertension and compensatory cardiac hypertrophy.

SHOJI et al. (1978 b) observed that G-PK and its stimulatory modulator are disproportionately elevated, relative to A-PK and its inhibitory modulator, in the fast-growing Morris hepatoma 3924 A when compared with the control or host liver. This finding is in line with an earlier report that the tissue concentration of cyclic GMP in this hepatoma is exceedingly high (THOMAS et al. 1973), thus further supporting the contention that imbalance in the effects of the cyclic nucleotides in favor of cyclic GMP appeares to be crucial for cell proliferation, tissue growth and neoplasia (for a review see GOLDBERG et al. 1975). A-PK and G-PK, and their modulators, along with other parameters in the cyclic nucleotide systems, may also be involved in other physiologic or pathologic processes or conditions, such as in blood pressure regulation and arteriovenous anastomosis (KUO et al. 1977); in the heart desensitized to catecholamines (TSE et al. 1978, 1979); in experimental diabetes (WALSH and ASHBY 1973; KUO 1975 b); and in the enlarged lung secondary to injury caused by butylated hydroxytoluene (KUO et al. 1978 c).

Based on the above studies, it is clear that A-PK and G-PK and their modulators are potential sites of bioregulation, thus profoundly influencing the overall effects of cyclic AMP and cyclic GMP. These changes can occur independent of, or in concert with, other parameters in the cyclic nucleotide systems, such as the receptors, cyclases, and phosphodiesterases.

F. Concluding Remarks

Although A-PK and G-PK appear to be homologous enzymes having many characteristics in common, a few fundamental differences do exist; the most notable ones being the mechanisms of activation of the two enzymes by their respective cyclic nucleotide, and the effects of the modulators and polydeoxyribonucleotides on phosphorylation of histone subfractions. One of the future directions of research would be to examine whether or not intracellular activation (or dissociation, if any) of G-PK is associated with an increased cyclic GMP level secondary to stimulation by certain physiological agents, such as acetylcholine. If so, by analogy to the activation of A-PK by certain cyclic AMP-elevating agents, this observation

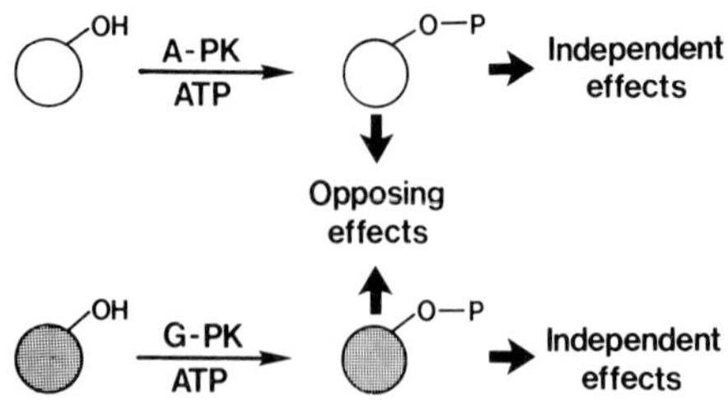

Fig. 8. A postulated mechanism by which A-PK and G-PK mediate the presumed independent or opposing effects of cAMP and cGMP by phosphorylating separate proteins

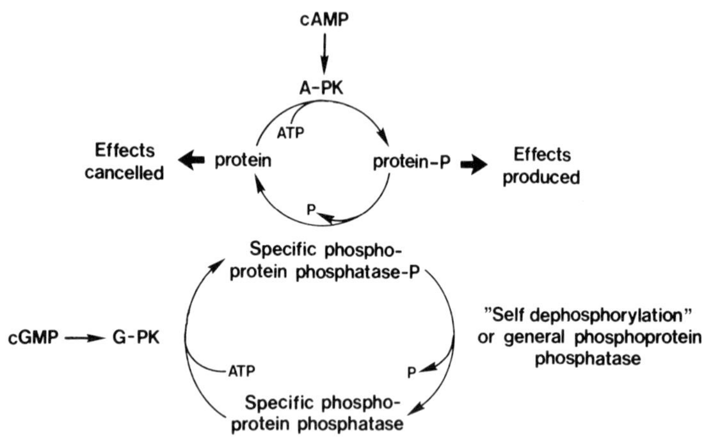

Fig. 9. A postulated mechanism by which A-PK and G-PK mediate the presumed opposing effects of cAMP and cGMP by phosphorylating separate proteins as in Fig. 8. In this particular case, the substrate protein for G-PK is a phosphoprotein phosphatase

would provide evidence to further support the notion that G-PK is the mediator of the effects of cyclic GMP. Elucidation of biochemical and physio-pathologic roles of G-PK, in relation to those of A-PK, would represent a new challenge. This would probably be best explored by seeking and identifying the endogenous substrate proteins and by establishing modifications in their properties upon phosphorylation by G-PK, very much like the procedures employed by Greengard and associates particularly for the problems concerning the A-PK system (Greengard 1978 a, b). This is an urgent research goal in view of the fact that not a single G-PK substrate protein with a defined physiological function has been identified. Lastly, similar information should be obtained for the recently described calcium-dependent protein kinase (Nishizuka et al. 1978, 1979) and the so-called cyclic nucleotide-independent protein kinase, so that a more complete picture pertaining to the biological processes regulated by these four distinct classes of multifunctional protein kinases may be synthesized.

One important question that remains to be answered is how cyclic AMP and cyclic GMP, via the phosphorylating action of A-PK and G-PK, bring about their

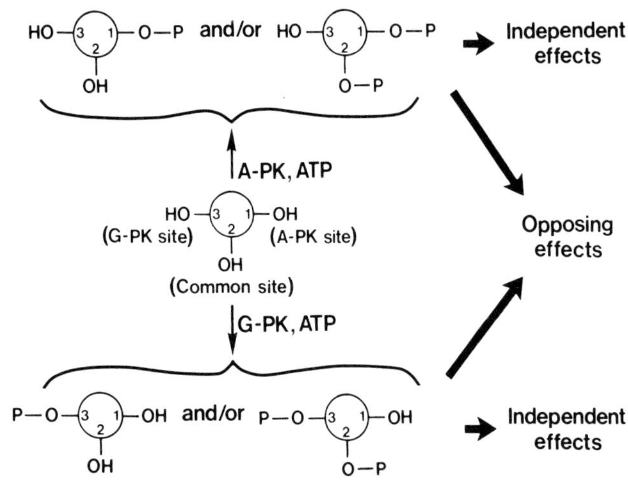

Fig. 10. A postulated mechanism by which A-PK and G-PK mediate the presumed independent or opposing effects of cAMP and cGMP by phosphorylating separate intramolecular sites in a given substrate protein

diverse and often opposing effects. The most obvious possibility is that A-PK and G-PK are phosphorylating separate proteins of different functions (Fig. 8). One special situation in this case is that the substrate protein for G-PK is a phosphoprotein phosphatase, with its activation dependent upon phosphorylation by G-PK; this G-PK-activated enzyme could dephosphorylate the active phosphoprotein produced by A-PK, and could attenuate or eliminate the effect brought about by cyclic AMP or A-PK (Fig. 9). The specific sites of phosphorylation by protein kinases, in turn, may be present within the same protein (i.e. intramolecular specificity), as in the case of Ser-32 and Ser-36 in histone H 2 B (NISHIZUKA et al. 1979; GLASS and KREBS 1979). It is conceivable, therefore, that phosphorylation of specific serine residue(s) by A-PK or G-PK would impart different, and possibly opposite, properties upon the same protein molecule. This mechanism is depicted in Fig. 10.

Acknowledgements. The original work from the authors' laboratory was supported by USPHS grants HL-15696, CA-23391 and T32-GM-07594.

References

Ashby CD, Walsh DA (1972) Characterization of the interaction of a protein inhibitor with adenosine 3′:5′-monophosphate-dependent protein kinases. I. Interaction with the subunit of the protein kinase. J Biol Chem 247:6637–6642
Ashby CD, Walsh DA (1973) Characterization of the interaction of a protein inhibitor with adenosine 3′:5′-monophosphate-dependent protein kinases. II. Mechanism of action with the holoenzyme. J Biol Chem 248:1255–1261
Blumenthal DK, Stull JT, Gill GN (1978) Phosphorylation of cardiac troponin by guanosine 3′:5′-monophosphate-dependent protein kinase. J Biol Chem 253:334–336
Boehme DH, Hosecki R, Marks N (1978) Protein phosphorylation in human synaptosomal membranes. Evidence for the presence of substrates for cyclic guanosine 3′:5′-monophosphate-dependent protein kinases. Brain Res Bull 3:697–700

Buss JE, McCune RW, Gill GN (1979) Comparison of cyclic nucleotide binding to adenosine 3′:5′-monophosphate and guanosine 3′:5′-monophosphate-dependent protein kinases. J Cyclic Nucleotide Res 5:225–237
Carnegie PR, Dunkley PR, Kemp BE, Murray AW (1974) Phosphorylation of selected serine and threonine residues in myelin basic protein by endogenous and exogenous protein kinases. Nature 249:147–150
Casnellie JE, Greengard P (1974) Guanosine 3′:5′-monophosphate-dependent phosphorylation of endogenous substrate proteins in membranes of mammalian smooth muscle. Proc Natl Acad Sci USA 71:1891–1895
Chihara-Nakashima M, Hashimoto E, Nishizuka Y (1977) Intrinsic activity of guanosine 3′:5′-monophosphate-dependent protein kinase similar to adenosine 3′:5′-monophosphate-dependent protein kinase. II. Phosphorylation of ribosomal proteins. J Biochem (Tokyo) 81:1863–1867
Cole HA, Perry SV (1975) The phosphorylation of troponin I from cardiac muscle. Biochem J 149:525–533
Corbin JD, Reimann EM, Walsh DA, Krebs EG (1970) Activations of adipose lipase by skeletal muscle cyclic adenosine 3′:5′-monophosphate-dependent protein kinase. J Biol Chem 245:4849–4851
Corbin JD, Keely SL, Soderling TR, Park CR (1975) Hormonal regulation of adenosine 3′:5′-monophosphate-dependent protein kinase. Adv Cyclic Nucleotide Res 5:265–279
de Jonge HR (1976) Cyclic nucleotide-dependent phosphorylation of intestinal epithelium proteins. Nature 262:590–593
de Jonge HR, Rosen OM (1977) Self-phosphorylation of cyclic guanosine 3′:5′-monophosphate-dependent protein kinase from bovine lung. J Biol Chem 252:2780–2783
Demaille JG, Peters KA, Fischer EH (1977) Isolation and properties of the rabbit skeletal muscle protein inhibitor of adenosine 3′:5′-monophosphate-dependent protein kinases. Biochemistry 16:3080–3086
Dill WL, Beavo JA, Bechtel PJ, Krebs EG (1975) Purification of rabbit skeletal muscle protein kinase regulatory subunit using cyclic adenosine 3′:5′-monophosphate affinity chromatography. Biochem Biophys Res Commun 62:70–77
Donnelly TE, Kuo JF, Reyes PL, Liu YP, Greengard P (1973a) Protein kinase modulator from lobster tail muscle. I. Stimulatory and inhibitory effects of the modulator on phosphorylation of substrate proteins by guanosine 3′:5′-monophosphate-dependent and adenosine 3′:5′-monophosphate-dependent protein kinases. J Biol Chem 248:190–198
Donnelly TE, Kuo JF, Miyamoto E, Greengard P (1973b) Protein kinase modulator from lobster tail muscle. II. Effects of the modulator on holoenzyme and catalytic subunit of guanosine 3′:5′-monophosphate-dependent and adenosine 3′:5′-monophosphate-dependent protein kinases. J Biol Chem 248:199–203
England PJ (1976) Studies on the phosphorylation of the inhibitory subunit of troponin during modification of contraction in perfused rat heart. Biochem J 160:295–304
England PJ (1977) Phosphorylation of the inhibitory subunit of troponin in perfused hearts of mice deficient in phosphorylase kinase. Evidence for phosphorylation by adenosine 3′:5′-monophosphate-dependent protein kinase in vivo. Biochem J 168:307–310
Engström L (1978) The regulation of liver pyruvate kinase by phosphorylation-dephosphorylation. Curr Top Cell Reg 13:29–51
Ferrendelli JA, Steiner AL, McDougal DR, Kipnis DL (1970) The effect of oxotremorine and atropine on cGMP and cAMP levels in mouse cerebral cortex and cerebellum. Biochem Biophys Res Commun 41:1061–1067
Flockerzi V, Speichermann N, Hofmann F (1978) A guanosine 3′:5′-monophosphate-dependent protein kinase from bovine heart muscle. J Biol Chem 253:3395–3399
Gill GN (1977) A hypothesis concerning the structure of cAMP- and cGMP-dependent protein kinase. J Cyclic Nucleotide Res 3:153–162
Gill GN, Walton GN (1979) Assay for cyclic nucleotide-dependent protein kinases. Adv Cyclic Nucleotide Res 10:93–106
Gill GN, Holdy KE, Walton GM, Kanstein CB (1976) Purification and characterization of 3′:5′-cyclic GMP-dependent protein kinase. Proc Natl Acad Sci USA 73:3918–3922

Gill GN, Walton GM, Sperry PJ (1977) Guanosine 3′:5′-monophosphate-dependent protein kinase from bovine lung. Subunit structure and characterization of the pure enzyme. J Biol Chem 252:6443–6449

Glass DB, Krebs EG (1979) Comparison of the substrate specificity of adenosine 3′:5′-monophosphate and guanosine 3′:5′monophosphate-dependent protein kinases. Kinetic studies using peptides corresponding to phosphorylation sites in histone H 2 B. J Biol Chem 254:9728–9738

Glass DB, Krebs EG (1980) Protein phosphorylation catalyzed by cyclic AMP-dependent and cyclic GMP-dependent protein kinases. Ann Rev Pharmacol Toxicol 20:363–388

Goldberg ND, Haddox MK, Nicol SE, Glass DB, Sanford CH, Kuehl FA, Estensen R (1975) Biological regulation through opposing influences of cyclic GMP and cyclic AMP: The Yin-Yang hypothesis. In: Drummond GI, Greengard P, Robison GA (eds) Advances in cyclic nucleotide research, vol 5. Raven Press, New York, pp 307–330

Greengard P (1978 a) Cyclic nucleotides, phosphorylated proteins and neuronal function. Distinguished lecture series of the Society of General Physiologists, vol 1. Raven Press, New York, pp 124

Greengard P (1978 b) Phosphorylated proteins as physiological effectors. Science 199:146–152

Greengard P, Kuo JF (1970) On the mechanism of action of cyclic AMP. In: Greengard P, Costa E (eds) Role of cyclic AMP in cell function. Advances in biochemical psychopharmacology, vol 3. Raven Press, New York, pp 287–306

Hashimoto E, Takeda M, Nishizuka Y (1975) Phosphorylated sites of calf thymus histone H 2 B by adenosine 3′:5′-monophosphate-dependent protein kinase from silkworm. Biochem Biophys Res Commun 66:547–555

Hashimoto E, Takeda M, Nishizuka Y, Hamana K, Iwai K (1976) Studies on the sites in histone phosphorylated by adenosine 3′:5′-monophosphate-dependent and guanosine 3′:5′-monophosphate-dependent protein kinases. J Biol Chem 251:6287–6293

Hashimoto E, Kuroda Y, Ku Y, Nishizuka Y (1979) Stimulation by polydeoxyribonucleotide of histone phosphorylation by guanosine 3′:5′-monophosphate-dependent protein kinase. Biochem Biophys Res Commun 87:200–206

Hofmann F, Sold G (1972) A protein kinase activity from rat cerebellum stimulated by guanosine 3′:5′-monophosphate. Biochem Biophys Res Commun 49:1100–1107

Huttunen JK, Steinberg D (1971) Activation and phosphorylation of purified adipose tissue hormone-sensitive lipase by cyclic AMP-dependent protein kinase. Biochim Biophys Acta 239:411–427

Inoue M, Kishimoto A, Takai Y, Nishizuka Y (1976) Guanosine 3′:5′-monophosphate-dependent protein kinase from silkworm: Properties of a catalytic fragment obtained by limited proteolysis. J Biol Chem 251:4476–4478

Kemp BE (1979) Phosphorylation of synthetic peptide analogs of rabbit cardiac troponin inhibitory subunit by the cyclic AMP-dependent protein kinase. J Biol Chem 254:2638–2642

Khoo JC, Gill GN (1979) Comparison of cyclic nucleotide specificity of adenosine 3′:5′-monophosphate-dependent and guanosine 3′:5′-monophosphate-dependent protein kinases. Biochim Biophys Acta 584:21–32

Khoo JC, Aquino AA, Steinberg D (1974) The mechanism of activation of hormone sensitive lipase in human adipose tissue. J Clin Invest 53:1124–1131

Khoo JC, Steinberg D, Huang JJ, Vagelos PR (1976) Triglyceride, diglyceride, monoglyceride, and cholesterol ester hydrolase in chicken adipose tissue activated by adenosine 3′:5′-monophosphate-dependent protein kinase. J Biol Chem 251:2882–2890

Khoo JC, Sperry PJ, Gill GN, Steinberg D (1977) Activation of hormone-sensitive lipase and phosphorylase kinase by cyclic GMP-dependent protein kinase. Proc Natl Acad Sci USA 74:4843–4847

Kobayashi R, Fang VS (1976) Studies on cyclic GMP-dependent protein kinase properties by blue dextran-sepharose chromatography. Biochem Biophys Res Commun 69:1080–1087

Krebs EG (1972) Protein kinases. Curr Topics Cell Regul 5:99–133

Krebs EG, Beavo JA (1979) Phosphorylation-dephosphorylation of enzymes. Ann Rev Biochem 48:923–959

Kuo JF (1974) Guanosine 3′:5′-monophosphate-dependent protein kinases in mammalian tissues. Proc Natl Acad USA 71:4037–4041

Kuo JF (1975a) Changes in relative levels of guanosine 3′:5′-monophosphate-dependent and adenosine 3′:5′-monophosphate-dependent protein kinases in lung, heart, and brain of developing guinea pigs. Proc Natl Acad Sci USA 72:2256–2259

Kuo JF (1975b) Changes in activities of modulators of cyclic AMP-dependent and cyclic GMP-dependent protein kinases in pancreas and adipose tissue from alloxan-induced diabetic rats. Biochem Biophys Res Commun 65:1214–1220

Kuo JF (1980) Cyclic nucleotide-dependent protein kinases: An overview. In: Vapaatalo H (ed) Progress in pharmacology, vol 4. Gustav Fischer, Stuttgart, New York, pp 21–30

Kuo JF, Greengard P (1969) Cyclic nucleotide-dependent protein kinases. IV. Widespread occurrence of adenosin 3′:5′-monophosphate-dependent protein kinase in various tissues and phyla of the animal kingdom. Proc Natl Acad Sci USA 64:1349–1355

Kuo JF, Greengard P (1970a) Cyclic nucleotide-dependent protein kinases. VI. Isolation and partial purification of a protein kinase activated by guanosine 3′:5′-monophosphate. J Biol Chem 245:2493–2498

Kuo JF, Greengard P (1970b) Cyclic nucleotide-dependent protein kinases. VII. Comparison of various histones as substrate for adenosine 3′:5′-monophosphate-dependent and guanosine 3′:5′-monophosphate-dependent protein kinases. Biochim Biophys Acta 212:434–440

Kuo JF, Greengard P (1974) Purification and characterization of cyclic GMP-dependent protein kinase. In: Hardman JF, O'Malley WO (eds) Methods in enzymology, vol XXXVIII/c. Academic Press, New York, pp 329–350

Kuo WN, Kuo JF (1976) Isolation of stimulatory modulator of guanosine 3′:5′-monophosphate-dependent protein kinase from mammalian heart devoid of inhibitory modulator of adenosine 3′:5′-monophosphate-dependent protein kinase. J Biol Chem 251:4283–4286

Kuo JF, Krueger BK, Sanes JR, Greengard P (1970) Cyclic nucleotide-dependent protein kinases. V. Preparations and properties of adenosine 3′:5′-monophosphate-dependent protein kinase from various bovin tissues. Biochim Biophys Acta 212:79–91

Kuo JF, Wyatt GR, Greengard P (1971) Cyclic nucleotide-dependent protein kinases. IX. Partial purification and some properties of guanosine 3′:5′-monophosphate-dependent and adenosine 3′:5′-monophosphate-dependent protein kinases from various tissues and species of arthropoda. J Biol Chem 246:7159–7167

Kuo JF, Lee TP, Reyes PL, Walton KG, Donnelly TE, Greengard P (1972) Cyclic nucleotide-dependent protein kinases. X. An assay method for the measurement of guanosine 3′:5′-monophosphate in various biological materials and study of agents regulating its levels in heart and brain. J Biol Chem 247:16–22

Kuo JF, Miyamoto E, Reyes PL (1974) Activation and dissociation of adenosine 3′:5′-monophosphate-dependent and guanosine 3′:5′-monophosphate-dependent protein kinases by various cyclic nucleotide analogs. Biochem Pharmacol 23:2011–2021

Kuo JF, Kuo WN, Shoji M, Davis CW, Seery VL, Donnelly TE (1976a) Purification and general properties of guanosine 3′:5′-monophosphate-dependent protein kinase from guinea pig fetal lung. J Biol Chem 251:1759–1766

Kuo JF, Patrick JG, Seery VL (1976b) Subunit structure of cyclic GMP-dependent protein kinase from guinea pig fetal lung: Dissociation of holoenzyme by cyclic GMP and histone. Biochem Biophys Res Commun 72:996–1002

Kuo JF, Kuo WN, Shoji M (1976c) Regulation by cyclic GMP and stimulatory protein kinase modulator of cyclic GMP-dependent protein kinase from brain and other tissues. In: Costa E, Giaconbini E, Paoletti R (eds) First and second messengers-New vistas, Advances in biochemical psychopharmacology, vol 15. Raven Press, New York, pp 379–389

Kuo WN, Shoji M, Kuo JF (1976d) Stimulatory modulator of guanosine 3′:5′-monophosphate-dependent protein kinase from mammalian tissues. Biochim Biophys Acta 437:142–149

Kuo JF, Davis CW, Tse J (1976e) Depressed cardiac cyclic GMP-dependent protein kinase in spontaneously hypertensive rats and its further depression by guanethidine. Nature 261:335–336

Kuo JF, Malveaux EJ, Patrick JG, Davis CW, Kuo WN, Pruitt A (1977) Cyclic GMP-dependent and cyclic AMP-dependent protein kinases, protein kinase modulators and phosphodiesterases in arteries and veins of dogs. Distribution and effects of arteriovenous fistula and arterial occlusion. Biochem Biophys Acta 497:785–796

Kuo JF, Shoji M, Kuo WN (1978a) Molecular and physiopathologic aspects of mammalian cyclic GMP-dependent protein kinase. Ann Rev Pharmacol Toxicol 18:341–355

Kuo JF, Shoji M, Kuo WN (1978b) Mammalian cyclic GMP-dependent protein kinase and its stimulatory modulator. In: George WJ, Ignarro LJ (eds) Advances in cyclic nucleotide research, vol 9. Raven Press, New York, pp 199–207

Kuo JF, Brackett NL, Stubbs JW, Shoji M, Helfman DM (1978c) Involvements of cyclic nucleotide systems in enlarged mice lungs produced by butylated hydroxytoluene (BHT). Biochem Pharmacol 27:1671–1675

Kuroda Y, Hashimoto E, Nishizuka Y (1976) Phosphorylated sites of calf thymus H2B histone by adenosine 3′:5′-monophosphate-dependent protein kinase from bovine cerebellum. Biochem Biophys Res Commun 71:629–635

Kuroda Y, Hashimoto E, Ku Y, Nishizuka Y (1979) A comment on the functional specificity of cyclic AMP-dependent and cyclic GMP-dependent protein kinases. J Biochem (Tokyo) 85:1099–1101

Langan T (1969) Phosphorylation of liver histone following the administration of glucagon and insulin. Proc Natl Acad Sci USA 64:1276–1283

Lincoln TM, Corbin JD (1977) Adenosine 3′:5′-monophosphate- and guanosine 3′:5′-monophosphate-dependent protein kinases: Possible homologous proteins. Proc Natl Acad Sci USA 74:3239–3243

Lincoln TM, Corbin JD (1978a) On the role of cAMP- and cGMP-dependent protein kinases in cell function. J Cyclic Nucleotide Res 4:3–14

Lincoln TM, Corbin JD (1978b) Purified cyclic GMP-dependent protein kinase catalyzes the phosphorylation of cardiac troponin inhibitory subunit (TN-I). J Biol Chem 253:337–339

Lincoln TM, Hall CL, Park CR, Corbin JD (1976) Guanosine 3′:5′-cyclic monophosphate-binding protein in rat tissues. Proc Natl Acad Sci USA 73:2559–2563

Lincoln TM, Dills WL, Corbin JD (1977) Purification and subunit composition of guanosine 3′:5′-monophosphate-dependent protein kinase from bovine lung. J Biol Chem 252:4269–4275

Lincoln TM, Flockhart DA, Corbin JD (1978) Studies on the structure and mechanism of action of guanosine 3′:5′-monophosphate-dependent protein kinase. J Biol Chem 253:6002–6009

Mackenzie CW, Donnelly TE (1979) Variable dependence on protein kinase stimulatory modulator for cyclic GMP stimulation of histone phosphorylation by rat liver cyclic GMP-dependent protein kinase. Biochem Biophys Res Commun 88:462–469

McCune RW, Gill GN (1979) Positive cooperativity in guanosine 3′:5′-monophosphate binding to guanosine 3′:5′-monophosphate-dependent protein kinase. J Biol Chem 254:5083–5091

Miyamoto E, Kuo JF, Greengard P (1969) Cyclic nucleotide-dependent protein kinases. III. Purification and properties of adenosine 3′:5′-monophosphate-dependent protein kinase from bovine brain. J Biol Chem 244:6395–6402

Miyamoto E, Petzgold GL, Kuo JF, Greengard P (1973) Dissociation and activation of adenosine 3′:5′-monophosphate-dependent and guanosine 3′:5′-monophosphate-dependent protein kinases by substrate proteins. J Biol Chem 248:179–189

Murad F, Manganiello V, Vaughan M (1971) A simple sensitive protein binding assay for guanosine 3′:5′-monophosphate. Proc Natl Acad Sci USA 68:736–739

Murofushi H (1974) Protein kinases in *Tetrahymena* cilia. II. Partial purification and characterization of adenosine 3′:5′-monophosphate-dependent protein kinases. Biochim Biophys Acta 370:130–139

Nakazawa K, Sano M (1975) Partial purification and properties of guanosine 3′:5′-monophosphate-dependent protein kinase from pig lung. J Biol Chem 250:7415–7419
Nimmo HG, Cohen P (1977) Hormonal control of protein phosphorylation. In: Greengard P, Robison GA (eds) Advances in cyclic nucleotide research, vol 8. Raven Press, New York, pp 145–266
Nishiyama K, Katakami H, Yamamura H, Takai Y, Shimomura R, Nishizuka Y (1975) Functional specificity of guanosine 3′:5′-monophosphate-dependent and adenosine 3′:5′-monophosphate-dependent protein kinases from silkworm. J Biol Chem 250:1297–1300
Nishizuka Y, Takai Y, Kishimoto A, Hashimoto E, Inoue M, Yamamoto M, Criss WE, Kuroda Y (1978) A role of calcium in the activation of a new protein kinase system. In: George WJ, Ignarro LJ (eds) Advances in cyclic nucleotide research, vol 9. Raven Press, New York, pp 209–220
Nishizuka Y, Takai Y, Hashimoto E, Kishimoto A, Kuroda Y, Sakai K, Yamamura H (1979) Regulatory and functional compartment of three multifunctional protein kinase systems. Mol Cell Biochem 23:153–165
Ram JL, Erlich YH (1978) Cyclic GMP-stimulated phosphorylation of membrane-bound protein from nerve roots of *Aplysia californica*. J Neurochem 30:487–491
Ramseyer J, Kanstein CB, Walton GM, Gill GN (1976) The use of affinity chromatography in purification of cyclic nucleotide receptor proteins. Biochim Biophys Acta 446:358–370
Riou JP, Claus TH, Flockhart DA, Corbin JD, Pilkis SJ (1977) In vivo and in vitro phosphorylation of rat liver fructose-1,6-biphosphatase. Proc Natl Acad Sci USA 74:4615–4619
Rochette-Egly C, Castagna M (1978) Evidence for a role of sulfhydryl groups in catalytic activity and subunit interaction of the cyclic GMP-dependent protein kinase from silkworm. Biochim Biophys Acta 526:107–115
Rubin CS, Rosen OM (1975) Protein phosphorylation. Ann Rev Biochem 44:831–887
Schlichter DJ, Casnellie JE, Greengard P (1978) An endogenous substrate for cGMP-dependent protein kinase in mammalian cerebellum. Nature 273:61–62
Shlatz LJ, Kimberg DV, Cattieu KA (1978) Cyclic nucleotide-dependent phosphorylation of rat intestinal microvillus and basal-lateral membrane proteins by an endogenous protein kinase. Gastroenterology 75:838–846
Shlyapnikov SV, Arutyunyan AA, Kurochkin SN, Memelova LV, Nesterova MV, Sashcenko LP, Severin ES (1975) Investigation of the sites phosphorylated in lysine-rich histone by protein kinase from pig brain. FEBS Lett 53:316–319
Shoji M, Patrick JG, Davis CW, Kuo JF (1977a) Guanosine cyclic monophosphate-dependent protein kinase from fetal calf heart. Purification, general properties and catalytic subunit. Biochem J 161:213–221
Shoji M, Patrick JG, Tse J, Kuo JF (1977b) Studies on the cyclic GMP-dependent protein kinase from bovine aorta. Possible existence of a catalytic subunit. J Biol Chem 252:4347–4353
Shoji M, Brackett NL, Tse J, Shapira R, Kuo JF (1978a) Molecular properties and mode of action of homogeneous preparation of stimulatory modulator of cyclic GMP-dependent protein kinase from the heart. J Biol Chem 253:3427–3434
Shoji M, Morris HP, Davis CW, Brackett NL, Kuo JF (1978b) Modified cyclic nucleotide systems in Morris hepatoma 3924A favoring expression of cyclic GMP effect. Biochim Biophys Acta 500:419–424
Soderling TR, Hickenbottom JP, Reiman EM, Hunkeler FL, Walsh DA, Krebs EG (1970) Inactivation of glycogen synthetase and activation of phosphorylase kinase by muscle adenosine 3′:5′-monophosphate-dependent protein kinase. J Biol Chem 245:6317–6328
Steiner AL, Parker CW, Kipnis DM (1970) The measurement of cyclic nucleotides by radioimmunoassay. Adv Biochem Psychopharmacol 3:89–111
Stull JT, Buss JE (1977) Phosphorylation of cardiac troponin by cyclic adenosine 3′:5′-monophosphate-dependent protein kinase. J Biol Chem 252:851–857
Takai Y, Nishiyama K, Yamamura H, Nishizuka Y (1975) Guanosine 3′:5′-monophosphate-dependent protein kinase from bovine cerebellum. Purification and characterization. J Biol Chem 250:4690–4695

Takai Y, Nakaya S, Inoue M, Kishimoto A, Nishiyame K, Yamamura H, Nishizuka Y (1976) Comparison of mode of activation of guanosine 3′:5′-monophosphate-dependent and adenosine 3′:5′-monophosphate-dependent protein kinases from silkworm. J Biol Chem 251:1481–1487
Takai Y, Kishimoto A, Iwasa Y, Kawahara Y, Mori T, Nishizuka Y (1979) Calcium-dependent activation of multifunctional protein kinase by membrane phospholipids. J Biol Chem 254:3692–3695
Thomas EW, Murad F, Looney WB, Morris HP (1973) Adenosine 3′:5′-monophosphate and guanosine 3′:5′-monophosphate concentration in Morris hepatomas of different growth rates. Biochim Biophys Acta 297:564–567
Tse J, Brackett NL, Kuo JF (1978) Alterations in activities of cyclic nucleotide systems and beta-adrenergic receptor-mediated activation of cyclic AMP-dependent protein kinase during progression and regression of cardiac hypertrophy induced by isoproterenol. Biochim Biophys Acta 542:399–412
Tse J, Powell JR, Baste CA, Priest RE, Kuo JF (1979) Isoproterenol-induced cardiac hypertrophy: Modifications of beta-adrenergic receptor, adenylate cyclase and ventricular contraction. Endocrinology 105:246–255
Van Leemput-Coutrez M, Canus J, Christophe J (1973) Cyclic nucleotide-dependent protein kinases of the rat pancreas. Biochem Biophys Res Commun 54:182–190
Wallach D, Davies PJA, Pastan I (1977) Cyclic AMP-dependent phosphorylation of filamin in mammalian smooth muscle. J Biol Chem 253:4739–4745
Walsh DA, Krebs EG (1973) Protein kinases. Enzymes 8:555–581
Walsh DA, Ashby CD (1973) Protein kinases: Aspects of their regulation and diversity. Recent Prog Horm Res 29:329–359
Walsh DA, Perkins JP, Krebs EG (1968) An adenosine 3′:5′-monophosphate-dependent protein kinase from rabbit skeletal muscle. J Biol Chem 243:3763–3765
Walsh DA, Ashby CD, Gonzales C, Calkins D, Fischer EH, Krebs EG (1971) Purification and characterization of a protein inhibitor of adenosine 3′:5′-monophosphate-dependent protein kinase. J Biol Chem 246:1977–1985
Yamamoto M, Takai Y, Hashimoto E, Nishizuka Y (1977) Intrinsic activity of guanosince 3′:5′-monophosphate-dependent protein kinase similar to adenosine 3′:5′-monophosphate-dependent protein kinase. I. Phosphorylation of histone fractions. J Biochem (Tokyo) 81, 1857–1863
Yamamoto N, Criss WE, Takai Y, Yamamura H, Nishizuka Y (1979) A hepatic soluble cyclic nucleotide-dependent protein kinase. Stimulation by basic polypeptides. J Biol Chem 254:5049–5052

Note Added in Proof

More recently, cyclic GMP-dependent protein phosphorylation in intact medial tissue and intact cells (Casnellie et al., 1980), and subcellular localization of G-PK and its substrates (Ives et al., 1980) in vascular smooth muscle, have been reported. A G-PK substrate protein, termed G-substrate ($M_r = 23{,}000$), from rabbit cerebellum has been purified to apparent homogeneity and characterized (Aswad and Greengard, 1981a), and its phosphorylation kinetics (Aswad and Greengard, 1981b) and amino acid sequences at the two phosphorylation sites (Aitken et al., 1981) have been determined. Other recent progresses made on G-PK also include its immunohistochemical localization in mammalian brain (Lohmann et al., 1981); detailed studies on its autophosphorylation (Foster et al., 1981); regulation of its activity by stimulatory modulator (Walton and Gill, 1980) and by histones and poly-arginine (Walton and Gill, 1981); its structural analysis by limited proteolysis (Monken and Gill, 1980); sites of phosphorylation of R subunit of type I A-PK (Hashimoto et al., 1981) and histone H1 (Zeilig et al., 1981) catalyzed by the enzyme.

Additional References

Aitken A, Bilham T, Cohen P, Aswad DW, Greengard P (1981) A specific substrate from rabbit cerebellum for guanosine 3′:5′-monophosphate-dependent protein kinase. III. Amino acid sequences at the two phosphorylation sites. J Biol Chem 256:3501–3506

Aswad DW, Greengard P (1981 a) A specific substrate from rabbit cerebellum for guanosine 3′:5′-monophosphate-dependent protein kinase. I. Purification and characterization. J Biol Chem 256:3487–3493

Aswad DW, Greengard P (1981 b) A specific substrate from rabbit cerebellum for guanosine 3′:5′-monophosphate-dependent protein kinase. II. Kinetic studies on its phosphorylation by guanosine 3′:5′-monophosphate-dependent and adenosine 3′:5′-monophosphate-dependent protein Kinases. J Biol Chem 256:3494–3500

Casnellie JE, Ives HE, Jamieson D, Greengard P (1980) Cyclic GMP-dependent protein phosphorylation in intact medial tissue and intact cells from vascular smooth muscle. J Biol Chem 255:3770–3776

Foster JL, Guttmann J, Rosen OM (1981) Autophosphorylation of cGMP-dependent protein kinase. J Biol Chem 256:5029–5036

Hashimoto E, Takio K, Krebs EG (1981) Studies on the site in the regulatory subunit of Type I cAMP-dependent protein kinase phosphorylated by cGMP-dependent protein kinase. J Biol Chem 256:5604–5607

Ives HE, Casnellie JE, Greengard P, Jamieson JD (1980) Subcellular localization of cyclic GMP-dependent protein kinase and its substrates in vascular smooth muscle. J Biol Chem 255:3777–3785

Lohmann SM, Walter U, Miller PE, Greengard P, De Camilli P (1981) Immunohistochemical localization of cyclic GMP-dependent protein kinase in mammalian brain. Proc Natl Acad Sci USA 78:653–657

Monken CE, Gill GN (1980) Structural analysis of cGMP-dependent protein kinase using limited proteolysis. J Biol Chem 255:7067–7070

Walton GM, Gill GN (1980) Protein effects on the activity of guanosine 3′:5′-monophosphate-dependent protein kinase. J Biol Chem 255:1603–1609

Walton GM, Gill GN (1981) Regulation of cyclic nucleotide-dependent protein kinase activity by histones and poly(L-arginine). J Biol Chem 256:1681–1688

Zeilig CE, Langan TE, Glass DB (1981) Sites in histone HI selectively phosphorylated by guanosine 3′:5′-monophosphate-dependent protein kinase. J Biol Chem 256:994–1001

CHAPTER 13

Calcium-Dependent Protein Phosphorylation

H. Schulman

Overview

In recent years there has been renewed appreciation of the importance of calcium in various modes of cell activation. It is becoming apparent that one can no longer separate the individual contributions of calcium and of the cyclic nucleotides in regulating the cellular responses of a variety of neuronal and hormonal stimuli. Although the actions of calcium do not appear to be mediated by a universal biochemical mechanism, this review considers the possibility that certain of the physiological effects of calcium may be mediated or modulated by protein phosphorylation.

The mechanisms involved in regulation of liver carbohydrate metabolism by catecholamines, vasopressin, and angiotensin are assessed. The role of α-adrenergic and β-adrenergic receptors in mediating the response to catecholamines appears to be species dependent. In rat liver, activation of the cAMP-dependent protein kinase does not appear to be necessary for stimulating glycogenolysis by epinephrine. The β-adrenergic receptor antagonist, propranolol, has no effect on epinephrine-stimulated glycogenolysis, while abolishing epinephrine-stimulated increases in cAMP. Conversely, phenoxybenzamine, an α-adrenergic receptor antagonist, blocks the effect of epinephrine on glycogenolysis without any effect on cAMP levels. cAMP-independent pathways are also suggested by studies on regulation of carbohydrate metabolism by vasopressin and angiotensin II. There is no evidence for the involvement of cGMP in any of these systems. Indeed, there is strong evidence that increased calcium levels resulting either from influx or from redistribution of intracellular stores may regulate glycogenolysis in these systems. Some of the effects of these hormones may result from activation of calcium-dependent protein kinases.

Calcium-dependent protein phosphorylation has now been demonstrated in several intact systems. These include synaptosomes, brain slices, mast cells, and isolated hepatocytes. Studies on synaptosomes, or pinched off nerve endings, indicate that depolarization by various agents leads to a calcium-dependent increase in the incorporation of $[^{32}P]O_4$ into several proteins. Such calcium-dependent protein phosphorylation may be involved in neurotransmitter synthesis or release. The possibility that release processes may in general be associated with calcium-stimulated changes in protein phosphorylation is strengthened by the studies showing changes in protein phosphorylation in mast cells stimulated by the secretagogue 48/80.

Studies with isolated hepatocytes demonstrate that although the actions of catecholamines, vasopressin, and angiotensin II on carbohydrate metabolism do not involve changes in cyclic nucleotide levels, stimulation of the cells by these hormones is accompanied by increased incorporation of $[^{32}P]O_4$ into various proteins, including pyruvate kinase and phosphorylase. Most of the proteins whose phosphorylation is regulated by these hormones are also regulated by glucagon, whose action is mediated by cAMP-dependent phosphorylation.

Four calcium dependent protein kinases are examined in detail. The effects of calcium on protein phosphorylation in synaptosomes may be mediated by a membrane-bound calcium-dependent protein kinase. This protein kinase requires an endogenous cytosolic, heat-stable protein for activity. This protein is shown to be identical to calmodulin, a ubiquitous calcium-binding protein, known to be involved in regulation of a variety of calcium-dependent processes. Authentic calmodulin and the endogenous protein kinase activator have identical physico-chemical characteristics, including activation of protein phosphorylation, sensitivity to calcium, and amino acid composition.

Protein I, one of the proteins phosphorylated in intact synaptosomes, is also phosphorylated in membrane fractions. Its phosphorylation in intact and lysed synaptosomes is regulated by cAMP and by calcium at distinct sites on the protein. In intact synaptosomes, 8-Br-cAMP stimulates phosphorylation of the cAMP-regulated site(s) while reducing phosphorylation of the calcium-regulated site(s). Thus, some of the actions of these two second messengers on neuronal function may intersect at the level of protein phosphorylation.

The calmodulin- and calcium-dependent protein phosphorylation system is not confined to brain but has a widespread distribution in membranes from a variety of rat tissues. Each tissue displays a tissue specific array of endogenous substrate proteins, consistent with the possibility that this phosphorylation system regulates specific effects of calcium in each of the tissues.

Myosin light chain kinase from skeletal muscle, smooth muscle, and a variety of nonmuscle cells, including platelets, baby hamster kidney cells, and brain cells is a calcium-dependent protein kinase which also requires calmodulin for activity. In vivo studies indicate that the state of phosphorylation of the P-light chain of myosin is changed during muscular activity. In smooth muscle, phosphorylation of myosin light chains can be directly correlated with ATPase activity and tension development. The function of myosin light chain kinase in skeletal muscle is not well understood. Phosphorylation of myosin light chain kinase by a cAMP-dependent protein kinase reduces its activity, providing a locus for interaction of calcium and cAMP in regulating muscle contraction.

Phosphorylase kinase, long known to be a calcium-dependent enzyme, has recently been shown to contain calmodulin as one of its intrinsic subunits. Unlike other calmodulin-dependent enzymes, the calmodulin in phosphorylase kinase does not dissociate from the enzyme complex in the absence of calcium. However, each tetrameric enzyme has four additional calmodulin binding sites which reversibly interact with calmodulin, leading to additional stimulation of the enzyme. Recent studies indicate that glycogen synthase may also be a substrate for phosphorylase kinase. Thus, this enzyme may account for some of the calcium-dependent phosphorylation observed in liver and other tissues. Regulation of phosphorylase kinase in vivo and in vitro by calcium and by cAMP is discussed.

A protein kinase with broad specificity has recently been identified in a variety of mammalian tissues. This cAMP-independent protein kinase is activated by calcium via two independent mechanisms. At relatively high calcium concentrations a neutral protease is able to convert the proenzyme of this protein kinase to a smaller active form. At lower calcium concentrations this enzyme can be activated without proteolysis if a membrane-associated factor or one of several purified phospholipids is present. The activated enzyme can phosphorylate phosphorylase kinase, glycogen synthase, and various histones.

The physiological and biochemical studies on intact tissue suggest the presence of cAMP-independent, calcium-dependent protein phosphorylation. The four calcium-dependent protein kinases discussed above may be involved in mediating or modulating some calcium-dependent physiological processes. The action of cAMP-dependent and calcium-dependent protein kinases on the same substrate proteins or on functionally related proteins may provide an interface for the interaction of these two second messenger systems.

A. Introduction

Publication of this volume attests to the great interest by scientists from diverse fields in the physiological and biochemical effects of cyclic nucleotides in various tissues. The inclusion of a chapter on calcium-dependent protein phosphorylation can be attributed to the increasing number of biological systems where an interaction between calcium and the cyclic nucleotides has been shown to occur. Indeed, it would have been surprising to find these second messenger systems and their elaborate regulatory components functioning as independent entities. Scientific appreciation of the importance of calcium in regulating various cellular functions has undergone dramatic reversals.

Early work on nerve and muscle physiology implicated calcium ion as an important intracellular messenger (Harvey and Macintosh 1940; Heilbrunn 1952; Luttgau and Niedergerke 1958). Subsequent biochemical studies demonstrated that the effect of calcium on actomyosin ATPase is the critical feature of the molecular basis of action of this cation in skeletal muscle (Huxley 1964). There was also strong evidence for a role of calcium in excitation-secretion coupling, including the stimulus-coupled release of neurotransmitters and peptide hormones (Katz and Miledi 1967, 1969; Douglas 1968). Interest in calcium action waned following the elegant studies of Sutherland and his colleagues demonstrating the involvement of cyclic AMP in the action of many hormones. These studies led to the hypothesis that cAMP is a ubiquitous "second messenger," the link between many extracellular messengers and control of intracellular events (Sutherland et al. 1968; Robison et al. 1971). Subsequently, it was found that a cAMP-dependent protein kinase in muscle mediated the breakdown of glycogen (WALSH et al. 1968) and that cAMP-dependent protein kinase activity was widespread in nature (KUO and GREENGARD 1969). It was proposed that the diverse biological effects of cAMP in various tissues might be mediated by regulating protein kinase activity (KUO and GREENGARD 1969). The appealing feature of this protein kinase hypothesis was that it provided a mechanism by which a single effector molecule, cAMP, could elicit such a diversity of physiological and biochemical effects. According to this hypothesis, cAMP need only interact with one intracellular receptor, the regulatory subunit of the protein kinase; the specificity of the action of cAMP would be determined by the nature of the substrates for cAMP-dependent protein kinase present in the various tissues. More than 21 enzymes, involved in a diverse array of cellular processes, have been shown to be substrates for cAMP-dependent protein kinases and for phosphoprotein phosphatases (KREBS and BEAVO 1979). Studies, both in vitro and in vivo, have clearly established protein phosphorylation as one of the principal mechanisms for hormonal regulation of many biological processes.

In recent years there has been renewed appreciation of the importance of calcium in various modes of cell activation. In many systems, the role of calcium as a second messenger is no less important than the role of cyclic nucleotides and it is apparent that their actions are often interrelated (BERRIDGE 1975; RASMUSSEN and GOODMAN 1977). A major apparent difference between the actions of calcium and cAMP is that, unlike cAMP, whose action in eukaryocit cells seems to be mediated exclusively by protein phosphorylation, several classes of enzyme appear to mediate the effects of calcium. This review considers the possibility, however, that although there is not a single universal mechanism for calcium action, certain of the physiological effects of calcium may also be mediated by protein phosphorylation.

It has been known for many years that a variety of biological systems contain protein kinases which are not responsive to fluctuating levels of cAMP. These kinases have often been referred to as cAMP-independent protein kinases (RUBIN and ROSEN 1975). Recent studies have indicated that many classes of regulatory agents that do not elicit changes in cyclic nucleotide levels can affect the phosphorylation of specific proteins in their target cells (GREENGARD 1978a, b). It has been suggested that phosphorylation of specific proteins is a final common pathway for some of the actions of diverse stimuli (GREENGARD 1978a).

In this review, I have elaborated on some recent findings on the role of calcium in regulation of a variety of cellular processes. Attention is focused primarily on both direct and indirect evidence suggesting that calcium-dependent protein phosphorylation may be involved in such regulation. The supporting evidence is presented at three levels. The first deals with the demonstration that carbohydrate metabolism can be regulated by agents that elevate intracellular calcium concentrations in the absence of significant changes in cyclic nucleotide levels (Sect. B). The second involves studies demonstrating changes in the state of phosphorylation of various substrate proteins elicited by the action of depolarizing agents and calcium on brain tissue, by the action of secretagogues and calcium on mast cells, and by the action of several hormones and calcium on liver cells (Sect. C). Finally, I have detailed recent biochemical studies on four calcium-dependent protein kinases and their possible cellular functions (Sects. D, E, F, G, and H). The four enzymes are a membrane-bound, calmodulin-, and calcium-dependent protein kinase (Sects. D and E); myosin light chain kinase (Sect. F); phosphorylase kinase (Sect. G); and a soluble, phospholipid- and calcium-dependent protein kinase (Sect. H). It is concluded that it is via these and/or similar protein kinases that the biological effects of hormones and agents that utilize calcium as an intracellular second messenger may be mediated or modulated. In addition, interaction between cyclic nucleotides and calcium, and thus between agents that utilize these second messengers, may occur at the level of protein phosphorylation (Sect. I).

B. Role of Calcium in Regulation of Carbohydrate Metabolism

The concept of cAMP as a second messenger arose from experiments dealing with the effects of catecholamines on hepatic glycogenolysis (SUTHERLAND et al. 1968; ROBISON et al. 1971). Ironically, recent studies of liver carbohydrate metabolism indicate that, in some animal species, cAMP may not mediate the effects of catecholamines on either glycogenolysis or gluconeogenesis and suggest the existence of alternate, cAMP-independent, pathways for phosphorylation of metabolic enzymes. These findings are described in detail in the review by John Exton (this volume) and will only be summarized here.

Glucagon is an example of a hormone whose physiological effects can be completely accounted for by the action of cAMP and cAMP-dependent protein kinase (ROBISON et al. 1971). The effects of epinephrine, however, appear to be mediated by different mechanisms in different tissues and species. In dog liver, it is believed to act by stimulation of a β-adrenergic receptor-linked adenylate cyclase system (HIMMS-HAGEN 1967). In mouse, guinea pig, cat, rabbit, and human liver, both α-receptors and β-receptors may be involved. The relative contributions of the α-receptor and β-receptor components appear to differ among species and has not been carefully examined in many species.

Although epinephrine stimulates both cAMP production and glycogenolysis in isolated rat hepatocytes, the effect on glycogenolysis is not mediated by the rise in cAMP levels. Elevation of cAMP elicited by epinephrine is completely abolished by the β-adrenergic receptor antagonist, propranolol, without any effect on glycogenolysis. Conversely, phenoxybenzamine, an α-adrenergic receptor antagonist, abolishes the effect of epinephrine on glycogenolysis without any effect on cAMP levels. Although, β-adrenergic receptor agonists stimulate glycogenolysis in rat liver, with accompanying changes in cAMP, they produce less dramatic responses than are obtained with α-adrenergic receptor agonists (CHERRINGTON et al. 1977). In addition, epinephrine, while stimulating carbohydrate metabolism, does so without activating the cAMP-dependent protein kinase (CHERRINGTON et al. 1976; BIRNBAUM and FAIN 1977). Thus, it appears that in rat liver the control of glycogenolysis by epinephrine is a cAMP-independent process (SHERLINE et al. 1972; TOLBERT et al. 1973; KNEER et al. 1974; OSBORN 1975; HUTSON et al. 1976; CHAN and EXTON 1977). The effects of epinephrine on gluconeogenesis (TOLBERT et al. 1973), glycogen synthase (HUTSON et al. 1976), pyruvate kinase (CHAN and EXTON 1977), and mitochondrial pyruvate carboxylation (GARRISON and BORLAND 1979) may also be regulated by activation of α-receptors. Thus, although the α-receptor-mediated effects of epinephrine on rat liver metabolism are similar to those seen by activation of β-receptors in other species, they occur by a cAMP-independent mechanism that does not involve activation of the cAMP-dependent protein kinase.

Carbohydrate metabolism is also regulated by vasopressin (KIRK and HEMS 1974; KEPPENS and DEWULF 1975), angiotensin II (KEPPENS and DEWULF 1976), and neuronal stimulation (SHIMAZU and AMAKAWA 1968) via mechanism(s) that do not involve changes in cAMP levels. There is no evidence for the involvement of cGMP in any of these systems (POINTER et al. 1976).

The cAMP-independent mechanisms involved in regulating carbohydrate metabolism have not been determined but there is considerable evidence favoring a role of calcium as a second messenger in many systems. It has been shown, for example, that angiotensin, vasopressin and the α-receptor agonist, phenylephrine, require the presence of extracellular calcium to elicit their effects on glycogenolysis (KEPPENS et al. 1977). Stimulation of both calcium influx and glycogenolysis by either phenylephrine or angiotensin was blocked by the appropriate antagonists. In addition, the stimulation of glycogenolysis was mimicked by the divalent cation ionophore, A 23187, in the presence of calcium (KEPPENS et al. 1977). The lack of a requirement for extracellular calcium for the action of a particular hormone does not eliminate calcium as the potential second messenger. The source of calcium may be either the extracellular milieu or intracellular stores (RASSMUSSEN and GOODMAN 1977). In a variety of systems there is a pronounced calcium efflux from cells following stimulation by hormones indicating that calcium is being released from intracellular stores (RASMUSSEN and GOODMAN 1977). In some systems both influx and redistribution of intracellular stores of calcium may occur (CHEN et al. 1978; BLACKMORE et al. 1978). It has recently been reported that mitochondria constitute the source of calcium which is released intracellularly upon addition of norepinephrine to hepatocytes (BABCOCK et al. 1979).

Studies in liver (SHIMAZU and AMAKAWA 1975; KHOO and STEINBERG 1975) and muscle (BROSTROM et al. 1971) suggest that phosphorylase kinase is a calcium-dependent enzyme (see Sect. G). Although it is possible that direct stimulation of phosphorylase kinase by calcium accounts for some of the effects of epinephrine (e.g., activation of phosphorylase), such stimulation cannot account for the effects of epinephrine on pyruvate kinase, gluconeogenesis, mitochondrial pyruvate carboxylation or amino acid transport (PARIZA et al. 1977). Thus, alternate mechanisms must exist for mediating the various effects of epinephrine and other hormones. Interestingly, most enzymes regulated by the calcium-dependent pathway(s), as well as those regulated by the cAMP-dependent pathway, are known to be regulated by cycles of phosphorylation and dephosphorylation. The appealing feature of control via chemical modification, such as phosphorylation, and the similarity of effects elicited by the calcium-dependent and calcium-independent mechanisms suggest that calcium may stimulate protein phosphorylation systems that are distinct from the cAMP-dependent protein kinases. Such calcium-regulated protein phosphorylation may mediate or modulate some of the actions of angiotensin, vasopressin, and the α-receptor agonists.

C. Role of Calcium in Regulation of Protein Phosphorylation in Intact Systems

I. Calcium-Dependent Protein Phosphorylation in Brain

1. Phosphorylation in Intact Synaptosomes

The important role of calcium in the physiology of the nervous system is well documented (BAKER 1972; RUBIN 1972). Stimulation of nerve cells leads to a wave of depolarization traveling down the axon which, at the nerve terminal, causes an influx of calcium and the subsequent release of neurotransmitters (BAKER et al. 1971). There is also evidence that calcium regulates the activity of neurotransmitter-synthesizing enzymes (e.g., PATRICK and BARCHAS 1974) and the electrical excitability of nerve cells (FRANKENHAEUSER 1957; KELLY et al. 1969). The biochemical mechanisms underlying these and other physiological effects of calcium have not yet been elucidated.

The possibility that one or more of the effects of calcium in the presynaptic nerve terminal might be mediated or modulated through the phosphorylation of specific proteins was examined in studies on intact synaptosomes (KRUEGER et al. 1976, 1977). Synaptosomes are pinched-off nerve endings produced by homogenization of brain tissue (GRAY and WHITTAKER 1962; RODRIGUEZ DE LORES ARNAIZ et al. 1967). They retain many morphological and functional characteristics of the nerve endings including the ability to synthesize ATP and to release neurotransmitters in a calcium-dependent manner. In these studies, a synaptosomal fraction (P 2) from rat cerebral cortex was preincubated with $[^{32}P]O_4$ of high specific activity for 30 min in order to label the intrasynaptosomal pool of ATP (KRUEGER et al. 1977). The synaptosomes were then incubated for 30 s in the absence or in the presence of veratridine or a high concentration of potassium, either with or without calcium present in the medium. Both veratridine and high potassium depolarize synaptosomal membranes and cause an influx of calcium from the medium (BLAUSTEIN 1975). The incubation was then terminated by addition of sodium dodecyl sulfate (SDS), and the state of phosphorylation of the synaptosomal proteins analyzed by SDS-polyacrylamide gel electrophoresis and autoradiography. An autoradiograph illustrating the incorporation of $[^{32}P]O_4$ into proteins of a crude synaptosomal fraction is shown in Fig. 1. During the 30 min preincubation, about 20 polypeptide bands incorporated $[^{32}P]O_4$. Following the 30 s treatment with either veratridine (100 μM) or high K^+ (60 mM) in the presence of 1 mM calcium, there was a change in the state of phosphorylation of several polypeptides. The major change seen was a marked increase of $[^{32}P]O_4$ incorporation into two polypeptides with apparent molecular weights of 86,000 and 80,000. These polypeptides, designated Ia and Ib, are also substrates for a membrane-bound cAMP-dependent protein kinase in brain and are collectively referred to as Protein I (UEDA and GREENGARD 1977). There was also a significant increase in incorporation of $[^{32}P]O_4$ into three polypeptides in the 50,000–60,000 Dalton range. Two other polypeptides of approximately 90,000 Daltons exhibited a decreased amount of $[^{32}P]O_4$ incorporation in the presence of either veratridine or high potassium. Depolarization of the synaptosomes had no effect on phosphorylation in the absence of calcium. Further purification of the synaptosomal preparation indicated that indeed the observed

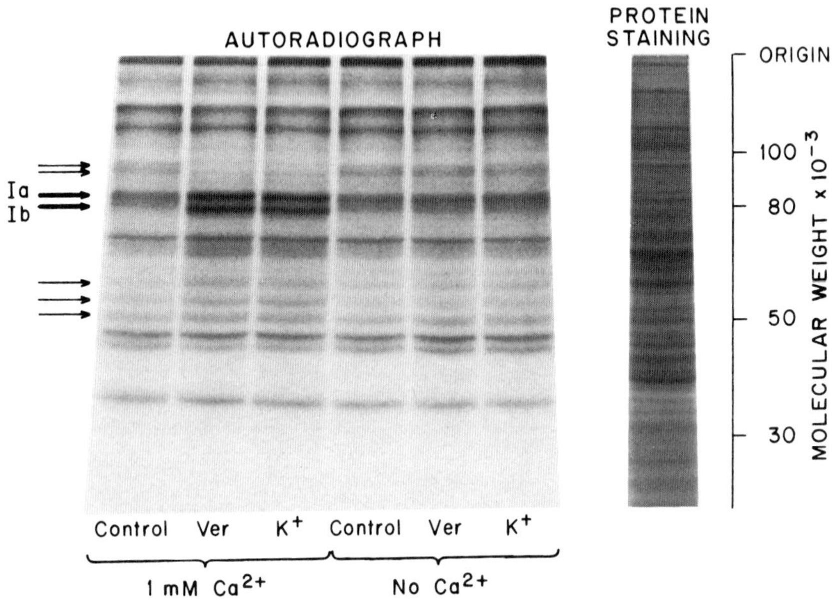

Fig. 1. Effect of veratridine and of high potassium, in the absence and presence of Ca^{2+}, on the phosphorylation of endogenous proteins in crude synaptosomal preparation from rat cerebral cortex. The synaptosome fraction was preincubated with ^{32}Pi for 30 min in the absence of calcium. Aliquots of this suspension were then incubated for 30 s in the absence (Control) or presence of 100 μ*M* veratridine (Ver) or 60 m*M* K^+ (K^+); 1 m*M* Ca^{2+} was present where indicated. The incubation was terminated by the addition of sodium dodecyl sulfate (SDS), and an aliquot of each sample subjected to SDS-polyacrylamide gel electrophoresis and autoradiography. The molecular weight scale was generated by determining the position of marker proteins of known molecular weight. The positions of Proteins Ia and Ib are indicated by bold arrows to the left of the autoradiograph. Light arrows indicate the positions of other protein bands whose phosphorylation was inhibited (upper two arrows) or stimulated (lower three arrows) by veratridine or high K^+. (From KRUEGER et al. 1977)

phosphorylation had occurred in intact synaptosomes. The results support the possibility that phosphoproteins may be involved in the regulation of certain presynaptic nerve terminal functions.

Protein I is the most prominent endogenous substrate for cAMP-dependent protein kinase in synaptic membrane preparations from brain (GREENGARD 1978 b). Although very little is known about its function, it has interesting biological properties which suggest an important role in the physiology of the synapse. In addition, its regulation by both cAMP and calcium make it a useful tool in analyzing the action and interaction of these second messengers. Subcellular fractionation, immunocytochemical studies, and lesion studies on rat brain indicate that Protein I is localized primarily in axon terminals at certain types of synapses, where it appears to be associated with synaptic vesicles (UEDA et al. 1979; BLOOM et al. 1979; DECAMILLI et al. 1979, SIEGHART et al. 1978 a). A good correlation between the ontogeny of synapse formation and the appearance of Proteins Ia and Ib in the brains of developing rats and guinea pigs has provided further evidence

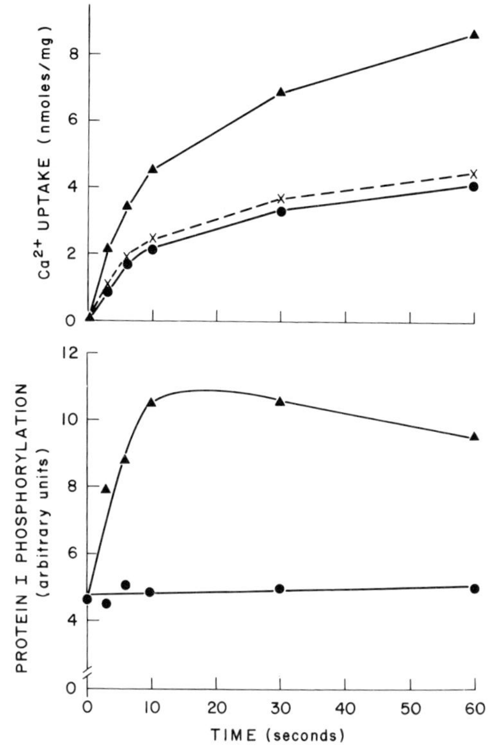

Fig. 2. Time course of $^{45}Ca^{2+}$ uptake (top) and Protein I phosphorylation (bottom) in a crude synaptosome preparation. Incubations were carried out in the absence (●) or presence (▲) of 100 µ*M* veratridine, which was added at zero time. The tracer amount of $^{45}Ca^{2+}$ was added together with the veratridine for the measurements of Ca^{2+} uptake. The accumulation of $^{45}Ca^{2+}$ in the absence of veratridine is attributable to an exchange of labeled for unlabeled Ca^{2+}, and not to a net accumulation of Ca^{2+}. The increment in Ca^{2+} uptake caused by veratridine is shown by the broken curve (×---×). (From KRUEGER et al. 1977)

that Proteins Ia and Ib are associated with synaptic structures (LOHMANN et al. 1978).

In addition to stimulating phosphorylation of various proteins, depolarization with veratridine or with high concentrations of potassium stimulated calcium influx into synaptosomes (KRUEGER et al. 1977). In the presence of calcium, A 23187 was shown to mimic the action of the depolarizing agents on both calcium influx and phosphorylation. Tetrodotoxin, which selectively inhibits the depolarizing effects of veratridine (ULBRICHT 1969), abolished the effects of veratridine on both calcium influx and Protein I phosphorylation while the action of high potassium and A 23187 plus calcium were unaffected.

The time-couse of [$^{45}Ca^{++}$]-uptake into synaptosomes and of Protein I phosphorylation, in the absence and in the presence of veratridine, is shown in Fig. 2. The veratridine-independent component of [$^{45}Ca^{++}$]-uptake is due to an exchange of labeled for unlabeled calcium rather than to a net accumulation of calcium by the synaptosomes. The veratridine-dependent net accumulation of calcium is indicated by the broken curve (Fig. 2, top). This component of uptake reaches its half-maximal level within 10 s after addition of veratridine. This calcium influx is sufficient to maximally stimulate Protein I phosphorylation (Fig. 2, bottom).

Despite the fact that Protein I is a substrate for the cAMP-dependent protein kinases (see below), the calcium-stimulated phosphorylation of Protein I in intact synaptosomes does not appear to be a consequence of some indirect effect of calcium on cyclic nucleotide levels. Neither veratridine nor 60 mM K^+, under conditions which stimulated phosphorylation of Protein I, had any significant effect on the level of cAMP in the preparation, either in the absence or in the presence of 3-isobutyl-1-methylxanthine (IBMX), a potent inhibitor of cyclic nucleotide phosphodiesterase (KRUEGER et al. 1977). Although 8-bromo-cAMP (but not 8-bromo-cGMP), stimulated phosphorylation of Protein I in intact synaptosomes, the site specificities of phosphorylation elicited by calcium and by 8-bromo-cAMP were different (Sect. D) (HUTTNER and GREENGARD 1979).

2. Phosphorylation in Intact Brain Tissue

The capacity of extracted Protein I to be phosphorylated in vitro by added cAMP-dependent protein kinase is a measure of vacant phosphorylation sites, and, therefore, provides an index of the state of phosphorylation of Protein I in vivo at the time of extraction. Regulation of the state of phosphorylation of Protein I in whole brains of mice in vivo has recently been demonstrated (STROMBOM et al. 1979). Several central nervous system depressants, including phenobarbitol, chloral hydrate, and urethane, administered to mice intraperitoneally, decreased the state of phosphorylation of Protein I. Conversely, the convulsant drugs pentylene-tetrazole and picrotoxin each increased the state of phosphorylation of Protein I.

Protein I phosphorylation has also been shown to be altered by depolarizing agents and cyclic nucleotides in intact slices of rat cerebral cortex incubated in Krebs-Ringer buffer (FORN and GREENGARD 1978). In control slices, Protein I existed almost entirely in the dephosphorylated form. Upon addition of veratridine, high potassium, 8-Br-cAMP, or IBMX, there was a rapid and marked increase in the level of phosphorylation of Proteins Ia and Ib.

These studies indicate that phosphorylation of Protein I at the site that is phosphorylated by the cAMP-dependent protein kinase can be regulated in synaptosomes, in brain slices, and in vivo by agents that mimic neuronal activity, supporting the possibility that it may be involved in calcium-dependent physiological processes at nerve terminals. In addition, these studies suggest the existence of a calcium-dependent, cAMP-independent, mechanism for regulating protein phosphorylation.

II. Calcium-Dependent Protein Phosphorylation in Mast Cells

Recently, the possibility that release processes may in general be associated with calcium-stimulated changes in protein phosphorylation was investigated using rat peritoneal mast cells (SIEGHART et al. 1978b). In experiments analogous to those with intact synaptosomes, mast cells were preincubated with $[^{32}P]O_4$ and then exposed to the classic mast cell secretagogue, 48/80, or to the calcium ionophore A 23187. A rapid phosphorylation of several polypeptides with molecular weights of 68,000, 59,000, and 42,000 Daltons accompanied secretion of granules by these cells. With 48/80, both degranulation and phosphorylation were stimulated by external calcium. In addition, at a later stage in the course of the mast cell response to 48/80, as secretion terminated, phosphorylation of a 78,000 Dalton polypeptide was seen to increase. Interestingly, the phosphorylation of this protein was selectively stimulated by the antiallergic drug cromolyn (cromoglycate), which simultaneously inhibited histamine release in the mast cell preparation (THEOHARIDES et al. 1980). The effect of cromolyn in stimulating the phosphorylation of this peptide paralleled its effect in inhibiting histamine release under a variety of experimental conditions. These results suggest a possible association between calcium-dependent protein phosphorylation and the initiation and termination of the secretory response in mast cells.

III. Regulation of Protein Phosphorylation in Isolated Hepatocytes

The possibility that stimulation of carbohydrate metabolism in rat hepatocytes is mediated or modulated by a protein phosphorylation system that is independent of cyclic nucleotides has recently been tested (GARRISON 1978). Techniques similar to those discussed above for intact synaptosomes and mast cells were used to examine the effect of various agents on protein phosphorylation. Glucagon, whose action in liver is mediated by cAMP (ROBISON et al. 1971), and α-adrenergic agonists, whose action is mediated by cAMP-independent mechanism(s) (see Sect. B), were compared. Cells prelabeled with $[^{32}P]O_4$ of high specific activity for 45 min incorporated $[^{32}P]O_4$ into about 27 polypeptide bands. Addition of glucagon or dibutyryl cAMP for a subsequent 3–6 min led to an increased incorporation of $[^{32}P]O_4$ into 12 of these 27 bands. The degree of stimulation varied, with maximal stimulation being about 600% of control. Polypeptide bands with similar molecular weights could be phosphorylated in vitro by addition of a cAMP-dependent protein kinase to hepatocyte cytosol. This is not unexpected, since glucagon action is believed to be mediated by cAMP. However, treatment of the hepatocytes with norepinephrine, epinephrine, or the selective α-adrenergic agonist phenylephrine in the presence of propranolol resulted in stimulation of the α-receptors with an accompanying change in the state of phosphorylation of 10 of the 12 bands whose phosphorylation was stimulated by glucagon and dibutyryl cAMP (Table 1). Although it is likely that some bands consist of several different phosphoproteins, the striking overlap in molecular weights suggests that the various agents are stimulating the phosphorylation of the same proteins. Because of the presence of propranolol, a selective β-adrenergic receptor antagonist, stimulation by the catecholamines is limited to activation of the α-adrenergic receptor. Consistent with the specificity of these agents, addition of glucagon led to increased cAMP levels whereas addition of norepinephrine, epinephrine, or phenylephrine, in the presence of 20 μM propranolol, did not alter control levels of cAMP. In addition, when the α-adrenergic antagonists ergotamine (20 μM) (Table 1) or phentolamine (20 μM) were included in the incubation medium, the effects of the α-receptor agonists on protein phosphorylation were abolished.

Phosphorylation of 4 polypeptide bands was stimulated by addition of A 23187 plus calcium (GARRISON 1978). This effect was absolutely dependent on the pres-

Table 1. A comparison of the effect of maximal doses of glucagon, norepinephrine and the calcium ionophore A23187 on the phosphorylation of liver cell supernatant proteins

Molecular weight	Magnitude of the phosphorylation change caused by various agonists: 10 μM Glucagon	20 μM Propranolol + 10 μM Norepinephrine	Norepinephrine, Propranolol +20 μM Ergotamine	10 μM A23187	Possible band identity
147,000	136±11	113± 6	102±3		
123,000	154±13	102± 2	100±1		
93,000	195±22	169±18	106±3	193±20	Phosphorylase
83,000	293±44	169±16	102±3		Glycogen synthase
77,000	357±34	162±17	100±3		
61,000	429±16	255±15	113±9	243±17	Pyruvate kinase
56,000	408±34	239±37	100±2		
52,000	596±64	225±14	109±6	183±17	
49,000	167± 8	142± 5	106±5	155±7	
46,000	192±24	140±12	108±6		
44,000	202±52	142±20	100±2		
41,000	133± 8	121± 7	106±3		

Isolated hepatocytes were incubated with $[^{32}P]O_4$ for 45 min and then stimulated with the various agents for 3–5 min (glucagon, propranolol plus norepinephrine, and A23187) or 6–15 min (norepinephrine plus propranolol plus ergotamine). The mobilities of the various polypeptide bands and the percentage of change in phosphorylation caused by the various agonists were compiled from autoradiographs of six experiments. The molecular weight of each band was estimated from a standard curve using molecular weight standards. The magnitude of the change in phosphorylation of each band was calculated from peak heights of optical density scans of the autoradiographs. The suggested band identities presented in Column 6 are tentative. (Adapted from GARRISON 1978)

ence of calcium in the medium suggesting that the ionophore was stimulating protein phosphorylation by increasing calcium influx. Each of these bands correspond in molecular weight to one of the bands stimulated by the α-adrenergic agonists (Table 1).

Tentative identification of some of the phosphoproteins has been made. Based on molecular weights of enzymes known to be stimulated in liver by α-receptor agonists, it was suggested that two of the regulated polypeptides, with molecular weights of 93,000 and 61,000 Daltons, may correspond to subunits of phosphorylase and pyruvate kinase, respectively. For example, stimulation of phosphorylase activity by increasing amounts of either glucagon or norepinephrine plus propranolol correlated with stimulation of $[^{32}P]O_4$ incorporation into the polypeptide band with a molecular weight of 93,000 (GARRISON 1978). Based on several studies on hormonal regulation of protein phosphorylation in rat hepatocytes, tentative designation can be made of several additional polypeptides listed in Table 1. A variety of evidence suggest that the 220,000 Daltons polypeptide may be acetyl coenzyme A carboxylase (WITTERS et al. 1979), the 123,000 Dalton polypeptide may be ATP-citrate lyase (ALEXANDER et al. 1979), and the 77,000 Dalton polypeptide may be phospofructokinase (CASTANO et al. 1979; KAGIMOTO and UYEDA 1979).

A recent study demonstrated that the action of angiotensin II and vasopressin on isolated rat hepatocytes may also be mediated by cAMP-independent protein phosphorylation (GARRISON et al. 1979). Treatment of pre-labeled cells with these hormones for 3 min led to increased incorporation of $[^{32}P]O_4$ into 10–12 cytosolic proteins with the same molecular weights as the proteins whose phosphorylation was stimulated by glucagon. The effects of angiotensin II, but not of vasopressin, could be greatly attenuated by the angiotensin antagonist, saralasin, indicating the involvement of a specific angiotensin II receptor. Two of the 10–12 phosphoproteins, with molecular weights of 93,000 and 61,000, were identified as phosphorylase (by affinity chromatography) and pyruvate kinase (by immunoprecipitation), respectively.

The effect of angiotensin II and vasopressin on incorporation of $[^{32}P]O_4$ into the various polypeptide bands was completely abolished when calcium was omitted from the incubation medium. Their effects on the enzymatic activities of phosphorylase, glycogen synthase and pyruvate kinase were also abolished, consistent with the possible relationship between phosphorylation of these enzymes and their state of activation. The action of glucagon did not require extracellular calcium. The polypeptide bands whose phosphorylation was stimulated by norepinephrine were not all affected by omission of calcium from the medium (GARRISON et al. 1979). Incorporation of $[^{32}P]O_4$ into the polypeptide bands with molecular weights of 83,000, 61,000, 56,000, and 52,000 did not require extracellular calcium. Inhibition of glycogen synthase (83,000 Daltons) and pyruvate kinase (61,000) also occurred in the absence of calcium, consistent with the incorporation of $[^{32}P]O_4$ into polypeptide bands with similar molecular weights. The differential requirement for extracellular calcium may indicate that more than one mechanism is involved in mediating the effects of the catecholamines. Alternatively, it is possible that the slight elevation of cAMP reported to be produced by α-receptor agonists in the absence of calcium (CHAN and EXTON 1977) may somehow lead to selective cAMP-dependent phosphorylation of the 4 polypeptides.

Activation of phosphorylase kinase may account for several of the effects of α-adrenergic agonists (e.g., their effect on phosphorylase and glycogen synthase) (see Sect. G). It would be worthwhile to examine its action on the other targets of epinephrine action. The only plausible explanation for increased $[^{32}P]O_4$ incorporation into polypeptide bands whose corresponding enzymatic activity was also shown to change is that a net incorporation of phosphate occurred. An increase in the total phosphate content of a protein may be elicited either by inhibiting a phosphoprotein phosphatase or by stimulating a protein kinase. Either scheme suggests the involvement of cAMP-independent protein kinase.

D. Calmodulin- and Calcium-Dependent Protein Kinase in Biological Membranes

There is considerable evidence for the existence of calcium-dependent protein phosphorylation in a variety of tissues. By analogy with the cAMP-dependent protein kinase system, one may expect such a protein kinase to be sensitive to calcium ion in the physiological range of concentrations, to be activated by calcium in a reversible manner, and to have broad specificity for a variety of protein substrates. This review analyzes four calcium-dependent protein phosphorylation systems which meet these criteria to various extents. Although none of these may ultimately be found to be a general protein kinase mediating the intracellular effects of calcium on a large number of substrates, the study of these enzymes will help to illustrate and elucidate possible schemes for regulation of protein kinase activity by calcium. Each of the four protein kinases may mediate some set of restricted actions of calcium. Furthermore, the action of these protein kinases illustrates biochemical schemes which may underly the physiological interactions of cyclic nucleotides and calcium.

I. Calcium-Dependent Protein Phosphorylation in Isolated Brain Membranes

The finding of calcium-stimulated changes in protein phosphorylation in intact synaptosomes incubated with $[^{32}P]O_4$ and the absence of detectable increases in cyclic nucleotide levels accompanying these changes suggested the presence of a cyclic nucleotide-independent protein phosphorylation system in brain. Evidence concerning the molecular mechanism by which calcium entry might stimulate phosphorylation of synaptosomal proteins has come recently from experiments using lysed synaptosomes (SCHULMAN and GREENGARD 1978 a). In these experiments, synaptosomes obtained from rat cerebral cortex were lysed by hypo-osmotic shock and incorporation of $[^{32}P]O_4$ from $[\gamma\text{-}^{32}P]$ATP into protein analyzed by SDS-polyacrylamide gel electrophoresis and autoradiography. The effect of calcium on endogenous protein phosphorylation in a synaptosomal lysate is shown in Fig. 3 (lanes 1 and 2). Calcium markedly stimulated $[^{32}P]O_4$ incorporation into many polypeptides in the lysed synaptosome preparation that contained membranes (synaptic membranes plus synaptic vesicles) and synaptosomal cytoplasm. Large increases in net $[^{32}P]O_4$ incorporation were seen in polypeptides of about 51,000 and 62,000 Daltons. Proteins Ia and Ib were among the polypeptide bands that showed calcium-dependent phosphorylation. Similar patterns of stimulation by calcium have been described by DELORENZO and FREEDMAN (1977). Preparations of brain tissue examined under different conditions also indicate the presence of calcium-dependent protein phosphorylation (HERSHKOWITZ 1978; BROWNING et al. 1979). The study detailed above suggests that the calcium-stimulated phosphorylation seen in intact synaptosomes (Fig. 1) may be a direct consequence of the activation of a calcium-sensitive protein kinase (Fig. 3) by calcium influx. Studies employing kainic acid to produce neuronal specific lesions in rat striatum suggest that both of these phosphorylation systems have a predominantly neuronal localization (SIEGHART et al. 1980).

The calcium-dependent phosphorylation observed in lysed synaptosomes was lost on preparation of cytoplasm-free membranes (Fig. 3, lanes 3 and 4). This phenomenon was not due to inactivation of the calcium-dependent protein kinase activity during experimental manipulations as it could be regained by addition of an amount of boiled (Fig. 3, lanes 5 and 6) or unboiled synaptosomal cytoplasm com-

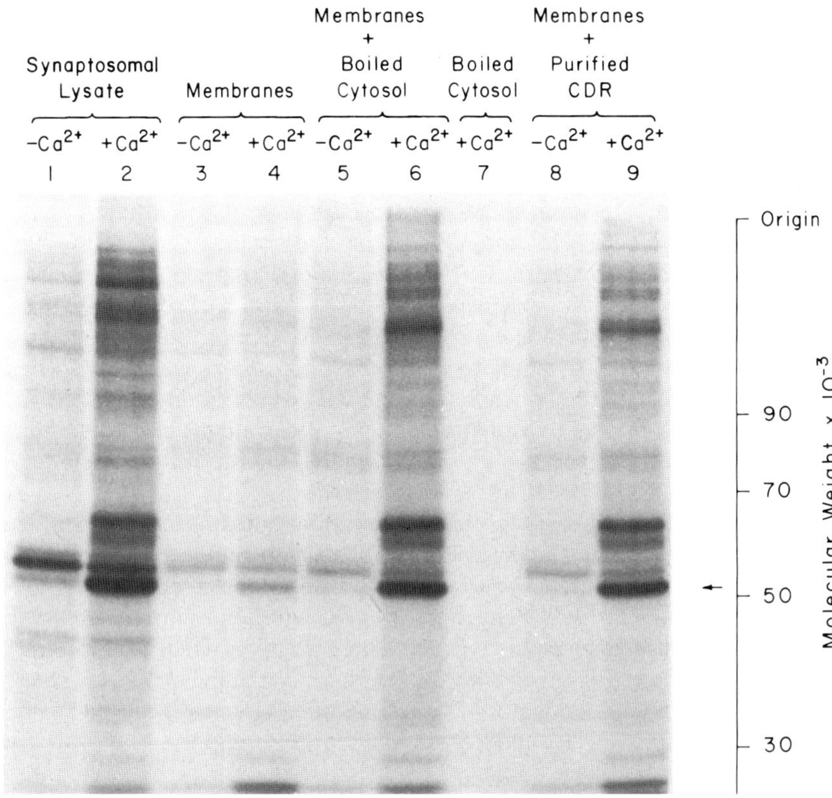

Fig. 3. Effect of calcium, synaptosomal cytoplasm, and purified CDR (calmodulin) on endogenous protein phosphorylation of brain membranes. Synaptosomes obtained from rat cerebral cortex were subjected to osmotic shock by homogenization in 10 volumes of ice-cold water (15 ml) and kept on ice for 30 min to optimize lysis. Total membranes were obtained from lysed synaptosomes by centrifugation at 150,000 $\times g$ for 30 min. The released synaptosomal cytosol was removed, heated at 95°–100 °C for 5 min, and used as the source of activator (boiled cytosol). The pellet was resuspended in 15 ml of 5 mM Tris-HCl (pH 7.0), centrifuged as above, and again resuspended in 15 ml of 5 mM Tris-HCl (pH 7.0). This resuspended pellet served as the source of activator-deficient membranes. The reaction mixture for assay of endogenous protein phosphorylation (final volume, 100 µl) contained: 50 mM PIPES buffer (pH 7.0), 10 mM $MgCl_2$, 1 mM dithioerythritol, 0.2 mM EGTA (minus calcium) or 0.2 mM EGTA + 0.5 mM $CaCl_2$ (plus calcium), 5 µM [γ-^{32}P]ATP (5.2×10^4 cpm/pmole), and, where indicated, lysed synaptosomes (50 µg protein), membranes (34 µg protein), boiled cytosol (14 µg protein), or purified CDR (0.25 µg protein). Incubation was carried out for 10 s at 30 °C, the reaction terminated by the addition of SDS, and an aliquot of the sample was analyzed for protein phosphorylation by SDS-polyacrylamide gel electrophoresis and autoradiography. The arrow indicates the 51,000-dalton phosphoprotein used for quantitative measurements of stimulation of [^{32}P]O_4 incorporation into membrane protein. (From SCHULMAN and GREENGARD 1978a)

mensurate with the amount of membrane protein present. This sample of boiled cytosol which conferred calcium-dependent phosphorylation on cytosol-free membranes did not exhibit any endogenous phosphorylation when incubated in the absence of membranes (Fig. 3, lane 7), nor did it show any protein kinase activity using histone, protamine, or casein as substrate. Thus, the calcium-dependent phosphorylation in synaptic membranes required a soluble heat stable factor present in synaptosomal cytosol.

A common feature of calcium-dependent processes that have been examined at the biochemical level is the involvement of a calcium binding protein (KRETSINGER 1976). A multifunctional calcium-binding protein (calmodulin; calcium-dependent regulator or CDR) has been shown to regulate the activity of a variety of enzymes, including cyclic nucleotide phosphodiesterase (CHEUNG 1970; KAKIUCHI and YAMAZAKI 1970) and a detergent solubilized preparation of adenylate cyclase (BROSTROM et al. 1975; CHEUNG et al. 1975) from mammalian brain (for a review see WOLFF and BROSTROM 1979; WANG and WAISMAN 1979; CHEUNG, this volume). Indeed, a preparation of authentic phosphodiesterase activator restored calcium-dependent protein phosphorylation to washed synaptosomal membranes (Fig. 3, lanes 8 and 9). The pattern of phosphorylation obtained on addition of boiled synaptosomal cytoplasm was indistinguishable from that obtained on addition of CDR (compare Fig. 3 lanes 6 and 9), suggesting that the two factors are functionally equivalent.

Stimulation of endogenous protein phosphorylation in synaptic membrane fractions was dependent on the presence of both calcium and either the cytosol factor or CDR. Thus, calcium in the absence of boiled cytosol and boiled cytosol in the absence of calcium were ineffective, whereas addition of both calcium and boiled cytosol restored phosphorylation (Fig. 3, compare lanes 3, 4, 5, and 6). Analogous results were obtained with CDR (Fig. 3, compare lanes 3, 4, 8, and 9). This endogenous kinase activator, like CDR, was extremely heat stable, nondialyzable, resistant to DNase and RNase, and sensitive to trypsin (SCHULMAN and GREENGARD 1978 a).

II. Identification of Protein Kinase Activator as Calmodulin

To determine the number and nature of endogenous protein kinase activators in brain, a preparation of cytosol from bovine cerebral cortex was fractionated and column fractions assayed for the ability to stimulate calcium-dependent protein phosphorylation in membrane preparations from rat brain (SCHULMAN and GREENGARD 1978 b). Throughout the purification, only a single peak of kinase activator was found, suggesting that only one protein in the cytosol is able to reconstitute calcium-dependent protein kinase activity. When the same column fractions were examined for the presence of calmodulin by assaying their ability to activate calmodulin-depleted phosphodiesterase, the presence of calmodulin coincided with the kinase activator peak. The yields of the two activities were nearly identical at each step of the purification and both copurified with an 18,000 Dalton polypeptide as revealed by SDS-polyacrylamide gel electrophoresis, thus suggesting that the endogenous activator of the calcium-dependent protein phosphorylation system is calmodulin.

The protein kinase activator is indistinguishable from authentic calmodulin based on a number of criteria.

As shown in Fig. 4, stimulation of protein phosphorylation by calmodulin and by kinase activator is identical. Synaptic membrane fractions washed free of cytosol were nearly devoid of calcium-dependent protein phosphorylation (Fig. 4,

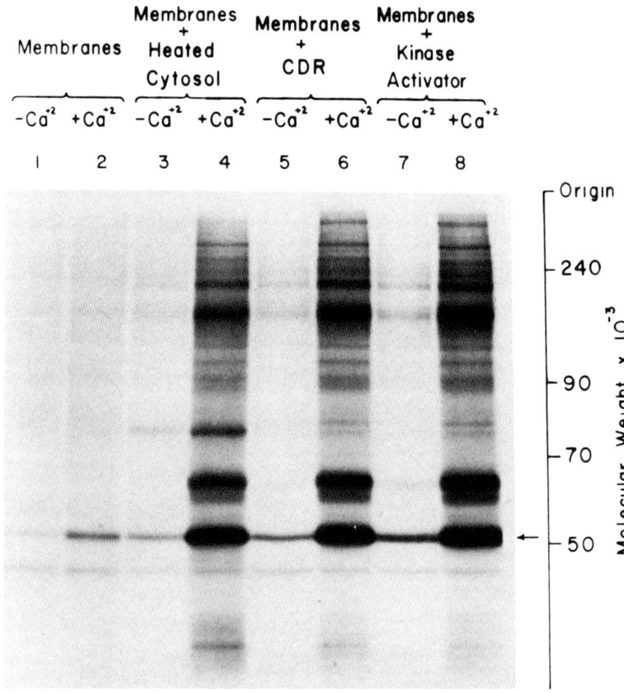

Fig. 4. Autoradiograph illustrating endogenous calcium-dependent protein phosphorylation of brain membranes, and activation by heated cytosol, purified CDR, or purified kinase activator. Cytosol-free membranes (34 μg of protein) were incubated in the standard reaction mixture containing 0.2 m*M* EGTA (minus calcium) or 0.2 m*M* EGTA + 0.5 m*M* $CaCl_2$ (plus calcium), 5 μ*M* [γ-^{32}P]ATP (5×10^4 cpm/pmol), and, where indicated, heated cytosol (14 μg protein), purified CDR (0.25 μg of protein) or purified kinase activator (0.25 μg of protein). The arrow indicates the 51,000 Dalton phosphoprotein used for quantitative measurements of stimulation of [^{32}P]O_4 incorporation into membrane protein. (From SCHULMAN and GREENGARD 1978 b)

lanes 1 and 2). Upon addition of authentic calmodulin, prepared from bovine cerebral cortex by the method of TEO et al. (1973) (Fig. 4, lanes 5 and 6) or upon addition of the protein kinase activator (Fig. 4, lanes 7 and 8), the same pattern of phosphorylation was obtained. The pattern with calmodulin and with kinase activator was qualitatively and quantitatively similar (Fig. 4, compare lanes 6 and 8).

Authentic calmodulin and protein kinase activator migrated identically, each as a single band on nondenaturing polyacrylamide gel electrophoresis (SCHULMAN and GREENGARD 1978 b). When gel slices were assayed, each preparation was found to have only a single peak of either protein kinase activator or phosphodiesterase activator with a mobility corresponding to the band of protein staining. On SDS-polyacrylamide gel electrophoresis, both calmodulin and the protein kinase activator migrated as homogeneous polypeptides. The two proteins comigrated and had an apparent molecular weight of 18,000 Daltons. A more complete biochemical characterization using criteria such as ultraviolet spectrum, isoelectric focusing and peptide mapping after partial proteolysis indicated that the two proteins, desig-

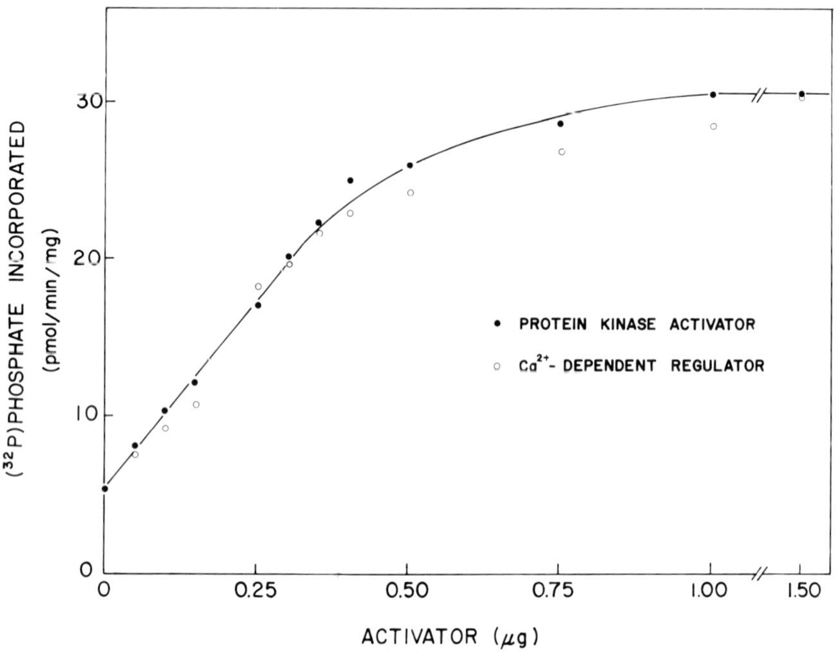

Fig. 5. Activation of calcium-dependent protein phosphorylation by purified kinase activator and by authentic calmodulin. Cytosol-depleted membranes were incubated with various amounts of purified kinase activator (●) or calmodulin (○) in the presence of calcium under standard conditions, and assayed for calcium-dependent protein phosphorylation of the 51,000 Dalton substrate protein. [^{32}P]O_4 incorporation into several other polypeptides was measured quantitatively in the same way and showed a similar response to the two factors. (From SCHULMAN and GREENGARD 1978 b)

nated as authentic calmodulin and protein kinase activator, are, in fact, identical. Indeed, analysis of the amino acid composition of the protein kinase activator shows that, within experimental error, it is identical to the amino acid composition of calmodulin reported in the literature (SCHULMAN et al. 1980).

III. Characteristics of Protein Kinase Activity

Calmodulin activates calcium-dependent protein phosphorylation in a dose-dependent manner. Authentic calmodulin or protein kinase activator stimulated [^{32}P]O_4 incorporation into the 51,000 Dalton polypeptide (see arrow, Figs. 3 and 4) with equal efficiency (Fig. 5). Each protein caused about a 6-fold stimulation, with half-maximal stimulation occurring at about 0.25 μg of protein.

A physiological role for this calcium-dependent phosphorylation system would dictate that the kinase be sensitive to low concentrations of calcium. It has been estimated that the basal cytoplasmic concentration of free Ca^{2+} in eukaryotic cells is 0.01 μM to 1 μM. Following cell stimulation, free Ca^{2+} concentration may be elevated to between 1 μM and 10 μM (KRETSINGER 1976, 1979). Indeed, as with other calmodulin-mediated processes, endogenous protein phosphorylation is sen-

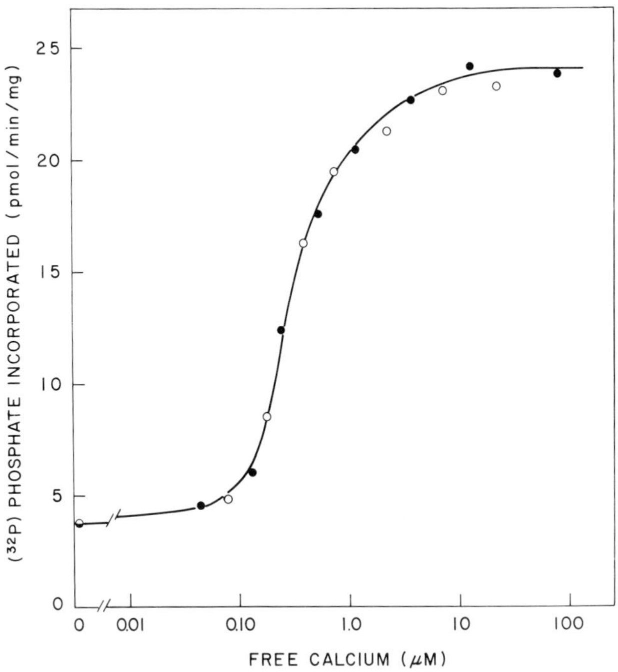

Fig. 6. Activation by calcium of protein phosphorylation, with either purified kinase activator or authentic calmodulin as the source of activator. Activator-depleted membranes were incubated with either 1 μg of kinase activator (●) or 1 μg of calmodulin (○) under standard conditions at pH 7.1. Free Ca^{2+} concentration was varied by use of Ca^{2+}/EGTA buffer containing 0.4 m*M* EGTA. Free Ca^{2+} concentration was determined by using an apparent binding constant for Ca^{2+}-EGTA of $7.61 \times 10^6\ M^{-1}$ (PORTZEHL et al. 1964). (From SCHULMAN and GREENGARD 1978 b)

sitive to physiologically relevant levels of calcium. The effect of various concentrations of calcium on protein phosphorylation is shown in Fig. 6. With either preparation of activator, half-maximal stimulation occurred at a free Ca^{2+} concentration of about 0.3 μ*M*. This is consistent with the findings that, under various conditions, half-maximal activation of calmodulin-dependent phosphodiesterase and binding of calcium to calmodulin in binding studies occur in the range of 0.2 μ*M* to 3 μ*M* free Ca^{2+} (TEO and WANG 1973; WOLFF et al. 1977). In vivo this phosphorylation system would then be expected to undergo dynamic changes in activity, cycling between a largely "on" and a largely "off" state.

IV. Function of Calcium-Dependent Protein Phosphorylation in Brain

Depolarization of intact synaptosomes with either veratridine or high potassium has been shown to stimulate a calcium-dependent release of neurotransmitters (BLAUSTEIN et al. 1972; BLAUSTEIN 1975), to regulate neurotransmitter synthesis (PATRICK and BARCHAS 1974), and to stimulate phosphorylation of Protein I (Fig. 1). It is possible that phosphorylation of Protein I, a vesicle-associated protein, and of other synaptosomal proteins, mediates or modulates some of the presynaptic effects of calcium, such as neurotransmitter synthesis and release. The calmodulin- and calcium-dependent protein kinase may have a role in such calcium-dependent processes. The possibility that calcium-dependent protein phosphorylation may regulate neurotransmitter synthesis is suggested by recent studies on activation of tryptophan hydroxylase, the enzyme that regulates synthesis of serotonin (KUHN et al. 1978; YAMAUCHI and FUJISAWA 1979).

The findings that depolarization-induced calcium influx regulates protein phosphorylation in intact synaptosomes (KRUEGER et al. 1976, 1977) and that synaptosomal membranes contain a calmodulin- and calcium-dependent protein kinase (SCHULMAN and GREENGARD 1978a, b) have recently been confirmed (DELORENZO et al. 1979). In addition, the latter investigators reported that release of norepinephrine from purified synaptic vesicles occurs simultaneously with activation of protein phosphorylation by calmodulin plus calcium. However, their suggestion that this calcium-dependent phosphorylation of vesicle-bound protein mediates neurotransmitter release from nerve terminals may not be warranted. The "release" of neurotransmitter which they measured was dependent on calcium and calmodulin but occurred in a vesicle preparation apparently in the absence of presynaptic plasma membrane (DELORENZO and FREEDMAN 1978), a necessary component of the physiological release process (DOUGLAS 1968). Their observations may represent an effect of calcium, magnesium, calmodulin, and ATP on vesicle integrity or permeability rather than representing a specific parameter of neurotransmitter release. Similar objections to the conclusions of these investigators have recently been raised (WOLFF and BROSTROM 1979).

V. Regulation of Protein I Phosphorylation by cAMP and by Calcium

1. Phosphorylation of Protein I by cAMP and Calcium in Intact and Lysed Synaptosomes

Studies using two phosphorylation systems (intact synaptosomes and synaptosomal membranes), indicated that two polypeptides with mobilities on SDS polyacrylamide gel electrophoresis similar to those of Proteins Ia and Ib were phosphorylated (KRUEGER et al. 1977; SCHULMAN and GREENGARD 1978a). Recent studies have indicated that these two calcium-regulated phosphoproteins are, in fact, identical to Proteins Ia and Ib (SIEGHART et al. 1979). Both sets of proteins were extractable from membrane under similar acid-extraction conditions, had similar molecular weights in several SDS polyacrylamide gel electrophoresis systems, had similar isoelectric points, and yielded phosphopeptides of similar if not identical molecular weights after partial proteolysis with a variety of proteases.

2. Tryptic Fingerprints of Phosphorylated Protein I

The relationship of cAMP- and calcium-regulated phosphorylation of Protein I was recently investigated in intact as well as in lysed synaptosome preparations from rat brain (HUTTNER and GREENGARD 1979). In order to compare the phosphorylation site(s) that were regulated by cAMP and calcium, phosphorylated Protein I was purified under conditions which did not alter its state of phosphorylation and analyzed by peptide mapping. After phosphorylation in intact and lysed synaptosomes Protein I was denatured with SDS and urea and purified by non-equilibrium pH gradient electrophoresis. Proteins Ia and Ib extracted from such a gel and analyzed on SDS polyacrylamide gels appear as major proteins by Coomassie blue staining, and as the only $[^{32}P]O_4$-labeled proteins (HUTTNER and GREENGARD 1979).

The two dimensional separation of phosphopeptides derived from purified Protein Ib after exhaustive digestion with trypsin is shown in Fig. 7. Five phosphopeptides, designated peptides 1–5 were observed with Protein Ib obtained from both intact and lysed synaptosomes. 8-Br-cAMP selectively stimulated the phosphorylation of peptide 1, both in intact and in lysed synaptosomes. This effect was not dependent on the presence of calcium. Veratridine-induced calcium influx into intact synaptosomes or the addition of calcium to lysed synaptosomes stimulated phosphorylation of peptide 1, as did 8-Br-cAMP, but in addition stimulated the phosphorylation of peptides 2–5. The effect of veratridine on the phosphorylation of peptides 1–5, as well as a similar effect obtained with potassium, was dependent on the presence of external calcium. In intact synaptosomes, 8-Br-cAMP reduced the amount of radioactive phosphate incorporated into peptides 2–5. The inhibition of phosphate incorporation into some peptides simultaneously with the stimulation of phosphate incorporation into another peptide elicited by 8-Br-cAMP in intact synaptosomes explains the inability to detect significant effects of 8-Br-cAMP on the total incorporation of phosphate into undigested Protein I (KRUEGER et al. 1977; HUTTNER and GREENGARD 1979).

These studies indicate that there are multiple phosphorylation sites in Protein I and that the sites are differentially regulation by cAMP and by calcium. The data are consistent with the possibility that cAMP stimulates the phosphorylation of only one site in Protein I (represented by peptide 1), whereas calcium stimulates the phosphorylation not only of this site, but also of additional site(s) (represented by peptides 2–5). In addition, cAMP reduces calcium-stimulated phosphorylation of some sites (represented by peptides 2–5).

Multiple sites of phosphorylation may be a common feature of enzymes and proteins under complex regulation (COHEN 1978) (see Sect. G). The finding that cAMP and calcium cause differential regulation of multiple-site phosphorylation of the same neuronal protein, Protein I, is of interest. Interaction between cAMP and calcium at the level of phosphorylation of Protein I may provide a molecular basis for physiological interactions between these two intracellular second messengers in those synaptic terminals containing this protein.

E. Widespread Occurrence of Calmodulin- and Calcium-Dependent Protein Phosphorylation in Biological Membranes

Calcium ion has been implicated in important regulatory functions in a variety of tissues (RASMUSSEN and GOODMAN 1977). For most of these functions, as for the neuronal functions discussed earlier, little is known about the biochemical basis for the action of calcium in these tissues. In light of the finding of calmodulin- and cal-

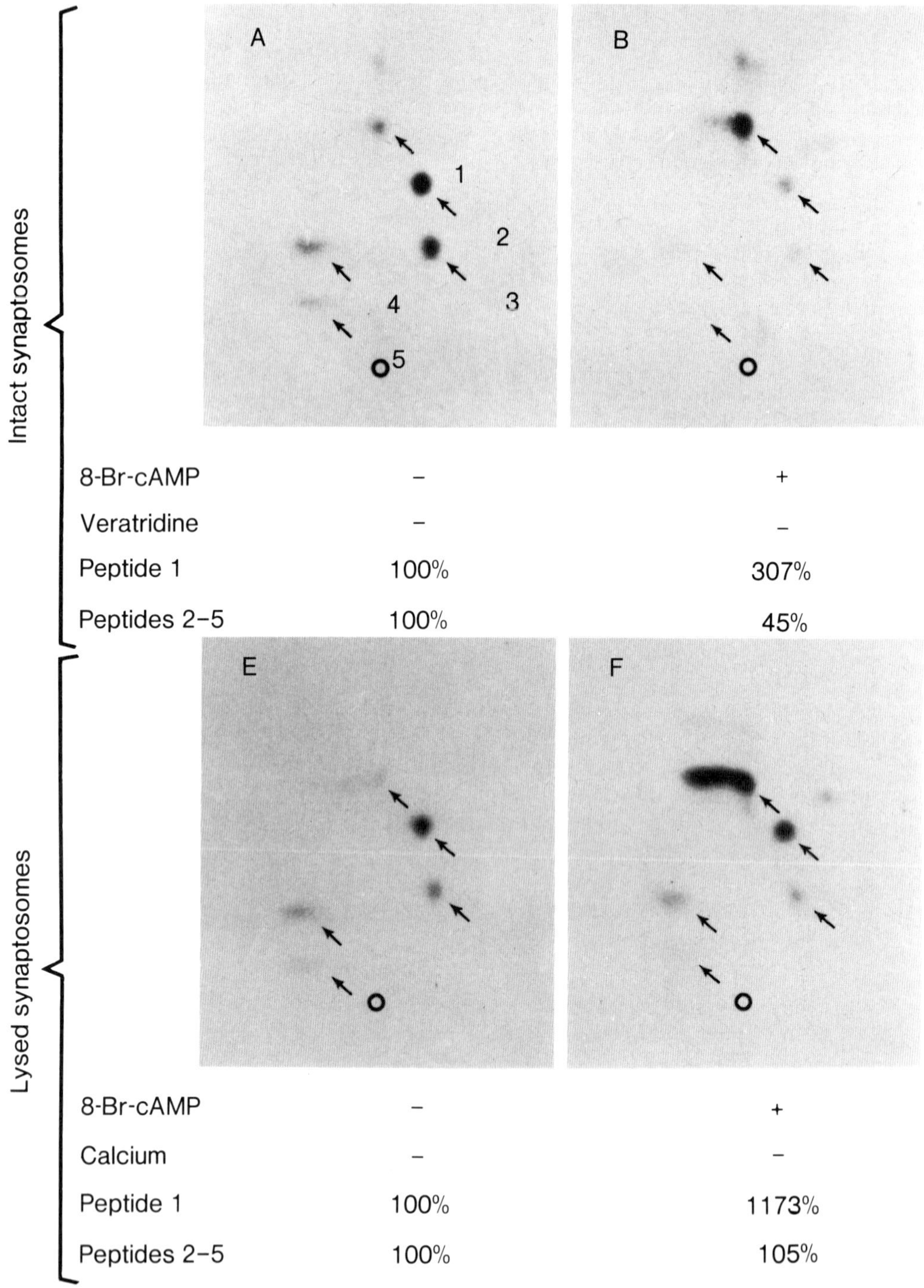

Fig. 7. Autoradiographs showing tryptic fingerprints of purified phosphorylated Protein Ib. Protein I that had been phosphorylated, in intact or lysed synaptosomes, was subjected to non-equilibrium pH gradient electrophoresis on polyacrylamide gels. The Protein I region of the gel was cut out and subjected to SDS-polyacrylamide gel electrophoresis. Proteins Ia and Ib were separately cut out of the gel and subjected to exhaustive digestions by trypsin. Tryptic fingerprints of purified Protein Ia and Protein Ib were qualitatively similar. Phos-

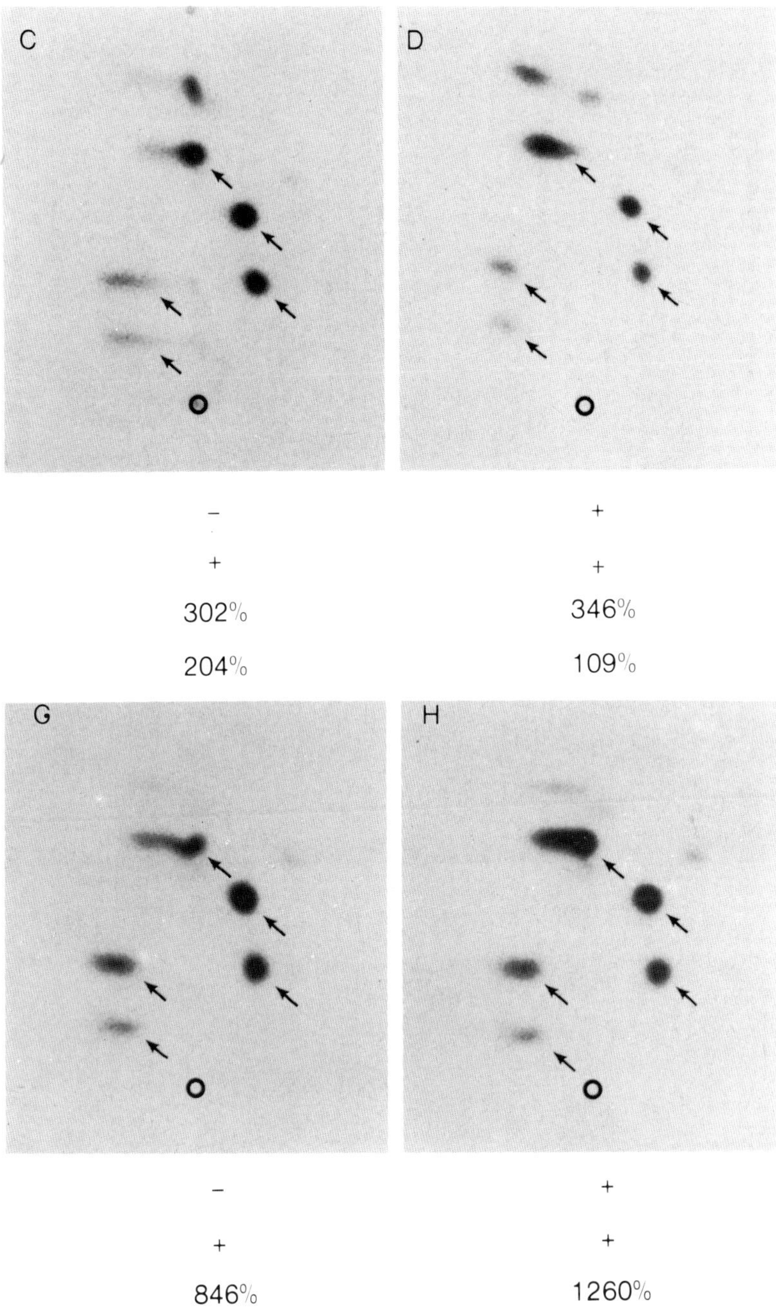

phopeptides derived from Protein Ib were spotted on cellulose plates (○ designates the origin) and were separated in two dimensions, first by electrophoresis in the horizontal dimension (negative pole left, positive pole right) and then by ascending chromatography in the vertical dimension. In each autoradiograph, the five arrows indicate peptides 1–5, as designated in panel *A*. The amount of phosphate incorporation into peptide 1 and into peptides 2 through 5 under various incubation conditions is given for both intact and lysed synaptosomes as percent of control. (From HUTTNER and GREENGARD 1979)

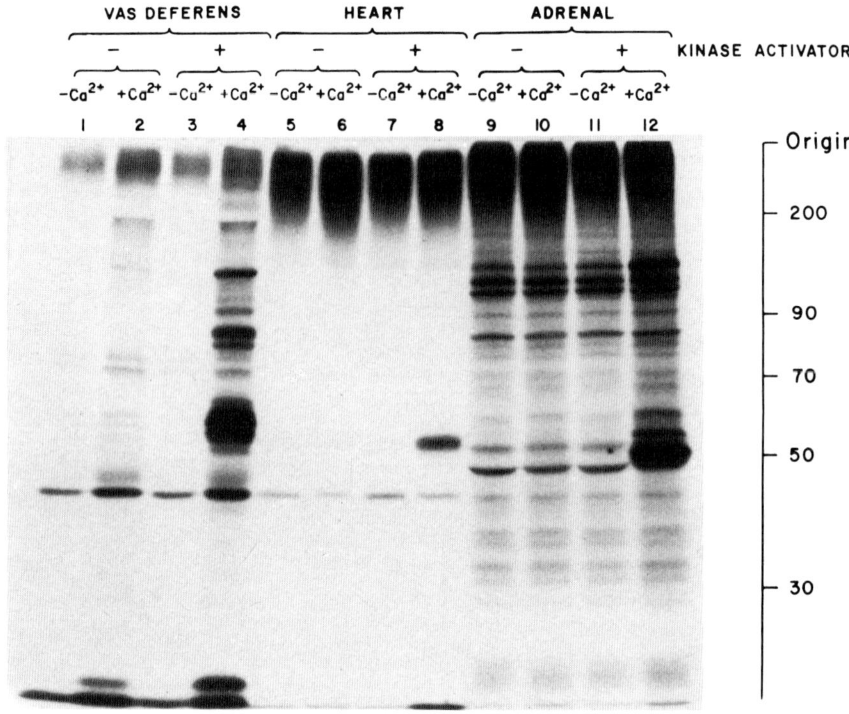
VAS DEFERENS
HEART
ADRENAL
KINASE ACTIVATOR
−Ca2+ +Ca2+ −Ca2+ +Ca2+ −Ca2+ +Ca2+ −Ca2+ +Ca2+ −Ca2+ +Ca2+ −Ca2+ +Ca2+
1 2 3 4 5 6 7 8 9 10 11 12
Origin
200
90
70
50
30
Molecular Weight x 10−3

a

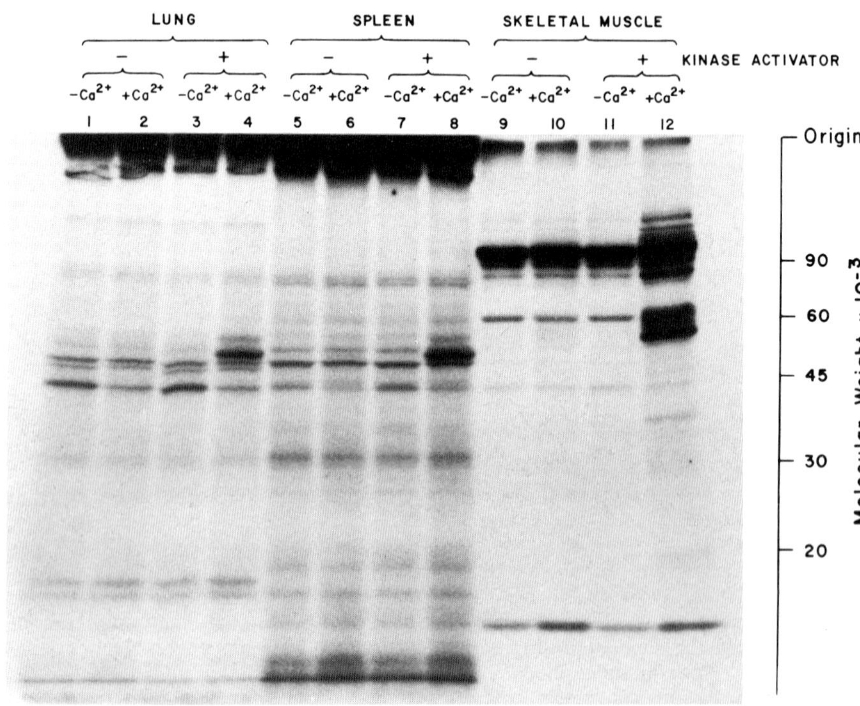
LUNG
SPLEEN
SKELETAL MUSCLE
KINASE ACTIVATOR
−Ca2+ +Ca2+ −Ca2+ +Ca2+ −Ca2+ +Ca2+ −Ca2+ +Ca2+ −Ca2+ +Ca2+ −Ca2+ +Ca2+
1 2 3 4 5 6 7 8 9 10 11 12
Origin
90
60
45
30
20
Molecular Weight x 10−3

b

cium-dependent protein phosphorylation in brain membranes, it was of interest to determine whether the action of calcium on membrane function in other tissues might also be mediated by such a mechanism (SCHULMAN and GREENGARD 1978 b). Analysis of endogenous protein phosphorylation in total membrane fractions from six representative rat tissues (lung, spleen, skeletal muscle, vas deferens, heart, and adrenal) is shown in Fig. 8. A total membrane fraction, washed free of cytosol and calmodulin, was prepared from each of the tissues and then assayed for endogenous calcium-dependent phosphorylation in either the absence or the presence of kinase activator purified from bovine brain. It can be seen that membranes from each of the six tissues contained a calmodulin- and calcium-dependent protein phosphorylation system. For example, in membranes from vas deferens, neither calcium alone nor protein kinase activator alone was effective, whereas addition of both calcium and protein kinase activator stimulated protein phosphorylation (Fig. 8 above, compare lanes 1–4). Each tissue displayed a different pattern of calcium-stimulated protein phosphorylation that was distinct from the cAMP-dependent phosphorylation seen in the same tissue. The conditions used to assay calcium-dependent protein phosphorylation in various tissues in this study were those that had been standardized for brain membranes and were not optimized for any other tissue. It is therefore possible that substrates for the calcium-dependent kinase, in addition to those detected here, exist in those membranes. The analysis does demonstrate, however, a widespread distribution of calmodulin- and calcium-dependent protein phosphorylation in membranes from various tissues. Each tissue differs in the number and molecular weight of the phosphoprotein substrates for this enzyme as well as in the proportion of proteins whose phosphorylation is regulated by calcium. For the vast majority of proteins, calcium-dependent phosphorylation also required the presence of calmodulin. The effect of calcium and calmodulin on the new class of protein kinase, the membrane-bound calmodulin- and calcium-dependent protein kinase, provides a possible mechanism by which calcium may directly regulate diverse biochemical processes.

F. Myosin Light Chain Kinase in Skeletal Muscle, Smooth Muscle, and Non-Muscle Cells

Tension development in mammalian muscle was one of the earliest biological processes shown to be regulated by calcium. At least three distinct types of muscle contractile mechanisms have been identified. They are exemplified by vertebrate skeletal muscle, vertebrate smooth muscle, and molluscan striated muscle (EBASHI et al. 1978). All three types of muscle appear to develop tension by ATP-dependent processes involving cyclic interactions between the thick and thin filaments of

◀ **Fig. 8 a and b.** Autoradiograph illustrating endogenous calmodulin- and calcium-dependent phosphorylation of membrane proteins from various rat tissues. Total membrane fractions, washed free of calmodulin were assayed for endogenous calcium-dependent phosphorylation in either the absence or the presence of 1 μg of calmodulin (kinase activator purified from bovine brain). The tissue studied, and the amounts of membrane protein applied to each lane were: lung, 7.7 μg; spleen, 11.4 μg; skeletal muscle, 5.0 μg; vas deferens, 5.4 μg; and adrenal, 6.8 μg. (From SCHULMAN and GREENGARD 1978 b) muscle myosin and actin, triggered by elevation of intracellular calcium. The basic difference among these three systems resides in the nature of the regulatory components that control their activity. Regulation by calcium could in principle occur via an effect on either the thick or thin filament. In skeletal muscle, the best studied of the three systems, contraction appears to result largely from a thin filament-regulated system (MURRAY and WEBER 1973). In the resting state, troponin, located on the thin filaments, inhibits actin-myosin interactions and the Mg^{2+}-ATPase associated with actomyosin. In the presence of calcium, or in the absence of the regulatory components, this inhibition is relieved and the actomyosin ATPase is maximally active.

Smooth muscle actomyosin ATPase is relatively inactive either in the absence of calcium or in the absence of the regulatory components. Two mechanisms have been proposed to explain activation of actomyosin ATPase in smooth muscle. The most widely accepted mechanism implicates a myosin-linked regulatory mechanism. It has been suggested that regulation of smooth muscle ATPase is mediated by phosphorylation-dephosphorylation of one of the light chains of myosin, the P-light chain, by myosin light-chain kinase and myosin light chain phosphatase. It has also been proposed, however, that an 80,000 Daltons protein and tropomyosin are sufficient to reconstitute a calcium-sensitive actomyosin (HIRATA et al. 1977). These same investigators found that the actomyosin ATPase activity of vertebrate smooth muscle does not correlate with the extent of phosphorylation of the P-light chain (MIKAWA et al. 1977).

Molluscan striated muscle does not appear to require troponin for tension development (SZENT-GYORGYI et al. 1973) although it contains nonstoichiometric amount of troponin. Control of ATPase activity of actomyosin from molluscan muscle is dependent on the presence of one of the light chains of myosin (KENDRICK-JONES et al. 1970). This light chain is a calcium-binding protein which functions by inhibiting ATPase activity and muscle tension development in the absence of calcium. Phosphorylation does not appear to be involved in the activation of ATPase by calcium in striated molluscan muscle.

I. Myosin Light Chain Kinase in Skeletal Muscle

Analysis of purified skeletal muscle myosin indicates that it consists of two heavy chains, each with a molecular weight of 200,000, and four light chains of which there are three types. Each myosin molecule contains two 180,000 Dalton light chains and one each of about 15,000 and 22,000 Dalton light chains (PERRIE et al. 1972). The 18,000 Dalton light chains were found to migrate as two bands on polyacrylamide gels run in the presence of 8 *M* urea (PERRIE and PERRY 1970). The molar ratio of these bands was dependent on the concentration of muscle mince used in the initial extract. It was subsequently shown that these two polypeptides had identical amino acid compositions but that one of them contained 1 mol of phosphate per mol polypeptide (PERRIE et al. 1973). This same light chain has been referred to in the literature as MI_3 (PERRIE and PERRY 1970), g_2 light chain (YAGI et al. 1975), DTNB light chain (WEEDS 1969; GAZITH et al. 1970), P-light chain (FREARSON and PERRY 1975), or referred to by its molecular weight (e.g., the 18,000 Dalton light chain). The term P-light chain will be used in this review since the phosphorylation of light chains from different sources will be described. It was shown that conversion of the dephosphorylated P-light chain to the phosphorylated P-light chain was catalyzed by an endogenous protein kinase present in sarcoplasm (PERRIE et al. 1973). The amino acid sequence neighboring the phosphoserine in myosin light chain was different from the site of phosphorylation in phosphorylase *b* that is regulated by phosphorylase kinase and the primary sites of phosphorylation that are regulated by the cAMP-dependent protein kinase suggesting that this is a distinct enzyme (NIMMO and COHEN 1977). An enzyme with the appropriate substrate specificity was partially purified from rabbit white skeletal muscle and designated as myosin light-chain kinase (PIRES et al. 1974).

Myosin light chain kinase from rabbit white muscle has been purified to apparent homogeneity (PIRES and PERRY 1977; YAZAWA and YAGI 1978; WANG and WAISMAN 1979). The enzyme is calcium-dependent and highly specific for the P-light chain. No significant incorporation of phosphate into the 15,000 or 22,000 Dalton light chains from skeletal muscle or the EDTA-light chain of the molluscan adductor muscle was detectable. The P-light chains of cardiac and red skeletal muscle as well as smooth muscle were effective substrates (PIRES and PERRY 1977). Phosphorylase kinase, glycogen phosphorylase, glycogen synthase, troponin-I, casein, and protamine were not good substrates (PIRES and PERRY 1977; WANG and WAISMAN 1979). A variety of histones were phosphorylated by the myosin light chain kinase, but their K_m values were about 20-fold higher than the K_m for the P-light chain (WANG and WAISMAN 1979). The enzyme showed an absolute dependence on the presence of calcium, although there are differences in published K_a's for calcium. Half-maximal activation of the kinase was reported to occur at 0.5 μM calcium (PIRES and PERRY 1977) and 10 μM calcium (YAZAWA and YAGI 1978). There are also differences reported for the K_m for the P-light chain with values of 0.1–0.2 mM (PIRES and PERRY 1977) and of 0.024 mM (YAZAWA and YAGI 1978) reported. The K_m for ATP is in the range of 0.2–0.4 mM. This is considerably higher than the K_m obtained with the cAMP-dependent protein kinases (NIMMO and COHEN 1977).

Several molecular weights have been reported for the catalytic subunit of myosin light chain kinase from skeletal muscle but there are indications that these differences may be due to proteolysis. Molecular weights have been reported to be 77,000 (PIRES and PERRY 1977), 80,000 (YAZAWA and YAGI 1978) or 95,000 Daltons (WANG and WAISMAN 1979). Subfragments of myosin light chain kinase possessing enzymatic activity can be obtained upon storage of the larger enzymes, indicating that perhaps even the myosin light chain kinase with the highest reported molecular weight may already represent a product of proteolysis occurring during purification (PIRES and PERRY 1977).

Myosin light chain kinase activity has recently been shown to require the presence of two proteins (YAZAWA and YAGI 1977; BARYLKO et al. 1978). The smallest protein was found to be a calcium-binding protein and was shown to be identical to calmodulin (YAGI et al. 1978; YAZAWA and YAGI 1978; BARYLKO et al. 1978). The calcium-binding protein and authentic calmodulin from bovine brain had identical cross-reactivity in stimulating the calmodulin-depleted phosphodiesterase and the myosin light chain kinase. Both proteins demonstrated equal sensitivity to activation by calcium, identical UV absorption difference spectra with or without calcium, and identical mobilities on SDS-polyacrylamide gel electrophoresis (YAGI et al. 1978; YAZAWA et al. 1978). Thus, as with the membrane-bound protein phosphorylation system (Sect. D), myosin light chain kinase appears to require both calcium and calmodulin for activity.

The relationship between the calcium-dependent myosin light chain kinase discovered by PERRY and his associates (PIRES and PERRY 1977) and the calmodulin- and calcium-dependent myosin light chain kinase (YAGI et al. 1978) is not certain. The major difference between the two enzymes is that the enzyme isolated by PERRY and his associates (PIRES and PERRY 1977) does not require calmodulin. However, because of the similarity in the molecular weight of calmodulin and the P-light chain, and because of the presence of calmodulin in muscle, it is possible that some calmodulin was present as a contaminant of the light chain preparation used by PERRY and his associates. The presence of calmodulin as a contaminant of the substrate preparation would make it appear as if the larger catalytic subunit was sufficient for myosin light chain kinase activity. It is therefore likely that the two groups are describing the same enzyme but this suggestion must await re-examination of the calmodulin-independent preparation.

Myosin light chain phosphatase has been purified from rabbit white skeletal muscle (MORGAN et al. 1976). The enzyme was composed of a single component with an apparent molecular weight of 70,000. It does not require calcium for activity and is highly specific for the P-light chain. The presence of both myosin light chain kinase and phosphatase activity in muscle is consistent with the possibility that the P-light chains undergo cycles of phosphorylation and dephosphorylation in vivo (see Sect. F.IV).

II. Myosin Light Chain Kinase in Smooth Muscle

1. Myosin Light Chain Kinase

A regulatory role for protein phosphorylation is more widely accepted for mammalian smooth muscle than for striated skeletal muscle (see Sect. F.IV). Myosin light chain kinase activity has been reported in smooth muscle obtained from several sources including chicken gizzard (AKSOY et al. 1976; IKEBE et al. 1977; SOBIESZEK 1977), rabbit uterus (FREARSON et al. 1976), guinea pig vas deferens (CHACKO et al. 1977), and turkey gizzard (SOBIESZEK 1977; ADELSTEIN et al. 1978). In each case it is the appropriate light chain which is phosphorylated. Smooth muscle tissue also contains P-light chain phosphatase activity, in accordance with a possible regulatory role for this phosphorylation system (FREARSON et al. 1976; CHACKO et al. 1977; IKEBE et al. 1977).

Myosin light chain kinase activity requires the presence of two components (DABROWSKA et al. 1977). The enzyme has been purified to apparent homogeneity from chicken gizzard (DABROWSKA et al. 1978) and turkey gizzard (ADELSTEIN et al. 1978). An extensive biochemical characterization of the enzyme has not been reported although the amino acid composition of the catalytic subunit has been published (DABROWSKA et al. 1977). The molecular weights of the two components have been reported to be 105,000 and 17,000 for the enzyme from chicken gizzard (DABROWSKA et al. 1977) and 125,000 for the catalytic subunit of the enzyme from

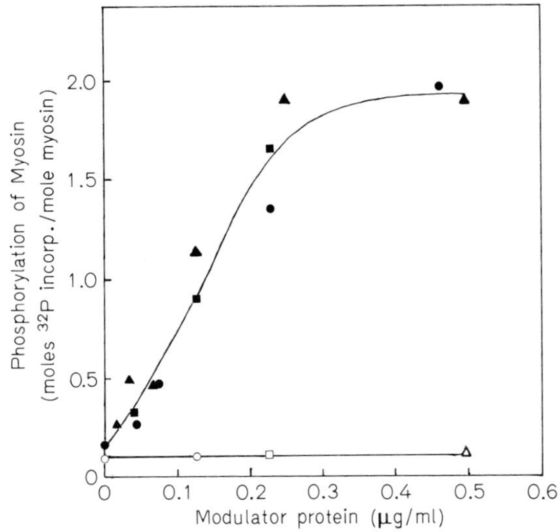

Fig. 9. Effect of the 17,000 Dalton myosin light chain kinase subunit from chicken gizzard and calmodulin from adrenal medulla and brain on the phosphorylation of gizzard myosin at a constant level of the 105,000 Dalton subunit. Assay conditions: 20 m*M* Tris HCl (pH 7.6) containing 10 m*M* $MgCl_2$, 50 m*M* KCl, 2.5 m*M* ATP, 0.46 mg gizzard myosin, 0.26 mg skeletal muscle actin, 0.075 mg skeletal muscle troponin, and 6.4 μg gizzard kinase 105,000 Dalton subunit in a reaction volume of 1 ml. The amount of the 17,000 Dalton subunit (●, ○) and calmodulin from adrenal medulla (▲, △) and from brain (■, □) was varied as shown. Open symbols indicate the inclusion of 1 m*M* EGTA. (Adapted from DABROWSKA et al. 1978)

turkey gizzard (ADELSTEIN et al. 1978). Based on the earlier work with myosin light chain kinase from skeletal muscle, proteolysis of the smooth muscle enzyme during purification should be carefully examined as a potential source of differences found among the various preparations.

The 17,000 Dalton component required for myosin light chain kinase activity has been shown to be calmodulin (DABROWSKA et al. 1978; ADELSTEIN et al. 1978). As shown in Fig. 9, at a constant level of the 105,000 Dalton subunit $[^{32}P]O_4$ incorporation into myosin was stimulated equally by the 17,000 Dalton subunit and by calmodulin isolated from either brain or adrenal medulla. Calmodulin from bovine brain and the low molecular weight activator in chicken gizzard smooth muscle were also equally effective in activating actomyosin ATPase and calmodulin-depleted phosphodiesterase (DABROWSKA et al. 1978). The two proteins had similar ultraviolet absorption spectra, similar mobilities on urea-polyacrylamide gel electrophoresis, similar changes in tyrosine fluorescence as a function of Ca^{2+} concentration, and similar amino acid compositions (DABROWSKA et al. 1978). Thus, myosin light chain kinase from both smooth and striated muscle is activated by calmodulin.

Myosin light chain kinase activity was dependent on physiological concentrations of free Ca^{2+}, with K_a for calcium being approximately 3 μ*M* (AKSOY et al. 1976). The stoichiometry of phosphorylation is similar to that of myosin light chain kinase from skeletal muscle, with 1 mol phosphate incorporated per mol

P-light chain (Fig. 9) (DABROWSKA et al. 1978). The substrate specificity of myosin light chain kinase from smooth muscle has not been carefully examined although preliminary indication is that only the P-light chain from either smooth or skeletal muscle is phosphorylated (FREARSON et al. 1976). The amino acid sequence near the phosphorylation site has not been reported.

2. Relationship Between cAMP and Calcium

Preliminary experiments indicate that cAMP may regulate the activity of myosin light chain kinase. Myosin light chain kinase from turkey gizzard is a substrate for the cAMP-dependent protein kinase from bovine heart (ADELSTEIN et al. 1978). Up to 1 mol of phosphate is incorporated per mol of myosin light chain kinase catalytic subunit. The rate of phosphorylation of myosin light chain kinase is approximately one-third the rate of histone II A phosphorylation. Phosphorylation occurred at a specific site on the enzyme and led to a 50% reduction in the rate at which the enzyme phosphorylates P-light chain from smooth muscle. Whether a similar effect occurs with intact myosin as substrate was not determined.

Phosphorylation of myosin light chain kinase by the cAMP-dependent protein kinase may represent another locus for the interaction of cAMP and calcium in regulation of biological process. Indirect effects of cAMP on muscle function have been characterized (BERRIDGE 1975). These include the general effects of cAMP on the energy state of cells as well as the effect of cAMP on calcium uptake by sarcoplasmic reticulum. Enhanced sequestration of calcium, stimulated by cAMP, has been proposed to explain the inhibition of tension development by cAMP. The findings of ADELSTEIN and associates (ADELSTEIN et al. 1978) suggest an additional mechanism to explain this observation. Thus, lowered cAMP levels (which may result from calcium activation of phosphodiesterase) may facilitate activation of myosin light chain kinase. Conversely, increased cAMP levels would attenuate the effect of calcium on myosin light chain kinase activity and hence of actomyosin ATPase and tension development (see below).

III. Myosin Light Chain Kinase in Non-Muscle Cells

The components of the contractile machinery, myosin, and actin, are not restricted to muscle cells. Interactions between these components may occur in non-muscle cells were they may be involved in diverse functions such as cytokinesis, mitosis, motility, secretion, and endocytosis (ADELSTEIN 1978). It appears that regulation of P-light chain phosphorylation occurs in non-muscle cells as well as in muscle cells. Myosin light chain kinase has been detected in human blood platelets (ADELSTEIN and CONTI 1975; DANIEL and ADELSTEIN 1976; DABROWSKA and HARTSHORNE 1978; HATHAWAY and ADELSTEIN 1979), on rabbit alveolar macrophages (TROTTER and ADELSTEIN 1979), bovine brain (DABROWSKA and HARTSHORNE 1978), and in baby hamster kidney (BHK-21) cells (YERNA et al. 1979). The enzyme from human blood platelets (HATHAWAY and ADELSTEIN 1979) and from BHK-21 cells (YERNA et al. 1979) has been purified to apparent homogeneity. There is general agreement that the catalytic subunit has a molecular weight of 105,000 Daltons.

Early studies (ADELSTEIN and CONTI 1975; DANIEL and ADELSTEIN 1976) indicated that the platelet enzyme was not stimulated by calcium. The catalytic subunit was found to have a molecular weight of 83,000 (DANIEL and ADELSTEIN 1976). It would now appear that this was an artifactual consequence of proteolysis (HATHAWAY and ADELSTEIN 1979). Myosin light chain kinase from smooth muscle can be converted from a calcium-sensitive to a calcium-insensitive form by incubation with a mammary tissue actomyosin preparation (BREMEL and SHAW 1978). The presence of proteases which can convert myosin light chain kinase to a calcium-insensitive form should be examined whenever a calcium-insensitive myosin light chain kinase is obtained.

There is also general agreement that myosin light chain kinase from nonmuscle cells requires calmodulin for activity (DABROWSKA and HARTSHORNE 1978; HATHAWAY and ADELSTEIN 1979; YERNA et al. 1979). The possible requirement for calmodulin was not investigated with the macrophage myosin light chain kinase (TROTTER and ADELSTEIN 1979). The myosin light chain phosphorylation activity in these non-muscle systems does not appear to be a calmodulin- and calcium-dependent protein kinase with broad substrate specificity. The enzyme was specific for the P-light chain. Other known substrates for protein kinases, such as α-casein, histones, phosphorylase *b*, protamine and phosvitin were not efficient substrates (HATHAWAY and ADELSTEIN 1979; YERNA et al. 1979). Preliminary kinetic studies indicate that the enzyme from non-muscle cells is similar to the enzyme from smooth muscle, with a K_m for ATP of 121 μM and K_m for light chain of 18 μM (HATHAWAY and ADELSTEIN 1979). Based on the requirement for calmodulin it is likely that this enzyme is also sensitive to physiological levels of free calcium.

IV. Mechanism of Activation of Myosin Light Chain Kinase

Myosin light chain kinases from the various cell types have similar physical and biochemical characteristics. It has not been determined whether differences among them reflect species differences or whether several types of myosin light chain kinases exist within any one species.

A common feature of most of the myosin light chain kinases from skeletal muscle, smooth muscle and non-muscle cells in the requirement for calcium and calmodulin. Calmodulin functions as an activator of the enzyme in the presence but not in the absence of calcium. Calcium is known to cause conformational changes in the structure of calmodulin (YAZAWA et al. 1978; WOLFF and BROSTROM 1979). The dose-response for activation of skeletal muscle myosin light chain kinase by calcium correlates well with the dose response for changes in the ultraviolet absorption spectrum of calmodulin elicited by calcium (YAGI et al. 1978). Myosin light chain kinase from skeletal muscle, smooth muscle, and non-muscle cells forms a reversible, calcium-dependent complex with affinity columns containing covalently bound calmodulin (YAZAWA and YAGI 1978; ADELSTEIN et al. 1978; DABROWSKA and HARTHSHORNE 1978). Thus, it would appear that increased intracellular calcium leads to the formation of a calmodulin-calcium complex which then binds to and activates the myosin light chain kinase. Such a mechanism is typical of most calmodulin mediated reactions (WANG and WAISMAN 1979).

V. Physiological Function of P-Light Chain Phosphorylation

1. Phosphorylation in Smooth Muscle and Non-Muscle Cells

There is a large body of evidence supporting the idea that tension development in smooth muscle and actomyosin ATPase in non-muscle cells is regulated by the phosphorylation and dephosphorylation of the P-light chain (ADELSTEIN et al. 1977). Calcium sensitivity in these tissues and cells therefore resides on myosin rather than on actin. The first demonstration of an effect of phosphorylation on muscle function was the experiment showing an increase in actin-activated ATPase activity resulting from phosphorylation of platelet myosin (ADELSTEIN and CONTI 1975). Dephosphorylation of the phosphorylated P-light chain with bacterial alkaline phosphatase reduced the ATPase activity. Thus, both unphosphorylated myosin and dephosphorylated myosin had low ATPase activities. A general cause and effect relationship between phosphorylation and ATPase activity has since been demonstrated in many systems, including proliferative myoblasts (SCORDILIS and ADELSTEIN 1977), chicken gizzard (AKSOY et al. 1976; SOBIESZEK 1977; IKEBE et al. 1977; GORECKA et al. 1976), guinea pig vas deferens (CHACKO et al. 1977) rabbit alveolar macrophages (TROTTER and ADELSTEIN 1979) and baby hamster kidney cells (YERNA et al. 1979). The response of ATPase activity and phosphate incorporation into gizzard P-light chains to increasing concentrations of calcium correlated well (AKSOY et al. 1976). In addition, desensitization of the ATPase activity by removal of a protein fraction from actomyosin also reduced the extent of phosphorylation of the P-light chains. A direct relationship between actin-activated ATPase and myosin phosphorylation was demonstrated using both actomyosin and myofibrils from chicken gizzard (SOBIESZEK 1977). A similar relationship between phosphorylation and actin-activated ATPase was demonstrated using a homogeneous preparation of myosin light chain kinase (CHACKO et al. 1977; DABROWSKA et al. 1977).

Two additional experiments demonstrated a relationship between phosphorylation of P-light chains and myosin function. Actomyosin threads were prepared from purified platelet myosin and purified muscle actin. A tensiometer was used to measure tension development (LEBOWITZ and COOKE 1978). A linear relationship between the extent of phosphorylation of myosin and isometric tension was obtained. Another study demonstrated that both collagen- and A 23187-stimulated secretion of serotonin from human blood platelets was accompanied by increases in P-light chain phosphorylation (HASLAM and LYNHAM 1977).

The only area of disagreement concerning the role of phosphorylation in smooth muscle and non-muscle cells is whether this is the only calcium-dependent mode of regulation. Whereas one study on guinea pig vas deferens indicated that maximal ATPase activity required both P-light chains and calcium (CHACKO et al. 1977) other studies on chicken gizzard (SOBIESZEK 1977; IKEBE 1977; SHERRY et al. 1978), pig stomach (SMALL and SOBIESZEK 1977), and skinned rabbit ileum fibers (CASSIDY et al. 1978) indicated that the P-light chains, in the absence of calcium, are sufficient for maximal ATPase activity. Experiments aimed at resolving this question in smooth muscle from chicken gizzard all indicate that phosphorylation is the only calcium-dependent process regulating ATPase activity (SHERRY et al. 1978). In a system containing both kinase and phosphatase, the rate of removal of phosphate from a maximally phosphorylated myosin was directly correlated with the rate of inhibition of actin-activated ATPase following removal of calcium. When ATP was replaced by adenosine 5′-O-(3-thiotriphosphate), the resultant thiophosphorylated P-light chains could not be readily dephosphorylated, trapping myosin in the thiophosphorylated state. As the amount of thiophosphorylation increased the sensitivity of the system to removal of calcium decreased (SHERRY et al. 1978).

These experiments indicate that dephosphorylation of myosin is a prerequisite for relaxation of muscle tension and that removal of calcium from its binding sites on myosin cannot counteract the presence of phosphorylated P-light chains. It therefore appears that direct calcium binding to the P-light chain, which has been shown to be involved in regulation of molluscan actomyosin ATPase (KENDRICK-JONES 1974), is not an important feature of actomyosin activity in gizzard smooth muscle.

2. Phosphorylation in Skeletal Muscle

The role of myosin phosphorylation in skeletal muscle is not well understood. It is widely accepted that the prominent feature of tension development in mammalian skeletal muscle is the release of an inhibitory state occurring upon the binding of calcium to troponin C (EBASHI et al. 1978). Unlike the situation in smooth muscle and non-muscle cells, phosphorylation of the P-light chain has no effect on actomyosin ATPase activity (MORGAN et al. 1976). Because the binding of calcium to troponin occurs at lower concentrations than its activation of skeletal muscle myosin light chain kinase, it has been suggested that in vivo myosin light chain kinase might be activated after maximal activity of actomyosin ATPase is achieved (YAZAWA and YAGI 1978).

The P-light chains of skeletal muscle do undergo changes in their state of phosphorylation in vivo. Analysis of the P-light chains during a single tetanus of frog sartorius muscle revealed an increased of 85%–90% in their phosphate content (BARANY and BARANY 1977). The P-light chain was the only protein to show increased incorporation of phosphate. This increase was significant, corresponding to about 0.4 mol phosphate per mol light chain. Tetanic stimulation and caffeine-induced contracture had similar effects on phosphate incorporation.

Tetanic electrical stimulation of rabbit white gracilis muscle increased phosphorylation of P-light chain from 0.5 mol phosphate per mol light chain at rest to 0.9 mol phosphate per mol light chain (STULL and HIGH 1977). In the resting muscle, troponin contained 1 mol phosphate and this amount was not changed by tetanic stimulation. This suggests that some inotropic responses correlate better with increased phosphorylation of myosin than of troponin. Differences in the state of phosphorylation of the P-light chains at rest between fast twitch muscle and cardiac muscle may reflect differences in the kinase to phosphatase ratio in these types of muscle (FREARSON et al. 1976).

Isometric tension development in rat extensor digitorum longus muscle led to an increase in the extent of phosphorylation of myosin from a pre-tetanic value of 0.1 mol phosphate per mol P-light chain to a value of 0.7 mol phosphate 7 s later (MANNING and STULL 1979). During a 1 s tetanic contracture the extent of phosphorylation of the P-light chains did not increase. Thus, phosphorylation of the P-light chains is not an obligatory step for contraction of skeletal muscle, consistent with the finding that the actomyosin ATPase activity is not dependent on the extent of phosphorylation of the P-light chains. Following tetanic stimulation the P-light chains remained in the phosphorylated state with phosphate content decreasing slowly. In contrast, the state of phosphorylation of another phosphoprotein, phosphorylase, rapidly declined to control levels following tetanic stimulation. The expected increase in peak twitch tension which normally follows tetanic stimulation was also observed and was found to decline exponentially at a rate similar to the dephosphorylation of light chains. Thus phosphorylation of the P-light chains does not correlate with the contractile state of the muscle, either during initiation of contraction or during the post-tetanic decrease in contraction, but does correlate with the potentiation of peak twitch tension seen in post-tetanic muscle.

G. Phosphorylase Kinase

Phosphorylase kinase plays a key role in control of glycogenolysis in mammalian muscle. The enzyme is regulated by both calcium and cAMP and thus mediates both neuronal and endocrine effects on glycogenolysis. At pH 6.8 the enzyme can be activated 50- to 100-fold as a consequence of phosphorylation by a cAMP-dependent protein kinase (COHEN 1973). Phosphorylase kinase also undergoes a slow self-phosphorylation or autophosphorylation in the presence of calcium and ATP (WALSH et al. 1971). The enzyme, activated by either mechanism as well as the unphosphorylated form, has an absolute requirement for calcium (BROSTROM et al. 1971; YEAMAN and COHEN 1975).

I. Regulation of Phosphorylase Kinase by Calmodulin

Initial studies on purified phosphorylase kinase demonstrated that the enzyme was composed of three types of subunits designated α, β, and γ with molecular weights of 145,000, 128,000, and 45,000 respectively (HAYAKAWA et al. 1973; COHEN 1973). Two isozymes of phosphorylase kinase have been identified and the corresponding kinases are designated as $(\alpha\beta\gamma)_4$ and $(\alpha'\beta\gamma)_4$ (COHEN 1974; JENNISSEN and HEILMEYER 1974). The α' subunit is slightly smaller than the α subunit (COHEN 1973). These isozymes are differentially distributed in fast and slow-twitch muscles. Phosphorylase kinase in rabbit psoas major muscle, which has predominantly fast-twitch fibers, is composed of 95%–98% $(\alpha\beta\gamma)_4$ isozyme (BURCHELL et al. 1976) with a molecular weight of 1,300,000 (HAYAKAWA et al. 1973).

Recently, a fourth subunit, designated δ, with a molecular weight of 17,000, was shown to be an integral component of phosphorylase kinase (COHEN et al. 1978). This subunit was present in the earlier preparations of the enzyme. However, because it is much smaller than the other subunits and it is stained poorly by Coomassie blue, it migrated as a faint band with the bromphenol blue dye front on SDS-polyacrylamide gels used to analyze the enzyme and was therefore not detected.

The δ-subunit was shown to be similar if not identical to calmodulin from brain (COHEN et al. 1978). The physical characteristics of the two proteins, including

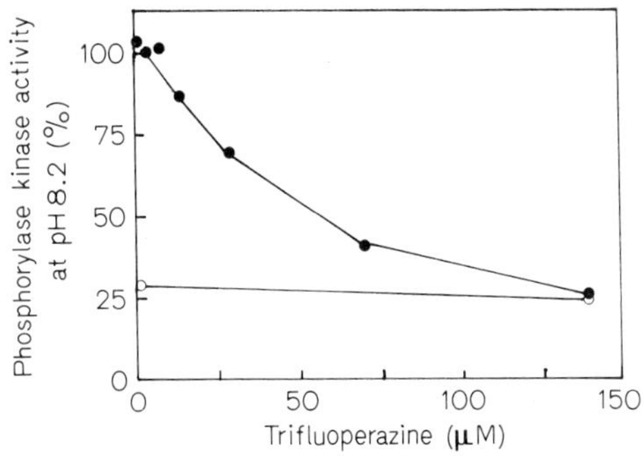

Fig. 10. Effect of added calmodulin and trifluoperazine on the activity of phoshorylase kinase at pH 8.2. Nonactivated phosphorylase kinase (0.14 μg/ml) was assayed in the presence of calcium, and in the absence (○) and presence (●) of calmodulin (6 μg/ml). In the absence of calcium there was essentially no activity either with or without saturating concentrations of calmodulin. 100% activity correspond to the activity in the presence of saturating concentrations of calmodulin. Under these assay conditions 5% of the phosphorylase *b* (0.15 mmol) was converted to phosphorylase *a* per min. (Adapted from SHENOLIKAR et al. 1979)

amino acid composition, the presence of trimethyllysine, ultraviolet absorption spectrum, electrophoretic mobility on polyacrylamide gels, and heat stability were indistinguishable. Functional characteristics, such as the ability to activate calmodulin-depleted phosphodiesterase and myosin light chain kinase as well as the ability to form a complex with troponin I and troponin T were also indistinguishable. About 0.94 mol calmodulin was bound per mol α subunit suggesting that $(\alpha\beta\gamma\delta)_4$ is the appropriate stoichiometry and composition of the complex (COHEN et al. 1978).

Unlike all other calmodulin-stimulated enzymes or biochemical processes, the calmodulin molecule in phosphorylase kinase does not dissociate in the absence of calcium (COHEN et al. 1978). Therefore, the molecular mechanism for activating phosphorylase kinase is different from the mechanism proposed for other calmodulin-dependent processes. Calmodulin could not be dissociated from the complex even in the presence of 8 *M* urea, provided that calcium was present (SHENOLIKAR et al. 1979).

Calmodulin appears to have a dual role in regulating phosphorylase kinase. Aside from calmodulin present as the δ subunit, addition of calmodulin to phosphorylase kinase further stimulated the enzyme 2- to 7-fold in the presence of calcium (SHENOLIKAR et al. 1979). Half-maximal activation was obtained at a molar ratio of camodulin to phosphorylase kinase of 10–1. The activity either with or without added calmodulin was absolutely dependent on calcium. Activation of phosphorylase kinase by the second calmodulin appears to be more typical of calmodulin action. Several experiments demonstrated that a dissociable complex could be formed between phosphorylase kinase $(\alpha\beta\gamma\delta)_4$, and additional calmodulin. This interaction was distinct from the interaction of the integral δ subunit with the other three subunits (SHENOLIKAR et al. 1979). For example, troponin-I, which is known to form a complex with calmodulin (AMPHLETT et al. 1976), had a differential effect on the integral δ subunit and on added calmodulin. Thus, troponin-I inhibited the calmodulin-stimulated activity while not affecting the enzyme without added calmodulin. A calmodulin binding protein, known to inhibit the activation of phosphodiesterase by calmodulin (KLEE and KRINKS 1978), was shown to block activation of phosphorylase kinase by the second molecule of calmodulin while not affecting the kinase in the absence of calmodulin (COHEN et al. 1979). As shown is Fig. 10, trifluoperazine, a psychotropic drug which has been shown to inhibit calmodulin-stimulated phosphodiesterase (LEVIN and WEISS 1976) by forming a complex with calmodulin (LEVIN and WEISS 1976), had only a small effect on phosphorylase kinase activity in the absence of added calmodulin. However, the 2- to 7-fold stimulation obtained by addition of calmodulin was completely abolished by trifluoperazine (Fig. 10). Finally, calmodulin-linked Sepharose has been shown to adsorb phosphorylase kinase in the presence of calcium suggesting that the enzyme forms a complex with a second molecule of calmodulin (COHEN et al. 1979).

II. Activation of Phosphorylase Kinase by Troponin C

Troponin C and calmodulin, although distinct proteins, are similar in both structure and function. The two proteins have a large amount of amino acid sequence homology (WATTERSON et al. 1976). Calmodulin-depleted phosphodiesterase was activated by troponin C to the same extent as that obtained with calmodulin, although calmodulin was 600-fold more effective (DEDMAN et al. 1977). Troponin C also activated phosphorylase kinase to the same extent as calmodulin, although it was 200-fold less effective than calmodulin (COHEN et al. 1979). The troponin complex, consisting of troponin-T, troponin-I, and troponin C, can also activate phosphorylase kinase, but does so at 1% the efficiency of calmodulin. Troponin C and the second calmodulin molecule are probably interacting on the same site since their effects were not additive (COHEN et al. 1979).

The ability of calmodulin and troponin to bind several different proteins may provide a mechanism for the formation of multienzyme complexes in vivo. For example, the δ subunit of phosphorylase kinase, although tightly bound to the enzyme, can activate myosin light chain kinase (COHEN et al. 1978). Presumably, under appropriate conditions, in the presence of calcium, a complex of myosin light chain kinase and phosphorylase kinase can be formed. This suggests that phosphorylase kinase may complex with other enzymes via a common activator, calmodulin. Likewise, the ability of the troponin complex to substitute for calmodulin in the activation of phosphorylase kinase suggests that, in the presence of calcium, phosphorylase kinase may form a complex with troponin and, thereby, with the thin filaments. In support of this possibility, it has been shown that, in vitro, phosphorylase kinase as well as phosphorylase phosphatase, glycogen phosphorylase and glycogen synthase, but not the bulk of cellular protein, associate with myofibrils in a calcium-dependent manner (COHEN et al. 1979). Because of the high concentration of phosphorylase kinase and the relatively low concentration of calmodulin in muscle, about 35% of total calmodulin is present as the δ subunit. This leaves only 65% of calmodulin available for activation of the kinase, assuming calmodulin does not bind to any other enzymes. Even with this unlikely assumption troponin would be present in 25-fold excess over calmodulin and would be present at sufficient concentration to activate phosphorylase kinase (COHEN et al. 1979).

III. Regulation of Phosphorylase Kinase in Vitro

A full analysis of the regulation of phosphorylase kinase is beyond the scope of this review and the reader should refer to recent reviews on this subject (NIMMO and COHEN 1977; COHEN 1978; COHEN 1980). Possible interactions between cAMP and calcium are described below.

1. Regulation by Calcium

Both the phosphorylated and unphosphorylated forms of phosporylase kinase require calcium (BROSTROM et al. 1971; YEAMAN and COHEN 1975). It has not been determined which subunit(s) is(are) responsible for calcium-sensitivity. However, based on the finding that the δ subunit is calmodulin, it is likely that calcium must interact with and regulate at least this subunit. Binding studies indicate that both the number of sites and the affinity of calcium for sites on phosphorylase kinase are dependent on the conditions used in the determination. In general, high affinity binding sites have activation constants ranging from 0.1 μM to 1.0 μM (BROSTROM et al. 1971; KILIMANN and HEILMEYER 1977). In a buffer of high ionic strength (e.g., 180 mM NH_4Cl) a single set of 8 binding sites with a K_a of 5.5×10^{-7} M was found per mol of non-activated phosphorylase kinase. In the presence of Mg^{2+} the apparent affinity of calcium for the enzyme was reduced to 3×10^{-6} M. At low ionic strength (1 mM sodium glycerophosphate, pH 6.8), 16 high affinity sites were detected; 12 with a K_a of 1.5×10^{-8} M and 4 with a K_a of 6×10^{-7} M. Since there are 4 mol of calmodulin per mol enzyme, each with 4 sites, all of the binding sites may reside on calmodulin. There is general agreement that calmodulin has 4 high affinity calcium-binding sites per mol protein although there is some disagreement about the binding constants. One study reported 2 classes of binding site in the presence of Mg^{2+}; 3 sites per molecule bound calcium with a K_a of 2×10^{-7} M and 1 site with a K_a of 1×10^{-6} M (WOLFF et al. 1977). However, another study reported that all 4 sites bound calcium equally, with a K_a of 2.5×10^{-6} M in the presence of 100 mM KCl (DEDMAN et al. 1977). Thus, although the stoichiometry is consistent with the subunit being the sole calcium-binding subunit, the results are inconclusive. Binding of calcium to freely soluble calmodulin may differ from binding of calcium to calmodulin bound to phosphorylase kinase (the δ subunit).

As indicated earlier, a consequence of calcium binding in the presence of ATP is the autophosphorylation of phosphorylase kinase. Maximal autophosphorylation was shown to result in an activity which is twice that obtained using cAMP-dependent protein kinase. About 28–36 mol phosphate are incorporated per $(\alpha\beta\gamma\delta)_4$ (WANG et al. 1976). The sites phosphorylated by the calcium-dependent autophosphorylation did not substantially overlap with the sites phosphorylated by the cAMP-dependent protein kinase, as prior phosphorylation by the cAMP-dependent protein kinase did not reduce subsequent phosphate incorporation via autophosphorylation. Only the α and β subunit were phosphorylated, with about 70% of the phosphate incorporated into the α-subunit. Since this study was performed prior to the discovery of the δ subunit, it is not known whether any phosphate is incorporated into the δ subunit. Calmodulin is not believed to be a phosphoprotein, however. It was not determined which phosphorylation site(s) is(are) associated with activation of the enzyme. The physiological role of autophosphorylation is not known. Tetanic electrical stimulation of rabbit gracilis muscle did not lead to appreciable activation of phosphorylase kinase suggesting that autophosphorylation did not occur during this type of muscle stimulation (STULL and MAYER 1971).

2. Regulation by cAMP

Phosphorylase kinase can be activated at pH 6.8 by phosphorylation catalyzed by the cAMP-dependent protein kinase under conditions which substantially reduce contribution of autophosphorylation (COHEN 1973; HAYAKAWA et al. 1973). Maximal phosphorylation lead to the incorporation of 1 mol phosphate into each of the α and β subunits with no incorporation into the γ or presumably the δ subunit. Analysis of the time course of phosphorylation indicated that the β subunits were phosphorylated faster than the α subunits, which showed a slight lag. The temporal increase in phosphorylase kinase activity closely paralleled the incorporation of phosphate into the β subunit (COHEN 1973). Conversely, upon addition of a phosphatase, the decrease in activity paralleled the decrease in phosphate content of the β subunit (COHEN and ANTONIW 1973). Phosphorylation of the α subunits at later stages in the reaction was not accompanied by a corresponding increase in activity. In fact, phosphorylation of the α subunit controls the rate of dephosphorylation of the β subunit and thereby regulates the rate of inactivation of the enzyme (COHEN and ANTONIW 1973).

IV. Regulation of Phosphorylase Kinase in Vivo

Although numerous enzymes have been demonstrated in vitro to undergo changes in activity upon phosphorylation, there have been very few systems where the state of phosphorylation of such enzymes was shown to change in vivo in response to some physiological stimulus. One problem in demonstrating such changes is that the state of phosphorylation of only a subset of multiple phosphorylation sites may change during any particular stimulus (e.g., see Sect. D.V). The solution to this problem is to isolate and characterize specific sites of phosphorylation and determine changes in the state of phosphorylation at each site. When such an analysis was performed on phosphorylase kinase in control and epinephrine treated animals (YEAMAN and COHEN 1975), it was found that, although there was only a small change in total phosphate content of the enzyme, each of the two sites known to be phosphorylated by the cAMP-dependent protein kinase had increased levels of phosphate in response to epinephrine. Neither site was phosphorylated in the absence of epinephrine. However, after hormone treatment, the extent of phosphorylation of the site on the α subunit was 50% while the extent of phosphorylation of the site on the β subunit was 40%. Thus, most of the phosphate on the enzyme was present on sites other than those regulated by the cAMP-dependent protein kinase. This explains why studies on the effect of epinephrine on total phosphate content (MAYER and KREBS 1970) detected only a change of 20%. As was the case for Protein I (see Sect. D.V), the same sites on phosphorylase kinase that are phosphorylated in vitro are also phosphorylated in vivo. Since the calcium-dependent autophosphorylation leads to phosphorylation of distinct sites, this experiment demonstrates that in rabbit skeletal muscle regulation of phosphorylase kinase activity by epinephrine is mediated by cAMP and cAMP-dependent protein phosphorylation. Phosphorylation of these two specific sites on the α and β subunits fully accounted for the increase in total phosphate content of the enzyme following hormone treatment (YEAMAN and COHEN 1975).

Similar biochemical characterization of the sites phosphorylated by autophosphorylation has not been performed. It would be interesting to determine whether phosphorylation of some of these sites is increased by electrical stimulation of muscle and whether it affects, or is affected by, simultaneous treatment with adrenaline.

There is some indication that phosphorylase kinase may phosphorylate several enzymes and proteins other than phosphorylase *b*. Casein (DELANGE et al. 1968), troponin I (STULL et al. 1972), troponin T (PERRY and COLE 1974), a 40,000 Dalton protein in synaptic plasma membranes (BROWNING et al. 1979) and glycogen synthase (DEPAOLI-ROACH et al. 1979; SODERLING et al. 1979 b; WALSH et al. 1979; EMBI et al. 1979) have been shown to be phosphorylated by phosphorylase kinase at significant rates. Distinct phosphorylation sites on skeletal muscle glycogen synthase, regulated by the cAMP-dependent protein kinase and a cAMP-independent protein kinase, have been identified (NIMMO and COHEN 1974; SODERLING et al. 1977). The cAMP-independent protein kinase has recently been shown to be calmodulin dependent (RYLATT et al. 1979; SRIVASTAVA et al. 1979). In fact, there is good evidence suggesting that this protein kinase is identical to phosphorylase kinase (DEPAOLI-ROACH et al. 1979; SODERLING et al. 1979 b; WALSH et al. 1979; EMBI et al. 1979). The enzyme phosphorylates a unique serine residue 7 amino acids from the N-terminus of glycogen synthase and results in a decrease in activity in the absence of glucose-6-P (SODERLING et al. 1979 a; RYLATT and COHEN 1979; EMBI et al. 1979). The amino acid sequence surrounding the phosphoserine is similar but not identical to the sequence surrounding the phosphoserine in phosphorylase. The rate of phosphorylation of phosphorylase *b* is several-fold faster than the rate of phosphorylation of glycogen synthase *a* (DEPAOLI-ROACH et al. 1979; SODERLING et al. 1979 b; EMBI et al. 1979). The relative contribution of the cAMP-dependent kinase and phosphorylase kinase towards regulation of glycogen synthase under various metabolic states has not been determined. It has recently been shown, however, that regulation of glycogen synthase by insulin, in mice which lack phosphorylase kinase because of a genetic defect, is normal, suggesting that the action of insulin on glycogen synthase is mediated via still another protein kinase (LEMARCHAND-BRUSTEL et al. 1979).

Studies of the substrate specificity of phosphorylase kinase indicate that this enzyme may be directly involved in regulation of muscle contraction and of glycogen synthesis in addition to its recognized function in glycogenolysis. This suggests that other metabolic enzymes may also be substrates for phosphorylase kinase. In numerous studies, however, its substrate specificity has been shown to be rather limited and one cannot describe this enzyme as a general calcium-dependent protein kinase.

Clearly the regulation of phosphorylase kinase is complex. The presence of several phosphorylase kinase phosphatases and inhibitory proteins for some phosphatases, which are themselves phosphoproteins (COHEN 1978), add additional levels of complexity that have not been elaborated here. Phosphorylase kinase provides another example of the interaction between cAMP and calcium. In resting rabbit skeletal muscle, phosphorylase kinase exists predominantly in the non-activated state. Stimulation of the muscle, resulting in elevated intracellular calcium, would stimulate phosphorylase kinase activity. The effect of epinephrine on glycogenolysis is presumably due to phosphorylation of the enzyme by the cAMP-dependent protein kinase. This phosphorylated enzyme still requires calcium for activity, providing a dependence of cAMP stimulation on the presence of calcium. Conversely, although the kinetics of activation via calcium-dependent autophosphorylation show a lag period, prior phosphorylation of the enzyme by the cAMP-dependent protein kinase eliminates this lag and may indicate that the calcium-dependent autophosphorylation is indirectly regulated by cAMP (WANG et al. 1976).

H. Phospholipid- and Calcium-Dependent Protein Kinase

A novel cAMP-independent protein kinase with broad specificity has recently been identified in a variety of mammalian tissues (for review see NISHIZUKA et al. 1978, 1979). This protein kinase is activated by calcium via two independent mechanisms. At relatively high calcium concentrations a neutral protease isolated from liver or brain is able to convert the proenzyme of this protein kinase to a smaller active form (TAKAI et al. 1977a, b; INOUE et al. 1977). At lower calcium concentrations this enzyme can be activated without proteolysis if a membrane-associated factor or one of several purified phospholipids is present (TAKAI et al. 1979a, b, c).

I. Irreversible Activation of Protein Kinase C by Calcium-Dependent Protease

A calcium-activated protein kinase system consisting of proenzyme (protein kinase C) and the protease-activated enzyme (protein kinase M) was found in a variety of mammalian tissues, including liver, brain, kidney, heart, skeletal muscle, and lung (NISHIZUKA et al. 1978). When extracted from these tissues, the activated form of the enzyme is present in low amounts, with most of it existing as the proenzyme. It was found that upon limited proteolysis with a calcium-dependent neutral protease (or other proteases), protein kinase C was converted to protein kinase M (TAKAI et al. 1977b; INOUE et al. 1977). At this point it is not known whether protein kinase M obtained from different tissues and protein kinase C obtained from different tissues are identical, but they appear to have similar properties. Protein kinase C was relatively inactive toward most substrates, but could phosphorylate protamine (TAKAI et al. 1977b). Protein kinase M had a broad substrate specificity, with calf thymus histone, protamine, phosphorylase kinase, glycogen synthase, but not casein or phosvitin serving as relatively good substrates (INOUE et al. 1977; KISHIMOTO et al. 1978; YAMAMOTO et al. 1978).

Proteolytic activation of protein kinase C could be catalyzed by a calcium-dependent protease or trypsin, although the calcium-dependent protease resulted in a more active protein kinase M. This protease is present in both liver and brain from rat (INOUE et al. 1977). It is most active at neutral pH and is absolutely dependent on the presence of calcium. The physiological significance of this mechanism for activating protein kinase C is questionable. The protease requires rather high levels of calcium for activity, with 2 m*M* calcium providing maximal stimulation. Protein kinase C has a molecular weight of 77,000 Daltons and is converted by the calcium-dependent protease to the 51,000 Dalton protein kinase M (YAMAMOTO et al. 1978). Once activated by proteolysis the enzyme no longer requires calcium for activity.

Protein kinase C (and protein kinase M) is distinct from both the cAMP- and cGMP-dependent protein kinases present in the same tissues (NISHIZUKA et al. 1979). Protein kinase C and protein kinase M are neither stimulated nor inhibited by either cAMP or cGMP. The physical and kinetic properties of these enzymes, such as pH optimum, molecular weight, metal ion requirement and K_m for ATP and histone, are significantly different. Protein kinase C and protein kinase M are not inhibited by the inhibitor protein of cAMP-dependent protein kinase (YAMAMOTO et al. 1978).

Protein kinase M appears to phosphorylate several substrates. A variety of histones can serve as substrates. Interestingly, Ser 38 in histone H 1 and Ser 32 and Ser 36 in histone H 2 B, the same seryl residues phosphorylated by the cAMP-dependent protein kinase, are also phosphorylated by protein kinase M. However, the cAMP-dependent protein kinase and protein kinase M differ in the relative rates of phosphorylation of the various sites.

Protein kinase M, but not protein kinase C, can phosphorylate phosphorylase kinase (KISHIMOTO et al. 1977). Both the α and β subunits of phosphorylase kinase are phosphorylated. This phosphorylation is accompanied by activation of phosphorylase kinase and parallels incorporation of phosphate into the β subunit. The enzyme is equally effective with either calf thymus whole histone or phosphorylase kinase as substrates (KISHIMOTO et al. 1978). Protein kinase M is also able to inactivate glycogen synthase. Although, it is equally as effective as the cAMP-dependent protein kinase in activating phosphorylase kinase, it is not as effective as the cAMP-dependent protein kinase in the inactivation of glycogen synthase (KISHIMOTO et al. 1978).

II. Reversible Activation of Protein Kinase C by Calcium and Membrane Factor

Another, more appealing mechanism for activation of protein kinase C has recently been reported (TAKAI et al. 1979 a, b, 1980). Protein kinase C can be activated, without a prior proteolytic conversion to protein kinase M, by addition of a membrane-bound factor. Unlike the mechanism described above, this mode of activation is reversible and requires relatively low concentrations of calcium.

Protein kinase C, partially purified from rat brain, was activated by a Triton X-100 extract of brain membranes in the presence of calcium (TAKAI et al. 1979 a) (Table 2). Activation could be reversed by addition of EGTA and reactivated by subsequent addition of excess calcium. As indicated in Table 2, activation of protein kinase C by membrane factor was insensitive to the presence of leupeptin, a protease inhibitor, whereas activation by the calcium-dependent protease was substantially reduced by leupeptin. Protein kinase C activated by the membrane factor

Table 2. Activation of protein kinase C by membrane factor and calcium-protease

Experiment	Reaction system	Histone phosphorylation % of control
1	Complete	100
	-Protein kinase C	0
	-Membrane factor	3.0
	-Calcium	5.5
	+Leupeptin	100
2	Complete	100
	-Protein kinase C	0.2
	-Calcium-protease	2.3
	-Calcium	1.6
	+Leupeptin	6.8

In experiment 1, 0.35 µg of protein kinase was activated by 1.3 µg (as protein) of membrane factor and phosphorylation of histone Hl assayed for 3 min at 30 °C. Where indicated calcium was present at 0.5 m*M* and leupeptin at 100 µg/ml. In experiment 2, protein kinase C was first activated with 4 µg of calcium-protease in the presence of 2 m*M* calcium for 13 min at 30 °C. Where indicated, leupeptin was present at 100 µg/ml. Subsequently, 0.35 µg of protein kinase was removed from the activation mixture and assayed as above, but in the absence of membrane factor and calcium. Calcium did not stimulate or inhibit the phosphorylation reaction catalyzed by the proteolytically activated enzyme. In experiments 1 and 2, 100% of control respresents 17,900 cpm and 16,800 cpm of acid-precipitable radioactivity, respectively. (Adapted from TAKAI et al. 1979a)

and protein kinase C activated by the calcium-dependent protease were equally effective in the phosphorylation of histone H 1. Addition of membranes to protein kinase C in the presence of calcium was also effective in activating it (TAKAI et al. 1979b). The activating factor was present largely in the particulate fraction in a large variety of tissues. The active factor in the Triton X-100 extract was shown to be phospholipid (TAKAI et al. 1979a). Several phospholipids could substitute for the membrane factor. The relative activities were in the order of phosphatidylinositol = phosphatidylserine > phosphatidic acid = diphosphatidylglycerol > phosphatidylethanolamine > phosphatidylcholine. Activation of the enzymic via this mechanism was an order of magnitude more sensitive to calcium than proteolytic activation. The K_a value for calcium in the phospholipid-dependent activation was less than 50 µ*M* (TAKAI et al. 1979a) and this effect was specific for calcium ion. Other divalent cations were ineffective except for Sr^{2+} which, at concentrations up to 0.3 m*M* was only 5% as effective as calcium. It is not known whether a calcium-binding subunit is involved in regulation of the enzyme although calmodulin could not substitute for phospholipid in the activation. Recent studies indicate a dual role for lipids in activation of protein kinase C (TAKAI et al. 1979c). Addition of a small amount of diacylglycerol markedly decreased the calcium and phospholipid concentrations required for maximal activation of protein kinase C. Diacylglycerol alone, in the presence of calcium, showed little or no stimulation of protein kinase activity. The effect of diacylglycerol could only be seen in the presence of phospholipid and was most pronounced when phosphatidylserine was used as the phospholipid. In the presence of phosphatidylserine, diacylglycerol increased V_{max} and lowered the activation constant for both calcium and phosphatidylserine. This effect of the diacylglycerol was specific for those species containing an unsaturated fatty acid at the 2-position (TAKAI et al. 1979c). In the presence of phosphatidylserine (2 μg), diolein (0.1 μg) lowered the K_a for calcium from 50 μ*M* to 2 μ*M* and activated the protein kinase whereas disaturated species had little or no effect on the K_a for calcium or on protein kinase activity. Determination of the molecular mechanism of this activation by calcium, diacylglycerol and phospholipid will have to await further purification and characterization of the enzyme.

Protein kinase C activated by calcium plus phospholipid shows substrate specificity which differs somewhat from that of protein kinase M. In the presence of calcium, histone H 1 is the preferred histone substrate by both enzymes. However, the ratio of H 1/H 2 B is higher with the enzyme activated by phospholipid and calcium than with the enzyme activated by proteolysis. In addition, the degree of stimulation of protein kinase C by phospholipid and calcium is different for the various histones, with histone H 1 phosphorylation being most stimulated and histone H 2 B phosphorylation least stimulated. Protein kinase M was 5-fold more active than the reversibly-activated protein kinase C when phosphorylase kinase was used as the substrate.

It seems fortuitous that both modes of activation of protein kinase C require calcium. The irreversible nature of the activation by the neutral protease and the requirement for "non-physiological" calcium levels (2–3 m*M*) make it unlikely that such activation plays a regulatory role in vivo. Reversible activation of protein kinase C by membranes and low calcium concentrations is a much more attractive mechanism. It has been demonstrated that the partially purified enzyme becomes bound to synaptic membranes and is activated in the presence of calcium (TAKAI et al. 1979a). This suggests that the membrane or a membrane factor is necessary because of its effect on the kinase, rather than its effect on the substrate. Further characterization of the system will be necessary to determine the mechanism of activation of the enzyme and its possible function in mediating or modulating some of the actions of calcium on membrane-associated processes.

Studies in numerous tissues have demonstrated an enhanced turnover of phosphatidylinositol in response to stimulation of a variety of receptors including α-adrenergic receptors, muscarinic cholinergic receptors and histaminergic (H_1) receptors (MICHELL 1975). This "phospholipid effect" involves an increase in unsaturated diacylglycerol produced by hydrolysis of phosphadidylinositol and is accompanied by calcium-influx into the stimulated cell. The finding of a protein kinase which requires both calcium and unsaturated diacylglycerol suggests a unique link between stimulation of cells by various extracellular agents and mediation of some of their physiological effects within the cell.

I. Concluding Comments

Several classes of enzyme appear to mediate the intracellular effects of calcium. A common feature of these enzymes is the involvement of calcium-binding proteins which serve as intracellular receptors for calcium. In recent years, one particular receptor, calmodulin, has received considerable attention because of its multifunctional properties. Since its initial discovery as an activator of cyclic nucleotide phosphodiesterase (CHEUNG 1970; KAKIUCHI and YAMAZAKI 1970), calmodulin has been found to affect many enzymes and cellular processes including certain adenylate cyclases (BROSTROM et al. 1975; CHEUNG et al. 1975), a membrane-bound calcium-dependent protein kinase (SCHULMAN and GREENGARD 1978 b), myosin light chain kinase (DABROWSKA et al. 1978; YAGI et al. 1978), phosphorylase kinase (COHEN et al. 1978; SHENOLIKAR et al. 1979), Ca^{2+} transport in erythrocytes (GOPINATH and VINCENZI 1977; JARRET and PENNISTON 1977), microtubule polymerization-depolymerization (MARCUM et al. 1978), plant NAD kinase (ANDERSON and CORMIER 1978) and platelet phospholipase A_2 (WONG and CHEUNG 1979).

The active form of Ca^{2+} for those enzymes and processes that are regulated by calmodulin appears to be the Ca^{2+}-calmodulin complex. Except for the δ subunit of phosphorylase kinase, calmodulin functions as a factor that associates with the enzymes that it regulates only in the presence of Ca^{2+}. An appealing feature of the mediation of hormonal responses via production of cAMP and activation of cAMP-dependent protein kinase is the enormous amplification of the original extracellular signal that it provides. Regulation of protein phosphorylation by calcium also results in an amplification of the original extracellular signal. The Ca^{2+}-calmodulin complex can regulate cAMP-dependent protein phosphorylation indirectly because of its effect on cAMP levels, that is, by its regulation of adenylate cyclase and cyclic nucleotide phosphodiesterase activities. Synthesis and degradation of cGMP may also be regulated by the Ca^{2+}-calmodulin complex, thereby indirectly affecting the activity of cGMP-dependent protein kinase. Thus, the Ca^{2+}-calmodulin complex may indirectly control numerous biochemical processes. The effect of the Ca^{2+}-calmodulin complex on phosphorylase kinase, myosin light chain kinase, and the newly discovered class of protein kinase, the membrane-bound, calmodulin- and calcium-dependent protein kinase, provides not only a mechanism for direct amplification of an extracellular signal but also a mechanism by which calcium may regulate diverse biochemical processes.

Experiments with rat hepatocytes suggest that glycogenolysis and gluconeogenesis can be regulated by both cAMP and calcium. Based on the conservative nature of biochemical evolution, it would not be surprising to find the same common pathway, regulation of metabolic enzymes by protein phosphorylation, being used by both second messengers. The nature and number of the protein kinases which mediate certain actions of calcium in liver and other tissues remain to be determined. Analysis of the substrate specificity of phosphorylase kinase and myosin light chain kinase indicate that neither enzyme can be termed a general calcium-dependent protein kinase, that is, one with broad substrate specificity. In contrast to cAMP action, which is mediated by only two related cAMP-dependent protein kinases, with broad specificity, it may turn out that the effects of calcium on protein phosphorylation are mediated by several distinct calcium-dependent protein kinases, each with a limited and non-overlapping substrate specificity. The other two calcium-dependent protein kinases detailed in this review appear to have both broad substrate specificity and a widespread tissue distribution. The requirement of protein kinase C for phospholipid and diacylglycerol, cellular constituents which are present in cell membranes and not in cell cytosol, suggests that this enzyme, like the membrane-bound calmodulin- and calcium-dependent protein kinase may function to regulate only membrane-bound substrate proteins. By analogy with the cAMP-dependent protein phosphorylation system and based on evidence for calcium-stimulated phosphorylation of soluble enzymes, one may expect to find additional, soluble, calcium-dependent protein kinases. The four protein kinases analyzed above provide examples of the molecular basis for the regulation of the postulated protein kinases by calcium as well as the basis for how the action of such protein kinases may affect the function of their substrate proteins.

One of the conclusions that has emerged from analysis of the four calcium-dependent protein kinases is that each of the two second messengers, calcium and cAMP, operates within the context of the presence of the other. Calcium and cAMP interact biochemically at a minimum of three levels. First, each can regulate the intracellular concentration of the other. Calcium affects cAMP levels by its regulation of adenylate cyclase and cyclic nucleotide phosphodiesterase (WANG and WAISMAN 1979). Conversely, cAMP can both increase and decrease intracellular levels of calcium (BERRIDGE 1975). cAMP can elevate free Ca^{2+} by stimulating calcium influx from the medium and by stimulating release of calcium from intracellular stores. cAMP can reduce free Ca^{2+} by stimulating uptake into intracellular stores. Second, these two second messengers can interact at the level of the enzymes that mediate their actions. For example, the cAMP-dependent protein kinase, which mediates the effects of cAMP in eukaryotes, is inhibited by calcium (NIMMO and COHEN 1977). Conversely, cAMP, via cAMP-dependent protein phosphorylation, has been shown to inhibit myosin light chain kinase, which mediates certain of the actions of calcium in muscle (ADELSTEIN et al. 1978). Third, calcium and cAMP can interact directly at the level of the cellular process that each regulates. Depending on the magnitude and the direction of their individual actions, their combined effects may be additive, synergistic, antagonistic, or independent. Phosphorylase kinase and Protein I are just two examples of proteins whose phosphorylation is regulated by both cAMP and calcium. Thus, in some instances, the two second messengers may interact at the level of their final common pathway, protein phosphorylation. Phosphorylation of protein I by the cAMP-dependent protein kinase reduces phosphorylation at another site on the same protein by a calcium-dependent protein kinase and may provide an example of antagonistic interaction between calcium and cAMP (HUTTNER and GREENGARD 1979). Regulation of phosphorylase kinase by cAMP and calcium provides several examples of additive or synergistic interaction between these second messengers. Thus, calcium greatly stimulates the activity of phosphorylase kinase in both the phosphorylated and unphosphorylated state and cAMP-dependent phosphorylation of the enzyme reduces the lag seen in calcium-dependent autophosphorylation of the enzyme. In addition, cAMP and cAMP-dependent protein kinase regulate the activity of an inhibitor of a phosphoprotein phosphatase which reverses the actions of calcium and cAMP on phosphorylase kinase (NIMMO and COHEN 1977).

Clearly, one can no longer discuss the regulation of cellular processes by cAMP and calcium independently of each other or of other possible second messengers. Protein phosphorylation is one area of research where the interaction of cAMP and calcium can be examined at the biochemical level. There is every indication that new and exciting findings that will enhance our understanding of numerous cellular processes will continue to emerge from studies of the various modes of regulation of protein phosphorylation.

Acknowledgements. I am grateful to Drs. F.E. Bloom, P. Cohen, J.G. DeMaille, W. Douglas, J.C. Garrison, P. Greengard, W.B. Huttner, P. Siekevitz, Y. Nishizuka, and J.H. Wang for making available material prior to publication and for helpful discussions. I wish to thank Dr. Gordon Ringold and Mr. Peter Nose for careful reading of the manuscript and Ms. Karen Benight for typing the manuscript. I am supported by Grant No. MH 32752 awarded by the National Institutes of Mental Health.

References

Adelstein RS (1978) Myosin phosphorylation, cell motility, and smooth muscle contraction. Trends Biochem Sci 3:27–30

Adelstein RS, Chacko S, Scordilis SP, Barylko B, Conti MA, Trotter JA (1977) Calcium binding proteins and calcium functions. In: Wassermann R et al. (eds). Elsevier, North Holland, pp 251–261

Adelstein RS, Conti MA (1975) Phosphorylation of platelet myosin increases actin-activated myosin ATPase activity. Nature 256:597–598

Adelstein RS, Conti MA, Hathaway DR, Klee CB (1978) Phosphorylation of smooth muscle myosin light chain kinase by the catalytic subunit of adenosine 3′:5′-monophosphate-dependent protein kinase. J Biol Chem 253:8347–8350

Aksoy MO, Williams D, Sharkey EM, Hartshorne DJ (1976) A relationship between Ca^{2+} sensitivity and phosphorylation of gizzard actomyosin. Biochem Biophys Res Commun 69:35–41

Alexander MC, Kowaloff EM, Witters LA, Dennihy DT, Avruch J (1979) Purification of a hepatic 123,000-dalton hormone-stimulated ^{32}P-peptide and its identification as ATP-citrate lyase. J Biol Chem 254:8052–8056

Amphlett GW, Vanaman TC, Perry SV (1976) Effects of the troponin C-like protein from bovine brain (brain modulator protein) on the Mg^{2+}-stimulated ATPase of skeletal muscle actomyosin. FEBS Lett 72:163–167

Anderson JM, Cormier MJ (1978) Calcium-dependent regulator of NAD kinase. Biochem Biophys Res Commun 84:595–602

Babcock DF, Chen J-LJ, Yip BP, Lardy HA (1979) Evidence for mitochondrial localization of the hormone-responsive pool of Ca^{2+} in isolated hepatocytes. J Biol Chem 254:8117–8120

Baker PF (1972) Transport and metabolism of calcium ions in nerve. Prog Biophys Mol Biol 24:177–223

Baker PF, Hodgkin AL, Ridgway EB (1971) Depolarization and calcium entry in squid giant axons. J Physiol (Lond) 218:709–755

Barany K, Barany M (1977) Phosphorylation of the 18,000-dalton light chain of myosin during a single tetanus of frog muscle. J Biol Chem 252:4752–4754

Barylko B, Kuznicki J, Drabikowski W (1978) Identification of Ca^{2+}-binding subunit of myosin light chain kinase from skeletal muscle with modulatory protein. FEBS Lett 90:301–304

Berridge MJ (1975) The interaction of cyclic nucleotides and calcium in the control of cellular activity. In: Greengard P, Robison GS (eds) Advances in cyclic nucleotide research, vol 6. Raven Press, New York, pp 1–98

Birnbaum MJ, Fain JN (1977) Activation of protein kinase and glycogen phosphorylase in isolated rat liver cells by glucagon and catecholamines. J Biol Chem 252:528–535

Blackmore PF, Brumley FT, Marks JL, Exton JH (1978) Studies on α-adrenergic activation of hepatic glucose output. Relationship between α-adrenergic stimulation of calcium efflux and activation of phosphorylase in isolated rat liver parenchymal cells. J Biol Chem 253:4851–4858

Blaustein MP (1975) Effects of potassium, veratridine, and scorpion venom on calcium accumulation and transmitter release by nerve terminals in vitro. J Physiol (Lond) 247:617–655

Blaustein MP, Johnson EM Jr, Needleman P (1972) Calcium-dependent norepinephrine release from presynaptic nerve endings in vitro. Proc Natl Acad Sci USA 69:2237–2240
Bloom FE, Ueda T, Battenberg E, Greengard P (1979) Immunocytochemical localization in synapses of Protein I, an endogenous substrate for protein kinases in mammalian brain. Proc Natl Acad Sci USA 76:5982–5986
Bremel RD, Shaw ME (1978) Conversion of calcium-sensitive myosin light chain kinase to a calcium-insensitive form. FEBS Lett 88:242–246
Brostrom CO, Huang Y-C, Breckenridge B McL, Wolff DJ (1975) Identification of a calcium-binding protein as a calcium-dependent regulator of brain adenylate cyclase. Proc Natl Acad Sci USA 72:64–68
Brostrom CO, Hunkeler FL, Krebs (1971) The regulation of skeletal muscle phosphorylase kinase by Ca^{2+}. J Biol Chem 246:1961–1967
Browning M, Bennett W, Lynch G (1979) Phosphorylase kinase phosphorylates a brain protein which is influenced by repetitive synaptic activation. Nature 278:273–275
Burchell A, Cohen PTW, Cohen P (1976) Distribution of isoenzymes of the glycogenolytic cascade in different types of muscle fiber. FEBS Lett 67:17–22
Cassidy P, Hoar PE, Kerrick WGL (1978) Tension development in rabbit ileum smooth muscle skinned by *Staphylococcal* α-toxin: Activation by Ca^{2+} and Sr^{2+}, irreversible activation in the presence of ATPγS. Biophys J 21:44a
Castano JG, Nieto A, Feliu JE (1979) Inactivation of phosphofructokinase by glucagon in rat hepatocytes. J Biol Chem 254:5576–5579
Chacko S, Conti MA, Adelstein RS (1977) Effect of phosphorylation of smooth muscle myosin on actin activation and Ca^{2+} regulation. Proc Natl Acad Sci USA 74:129–133
Chan TM, Exton JH (1977) α-Adrenergic-mediated accumulation of adenosine 3′:5′ monophosphate in calcium-depleted hepatocytes. J Biol Chem 252:8645–8651
Chen J-LJ, Babcock DF, Lardy HA (1978) Norepinephrine, vasopressin, glucagon, and A 23187 induce efflux of calcium from an exchangeable pool in isolated rat hepatocytes. Proc Natl Acad Sci USA 75:2234–2238
Cherrington AD, Assimacopoulos FD, Harper SC, Corbin JD, Park CR, Exton JH (1976) Studies on the α-adrenergic activation of hepatic glucose output. J Biol Chem 251:5209–5218
Cherrington AD, Hundley RF, Dolgin S, Exton JH (1977) Studies of the role of beta-adrenergic receptors in the activation of phosphorylase in rat hepatocytes by catecholamines. J Cyclic Nucleotide Res 3:263–273
Cheung WY (1970) Cyclic 3′,5′-nucleotide phosphodiesterase. Demonstration of an activator. Biochem Biophys Res Commun 38:533–538
Cheung WY, Bradham LS, Lynch TJ, Lin YM, Tallant EA (1975) Protein activator of cyclic 3′:5′-nucleotide phosphodiesterase of bovine or rat brain also activates its adenylate cyclase. Biochem Biophys Res Commun 66:1055–1062
Cohen P (1973) The subunit structure of rabbit skeletal muscle phosphorylase kinase, and the molecular basis of its activation reactions. Eur J Biochem 34:1–14
Cohen P (1974) The role of phosphorylase kinase in nervous and hormonal control of glycogenolysis in muscle. Biochem Soc Sym 39:51–73
Cohen P (1978) The role of cyclic-AMP-dependent protein kinase in the regulation of glycogen metabolism in mammalian skeletal muscle. Curr Top Cell Regul 14:117–196
Cohen P (1980) The role of calmodulin and troponin in the regulation of phosphorylase kinase from mammalian skeletal muscle. In: Cheung WY (ed) Calcium-binding proteins as cellular regulators. Academic Press, New York, in press
Cohen P, Antoniw JF (1973) The control of phosphorylase kinase phosphatase by "second site phosphorylation"; a new form of enzyme regulation. FEBS Lett 34:43–47
Cohen P, Burchell A, Foulkes JG, Cohen PTW, Vanaman TC, Nairn AC (1978) Identification of the Ca^{2+}-dependent modulator protein as the fourth subunit of rabbit skeletal muscle phosphorylase kinase. FEBS Lett 92:287–293 (1978)
Cohen P, Picton C, Klee CB (1979) Activation of phosphorylase kinase from rabbit skeletal muscle by calmodulin and troponin. FEBS Lett 104:25–30
Dabrowska R, Aromatorio D, Sherry JMF, Hartshorne DJ (1977) Composition of the myosin light chain kinase from chicken gizzard. Biochem Biophys Res Commun 78:1263–1271

Dabrowska R, Hartshorne DJ (1978) A Ca^{2+}- and modulator-dependent myosin light chain kinase from non-muscle cells. Biochem Biophys Res Commun 85:1352–1359

Dabrowska R, Sherry JMF, Aromatorio DK, Hartshorne DJ (1978) Modulator protein as a component of the myosin light chain kinase from chicken gizzard. Biochemistry 17:253–258

Daniel JL, Adelstein RS (1976) Isolation and properties of platelet myosin light chain kinase. Biochemistry 15:2370–2377

De Camilli P, Ueda T, Bloom FE, Greengard P (1979) Widespread distribution of Protein I in the central and peripheral nervous system. Proc Natl Acad Sci USA 76:5977–5981

Dedman JR, Potter JD, Means AR (1977) Biological cross-reactivity of rat testis phosphodiesterase activator protein and rabbit skeletal muscle troponin-C. J Biol Chem 252:2437–2440

DeLange RJ, Kemp RG, Riley WD, Cooper RA, Krebs EG (1968) Activation of skeletal muscle phosphorylase kinase by adenosine triphosphate and adenosine 3′,5′-monophosphate. J Biol Chem 243:2200–2208

DeLorenzo RJ Freedman SD (1977) Calcium-dependent phosphorylation of synaptic vesicle proteins and its possible role in mediating neurotransmitter release and vesicle function. Biochem Biophys Res Commun 77:1036–1043

DeLorenzo RJ, Freedman SD (1978) Calcium-dependent neurotransmitter release and protein phosphorylation in synaptic vesicles. Biochem Biophys Res Commun 80:183–192

DeLorenzo RJ, Freedman SD, Yohe WB, Maurer SC (1979) Stimulation of Ca^{2+}-dependent neurotransmitter release and presynaptic nerve terminal protein phosphorylation by calmodulin and a calmodulin-like protein isolated from synaptic vesicles. Proc Natl Acad Sci USA 76:1838–1842

DePaoli-Roach AA, Roach PJ, Larner J (1979) Rabbit skeletal muscle phosphorylase kinase. Comparison of glycogen synthase and phosphorylase as substrates. J Biol Chem 254:4212–4219

Douglas WW (1968) Stimulus-secretion coupling: the concept and clues from chromaffin and other cells. Br J Pharmacol 34:451–474

Ebashi S, Mikawa T, Hirata M, Nononura Y (1978) The regulatory role of calcium in muscle. Ann NY Acad Sci 307:451–461

Embi N, Rylatt DB, Cohen P (1979) Glycogen synthase kinase-2 and phosphorylase kinase are the same enzyme. Eur J Biochem 100:339–347

Forn J, Greengard P (1978) Depolarizing agents and cyclic nucleotides regulate the phosphorylation of specific neuronal proteins in rat cerebral cortex slices. Proc Natl Acad Sci USA 75:5195–5199

Frankenhaeuser B (1957) The effect of calcium on the myelinated nerve fibre. J Physiol (Lond) 137:245–260

Frearson N, Focant BWW, Perry SV (1976) Phosphorylation of a light chain component of myosin from smooth muscle. FEBS Lett 63:27–32

Frearson N, Perry SV (1975) Phosphorylation of the light-chain components of myosin from cardiac and red skeletal muscles. Biochem J 151:99–107

Garrison JC (1978) The effects of glucagon, catecholamines, and the calcium ionophore A 23187 on the phosphorylation of rat hepatocyte cytosolic proteins. J Biol Chem 253:7091–7100

Garrison JC, Borland MK (1979) Regulation of mitochondrial pyruvate carboxylation and gluconeogenesis in rat hepatocytes via an α-adrenergic, adenosine 3′:5′-monophosphate-independent mechanism. J Biol Chem 254:1129–1133

Garrison JC, Borland MK, Florio VA, Twible DA (1979) The role of calcium ion as a mediator of the effects of antiotensin II, catecholamines, and vasopressin on the phosphorylation and activity of enzymes in isolated hepatocytes. J Biol Chem 254:7147–7156

Gazith J, Himmelbarb S, Harrington WF (1970) Studies on the subunit structure of myosin. J Biol Chem 245:15–22

Gopinath RM, Vincenzi FF (1977) Phosphodiesterase protein activator mimics red blood cell cytoplasmic activator (Ca^{2+}-Mg^{2+}) ATPase. Biochem Biophys Res Commun 77:1203–1209

Gorecka A, Aksoy MO, Hartshorne DJ (1976) The effect of phosphorylation of gizzard myosin on actin activation. Biochem Biophys Res Commun 71:325–331

Gray EG, Whittaker VP (1962) The isolation of nerve endings from brain: An electronmicroscopic study of cell fragments derived by homogenization and centrifugation. J Anat 96:79–88

Greengard P (1978a) Phosphorylated proteins as physiological effectors. Science 199:146–152

Greengard P (1978b) Cyclic nucleotides, phosphorylated proteins, and neuronal function. Raven Press, New York

Harvey AM, MacIntosh FC (1940) Calcium and synaptic transmission in a sympathetic ganglion. J Physiol (Lond) 97:408–416

Haslam RJ, Lynham JA (1977) Relationship between phosphorylation of blood platelet proteins and secretion of platelet granule constituents. I. Effects of different aggregating agents. Biochem Biophys Res Commun 77:714–722

Hathaway DR, Adelstein RS (1979) Human platelet myosin light chain kinase requires the calcium-binding protein calmodulin for activity. Proc Natl Acad Sci USA 76:1653–1657

Hayakawa T, Perkins JP, Krebs EG (1973) Studies on the subunit structure of rabbit skeletal muscle phosphorylase kinase. Biochemistry 12:574–580

Heilbrunn LV (1952) An outline of general physiology, 3rd ed. W.B. Saunders Co, Philadelphia

Hershkowitz M (1978) Influence of calcium on phosphorylation of a synaptosomal protein. Biochim Biophys Acta 542:274–283

Himms-Hagen J (1967) Sympathetic regulaton of metabolism. Pharmacol Rev 19:367–461

Hirata M, Mikawa T, Nonomura Y, Ebashi S (1977) Ca^{2+} regulaton in vascular smooth muscle. J Biochem (Tokyo) 82:1793–1796

Hutson NJ, Brumley FT, Assimacopoulos FD, Harper SC, Exton JH (1976) Studies on the α-adrenergic activation of hepatic glucose output. I. Studies on the α-adrenergic activation of phosphorylase and gluconeogenesis and inactivation of glycogen synthase in isolated rat liver parenchymal cells. J Biol Chem 251:5200–5208

Huttner WB, Greengard P (1979) Multiple phosphorylation sites in Protein I and their differential regulation by cyclic AMP and calcium. Proc Natl Acad Sci USA 76:5402–5406

Huxley AF (1964) Muscle. Ann Rev Physiol 26:131–152

Ikebe M, Onishi H, Watanabe S (1977) Phosphorylation and dephosphorylation of a light chain of the chicken gizzard myosin molecule. J Biochem (Tokyo) 82:299–302

Inoue M, Kishimoto A, Takai Y, Nishizuka Y (1977) Studies on a cyclic nucleotide-independent protein kinase and its proenzyme in mammalian tissues. Proenzyme and its activation by calcium-dependent protease from rat brain. J Biol Chem 252:7610–7616

Jarrett HW, Penniston JT (1977) Partial purification of the Ca^{2+}-Mg^{2+} ATPase activator from human erythrocytes: Its similarity to the activator of 3′:5′-cyclic nucleotide phosphodiesterase. Biochem Biophys Res Commun 77:1210–1216

Jennissen HP, Heilmeyer LMG (1974) Multiple forms of phosphorylase kinase in red and white skeletal muscle. FEBS Lett 42:77–80

Kagimoto T, Uyeda K (1979) Hormone-stimulated phosphorylation of liver phosphofructokinase in vivo. J Biol Chem 254:5584–5587

Kakiuchi S, Yamazaki R (1970) Calcium-dependent phosphodiesterase activity and its activating factor (PAF) from brain: Studies on cyclic 3′:5′-nucleotide phosphodiesterase (III). Biochem Biophs Res Commun 41:1104–1110

Katz B, Miledi R (1967) A study of synaptic transmission in the absence of nerve impulses. J Physiol (Lond) 192:407–436

Katz B, Miledi R (1969) Tetrodotoxin-resistant electrical activity in presynaptic terminals. J Physiol (Lond) 203:459–487

Kelly JS, Krnjevic K, Somjen GJ (1969) Divalent cations and electrical properties of cortical cells. J Neurobiol 1:197–208

Kendrick-Jones J (1974) Role of myosin light chains in calcium regulation. Nature 244:631–634

Kendrick-Jones J, Lehman W, Szent-Gyorgyi AG (1970) Regulation in molluscan muscles. J Molec Biol 54:313–326

Keppens S, DeWulff H (1975) The activation of liver glycogen phosphorylase by vasopressin. FEBS Lett 51:29–32
Keppens S, DeWulff H (1976) The activation of liver glycogen phosphorylase by angiotensin II. FEBS Lett 68:279–282
Keppens S, Vandenheede JR, DeWulff H (1977) On the role of calcium as second messenger in liver for the hormonally induced activation of glycogen phosphorylase. Biochim Biophys Acta 496:448–457
Khoo JC, Steinberg D (1975) Stimulation of rat liver phosphorylase kinase by micromolar concentrations of Ca^{2+}. FEBS Lett 57:68–72
Kilimann M, Heilmeyer LMG, Jr (1977) The effect of Mg^{2+} on the Ca^{2+}-binding properties of non-activated phosphorylase kinase. Eur J Biochem 73:191–197
Kirk CJ, Hens DA (1974) Hepatic action of vasopressin: Lack of a role for adenosine-3′,5′-cyclic monophosphate. FEBS Lett 47:128–131
Kishimoto A, Mori T, Takai Y, Nishizuka Y (1978) Comparison of calcium-activated, cyclic nucleotide-independent protein kinase and adenosine 3′:5′-monophosphate-dependent protein kinase as regards the ability to stimulate glycogen breakdown in vitro. J Biochem (Tokyo) 84:47–53
Kishimoto A, Takai Y, Nishizuka Y (1977) Activation of glycogen phosphorylase kinase by a calcium-activated cyclic nucleotide-independent protein kinase system. J Biol Chem 252:7449–7452
Klee CB, Krinks MH (1978) Purification of cyclic 3′,5′-nucleotide phosphodiesterase inhibitory protein by affinity chromatography on activator protein coupled to Sepharose. Biochemistry 17:120–126
Kneer NM, Bosch AL, Clark MG, Lardy HA (1974) Glucose inhibition of epinephrine stimulation of hepatic gluconeogenesis by blockade of the α-receptor function. Proc Natl Acad Sci USA 71:4523–4527
Krebs EG, Beavo JA (1979) Phosphorylation-dephosphorylation of enzymes. Ann Rev Biochem 48:923–959
Kretsinger RH (1976) Calcium-binding proteins. Ann Rev Biochem 45:239–266
Kretsinger RH (1979) The informational role of calcium in the cytosol. In: Greengard P, Robison GA (eds) Advances in cyclic nucleotide research, vol 11. Raven Press, New York, pp 1–26
Krueger BK, Forn J, Greengard P (1976) Calcium-dependent protein phosphorylation in rat brain synaptosomes. Abstracts of the Sixth Annual Meeting of the Society of Neuroscience, p 1007
Krueger BK, Forn J, Greengard P (1977) Depolarization-induced phosphorylation of specific proteins, mediated by calcium ion influx, in rat brain synaptosomes. J Biol Chem 252:2764–2773
Kuhn DM, Vogel RL, Lovenberg W (1978) Calcium-dependent activation of tryptophan hydroxylase by ATP and magnesium. Biochem Biophys Res Commun 82:759–766
Kuo JF, Greengard P (1969) Cyclic nucleotide-dependent protein kinases. IV. Widespread occurrence of adenosine 3′,5′-monophosphate-dependent protein kinase in various tissues and phyla of the animal kingdom. Proc Natl Acad Sci USA 64:1349–1355
Lebowitz EA, Cooke R (1978) Contractile properties of actomyosin from human blood platelets. J Biol Chem 253:5443–5447
Levin RM, Weiss B (1976) Mechanism by which psychotropic drugs inhibit adenosine cyclic 3′,5′-monophosphate phosphodiesterase of brain. Molec Pharmacol 12:581–589
Levin RM, Weiss B (1977) Binding of the trifluoperazine to the calcium-dependent activator of cyclic nucleotide phosphodiesterase. Molec Pharmacol 13:690–697
Le Marchand-Brustel Y, Cohen PTW, Cohen P (1979) Insulin activates glycogen synthase in phosphorylase kinase deficient mice. FEBS Lett 105:235–238
Lohmann SM, Ueda T, Greengard P (1978) Ontogeny of synaptic phosphoproteins in brain. Proc Natl Acad Sci USA 75:4037–4041
Luttgau HC, Niedergerke R (1958) The antagonism between Ca and Na ions on the frog's heart. J Physiol (Lond) 143:486–505
Manning DR, Stull JT (1979) Myosin light chain phosphorylation and phosphorylase *a* activity in rat extensor digitorum longus muscle. Biochem Biophys Res Commun 90:164–170

Marcum JM, Dedman JR, Brinkley BR, Means AR (1978) Control of microtubule assembly-disassembly by calcium-dependent regulator protein. Proc Natl Acad Sci USA 75:3771–3775

Mayer SE, Krebs EG (1970) Studies on the phosphorylation and activation of skeletal muscle phosphorylase and phosphorylase kinase in vivo. J Biol Chem 245:3153–3160

Michell RH (1975) Inositol phospholipids and cell surface receptor function. Biochim Biophys Acta 415:81–147

Mikawa T, Nonomura Y, Ebashi S (1977) Does phosphorylation of myosin light chain have direct relation to regulation in smooth muscle? J Biochem (Tokyo) 82:1789–1991

Morgan M, Perry SV, Ottaway J (1976) Myosin light-chain phosphatase. Biochem J 157:687–697

Murray J, Weber A (1973) Molecular control mechanisms in muscle contraction. Physiol Rev 53:612–673

Nimmo HG, Cohen P (1974) Glycogen synthetase kinase 2 (GSK 2). The identification of a new protein kinase in skeletal muscle. FEBS Lett 47:162–166

Nimmo HG, Cohen P (1977) Hormonal control of protein phosphorylation. In: Greengard P, Robison GA (eds) Advances in cyclic nucleotide research, vol 8. Raven Press, New York, pp 145–266

Nishizuka Y, Takai Y, Hashimoto E, Kishimoto A, Kuroda Y, Kuniyasu S, Yamamura H (1979) Regulatory and functional compartment of three multifunctional protein kinase systems. Molec Cell Biochem 23:153–165

Nishizuka Y, Takai Y, Kishimoto A, Hashimotom E, Inoue M, Yamamoto M, Criss WE, Kuroda Y (1978) A role of calcium in the activation of a new protein kinase system. In: George WJ, Ignarro LJ (eds) Advances in cyclic nucleotide research, vol 9. Raven Press, New York, pp 209–220

Osborn D (1975) Comparison of the effects of selective α- and β-receptor agonists on intracellular cyclic AMP levels and glycogen phosphorylase activity in guinea-pig liver. Br J Pharmacol 55:286P–287P

Pariza MW, Butcher FR, Becker JE, Potter VR (1977) 3′:5′-cyclic AMP: Independent induction of amino acid transport by epinephrine in primary cultures of adult rat liver cells. Proc Natl Acad Sci USA 74:234–237

Patrick RL, Barchas JD (1974) Stimulation of synaptosomal dopamine synthesis by veratridine. Nature 250:737–739

Perrie WT, Perry SV (1970) An electrophoretic study of the low molecular weight components of myosin. Biochem J 119:31–38

Perrie WT, Smillie LB, Perry SV (1972) A phosphorylated light-chain component of myosin. Biochem J 128:105P–106P

Perrie WT, Smillie LB, Perry SV (1973) A phosphorylated light chain of myosin from skeletal muscle. Biochem J 135:151–164

Perry SV, Cole HA (1974) Phosphorylation of troponin and the effects of interactions between the components of the complex. Biochem J 141:733–743

Pires EMV, Perry SV (1977) Purification and properties of myosin light-chain kinase from fast skeletal muscle. Biochem J 167:137–146

Pires EMV, Perry SV, Thomas MAW (1974) Myosin light-chain kinase, a new enzyme from striated muscle. FEBS Lett 41:292–296

Pointer RH, Butcher FR, Fain JN (1976) Studies on the role of cyclic guanosine 3′:5′-monophosphate and extracellular Ca^{2+} in the regulation of glycogenolysis in rat liver cells. J Biol Chem 251:2987–2992

Porlzehl H, Caldwell PC, Ruegg JC (1964) The dependence of contraction and relaxation of muscle fibres from the crab *maia squinado* on the internal concentration of free calcium ions. Biochim Biophys Acta 79:581–591

Rasmussen H, Goodman DBP (1977) Relationships between calcium and cyclic nucleotides in cell activation. Physiol Rev 57:421–509

Robison GA, Butcher RW, Sutherland EW (1971) Cyclic AMP. Academic Press, New York

Rodriguez de Lores Arnaiz G, Alberici M, DeRobertis E (1967) Ultrastructural and enzymatic studies of cholinergic and noncholingeric synaptic membranes isolated from brain cortex. J Neurochem 14:215–225

Rubin RP (1972) The role of calcium in the release of neurotransmitter substances and hormones. Pharmacol Rev 22:389–428
Rubin CS, Rosen OM (1975) Protein phosphorylation. Annu Rev Biochem 44:831–887
Rylatt DB, Cohen P (1979) Amino acid sequence at the site on rabbit skeletal muscle glycogen synthase phosphorylated by the endogenous glycogen synthase kinase-2 activity. FEBS Lett 98:71–75
Rylatt DB, Embi N, Cohen P (1979) Glycogen synthase kinase-2 from rabbit skeletal muscle is activated by the calcium-dependent regulator protein. FEBS Lett 98:76–80
Schulman H, Greengard P (1978 a) Stimulation of brain membrane protein phosphorylation by calcium and an endogenous heat-stable protein. Nature 271:478–479
Schulman H, Greengard P (1978 b) Ca^{2+}-dependent protein phosphorylation system in membranes from various tissues, and its activation by "calcium-dependent regulator". Proc Natl Acad Sci USA 75:5432–5436
Schulman H, Huttner WB, Greengard P (1980) Calcium-dependent protein phosphorylation in mammalian brain and other tissues. in Ca^{2+}-binding proteins as cellular regulators. Cheung WY (ed) Raven Press, 219–252
Scordilis SP, Adelstein RS (1977) Myoblast myosin phosphorylation is a prerequisite for actin-activation. Nature 268:558–560
Shenolikar S, Cohen PTW, Cohen P, Nairn AC, Perry SV (1979) The role of calmodulin in the structure and regulation of phosphorylase kinase from rabbit skeletal muscle. Eur J Biochem 100:329–337
Sherline P, Lynch A, Glinsmann Wh (1972) Cyclic AMP and adrenergic receptor control of rat liver glycogen metabolism. Endocrinology 91:680–690
Sherry JMF, Gorecka A, Aksoy MO, Dabrowska R, Hartshorne DJ (1978) Roles of calcium and phosphorylation in the regulation of the activity of gizzard myosin. Biochemistry 17:4411–4418
Shimazu T, Amakawa A (1968) Regulation of glycogen metabolism in liver by the autonomic nervous system. III. Differential effects of sympathetic-nerve stimulation and of catecholamines on liver phosphorylase. Biochim Biophys Acta 165:349–356
Shimazu T, Amakawa A (1975) Regulation of glycogen metabolism in liver by the autonomic nervous system. VI. Possible mechanism of phosphorylase activation by the splanchnic nerve. Biochim Biophys Acta 385:242–256
Sieghart W, Forn J, Greengard P (1979) Ca^{2+} and cyclic AMP regulate phosphorylation of same two membrane-associated proteins specific to nerve tissue. Proc Natl Acad Sci USA 76:2475–2479
Sieghart W, Forn J, Schwarcz R, Coyle JT, Greengard P (1978 a) Neuronal localization of specific brain phosphoproteins. Brain Res 156:345–350
Sieghart W, Schulman H, Greengard P (1980) Neuronal localization of Ca^{2+}-dependent protein phosphorylation in brain. J Neurochem, in press
Sieghart W, Theoharides TC, Alper SL, Douglas WW, Greengard P (1978 b) Calcium-dependent protein phosphorylation during secretion by exocytosis in the mast cell. Nature 275:329–331
Small JV, Sobieszek A (1977) Ca-regulation of mammalian smooth muscle actomyosin via a kinase-phosphatase-dependent phosphorylation and dephosphorylation of the 20,000-M_r light chain of myosin. Eur J Biochem 76:521–530
Sobieszek A (1977) Ca-linked phosphorylation of a light chain of vertebrate smooth muscle myosin. Eur J Biochem 73:477–483
Soderling TR, Jett MF, Hutson NJ, Khatra BS (1977) Regulation of glycogen synthase. Phosphorylation specificities of cAMP-dependent and cAMP-independent kinases for skeletal muscle synthase. J Biol Chem 252:7517–7524
Soderling TR, Sheorain VS, Ericsson LH (1979 a) Phosphorylation of glycogen synthase by phosphorylase kinase. Stoichiometry, specificity and site of phosphorylation. FEBS Lett 106:181–184
Soderling TR, Srivastava AK, Bass MA, Khatra BS (1979 b) Phosphorylation and inactivation of glycogen synthase by phosphorylase kinase. Proc Natl Acad Sci USA 76:2536–2540

Srivastava AK, Waisman DM, Brostrom CO, Soderling TR (1979) Stimulation of glycogen synthase phosphorylation by calcium-dependent regulator protein. J Biol Chem 254:583–586
Strombom U, Forn J, Dolphin AC, Greengard P (1979) Regulation of the state of phosphorylation of specific neuronal proteins in mouse brain by in vivo administration of anesthetic and convulsant agents. Proc Natl Acad Sci USA 76:4687–4690
Stull JT, Brostrom CO, Krebs EG (1972) Phosphorylation of the inhibitor component of troponin by phosphorylase kinase. J Biol Chem 247:5272–5274
Stull JT, High CW (1977) Phosphorylation of skeletal muscle contractile proteins in vivo. Biochem Biophys Res Commun 77:1078–1083
Stull JT, Mayer SE (1971) Regulation of phosphorylase activation in skeletal muscle in vivo. J Biol Chem 246:5716–5723
Sutherland EW, Robison GS, Butcher RW (1968) Some aspects of the biological role of adenosine 3′,5′-monophosphate (cyclic AMP). Circulation 37:279–306
Szent-Gyorgyi AG, Szentkiralyi EM, Kendrick-Jones J (1973) The light chains of scallop myosin as regulatory subunits. J Mol Biol 74:179–203
Takai Y, Kishimoto A, Inoue M, Nishizuka Y (1977a) Studies on a cyclic nucleotide-independent protein kinase and its proenzyme in mammalian tissues. Purification and characterization of an active enzyme from bovine cerebellum. J Biol Chem 252:7603–7609
Takai Y, Kishimoto A, Iwasa Y, Kawahara Y, Mori T, Nishizuka Y (1979a) Calcium-dependent activation of a multifunctional protein kinase by membrane phospholipids. J Biol Chem 254:3692–3695
Takai Y, Kishimoto A, Iwasa Y, Kawahara Y, Mori T, Nishizuka Y, Tamura A, Fujii T (1979b) A role of membranes in the activation of a new multifunctional protein kinase system. J Biochem (Tokyo) 86:575–578
Takai Y, Kishimoto A, Kikkawa U, Mori T, Nishizuka Y (1979c) Unsaturated diacylglycerol as a possible messenger for the activation of calcium-activated, phospholipid dependent protein kinase. Biochem Biophys Res Commun 91:1218–1224
Takai Y, Yamamoto M, Inoue M, Kishimoto A, Nishizuka Y (1977b) A proenzyme of cyclic nucleotide-independent protein kinase and its activation by calcium-dependent neutral protease from rat liver. Biochem Biophys Res Commun 77:542–550
Teo TS, Wang JH (1973) Mechanism of activation of a cyclic adenosine 3′:5′-monophosphate phosphodiesterase from bovine heart by calcium ions. Identification of the protein activator as a Ca^{2+}-binding protein. J Biol Chem 248:5950–5955
Teo TS, Wang TH, Wang JH (1973) Purification and properties of the protein activator of bovine heart cyclic adenosine 3′,5′-monophosphate phosphodiesterase. J Biol Chem 248:588–595
Theoharides TC, Sieghart W, Greengard P, Douglas WW (1980) Antiallergic drug cromolyn may inhibit histamine secretion by regulating phosphorylation of a mast cell protein. Science 207:80–82
Tolbert MEM, Butcher FR, Fain JN (1973) Lack of correlation between catecholamine effects on cyclic adenosine 3′:5′-monophosphate and gluconeogenesis in isolated rat liver cells. J Biol Chem 248:5686–5692
Trotter JA, Adelstein RS (1979) Macrophage myosin. Regulation of actin-activated ATPase activity by phosphorylation of the 20,000alton light chain. J Biol Chem 254:8781–8785
Ueda T, Berzins K, Cohen RS, Blomberg R, Siekevitz P, Greengard P (1979) Sub-cellular distribution in cerebral cortex of two proteins phosphorylated by a cAMP-dependent protein kinase. J Cell Biol 83:308–319
Ueda T, Greengard P (1977) Adenosine 3′:5′-monophosphate-regulated phosphoprotein system of neuronal membranes. I. Solubilization, purification, and some properties of an endogenous phosphoprotein. J Biol Chem 252:5155–5163
Ulbricht W (1969) The effect of veratridine on excitable membranes of nerve and muscle. Ergeb Physiol Biol Chem Exp Pharmakol 61:18–71
Walsh DA, Perkins JP, Brostrom CO, Ho ES, Krebs EG (1971) Catalysis of the phosphorylase kinase activation reaction. J Biol Chem 246:1968–1976
Walsh DA, Perkins JP, Krebs EG (1968) An adenosine 3′,5′-monophosphate dependent protein kinase from rabbit skeletal muscle. J Biol Chem 243:3763–3765

Walsh KX, Millikin DM, Schlender KK, Reimann EM (1979) Calcium-dependent phosphorylation of glycogen synthase by phosphorylase kinase. J Biol Chem 254:6611–6616

Wang JH, Stull JT, Huang TH, Krebs EG (1976) A study of the autoactivation of rabbit muscle phosphorylase kinase. J Biol Chem 251:4521–4527

Wang JH, Waisman DM (1979) Calmodulin and its role in the second messenger system. In: Current topics in cellular regulation, vol 15. Academic Press, New York, pp 47–109

Watterson DM, Harrelson WG, Jr, Keller PM, Sharief F, Vanaman TC (1976) Structural similarities between the Ca^{2+}-dependent regulatory proteins of 3′:5′-cyclic nucleotide phospodiesterase and actomyosin ATPase. J Biol Chem 251:4501–4513

Weeds AG (1969) Light chains of myosin. Nature 223:1362–1364

Witters LA, Kowaloff EM, Avruch J (1979) Glucagon regulation of protein phosphorylation. Identification of acetyl coenzyme A carboxylase as a substrate. J Biol Chem 254:245–248

Wolff DJ, Brostrom CO (1979) Properties and functions of the calcium-dependent regulator protein. In: Greengard P, Robison GA (eds) Advances in cyclic nucleotide research, vol 11. Raven Press, New York, pp 27–88

Wolff DJ, Poirier PG, Brostrom CO, Brostrom MA (1977) Divalent cation binding properties of bovine brain Ca^{2+}-dependent regulator protein. J Biol Chem 252:4108–4117

Wong P-K, Cheung WY (1979) Calmodulin stimulates human platelet phospholipase A_2. Biochem Biophys Res Commun 90:473–480

Yagi K, Okamoto Y, Yazawa Y (1975) Low molecular weight components (g-chains) of myosin from rabbit skeletal muscle. Separation, amino acid compositions and contents in myosin. J Biochem (Tokyo) 77:333–342

Yagi K, Yazawa M, Kakiuchi W, Ohshima M, Uenishi K (1978) Identification of an activator protein for myosin light chain kinase as the Ca^{2+}-dependent modulator protein. J Biol Chem 253:1338–1340

Yamamoto M, Takai Y, Inoue M, Kishimoto A, Nishizuka Y (1978) Characterization of cyclic nucleotide-independent protein kinase produced enzymatically from its proenzyme by calcium-dependent neutral protease from rat liver. J Biochem (Tokyo) 83:207–212

Yamauchi T, Fujisawa H (1979) Activation of tryptophan 5-monooxygenase by calcium-dependent regulatory protein. Biochem Biophys Res Commun 90:28–35

Yazawa M, Kuwayama H, Yagi K (1978) Modulatory protein as a Ca^{2+}-dependent activator of rabbit skeletal myosin light-chain kinase. J Biochem (Tokyo) 84:1253–1258

Yazawa M, Yagi K (1977) A calcium-binding subunit of myosin light chain kinase. J Biochem (Tokyo) 82:237–289

Yazawa M, Yagi K (1978) Purification of modulator-deficient myosin light chain kinase by modulator protein-sepharose affinity chromatography. J Biochem 84:1259–1265

Yeaman SJ, Cohen P (1975) The hormonal control of activity of skeletal muscle phosphorylase kinase: Phosphorylation of the enzyme at two sites in vivo in response to adrenaline. Eur J Biochem 51:93–104

Yerna M-J, Dabrowska R, Hartshorne DJ, Goldman RD (1979) Calcium-sensitive regulation of actin-myosin interactions in baby hamster kidney (BHK-21) cells. Proc Natl Acad Sci USA 76:184–188

CHAPTER 14

Photoaffinity Labeling of Cyclic AMP-Dependent and Cyclic GMP-Dependent Protein Kinases

U. WALTER and P. GREENGARD

Overview

Photoaffinity labeling is a useful technique for the study of ligand-receptor interactions. Irradiation of chemically modified ligands with light produces highly reactive intermediates which are capable of forming covalent bonds with the receptors. Photoaffinity labeling has been used for many different ligands and has been especially successful for cyclic nucleotides. Several photoaffinity analogues of cAMP and cGMP are available. 8-Azido-cAMP (8-N_3-cAMP) has been the one most widely and successfully used. 8-N_3-cAMP binds to the regulatory subunit of a Type II cAMP-dependent protein kinase with a slightly lower affinity than cAMP and is able to label all cAMP-binding sites under saturating conditions. The data indicate that 8-N_3-cAMP is an effective analogue of cAMP and a very efficient photoaffinity label for cAMP-binding sites. Regulatory subunits of Type I and Type II cAMP-dependent protein kinases are the only proteins present in all tissues studied which are specifically labeled by 8-N_3-cAMP. Other minor labeled proteins appear to be proteolytic fragments of the two types of regulatory subunits. The finding that most, if not all, specific cAMP-receptor proteins are regulatory subunits of cAMP-dependent protein kinase strongly supports the hypothesis that cAMP-dependent protein kinases and their subunits mediate all the physiological effects of cAMP. Photoaffinity labeling has been used to detect and characterize proteolytic fragments of cAMP-dependent protein kinases and to study the regulation of the amounts of cAMP-dependent protein kinases and their subunits. Similarly, photoaffinity analogues of cGMP have been used to label cGMP-dependent protein kinases. Photoaffinity analogues of cyclic nucleotides are powerful tools for the investigation of cAMP- and cGMP-receptor proteins. Studies with these analogues should contribute substantially to an understanding of the mechanism of action of cyclic nucleotides as second messengers.

A. Introduction

I. Principles of Affinity Labeling and Photoaffinity Labeling

In 1962, SINGH, THORNTON, and WESTHEIMER introduced the use of photo-generated reagents with diazoacetyl chymotrypsin (SINGH et al. 1962). Since that initial report, photoaffinity labeling has emerged as an important technique in the study of ligand-binding molecules. Several excellent reviews on the principles and use of photoaffinity labeling in general have been published (KNOWLES 1972; CREED 1974; COOPERMAN 1976; BAYLEY and KNOWLES 1977; WOLD 1977; CHOWDHRY and WESTHEIMER 1979).

Affinity labeling and its subspecialty, photoaffinity labeling, are applicable in principle to all processes in which one molecule binds to another. In biology, these techniques have been applied most successfully to study the binding of ligands to their specific receptors which are primarily proteins. The ligands of interest include substrates, inhibitors, coenzymes, hormones, neurotransmitters, antigens, haptens, allosteric modulators and other regulatory substances. The principal goal of affinity labeling and photoaffinity labeling is the formation of a covalent bond between the ligand and the receptor. Such covalent bond formation is useful for the study of a wide range of problems, including the following:

a) Permanent activation or inactivation of an enzyme or receptor.

b) Identification and characterization of a receptor molecule.

c) Isolation and sequencing of the active binding site of the receptor molecule.

Although certain compounds may have some intrinsic photolability, photoaffinity labeling usually requires insertion of a photolabile group *via* chemical modification of a ligand. Such photolabile groups include diazocarbonyl-, diazomalonyl-, diazoacetyl-, and aryl azide residues. Irradiation of these chemically modified ligands with visible or ultraviolet light produces highly reactive intermediates, carbenes from diazoacyl derivatives and nitrenes from aryl azides. These highly reactive intermediates are able to form a stable covalent bond with amino acids at the binding site of the receptor molecule. Possible reactions of carbenes and nitrenes have been reviewed (KNOWLES 1972; BAYLEY and KNOWLES 1977). The advantage of photoaffinity labeling over classical affinity labeling is that the experimenter determines the time when the photolabile group is converted to a species of a very high chemical reactivity.

The success of photoaffinity labeling of receptor proteins depends on meeting several criteria:

a) The photoaffinity reagent should be readily synthesized and chemically stable. For biological studies, high stability in aqueous solutions is practically always required.

b) Photolysis should produce an active intermediate that reacts with amino acid residues of proteins.

c) The photoaffinity reagent should mimic the natural ligand, i.e. it should bind specifically to the binding site of the macromolecule.

d) Upon photolysis it should couple covalently to the binding site of the macromolecule with high specificity and efficiency.

e) The natural ligand should be able to prevent the light induced labeling of the receptor site by the photoaffinity analogue.

f) The light induced covalent bond should be stable under a wide variety of experimental conditions to allow for analysis.

Photoaffinity labeling has been successfully used to study the combining site of antibodies, to characterize transport proteins and to identify receptor proteins for acetylcholine, steroid hormones and peptide hormones (for a review see CHOWDHRY and WESTHEIMER 1979). In the last few years, photoaffinity analogues of cyclic nucleotides have been increasingly used to study receptor proteins for cAMP and cGMP, and that is the subject of this Chapter. It will be helpful for understanding this material to review briefly what is known about cAMP- and cGMP-receptor proteins. The reader is also referred to Chaps. 11 and 12 in this volume.

II. Brief Review of cAMP- and cGMP-Receptor Proteins

The discovery of a cAMP-dependent protein kinase (WALSH et al. 1968) led to the suggestion that cAMP may exert most or all of its physiological effects through such cAMP-dependent protein kinases (KUO and GREENGARD 1969). It was later found that, in *E. coli*, cAMP stimulates gene transcription *via* CRP (Cyclic AMP Receptor Protein) which is not a protein kinase (DE CROMBRUGGHE et al. 1971; PASTAN and PERLMAN 1972). For eukaryotes, however, the hypothesis (KUO and GREENGARD 1969) remains valid that cAMP-dependent protein kinases are the major if not the only receptor proteins for cAMP, and this hypothesis is supported by the photoaffinity labeling experiments reviewed in this Chapter. Two types of cAMP-dependent protein kinase, designated Type I and Type II, can be resolved by DEAE-cellulose chromatography, which differ in their tissue distribution, biochemical and immunological properties (CORBIN et al. 1975; HOFMANN et al. 1975; FLEISCHER et al. 1976; HOFMANN et al. 1977). These two cAMP-dependent protein kinases have identical catalytic but different regulatory subunits (the cAMP-binding unit). The regulatory subunit of the Type I protein kinase (R-I) has a molecular weight of 47,000, whereas the regulatory subunit of the Type II protein kinase (R-II) has a molecular weight of 55,000. Another major difference is autophosphorylation, namely the phosphorylation of the regulatory subunit by the catalytic subunit, which has been observed with the Type II but not with the Type I cAMP-dependent protein kinase. Although the R-I and R-II regulatory subunits are the best characterized cAMP-receptor proteins in eukaryotes, there are a few other proteins which are able to bind cAMP. Obviously, cAMP-degrading enzymes, the phosphodiesterases, must be capable of binding cAMP. Phosphofructokinase was reported to bind ADP, cAMP, and AMP in the μmolar range (KEMP and KREBS 1967).

A protein which binds cAMP and adenosine has been found in rabbit erythrocytes (YUH and TAO 1974) and mouse liver (UELAND and DØSKELAND 1977). This protein, which has a molecular weight of 45,000–48,000 on sodium dodecyl sulfate polyacrylamide-gels, has been identified recently as S-adenosylhomocysteine hydrolase (HERSHFIELD and KREDICH 1978; SAEBØ and UELAND 1978). However, it remains to be demonstrated that the binding of cAMP to phosphofructokinase and to S-adenosylhomocysteine hydrolase is of physiological significance.

A specific cGMP-dependent protein kinase was found initially in several invertebrate tissues (KUO and GREENGARD 1970; KUO et al. 1971). This enzyme has subsequently been found in a number of vertebrate tissues and has been purified to homogeneity from bovine lung (GILL et al. 1976; DEJONGE and ROSEN 1977; LINCOLN et al. 1977) and bovine heart (FLOCKERZI et al. 1978). This enzyme was shown to be a dimer consisting of two identical subunits with molecular weight of 74,000 (GILL et al. 1976) to 81,000 (LINCOLN et al. 1977). Although this cGMP-dependent protein kinase is the only well-characterized receptor for cGMP, there are a few

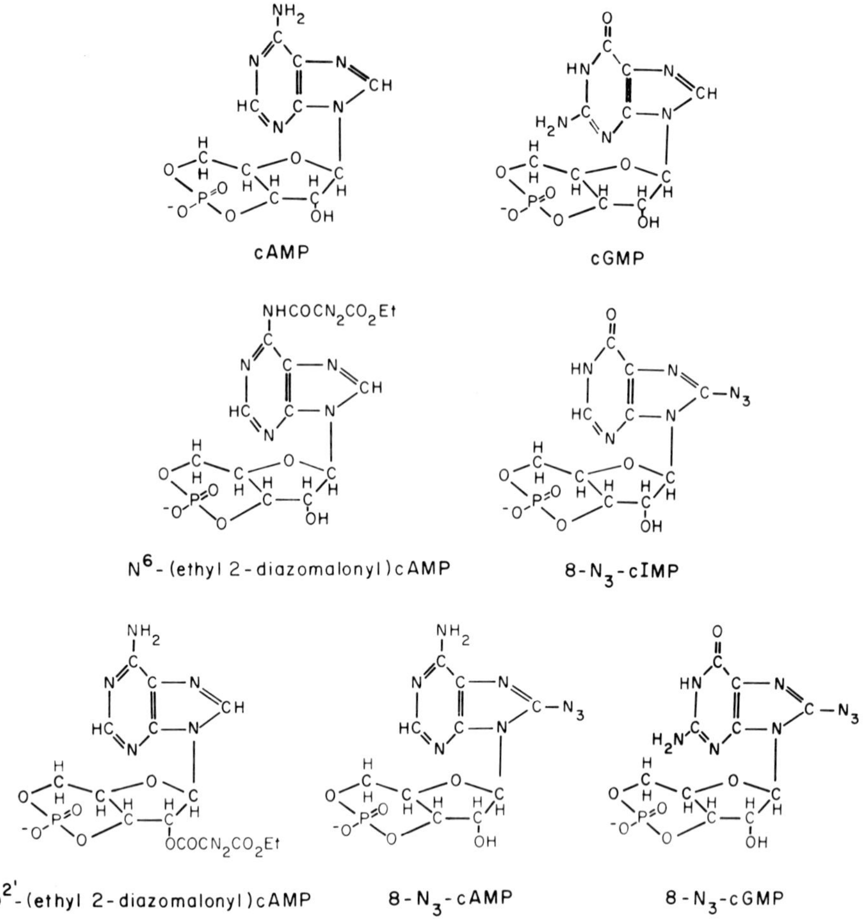

Fig. 1. Cyclic nucleotides and some of their photoaffinity analogues

reports that other cGMP-binding proteins may exist (LINCOLN et al. 1976; HAMET and COQUIL 1978).

It is the purpose of this Chapter to review how photoaffinity analogues of cAMP and cGMP can be used to identify and characterize specific cAMP- and cGMP-binding proteins.

B. Photoaffinity Analogues of cAMP and cGMP

I. Available Photoaffinity Analogues of cAMP and cGMP

Figure 1 shows the structures of cAMP, of cGMP, and of some of their analogues which have been successfully used in photoaffinity labeling experiments.

It has been shown that unmodified cAMP and cGMP may undergo photoincorporation into different receptors present in extracts from testis and adrenal cortex (ANTONOFF and FERGUSON 1974; ANTONOFF et al. 1976). The amount of cAMP and cGMP covalently incorporated was very low compared to that bound non-covalently. Also, an especially long exposure to UV-light was required.

Several different diazomalonyl derivatives of cAMP have been synthesized (BRUNSWICK and COOPERMAN 1971). $O^{2'}$-(Ethyl-2-diazomalonyl)cAMP was used for the photoaffinity labeling of rabbit muscle phosphofructokinase (COOPERMAN and BRUNSWICK 1973). N^6-(Ethyl-2-diazomalonyl)cAMP was used to label cAMP-binding proteins of erythrocyte membranes (RUBIN 1975). A single protein with molecular weight of 48,000 was labeled and identified as the regulatory subunit of a membrane-bound cAMP-dependent protein kinase. The labeling efficiency appeared to be rather low.

HALEY was the first to recognize that azide derivatives of nucleotides have the potential to be valuable photoaffinity analogues. He synthesized 8-N_3-ATP and demonstrated its ability to label ATP-binding sites of human red cell membranes (HALEY and HOFFMANN 1974). Subsequently, he synthesized 8-N_3-cAMP, which was found to label cAMP-binding sites of human red cell membranes (HALEY 1975) and of a partially purified cAMP-dependent protein kinase (POMERANTZ et al. 1975). 8-N_3-cAMP has been shown to be a valuable photoaffinity reagent as will be discussed later.

8-N_3-cIMP, closely related to both cGMP and cAMP, has been demonstrated to be effective as a photoaffinity label for cGMP- and cAMP-dependent protein kinases (CASNELLIE et al. 1978) and 8-N_3-cGMP has recently been synthesized and used as a photoaffinity label for a partially purified cGMP-dependent protein kinase (GEAHLEN et al. 1979).

II. Synthesis and Characterization of 8-N_3-[^{32}P]cAMP

8-N_3-cAMP is derived from cAMP *via* a two-step synthetic procedure (MUNEYAMA et al. 1971). cAMP is brominated to yield 8-bromo-cAMP which subsequently undergoes an azide exchange reaction to yield 8-N_3-cAMP. Detailed descriptions of the synthesis of the radioactive analogue, 8-N_3-[^{32}P]cAMP have been published (HALEY 1975; WALTER et al. 1977a; HALEY 1977). It is always important to check the purity and photolability of 8-N_3-[^{32}P]cAMP employing thin layer chromatography. 8-N_3-[^{32}P]cIMP can be quantitatively prepared from 8-N_3-[^{32}P]cAMP by a simple deamination reaction (CASNELLIE et al. 1978; CASNELLIE et al. 1980). 8-N_3-[^{32}P]cAMP and 8-N_3-[^{32}P]cIMP appear to be stable for at least 4 weeks when stored in methanol at -20 °C. 8-N_3-[^{32}P]cGMP is similarly prepared by a bromination and subsequent azide exchange reaction (GEAHLEN et al. 1979), although the yield and stability of this compound do not appear to be as good as for 8-N_3-[^{32}P]cAMP.

III. Procedures for Photoaffinity Labeling and Other Methods to Assay cAMP- and cGMP-Dependent Protein Kinases

Photoaffinity labeling experiments are performed using glass scrology plates, for example Pyrex spot plates. Solutions are placed into the individual wells and kept at 4 °C throughout the experiment. A typical reaction mixture (final volume, 0.1 ml) contains 50 m*M* sodium morpholino-ethanesulfonate (pH 6.2), 10 m*M* $MgCl_2$, 1 m*M* IBMX, 0.5 m*M* β-mercaptoethanol, 1 μ*M* 8-N_3-[^{32}P]cAMP (2–4 Ci/mmol) and various amounts of protein up to 200 μg. To detect specific labeling, incubations are usually carried out in the absence and presence of an excess of unlabeled cAMP, for example, 20 μ*M*. Preincubations are usually carried out for 60 min to allow reversible cyclic nucleotide binding to reach equilibrium. Although 8-N_3-cAMP is not highly sensitive to normal day light, it is recommended to keep the sample in the dark. After the preincubation period, the samples are irradiated for 10 min at 254 nm with a Mineralite UVS-11 hand lamp at a distance of 8 cm. Total covalent incorporation of 8-N_3-[^{32}P]cAMP can be measured by acid precipitaton, but this technique is not very useful, especially for complex protein mixtures, since individual proteins are not separated. For the separation of proteins after photolysis, each sample is mixed with 50 μl of a sodium dodecyl sulfate-containing "stop solution" and heated for 2 min at 100 °C. The entire sample is then subjected to polyacrylamide slab gel electrophoresis. After the electrophoresis is completed, the slab gels are stained for protein, destained and dried, and autoradiography is carried out. Radioactive bands localized by autoradiography can be cut out of the dried gel and counted by liquid scintillation spectrometry to determine the absolute amounts of radioactivity incorporated. The reader is referred elsewhere for more details (WALTER et al. 1977a, b). The method for photoaffinity labeling for the other azide derivatives of cyclic nucleotides is very similar to that described here for 8-N_3-cAMP.

Photoaffinity labeling of cAMP- and cGMP-dependent protein kinases is often compared with other useful assays for cyclic nucleotide-dependent protein kinases such as endogenous phosphorylation, histone kinase activity and cyclic nucleotide-binding capacity.

A typical reaction mixture for measurement of phoshorylation of endogenous proteins (final volume 0.1 ml) contains 50 m*M* morpholinoethanesulfonate (pH 6.2), 10 m*M* $MgCl_2$, 1 m*M* IBMX, various amounts of protein up to 200 μg, with and without 10 μ*M* cAMP. The sample is preincubated for 1 min at 30 °C, and the phosphorylation reaction is then initiated by the addition of 4 μ*M* [γ-^{32}P]ATP. (This concentration of ATP is not saturating for many reactions.) The reaction, carried out for 10 s to 1 min in a Dubnoff metabolic shaker, is terminated by the addition of 50 μl of a sodium dodecyl sulfate-containing stop solution, and the sample is heated for 2 min at 100 °C. The entire sample is then subjected to sodium dodecyl sulfate polyacrylamide slab gel electrophoresis and further processed as described for the photoaffinity labeling experiments. A detailed description of endogenous protein phosphorylation, as studied in this laboratory, has been published (RUDOLPH and KRUEGER 1979).

Enzyme activity of cAMP- and cGMP-dependent protein kinase is routinely determined by measuring the rate of incorporation of radioactive phosphate from [γ-^{32}P]ATP into histones in the presence and absence of cyclic nucleotides. Separation of the phosphohistone from ATP is usually accomplished by acid precipitation of histones onto filter paper disks or by specific adsorption of histones to phosphocellulose paper. Another rapid assay for cAMP- and cGMP-dependent protein kinases is provided by measuring the direct binding of radioactive cAMP

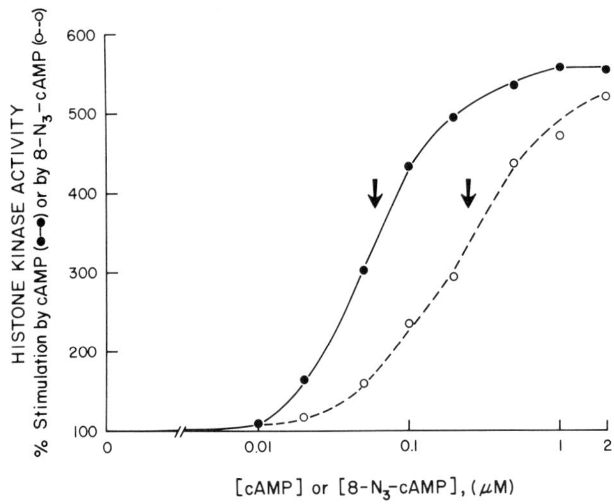

Fig. 2. Stimulation of the histone kinase activity of a Type II cAMP-dependent protein kinase by cyclic AMP and 8-N_3-cAMP. The enzyme was purified 800-fold from bovine heart. Half-maximal stimulation was obtained at 0.06 μM cAMP and at 0.25 μM 8-N_3-cAMP. (U. WALTER and P. GREENGARD unpublished)

or cGMP to the protein kinase. Bound cyclic nucleotide is separated from free cyclic nucleotide by filtration on Millipore filters. A detailed description of both the histone kinase assay and the binding assay has appeared (GILL and WALTON 1979).

C. Studies with Purified cAMP-Dependent Protein Kinase

I. Photoaffinity Labeling of Regulatory Subunit

A good photoaffinity reagent should be able to mimic its natural ligand reasonably well. It was recognized early that 8-N_3-cAMP is a good activator of cAMP-dependent protein kinase but a poor substrate for phosphodiesterases (MUNEYAMA et al. 1971). For the comparison of some binding and photoaffinity labeling properties of 8-N_3-cAMP, we used a well-characterized cAMP-receptor protein, a cAMP-dependent protein kinase from bovine heart. This enzyme, a Type II protein kinase, offers the advantage that the regulatory subunit can be measured not only by the cAMP-binding assay but also by autophosphorylation. 8-N_3-cAMP was able to replace cAMP in stimulating the histone kinase activity of the purified cAMP-dependent protein kinase from bovine heart (Fig. 2), although slightly higher concentrations of 8-N_3-cAMP were required. Thus, half-maximal stimulation of histone kinase activity was obtained at 0.06 μM cAMP and at 0.25 μM 8-N_3-cAMP. Figure 3 shows an experiment where the bovine heart protein kinase was examined for photoaffinity labeling by 8-N_3-[^{32}P]cAMP and for autophosphorylation. The regulatory subunit of the holoenzyme was maximally phosphorylated both in the absence and presence of cAMP since the catalytic subunit of this cAMP-dependent

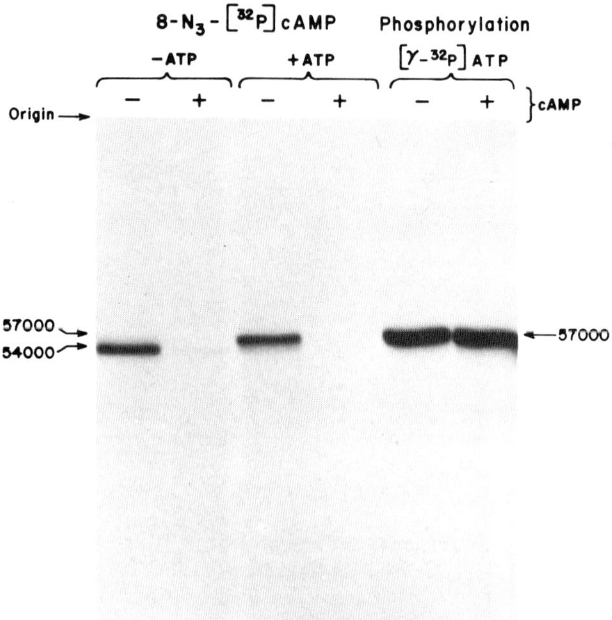

Fig. 3. Autoradiograph showing 8-N_3-[^{32}P]cAMP incorporation into and autophosphorylation of Type II cAMP-dependent protein kinase from bovine heart. Photoaffinity labeling was carried out using 1 μ*M* 8-N_3-[^{32}P]cAMP in the absence or presence of 1 m*M* ATP and of 20 μ*M* cAMP. Phosphorylation was carried out in the absence or presence of 10 μ*M* cAMP. Samples containing 0.5 μg of protein were analyzed by sodium dodecyl sulfate-polyacrylamide gel electrophoresis. The arrows indicate the apparent molecular weight of the radioactive bands. (Taken from WALTER and GREENGARD 1978, with permission)

protein kinase is able to phosphorylate the regulatory subunit both in the undissociated and dissociated state of the holoenzyme (RANGEL-ALDAO and ROSEN 1976). Using the photoaffinity labeling procedure, the enzyme covalently incorporated 8-N_3-[^{32}P]cAMP into the regulatory subunit, and this incorporation was completely prevented by an excess of cold cAMP. Prephotolysis of the 8-N_3-[^{32}P]cAMP also completely prevented the photoactivated incorporation (not shown). A small decrease in electrophoretic mobility of the regulatory subunit, detected by 8-N_3-[^{32}P]cAMP incorporation, was observed when the photoaffinity labeling was carried out in the presence of ATP, resulting in a shift of the apparent molecular weight of the regulatory subunit from 54,000 in the absence to 57,000 in the presence of ATP. This shift is due to the phosphorylation of the regulatory subunit (HOFMANN et al. 1975). The concentration-dependence of reversible binding of [^{3}H]cAMP and 8-N_3-[^{32}P]cAMP and the photoactivated incorporation of 8-N_3-[^{32}P]cAMP were compared using the bovine heart protein kinase (Fig. 4). In agreement with the results of the histone kinase assay, higher concentrations of 8-N_3-[^{32}P]cAMP than of [^{3}H]cAMP were required for half-maximal binding. Dissociation constants (K_d) were 0.09 μ*M* for [^{3}H]cAMP and 0.33 μ*M* for 8-N_3-[^{32}P]cAMP. The photoactivated incorporation of 8-N_3-[^{32}P]cAMP displayed a similar concentration dependence, half-maximal incorporation occurring at

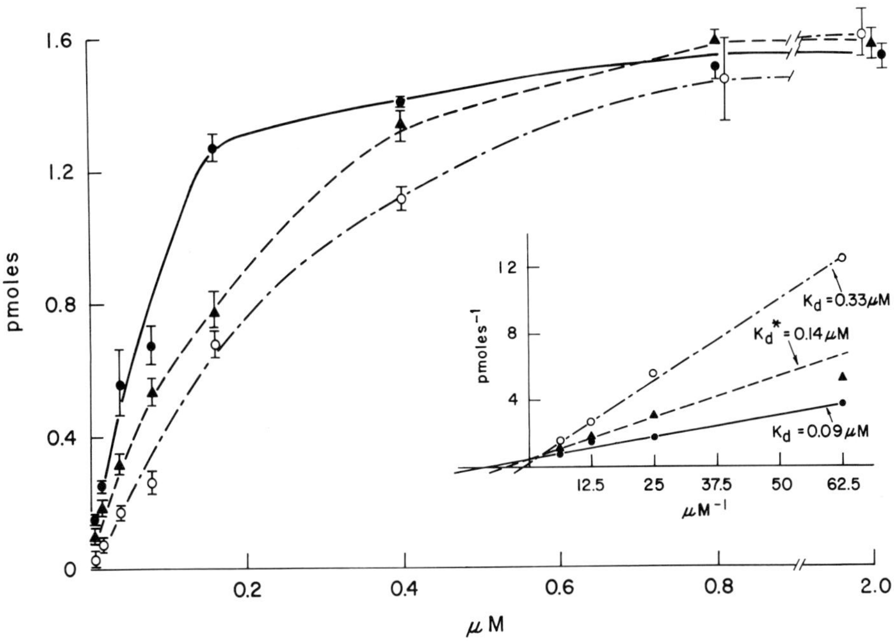

Fig. 4. Concentration-dependent photoactivated incorporation of 8-N_3-[^{32}P]cAMP (▲—▲) into, and reversible binding of [^{3}H]cAMP (●—●) or 8-N_3-[^{32}P]cAMP (○—○), to, Type II cAMP-dependent protein kinase from bovine heart. The protein kinase (0.5 μg of protein) was analyzed for reversible binding using the indicated concentrations of [^{3}H]cAMP or 8-N_3-[^{32}P]cAMP. The same amount of enzyme was analyzed for the photoactivated incorporation of 8-N_3-[^{32}P]cAMP using the indicated concentrations of 8-N_3-[^{32}P]cAMP. All results are expressed as pmoles of [^{3}H]cAMP or 8-N_3-[^{32}P]cAMP per 0.5 μg of enzyme protein. The mean value and standard deviation of triplicate determinations is shown. Dissociation constants (K_d) were calculated from double-reciprocal plots of the results (inset). *The K_d value given for 8-N_3-[^{32}P]cAMP incorporation represents the concentration of 8-N_3-[^{32}P]cAMP producing half-maximal incorporation, and is not a true dissociation constant. (Taken from Walter and Greengard 1978, with permission)

0.14 μ*M* 8-N_3-[^{32}P]cAMP. In addition, the results of Fig. 4 demonstrate that 8-N_3-[^{32}P]cAMP is able, under saturating conditions, to covalently label all available cAMP-binding sites of the protein kinase from bovine heart. These experiments illustrate that 8-N_3-cAMP, at least in the case of the cAMP-dependent protein kinase from bovine heart, is:

a) a suitable analogue of cAMP,

b) a true photoaffinity label for cAMP-binding sites,

c) an extremely efficient photoaffinity analogue (most or all available cAMP-binding sites are covalently labeled under appropriate conditions during one course of photolysis).

The last point was unexpected since most photoaffinity analogs, including other nucleotide photoaffinity analogues (Owens and Haley 1978), label at best only a fraction of the total binding sites. One explanation for the very efficient photoactivated incorporation of 8-N_3-cAMP is obviously the high affinity of the regulatory subunit for cAMP and 8-N_3-cAMP. In addition, the cyclic nucleotide may be bound in a rigid conformation, in which the nitrene formed during photolysis has little or no access to solvent and other solute molecules. This possibility has been suggested by experiments with the bovine heart protein kinase using a fluorescent photoaffinity analogue of cAMP, 8-azido-1,N^6-ethenoadenosine 3′,5′-cyclic monophosphate (DREYFUSS et al. 1978a). The observed fluorescent signal was blue-shifted and highly polarized upon binding of the ligand to the receptor, indicating that the cAMP-receptor sites are highly hydrophobic and immobilize the adenosine moiety of the cyclic nucleotide in a rather rigid conformation.

II. Photoaffinity and Affinity Labeling of Catalytic Subunit

The Type I cAMP-dependent protein kinase holoenzyme is known to bind ATP with high affinity (HOFMANN et al. 1975). Binding of ATP to the Type I protein kinase decreases the affinity of this enzyme for cAMP. 8-N_3-ATP was used to identify the high affinity binding site for ATP (HOPPE and FREIST 1979). Photoactivated incorporation of 8-N_3-ATP was observed only into the catalytic and not into the regulatory subunit. In addition, a loss of catalytic activity, proportional to the amount of 8-N_3-ATP incorporated, was found. These results strongly support the conclusion that the high affinity ATP-binding site of the Type I protein kinase holoenzyme is located on the catalytic subunit and not on the regulatory subunit. This agrees well with another study which reported that ATP does not bind to the free regulatory subunit of Type I protein kinase (SCHWECHHEIMER and HOFMANN 1977). Affinity labeling and inactivation of the catalytic subunit of cAMP-dependent protein kinase has also been achieved by the use of N^{α}-tosyl-L-lysine chloromethyl ketone (KUPFER et al. 1979) or by the use of 5′-p-fluorosulfonylbenzoyl-adenosine (HIXSON and KREBS 1979; ZOLLER and TAYLOR 1979).

D. Application of Photoaffinity Labeling to Crude Systems

One of the important applications of photoaffinity labeling experiments is to identify specific receptor proteins in crude extracts. Such an experiment is shown in Fig. 5, in which the photoactivated incorporation of 8-N_3-[^{32}P]cAMP into soluble proteins from rat heart, liver and uterus was examined. Of the many proteins present in such soluble fractions, only a few proteins incorporated 8-N_3-[^{32}P]cAMP to a significant extent. Two of them, with molecular weights of 47,000 and 54,000, were common to all tissues and specifically incorporated 8-N_3-[^{32}P]cAMP as indicated by the finding that this incorporation could be prevented by an excess of unlabeled cAMP. A few other proteins non-specifically incorporated 8-N_3-[^{32}P]cAMP as indicated by the observation that it occurred both in the absence and in the presence of an excess of cAMP. One of these non-specifically labeled proteins with a molecular weight of 67,000 is serum albumin (WALTER et al. 1977a).

It was found that the presence of a low concentration of β-mercaptoethanol (i.e. 0.5 mM) could prevent most non-specific labeling without affecting the specific labeling (WALTER and GREENGARD 1978; LOHMANN et al. 1978). This suggests that β-mercaptoethanol can act as a "scavenger" of free radicals generated during photolysis (RUOHO et al. 1973). Occasionally proteins with molecular weights

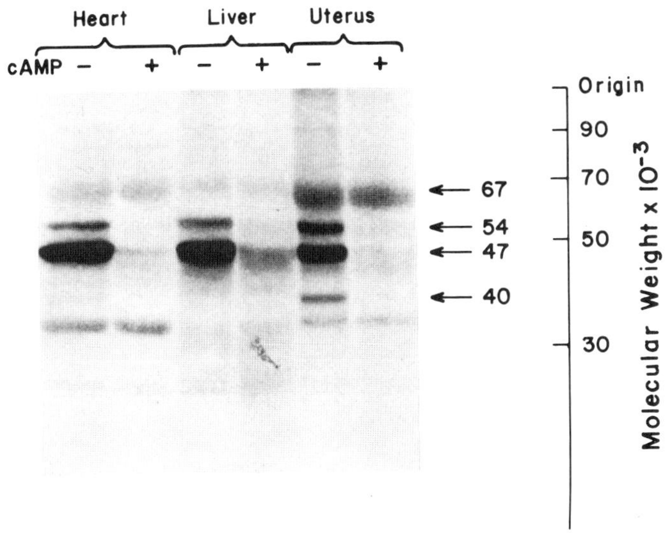

Fig. 5. Autoradiograph showing the photoactivated incorporation of 8-N_3-[^{32}P]cAMP into various protein bands of cytosol from heart, liver, and uterus of the rat. The photoactivated incorporation using 0.4 μ*M* 8-N_3-[^{32}P]cAMP was performed in the absence or presence of 10 μ*M* cAMP. Each channel contained 200 μg of protein. (Taken from WALTER et al. 1977a, with permission)

smaller than 47,000 and 54,000 were observed which also specifically incorporated 8-N_3-[^{32}P]cAMP. For example, such a protein, with a molecular weight of 40,000, was found in uterus (Fig. 5). A variety of evidence suggests that these lower molecular weight proteins are most likely the product of proteolysis of the 47,000 and 54,000 Dalton proteins. For example, treatment of cytosol fractions with trypsin prior to the photoaffinity experiments resulted in a dramatic decrease of the 54,000 Dalton protein and a concomitant increase of the 40,000 Dalton protein (U. Walter, unpublished experiments). Half-maximal, photoactivated incorporation was obtained at about 0.06 μ*M* 8-N_3-[^{32}P]cAMP in the case of the 47,000 Dalton protein and at about 0.2 μ*M* 8-N_3-[^{32}P]cAMP in the case of the 54,000 Dalton protein. Other experiments demonstrated that cAMP was very specific in preventing the photoactivated incorporation into soluble proteins of several tissues. When tested at a 20- to 30-fold higher concentration than 8-N_3-[^{32}P]-cAMP, several nucleotides and other substances including cGMP, AMP, ATP, and adenosine had little or no effect on the 8-N_3-[^{32}P]cAMP incorporation into any of the labeled proteins (WALTER et al. 1977a). Apart from cAMP, only cIMP was a potent inhibitor of the 8-N_3-[^{32}P]cAMP incorporation into the 47,000 Dalton protein. The failure of adenosine to inhibit any 8-N_3-[^{32}P]-cAMP incorporation suggested that the "adenosine-cAMP-binding protein" (SUGDEN and CORBIN 1976; UELAND and DØSKELAND 1977; HERSHFIELD and KREDICH 1978; YUH and TAO 1974) was either not present, undetectable or not a target for the photoaffinity labeling by 8-N_3-[^{32}P]cAMP.

In view of the molecular weights of the purified regulatory subunits of the Type I and Type II cAMP-dependent protein kinase (HOFMANN et al. 1975), it

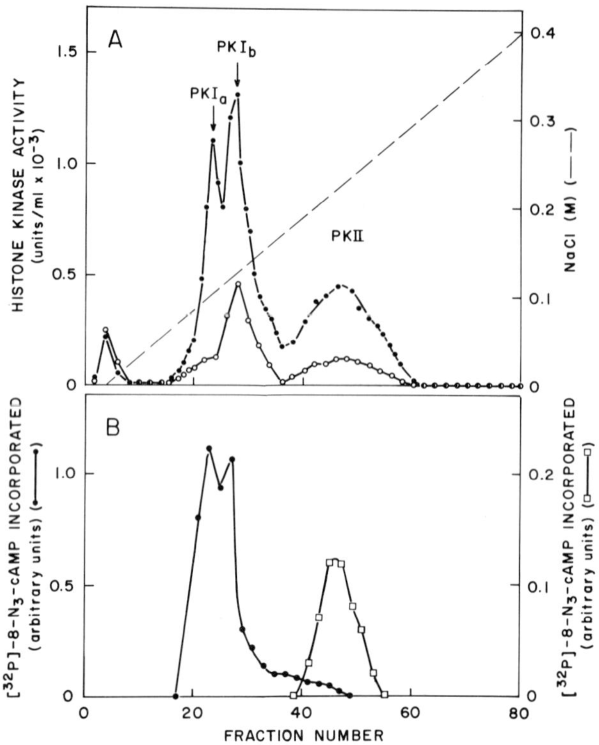

Fig. 6 A, B. DEAE-cellulose column chromatography of rat liver cytosol. Ten milliliters of a 100,000 × *g* supernatant of liver, containing 22.5 mg of protein/ml, were applied to a DEAE-cellulose column and the column eluted using a NaCl gradient as indicated. Fractions were 4.5 ml. **A** aliquots of 20 µl were assayed for histone kinase activity in the absence (○—○) or presence (●—●) of 10 µ*M* cAMP. *PK*, protein kinase. **B** aliquots of 70 µl were analyzed for the photoactivated incorporation of 8-N_3-[^{32}P]cAMP into the 47,000 (●—●) and into the 54,000 (□—□) molecular weight protein band. (Taken from WALTER et al. 1977 a, with permission)

seemed very likely that the 47,000 and 54,000 Dalton proteins specifically labeled by 8-N_3-[^{32}P]cAMP in the cytosol of various tissues represented the regulatory subunits of Type I and Type II cAMP-dependent protein kinase. This was substantiated by the finding that the 47,000 Dalton protein copurified with the Type I protein kinase (PK-I) and that the 54,000 Dalton protein copurified with the Type II protein kinase (PK-II) (Fig. 6; WALTER et al. 1977 a; WALTER et al. 1977 b) using DEAE-cellulose chromatography. The Type I protein kinase from liver was partially resolved by DEAE-cellulose chromatography into two peaks designated PK-Ia and PK-Ib, both of which contained the 47,000 Dalton protein. This indicates that the Type I protein kinase may have two subtypes of regulatory subunits which differ in their isoelectric point but not in their molecular weight. Indeed, this has been reported for neuroblastoma cells (PRASHAD et al. 1979). The photoactivated 8-N_3-[^{32}P]cAMP incorporation into the 47,000 Dalton protein and 54,000 Dalton protein of crude extracts was shown to be quantitative: a) the sum of the 8-N_3-[^{32}P]cAMP incorporation into the 47,000 and 54,000 Dalton proteins agreed well with the total amount of specific cAMP-binding sites as determined by reversible binding of cAMP (WALTER et al. 1977a); b) continuous replacement of photolysed material with fresh 8-N_3-[^{32}P]cAMP and subsequent photolysis did not result in a further incorporation of radioactivity into the 47,000 or 54,000 Dalton proteins (WALTER et al. 1977a); c) labeling of the purified regulatory subunit of the bovine heart protein kinase was not affected by the presence of crude cytosol or membrane fractions indicating that quantitative labeling can be obtained with crude fractions (LOHMANN et al. 1978).

The relative distribution of Type I and Type II cAMP-dependent protein kinase activity in several tissues agreed well with the relative distribution of the 47,000 and 54,000 Dalton proteins, respectively, which specifically incorporated 8-N_3-[^{32}P]cAMP (WALTER et al. 1977a), indicating a constant ratio of regulatory and catalytic subunit of the cAMP-dependent protein kinase holoenzyme. This is in agreement with another study in which equimolar concentrations of cAMP-binding capacity and cAMP-dependent histone kinase activity were found in several tissues (HOFMANN et al. 1977). That study, using an immunochemical approach, concluded that the total cAMP-binding capacity found in soluble extracts could be attributed to the regulatory subunits of the Type I and Type II cAMP-dependent protein kinases.

The photoaffinity labeling experiments described demonstrated that the two proteins with molecular weights of 47,000 and 54,000 which specifically incorporated 8-N_3-[^{32}P]cAMP represent the regulatory subunits of Type I and Type II cAMP-dependent protein kinase, R-I and R-II, respectively. Photoaffinity labeling with 8-N_3[^{32}P]cAMP can be used in crude extracts for the simultaneous and quantitative estimation of the amounts of these regulatory subunits, a task which otherwise could only be accomplished by time-consuming column chromatography or a detailed immunochemical approach.

E. Specific Examples of the Usefulness of the Photoaffinity Label 8-N_3-[^{32}P]cAMP

I. Distribution and Differences of Type I and Type II cAMP-Dependent Protein Kinases

In each of several tissues from three different species it was found that the regulatory subunit of the Type I protein kinases differed in molecular weight from that of the regulatory subunit of the Type II protein kinases. Furthermore, a comparison of photoaffinity labeling by 8-N_3-cAMP with endogenous phosphorylation allowed the conclusion that all Type II protein kinases are able to catalyze the phosphorylation of their regulatory subunit (autophosphorylation) whereas all Type I protein kinases examined do not catalyze such a reaction (WALTER et al. 1977b). Photoaffinity labeling experiments also helped to establish that the majority of Type II protein kinase is in the phosphorylated form in normal tissue extracts (RANGEL-ALDAO et al. 1979).

II. Proteolytic Fragments of cAMP-Dependent Protein Kinases

It is well known that cAMP-dependent protein kinases are very susceptable to proteolysis (NIMMO and COHEN 1977), and proteolysis is the most likely cause for the observation of multiple forms of cAMP-dependent protein kinases in many early publications. It had even been proposed that the Type I protein kinase is derived from the Type II protein kinase by limited proteolysis, since the Type I enzyme has an identical catalytic but smaller regulatory subunit than the Type II enzyme. However, three lines of evidence indicate conclusively that the Type I and Type II protein kinases are truly separate enzymes due to their different regulatory subunit: a) the regulatory subunits R-I and R-II differ fundamentally in their immunological properties (FLEISCHER et al. 1976; HOFMANN et al. 1977); b) peptide maps obtained by limited proteolysis of 8-N_3-[^{32}P]cAMP-labeled regulatory subunits R-I and R-II displayed completely different patterns (RANGEL-ALDAO et al. 1979); c) the two types of protein kinases and their regulatory subunits appear to be regulated differentially (RUSSELL 1978; WALTER et al. 1979).

However, specific cAMP-binding proteins with molecular weights smaller than those of the R-I and R-II regulatory subunits have frequently been observed (SUGDEN and CORBIN 1976; WALTER et al. 1977b; CORBIN et al. 1978; SRIVASTAVA and STELLWAGEN 1978; WEBER and HILZ 1978; IMASHUKU et al. 1979; POTTER and TAYLOR 1979a; WALTER et al. 1979). It is generally believed that these cAMP-binding proteins are proteolytic degradation products of the R-I and R-II regulatory subunit: a) aging of tissue extracts results in a decrease of the R-I and R-II regulatory subunit with a concomitant increase of multiple cAMP-binding proteins with smaller molecular weights; b) the presence of various types of known protease inhibitors during tissue extraction prevents or decreases the formation of such smaller cAMP-binding proteins; c) some of these smaller cAMP-binding proteins present in tissue extracts can be obtained by treatment of the R-I and R-II regulatory subunits with trypsin or other proteases.

Most of these fragments obtained by limited proteolysis of the R-I or R-II regulatory subunits retain a high affinity for cAMP but lose the ability to inhibit the catalytic subunit of the cAMP-dependent protein kinase (RANNELS and CORBIN 1979; POTTER and TAYLOR 1979b). Photoaffinity labeling by 8-N_3-[^{32}P]cAMP has been a powerful method for the analysis and detection of these proteolytic fragments. This technique should prove very helpful for the investigation of structural and functional domains of the cAMP-binding site of the two regulatory subunits. It should also be helpful for the investigation of whether these multiple proteolytic fragments have any physiological role.

III. Subcellular Distribution of cAMP-Dependent Protein Kinases and cAMP-Receptor Proteins

In most tissues the major portion of cAMP-dependent protein kinase is in the soluble fraction (HOFMANN et al. 1977). However, mammalian brain is a rich source both for soluble and for particulate cAMP-dependent protein kinase (MAENO et al. 1971; UNO et al. 1976; HOFMANN et al. 1977). Using 8-N_3-[^{32}P]cAMP, we therefore examined soluble and particulate fractions from brain of several species for the

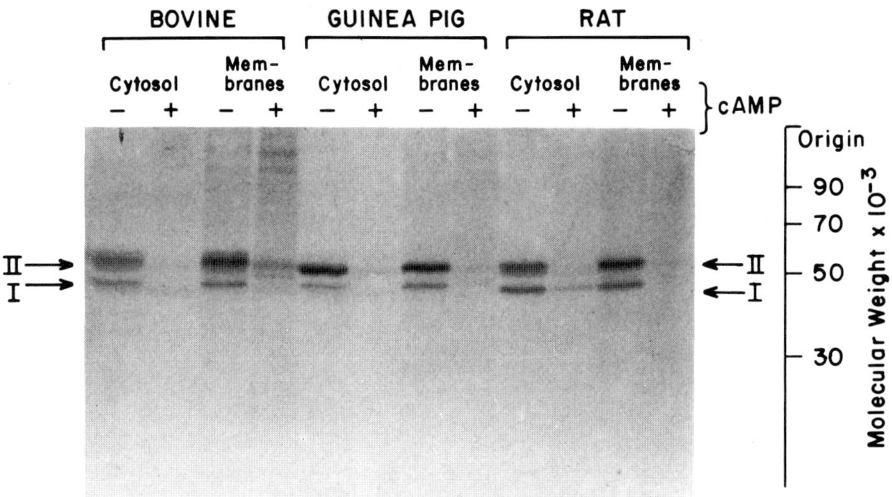

Fig. 7. Autoradiograph showing photoactivated incorporation of 8-N_3-[^{32}P]cAMP into cytosol and membrane fractions of cerebral cortex from bovine, guinea pig, and rat. The incubations were carried out using 1 μ*M* 8-N_3-[^{32}P]cAMP in the absence or presence of 20 μ*M* cAMP. Each channel contained 200 μg of protein. (Taken from WALTER et al. 1978, with permission)

presence of specific cAMP-receptor proteins (WALTER et al. 1978). Both cytosol and particulate fraction of rat, guinea pig, and bovine cerebral cortex were found to contain two specific cAMP-receptor proteins with apparent molecular weights of 47,000 and 52,000–55,000, R-I and R-II, respectively (Fig. 7). Subcellular fractionation revealed that these two cAMP-receptor proteins were present in all primary subfractions including the nuclear fraction, but highest specific activities were detected in the cytosol fractions and fractions rich in synaptic membranes. For both the cytosol and membrane fractions, the two cAMP-receptor proteins represented almost all of the proteins specifically labeled by 8-N_3-[^{32}P]cAMP, and were identified as the regulatory subunit of the Type I and Type II cAMP-dependent protein kinase (WALTER et al. 1977b; WALTER et al. 1978). The particulate cAMP-receptor proteins and cAMP-dependent protein kinases could be extracted with 0.1% Triton X-100 but not with salt up to 0.5 *M* KCl. As in the case of cytosol from different tissues, the particulate fraction of brain was found to contain two types of cAMP-dependent protein kinases which differed in their elution profile on DEAE-cellulose (WALTER et al. 1978). Also, as with soluble protein kinases, the two particulate protein kinases had regulatory subunits which could be distinguished by their molecular weight and by their ability to undergo autophosphorylation. These results indicate that in both cytosol and membrane fractions the Type I and Type II cAMP-dependent protein kinase differ from each other by the properties of their regulatory subunit. These results also indicate that the particulate Type I and Type II protein kinase appear to be remarkably similar to their soluble counterparts. Other investigators (RUBIN et al. 1979) demonstrated that the soluble and particulate Type II protein kinase of bovine brain could not be distinguished by a variety of biochemical and immunological methods.

Studies carried out with heart tissue of several species suggested that the Type II protein kinase is principally associated with the particulate fraction, whereas the Type I protein kinase is principally found in the soluble fraction (CORBIN and KEELY 1977; CORBIN et al. 1977). However, this difference in subcellular distribution does not seem to be the case for all tissues. Analysis by 8-N_3-[^{32}P]cAMP incorporation as well as by regular DEAE-cellulose chromatography demonstrated that the Type II protein kinase constitutes the major part of cAMP-dependent protein kinase activity in both cytosol and particulate fraction of brain (WALTER et al. 1977b; WALTER et al. 1978). Moreover, in brain the ratio of Type I to Type II protein kinase appears to be similar in the soluble and particulate fractions. In human erythrocytes the membrane-bound protein kinase was found to be a Type I protein kinase whereas the soluble enzyme was predominantly a Type II protein kinase, analyzed both by photoaffinity labeling as well as by DEAE-cellulose chromatography (DREYFUSS et al. 1978b).

IV. Regulation of cAMP-Dependent Protein Kinase and cAMP-Receptor Proteins

The quantitative nature of the photoaffinity labeling by 8-N_3-[^{32}P]cAMP allows not only the detection of cAMP-binding proteins but also the measurement of the Type I and Type II regulatory subunits. This has been used to study the regulation of cAMP-receptor proteins in neuroblastoma-glioma hybrid cells (WALTER et al. 1979). In many respects, these cells have properties of neurons (HAMPRECHT 1976). The presence of dibutyryl-cAMP in the growth medium causes these cells to reduce their rate of division and to produce long, neurite-like extensions (HAMPRECHT 1976). Moreover, neuroblastoma X glioma hybrid cells, after treatment with dibutyryl cAMP, are able to form cholinergic synapses with primary and continuous cultures of striated muscle cells (NELSON et al. 1976; NELSON et al. 1977; CHRISTIAN et al. 1977).

Neuroblastoma-glioma hybrid cells were found to contain high levels of cAMP-dependent protein kinase activity and high levels of the R-I and R-II regulatory subunit (WALTER et al. 1979). Treatment of the hybrid cells with dibutyryl cAMP increased the level of the R-I regulatory subunit 2- to 5-fold (see Fig. 8) but did not significantly affect the level of either the R-II regulatory subunit or the cAMP-dependent protein kinase activity. The effect of dibutyryl cAMP could be mimicked by prostaglandin E_1 and by 3-isobutyl-l-methylxanthine, both of which are known to raise cAMP levels in neuroblastoma-glioma hybrid cells. Further analysis by DEAE-cellulose chromatography (Fig. 9) and by gel filtration (WALTER et al. 1979) revealed that the R-I regulatory subunit was present in both control and dibutyryl-cAMP treated cells mainly as free regulatory subunit. In contrast, the R-II regulatory subunit was associated with the Type II protein kinase which constituted almost all cAMP-dependent protein kinase activity in these cells. This situation appears to differ from normal tissues in which equimolar levels of the catalytic and regulatory subunits of cAMP-dependent protein kinase have been found, suggesting that regulation of the level of catalytic and regulatory subunit is usually coordinated (CORBIN et al. 1977, HOFMANN et al. 1977; LOHMANN et al. 1978).

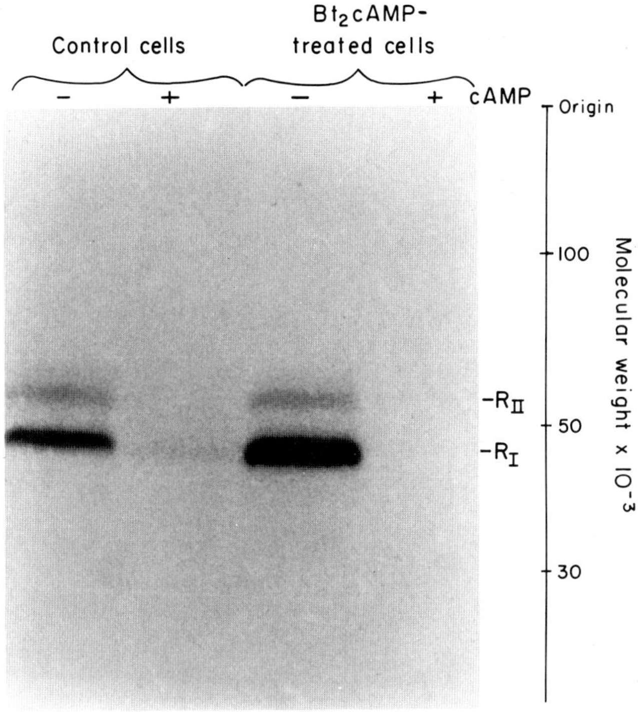

Fig. 8. Autoradiograph showing the effect of dibutyryl cAMP (Bt_2cAMP) treatment of neuroblastoma X glioma hybrid cells on the incorporation of 8-N_3-[^{32}P]cAMP into cytosol proteins. Cells were cultured for 3 days in the absence (control cells) or presence (Bt_2cAMP-treated cells) of 1 m*M* Bt_2cAMP. Cytosol (5 µg of protein) was analyzed for 8-N_3-[^{32}P]cAMP incorporation in the absence and presence of 20 µ*M* cAMP. R-I and R-II indicate the location of the two specific cAMP-receptor proteins. Quantitation of 8-N_3-[^{32}P]cAMP incorporation showed that cytosol from control cells contained 7.3 pmol of R-I per mg of protein and 2.1 pmol of R-II per mg of protein and that cytosol from Bt_2cAMP-treated cells contained 17.5 pmol of R-I per mg of protein and 2.5 pmol of R-II per mg of protein. (Taken from WALTER et al. 1979, with permission)

The extent to which the two types of cAMP-dependent protein kinase, which differ in their regulatory but not in their catalytic subunits, have separate physiological roles is not yet clear. It has been suggested that the two protein kinases have different roles in the regulation of the cell cycle and in differentiation (JUNGMANN and RUSSELL 1977; RUSSELL 1978). The possibility cannot be ruled out that the R-I and R-II regulatory subunits may have functions in addition to that of inhibiting the catalytic subunit of the protein kinase. The results obtained with neuroblastoma X glioma hybrid cells raise the possibility that free regulatory subunits could play a role in the regulation of cellular functions such as growth and differentiation. There is some evidence that the R-I and R-II regulatory subunits are present in nuclei of bovine liver (FRIEDMAN and CHAMBERS 1978). It is of interest to note that dibutyryl cAMP increased the level of both soluble and chromatin associated

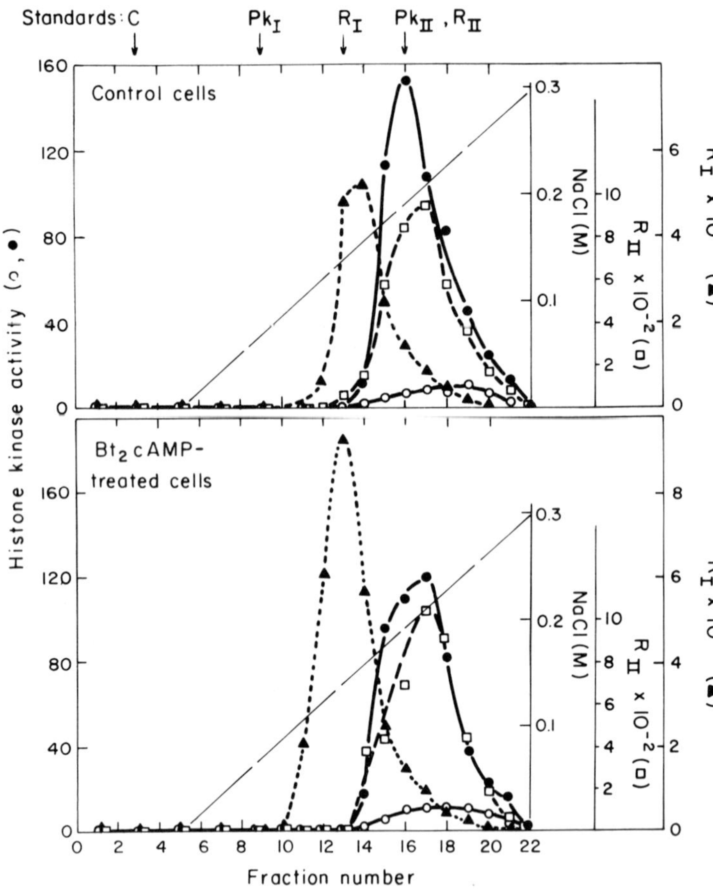

Fig. 9. DEAE-cellulose column chromatography of cytosol from control (Upper) and Bt_2cAMP-treated (Lower) neuroblastoma X glioma hybrid cells. Samples (1 ml) containing 1.8 mg of protein were applied to DEAE-cellulose columns (1.3 × 2.5 cm) that had been equilibrated with 10 m*M* Tris-HCl, pH 7.2/1 m*M* EDTA/50 units of Trasylol per ml. After washing with 10 ml of the same buffer, the enzymes were eluted from the column with a linear gradient of 0–0.3 *M* NaCl in the same buffer. Fractions of 2.4 ml were collected. Aliquots (50 µl) of each fraction were analyzed for histone kinase activity in the absence (○) and presence (●) of cAMP. The results are expressed as units/ml of eluate. Aliquots (80 µl) of each fraction were also analyzed for 8-N_3-[^{32}P]cAMP incorporation into cAMP receptor proteins R_I (▲) and R_{II} (□) under standard conditions. These results are expressed as cpm/ml of eluate. Purified components of cAMP-dependent protein kinases from bovine heart were used as standards: C, catalytic subunit; Pk_I, type I holoenzyme; R_I, type I regulatory subunit dimer; Pk_{II}, type II holoenzyme; R_{II}, type II regulatory subunit dimer. The arrows indicate the elution positions of these proteins used as standards. (Taken from WALTER et al. 1979, with permission)

R-I regulatory subunit in neuroblastoma cells (PRASHAD et al. 1979). Furthermore, ovariectomy and dibutyryl-cAMP treatment, which caused the regression of certain mammary tumors, also increased the amount of cAMP-binding protein in these tumors; this protein appeared to be the R-II regulatory subunit (BODWIN et al. 1978; CHO-CHUNG 1979). Further work is required to investigate the physiological role of the two types of cAMP-dependent protein kinases and their different regulatory subunits. Photoaffinity labeling with 8-N_3-[^{32}P]cAMP should prove to be an important tool for these investigations.

F. Photoaffinity Labeling of cGMP-Dependent Protein Kinase

Although cGMP was discovered as a physiologically occurring cyclic nucleotide in 1963 (ASHMAN et al. 1963), much less is known about its biological role and receptors than is the case for cAMP. A cGMP-dependent protein kinase has been purified from bovine lung (GILL et al. 1976; DEJONGE and ROSEN 1977; LINCOLN et al. 1977) and bovine heart (FLOCKERZI et al. 1978) and has been found to differ considerably from the two cAMP-dependent protein kinases. The enzyme has been shown to be a dimer composed of two identical subunits (molecular weight 74,000–81,000) each of which appears to contain cGMP-binding as well as catalytic activity. The success of the studies with 8-N_3-[^{32}P]cAMP suggested that a photoaffinity analogue of cGMP would be useful for the study of the distribution of cGMP-dependent protein kinase and in the search for other cGMP-binding proteins.

Since 8-N_3-cIMP has been shown to be a good activator of the cGMP-dependent protein kinase from lobster, and since 8-N_3-cIMP can easily be prepared from 8-N_3-cAMP (MILLER et al. 1973), we investigated 8-N_3-[^{32}P]cIMP as a potential photoaffinity analogue for cGMP-binding sites (CASNELLIE et al. 1978). Half-maximal activation (K_a) of a partially-purified cGMP-dependent protein kinase from bovine pulmonary arteries was obtained at 0.014 μM cGMP and 1.0 μM 8-N_3-cIMP. Furthermore, 8-N_3-cIMP could competitively inhibit the binding of cGMP to this enzyme with an inhibition constant (K_i) of 2.5 μM. The K_d for cGMP binding was 0.022 μM. These results indicated that 8-N_3-cIMP could bind to the cGMP-binding site of this protein kinase although with a 100-fold lower affinity than cGMP. When the photoactivated incorporation of 8-N_3-[^{32}P]cIMP into soluble extracts of bovine pulmonary arteries was examined, several proteins were found to incorporate radioactivity, including the R-I and R-II regulatory subunit (CASNELLIE et al. 1978). 1 μM cGMP but not 1 μM cAMP prevented the incorporation of 8-N_3-[^{32}P]cIMP into only one protein present in soluble extracts of bovine pulmonary arteries, and this protein had an apparent molecular weight of 74,000 (Fig. 10). Subsequent experiments showed that this protein represented the cGMP-dependent protein kinase as demonstrated by the photoactivated incorporation of 8-N_3-[^{32}P]cIMP into the purified cGMP-dependent protein kinase from bovine lung (Fig. 11). Although the labeling of the cGMP-dependent protein kinase by 8-N_3-[^{32}P]cIMP does not proceed with high efficiency (in contrast to the labeling of cAMP-dependent protein kinase by 8-N_3-[^{32}P]cAMP), this technique should allow the measurement of cGMP-dependent protein kinase in crude extracts if properly controlled for the efficiency of the labeling process.

Recently, the synthesis and use of 8-N_3-cGMP as a photoaffinity label for a partially-purified cGMP-dependent protein kinase was described (GEAHLEN et al. 1979). This compound has several kinetic properties that suggest that it could be an even better photoaffinity label for cGMP-dependent protein kinase than 8-N_3-

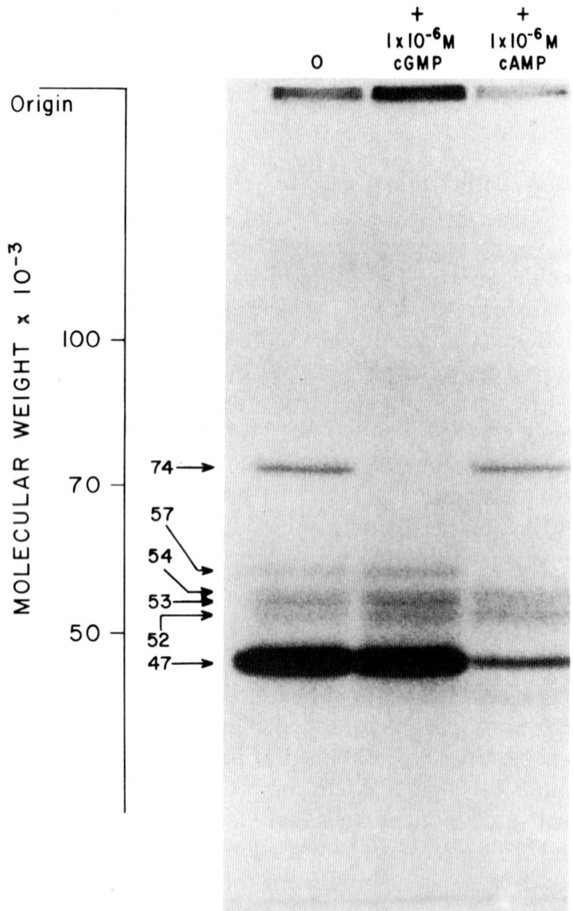

Fig. 10. Autoradiograph showing the photoactivated incorporation of 8-N_3-[^{32}P]cIMP into protein bands of cytosol from the medial layer of calf pulmonary arteries. The photoactivated incorporation was performed using 1 μM 8-N_3-[^{32}P]cIMP in the absence or presence of 1 μM cGMP or 1 μM cAMP. Each channel contained 80 μg of protein. (Taken from CASNELLIE et al. 1978, with permission)

cIMP. However, more information on the stability of 8-N_3-cGMP, on its use as a photoaffinity label in crude systems, and on the efficiency of the photolabeling process is required before a full evaluation of its potential can be made.

G. Problem Areas

Despite the success of using cyclic nucleotide analogues as photoaffinity labels, there are a few difficulties which are frequently encountered. Some of these difficulties are common to other photoaffinity labels. Non-specific labeling is often observed even with 8-N_3-cAMP, especially at concentrations necessary for saturating both the R-I and R-II regulatory subunit. It is therefore always necessary, for the

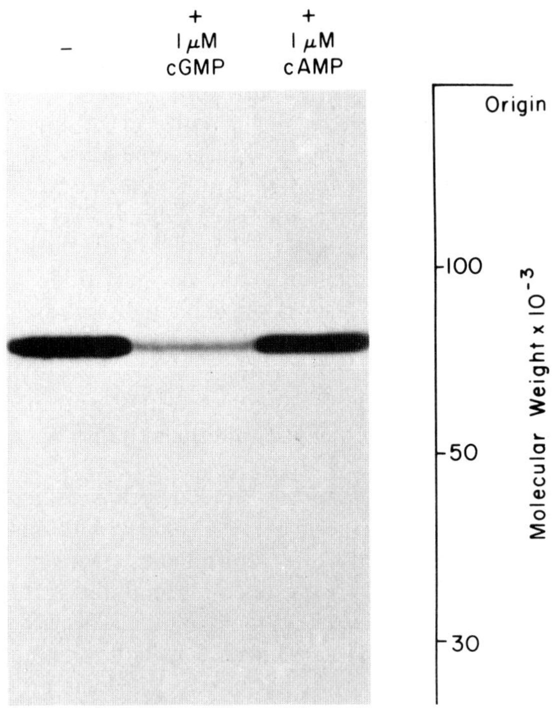

Fig. 11. Autoradiograph showing the photoactivated incorporation of 8-N_3-[^{32}P]cIMP into purified cGMP-dependent protein kinase from bovine lung. The photoactivated incorporation was performed using 1 μ*M* 8-N_3-[^{32}P]cIMP in the absence or presence of 1 μ*M* cGMP or 1 μ*M* cAMP. Each channel contained 0.28 μg of homogeneous cGMP-dependent protein kinase. (U. WALTER and P. GREENGARD, unpublished)

identification of cAMP-receptor proteins, to demonstrate that physiological concentrations of cAMP are able to prevent the photoaffinity labeling by 8-N_3-cAMP. Low concentrations of β-mercaptoethanol, a "scavenger" of free radicals (RUOHO et al. 1973), are very effective in reducing non-specific labeling without affecting specific labeling. However, at high concentrations, thiols, especially dithiols such as dithiothreitol and dithioerythritol, may react with 8-azido-adenosine derivatives to form the corresponding 8-amino-adenosine derivatives (CARTWRIGHT et al. 1976; STAROS et al. 1978).

Although the quantitative nature of the photoaffinity labeling by 8-N_3-[^{32}P]cAMP has been well documented (WALTER et al. 1977a, 1978; RANGEL-ALDAO et al. 1979; RUBIN et al. 1979), quantitative labeling should not be assumed for every experimental condition. For quantitative studies, the chemical and radiochemical purity of 8-N_3-[^{32}P]cAMP needs to be carefully controlled. Quantitative labeling requires saturating concentrations of the photoaffinity analogue. Substances which affect the affinity of the receptor protein obviously could affect the efficiency of the photoaffinity labeling. For example, ATP is known to affect the affinity of cAMP-dependent protein kinases for cAMP (HOFMANN et al. 1975) and to reduce the photoaffinity labeling of the membrane-bound Type I protein kinase of human erythrocytes (OWENS and HALEY 1978). However, in cases where quantitative incorporation is not obtained during one course of photoaffinity labeling, quantitative incorporation may be accomplished by several successive labeling procedures with fresh photoaffinity analogue. However, it should be kept in mind that increasing exposure to UV-light may cause some cross-linking of proteins.

Photoaffinity labeling may also be significantly affected by the levels of bound or free cyclic nucleotides present in crude tissue extracts. Gel filtration or dialysis removes only free cyclic nucleotides and not bound cyclic nucleotides (CORBIN et al. 1978). However, the measurement of total cyclic nucleotide binding sites in extracts with bound cyclic nucleotides is possible if conditions are used which allow the complete exchange of bound and free cyclic nucleotides (CORBIN et al. 1978; WALTER et al. 1979).

Another problem area already discussed is the effect of proteolysis on the pattern of specific cAMP-binding proteins. Conditions should be chosen which minimize proteolysis (WALTER et al. 1979).

The cAMP-receptor protein (CRP) of *E. coli*, which is not a cAMP-dependent protein kinase, was found to be labeled by 8-N_3-cAMP (AIBA and KRAKOW 1979). In contrast, no evidence is available that phosphodiesterases are labeled by 8-N_3-[^{32}P]cAMP. This may seem surprising since phosphodiesterases are capable of binding cAMP and seem to be present in most tissues (APPLEMAN et al. 1973). An explanation could be that phosphodiesterases usually have a much lower affinity for cyclic nucleotides than do the cyclic nucleotide-dependent protein kinases. Furthermore, 8-substituted derivatives of cyclic nucleotides such as 8-N_3-cAMP have been shown to be poor substrates for phosphodiesterases (MUNEYAMA et al. 1971). This raises the possibility that there may be cAMP-receptor proteins which are not labeled or are poorly labeled by 8-N_3-cAMP.

H. Conclusions

Most, if not all, specific cAMP-binding activity can be attributed to the two types of regulatory subunit of cAMP-dependent protein kinases, as demonstrated by analysis of cAMP-binding proteins using the photoaffinity label 8-N_3-cAMP. These results support the hypothesis that in eukaryotes all effects of cAMP are mediated by cAMP-dependent protein kinases. This conclusion is also supported by an immunochemical study which demonstrated that all specific cAMP-binding activity could be precipitated by antisera against the two types of regulatory subunit (HOFMANN et al. 1977). The experiments using 8-N_3-cAMP also support the concept of a widespread distribution of two types of cAMP-dependent protein kinases which differ in their regulatory subunit. There appears to be a considerable species and tissue variation in the relative amounts of the two types of protein kinases and in the extent to which they are soluble and membrane-bound. The quantitative nature of the photoactivated incorporation of 8-N_3-cAMP has proven useful for studies investigating the regulation of the level of the R-I and R-II regulatory subunits. It should also be possible to use this technique for studies investigating the occupancy of the cAMP-binding sites of the R-I and R-II regulatory subunits under various hormonal and metabolic conditions. These studies will hopefully contribute to answering the questions of whether the two types of cAMP-dependent protein kinases have different physiological roles and whether the regulatory subunits have functions in addition to that of inhibiting the catalytic subunit of the protein kinase. Photoaffinity analogues of cGMP should be helpful for studies on cGMP-binding proteins and their physiological role.

Acknowledgments. The work of our collaborators in the original research from this laboratory is gratefully acknowledged. The original work from this laboratory reviewed here was supported by grants from the United States Public Health Service and by a grant from the McKnight Foundation. Drs. Yoram Milner and Clive Palfrey provided valuable criticism of the manuscript. We also would like to thank Annette E. Gwardyak for her expert assistance in preparing the manuscript.

References

Aiba H, Krakow JS (1979) Photoaffinity labeling of the *E. coli* cyclic AMP receptor protein with 8-azido-adenosine 3′,5′-monophosphate. Fed Proc 38:231

Antonoff RS, Ferguson JJ (1974) Photoaffinity labeling with cyclic nucleotides. J Biol Chem 249:3319–3321

Antonoff RS, Ferguson JJ, Idelkope G (1976) Direct photoaffinity labeling of cyclic nucleotide binding proteins with guanosine-3′,5′-monophosphate. Photochem Photobiol 23:327–329

Appleman MM, Thompson WJ, Russell TR (1973) Cyclic nucleotide phosphodiesterases. Adv Cyclic Nucleotide Res 3:65–98

Ashman DF, Lipton R, Melicow MM, Price TD (1963) Isolation of adenosine 3′,5′-monophosphate and guanosine 3′,5′-monophosphate from rat urine. Biochem Biophys Res Commun 11:330–334

Bayley H, Knowles JR (1977) Photoaffinity labeling. Methods Enzymol 46:69–114

Bodwin JS, Clair T, Cho-Chung YS (1978) Inverse relation between estrogen receptors and cyclic adenosine 3′:5′-monophosphate-binding proteins in hormone-dependent mammary tumor regression due to dibutyryl cyclic adenosine 3′:5′-monophosphate treatment or ovariectomy. Cancer Res 38:3410–3413

Brunswick J, Cooperman BS (1971) Photo-affinity labels for adenosine 3′:5′-cyclic monophosphate. Proc Natl Acad Sci USA 68:1801–1804

Cartwright JS, Hutchinson DW, Armstrong VW (1976) The reaction between thiols and 8-azidoadenosine derivatives. Nucleic Acids Res 3:2331–2339

Casnellie JE, Schlichter DJ, Walter U, Greengard P (1978) Photoaffinity labeling of a guanosine 3′:5′-monophosphate-dependent protein kinase from vascular smooth muscle. J Biol Chem 253:4771–4776

Casnellie JE, Ives HE, Jamieson JD, Greengard P (1980) cGMP-dependent protein phosphorylation in intact medial tissue and isolated cells from vascular smooth muscle. J Biol Chem 255:3770–3776

Cho-Chung YS (1979) On the mechanism of cyclic AMP-mediated growth arrest of solid tumors. Adv Cyclic Nucleotide Res 12:in press

Chowdhry V, Westheimer FH (1979) Photoaffinity labeling of biological systems. Ann Rev Biochem 48:293–325

Christian CN, Nelson PG, Peacock J, Nirenberg M (1977) Synapse formation between two clonal cell lines. Science 196:995–998

Cooperman BS, Brunswick J (1973) On the photoaffinity labeling of rabbit muscle phosphofructokinase with $O^{2'}$-(ethyl-2-diazomalonyl)adenosine 3′:5′-cyclic monophosphate. Biochemistry 12:4079–4084

Cooperman BS (1976) Photoaffinity labeling of proteins and more complex receptors. In: Smith KC (ed) Aging, carcinogenesis and radiation biology. Plenum, New York, pp 315–340

Corbin JD, Keely SL, Park CR (1975) The distribution and dissociation of cyclic adenosine 3′,5′-monophosphate-dependent protein kinases in adipose, cardiac and other tissues. J Biol Chem 250:218–225

Corbin JD, Keely SL (1977) Characterization and regulation of heart adenosine 3′,5′-monophosphate-dependent protein kinase isozymes. J Biol Chem 252:910–918
Corbin JD, Sugden PH, Lincoln TM, Keely SL (1977) Compartmentalization of adenosine 3′:5′-monophosphate and adenosine 3′:5′-monophosphate-dependent protein kinase in heart tissue. J Biol Chem 252:3854–3861
Corbin JD, Sugden PH, West L, Flockhart DA, Lincoln TM, McCarthy D (1978) Studies on the properties and mode of action of the purified regulatory subunit of bovine heart adenosine 3′:5′-monophosphate-dependent protein kinase. J Biol Chem 253:3997–4003
Creed D (1974) Photochemical probes for biological interactions. Photochem Photobiol 19:459–462
DeCrombrugghe B, Chen B, Anderson W, Nissley P, Gottesman M, Perlman M, Pastan I (1971) LacDNA, RNA polymerase, cyclic AMP receptor protein, cyclic AMP, Lac repressor and inducer are the essential elements for controlled Lac transcription. Nature 231:139–142
DeJonge H, Rosen OM (1977) Self-phosphorylation of cyclic guanosine 3′:5′-monophosphate-dependent protein kinase from bovine lung. J Biol Chem 252:2780–2783
Dreyfuss G, Schwartz K, Blout ER, Barrio JR, Lin FT, Leonard NJ (1978a) Fluorescent photoaffinity labeling: Adenosine 3′,5′-cyclic monophosphate receptor sites. Proc Natl Acad Sci USA 75:1199–1203
Dreyfuss G, Schwartz KJ, Blout ER (1978b) Compartmentalization of cyclic AMP-dependent protein kinases in human erythrocytes. Proc Natl Acad Sci USA 75:5926–5930
Fleischer N, Rosen OM, Reichlin M (1976) Radioimmunoassay of bovine heart protein kinase. Proc Natl Acad Sci USA 73:54–58
Flockerzi V, Speichermann N, Hofmann F (1978) A guanosine-3′:5′-monophosphate-dependent protein kinase from bovine heart muscle. J Biol Chem 253:3395–3399
Friedman DL, Chambers DA (1978) Cyclic nucleotide-binding proteins detected by photoaffinity labeling in nucleus and cytoplasm of bovine liver. Proc Natl Acad Sci USA 75:5286–5290
Geahlen RL, Haley BE, Krebs EG (1979) Synthesis and use of 8-azido-guanosine 3′,5′-cyclic monophosphate as photoaffinity label for cyclic GMP-dependent protein kinase. Proc Natl Acad Sci USA 76:2213–2217
Gill GN, Holdy KE, Walton GM, Kanstein CB (1976) Purification and characterization of 3′:5′-cyclic GMP-dependent protein kinase. Proc Natl Acad Sci USA 73:3918–3922
Gill GN, Walton GM (1979) Assay of cyclic nucleotide-dependent protein kinase. Adv Cyclic Nucleotide Res 10:93–106
Haley BE (1975) Photoaffinity labeling of adenosine 3′,5′-cyclic monophosphate binding sites of human red cell membranes. Biochemistry 14:3852–3857
Haley BE (1977) Adenosine 3′,5′-cyclic monophosphate binding sites. Methods Enzymol 46:339–346
Haley BE, Hoffman JF (1974) Interactions of a photoaffinity ATP analog with cation-stimulated adenosine triphosphatases of human red cell membranes. Proc Natl Acad Sci USA 71:3367–3371
Hamet P, Coquil JF (1978) Cyclic GMP binding and cyclic GMP phosphodiesterase in rat platelets. J Cyclic Nucleotide Res 4:281–290
Hamprecht B (1976) Neuron models. Angew Chem Int Ed Engl 15:194–206
Hershfield MS, Kredich NM (1978) S-adenosylhomocysteine hydrolase is an adenosine-binding protein: A target for adenosine toxicity. Science 202:757–760
Hixson CS, Krebs EG (1979) Affinity labeling of catalytic subunit of bovine heart muscle cyclic AMP-dependent protein kinase by 5′-p-fluorosulfonylbenzoyl-adenosine. J Biol Chem 254:7509–7514
Hofmann F, Beavo JA, Bechtel PJ, Krebs EG (1975) Comparison of adenosine 3′:5′-monophosphate-dependent protein kinases from rabbit skeletal and bovine heart muscle. J Biol Chem 250:7795–7801
Hofmann F, Bechtel PJ, Krebs EG (1977) Concentrations of cyclic AMP-dependent protein kinase subunits in various tissues. J Biol Chem 252:1441–1447
Hoppe J, Freist W (1979) Localization of the high-affinity ATP site in adenosine-3′:5′-monophosphate-dependent protein kinase type I. Eur J Biochem 93:141–146

Imashuku S, Fossett MC, Green AA (1979) Characterization of adenosine cyclic 3′:5′-monophosphate-binding proteins in human neuroblastoma. Cancer Res 39:3006–3013

Jungmann RA, Russell DH (1977) Cyclic AMP, cyclic AMP-dependent protein kinase, and the regulation of gene expression. Life Sci 20:1787–1798

Kemp RG, Krebs EG (1967) Binding of metabolites by phosphofructokinase. Biochemistry 6:423–434

Knowles JR (1972) Photogenerated reagents for biological receptor-site labeling. Acc Chem Res 5:155–160

Kuo JF, Greengard P (1969) Cyclic nucleotide-dependent protein kinases. IV. Widespread occurrence of adenosine 3′,5′-monophosphate-dependent protein kinase in various tissues and phyla of the animal kingdom. Proc Natl Acad Sci USA 64:1349–1355

Kuo JF, Greengard P (1970) Cyclic nucleotide-dependent protein kinases. VI. Isolation and partial purification of a protein kinase activated by guanosine 3′,5′-monophosphate. J Biol Chem 245:2493–2498

Kuo JF, Wyatt GR, Greengard P (1971) Cyclic nucleotide-dependent protein kinases. IX. Partial purification and some properties of guanosine 3′,5′-monophosphate-dependent and adenosine 3′,5′-monophosphate-dependent protein kinases from various tissues and species of arthropoda. J Biol Chem 246:7159–7167

Kupfer A, Gani V, Jimenex JS, Shaltiel S (1979) Affinity labeling of the catalytic subunit of cyclic AMP-dependent protein kinase by N^{α}-tosyl-L-lysine chloromethyl ketone. Proc Natl Acad Sci USA 76:3073–3077

Lincoln TM, Hall CL, Park CR, Corbin JD (1976) Guanosine 3′:5′-cyclic monophosphate binding proteins in rat tissues. Proc Natl Acad Sci USA 73:2559–2563

Lincoln TM, Dills WL, Corbin JD (1977) Purification and subunit composition of guanosine 3′:5′-monophosphate-dependent protein kinase from bovine lung. J Biol Chem 252:4269–4275

Lohmann SM, Walter U, Greengard P (1978) Protein kinases in developing rat brain. J Cyclic Nucleotide Res 4:445–452

Maeno H, Johnson EM, Greengard P (1971) Subcellular distribution of adenosine 3′,5′-monophosphate-dependent protein kinase in rat brain. J Biol Chem 246:134–142

Miller JP, Boswell KH, Muneyama K, Simon LN, Robins RK, Shuman DA (1973) Synthesis and biochemical studies of various 8-substituted derivatives of guanosine 3′,5′-cyclic phosphate, inosine 3′,5′-cyclic phosphate, and xanthosine 3′,5′-cyclic phosphate. Biochemistry 12:5310–5319

Muneyama K, Bauer RJ, Shuman DA, Robins RK, Simon LN (1971) Chemical synthesis and biological activity of 8-substituted adenosine 3′,5′-cyclic monophosphate derivatives. Biochemistry 10:2390–2395

Nelson P, Christian C, Nirenberg M (1976) Synapse formation between clonal neuroblastoma X glioma hybrid cells and striated muscle cells. Proc Natl Acad Sci USA 73:123–127

Nelson PG, Christian CN, Daniels MP, Henkart M, Bullock P, Mullinax D, Nirenberg M (1977) Formation of synapses between cells of neuroblastoma X glioma hybrid clone and mouse myotubes. Brain Res 147:245–259

Nimmo HG, Cohen P (1977) Hormonal control of protein phosphorylation. Adv Cyclic Nucleotide Res 8:145–266

Owens JR, Haley BE (1978) Use of photoaffinity nucleotide analogs to determine the mechanism of ATP regulation of a membrane-bound, cAMP-activated protein kinase. J Supramol Struct 9:57–68

Pastan I, Perlman RL (1972) Regulation of gene transcription in E. coli by cyclic AMP. Adv Cyclic Nucleotide Res 1:11–16

Pomerantz AH, Rudolph SA, Haley BE, Greengard P (1975) Photoaffinity labeling of a protein kinase from bovine brain with 8-azido-adenosine 3′,5′-monophosphate. Biochemistry 14:3858–3862

Potter RL, Taylor SS (1979a) Relationship between structural domains and function in the regulatory subunit of cAMP-dependent protein kinases I and II from porcine skeletal muscle. J Biol Chem 254:2413–2418

Potter RL, Taylor SS (1979 b) The structural and functional domains of cAMP-dependent protein kinase I and II. Fed Proc 38:316

Prashad N, Rosenberg RN, Wischmeyer B, Ulrich C, Sparkman D (1979) Induction of adenosine 3′,5′-monophosphate-binding proteins by $N^6,O^{2'}$-dibutyryladenosine 3′,5′-monophosphate in mouse neuroblastoma cells. Analysis by two-dimensional gel electrophoresis. Biochemistry 18:2717–2725

Rangel-Aldao R, Rosen OM (1976) Mechanism of self-phosphorylation of adenosine 3′:5′-monophosphate-dependent protein kinase from bovine cardiac muscle. J Biol Chem 251:7526–7529

Rangel-Aldao R, Kupiec JW, Rosen OM (1979) Resolution of the phosphorylated and dephosphorylated cAMP-binding proteins of bovine cardiac muscle by affinity labeling and two-dimensional electrophoresis. J Biol Chem 254:2499–2508

Rannels SR, Corbin JD (1979) Characterization of small cAMP-binding fragments of cAMP-dependent protein kinases. J Biol Chem 254:8605–8610

Rubin CS (1975) Adenosine 3′:5′-monophosphate-regulated phosphorylation of erythrocyte membrane proteins. J Biol Chem 250:9044–9052

Rubin CS, Rangel-Aldao R, Sarkar D, Erlichman J, Fleischer N (1979) Characterization and comparison of membrane-associated and cytosolic cAMP-dependent protein kinases. J Biol Chem 254:3797–3805

Rudolph SA, Krueger BK (1979) Endogenous protein phosphorylation and dephosphorylation. Adv Cyclic Nucleotide Res 10:107–133

Ruoho AR, Kiefer H, Roeder PE, Singer S (1973) The mechanism of photoaffinity labeling. Proc Natl Acad Sci USA 70:2567–2571

Russell DH (1978) Type I cyclic AMP-dependent protein kinase as a positive effector of growth. Adv Cyclic Nucleotide Res 9:493–506

Saebø J, Ueland PM (1978) An adenosine 3′:5′-monophosphate-adenosine-binding protein from mouse liver. FEBS Lett 96:125–128

Schwechheimer K, Hofmann F (1977) Properties of regulatory subunit of cyclic AMP-dependent protein kinase (peak I) from rabbit skeletal muscle prepared by urea treatment of the holoenzyme. J Biol Chem 252:7690–7696

Singh A, Thornton ER, Westheimer FH (1962) The photolysis of diazoacetyl chymotrypsin. J Biol Chem 237:3006–3008

Srivastava AK, Stellwagen RH (1978) Presence of the sites for interacting with cyclic AMP and with catalytic subunit on small fragments of protein kinase regulatory subunit. J Biol Chem 253:1752–1755

Staros JV, Bayley H, Standring DN, Knowles JR (1978) Reduction of aryl azides by thiols: Implications for the use of photoaffinity reagents. Biochem Biophys Res Commun 80:568–572

Sugden PH, Corbin JD (1976) Adenosine 3′:5′-cyclic monophosphate-binding proteins in bovine and rat tissues. Biochem J 159:423–437

Ueland PM, Døskeland SO (1977) An adenosine 3′:5′-monophosphate-adenosine binding protein from mouse liver. J Biol Chem 252:677–686

Uno I, Ueda T, Greengard P (1976) Differences in properties of cytosol and membrane-derived protein kinases. J Biol Chem 251:2192–2195

Walsh DA, Perkins JP, Krebs EG (1968) An adenosine 3′,5′-monophosphate-dependent protein kinase from rabbit skeletal muscle. J Biol Chem 243:3763–3765

Walter U, Uno I, Liu AY-C, Greengard P (1977 a) Identification, characterization, and quantitative measurement of cyclic AMP receptor proteins in cytosol of various tissues using a photoaffinity ligand. J Biol Chem 252:6494–6500

Walter U, Uno I, Liu AY-C, Greengard P (1977 b) Study of autophosphorylation of isoenzymes of cyclic AMP-dependent protein kinases. J Biol Chem 252:6588–6590

Walter U, Greengard P (1978) Quantitative labeling of the regulatory subunit of Type II cAMP-dependent protein kinase from bovine heart by a photoaffinity analog. J Cyclic Nucleotide Res 4:437–444

Walter U, Kanof P, Schulman H, Greengard P (1978) Adenosine 3′:5′-monophosphate receptor proteins in mammalian brain. J Biol Chem 253:6275–6280

Walter U, Costa MRC, Breakefield XO, Greengard P (1979) Presence of free cyclic AMP receptor protein and regulation of its level by cyclic AMP in neuroplastoma-glioma hybrid cells. Proc Natl Acad Sci USA 76:3251–3255
Weber W, Hilz H (1978) Adenosine-3′:5′-monophosphate-binding proteins from bovine kidney. Eur J Biochem 83:215–225
Wold F (1977) Affinity labeling. An overview. Methods in Enzymol 46:3–14
Yuh K-CM, Tao M (1974) Purification and characterization of adenosine cyclic 3′,5′-monophosphate binding protein factors from rabbit erythrocytes. Biochemistry 13:5220–5226
Zoller MJ, Taylor SS (1979) Affinity labeling of the nucleotide binding site of the catalytic subunit of cAMP-dependent protein kinase using p-fluorosulfonyl-[^{14}C]benzoyl-5′-adenosine. Identification of a modified lysine residue. J Biol Chem 254:8363–8368

CHAPTER 15

Nuclear Protein Phosphorylation and the Regulation of Gene Expression

E. M. JOHNSON

Overview

The involvement of cyclic AMP-dependent protein kinase in the phosphorylation of chromosomal proteins is well known. Perhaps the best examples of nuclear proteins phosphorylated by this enzyme are histone H 1 and certain non-histone proteins, each of which is phosphorylated in vivo in response to agents that raise levels of cyclic AMP. Histone H 1 from calf thymus is phosphorylated at serine residue 38, and peptide analogs of this site have been used in studies of catalysis by the C-subunit of the kinase in vitro. Kinetic analyses indicate that phosphorylation proceeds by a random sequential interaction of the enzyme with ATP and peptide substrates. The data do not rule out the presence of a transient phosphoenzyme intermediate in catalysis, but they do indicate the formation of a ternary complex between enzyme, Mg-ATP and peptide which may be significant in the design of active site-directed inhibitors. Studies on the location and translocation of kinase subunits suggest a complex regulatory role for this enzyme in the nucleus. One consequence of histone phosphorylation could be an alteration of nucleosome structure leading to changes in functional activities over broad regions of the chromatin. However, at this point no functional significance can be attributed to any site-specific nuclear protein phosphorylation.

Newly-developed methods should aid in clarifying functional roles of protein kinase substrates. One method involves the isolation of cell mutants deficient in different aspects of phosphorylation. These can be used to assess the importance of phosphorylation in various cellular processes. Another new method involves the use of 5′-[γ-S]ATP as a protein kinase substrate. The thiophosphate group serves as an affinity probe for isolation of newly-phosphorylated proteins. This procedure allows examination of activity of a phosphorylated protein separated from its non-phosphorylated counterparts.

Numerous phosphoproteins have been implicated in reactions related to gene expression both at transcriptional and translational levels. In many cases it is difficult to consider separately cyclic AMP-dependent and -independent phosphorylations with regard to their ultimate roles. For example, translation may involve multiple phosphorylation of ribosomal proteins, not all responsive to cyclic AMP. Phosphorylation occurs on both ribosomal subunit proteins and initiation factors.

Recently, peptides encoded by viral nucleic acids have been shown to possess protein kinase activity. The pp 60^{src} protein specified by avian sarcoma virus is apparently a cyclic AMP-independent protein kinase. It is interesting that this protein is itself a phosphoprotein, and that phosphorylation at one site, a serine residue, is stimulated by cyclic AMP. These results point out the importance of protein phosphorylation to the process of cell transformation and emphasize the necessity for functional identification of protein kinase substrates.

A. Introduction

Post-synthetic modifications of protein primary structure, including phosphorylation, provide an intracellular mechanism for altering the activities of pre-existing proteins. Phosphorylation is thus a means by which the existing cell apparatus can exert its influence upon subsequent expression of the genetic program, either by affecting RNA synthesis at the stages of transcription or processing, or protein synthesis at the various stages of translation. For example, phosphorylation of chromosomal proteins is recognized as one possible mechanism through which extracellular signals can exert effects on genetic processes. Studies on phosphorylation of both histones and non-histone proteins have established temporal correlations of phosphorylation of individual proteins with various chromosomal activities, such as activation of chromatin for transcription or condensation of chromatin during mitosis. Studies on several protein kinases, now quite well characterized, have identified enzymes which phosphorylate specific proteins in vitro and which may be candidates to do so in vivo. The effects of cyclic AMP and cyclic GMP on certain of these kinases have been well documented. Despite a now detailed knowledge of sites of phosphorylation and enzymes responsible, a gap exists with regard to the functional significance of many phosphorylation events. However, several recent studies have been directed toward identification of in vivo kinase substrates and their functions. These include studies on somatic cell mutations affecting the protein kinase system and on methods for isolating newly-phosphorylated proteins.

The subject undertaken for this review is vast, and a comprehensive recitation of references is impractical. Fortunately, several recent reviews cover much of the early literature on chromosomal phosphoroteins (PUMO and KLEINSMITH 1978), histones and their modifications (JOHNSON and ALLFREY 1978; ISENBERG 1979), nuclear protein kinases (JUNGMANN et al. 1976; JOHNSON 1977), chromatin subunit structure and alterations (KORNBERG 1977; ALLFREY 1980), and phosphorylation in control of protein synthesis (OCHOA and DE HARO 1979). In the present review I discuss several developments (for example, kinase activity encoded by viral genes) which may enlarge the concept of phosphorylation in the regulation of gene expression. In some cases, phosphorylation or the kinases discussed may not be regulated by cyclic nucleotides. However, as is clear from the case of histone H 1, an assessment of both cyclic nucleotide dependent and independent phosphorylation sites is necessary for an understanding of the role of phosphorylation. I have also attempted to focus on recent methodology which may help to identify and characterize those phosphorylation events which are significant in terms of cellular function.

B. Nuclear Cyclic Nucleotide-Dependent Protein Kinase Activities

I. Cyclic AMP-Dependent Protein Kinase

1. Translocation to Nuclei

Experiments on rat ovarian nuclei showed that nuclei and chromatin from newborn rat ovaries possess low levels of protein kinase activity relative to preparations from sexually mature rats (DE ANGELO et al. 1973; JUNGMANN et al. 1976). On the basis of further experiments it was proposed that chorionic gonadotropin can, by raising intracellular cyclic AMP levels, induce translocation of cyclic AMP-dependent protein kinase from the cytoplasm to the nucleus of ovarian cells (JUNGMANN et al. 1974).

When calf ovary cytoplasmic protein kinase is incubated with either ovarian chromatin or intact nuclei in the presence of ^{3}H-cyclic AMP, the labeled cyclic nucleotide is bound to the chromatin. Chromatin protein kinase activity is also increased by this procedure, and this activity is cyclic AMP-independent, suggesting that free catalytic (C)-subunits of cyclic AMP-dependent protein kinase are translocated to the nuclei. This in vitro translocation of protein kinase subunits is apparently dependent on dissociation of the enzyme by cyclic AMP, and the phenomenon is temperature-sensitive, proceeding more rapidly at 30 °C than at 4 °C.

Cyclic AMP-binding protein from calf ovary cytosol bound to calf ovary chromatin more effectively than to chromatin from liver, adrenal, or uterine cells, and the presence of non-histone chromatin protein was important for maximal binding (JUNGMANN et al. 1975).

PALMER and co-workers (1974) found that nuclei isolated from glucagon-stimulated, perfused rat liver contain two to three times as much protein kinase activity as do nuclei from livers of control animals. Correspondingly, cyclic AMP-dependent protein kinase activity is decreased in cytoplasm extracted from the glucagon-stimulated liver. Addition of either the heat-stable protein kinase inhibitor of ASHBY and WALSH (ASHBY and WALSH 1972, 1973) or the protein kinase R-subunit can reduce the elevated nuclear protein kinase activity. These results were taken to indicate that, in response to increased cyclic AMP levels, cyclic AMP-dependent protein kinase molecules of the liver cytoplasm are translocated to the nucleus, either in holoenzyme form or as free catalytic subunits.

Not all translocation measured by in vitro subcellular distribution of subunits is likely to be of in vivo significance. One possible problem is that the R and C subunits of the kinase holoenzyme have acidic and basic isoelectric points, respectively, and may be expected to bind ionically to some subcellular components following homogenization. KEELY et al. (1975) perfused rat hearts with epinephrine or 1-methyl-3-isobutylxanthine, agents which raise cyclic AMP levels and activate cyclic AMP-dependent protein kinase. It was found that activated catalytic subunits are redistributed from the cytoplasm to various particulate binding sites, and that this translocation can be mimicked by adding cyclic AMP to heart homogenates. However, prior boiling, phospholipase or trypsin treatments of the particulate material do not alter the binding of C-subunits. C-subunit binding is prevented by 150 mM KCl. It was concluded that protein kinase activity translocated to particulate fractions may be overestimated when isolation of subcellular fractions is carried out under conditions of low ionic strength, since an ionic interaction of the C-subunit with subcellular binding sites could occur during homogenization of cells under such conditions. The authors suggested that in studies on translocation cell homogenization be conducted in media containing 0.10–0.15 *M* NaCl or KCl (KEELY et al. 1975).

In subsequent experiments by JUNGMANN et al. (1976) it was reported that, in ovaries from neonatal pigs, 8-p-chlorophenylthio cyclic AMP induced migration of intact cyclic AMP-dependent protein kinase from cytoplasm to nucleus. In this case, nuclei were isolated by a nonaqueous procedure, reducing possible salt effects during homogenization, and protein kinase was detected by activity measurements as well as by reaction with anti-cyclic AMP-dependent protein kinase antiserum. Thus binding of protein kinase subunits to nuclear components is not likely here to be due merely to non-specific ionic interactions.

In additional studies performed using rat adrenal medulla cells, it was observed that reserpine and carbamylcholine induce translocation of activated protein kinase C-subunits from cytoplasm to nucleus, and that these effects are correlated with an increase in cyclic AMP levels in the cells (COSTA et al. 1976; GUIDOTTI et al. 1978). In this study, homogenization in the presence of 0.15 *M* NaCl did not prevent binding of protein kinase C-subunits in the nuclei. The authors hypothesized that nuclear translocation of cyclic AMP-dependent protein kinase is involved in a transsynaptic induction of tyrosine-3-monooxygenase.

Translocation of cyclic AMP-dependent protein kinase to nuclei has also been observed after norepinephrine treatment of C6–2B glial tumor cells (SALEM and DE VELLIS 1976) and 8-p-chlorophenylthio cyclic AMP treatment of WI-38 cells (JUNGMANN et al. 1976).

Recent cytochemical experiments support the contention that nuclear translocation of cyclic AMP-dependent protein kinase accompanies increased cyclic AMP levels in rat liver cells (BYUS and FLETCHER 1979). Administration of dibutyryl- or 8-bromo-cyclic AMP to Reuber H35 cells results in a rapid appearance of kinase C-subunit in the nucleolus and nucleus, as detected using fluorescent-labelled, heat-stable protein kinase inhibitor.

Nuclear translocation of cyclic AMP-dependent protein kinase in response to hormonal signals is reminiscent of the nuclear migration of steroid hormone receptors (VEDECKIS et al. 1978), which is known to result in transcriptional activation of specific genes (ROYAL et al. 1979). In the case of cyclic AMP there is at this time no known eukaryotic gene which depends upon protein kinase for initiation of transcription. There is evidence that DNA binding by the kinase C-subunit displays a degree of selectivity (JOHNSON et al. 1975a) and that specific non-histone proteins serve as substrates for the kinase in chromatin (JOHNSON and ALLFREY 1972). Further experiments should help evaluate the functional significance of nuclear translocation.

2. Endogenous Nuclear Cyclic AMP-Dependent Protein Kinase

The ability of cyclic AMP to stimulate phosphorylation of certain nuclear proteins may not be dependent upon translocation of kinase subunits. In certain cells, cyclic

AMP-dependent protein kinase holoenzyme can be detected in the nucleus in the absence of prior manipulation of cyclic AMP levels (JOHNSON and ALLFREY 1972; JOHNSON et al. 1975a). This is readily evident in thymocytes, cells in which the nucleus occupies a large proportion of the total internal volume of the cell (JOHNSON et al. 1975a), but it is also true of bovine brain (MAENO et al. 1971) and rat liver (MATSUMURA and NISHIZUKA 1974; KISH and KLEINSMITH 1974). Calf thymus possesses one of the highest levels of cyclic AMP-dependent protein kinase of any tissue (POMERANTZ 1979; POMERANTZ et al. 1977), and both R and C subunits can be isolated from purified nuclei (JOHNSON et al. 1975a). In thymocytes, cyclic AMP-dependent protein kinase activity has been found in nuclei highly purified by non-aqueous procedures following rapid freezing in liquid nitrogen (JOHNSON 1977). These procedures greatly reduce any possibility of enzyme redistribution between cytoplasm and nucleus during isolation. Despite this presence of protein kinase enzymes, the level of labelled phosphate incorporation into nuclear protein in resting peripheral blood lymphocytes is very low (JOHNSON et al. 1973). However, when these lymphocytes are activated by concanavalin A, the level of overall phosphorylation increases nearly 8-fold within a few hours, and this increase is accompanied by a dramatic increase in the mass of proteins bound to DNA in the chromatin (JOHNSON et al. 1974). These results suggest that certain cellular signals can increase the activities of different kinase enzymes in nuclei, and allow the possibility of translocation. In this system, however, the relative contributions of translocated vs. activated endogenous protein kinase activities are difficult to distinguish.

II. Cyclic GMP-Dependent Protein Kinase

Cyclic GMP-dependent protein kinase differs from the cyclic AMP-dependent enzyme in several important structural details. Cyclic GMP activates the cyclic GMP-dependent enzyme without causing an apparent dissociation of the holoenzyme (GILL et al. 1976; LINCOLN et al. 1977; SHOJI et al. 1978). A catalytically-active subunit can be obtained, however, by incubating with both histone and a high concentration (100 μM) or cyclic GMP (MIYAMOTO et al. 1973). A catalytic polypeptide fragment of the holoenzyme can be obtained by limited proteolysis (LINCOLN et al. 1977). Cyclic GMP-dependent protein kinase may be composed of 2 identical protein subunits which undergo a structural alteration upon binding cyclic GMP (LINCOLN et al. 1978).

Cyclic GMP-dependent protein kinase is known to phosphorylate several histones (see JOHNSON and ALLFREY 1978). In addition, it has been observed that phosphorylation of chromosomal proteins in equine lymphocytes is stimulated by extracellular addition of agents which raise levels of cyclic GMP (E. M. JOHNSON and HADDEN 1975). Cyclic GMP binding to a subunit of partially purified RNA polymerase I has also been reported (L. D. JOHNSON and HADDEN 1975), raising the possibility that cyclic GMP is involved in ribosomal RNA synthesis or metabolism in the nucleolus. At this point, no specific polypeptide site on any functionally-identified protein is known to be phosphorylated in a cylcic GMP-dependent reaction in vivo.

III. Cytochemical Localizations of Protein Kinase Activities

Antibodies to conjugates of cyclic AMP and cyclic GMP have been used in cytochemical studies on the localization of binding proteins for these nucleotides. This technique is subject to limitations, one of which is that bound cyclic nucleotides may not always be reactive with antibodies. Nonetheless, data have been obtained which may help identify in vivo roles of the cyclic nucleotides, especially cyclic GMP. SPRUILL et al. (1978) have found that cyclic GMP, but not cyclic AMP, is distributed along the chromosomes of *D. melanogaster* salivary glands, as detected with a fluorescent antibody. This localization may reflect the tight binding of cyclic GMP to certain chromosomal proteins (JOHNSON et al. 1975b). Heat shock, which selectively activates several *Drosophila* genes, results in a dramatic increase in fluorescence intensity over specific chromosome loci. Locus 93 D displays the greatest cGMP-antibody fluorescence and is also transcriptionally active, as detected by ^{3}H-uridine autoradiography. These results are further evidence for a role for cyclic GMP in the modulation of certain types of RNA synthesis.

BYUS and FLETCHER (1979) have employed the heat-stable protein kinase inhibitor as a fluorescent probe for the free C-subunit in H 35 cell monolayers and rat liver cells. Most striking in these experiments is the localization of the C-subunit in the cell nucleolus following activation by dibutyryl cyclic AMP or 8-bromo cyclic AMP. MEDNIEKS et al. (1979) have used antiserum to type II protein kinase and a fluorescent-conjugated secondary antibody to localize protein kinase in lymphocytes. It was found that in bovine B-lymphoid cells the type II enzyme is located in the nucleus.

STEINER et al. (1978) have prepared antisera to both the C and R subunits of cyclic AMP-dependent protein kinase and used these along with a fluorescent secondary antibody to examine distribution of the subunits in regenerating rat liver. Their results show an apparent redistribution of both subunits to nuclei during regeneration. However, further experiments with purified antibodies of proven specificity are necessary to confirm this. The results are consistent with those of KALLOS (1977), who observed, using a method of photoaffinity labelling, binding of cytoplasmic cyclic AMP receptor protein in liver nuclei.

C. Histone Phosphorylation

Histones are among the best substrates for cyclic nucleotide-dependent protein kinases in vitro. Following early reports on the incorporation of phosphate into lysine-rich histones (KLEINSMITH et al. 1966; ORD and STOCKEN 1966; LANGAN and SMITH 1967), considerable progress has been made concerning structural and regulatory aspects of histone phosphorylation. Despite evidence that various histones can be phosphorylated by cyclic nucleotide-dependent protein kinases in vitro, evidence for in vivo cyclic nucleotide-dependent phosphorylation is presently convincing only for histone H 1. Thus, H 1 will be considered in greatest detail in this review. In the last few years some insight has been gained into the structural role of H 1 in the chromatin, allowing an assessment of possible effects of site-specific H 1 phosphorylations.

I. H 1 Phosphorylation

1. Cylic Nucleotide-Dependent H 1 Phosphorylation

H 1 is the largest of the five major histone classes, possessing a molecular weight of about 21,000, corresponding to 210–220 amino acids. Unlike the other major histones, H 1 displays considerable tissue and species heterogeneity. There exist multiple subfractions of H 1 in all mammalian sources examined. Early evidence that phosphorylation of H 1 from rat liver is catalyzed by cyclic AMP-dependent protein kinase has previously been reviewed by LANGAN (1973).

In vivo incorporation of ^{32}P-phosphate into rat liver H 1 can be stimulated by injection of hormones such as glucagon, or by dibutyryl cyclic AMP (LANGAN 1969 a, b, 1970, 1973; TAKEDA and OHGA 1973). In experiments performed by LANGAN and co-workers, histone H 1 phosphorylation was measured by isolation of a phosphorylated polypeptide obtained upon trypsin digestion of isolated H 1. An 8- to 20-fold enhancement of ^{32}P-phosphate incorporation into the specific H 1 fragment was observed upon injection of glucagon or dibutyryl cyclic AMP. Hydrocortisone and ACTH had no effect upon phosphorylation of liver H 1 (LANGAN 1969 b, 1970, 1971). In response to glucagon injection, enhanced H 1 phosphorylation was detectable after 15 min and was maximal after about 1 h. Phosphorylation was not influenced by injection of actinomycin D or cycloheximide. The possibility that cyclic AMP-mediated phosphorylation of H 1 might occur artifactually upon liver homogenization was dealt with in subsequent studies by WICKS and co-workers (1975), who extracted H 1 directly from cells with sulfuric acid, thus allowing little opportunity for histone phosphorylation to occur during homogenization. These studies indicate that the ability of various cyclic AMP analogs to induce tyrosine aminotransferase in Reuber H 35 hepatoma cells is correlated with the ability of these analogs to stimulate phosphorylation of endogenous H 1. All analogs which induce the enzyme also stimulate phosphorylation of a specific serine residue on H 1, although no causal relationsip between these events has been established.

At low enzyme concentrations in vitro, cyclic AMP-dependent protein kinase phosphorylates only one site on histone H 1. This site has been localized to a single serine residue corresponding to position 37 of rabbit thymus H 1, or ser 38 of calf thymus H 1 (LANGAN 1971; LANGAN et al. 1971). Results of SHLYAPNIKOV et al. (1975) agree with the earlier results of LANGAN (1971) indicating that isolated cyclic AMP-dependent protein kinase transfers only 1 mole of phosphate per mole of histone H 1 in vitro, and that the site of phosphorylation of calf H 1 is Ser 38. The amino acid sequence of the H 1 region containing this serine residue is known for both rabbit thymus and calf thymus H 1 fractions (RALL and COLE 1971). At higher kinase concentrations more than 1 H 1 site may be phosphorylated (ZEILIG et al. 1979). It has been reported that histone kinase preparations from several tissues can phosphorylate two distinct sites on the H 1 molecule. Separation of kinase enzyme activities from these tissues allowed the determination that cyclic AMP-dependent protein kinase catalyzes phosphorylation specifically of Ser 38, while a separate cyclic AMP-independent histone kinase catalyzes phosphorylation of a serine residue (shown in Fig. 1) which corresponds to Ser 105 of rabbit thymus H 1

```
                                                                        P*(G)
Ac-Ser-Glu-Ala-Pro-Ala-Glu-Thr-Ala-Ala-Pro-Ala-Pro-Ala-Glu-Lys-Ser-Pro-Ala-Lys-Lys-Lys-Lys-
    1                                       10                                      20

                                                                   Ser-P*‡
Ala-Ala-Lys-Lys-Pro-Gly-Ala-Gly-Ala-Ala-Lys-Arg-Lys-Ala-Ala-Gly-Pro-Pro-Val-Ser-Glu-Leu-Ile-
                                30                                      40

Thr-Lys-Ala-Val-Ala-Ala-Ser-Lys-Glu-Arg-Asn-Gly-Leu-Ser-Leu-Ala-Ala-Leu-Lys-Lys-Ala-Leu-Ala-
                    50                                      60

Ala-Gly-Gly-Tyr-Asp-Val-Glu-Lys-Asn-Asn-Ser-Arg-Ile-Lys-Leu-Gly-Leu-Lys-Ser-Leu-Val-Ser-Lys-
    70                                      80                                      90

                                                P‡
Gly-Thr-Leu-Val-Glu-Thr-Lys-Gly-Thr-Gly-Ala-Ser-Gly-Ser-Phe-Lys-Leu-Asn-Lys-Lys-Ala-Ala-Ser-
                                    100                                     110

                                                                                    P*(G)
Gly-Glu-Ala-Lys-Pro-Lys-Pro-Lys-Lys-Ala-Gly-Ala-Ala-Lys-Pro-Lys-Lys-Pro-Ala-Gly-Ala-Thr-Pro-
                    120                                     130

                                                            P*(G)
Lys-Lys-Pro-Lys-Lys-Ala-Ala-Gly-Ala-Lys-Lys-Ala-Val-Lys-Lys-Thr-Pro-Lys-Lys-Ala-Pro-Lys-Pro-
        140                                     150                                     160

                                                                                P*(G)
Lys-Ala-Ala-Ala-Lys-Pro-Lys-Val-Ala-Lys-Pro-Lys-Ser-Pro-Ala-Lys-Val-Ala-Lys-Ser-Pro-Lys-Lys-
                                        170                                     180

Ala-Lys-Ala-Val-Lys-Pro-Lys-Ala-Ala-Lys-Pro-Lys-Ala-Pro-Lys-Pro-Lys-Ala-Ala-Lys-Ala-Lys-Lys-
                                190                                     200

Thr-Ala-Ala-Lys-Lys-Lys-Lys-OH
            210
```

Fig. 1. Compilation of phosphorylation sites of histone H 1 residues observed in different cells. The sequence of rabbit thymus H 1 subfraction RTL 3 is depicted (COLE 1977). Serine 37 is a site of amino acid heterogeneity in rabbit H 1 fractions (LANGAN et al. 1971). In calf thymus, phosphorylation of the corresponding site occurs at serine 38. The symbol * indicates that modification of a site has been observed to occur in vivo, and † indicates that modification has been reported in vitro. The sites denoted by (G) correspond to growth-associated sites reported by LANGAN (1978)

(LANGAN 1971; LANGAN and HOHMANN 1975). It is not yet clear whether or not H 1 sites other than Ser 38 may be phosphorylated by this enzyme in vivo.

Ser 38 is a site of amino acid heterogeneity in H 1 molecules of certain species. For instance, H 1 fraction 3 from rabbit thymus has alanine in place of serine at position 37 (analogous to calf thymus Ser 38), while H 1 fraction 4 from rabbit thymus has serine at position 37 (LANGAN et al. 1971). This specific substitution of alanine for serine, occurring at a major site of histone phosphorylation, determines the substrate capacity of a class of H 1 molecules of a particular cell. It is interesting to contemplate the possible biological need for a cell to possess a class of H 1 molecules insensitive to stimulation by cyclic AMP.

Experiments of POMERANTZ et al. (1977) have shown that synthetic polypeptide analogs of the Ser 38 H 1 phosphorylation site can serve as good substrates for the purified C-subunit. Previous experiments of KEMP et al. (1976) and ZETTERQVIST et al. (1976) suggest a model for specificity in which vicinal arginine residues removed from a phosphorylatable serine strongly enhance the substrate capacity of a given polypeptide. Ser 38 in H 1 generally fits this model, and the peptide analog

Arg-Lys-Ala-Ser-Gly-Pro (H 1–6)

has a maximal velocity (V_m) of phosphorylation, under the conditions examined, approximately 2-fold higher than that of whole H 1 (POMERANTZ et al. 1977). The peptide

Arg-Arg-Lys-Ala-Ser-Gly-Pro (H 1–7)

has a V_m nearly 5-fold that of H 1–6, but the K_m of the enzyme for H 1–7 is twice as high as that for H 1–6. Thus, the effect of vicinal arginines is primarily on rate of catalysis rather than on substrate binding by the enzyme. This may result from an entropy-lowering, conformation-stabilizing effect of the arginine residues on the substrate polypeptide backbone.

POMERANTZ et al. (1977), in our laboratory, studied the mechanism of phosphorylation of the synthetic histone sequences by the C-subunit of cyclic AMP-dependent protein kinase. The kinetic data obtained argue strongly for a sequential interaction of enzyme with ATP and peptide substrates. Our results contradict those of SEVERIN et al. (1978), who report a stable phosphohistidine covalent intermediate in the transfer of phosphate from ATP to substrate catalyzed by pig brain protein kinase. We find that the C-subunit of the thymus protein kinase is autophosphorylated at multiple sites, but that this is most likely the result of phosphorylation of some C-subunit molecules by others. We find only very slight ATPase activity by the purified C-subunit, in the presence of the substrate analog

Leu-Arg-Arg-Ala-Ala-Leu-Gly.

This heptapeptide has Ala substituted for Ser at a phosphorylation site and binds to the catalytic site of the C-subunit. With either this analog or with no peptide substrate present, using α-^{32}P-ATP as a probe, we detect extremely low levels of ATP-ADP phosphohydrolase activity which are explained kinetically as merely catalysis of transfer of phosphate to H_2O, as though H_2O were substituting for substrate peptide at the catalytic site (POMERANTZ 1979). We have been unable to detect any phosphoenzyme species which are capable of transfering phosphate to a substrate polypeptide, as has been reported by SEVERIN et al. (1978), even under conditions expected to preserve phosphohistidine. The significance of the phosphohistidine enzyme group found by SEVERIN et al. (1978) is not clear since it is apparently obtained under conditions not conducive to maximum phosphotransferase activity. Our data do not entirely rule out a transient phosphoenzyme intermediate. The results do suggest that, rather than via a ping-pong mechanism, C-subunit catalysis proceeds via a random sequential mechanism similar to those of other ATP phosphotransferases such as *E. coli* galactokinase (GULBINSKY and CLELAND 1968), yeast hexokinase (DANENBERG and CLELAND 1975) and rabbit muscle creatine kinase (MORRISON and HEYDE 1972). The formation of a substrate-enzyme-ATP ternary intermediate in phosphorylation as proposed by POMERANTZ et al. (1977) could be an important consideration in the design of transition state analogs which might selectively inhibit phosphorylation by cyclic AMP-dependent protein kinase.

Cyclic GMP-dependent protein kinase also phosphorylates histone H 1. FLOCKERZI et al. (1978) found that a purified cyclic GMP-dependent protein kinase from heart muscle phosphorylates both H 1 and H 2 B, and that affinity of the enzyme is much higher for H 2 B. H 1 phosphorylation occurs preferentially at low (5 m*M*) Mg^{++} concentrations. No phosphorylation site was reported. ZEILIG et al. (1979) have reported that a cyclic GMP-dependent protein kinase from bovine lung phosphorylates more than 1 site on H 1. Ser 37 is the only H 1 N-terminal site phosphorylated by cyclic GMP-dependent protein kinase. In addition 2 C-terminal sites are phosphorylated, and these are distinct from growth-associated phosphorylation sites. It should be interesting to see whether phosphorylation of any of these H 1 sites is mediated by cyclic GMP in vivo.

2. Cyclic Nucleotide-Independent H 1 Phosphorylation

Numerous references attest to an enhanced phosphorylation of H 1 at specific points in the proliferative cycle of mammalian cells (BALHORN et al. 1972; BRADBURY et al. 1974; HOHMANN et al. 1976; DIXON et al. 1975; reviewed by JOHNSON and ALLFREY 1978). Early results will be considered here only as they relate to more recent data.

Experiments concerning H 1 phosphorylation during the cell cycle have been performed using the acellular slime mold *Physarum polycephalum*. The natural synchrony of nuclear division in this organism allows precise timing of events in the proliferative cycle. BRADBURY and co-workers have observed that a peak of H 1 phosphorylation occurs in late G 2, just preceding chromosome condensation and nuclear division (BRADBURY et al. 1973, 1974). It was hypothesized that H 1 phosphorylation at this point in the cycle may constitute a mechanism for initiating chromosome condensation (BRADBURY et al. 1974). It was reported that exogenously added histone kinase, incorporated through the easily permeable outer membrane of the *Physarum* synctium, could advance the onset of mitosis in synchronously growing nuclei (INGLIS et al. 1976).

Assessment of the biological function of H 1 phosphorylation at mitosis is complicated by differences in the timing of the onset of both phosphorylation and dephosphorylation in different systems. Studies on *Physarum* suggest that phosphorylation in late G 2 is reversed by dephosphorylation prior to metaphase. Studies in Chinese hamster cells suggest that if sodium bisulfite (or some equivalent phosphatase inhibitor) is used in isolation buffers, then high levels of H 1 and H 3 phosphorylation can be detected in metaphase (GURLEY et al. 1978; D'ANNA et al. 1978). JACKSON and colleagues (1976) have found that inhibition of phosphatase activity with $ZnCl_2$ in metaphase HTC cells prevents the dephosphorylation of H 1 which occurs as the cells proceed from M into G 1. It was found that chromosomes decondensed normally in G 1, although H 1 remained phosphorylated. These results argue against involvement of massive H 1 phosphorylation in maintenance of the metaphase condensed state but do not rule out the possibility that phosphorylation acts as an initiator of condensation.

Growth-associated H 1 phosphorylation occurs on sites other than Ser 38 (DIXON et al. 1975; LANGAN 1978) and apparently involves cyclic AMP-independent histone kinases (COMBER and TAYLOR 1974; LANGAN and HOHMANN 1975). DIXON and colleagues (1975) have reported that in developing trout testis, the H 1 sequence -Lys-Ser-Pro-Lys- may represent a recurrent phosphorylation site at the C-terminal end of the molecule. A generally similar result was found by LANGAN (1978) who reported sequences of several growth-associated phosphorylation sites in Ehrlich ascites H 1. These results are summarized in Fig. 1, in which all reported sites have been transposed to the rabbit RTL 3 H 1 subfraction (COLE 1977).

Histone phosphatases have been described which can dephosphorylate phospho-H 1 (MEISLER and LANGAN 1969; VOKAER et al. 1974). Presently little is known regarding in vivo site specificity or regulation of these enzymes.

II. Phosphorylation of Histones Other than H 1

The histones H 2 B, H 2 A, H 3, and H 4 constitute the core histones of nucleosomes in eukaryotes (see KORNBERG 1977; ISENBERG 1979). Of these histones, sequences of H 3 and H 4 are most strongly conserved through evolution while H 2 A and H 2 B show greater sequence divergence. All of these histones are phosphorylated, and related results have been reviewed recently (JOHNSON 1977; JOHNSON and ALLFREY 1978).

Histone H 2 B serves as a good substrate for cyclic AMP-dependent protein kinase in vitro, and under certain circumstances gives a higher maximum velocity of phosphorylation than does H 1 (JOHNSON and ALLFREY 1972; SHLYAPNIKOV et al. 1975).

Amino acid sequences at sites of phosphorylation of calf thymus H 2 B (MW = 13,800) by cyclic AMP-dependent protein kinase have been reported by FARAGO and co-workers (1975), using human lymphocyte protein kinase, by SHLYAPNIKOV and co-workers (1975), using pig brain protein kinase, and by HASHIMOTO and co-workers (1975), using silkworm protein kinase. FARAGO et al. (1975) isolated a tryptic phosphopeptide from H 2 B containing serine residues 36 and 38. Both SHLYAPNIKOV et al. (1975) and HASHIMOTO et al. (1975) determined that serine 36 of H 2 B is phosphorylated. HASHIMOTO and co-workers also reported phosphorylation of serine 32 by the protein kinase. KURODA and colleagues (1976) found that purified protein kinases from silkworm and bovine cerebellum each phosphorylated sites serine 32 and serine 36 of claf thymus H 2 B.

Recently GLASS and KREBS (1979) have found that synthetic H 2 B sequence analogs

Arg-Lys-Arg-Ser-Arg-Lys-Gln
32

and

Arg-Lys-Gln-Ser-Tyr-Ser-Val
36 38

can serve as substrates for both cyclic AMP- and cyclic GMP-dependent protein kinases. Neither enzyme was observed to phosphorylate Ser 38 in the latter peptide. The two enzymes differed in specificity in response to changes in Mg^{++} concentration or pH, but the significance of these changes in the nucleus is not cellar. EDLUND et al. (1977) have also employed synthetic H 2 B analogs to study specificity of cyclic GMP-dependent protein kinase. These workers did not observe significant phosphorylation of residues corresponding to Ser 32 or Ser 36 in hexapeptides, but they did detect phosphorylation of a decapeptide containing Ser residues 32, 36, and 38.

Histone H 3 (MW = 15,300 in calf thymus) is phosphorylated in cultured Chinese hamster cells (GURLEY and WALTERS 1973), in developing trout testis (DIXON et al. 1975), and in maturing avian erythroid cells (RUIZ-CARRILLO et al. 1976). Evidence obtained by DIXON and colleagues (1975) indicates that, in trout testis, H 3 may have one phosphorylation site per molecule. Studies performed using cultured Chinese hamster cells indicate that H 3 phosphorylation occurs at low levels throughout interphase, and that, just prior to mitosis, an intense burst of H 3 phosphorylation accompanies the previously described phosphorylation of H 1 (GURLEY et al. 1975). H 3 serves as a substrate for cyclic AMP-dependent protein kinase in vitro (CHEN and WALSH 1971; JOHNSON and ALLFREY 1972), although there is little evidence to suggest that this enzyme catalyzes in vivo H 3 phosphorylation. Recently, SHOEMAKER and CHALKLEY (1978) have isolated an H 3-specific histone kinase from nuclei of rat and calf thymus. The kinase reportedly phosphorylates H 3 at a single site which appears identical to the site phosphorylated during metaphase-associated H 3 phosphorylation. Further characterization of H 3 phosphorylation sites should aid in assessing possible mediation by regulatory agents.

Histone H 2 A (MW = 14,300 in calf thymus) is phosphorylated in vivo in regenerating rat liver (SUNG et al. 1971), during spermatid development in trout testis (MARUSHIGE et al. 1969), and in synchronously dividing Chinese hamster cells (GURLEY and WALTERS 1973). Phosphorylation of H 2 A in maturing avian erythrocytes persists after the phoshorylation of the other major histone classes has ceased (RUIZ-CARRILLO et al. 1976).

H 2 A serves as a substrate for cyclic AMP-dependent protein kinase in vitro (JOHNSON and ALLFREY 1972; SHLYAPNIKOV et al. 1975). SHLYAPNIKOV and coworkers have reported that a cyclic AMP-dependent protein kinase from pig brain has very high substrate specificity and phosphorylates only histones H 1, H 2 A, and H 2 B. In the case of H 2 A, the enzyme transfers no more than 1 mole of phosphate to 1 mole of histone. The site of H 2 A phosphorylation has been identified as serine 19 (SHLYAPNIKOV et al. 1975).

One interesting aspect of H 2 A is that its N-terminal amino acid is a serine residue which is at times both phosphorylated, on its -OH group, and acetylated, on its $-NH_2$ group (DIXON et al. 1975; see JOHNSON and ALLFREY 1978). Possible double modification of the same amino acid residue raises the question of whether there are coordinate controls of phosphorylating and acetylating enzymes.

The N-terminal serine of histone H 4 (MW = 10,200 in calf thymus) is phosphorylated in trout testis (DIXON et al. 1975) and in avian erythroid cells (RUIZ-CARRILLO et al. 1975). In trout testis, phosphorylation at this residue occurs following histone synthesis and may be concerned with altering DNA-histone interactions subsequent to insertion of H 4 into the chromatin (LOUIE et al. 1973). As mentioned above with regard to H 2A, the N-terminal serine of H 4 is subject to two modifications: phosphorylation and acetylation. Histone H 4 is a poor substrate for cyclic AMP-dependent protein kinase in vitro (JOHNSON and ALLFREY 1972). It has been reported that arginine-rich histones are effective substrates for cyclic GMP-dependent protein kinases isolated from different tissues (KUO 1974). MASARACCHIA et al. (1977) have reported the isolation from murine lymphosarcoma cells of two histone kinases specific for H 4. H 4-Kinase (I) was found to catalyze formation of 1.02 moles of phosphoserine per mole of H 4. The enzyme did not catalyze phosphorylation of a synthetic hepatapeptide corresponding to the N-terminal sequence of H 4, but it did catalyze phosphorylation of a synthetic peptide analog corresponding to H 4 residues 13–19 at serine 47. H 4-Kinase (II) phosphorylated the H 4 peptide containing serine 1.

Histone H 5 is a lysine-rich histone (MW = 20,900) which increases in proportion in the chromatin during maturation – and progressive inactivation – of avian erythrocytes. This protein is phosphorylated in erythrocytes (SUNG 1977) and serves as a substrate in vitro for cyclic AMP-dependent protein kinase (FASY et al. 1979). FASY and co-workers have observed that phosphorylation of H 5 in a cyclic AMP-dependent reaction reduces affinity of H 5 for the DNA.

III. Role of H 1 Phosphorylation in Altering Nucleosome Structure

Early observations indicated that H 1 phosphorylation could alter the circular dichroism of H 1-DNA complexes (ADLER et al. 1971) and the template activity of chromatin (WATSON and LANGAN 1973), although little could be deduced about molecular mechanisms involved. Although the structural role of H 1 in chromatin is still not entirely clear, recent work has revealed many details of the interaction of H 1 with the DNA and histones of nucleosomes. H 1 is presently believed to be involved in the organization of nucleosomes into higher order chromatin structures. It has been proposed that histone H 1 is associated with the linker DNA between 140 base-pair (bp) DNA segments protected by core histones (NOLL and KORNBERG 1977). H 1 is present at an average of 1 molecule per nucleosome and, when chemically cross-linked to other proteins in calf thymus nuclei, can be detected bound to core histone H 2A (GARRARD et al. 1979). Analysis of cross-links shows that the central, globular region of H 1 from residues 74–106 (see Fig. 1) is in contact with the central region of H 2A. The NH_2- and -COOH termini of H 1 most likely extend outside the nucleosome core and may bind 10 bp at each end of the core particle DNA, in keeping with data of NOLL and KORNBERG (1977) indicating that H 1 protects 20 bp in the 160 bp protected subunit of the nucleosome repeat. THOMA et al. (1979) have presented data indicating that H 1 is located in the region of entry and exit points of the DNA in a complete two-turn, 166 bp nucleosome particle. When higher-order chromatin structures are formed, these regions on neighboring nucleosomes may come closer together so that an H 1 polymer is formed in the center of the superhelical structures. Phosphorylation of H 1 might be involved in loosening such highly-packaged chromatin to convert inactive DNA regions to a form available for DNA or RNA polymerase enzymes.

FASY et al. (1979) have observed that phosphorylation of a single residue – Ser 38 – on H 1 alters the binding of the histone to DNA as measured by affinity chromatography. Phosphorylation reduces affinity more than would be expected for a simple charge alteration. Reduced H 1-DNA binding might be expected to alter higher order chromatin structure and perhaps to lead to unfolding of nucleosomes. It is now well documented that nucleosome structure is altered on active gene sequences (WEINTRAUB and GROUDINE 1976; GAREL and AXEL 1976; JOHNSON et al. 1978). During transcription of genes coding for ribosomal RNA, nucleosomes are altered in conformation, and fully extended 144 bp protected segments can be isolated from the transcribing sequences (JOHNSON et al. 1979). It is not yet known whether H 1 phosphorylation plays a role in activation of any specific gene. However, it is reasonable to assume that many of the observed transitions in nucleosome configuration are due to modifications of histone structure.

D. Non-Histone Protein Phosphorylation

Phosphate incorporation into proteins of the cell nucleus occurs primarily in non-histone proteins which are tightly associated with the chromatin (TENG et al. 1971; TAKEDA et al. 1971; KISH and KLEINSMITH 1974). The non-histone protein/DNA ratio is highest in chromatin actively engaged in the synthesis of RNA or DNA and lowest in chromatin that is synthetically inactive (KARN et al. 1974; JOHNSON et al. 1974; ALLFREY et al. 1974). During stimulation of equine lymphocytes to proliferate with concanavalin A, the non-histone protein/DNA ratio increases from a value of approximately 0.2 in resting cells to a value of nearly 0.8 within 6 h (JOHNSON et al. 1974). This increase in protein is accompanied by increases of several-fold in both protein kinase activity and total phosphate incorporation. It is known that phosphorylation of many of the nonhistone proteins is stimulated by cyclic AMP (JOHNSON and ALLFREY 1972; MATSUMURA and NISHIZUKA 1974) or cyclic GMP (JOHNSON and HADDEN 1975). However, inhibition of phosphorylation of certain non-histone proteins by cyclic AMP has also been reported (KISH and KLEINSMITH 1974; JOHNSON and HADDEN 1975), and this may be ascribed to stimulation by the cyclic nucleotide of a nuclear phosphoprotein phosphatase (DOKAS et al. 1978).

Many reviews have discussed functional aspects of the non-histone proteins and their phosphorylation (KLEINSMITH 1974; JOHNSON 1977; PUMO and KLEINSMITH 1978). The non-histone proteins are hypothesized to be involved in a variety of chromatin functions, and their phosphorylation is likely to modulate an array of processes, including selective gene transcription (KLEINSMITH et al. 1976) and the packaging or processing of newly-synthesized messenger RNA (KARN et al. 1977). The current emphasis on research in this area is through characterization of individual non-histone proteins and effects of phoshorylation.

I. Phosphorylation of Eukaryotic DNA-Dependent RNA Polymerases

Numerous references now document the phosphorylation of subunits of eukaryotic RNA polymerase II, the enzyme which in all or nearly all cells catalyzes the transcription of genes coding for messenger RNA (cf. CHAMBON 1975). In yeast an

RNA polymerase II subunit of 24,000 Daltons is selectively phosphorylated in vivo (BELL et al. 1977). KRANIAS et al. (1977) have reported that a 25,000 Dalton subunit of calf thymus RNA polymerase II is phosphorylated in vitro by homologous cyclic AMP-dependent protein kinase. In this case, activity of the polymerase was stimulated several-fold by the C-subunit of protein kinase in the presence of ATP. DAHMUS (1976) has reported phosphorylation of ascites tumor RNA polymerase II by homologous cyclic AMP-independent protein kinase.

RNA polymerase I is the enzyme which catalyzes transcription of ribosomal genes in eukaryotes (CHAMBON 1975). This enzyme is phosphorylated in rat liver (HIRSCH and MARTELO 1976), calf thymus (JUNGMAN et al. 1974) and yeast (BELL et al. 1977). RNA polymerase III, which transcribes genes including those coding for 5S ribosomal RNA (CHAMBON 1975) is also phosphorylated in yeast (BELL et al. 1977). Phosphorylated subunits are of 24,000 and 20,000 Daltons.

Phosphorylation of RNA polymerase may be part of a pleiotypic response to cyclic AMP which includes, in certain target cells, an alteration in chromatin structure accompanied by induction of synthesis of messenger RNAs coding for certain enzymes. Enzymes induced at the transcriptional level by cyclic AMP include phosphoenolpyruvate carboxykinase (IYNEDJIAN and HANSON 1977) and tyrosine aminotransferase (NOGUCHI et al. 1978) in rat liver. STEINBERG and COFFINO (1979) have shown, using 2-dimensional gel electrophoresis, that levels of several proteins in S 49 mouse lymphoma cells are regulated by cyclic AMP and cyclic AMP-dependent protein kinase, possibly through effects on levels of specific messenger RNAs.

Recently a nucleolar phosphoprotein of 70,000 Daltons has been isolated from the lower eukaryote *Physarum polycephalum* (KUEHN et al. 1979). This protein is phosphorylated in a polyamine-dependent reaction (ATMAR et al. 1978; KUEHN et al. 1979) and is tightly associated with chromatin containing the palindromic ribosomal genes of *Physarum*. The phosphoprotein stimulated rRNA synthesis 5-fold by RNA polymerase I within a rDNA-protein complex isolated from nucleoli (KUEHN et al. 1979). Although the phosphoprotein specifically bound several restriction fragments of the purified rDNA, the significance of binding with respect to transcription is not clear since these fragments are distant from the ribosomal RNA transcription units (CAMPBELL et al. 1979). At this point no purification of a cyclic AMP-dependent protein kinase from *Physarum* has been described. KUEHN (1972) has reported that activity of a cyclic AMP-inhibited protein kinase varies with cell cycle time.

II. Phosphorylation of HMG Proteins

The HMG (high mobility group) proteins are a limited class of non-histone proteins which are found associated with a small percentage of nucleosomes (GOODWIN et al. 1973; ALLFREY 1980). These proteins may play a role in functional transitions in nucleosome structure. HMG proteins are released from avian erythrocyte chromatin by DNase I treatment to selectively digest active globin genes (VIDALI et al. 1977; WEISBROD and WEINTRAUB 1979). HMG proteins 14 and 17 can be reconstituted with chromatin to confer DNase I sensitivity (WEISBROD et al. 1980). Recently, we have found that a calf thymus protein in this class, HMG-1, is phosphorylated by the C-subunit of cyclic AMP-dependent protein kinase (SUN et al. 1980). The regulatory consequences of such phoshorylation remain to be determined.

E. Phosphorylation of Ribosomal Proteins

Protein phosphorylation, as modulated by hormones and cyclic nucletides, may influence the expression of transcriptionally active genes through effects on protein synthesis at the level of translation. Phosphorylation of ribosomal proteins is enhanced by cyclic AMP in mouse adrenal cells (ROOS 1973), rat cerebral cortical cells (ROBERTS and ASHBY 1978) and rabbit reticulocytes (CAWTHORN et al. 1974), among others. A single protein of the 40 S subunit of reticulocyte ribosomes, denoted S 13, is phosphorylated by cyclic AMP-dependent protein kinase in vitro (TRAUGH and PORTER 1976). GRESSNER and WOOL (1976) have reported stimulation of phosphorylation of a rat liver small subunit ribosomal protein (MW = 31,000) by cyclic AMP in vivo. It has been suggested that ribosomal protein phosphorylation may be involved in attachment of the ribosome to messenger RNA, initiation factors or other components of the protein synthetic mechanism (KABAT 1970; TRAUGH and PORTER 1976; GRESSNER and WOOL 1976; ROBERTS and ASHBY 1978), but no specific mechanism is presently accepted.

Protein chain initiation is inhibited in heme-deficient reticulocyte lysates due to activation of a heme-regulated protein kinase which phosphorylates the α-subunit (MW = 38,000) of the initiation factor eIF-2 (KRAMER et al. 1976; LEVIN et al. 1979) DATTA et al. (1977) reported that the C-subunit of cyclic AMP-dependent protein kinase is latent in the presence of heme and may be activated during heme deficiency. However, more recent observations (LEVIN et al. 1979) suggest that the heme-regulated eIF-2 kinase is cyclic AMP-independent. Initiation factors eIF-3 and eIF-4 B are also phosphorylated in reticulocytes, and this may be of significance due to the possible participation of the latter in messenger RNA selection (BENNE et al. 1978).

F. Protein Phosphorylation and Viral Transformation

Recently it has been reported that the avian sarcoma virus (ASV) src gene product, pp 60^{src}, is capable of catalyzing protein phosporylation (COLLETT and ERIKSON 1978). Although the pp 60^{src} protein can phosphorylate IgG in immune precipitates (COLLETT and ERIKSON 1978; LEVINSON et al. 1978), some question remains as to whether kinase activity is the primary function of this protein in cells. MANESS et al. (1979) have reported that highly-purified pp 60^{src} from rat tumor cells can catalyze phosphorylation of α-casein in solution, strengthening the hypothesis that pp 60^{src} is a protein kinase in vivo. Evidence exists that the activity of the src gene product is cyclic nucleotide independent (ERIKSON et al. 1979). It is interesting, however, that the pp 60^{src} protein is itself phosphorylated at 2 sites, and that phosphorylation at one site (a serine residue) is cyclic AMP-responsive (COLLETT et al. 1979). Cyclic AMP has been observed to amplify replication of RNA tumor virus-like particles in Chinese hamster ovary cells (TIHON and GREEN 1973). It has been found that pp 60^{sarc}, the product of a gene closely homologous to that for pp 60^{src}, is present at low levels in normal, non-transformed cells and is itself a protein kinase (COLLETT et al. 1979). These results raise questions about the role of pp 60^{src} in cell transformation. An interesting characteristic of the src gene product is that it apparently phosphorylates tyrosine residues on proteins (HUNTER and SEFTON 1980), as has also recently been reported for a protein kinase in polyoma T-antigen immunoprecipitates (ECKHART et al. 1979). Genes of transforming viruses other than ASV may also code for protein kinase activities. For example, thermolabile protein kinase molecules have been found associated with a temperature-sensitive murine sarcoma virus pseudotype (SEN et al. 1979). (See Cold Spring Harbor Symposia on Quantitative Biology, Vol 44, 1979 for several references to protein kinases of transforming viruses.) It will be interesting to determine the relationship of these protein kinases to overall mechanisms of cell transformation and oncogenesis.

G. Functional Aspects of Protein Kinase Substrates: A Procedure for Isolating Newly-Phosphorylated Polypeptides

In order to fully understand the biological role of any protein kinase it is necessary to identify natural substrates for the enzyme in cells and to determine the effects of phosphorylation on these substrates' activities. Criteria for determining whether phosphorylation has a regulatory influence on the activity of a given protein have previously been described (JOHNSON 1977). It is most important to be able to measure whether activity of a protein phosphorylated at a specific site differs from activity of the non-phosphorylated protein species. At this point there are no methods in general use for separating phosphorylated from non-phosphorylated polypeptides. Recently, however, a method has been described by SUN et al. (1980) which allows separation of proteins containing a single newly-added phosphate moiety from their counterparts lacking this moiety. The method makes use of the fact that the ATP analog adenosine 5′-[γ-s]triphosphate, first synthesized by GOODY and ECKSTEIN (1971), serves as a phosphate donor in reactions catalyzed by various protein kinases (GRATECOS and FISCHER 1974; MORGAN et al. 1976). The thiophosphate group, when transferred to proteins, acts as an affinity probe and allows their separation on columns containing mercury bound to a solid support (RUIZ-CARRILLO and ALLFREY 1973), as summarized in Fig. 2. SUN et al. (1980) have reported that thiophosphorylated histone H 1 is quantitatively retained on an organomercurial column while phosphorylated H 1 is not. Proteins retained on the organomercurial column can be eluted with dithiothreitol (DTT). It was also found that H 1 is thiophosphorylated and retained on an organomercurial column when cells in culture are incubated with thiophosphate. An added advantage of using thiophosphate to study protein phosphorylation may be that thiophosphorylated proteins are more resistant than their phosphorylated counterparts to the action of certain phosphatases (GRATECOS and FISCHER 1974).

One possible complication during affinity chromatography of thiophosphorylated proteins is that many proteins possess reduced cysteinyl-SH groups which would be bound to the organomercurial column. (Note in Fig 1 that H 1 has no Cys

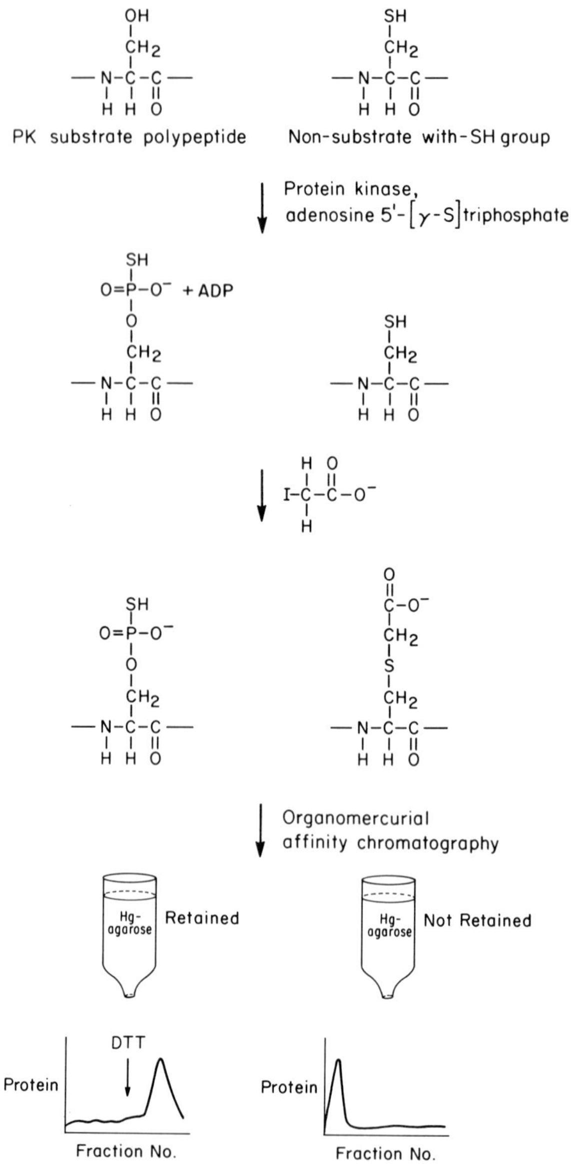

Fig. 2. Affinity chromatography of thiophosphorylated proteins on organomercurial columns. The diagram summarizes the procedure of Sun et al. (1980) for separating newly-phosphorylated protein kinase substrates from their non-phosphorylated or previously phosphorylated counterparts

residues.) The -SH groups can be carboxymethylated using iodoacetate or iodoacetamide under conditions in which thiophosphate -SH groups are not. It has been reported that thiophosphorylated histone H 3 (which contains a single Cys residue) is retained on the organomercurial column after reaction with iodoacetate while non-thiophosphorylated H 3 is not retained. Non-phosphorylated H 3 is retained on the column before carboxymethylation, but not after (see Fig. 2). Unfortunately, Cys -SH groups at many protein sites are not reactive with iodoacetate without prior denaturation. Thus, it would be premature to assume that the system described for affinity purification will be equally amenable for all protein kinase substrates. However, in certain cases it may be feasible to isolate enzymatically active, newly-phosphorylated proteins and to measure their activities compared with non-phosphorylated forms. For example, it should be feasible to study the contribution of selectively-phosphorylated and non-phosphorylated H 1 species to transitions in nucleosome structure.

H. Protein Phosphorylation as a Means of Altering the Genetic Program: The Problem of Specificity

One theoretical question is frequently posed concerning the functional significance of histone phosphorylation at certain sites, particularly as catalyzed by cyclic AMP-dependent protein kinase: Is it possible for this phoshorylation to mediate changes in functional states of some genes but not others? The issue in this instance is how phosphorylation of histones, present on virtually all gene sequences, mediated by a protein kinase, present in virtually all animal cells, can alter activity (e.g. transcription) at specific gene loci. The answer to this question is that it is certainly possible for histone phosphorylation (at least at certain sites) to mediate changes in functional states of specific genes, even though such phosphorylation may not itself be the primary specificity determinant. It may be conceptually convenient to view histone phosphorylation, and other histone modifications as well, as processes can alter the state of packaging of large – but discrete – segments of the genome to render them either more or less accessible to enzymes which react more specifically with DNA sequences. One may ask, to what degree does histone phosphorylation involve discrete segments of the genome? In the case of phosphorylation of Ser 38 in rat liver in response to hormones or cyclic AMP, evidence indicates that as much as 1% of the nuclear H 1 is phosphorylated (LANGAN 1973). This is greater than the percent of the total genome which could be transcriptionally activated by these agents, but obviously does not include H 1 of most nucleosomes. (Another modification, histone acetylation, is believed to involve transcriptionally active chromatin regions [cf. ALLFREY 1980]. Thus, although the histones themselves probably possess little or no capacity to recognize specific base sequences (see KORNBERG 1977; ISENBERG 1979), their modification may involve enzymes with at least a limited capacity to differentiate among chromatin regions. The basis for this limited specificity is not fully understood. A degree of selectivity in DNA binding has been attributed to the C subunit of calf thymus cyclic AMP-dependent protein kinase (JOHNSON et al. 1975a). At some level the selective transcription of genes must depend upon specific base sequence recognition by proteins. Alternation of chromatin structure of genomic segments by histone modifications could predispose these segments to react with proteins capable of sequence recognition. In this way histone phoshorylation, although not the primary determinant of specific DNA sequence recognition, could create a favorable chromatin configuration for sequence recognition by more selective proteins.

Over the past several years it has become most clear that protein phosphorylation is a very broadly employed mechanism in cells for altering structures of numerous proteins in response to different hormones (see GREENGARD 1978). Unfortunately, much of the early work on phosphoproteins consisted of identifying (perhaps enumerating is a more appropriate word) proteins which are phosphorylated, either in vivo or as protein kinase substrates in vitro. Today it would be reasonable to consider a protein which cannot be phosphorylated under any circumstances as a very rare protein. Many early demonstrations of protein phosphorylation must thus be reassessed in terms of functional significance. In considering the non-histone chromatin phosphoproteins, attempts must be made to isolate in native form individual proteins with defined physical and chemical characteristics, the activities of which can be sought using systems comprising purified components including specific DNA sequences. In considering enzymes of known activity, such as RNA polymerases, attempts must be made to demonstrate that phosphorylation influences not only enzyme activity, but also correct transcription of RNA from a well-defined gene sequence.

Studies on the significance of histone phosphorylation involve problems of an unusual nature, since the "activity" of these proteins is the structural organization of chromatin. Perhaps a promising approach to these problems would involve the use of cellular mutants lacking specific protein kinase activities (STEINBERG et al. 1978; STEINBERG and COFFINO 1979; LECAM et al. 1979). In such a system it may be possible to identify nucleosome or supernucleosome structural features which are sensitive to the disruption of histone phosphorylation

I. Conclusions

Protein phosphorylation may influence gene expression at various levels including those of chromosomal configuration, transcription, RNA processing and transport, ribosome configuration and binding to RNA, or translation. An understanding of the processes affected by phosphorylation will involve determination of the functional activities of many proteins phosphorylated in cells. This will require the isolation and biochemical characterization of different protein kinase substrates in native form in both phosphorylated and non-phosphorylated states.

Acknowledgement. Work was supported by the NIH (GM 26170). I thank Drs. V.G. Allfrey, A.H. Pomerantz, and I.Y.-C. Sun for experimental contributions and stimulating discussion.

References

Adler AJ, Schaffhausen B, Langan TA, Fasman GD (1971) Altered conformational effects of phosphorylated lysine-rich histone (f-1) in f-1-deoxyribonucleic acid complexes: Circular dichroism and immunological studies. Biochemistry 10:909–913

Allfrey VG (1980) Molecular aspects of the regulaton of eukaryotic transcription. Nucleosomal proteins and their post-synthetic modifications in the control of DNA conformation and template function. In: Goldstein L, Prescott DM (eds) Cell biology: a comprehensive treatise, vol 3. Academic Press, New York, pp 347–437

Allfrey VG, Inoue A, Karn J, Johnson EM, Vidali G (1974) Phoshorylation of DNA-binding nuclear acidic proteins and gene activation in the HeLa cell cycle. Cold Spring Harbor Symp Quant Biol 38:785–801

Ashby CD, Walsh DA (1972) Characterization of the interaction of a protein inhibitor with adenosine 3′,5′-monophosphate-dependent protein kinases. J Biol Chem 247:6637–6642
Ashby CD, Walsh DA (1973) Characterization of the interaction of a protein inhibitor with cyclic AMP-dependent protein kinases. II. Mechanism of action with the holoenzyme. J Biol Chem 248:1255–1261
Atmar VJ, Daniels GR, Kuehn GD (1978) Polyamine stimulation of phosphorylation of nonhistone acidic protein in nuclei and nucleoli from *Physarum polycephalum*. Eur J Biochem 90:29–37
Balhorn R, Chalkley R, Granner D (1972) Lysine-rich histone phosphorylation: A positive correlation with cell replication. Biochemistry 11:1094–1098
Bell GI, Valenzuela P, Rutter WJ (1977) Phosphorylation of yeast DNA-dependent RNA polymerases in vivo and in vitro. J Biol Chem 252:3082–3091
Benne R, Edman J, Traut RR, Hershey JWB (1978) Phosphorylation of eukaryotic protein synthesis initiation factors. Proc Natl Acad Sci USA 75:108–112
Bradbury EM, Inglis RJ, Matthews HR, Sarner N (1973) Phosphorylation of very lysine-rich histone in *Physarum polycephalum*. Eur J Biochem 33:131–139
Bradbury EM, Inglis RJ, Matthews HR, Sarner N (1974) Molecular basis of control of mitotic cell division in eukaryotes. Nature 249:553–556
Byus CV, Fletcher WH (1979) Intracellular localization of cAMP-dependent protein kinase in cultured Reuber H 35 cells and rat liver. J Cell Biol 83:421 a
Campbell GR, Littau VC, Melera PW, Allfrey VG, Johnson EM (1979) Unique sequence arrangement of ribosomal genes in the palindromic rDNA molecule of Physarum polycephalum. Nucleic Acids Res 6:1435–1447
Cawthorn ML, Bitte LF, Krystosek A, Kabat D (1974) Effect of cyclic adenosine 3′:5′-monophosphate on ribosomal protein phosphorylation in reticulocytes. J Biol Chem 249:275–278
Chambon P (1975) Eukaryotic nuclear RNA polymerases. Ann Rev Biochem 44:613–638
Chen L, Walsh DA (1971) Multiple forms of hepatic adenosine 3′:5′-monophosphate-dependent protein kinase. Biochemistry 10:3614–3621
Cole RD (1977) Special features ofthe structure of H 1 histones. In: T'so POP (ed) Molecular biology of the mammalian genetic apparatus. North-Holland Publ, Amsterdam, pp 93–103
Collett MS, Erikson RL (1978) Protein kinase activity associated with the avian sarcoma virus src gene product. Proc Natl Acad Sci USA 75:2021–2024
Collett MS, Erikson E, Erikson RL (1979) Structural analysis of avian sarcoma virus transforming protein: sites of phosphorylation. J Virol 29:770–781
Collett MS, Erikson E, Purchio AF, Brugge JS, Erikson RL (1979) A normal cell protein similar in structure and function to the avian sarcoma virus transforming gene product. Proc Natl Acad Sci USA 76:3159–3163
Comber HJ, TaylorDM (1974) Changes in histone phosphorylation and adenosine 3′:5′-cyclic monophosphate during the initiation of deoxyribonucleic acid synthesis and mitosis in rat kidney. Biochem Soc Trans 2:74–76
Costa E, Kurosawa A, Guidotti A (1976) Activation and nuclear translocation of protein kinase during transsynaptic induction of tyrosine 3-monooxygenase. Proc Natl Acad Sci USA 73:1058–1062
Dahmus ME (1976) Stimulation of ascites tumor RNA polymerase II by protein kinase. Biochemistry 15:1821–1829
Danenberg KD, Cleland WW (1975) Use of chromium adenosine triphosphate and lyxose to elucidate the kinetic mechanism and coordination state of the nucleotide substrate for yeast hexokinase. Biochemistry 14:28–39
D'Anna JA Jr, Gurley LR, Deaven LL (1978) Dephosphorylation of histones H 1 and H 3 during the isolation of metaphase chromosomes. Nucleic Acids Res 5:3195–3207
Datta A, de Haro C, Sierra JM, Ochoa S (1977) Role of 3′:5′-cyclic-AMP-dependent protein kinase in regulation of protein synthesis in reticulocyte lysates. Proc Natl Acad Sci USA 74:1463–1467
De Angelo AB, Lee PC, Jungmann RA (1973) Ovarian cyclic AMP-binding protein and protein kinase activities during postnatal development of the rat. Am Zool 13:1285

Dixon GH, Candido EPM, Houda BM, Louie AJ, MacLeod AR, Sung MT (1975) The biological roles of post-synthetic modifications of basic nuclear proteins. In: Fitzsimmons DW, Wolstenholme GEW (eds) The structure and function of chromatin. CIBA Symposium 28. Elsevier, Amsterdam, pp 229–258
Dokas L, Rittschof D, Kleinsmith LJ (1978) Effects of cyclic adenosine 3′,5′-monophosphate on phosphoprotein kinase and phosphatase fractions prepared from rat liver nuclei. Arch Biochem Biophys 191:578–589
Eckhart W, Hutchinson MA, Hunter T (1979) An activity phosphorylating tyrosine in polyoma T-antigen immunoprecipitates. Cell 18:925–933
Edlund B, Zetterqvist O, Ragnarsson U, Engström L (1977) Phosphorylation of synthetic peptides by (^{32}P)ATP and cyclic GMP-stimulated protein kinase. Biochem Biophys Res Commun 79:139–144
Erikson RL, Collett MS, Erikson E, Purchio AF (1979) Evidence that the avian sarcoma virus transforming gene product is a cyclic AMP-independent protein kinase. Proc Natl Acad Sci USA 76:6260–6264
Farago A, Romhanyi T, Antoni F, Takats A, Fabian F (1975) The phoshorylated site of calf thymus F2b histone by the cyclic AMP-dependent protein kinase. Nature 254:88
Fasy TM, Inoue A, Johnson EM, Allfrey VG (1979) Phosphorylation of H1 and H5 histones by cyclic AMP-dependent protein kinase reduces DNA binding. Biochim Biophys Acta 564:322–334
Flockerzi V, Speichermann N, Hofmann F (1978) A guanosine 3′:5′-monophosphate-dependent protein kinase from bovine heart muscle. J Biol Chem 253:3395–3399
Garel A, Axel R (1976) Selective digestion of transcriptionally active ovalbumin genes from oviduct nuclei. Proc Natl Acad Sci USA 73:3966–3970
Garrard WT, Boulikas T, Wiseman JM (1979) Points of contact between histone H1 and the histone octamer. J Cell Biol 83:146a
Gill GN, Holdy KE, Walton GM, Kanstein CB (1976) Purification and characterization of 3′:5′-cyclic GMP-dependent protein kinase. Proc Natl Acad Sci USA 73:3918–3922
Glass DB, Krebs EG (1979) Comparison of the substrate specificity of adenosine 3′:5′-monophosphate- and guanosine 3′:5′-monophoshate-dependent protein kinases. J Biol Chem 254:9728–9738
Goodwin GH, Sanders C, Johns EW (1973) A new group of chromatin-associated proteins with a high content of acidic and basic amino acids. Eur J Biochem 38:14–19
Goody RS, Eckstein F (1971) Thiophosphate analogs of nucleoside di- and triphosphates. J Am Chem Soc 93:6252–6257
Gratecos D, Fischer EH (1974) Adenosine 5′-O(3-thiotriposphate) in the control of phoshorylase activity. Biochem Biophys Res Commun 58:960–967
Greengard P (1978) Cyclic nucleotides, phosphorylated proteins and neuronal function. Raven Press, New York
Gressner AM, Wool IG (1976) Influence of glucagon and cyclic adenosine 3′:5′-monophosphate on the phosphorylation of rat liver ribosomal protein S6. J Biol Chem 251:1500–1504
Guidotti A, Chuang DM, Hollenbeck R, Costa E (1978) Nuclear translocation of catalytic subunits of cytosol cAMP-dependent protein kinase in the transsynaptic induction of medullary tyrosine hydroxylase. Adv Cyclic Nucleotide Res 9:185–197
Gulbinsky JS, Cleland WW (1968) Kinetic studies of E. coli galactokinase. Biochemistry 7:566–575
Gurley LR, D'Anna JA, Barham SS, Deaven LL, Tobey RA (1978) Histone phosphorylation and chromatin structure during mitosis in Chinese hamster cells. Eur J Biochem 84:1–15
Gurley LR, Walters RA (1973) Evidence from triton X-100 polyacrylamide gel electrophoresis that histone f2a2, not f2b, is phosphorylated in Chinese hamster cells. Biochem Biophys Res Comm 55:697–703
Gurley LR, Walters RA, Tobey RA (1975) Sequential phosphorylation of histone subfractions in the Chinese hamster cell cycle. J Biol Chem 250:3936–3944
Hashimoto E, Takeda M, Nishizuka Y (1975) Phosphorylated sites of calf thymus histone H2B by adenosine 3′,5′-monophosphate-dependent protein kinase from silkworm. Biochem Biophys Res Comm 66:547–555

Hirsch J, Martelo OJ (1976) Phosphorylation of rat liver ribonucleic acid polymerase I by nuclear protein kinases. J Biol Chem 251:5408–5413
Hohmann P, Tobey RA, Gurley LR (1976) Phosphorylation of distinct regions of histone H 1: Relationship to the cell cycle. J Biol Chem 251:3685–3692
Hunter T, Sefton BM (1980) Transforming gene product of Rous sarcoma virus phoshorylates tyrosine. Proc Natl Acad Sci USA 77:1311–1315
Inglis RJ, Langan TA, Matthews HR, Hardie DG, Bradbury EM (1976) Advance of mitosis by histone phosphokinase. Exp Cell Res 97:418–425
Isenberg I (1979) Histones. In: Snell ES, Boyer PD, Meister A, Richardson CC (eds) Ann Rev Biochem, vol 48. Annual Reviews, Inc., Palo Alto, pp 159–191
Iynedjian PB, Hanson RW (1977) Increase in level of functional messenger RNA coding for phosphoenolpyruvate carboxykinase (GTP) during induction by cyclic adenosine 3′:5′-monophosphate. J Biol Chem 252:655–662
Jackson V, Shires A, Tanphaichitr N, Chalkley R (1976) Modifications to histones immediately after synthesis. J Mol Biol 104:471–483
Johnson EM (1977) Cyclic AMP-dependent protein kinase and its nuclear substrate proteins. In: Greengard P, Robison GA (eds) Adv Cyclic Nucleotide Res, vol 8. Raven Press, New York, pp 267–309
Johnson EM, Allfrey VG (1972) Differential effects of cyclic adenosine 3′,5′-monophosphate on phosphorylation of rat liver nuclear acidic proteins. Arch Biochem Biophys 152:786–794
Johnson EM, Allfrey VG (1978) Biochemical modifications of histone primary structure: phosphorylation and acetylation as related to chromatin conformation and function. In: Litwack G (ed) Biochemical actions of hormones. Academic Press, New York, pp 1–51
Johnson EM, Allfrey VG, Bradbury EM, Matthews HR (1978) Altered nucleosome structure containing DNA sequences complementary to 19S and 26S ribosomal RNA in Physarum polycephalum. Proc Natl Acad Sci 75:1116–1120
Johnson EM, Campbell GR, Allfrey VG (1979) Different nucleosome structures on transcribing and non-transcribing ribosomal gene sequences. Science 206:1192–1194
Johnson EM, Hadden JW (1975) Phosphorylation of lymphocyte nuclear acidic proteins: regulation by cyclic nucleotides. Science 187:1198–1200
Johnson EM, Hadden JW, Inoue A, Allfrey VG (1975a) DNA-binding by cyclic adenosine 3′,5′-monophosphate-dependent protein kinase from calf thymus nuclei. Biochemistry 14:3873–3883
Johnson EM, Inoue A, Crouse LJ, Allfrey VG, Hadden JW (1975b) Effects of cyclic GMP upon DNA binding by a calf thymus nuclear protein fraction. Biochem Biophys Res Commun 65:714–721
Johnson EM, Karn J, Allfrey VG (1974) Early nuclear events in the induction of lymphocyte proliferation by mitogens: Effects of concanavalin A on the phosphorylation and distribution of non-histone nuclear proteins. J Biol Chem 249:4990–4999
Johnson LD, Hadden JW (1975) Cyclic GMP and lymphocyte proliferation: Effects on DNA-dependent RNA polymerase I and II activities. Biochem Biophys Res Commun 66:1498–1505
Jungmann RA, Christensen ML, Schweppe JS, Mednieks MI, Spielvogel AM (1976) Cyclic AMP-mediated nuclear translocation of cytoplasmic cAMP-dependent protein kinase: identity of the nuclear and cytoplasmic enzymes. In: Abou Sabe M (ed) Cyclic nucleotides and the regulation of cell growth. Dowden, Hutchinson, Ross, Stroudsberg, Pa., pp 225–252
Jungmann RA, Hiestand PC, Schweppe JS (1974) Mechanism of action of gonadotropin. IV. Cyclic AMP-dependent translocation of ovarian cyclic AMP-binding protein and protein kinase to nuclear acceptor sites. Endocrinology 94:168–183
Jungmann RA, Lee S, DeAngelo AB (1975) Translocation of cytoplasmic protein kinase and cyclic adenosine monophosphate-binding protein to intracellular acceptor sites. Adv Cyclic Nucleotide Res 5:281–306
Kabat D (1970) Phosphorylation of ribosomal proteins in rabbit reticulocytes. Characterization and regulatory aspects. Biochemistry 9:4160–4175
Kallos J (1977) Photochemical attachment of cyclic AMP binding protein(s) to the nuclear genome. Nature 265:705–710

Karn J, Johnson EM, Vidali G, Allfrey VG (1974) Differential phosphorylation and turnover of nuclear acidic proteins during the cell cycle of synchronized HeLa cells. J Biol Chem 249:667–677
Karn J, Vidali G, Boffa LC, Allfrey VG (1977) Characterization of the non-histone nuclear proteins associated with rapidly labeled heterogeneous nuclear RNA. J Biol Chem 252:7307–7322
Keely SL, Jr, Corbin, JD, Park CR (1975) On the question of translocation of heart cyclic AMP-dependent protein kinase. Proc Natl Acad Sci USA 72:1501–1504
Kemp BE, Benjamini E, Krebs EG (1976) Synthetic hexapeptide substrates and inhibitors of 3′,5′-cyclic AMP-dependent protein kinase. Proc Natl Acad Sci USA 73:1038–1042
Kish VM, Kleinsmith LJ (1974) Nuclear protein kinases: Evidence for their heterogeneity tissue specificity, substrate specificity and differential responses to adenosine 3′,5′-monophosphate. J Biol Chem 249:750–760
Kleinsmith LJ (1974) Acidic nuclear phosphoproteins. In: Cameron IL, Jeter JR, Jr (eds) Acidic proteins of the nucleus. Academic Press, New York, pp 103–135
Kleinsmith LJ, Allfrey VG, Mirsky AE (1966) Phosphoprotein metabolism in isolated lymphocyte nuclei. Proc Natl Acad Sci USA 55:1182–1189
Kleinsmith LJ, Stein J, Stein G (1976) Dephosphorylation of non-histone proteins specifically alters the pattern of gene transcription in reconstituted chromatin. Proc Natl Acad Sci USA 73:1174–1178
Kornberg R (1977) Structure of chromatin. Ann Rev Biochem 46:931–954
Kramer G, Cimadivella M, Hardesty B (1976) Specificity of the protein kinase activity associated with the hemin-controlled repressor of rabbit reticulocyte. Proc Natl Acad Sci USA 73:3078–3082
Kranias EG, Schweppe JS, Jungmann RA (1977) Phosphorylative and functional modifications of nucleoplasmic RNA polymerase II by homologous adenosine 3′,5′-monophosphate-dependent protein kinase from calf thymus and by heterologous phosphatase. J Biol Chem 252:6750–6758
Kuehn GD (1972) Cell cycle variation in cyclic adenosine 3′,5′-monophosphate-dependent inhibition of a protein kinase from Physarum polycephalum. Biochem Biophys Res Commun 49:414–419
Kuehn GD, Affolter H-U, Atmar VJ, Seebeck T, Gubler U, Braun R (1979) Polyamine mediated phosphorylation of a nucleolar protein from Physarum polycephalum that mediates rRNA synthesis. Proc Natl Acad Sci USA 76:2541–2545
Kuo JF (1974) Guanosine 3′,5′-monophosphate-dependent protein kinase in mammalian tissues. Proc Natl Acad Sci USA 71:4037–4041
Kuroda Y, Hashimoto E, Nishizuka Y, Hamana K, Iwai K (1976) Phosphorylated sites of calf thymus H 2 B histone by adenosine 3′:5′-monophosphate-dependent protein kinase from bovine cerebellum. Biochem Biophys Res Commun 71:629–635
Langan TA (1969 a) Action of adenosine 3′,5′-monophosphate-dependent histone kinase in vivo. J Biol Chem 244:5763–5765
Langan TA (1969 b) Phosphorylation of liver histone following the administration of glucagon and insulin. Proc Natl Acad Sci USA 64:1276–1283
Langan TA (1970) Phosphorylation of histones in vivo under the control of cylcic AMP and hormones. In: Greengard P, Costa E (eds) Role of cyclic AMP in cell function. Raven Press, New York, pp 307–323
Langan TA (1971) Phosphorylation of separate sites in lysine-rich histone by cyclic AMP-dependent and independent protein kinases. Fed Proc 30:1089 (abstract)
Langan TA (1973) Protein kinases and protein kinase substrates. Adv Cyclic Nucleotide Res 3:99–153
Langan TA (1978) Methods for the assessment of site-specific histone phosphorylation. In: Stein G, Stein J, Kleinsmith LJ (eds) Methods in cell biology, vol 19. Academic Press, New York, pp 127–142
Langan TA, Hohmann P (1975) Analysis of phosphorylation sites in lysine-rich (H 1) histone: an approach to the determination of structural chromosomal protein functions. In: Stein GS, Kleinsmith LJ (eds) Chromosomal proteins and their role in the regulation of gene expression. Academic Press, New York, pp 113–125

Langan TA, Rall SC, Cole RD (1971) Variation in primary structure at a phosphorylation site in lysine-rich histones. J Biol Chem 246:1942–1948
Langan TA, Smith LK (1967) Phosphorylation of histones and protamines by a specific protein kinase from liver. Fed Proc 26:603 (abstract)
Le Cam A, Singh T, Roth C, Cabral F, Pastan I, Gottesman I (1979) Use of cyclic AMP-resistant CHO mutants to identify substrates of cyclic AMP-dependent protein kinase. J Cell Biol 83:245a
Levin D, Ernst V, London I (1979) Effects of the catalytic subunit of cyclic AMP-dependent protein kinase (type II) from reticulocytes and bovine heart muscle on protein phosphorylation and protein synthesis in reticulocyte lysates. J Biol Chem 254:7935–7941
Levinson AD, Oppermann H, Levintow L, Varmus HE, Bishop JM (1978) Evidence that the transforming gene of avian sarcoma virus encodes a protein kinase associated with a phosphoprotein. Cell 15:561–572
Lincoln TM, Dills WL, Jr, Corbin JD (1977) Purification and subunit composition of guanosine 3′:5′-monophosphate-dependent protein kinase from bovine lung. J Biol Chem 252:4269–4275
Lincoln TM, Flockhart DA, Corbin JD (1978) Studies on the structure and mechanism of activation of the guanosine 3′:5′-monophosphate-dependent protein kinase. J Biol Chem 253:6002–6009
Louie AJ, Candido EPM, Dixon GH (1973) Enzymatic modifications and their possible roles in regulating the binding of basic proteins to DNA and in controlling chromosomal structure. Cold Spring Harbor Symp Quant Biol 38:803–819
Maeno H, Johnson EM, Greengard P (1971) Subcellular distribution of adenosine 3′,5′-monophosphate-dependent protein kinase in rat cerebrum. J Biol Chem 246:134–142
Maness PF, Engeser H, Greenberg ME, O'Farrell M, Gall WE, Edelman GM (1979) Characterizaton of the protein kinase activity of avian sarcoma virus src gene product. Proc Natl Acad Sci USA 76:5028–5032
Marushige K, Ling V, Dixon GH (1969) Phosporylation of chromosomal basic proteins in maturing trout testis. J Biol Chem 244:5953–5958
Masaracchia RA, Kemp BE, Walsh DA (1977) Histone 4 phosphotransferase activities in proliferating lymphocytes. J Biol Chem 252:7109–7117
Matsumura S, Nishizuka Y (1974) Phosphorylation of endogenous hepatic proteins by adenosine 3′,5′-monophosphate-dependent protein kinase. J Biochem (Tokyo) 76:29–38
Mednieks MI, Lawellin D, Kawano-Valuskas L, Jungmann RA (1979) Immunocytochemical localization of cyclic AMP-dependent protein kinase isozymes in normal and leukemic lymphoid cells. J Cell Biol 83:126a
Meisler MH, Langan TA (1969) Characterization of a phosphatase specific for phosphorylated histones and protamines. J Biol Chem 244:4961–4968
Miyamoto E, Petzold GL, Kuo JF, Greengard P (1973) Dissociation and activation of adenosine 3′,5′-monophosphate-dependent and guanosine 3′,5′-monophosphate-dependent protein kinases by cyclic nucleotides and by substrate proteins. J Biol Chem 248:179–189
Morgan M, Perry SV, Ottaway J (1976) Myosin light-chain phosphatase. Biochem J 157:687–697
Morrison JF, Heyde E (1972) Enzymic phosphoryl group transfer. Ann Rev Biochem 41:29–54
Noguchi T, Diesterhaft M, Granner D (1978) Dibutyryl cyclcic AMP increases the amount of functional messenger RNA coding for tyrosine aminotransferase in rat liver. J Biol Chem 253:1332–1335
Noll M, Kornberg RD (1977) Action of micrococcal nuclease on chromatin and the location of histone H 1. J Mol Biol 109:393–404
Ochoa S, de Haro C (1979) Regulation of protein synthesis in eukaryotes. Ann Rev Biochem 48:549–580
Ord MG, Stocken LA (1966) Metabolic properties of histones from rat liver and thymus gland. Biochem J 98:888–897
Palmer WK, Castagna M, Walsh DA (1974) Nuclear protein kinase activity in glucagon-stimulated perfused rat livers. Biochem J 143:469–471

Pomerantz AH (1979) Studies on the reaction mechanism of the catalytic subunit of calf thymus cyclic AMP-dependent protein kinase using synthetic polypeptide substrates. PhD Thesis, The Rockefeller, University.

Pomerantz AH, Allfrey VG, Merrifield RB, Johnson EM (1977) Studies on the mechanism of phosphorylation of synthetic polypeptides by a calf thymus cyclic AMP-dependent protein kinase. Proc Natl Acad Sci USA 74:4261–4265

Pumo DE, Kleinsmith LJ (1978) Methods for the assessment of nonhistone phosphorylation (acid-stable, alkali-labile linkages). In: Stein G, Stein J, Kleinsmith LJ (eds) Methods in cell biology, vol 19. Academic Press, New York, pp 119–126

Rall SC, Cole RD (1971) Amino acid sequence and sequence variability of the amino terminal regions of lysine-rich histones. J Biol Chem 246:7175–7190

Roberts S, Ashby CD (1978) Ribosomal protein phosphorylation in rat cerebral cortex in vitro. J Biol Chem 253:288–296

Roos BA (1973) ACTH and cAMP stimulation of adrenal ribosomal protein phosphorylation. Endocrinology 93:1287–1293

Royal A, Garapin A, Cami B, Perrin F, Mandel JF, Le Meur M, Brégégègre F, Gannon F, Le Pennec JP, Chambon P, Kourilsky P (1979) The ovalbumin gene region: common features in the organization of three genes expressed in chicken oviduct under hormonal control. Nature 279:125–132

Ruiz-Carrillo A, Allfrey VG (1973) A method for the purification of histone fraction F3 by affinity chromatography. Arch Biochem Biophys 154:185–191

Ruiz-Carrillo A, Wangh LJ, Allfrey VG (1975) Processing of newly synthesized histone molecules. Science 190:117–128

Ruiz-Carrillo A, Wangh LJ, Allfrey VG (1976) Selective synthesis and modification of nuclear proteins during maturation of avian erythroid cells. Arch Biochem Biophys 174:273–290

Salem R, de Vellis J (1976) Protein kinase activity and cyclic AMP-dependent protein phosphorylation in subcellular fractions after norepinephrine treatment of glial cells. Fed Proc 35:296 (abstract)

Sen A, Todaro GJ, Blair DG, Robey WG (1979) Thermolabile protein kinase molecules in a temperature-sensitive murine sarcoma virus pseudotype. Proc Natl Acad Sci USA 76:3617–3621

Severin ES, Saschenko LP, Kochetkov SN, Kurochkin SN (1978) Structure and function of protein kinase from pig brain. Adv Cyclic Nucleotide Res 9:171–184

Shlyapnikov SV, Arutyunyan AA, Kurochkin SN, Memelova LV, Nesterova MV, Saschenko LP, Severin ES (1975) Investigation of the sites phosphorylated in lysine-rich histones by protein kinase from pig brain. FEBS Lett 53:316–319

Shoemaker CB, Chalkley R (1978) An H3 histone-specific kinase isolated from bovine thymus chromatin. J Biol Chem 253:5802–5807

Shoji M, Brackett NL, Tse J, Sapira R, Kuo JF (1978) Molecular properties and mode of action of homogeneous preparation of stimulatory modulator of cyclic GMP-dependent protein kinase from the heart. J Biol Chem 253:3427–3434

Spruill WA, Hurwitz DR, Lucchesi JC, Steiner AL (1978) Association of cyclic GMP with gene expression of polytene chromosomes of Drosophila melanogaster. Proc Natl Acad Sci USA 75:1480–1484

Steinberg RA, Coffino P (1979) Two-dimensional gel analysis of cyclic AMP effects in cultured S49 mouse lymphoma cells: Protein modifications, inductions, and repressions. Cell 18:719–733

Steinberg RA, Wetters TV, Coffino P (1978) Kinase-negative mutants of S49 mouse lymphoma cells carry a trans-dominant mutation affecting expression of cAMP-dependent protein kinase. Cell 15:1351–1361

Steiner AL, Koide Y, Earp HS, Bechtel PJ, Beavo JA (1978) Compartmentalization of cyclic nucleotides and cyclic AMP-dependent protein kinases in rat liver: Immunocytochemical demonstration. Adv Cyclic Nucleotide Res 9:691–705

Sun IY-C, Johnson EM, Allfrey VG (1980) Affinity purification of newly-phosphorylated protein molecules. J Biol Chem 255:742–747

Sung MT (1977) Phosphorylation and dephosphorylation of histone V (H 5): Controlled condensation of avian erythrocyte chromatin. Biochemistry 16:286–290
Sung MT, Dixon GH, Smithies O (1971) Phosphorylation and synthesis of histones in regenerating rat liver. J Biol Chem 246:1358–1364
Takeda M, Ohga Y (1973) Adenosine 3′,5′-monoposphate and histone phosphorylation during enzyme induction by glucagon in rat liver. J Biochem (Tokyo) 73:621–629
Takeda M, Yamamura H, Ohga Y (1971) Phosphoprotein kinases associated with rat liver chromatin. Biochem Biophys Res Commun 42:103–110
Teng CS, Teng CT, Allfrey VG (1971) Studies of nuclear acidic proteins: Evidence for their phosphorylation, tissue specificity, selective binding to DNA and selective effects on transcription. J Biol Chem 246:3597–3609
Thoma F, Koller T, Klug A (1979) Involvement of histone H 1 in the organization of the nucleosome and of the salt-dependent superstructures of chromatin. J Cell Biol 83:403–427
Tihon C, Green M (1973) Cyclic AMP-amplified replication of RNA tumor virus-like particles in Chinese hamster ovary cells. Nature (New Biol) 244:227–231
Traugh JA, Porter GG (1976) A comparison of ribosomal proteins from rabbit reticulocytes phosphorylated in situ and in vitro. Biochemistry 15:610–616
Vedeckis WV, Schrader WT, O'Malley BW (1978) The chick oviduct progesterone receptor. In: Litwack G (ed) Biochemical actions of hormones. Academic Press, New York, pp 321–372
Vidali G, Boffa LC, Allfrey VG (1977) Selective release of chromosomal proteins during limited DNAse I digestion of avian erythrocyte chromatin. Cell 12:409–415
Vokaer A, Iacobelli S, Kram R (1974) Phosphoprotein phosphatase activity associated with estrogen-induced protein in rat uterus. Proc Natl Acad Sci USA 71:4482–4486
Watson G, Langan TA (1973) Effects of F 1 histone and phosphorylated F 1 histone on template activity of chromatin. Fed Proc 32:588
Weintraub H, Groudine M (1976) Chromosomal subunits in active genes have an altered conformation. Science 193:848–856
Weisbrod S, Weintraub H (1979) Isolation of a subclass of nuclear proteins responsible for conferring DNase I-sensitive structure on globin chromatin. Proc Natl Acad Sci USA 76:631–635
Weisbrod S, Groudine M, Weintraub H (1980) Interaction of HMG 14 and 17 with actively-transcribed genes. Cell 19:289–301
Wicks WD, Koontz J, Wagner K (1975) Possible participation of protein kinase in enzyme induction. J Cyclic Nucleotide Res 1:49–58
Zeilig CE, Langan TA, Glass DB (1979) Phosphorylation of histone H 1 by cyclic GMP and cyclic AMP-dependent protein kinases. J Cell Biol 83:244 a
Zetterqvist Ö, Ragnarsson U, Humble E, Berglund L, Engström L (1976) The minimum substrate of cyclic AMP-stimulated protein kinase, as studied by synthetic peptides representing the phosphorylatable site of pyruvate kinase (type L) of rat liver. Biochem Biophys Res Commun 70:696–703

Subject Index

Compiled by Barbara Geller Nathanson, Ph. D.

The index covers this Volume only. Companion Volume 58, Part II, has a separate index. The *italic* numbers indicate pages where the corresponding entry is discussed in detail. For certain very general categories, such as *Cyclic AMP*, the reader should look under more specific entries, such as *Adenylate cyclase* or *Synthesis of cyclic nucleotides*

Acknowledgements: I wish to thank Kathleen Sullivan, Edward Hunnicutt, and Christopher Owen for their assistance in compiling this index

Handbook of Experimental Pharmacology

Continuation of "Handbuch der experimentellen Pharmacology"

Volume 16
Erzeugung von Krankheitszuständen durch das Experiment/Experimental Production of Diseases

Part 1
Blut (In Vorbereitung)

Part 2
Atemwege

Part 3
Heart and Circulation

Part 4
Niere, Nierenbecken, Blase

Part 5
Liver

Part 6
Schilddrüse
(In Vorbereitung)

Part 7
Zentralnervensystem

Part 8
Stütz- und Hartgewebe

Part 9
Infektionen I

Part 10
Infektionen II

Part 11 A
Infektionen III

Part 11 B
Infektionen IV

Part 12
Tumoren I

Part 13
Tumoren II

Part 14
Tumoren III
(In Vorbereitung)

Part 15
**Kohlenhydratstoffwechsel, Fieber/
Carbohydrate Metabolism, Fever**

Volume 17
Ions alcalino-terreux

Part 1
Systèmes isolés

Part 2
Organismes entires

Volume 18
Histamine and Anti-Histaminics

Part 1
Histamine. Its Chemistry. Metabolism and Physiological and Pharmacological Actions

Part 2
Histamine II and Anti-Histaminics

Volume 19
5-Hydroxytryptamine and Related Indolealkylamines

Volume 20: Part 1
Pharmacology of Fluorides I

Part 2
Pharmacology of Fluorides II

Volume 21
Beryllium

Volume 22: Part 1
Die Gestagene I

Part 2
Die Gestagene II

Volume 23
Neurohypophysical Hormones and Similar Polypeptides

Volume 24
Diuretica

Volume 25
Bradykinin, Kallidin and Kallikrein

Volume 26
H. Fischer: **Vergleichende Pharmakologie von Überträgersubstanzen in tiersystematischer Darstellung**

Volume 27
Anticoagulantinen

Springer-Verlag
Berlin
Heidelberg
New York

Handbook of Experimental Pharmacology

Continuation of "Handbuch der experimentellen Pharmakologie"

Springer-Verlag
Berlin
Heidelberg
New York